ENVIRONMENTAL HEALTH - PHYSICAL, CHEMICAL AND BIOLOGICAL FACTORS

FUNGI

TYPES, ENVIRONMENTAL IMPACT AND ROLE IN DISEASE

Environmental Health - Physical, Chemical and Biological Factors

Additional books in this series can be found on Nova's website under the Series tab.

Additional e-books in this series can be found on Nova's website under the e-book tab.

Environmental Science, Engineering and Technology

Additional books in this series can be found on Nova's website under the Series tab.

Additional e-books in this series can be found on Nova's website under the e-book tab.

ENVIRONMENTAL HEALTH - PHYSICAL, CHEMICAL AND BIOLOGICAL FACTORS

FUNGI

TYPES, ENVIRONMENTAL IMPACT AND ROLE IN DISEASE

ADOLFO PAZ SILVA
AND
MARÍA SOL
EDITORS

Nova Science Publishers, Inc.
New York

For permission to use material from this book please contact us:
Telephone 631-231-7269; Fax 631-231-8175
Web Site: http://www.novapublishers.com

Library of Congress Cataloging-in-Publication Data

Fungi : types, environmental impact, and role in disease / editors, Adolfo Paz Silva, Marma Sol.
p. cm.
Includes index.
ISBN 978-1-61942-671-9 (hardcover)
1. Fungi. 2. Fungi--Classification. 3. Fungi--Ecology. 4. Mycoses--Environmental aspects. 5. Mycology. I. Silva, Adolfo Paz. II. Sol, Marma.
QK603.F94 2011
579.5--dc23
2011051312

Published by Nova Science Publishers, Inc. † New York

Contents

Preface

Fungi are very interesting organisms actively involved in a great number of environmental processes. Both animals and humans can also be affected by fungi, some of them especially important among immunocompromised individuals. This book provides research on the different types, impact and role in disease of fungi. Some of the topics discussed include gut fungi; carbohydrate uptake and catabolism in fungi; candida species as causative agents in infections; effect of polyene antibiotics on the potential pathogenicity factors in saprophytic micromycetes; and ectomycorrhizal fungi and their role in symbiosis.

Chapter I - Gut fungi represent a special anaerobic group of microscopic eukaryotes inhabiting the digestive tract of ruminant as well as many non-ruminant herbivores. Every trait of these fungi is singular and exceptional. Morphologically, gut fungi are the only representatives of kingdom Fungi demanding anaerobiosis, and therefore, possessing hydrogenosomes instead of mitochondria. Genetically, anaerobic fungi represent organisms with the lowest G+C content ever found in Nature and with duplication of a variety of important hydrolytic genes, which is a trait not common among eukaryotes. Enzymologically, anaerobic fungi possess cellulosomes, the enzyme factories known only in some bacteria, carrying the most active (hemi)cellulases. These special features have always complicated the determination and phylogenetic position of anaerobic fungi. Taxonomical problems have been resolved recently by establishment of the separate phylum as a result of reconstruction of the large group of flagellated fungi called chytridiales. The new phylum comprises a single class (Neocallimastigomycetes), one order (Neocallimastigales) and a single family (Neocallimastigaceae), which covers six genera of anaerobic fungi including *Neocallimastix*, *Piromyces*, *Caecomyces*, *Orpinomyces*, *Anaeromyces* and *Cyllamyces*. All species of anaerobic gut fungi secrete a variety of hydrolytic enzymes, including cellulases, xylanases, mannases, esterases, glucosidases, and glucanases, which effectively hydrolyze plant biomass consisting mainly from cellulose and hemicellulose. These enzymes exhibit higher activities than commercial enzyme products derived from *Trichoderma reeseii* or *Aspergillus niger*. The capacity of extracellular plant cell-wall enzymes from anaerobic fungi to degrade cellulose, hemicellulose, and various lignocellulosic compounds indicates high potential feasibility of gut fungi hydrolases in commercial processes. Their experimental applications in livestock production, paper industry, biofuel and renewable energy production are discussed in this chapter.

Chapter II - Fungal species have evolved different strategies to obtain and degrade resources available in their natural environments and/or their hosts. In general, these compounds can act as a carbon source or signaling molecules indicating the presence of the host in the surroundings. These molecules have been extensively studied as they contribute to host specificity and determine regulatory aspects of the host-fungus interaction. The knowledge and understanding of the metabolic traits and regulatory aspects of host-derived metabolites represent an interesting subject for research leading to the identification of novel targets for fungal disease control and management. In this chapter, the most relevant findings on filamentous fungi carbohydrate metabolism are reviewed. This review also provides a comparative analysis of carbohydrate metabolism in symbiotic, pathogenic and saprophytic species including sugar uptake and further degradation through glycolysis and tricarboxylic acid pathway. Even though fungal species interacting with plants are the main subject of discussion, the human pathogens *Candida albicans* and *Aspergillus fumigatus* are also included in the analysis.

Chapter III - Ecological changes, human demographic modifications, international travel and commerce as well as fungal adaptation to the environment and its conditions such as antifungal drugs by genetic recombination cause emergence or re-emergence of infectious diseases. Cryptococcus species complex is a large genus, where the pathogenic species C. neoformans (var. neoformans and var. grubii) and C. gatti are the agents of cryptococcosis. This is a ubiquitous fungus, responsible for disease in both immunocompetent and immunocompromised hosts and causes approximately 600,000 deaths/yr in North America.

C. neoformans is ubiquitous in the environment due to dissemination by pigeons following migratory and trade routes, and C. gattii is restricted to tropical/subtropical regions because it is not associated to pigeon guano, but to some tree species. Simultaneously, an increase of prevalence has been reported in patients living with HIV/AIDS, receiving aggressive cancer therapy or solid organ transplant recipients.

Coccidioides immitis and C. posadasii are responsible for the most common respiratory mycoses. C. immitis is found in California and C. posadasii elsewhere. Although this disease is considered endemic, epidemic outbreaks have been documented, mostly related to droughts, as well as immigration of susceptible individuals. Immunosuppressed patients are the most severely affected, with various clinical manifestations. Some associated factors have been identified regarding the epidemics, while new approaches to treatment and prevention are being applied.

Sporotrichosis is a subcutaneous mycosis caused by Sporothrix species complex fundamentally transmitted by traumatic inoculation with contaminated plants, wood, and other vegetable debris. Cases inoculated by animals, particularly cats, have been documented. This is sometimes considered an occupational disease because it commonly presents in gardeners and field workers, but epidemic outbreaks have occurred in several countries affecting people with different occupations, regardless of their immune status.

Chapter IV - The intent of this article is to review recent literature on the role of *Candida albicans* in infection. Candidemia, invasive candidiasis in neonates, disseminated candidiasis, invasive candidiasis in critically ill patients, Candida cerebrospinal fluid infection following neurosurgery, Candida cerebral abscess in premature infants. Hematogenous dissemination of fungus of the eyes manifested as chorioretinitis or endophthalmitis, oropharyngeal candidiasis, Candida endocarditis, Candida pneumonia in immunocompromised patients, intra-abdominal pancreatic fungal infection, recurrent vulvovaginal candidasis, Candida interdigital ulcers, Candida interdigital ulcers, congenital cutaneous candidiasis, and burn wound infection are discussed.

Chapter V - The influence of polyene antibiotics on formation by saprophytic micromycetes of such pathogenicity factors as pigments, exopolysaccharides, phospholipases, and catalase was examined. The tested polyenes included medicinal antibiotics amphotericin B (heptaene) and nystatin (tetraene), as well as hexaene antibiotic 30 which was specially isolated with a view to its use for protection against biodeterioration. It was shown that that all the polyene antibiotics tested promote the formation of exopolysaccharides and phospholipases. The effects exerted on the catalase activity range from promotive to inhibitory, depending on the specific test culture. However, by contrast to chemical fungicides, polyenes inhibit synthesis of pigments in all the saprophytic micromycetes tested. It was demonstrated that, when the polyene antibiotics are combined with surfactants, the fungicidal concentrations of the polyenes decrease by nearly an order of magnitude. Combined agents comprised of the polyenes and surfactants also suppress formation of pathogenicity factors (pigments, exopolysaccharides, phospholipases, and catalase) in saprophytic fungi. These findings suggest that non-medicinal antifungal polyene antibiotics, taken both alone and in combination with surfactants, offer promise for protection of industrial materials against mycodeterioration.

Chapter VI - Mycorrhiza is the most ancient and widespread mutualistic symbiotic association between plants and soil borne fungi. Mycorrhizae *sensu lato* are dual organs of absortion formed by mutualistic symbiotic fungi inhabiting healthy absorbing organs such as roots, rhizomes or thalli of most terrestrial plants and also many aquatics and epiphytes. In nature the soil nutrient acquisition of plants occurs to great extent via these mycorrhizal organs. It is now widely accepted that mycorrhizae would have played a fundamental role in the colonization of the land by plants functioning as nutrient absorbing structures even before the appearance of roots. Mycorrhizae are traditionally classified into categories on the basis of their fungal partners and macro- and microscopical anatomical structures of the dual organs. Among them ectomycorrhiza (ECM) is typically formed between the fine roots of trees of boreal and temperate forest and fungal species usually belonging to Basidiomycota and Ascomycota. In this chapter we discuss the evolutionary origin of ECM, some of its morphological and anatomical characters, the ecological significance of ECM and some of the molecular traits related to establishment and functioning of ECM structures between fungi and their plant hosts.

Chapter VII - Two experiments were designed to investigate the effects of soil disturbance and organic P (Na-phytate and lecithin) on mycorrhiza formation and symbiotic efficiency of AMF associated with maize in greenhouse. The results showed that P forms did not affect the root colonization of BEG167 and *Glomus versifome* (*G.v*) with maize. *G.v* produced more extraradical fungal biomass than BEG167 when KH_2PO_4 and lecithin were applied. The treatments with lecithin and KH_2PO_4 had higher maize biomass than those supplied with Na-phytate. BEG167 and *G.v* differed in maize biomass contribution when Na-phytate was applied. Soil disturbance reduced significantly root colonization, and significantly or slightly decreased shoot biomass and P uptake of maize colonized by BEG141. However, mycorrhiza development and symbiotic efficiency of *G.v* were not influenced by soil disturbance. On cellulose acetate membrane, numerous regeneration hyphae with anastomosis structure, cluster structure and dichotomous structure were observed. The results suggested that rapid recovery of extraradical hyphae of AMF might compensate the P contribution loss caused by soil disturbance and buffer the decrease of propagule reserve in soil ecosystem.

Chapter VIII - The interest in maintaining and increasing biodiversity has increased worldwide since The Treaty on Biological Diversity of Rio de Janeiro in 1992. This century, awareness of sustainability has grown too. The awareness of biodiversity and sustainability is not only focused on the above-ground environment, also the below-ground soil environment is considered. The soil environment occurring globally provides services that humans draw from nature. The so-called ecosystem services include products like food and processes like nutrient transfer. Numerous people consider the ecosystem services as public benefits contributing to welfare for free and infinitely available. The ecosystem services have considerable economic value and the consideration on this value is growing. However, the services provided by the soil can be used within certain conditions for sustainability reducing the risks of irreversible changes. Durable and sustainable soil ecosystem services have more economic value, certainly on the long term, than those services used for welfare without any consideration for the environment and biodiversity. Sustainable soil ecosystem services require a good soil quality integrating the physical, chemical, and biological components of soils and their interactions. This implies elimination of soil degradation, reduction of input of chemical fertilizers and crop protection agents as well as decrease of dehydration of soil and management of beneficial soil organisms. A good soil quality determines agricultural sustainability, environmental quality and, consequently, plant, animal and human health. For such good soil quality, knowledge is needed on the contribution of soil-organisms to durable and sustainable soil. Mycorrhizal fungi are an important group of the soil-organisms being beneficial to sustainable soils. These soil fungi live in symbiosis with the great majority of plants, trees and crops. Evidence is growing that mycorrhizal fungi enhance the plant health and crop yield. The efficacy of mycorrhizal fungi vary with abiotic factors including soil type and structure. In this paper, the efficacy of mycorrhizal fungi in relation to abiotic factors will be presented based on recent data. Preconditions on the soil characteristics for sustainable soils of good quality and for optimal development of mycorrhizal fungi are provided. The economic value of optimal mycorrhizal fungal development is shown and related to soil ecosystem services. European policy makers are working on legislation for soil protection and management. The European Commission has proposed a Soil Framework Directive to address soil degradation across the European Union, and to repair the damage that has already been done. Indicators for soil quality could contribute to monitoring the condition of soils when reflected in standards, like soil quality metrics. Mycorrhizal fungi are indicative for changes in soils being sensitive to changes in the physical-chemical conditions of the soil. In this paper, it is shown that mycorrhizal fungi are reliable and sensitive indicators for disturbances in the soils to be used for the development of soil quality metrics. Different soil quality metrics with different categories ranging from very bad to very good are described based on a considerable number of root samples. The applicability of the soil quality metrics is discussed and their potential usefulness for the legislations for protection of soil quality is shown.

Chapter IX - Understanding the environmental behaviour of radionuclides in forest ecosystems is of great importance to ensure an adequate level of radioprotection. The accumulation of radionuclides by fungi and the process of their transfer from the soil play a significant part in this behaviour because of the ability of these organisms to accumulate heavy metals, including most radionuclides. As a consequence of radioactive fallout and nuclear accidents, anthropogenic radionuclides have entered forest ecosystems, with radiocæsium being among the most hazardous in the medium-to-long term. As a general rule, fungi are able to accumulate radiocæsium very efficiently. However, there is a large range of variation, about seven orders of magnitude, in its content as reported in the literature. In areas heavily affected by a radioactive deposition, the consumption of mushrooms by man or animals contributes to increasing their body's radiocæsium content. There have been fewer studies of radionuclides other than radiocæsium, but on occasions they can also be of importance. In areas with low or almost negligible levels of anthropogenic radioactive contamination, the content of radionuclides belonging to the natural radioactive decay series, such as ^{210}Po and ^{210}Pb, contributes significantly to the internal dose of the general population. The uptake of naturally occurring radionuclides also shows a great range of variation, although not as wide as that observed for radiocæsium. Not all species of fungi accumulate radionuclides at the same level. There is also a difference based on the fungus's nutritional mechanism, with mycorrhizae generally having a greater radiocæsium content than saprophytes. But there are also other factors that influence the accumulation and the transfer of radionuclides from the soil, such as the quantity of radionuclides deposited as a consequence of the fallout event, the chemical form of those radionuclides, and the soil's properties (pH, and organic matter and clay contents), among others.

Chapter X - Fungi play vital role in nutrient cycling by achieving process of decomposition of organic matter and breakdown of complex compounds in nature. This role of fungi coupled with their fundamental strategy of adaptation to various environmental factors may be used for designing systems to enable the elimination of biopolymers and man-made xenobiotics from the environment via biosorption. Soil contaminants like heavy metals, radionuclides, polyaromatic hydrocarbons-PAHs and chemicals used in agriculture that are toxic and carcinogenic agents could be diminished or removed by activity of fungal exoenzymes. In this review, special attention was paid to lignolityc enzyme system expressed by white-rot fungi recognized as successful agent in bioremediation of a large variety of chemicals that are, like lignin, relatively long lived in the environment. Furthermore, fungi are very important in natural cycling of metal ions due to their great accumulation potential for heavy metals (Pb, Hg, Cd) and radionuclides (^{137}Cs), implicating them as good bioindicators of the pollution in urban and industrial areas and in contaminated forest ecosystems. These processes in macrofungi are influenced by environmental factors like metal concentrations of soil and substrate, pH, organic matter and contamination by atmospheric deposition as well as fungal factors like fungal structure, biochemical composition, decomposition activity, development of mycelium and sporocarps or portion of fruiting body. Concentration of radionuclides in fungi is determined by the amount of radioactivity precipitation, concentration of stable - non radioactive or analogous element in soil, soil characteristics (mineral composition, pH) and its taxonomic and nutritional identity. Macrofungi maintain ecological balances used as bioindicators or as remediation agents of contaminated environment. Also, edibility and medicinal properties are of a great importance for humans while majority of edible and medicinal mushrooms can accumulate high amounts of heavy metals and other environmental pollutants. This chapter summarizes relevant biological features of fungi (position in tree of life, nutritional strategy, enzyme systems), especially lignicolous macrofungi (mostly white-rot), as a basic tool for resolving physiological, ecological and biotechnological potential of fungi in changing polluted environment thus restoring the natural environment.

Chapter XI - The reduction of vegetation cover has considerably increased. A lot of efforts are being undertaken to repair degraded lands. In African countries, exotic plants were established in this purpose to rehabilitate affected ecosystems. These exotic plantations play a key role in improving soils fertility partly because they are able to establish a symbiotic relationship with mycorrhizal fungi which promote plant development and allow better resistance of plant to stresses. Some of these reforestation programs were not successful in certain zones but this situation is overcome when these exotic trees are transferred together with appropriated mycorrhizal fungi from their zone of origin. Nevertheless, furthers studies carried out with exotic trees also showed that inoculation with local or more specific mycorrhizal fungi strains can strongly stimulate plant development. Therefore, it is important to find mycorrhizal fungi strains that are more beneficial for the establishment of exotic trees in foreign areas. In this chapter, we will expose i) the major reforestation programs carried out in arid and semi-arid African zones with exotic trees, discuss ii) the role of mycorrhizal symbiosis in their establishment and iii) in the mitigation of negative influence of these exotic trees in soil microbiota.

Chapter XII - Endophytic fungi are known to be abundant colonizers of the internal tissues of forest trees, but their ecological functions are still largely unknown. Recent studies indicate that endophytes may associate with tree's resistance and tolerance properties, and they are thus potential bio-agents that could be utilized in sustainable forest protection and management. To gain a better understanding of the endophytes' potential role in shaping forest health we need more evidence in the form of ecological studies of endophyte communities, in various tissues, across space, and time. The recent advances in molecular methods have given us new and effective tools to obtain such data. Studies of endophyte functions are further facilitated with the development of new high through-put screening methods for substrate use and competitive ability. Fungi are known as chemical factories of natural compounds with biological properties. Beside their potential as antagonists against pests and diseases, the tree-associated endophytic fungi therefore also appear as an emerging source of novel biomolecules for industrial or clinical applications outside forestry. This chapter presents some of the current methodological approaches that are likely to be valuable in studies on endophyte diversity in forest trees, and discusses the goals and impacts of the studies that aim at disentangling the beneficial potential of fungal endophytes in trees. A new concept, bioactive symbiosis, is suggested as a general framework for these studies.

Chapter XIII - Fungal diseases influence the consequences of competition between host and non-host plants and their co-existence, leading to changes in the species richness and diversity in forest ecosystems in which the hosts are embedded. Long-term studies of forest communities infested by serious fungal diseases have demonstrated the crucial importance of fungal pathogens in the maintenance of forest biodiversity. Here, I review long-term dynamics of plant populations and communities infested by fungal diseases and discuss the roles of fungal pathogens in the maintenance of the biodiversity of forest communities. Fungal diseases causing the death of host tissues at the shoot level, such as zonate leaf blight of dogwood and Swiss needle cast of Douglas-fir, have been shown to cause short-term changes in the shoot dynamics and primary productivity of individual host trees. However, the impacts of these diseases on long-term dynamics of host populations and forest communities remain to be investigated. In contrast, fungal diseases causing severe mortality of host populations, such as dogwood anthracnose, beech bark disease, Dutch elm disease, chestnut blight, phytophthora dieback, and laminated root rot, have been demonstrated to significantly impact the diversity and long-term dynamics of forest communities. Infestation by these fungal diseases leads to a decline of the host populations and replacement by other plant species, which can result in an increase, decrease, or undetectable changes in the number of plant species in forest communities. The response of forest diversity to diseases is variable and appears to depend on the virulence of fungal pathogens, relative abundance of hosts in forest communities, and environmental factors.

Chapter XIV - Fungi are a wide group of organisms, being free-living or parasitic on both plants and animals. Some of them can infect insects and nematodes, being these last ones called nematophagous or nematode trapping fungi (NTF).This behaviour can have practical applications, namely in the field of biological control of gastrointestinal nematodes, since they kill their exogenous larval stages. There is still a gap in the knowledge of their mode of action and their use in regular deworming programs associated with anthelmintics. The purposes of this chapter are to enrol the most important groups of fungi used in parasite control (namely the NTF), the way they interact with parasites and how they can be used in parasite control programs along the time. As an example of their practical application, data concerning the use of the nematophagous fungi *Duddingtonia flagrans* in the control of horse strongylosis in Portugal will be described. Field trials were performed in horse stud farms near Lisbon, evaluating the capacity of *D. flagrans* to reduce strongyle infection. Field work involved always two groups of horses with mixed strongyle infection: one receiving a daily oral dose of *D. flagrans* (D), while the control one (C) received a similar amount of food without fungus. The L3 were reduced significantly in faecal cultures and pastures of D groups, being more effective in warmer seasons, in both pasture and stabled horses. Therefore, *D. flagrans* can be easily associated with the classic control methods of strongylosis in a horse management system involving pasture and stabled animals.

Chapter XV - The capacity of microfungi known as 'nematophagous fungi' to reduce the numbers of parasitic nematodes is anything but a new subject. However, the research into the actual development of these fungi for biological control purposes and their practical application in veterinary medicine has made inroads only during the last two decades. This chapter intends to shed light on those inroads, focusing on the application of biological control of parasites of veterinary importance. Examples of parasite-destroying fungi other than nematophagous ones are also mentioned in the chapter. The four functional categories of nematophagous fungi are nematode-trapping, endoparasitic, egg- and cyst-parasitic fungi, and toxin-producing fungi. Although species of all categories have shown capabilities to act as biological control agents at different levels, the first group has been more extensively studied for that purpose to this date. However, the field for using the second and third groups remains wide open. While the last category is briefly mentioned and exemplified, this chapter focuses on the first three categories, including their description, biology, and behaviour in the environment. The research into using nematophagous fungi to control parasitic nematodes of veterinary importance has not only elucidated issues such as which fungal species are the most promising candidates and their individual nematode suppressive effect under controlled conditions, but has also raised problematic questions ranging from the use of the fungi in different geographical and climate regions to the formulation, commercialisation and regulatory aspects of their use. All these and the future perspectives of this exciting field of knowledge will be covered in this chapter.

Chapter XVI - Grazing livestock systems ensure nutrition of the animals, provide high quality food (milk, meat) and limit the presence of unwanted vegetation in an ecologically sustainable procedure. Pasturing practices can control grass competition for moisture, nutrients, and sunlight, thereby enhancing tree growth and the benefits from timber production. The presence of grazing animals provides organic fertilization so livestock manure recycles nutrients to trees and forage. Productivity of grazing animals can be lessened by pathogens as the gastrointestinal parasites. In infected animals, eggs from several nematode genera are passed in the feces and develop in the soil until the infective third larval stage (L3). Animals become infected when feedind the grass with the L3. Several chemical compounds exhibiting anthelmintic activity are commercially available for their control and frequently employed in farm animals. Some troubles related to low efficacy and/or emergence of anthelmintic resistance (AR) in different animal species has been widely reported. By considering the AR, together with chemotherapy affects only to the parasites in the animals, the importance for certain measures on the environment to reduce the presence of free living and infective stages to prevent the infection of the animals is emphasized. *Duddingtonia flagrans* is a nematode-trapping fungus very useful to capture nematodes from mycelia, due to its ability to produce a large quantity of chlamydospores able to resist the passage through the gastrointestinal tract of ruminants and monogastrics once they are orally administered into animals. The possibilities for reducing the frequence of administration of dewormers to livestock by using *D. flagrans* chlamydospores in horses under a pasturing management were evaluated. Equines received ivermectin and/or fungal spores. Chemotherapy was very successful, but the presence of infective stages in the environment conferred a marked temporal solution. When using chlamydospores of the *D. flagrans* nematode-trapping fungus, a high reduction of the infective stages (L3) from the pasture was noted. It is concluded that chemotherapy and the administration of spores prevents the infection in grazing horses and helps to decrease the frequency of deworming.

Chapter XVII - The gastrointestinal nematodes infections are the most frequent pathological processes in the small ruminant production systems and are considered to be the mainspring productive loss in small ruminants production systems based on pasture globally. For this reason, the implementation of control methods to support the parasitic population inside a level compatible with the economic production has been consolidated as a routine practice in the managing of livestock productions. During the last fifty years the control of the parasitic diseases and concretely the gastrointestinal nematodes, it has been based almost exclusively on the systematic use of antiparasitic drugs. Nevertheless as consequence of the improper or inadequate use of the chemical products, the anthelmintic resistance has stopped being an academic curiosity to turn into the principal threat for the developments of small ruminants throughout the world. The increasing prevalence and severity of anthelmintic resistance together with increased market demand for ecological or green animal products, has led to the search for alternative methods of parasite control. The new methods of gastrointestinal nematodes control should minimize the use of drugs allow to obtain quality products, guarantee the animal welfare and reduce the environment residue, contributing to the sustainability of the livestock systems.

Chapter XVIII - Overall helminthic infections are an old problem and a serious barrier for cattle ranching. Typically, large portion of nematodes in fecal environment passes from egg to larval stages, reaching the infective form. Nematodiosis control in general is accomplished through drugs known as anthelmintics. This control has been secular, i.e., it is culturally held in all parts of the world for long time. However, except experiences gained in the past, currently a major problem has been its indiscriminate and by wrong way use, which has been extensively described as a major responsible for parasitic resistance installation. Thus, the uses of nematophagous fungi have been tested under controlled conditions and in natural conditions to help fighting parasitic nematodes. They are divided into three groups, endoparasites, predators, and ovicidal. In practice, the "biological control" should be understood as the use of natural antagonists, in these particular case nematophagous fungi, to reduce the target organ, which is the nematode, to an acceptable level. For over 20 years, in South America, Brazil, Professor Jackson Victor de Araújo from Universidade Federal de Viçosa (Federal University of Viçosa) led studies shown that *Duddingtonia flagrans* species destroys nematodes populations of domestic animals in laboratory and natural tests. The future prospects for using a biological control agent consider basically be virulent, be efficient, present easy and cheap production, not to change the farmer's usual cultural practices for their implementation, be compatible with chemicals, be able to quickly colonize and persist in the soil, be effective for several host species to increase the interest of industrial companies to develop it commercially, not present deleterious effects on other organisms than the target, and maintain viability during storage without special care. Although a biocontrol agent with all these attributes is not yet available, there are researches for this purpose.

Chapter XIX - Ruminant parasitic gastrointestinal nematodes (GIN) are organisms responsible for severe economic losses affecting grazing ruminant industry all over the world. The most common method of control of these parasites is through the regular use of antiparasitic treatments in the form of commercial anthelmintic (AH) drugs. Nevertheless, the effectiveness of the AH products is dramatically declining due to the development of resistant worms that can survive the AH treatments. As a result, animals keep showing clinical signs of nematodosis and low productivity even after treatment with conventional AH drugs. Additionally, the use of conventional AH drugs have other disadvantages. The AH molecules can remain as residual contaminants in animal products intended for human consumption. Besides, some active molecules can be eliminated in the feces of treated animals and reach the soil leaving beneficial organisms of the soil under risk of destruction. There is an extraordinary group of microfungi living in the soil that is classified as natural antagonist of nematodes. The fungi in this group are able to form trapping devices to capture and destroy infective larvae of ruminant parasitic nematodes before they leave the feces and move to the field. Thus, they can act as biological control agents against the free living stages of GIN. In this chapter a general overview of nematophagous fungi and the perspectives of their use as tools for the biological control of GIN is presented. Special emphasis is directed to the results obtained by the group of Mexican researchers working with this interesting group of fungi mainly on its biotechnological potential for the control of GIN. Its successful application may help to reduce the risk of parasitic diseases affecting grazing ruminants around the world.

Chapter XX - White-rot fungi (WRF) are essentially basidiomycetes that exhibit singular biochemical abilities and potential applications. These fungi are efficient carbon recyclers in nature, can degrade lignin completely to carbon dioxide, act as ecological indicators and are important components of ecosystems. Their unique ability to degrade lignin is due to their capability to synthesize a set of oxidative (ligninolytic) extracellular enzymes, mainly peroxidases and laccases. The ligninolytic enzymes of WRF have broad substrate specificity and have been implicated in the transformation and mineralization of organopollutants, including pesticides, poly-aromatic hydrocarbons and dyes. The most studied WRF is

Phanerochaete chrysosporium. Also quite extensively studied are species of the genera *Trametes*, *Bjerkandera*, *Ganoderma* and *Pleurotus*. The biotechnological potential of ligninolytic enzymes has stimulated the search for other promising producers. The WRF biodiversity in temperate environments has been more widely studied than in tropical environments. This review focus especially on the biodiversity of WRF of tropical environments, mainly of Brazilian biomes, their impact on rain forest ecosystems, their economical impact as plant pathogens, their possible uses in green chemistry and the state of the art of research on this exceptional group of fungi.

Chapter XI - White-rot fungi (WRF), the most frequent wood-rotting organisms, are so far unique in their ability to completely degrade all components of our principal renewable resource, lignocellulose. They produce an impressive array of enzymes (cellulases, hemicellulases, pectinases and ligninases) with potential in a wide range of biotechnological applications. WRF and their ligninolytic enzymes (such as laccase and Mn-peroxidase) are able to degrade many pollutants. In pulp and paper manufacturing they are applied in biopulping, biobleaching and effluent treatment. Recently, there has been a growing interest in studying the ligninolytic enzymes of a wider array of WRF, with the expectation of finding better lignin-degrading systems for use in various biotechnological applications. To be able to use the white-rotters in these processes it is necessary to learn how their enzymes are secreted, how they operate and under what conditions they are active. Our current research is focused on these subjects and in the screening of new isolates for their ability to degrade xenobiotics, in the search of more efficient strains. Reducing the cost for enzyme production is still needed in order to develop enzymatic treatment processes for different industrial and environmental applications, competitive with other treatment technologies. At present, one of the objectives of our work is reducing this cost by means of process optimization using statistical experimental designs, and the use of cheaper growth substrates such as agricultural and food wastes. We identified new strains of *Trametes versicolor, Trametes villosa, Stereum hirsutum* and *Fomes sclerodermeus* which demonstrated an outstanding ability in dye degradation. *Trametes trogii* BAFC 463 almost completely removed several dyes as well as high priority pollutants such as polycyclic aromatic hydrocarbons and polychlorinated biphenyls. *Pycnoporus sanguineus* BAFC 2126 proved to be a potential candidate for use in softwoods biopulping processes. Efficient bleaching of loblolly pine kraft pulp could be achieved in a chlorine-free sequence, with culture supernatants from *T. trogii*, followed by a peroxide stage under mild conditions. Its high potential purified laccase, was recently applied as biocatalyst for oxygen electro-reduction in biofuel cells. With our results we wish to contribute to expand the knowledge on this important group of fungi and their biotechnological applications.

Chapter XII - Environmental pollution with hazardous xenobiotic compounds has become one of the major ecological problems world-wide. In addition, the implementation of more stringent environmental regulations, especially in the most developed countries, has impelled the necessity of finding innovative and environmentally-friendly treatment technologies to complement or substitute the conventional ones. White-rot fungi are the only micro-organisms able to degrade the whole wood components and are by far the most efficient ligninolytic micro-organism described to date. This ability to degrade lignin is related to the secretion of a non-specific and non-stereo selective extracellular enzymatic system. This enzymatic system consists mainly of oxidative enzymes like lignin peroxidases (LiP; E.C. 1.11.1.14), manganese-dependent peroxidases (MnP; E.C.1.11.1.13), laccases (EC

1.10.3.2) and hydrogen peroxide-generating oxidases. The same unique non-specific mechanisms that give these fungi the ability to degrade lignin also allow them degrading a wide range of recalcitrant pollutants such as polycyclic aromatic hydrocarbons (PAHs), pesticides, fuels, alkanes, polychlorinated biphenyls (PCBs), explosives and synthetic dyes. This chapter deals with the developments made for the use of white-rot fungi in the treatment of pollutants and their future perspectives.

Section 1: General Knowledge on Fungi: Physiology and Pathogenesis

In: Fungi: Types, Environmental Impact and Role in Disease
Editors: Adolfo Paz Silva and María Sol
ISBN: 978-1-61942-671-9

Chapter I

Gut Fungi: Classification, Evolution, Life Style and Application

***K. Fliegerova**[*1]**, J. Mrazek**[1] **and K. Voigt**[2]*

[1] Institute of Animal Physiology and Genetics,
Czech Academy of Sciences, v.v.i., Prague, Czech Republic

[2] Jena Microbial Research Collection at the Leibniz Institute for
Natural Product Research and Infection Biology and the University of Jena,
Institute of Microbiology, Jena, Germany

Abstract

Gut fungi represent a special anaerobic group of microscopic eukaryotes inhabiting the digestive tract of ruminant as well as many non-ruminant herbivores. Every trait of these fungi is singular and exceptional. Morphologically, gut fungi are the only representatives of kingdom Fungi demanding anaerobiosis, and therefore, possessing hydrogenosomes instead of mitochondria. Genetically, anaerobic fungi represent organisms with the lowest G+C content ever found in Nature and with duplication of a variety of important hydrolytic genes, which is a trait not common among eukaryotes. Enzymologically, anaerobic fungi possess cellulosomes, the enzyme factories known only in some bacteria, carrying the most active (hemi)cellulases. These special features have always complicated the determination and phylogenetic position of anaerobic fungi. Taxonomical problems have been resolved recently by establishment of the separate phylum as a result of reconstruction of the large group of flagellated fungi called chytridiales. The new phylum comprises a single class (Neocallimastigomycetes), one order (Neocallimastigales) and a single family (Neocallimastigaceae), which covers six genera of anaerobic fungi including *Neocallimastix*, *Piromyces*, *Caecomyces*, *Orpinomyces*, *Anaeromyces* and *Cyllamyces*. All species of anaerobic gut fungi secrete a variety of hydrolytic enzymes, including cellulases, xylanases, mannases, esterases, glucosidases, and glucanases, which effectively hydrolyze plant biomass consisting mainly from cellulose and hemicellulose. These enzymes exhibit higher activities than commercial enzyme products derived from *Trichoderma reeseii* or *Aspergillus niger*. The

[*] E-mail address: fliegerova@iapg.cas.cz.

capacity of extracellular plant cell-wall enzymes from anaerobic fungi to degrade cellulose, hemicellulose, and various lignocellulosic compounds indicates high potential feasibility of gut fungi hydrolases in commercial processes. Their experimental applications in livestock production, paper industry, biofuel and renewable energy production are discussed in this chapter.

1. Introduction

J. W. Foster declared in 1949 the opinion, that molds are highly oxidative and unable to metabolize carbohydrates anaerobically [1]. This statement known as mycological dogma has been probably the main reason of relatively late discovery of anaerobic fungi. Flagellated microorganisms observed in rumen fluid already at the beginning of last century were wrongly considered as protozoans [2, 3] and it took more than sixty years to determine them as fungi. The discovery, identification and early exploration of these microorganisms is closely related to the personality of the British scientist Colin Orpin, who recognized these cells as anaerobic fungi, described their major stages of life cycle [4], discovered chitin as the main structural polysaccharide of their cell walls [5], and classified the first species of anaerobic fungi [6-8].

Table 1. Survey of known strains of anaerobic fungi

Genus	Species	References
Neocallimastix	*frontalis*	[16]
Neocallimastix	*hurleyensis*	[17]
Neocallimastix	*patriciarum*	[8]
Neocallimastix	*variabilis*	[18]
Anaeromyces	*elegans*	[19]
Anaeromyces	*mucronatus*	[20]
Caecomyces	*communis**	[6]
Caecomyces	*equi*	[21]
Caecomyces	*sympodialis*	[22]
Cyllamyces	*aberensis*	[23]
Orpinomyces	*bovis*	[24]
Orpinomyces	*intercalaris*	[25]
Orpinomyces	*joyonii***	[26]
Piromyces	*citronii*	[27]
Piromyces	*communis*	[21, 24]
Piromyces	*dumbonica*	[28]
Piromyces	*mae*	[28]
Piromyces	*minutus*	[29]
Piromyces	*polycephalus*	[30]
Piromyces	*rhizinflata*	[31]
Piromyces	*spiralis*	[32]

* Originally described as *Sphaeromonas communis.*

** Originally described as *Neocallimastix joyonii.*

The taxonomical shift of anaerobic fungi from Phycomycetes [4] to Chytridiomycetes [9-14] has been driven both by their extraordinary morphology and rDNA composition. The final determination as Neocallimastigomycetes has been achieved by the modern methods of molecular biology based on multi-gene phylogeny of DNA sequences, mainly originating from the nuclear ribosomal RNA operon [15].

Thus thirty-two years after the discovery, the anaerobic fungi are classified into a separate phylum, the Neocallimastigomycota, which encompasses one class (Neocallimastigomycetes), one order (Neocallimastigales) and a single family, the Neocallimastigaceae [15]. Inside this family six genera of anaerobic fungi including *Neocallimastix*, *Piromyces*, *Caecomyces*, *Orpinomyces*, *Anaeromyces* and *Cyllamyces* have been determined up to now (Table 1).

However, recent research dealing with diversity of animal gut fungi indicates that the number of 6 genera and 21 species of anaerobic fungi is neither terminal nor most likely confined to ruminant and large herbivores only [33, 34].

2. Role of Gut Fungi

Gut fungi are symbiotic microorganisms occupying a unique ecological niche, the anaerobic environment of the rumen and gastrointestinal tract of large herbivores. In the digestive tract, fungi contribute, in association with bacteria and protozoa, to hydrolyzation of diet fiber resulting in the production of end fermentation products (volatile fatty acids) that can be utilized by host animal as source of nutrition. The complete set of enzymes necesary for plant cell-wall degradation provides anaerobic fungi to penetrate plant cell walls, access fermentable substrates, degrade and weaken plant tissue and reduce the size of diet particles. Released zoospores attach to the feed fragments, encyst them and germinate to produce a fungal thallus, which is composed of rhizoids and the sporangium. The rhizoid system can be highly branched and is able to penetrate deeply into stems and thus possibly aids to decompose substrate also mechanically [35]. Due to the capability to colonize the recalcitrant plant components like sclerenchyma and vascular system, gut fungi are known to decompose lignin-containing plant walls, even if they do not utilize phenolics or lignin [36]. Thus, the role of gut fungi seems to be lucid. However, the factual state, number, representation, distribution and the extent of their participation in the fermentation processes has not been yet fully clarified.

A study of Lee et al. [37] suggested that the contribution of the rumen fungi to cell wall degradation may greatly exceed the fermentation part of bacteria. It is believed that gut fungi are the very first colonizer of diet fiber, that enzymatically as well as physically open up the plant tissue for surface-acting bacteria. Rumen bacteria attach the phytomass more quickly; however, they preferentially colonize sites of damaged plant material. The anaerobic fungi and bacteria are probably in a complicated relation with each other, which may even be contraproductive. The inhibition of rumen fungi by some cellulolytic strains of *Ruminococcus albus* [38], *R. flavefaciens* [39-41], and *Butyrivibrio fibrisolvens* [42] was demonstrated by several *in vitro* studies. On the other hand, these results cannot be generalized because the type of interaction with fibrolytic bacteria depends on the content and diversity of the anaerobic fungal genera. Moreover, the interaction of anaerobic fungi with nonfibrolytic bacteria was found to be beneficial and a close positive association of gut fungi with methanogens is well known [43, 44]. The syntrophic co-cultures with *Archaea* increase amount of fungal biomass and markedly improve rate and extent of fungal cellulolysis and xylanolysis [45].

3. Special Features of Anaerobic Fungi

Anaerobic fungi differ considerably from all known members of the kingdom Fungi in many characteristics. This deviation from aerobic fungi is very probably caused by the specificity of their biotope, which is digestive tract of herbivores. The most apparent exceptional feature is the ability to cope with anaerobiosis and the inability to cope with aerobiosis resulting in the fact that mitochondria were replaced by hydrogenosomes during secondary rearrangements at the cellular level. Rumen or gut fungi are the only fungi that do not only need no oxygen for their life cycle, but oxygen is toxic for them. Anaerobic fungi cannot „breathe“, and therefore do not require mitochondria with their whole machinery of the respiratory chain. The energetic center of anaerobic fungi is represented by hydrogenosomes. These organelles, under anoxic conditions, decarboxylate malate into acetate, CO_2, and H_2 with concomitant production of energy in the form of ATP [46, 47]. Hydrogenosomes of gut fungi are unique and followed probably an alternative way to adapt to anaerobic environments. The available functional and phylogenetic evidence allows the conclusion that the hydrogenosomes of anaerobic fungi, trichomonads and ciliates are substantially different, and that they evolved independently from each other. Yarlett and Hackstein [48] suggested the hypothesis that the hydrogenosomes of anaerobic fungi evolved from fungal mitochondria, the hydrogenosomes of trichomonads from a hydrogen-producing protomitochondrion-like ancestral organelle, and the hydrogenosomes of *Nyctotherus ovalis* (hydrogenosome containing their own DNA) from the mitochondria of an aerobic ciliate retaining substantial feature of a classical mitochondrion. On the other hand, Benchimol et al. [49] demonstrated a sharp similarity between hydrogenosomes of anaerobic fungi and trichomonad protozoa based on observations of this organelle division. In consequence, Benchimol [50] suggests that hydrogenosomes are homologous organelles in unrelated species, weakens the hypothesis of their polyphyletic origin, and reinforces the hypothesis that fungal and trichomonad hydrogenosomes are derived from an ancestral endosymbiont preceded by a singular event of endosymbiosis during the course of evolution.

A phylogenetic tree which reconstructs the evolution of anaerobic gut fungi in relation to other related aerobic fungal lineages is presented in figure 1. Anaerobic gut fungi (marked in the rectangle) appear as monophyletic group (BP = 100 %).

Within this group the *Orpinomyces* clade contains the monocentric anaerobic species *Neocallimastix frontalis* and *Piromyces communis* (BP = 68 %), whereas *Anaeromyces* forms a well supported monophylum (BP = 78 %). The aerobic chytrids (indicated in bold font) forming a single opisthokont flagellum at the end of their zoospores appear to be the closest phylogenetic ancestors of the anaerobic gut fungi whose zoospores form more than a single flagellum. Thus, the presence of flagellum appears to be an important synapomorphic character for the group of zoosporic fungi as a whole taxon. The type of flagellation will be elucidated in the following section.

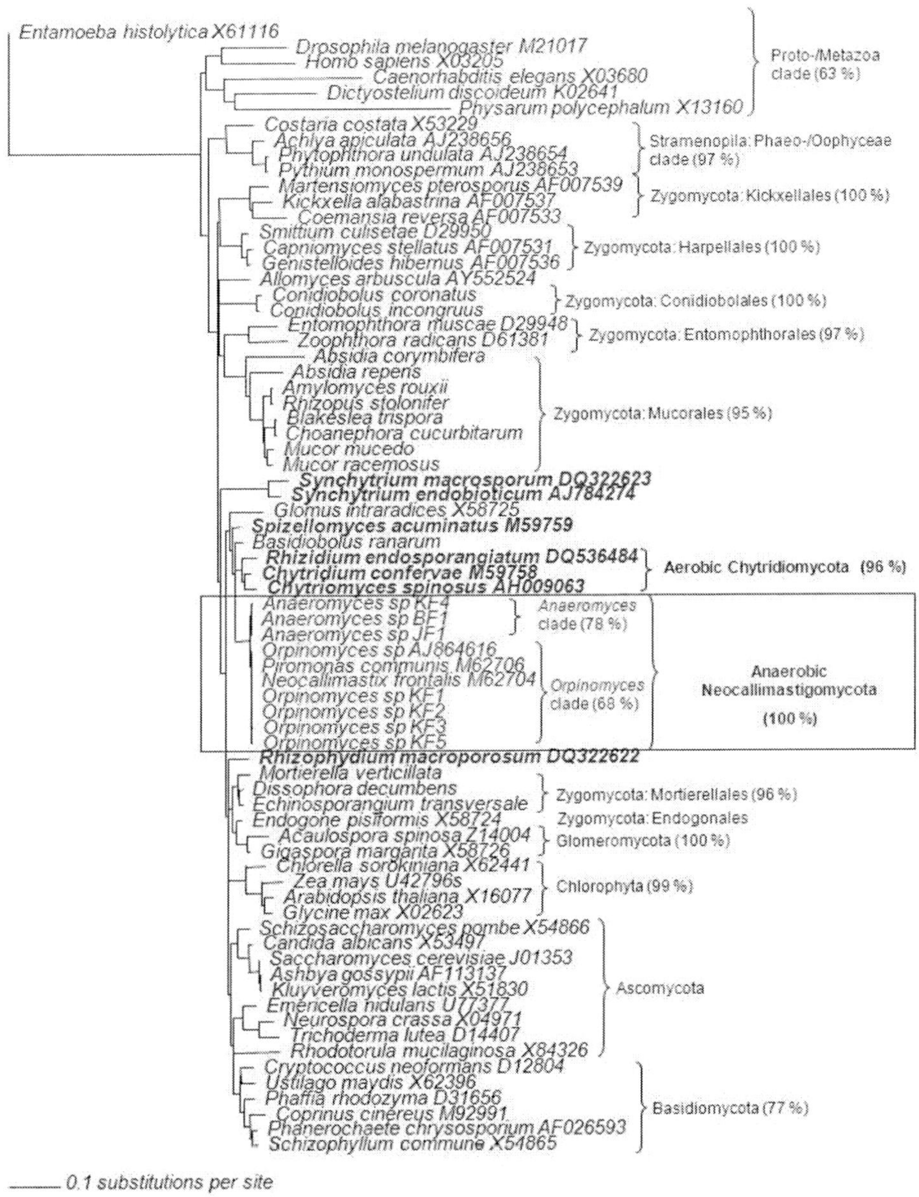

Figure 1. Neighbor-Joining phylogenetic tree of 589 aligned nucleotide characters of the 5′part of the 18S rDNA of 73 taxa. Numbers behind taxon names represent accession numbers as obtained from GenBank at www.ncbi.nlm.nih.gov. Numbers behind clade designations indicate bootstrap proportions (BP) of 1,000 neighbor-joining replicates of Jukes-Canor distances for measurement of the clade stability of each node in the phylogenetic tree.

Polyflagellate zoospores are another exceptional morphological feature of anaerobic fungi. Flagellate zoospores are responsible for asexual reproduction of Chytridiomycota, Blastocladiomycota and Neocallimastigomycota, however the first two phyla produce exclusively uniflagellate zoospores and only two genera of gut fungi produce polyflagellate motile cells. This singular feature is found in polycentric *Orpinomyces* and monocentric *Neocallimastix* species. Their zoospores have 7 – 20 flagella inserted in two rows. Zoospores of other anaerobic fungal genera are uniflagellate, however sometimes bi- and tri-flagellate cells can be observed. Flagellate fungi are assumed to form early-diverging clades within the kingdom Fungi, because the simple aquatic forms with flagellate spores are believed to be the fungal ancestor [51]. The loss of flagellum was certainly an important evolutionary event leading to the diversification of terrestrial fungi. From this point of view, Neocallimastigomycota are extremely interesting because multiflagellate zoospores indicate the outstanding and extraordinary evolutionary history of anaerobic fungi, which is, to date, not well-understood yet.

Not only morphological, but also genetic singularities are harboured by anaerobic fungi in secrecy. The percentual representation of guanine and cytosine bases ranging from 5 to 20 % in the genomic DNA of gut fungi exhibits the lowest GC content ever reported in any organism. This extreme nucleotide bias is reflected in both the coding and non-coding regions of the genome, with codon usage tending towards more AT-rich codons [52]. The non-coding regions are even more AT-rich with many sections expected to be near or above 95 mol% AT content [53-56]. The codon bias resulting from the low G+C content leads to a depletion in amino acids coded by GC rich codons (e.g. arginine, proline, glycine) and an overabundance of amino acids coded by AT rich codons (e.g. lysine, phenylalanine, tyrosine) influencing thus the amino acid composition of encoded proteins [57].

The main attractive feature of anaerobic fungi is, however, their excellent enzymatic system, which is connected with unique ability to organize hydrolases in large multienzyme complexes called cellulosomes. The Neocallimastigomycota is the only known member of the kingdom Fungi possessing cellulosomes and this can explain their cellulolytic superiority over aerobic cellulolytic fungi. These exocellular enzyme complexes are extremely active and can degrade both amorphous and crystalline cellulose [58]. Research indicates that cellulosomes of anaerobic fungi are similar to the most investigated cellulosome of bacteria *Clostridium thermocellum*, but on the other hand some important differences have been already proven. The composition of fungal cellulosome is similar to bacterial one including the nonenzymatic scaffolding protein(s) with enzyme binding site called cohesins and variety of cellulosomal enzymes with dockerins, which interact with the cohesins in the scaffolding protein [59, 60]. Fungal noncatalytic dockerin domain has function similar to bacterial dockerins [61], but no sequence homology to clostridial dockerins has been found [62-66]. Moreover, so far no scaffolding polypeptide has been isolated from a cellulosome of any anaerobic fungus, and there is no detailed knowledge regarding cohesins in scaffolding peptides from gut fungi. Fungal cellulosomes have been described up to now for *Neocallimastix*, *Piromyces*, and *Orpinomyces* species and molecular biological studies indicate that enzymes associated with the fungal cellulosomes from these three genera are modular like those of anaerobic bacteria.

4. Amazing Enzymes of Anaerobic Fungi

Anaerobic fungi synthetize a variety of hydrolytic enzymes, including cellulases, xylanases, mannases, esterases, glucosidases, and glucanases, which effectively hydrolyze plant biomass consisting mainly from cellulose and hemicellulose, the most abundant polysaccharides in the biosphere. Gut fungi decompose complicated structure of cellulose chain by several types of enzymes including endoglucanase (cellulase), which splits β-1,4-glycosidic bonds randomly within the cellulose chain, exoglucanases (cellobiosidase or cellobiohydrolases), which cleave cellobiose from the ends of the cellulose chain, and β-glucosidases (cellobiases), which convert cellobiose and other low molecular mass cellodextrins into glucose. Cellulolytic hydrolases of anaerobic fungi pertain according to amino acid sequences, hydrophobic clusters, and stereochemistry to the several different glycoside hydrolase (GH) families. β-glucosidases represent usually family GH1 or sometimes GH3, exoglucanases are registered in GH6. Endoglucanases are generally of family GH5, but some are classified in family 6. For *Piromyces* genus unusual cellulases of family 9 [54, 55], 45 [67-69] and 48 [70] have been discovered. Cellulases of family 5 from anaerobic fungi are intronless and their catalytic domains show homology with those of GH from several anaerobic bacteria. It has been suggested that the family 5 cellulase gene of anaerobic fungi was originally transferred from a ruminal bacterium and subsequently underwent gene duplication [52]. Cellulases of family 6 from anaerobic fungi are often intron-containing and this finding is considered to be a good evidence for fungal (eukaryotic) origin of this gene.

Also enzymatic conversion of xylan, the principal type of hemicellulose, to its monomeric components requires the participation of several enzymes, which are effectively secreted by anaerobic fungi. The main hydrolase is xylanase (endoxylanase) cleaving β-1, 4-xylosidic linkage randomly within the main xylan chain and xylosidase, which removes successive D-xylose residues from the non-reducing termini. The ester bonds on xylan and the side chains are hydrolyzed by feruloyl acid esterases, acetylxylan esterases, arabinases, or α-glucuronidases depending on the type of branching group. Xylanases of anaerobic fungi belong usually into GH family 11, rarely to family 10. Other types of hemicellulases like licheninase (1, 3-1, 4-β-D-glucan 4-glucanohydrolase) and mannanase are registered in GH families 16 and 26, respectively. Esterases of gut fungi are also very various pertaining to carbohydrate esterase (CE) families 1, 2, 3 and 6.

In connection with xylan degradation, the fate of pentose's yielding from the hydrolysis of hemicellulose should be mentioned. Also in this metabolic conversion the anaerobic fungi have chosen the unusual way, which has been elucidated only recently in *Piromyces* sp. [71]. Enzymatic and molecular analysis of this gut fungus provided evidence that xylose is converted to xylulose by xylose isomerase and xylulose is phosphorylated by D-xylulokinase to xylulose-5-phosphate, which is a key intermediate in pentose metabolism. This metabolic route resembles situation in bacteria, because in the majority of yeasts and fungi xylose is converted via xylose reductase and xylitol dehydrogenase to xylulose, which is subsequently phosphorylated [72, 73]. Xylose isomerase from several strains of anaerobic fungi has been intensively studied with respect to its biotechnological exploitation in ethanol production as mentioned below in chapter regarding application.

All genera of anaerobic fungi degrade cellulose and hemicellulose, but the range of their hydrolysation depends on the degrees of crystallinity. Moreover gut fungi exhibit genus-dependent differences in their activity towards (hemi)celluloses and recalcitrant substrates. *Neocallimastix* and *Piromyces* sp. are supposed to be the most active genera [74-76], while *Caecomyces* sp. the least effective genus [77, 78]. Except for this set of (hemi-)cellulolytic enzymes, anaerobic fungi secrete numbers of other less important enzymes including disaccharidases, amylases and amyloglycosidases, pectin lyase, chitinase and proteases. The range of enzymes was found to be similar for anaerobic fungi isolated from the foregut and hindgut of nonruminant herbivores and for strains from forestomach of ruminants [79]. Some of these enzymes act individually and are free in solution, whereas others are constituents of large (hemi)cellulase multienzyme complexes mentioned above as cellulosomes. Enzymes organized in cellulosomal complexes containing endo- and exo-cellulases, hemicellulases and esterases have been described for *Piromyces*, *Orpinomyces* and *Neocallimastix* species. Another remarkable feature of anaerobic fungi insist in containing more than one copy of genes encoding (hemi)-cellulases in their genome. This gene duplications result in clusters of almost identical genes arranged head-to-tail on the genome. In contrast to other eukaryotes, this phenomenon appears frequently in anaerobic fungi [55]. Genome of anaerobic fungi thus can serve as an excellent model for study of horizontal gene transfer. Sequence comparison of broad range of hydrolases produced by anaerobic fungi including catalytic as well as non-catalytic domains indicates both bacterial and eukaryotic (fungal or even animal) origin of these enzymes.

5. Potential Application

The broad range of potent polysaccharide-degrading enzymes of anaerobic fungi has always attracted the attention of scientific and biotechnological community. The ambitions of exploitation of the effective hydrolytic facilities of gut fungi entered several fields, especially food, animal feed, brewery, textile, paper and biofuel industries. The applications naturally face many problems and the difficulties to set up continuous-flow cultures to produce anaerobic fungal enzymes represent one of the biggest limitation. Processes using immobilized growing cells seem to be more promising than traditional fermentations with free cells. The effort to immobilize anaerobic fungal mycelium and/or zoospores in alginate beads has been recently reported by several authors for monocentric *Piromyces* sp. [80, 81] and *Caecomyces* sp. [82] and polycentric *Orpinomyces* [80, 81] and *Anaeromyces* sp. [83]. These studies suggest that immobilization may be more suited for polycentric anaerobic fungi, because mycelial growth on the surface observed for monocentric *Piromyces* is undesirable effect leading to the growth in culture liquor and represents major problem that has to be overcome before immobilized fungi can find wide use in the fermentation industry. Also immobilization of enzymes has not been without problems. Immobilized proteins had difficulties to accommodate small substrate and had a lower rate of hydrolysis as has been shown by Mesta et al. [84] trying to immobilize chimeric endoxylanase from *N. frontalis* on metal-chelate matrix. Interesting results have been achieved during the expression of recombinant xylanase gene from *Neocallimastix patriciarum* in Canola seeds [85]. Oil-bodies extract from transgenic seed exhibited xylanase activity and immobilized enzyme could be recycled by floatation several times without loss of activity. Another experimental streams directed to enhanced xylose and cellulose degradation are represented by efforts to induce changes in catalytic domain of enzymes of anaerobic fungi and/or to produce hydrolytic enzymes of anaerobic fungi in microorganisms already used in commercial processes.

The livestock production is the industrial field with high interest in application of fibrolytic enzymes of anaerobic fungi to improve the efficiency of feed utilization. The possible use of rumen fungi as probiotics has been already studied on sheep, buffalos, and cows [86-89]. Direct supplementation of *Orpinomyces* sp., *Piromyces* sp. or *Anaeromyces* sp. cultures improved animal growth rate and feed efficiency. Administration of fungal cultures increased digestibility of cell-wall contents and crude protein, rumen fermentation parameters were improved, and the average daily weight gain was significantly higher. On the other hand the application of anaerobic fungal enzyme has not induced the response obtained with administration of live fungal culture. Another research is directed to use the ruminal fungi features for non-ruminant animals, especially economically important swines and poultry. Cereals, the main part of monogastric feed, contain in the cell walls hardly digestible nonstarch polysaccharides like β-glucan in barley and wheat, and arabinoxylans in rye and oats. The antinutritive effect of these compounds for monogastric animals is well known and unwanted [90]. Anaerobic fungi are able to split these hemicellulosic complexes and it has been already proved that application of their glycoside hydrolases to barley or wheat diets increases the growth of animals. Addition of fungal β-glucanase to the basic barley diet resulted even in 25% increase in growth of broiler chicks [91]. The high-priced production of enzymes from anaerobic fungi has induced the genetic engineering attempts to introduce xylanase and cellulase genes of gut fungi into the genome of bacterium *Lactobacillus reuteri*, which is natural component of microflora of broiler digestive tract [92, 93].

The great interest in enzymes of anaerobic fungi has been manifested by paper industry. Fungal cellulases and mainly xylanases are desired as environmental friendly substances to reduce the amount of alkalies and chlorines used for treatment of paper pulp. Xylanases of anaerobic fungal strain *Orpinomyces* have decreased the lignin content in paper stuff, increased the brightness of paper and positively affected the further bleaching with ozone [94]. Cellulases of the same genus have been used for deinking of recycled mixed office paper and significantly reduced both dirt count and residual ink area and increased brightness of recycled paper [95]. The pulp bleaching entails high temperatures and high pH and therefore genetic engineering aspires to construct thermophilic xylanase stable in alkaline environments. For this reason the catalytic domain of a xylanase from *Neocallimastix patriciarum* has been changed to be more alkalophilic [96] through directed evolution using error-prone PCR. The resulting composite mutant xylanase was alkalic tolerant maintaining 25% of its activity at pH 9. Moreover the mutant xylanase was more thermostable at $60^{O}C$ [97].

Enzymes of anaerobic fungi are involved also in tests dealing with fermentative production of ethanol from lignocellulosic residues. The wild - type of *Saccharomyces cerevisiae*, which is the most used industrial microbial strain for ethanol production, cannot metabolize xylose and arabinose, two important pentoses resulting from hemicellulose hydrolysis. The genetic engineering is therefore concentrated to modify this yeast to be able to metabolize xylan or at least monosaccharide xylose. Efforts with incorporation of rumen fungal xylanases [98, 99] or cellobiohydrolases [100] into different kinds of aerobic fungi have not been surprisingly as successful as introduction of xylose isomerase, the pentose cycle enzyme of gut fungi. Strain of *Sacharomyces cerevisiae* genetically modified to change xylose into xylulose using the isomerase of anaerobic fungi *Piromyces* and/or *Orpinomyces* [101-103] represents at this time the most promising species for industrial production of ethanol [104].

The unworn involvement of rumen fungi into the biogas production has been inspired by growing demands for renewable energy. The possibility of *Anaeromyces* and *Piromyces* strains to integrate into biogas-producing anaerobic sludge bacterial ecosystem, to improve degradation of substrate polysaccharides and consequently to influence methane production has been already tested in laboratory conditions [105]. Even if small batch cultures have indicated positive influence of anaerobic fungi on biogas production, the application of this approach in commercial full-scale biogas does not seems feasible at this moment. However, the detection of gut fungi in landfill samples published by Lockhart et al. [106], the only report of the presence of anaerobic fungi outside the gastrointestinal animal tract, indicates the unraveled and unsuspected opportunities of these special, exceptional, but not easily graspable fungi with excellent enzymatic properties.

Acknowledgments

This work was supported by the Czech Science Foundation (project no. P503/10/P394) and the Deutsche Forschungsgemeinschaft (project no. Vo772/7-1). The research on anaerobic fungi is part of 'Ruminomics' project (no. 289319) supported by the EC's Seventh Framework Programme: Food, Agriculture, Fisheries and Biotechnology.

References

[1] Foster, J.W. *Chemical activities of the fungi.* New York: Academic Press; 1949.

[2] Liebetanz, E. (1910). Die parasitischen protozoen der wiederkauermagens. *Archiv für Protistenk,* 19, 19-90.

[3] Braune, R. (1913). Untersuchungen über die in Wiederkauermagen vorkommenden Protozoen. *Archiv für Protistenk,* 32, 111-170.

[4] Orpin, C.G. (1975). Studies on the rumen flagellate *Neocallimastix frontalis. Journal of General Microbiology,* 91, 249-62.

[5] Orpin, C.G. (1977). The occurrence of chitin in the cell walls of the rumen organisms *Neocallimastix frontalis, Piromonas communis* and *Sphaeromonas communis. Journal of General Microbiology,* 99, 215-218.

[6] Orpin, C.G. (1976). Studies on the rumen flagellate *Sphaeromonas communis. Journal of General Microbiology,* 94, 270-80.

[7] Orpin, C.G. (1977). The rumen flagellate *Piromonas communis*: its life, history and invasion of plant material in the rumen. *Journal of General Microbiology,* 99, 107-117.

[8] Orpin, C.G. and Munn, E.A. (1986). *Neocallimastix patriciarum*: new member of the Neocallimasticaceae inhabiting the sheep rumen. *Transactions of the British Mycological Society,* 86, 178-181.

[9] Barr, D.J.S. (1980). An outline for the classification of the Chytridiales, and for a new order, the Spizellomycetales. *Canadian Journal of Botany,* 58, 2380-2394.

[10] Barr, D.J.S. (1988). How modern systematics relates to the rumen fungi. *BioSystem,* 21, 351-356.

[11] Dore, J. and Stahl, D.A. (1991). Phylogeny of anaerobic rumen Chytridiomycetes inferred from small subunit ribosomal RNA sequence comparison. *Canadian Journal of Botany,* 69, 1964-1971.

[12] Bowman, B.H., Taylor, J.W., Brownlee, A.G., Lee, J., Lu, S.D. andWhite, T.J. (1992). Molecular evolution of the fungi: relationship of the Basidiomycetes, Ascomycetes, and Chytridiomycetes. *Molecular Biology and Evolution,* 9, 285-296.

[13] Li, J. and Heath, J.B. (1992). The phylogenetic relationship of the anaerobic chytridiomycetous gut fungi (Neocallimasticaceae) and the Chytridiomycota. I. Cladistic analysis of rRNA sequences. *Canadian Journal of Botany,* 70, 1738-1746.

[14] Li, J., Heath, J.B. and Packer, L. (1993). Phylogenetic relationships of the anaerobic chytridiomycetous gut fungi (Neocallimasticaceae) and the Chytridiomycota. II. Cladistic analysis of structural data and description of Neocallimasticales ord. nov. *Canadian Journal of Botany,* 71, 393-407.

[15] Hibbett, D. S., Binder, M., Bischhoff, J. F. and 64 others. (2007). A higher-level phylogenetic classification of the Fungi. *Mycological Research,* 111, 509-547.

[16] Heath, I.B., Bauchop, T. and Skipp, R.A. (1983). Assignement of the rumen anaerobe *Neocallimastix frontalis* to the Spizellomycetales (Chytridiomycetes) on the basis of its polyflagellate zoospore ultrastructure. *Canadian Journal of Botany,* 61, 295-307.

[17] Webb, J. and Theodorou, M.K. (1991). *Neocallimastix hurleyensis* sp. nov., an anaerobic fungus from the ovine rumen. *Canadian Journal of Botany,* 69, 1220-1224.

[18] Ho, Y.W., Barr, D.J.S., Abdullah, N., Jalaludin, S. and Kudo, H. (1993). *Neocallimastix variabilis,* a new species of anaerobic fungus from the rumen of cattle. *Mycotaxon,* 46, 241-258.

[19] Ho, Y.W., Barr, D.J.S., Abbdulah, N., Jalaludin, S. and Kudo, H. (1993). *Anaeromyces,* an earlier name for *Ruminomyces. Mycotaxon,* 47, 283 – 284.

[20] Breton, A., Bernalier, A., Dusser, M., Fonty, G., Gaillard-Martinie, B. and Guillot, J. (1990). *Anaeromyces mucronatus* nov. gen. nov. sp. A strictly anaerobic rumen fungus with polycentric thallus. *FEMS Microbiology Letters,* 70, 177-182.

[21] Gold, J.J., Heath, I.B. and Bauchop, T. (1988). Ultrastructural description of a new chytrid genus of caecum anaerobe, *Caecomyces equi* gen. nov., sp. nov., assigned to the Neocallimasticaceae. *BioSystems, 2*1, 403-415.

[22] Chen, Y.C., Tsai, S.D., Cheng, H.L., Chien, C.Y., Hu, C.Y. and Cheng, T.Y. (2007). *Caecomyces sympodialis* sp. nov., a new rumen fungus isolated from *Bos indicus. Mycologia,* 99, 125-130.

[23] Ozkose, M., Thomas, B.J., Davies, D.R., Griffith, G.W. and Theodorou, M.K. (2001). *Cyllamyces aberensis* gen. nov., a new anaerobic gut fungus with branched sporangiophores isolated from cattle. *Canadian Journal of Botany,* 79, 666-673.

[24] Barr, D.J.S., Kudo, H., Jakober, K.D. and Cheng, K.J. (1989). Morphology and development of rumen fungi: *Neocallimastix* sp., *Piromyces communis,* and *Orpinomyces bovis* gen. nov., sp. nov. *Canadian Journal of Botany,* 67, 2815-2824.

[25] Ho, Y.W., Abdullah, N. and Jalaludin, S. (1994). *Orpinomyces intercalaris,* a new species of polycentric anaerobic fungus from cattle. *Mycotaxon,* 50, 139-150.

[26] Breton, A., Bernalier, A., Bonnemoy, F., Fonty, G., Gaillard, B. and Gouet, P. (1989). Morphological and metabolic characterization of a new species of strictly anaerobic rumen fungus: *Neocallimastix joyonii. FEMS Microbiology Letters,* 58, 309-314.

[27] Gaillard-Martinie, B., Breton, A., Dusser, M. and Julliand, V. (1995). *Piromyces citronii* sp. nov., a strictly anaerobic fungus from the equine cecum: a morpological, metabolic, and ultrastructural study. *FEMS Microbioogy Letters,* 130, 321-326.

[28] Li, J., Heath, J.B. and Bauchop, T. (1990). *Piromyces mae* and *Piromyces dumbonica*, two new species of uniflagellate anaerobic chytridiomycete fungi from the hindgut of the horse and elephant. *Canadian Journal of Botany,* 68, 1021-1033.

[29] Ho, Y.W., Barr, D.J.S., Abdullah, N., Jalaludin, S. and Kudo, H. (1993). A new species of *Piromyces* from the rumen of deer in Malaysia. *Mycotaxon,* 47, 285-293.

[30] Chen, Y.C. and Hseu, R.S. (2002). *Piromyces polycephalus* (Neocallimastigaceae), a new rumen fungus. *Nova Hedwigia,* 75, 409-414.

[31] Breton, A., Dusser, M., Gaillard-Martinie, B., Guillot, J., Millet, L. and Prensier, G. (1991). *Piromyces rhizinflata* nov. sp., a strictly anaerobic fungus from feces of the saharan ass: a morphological, metabolic and ultrastructural study. *FEMS Microbiology Letters,* 82, 1-8.

[32] Ho, Y.W., Barr, D.J.S., Abdullah, N., Jalaludin, S. and Kudo, H. (1993). *Piromyces spiralis*, a new species of anaerobic fungus from the rumen of goat. *Mycotaxon,* 48, 59-68.

[33] Nicholson, M.J., McSweeney, C.S., Mackie, R.I., Brookman, J.L. and Theodorou, M.K. (2010). Diversity of anaerobic gut fungal populations analyzed using ribosomal ITS1 sequences in faeces of wild and domesticated herbivores. *Anaerobe,* 16, 66-73.

[34] Liggenstoffer, A.S., Youseff, N.H., Couger, M.B. and Elshaded, M.S. (2010). Phylogenetic diversity and community structure of anaerobic gut fungi (phylum Neocallimastigomycota) in ruminant and non-ruminant herbivores. *ISME Journal,* 4, 1225-1235.

[35] Orpin, C.G. Anaerobic fungi: taxonomy, biology and distribution in nature. In: Mountford D, Orpin C editors. *Anaerobic fungi: biology, ecology and function.* New York, USA: Marcel Dekker Inc.; 1994; 1- 46.

[36] Akin, DE. Ultrastructure of plant cell-wals degraded by anaerobic fungi. In: Mountford D, Orpin C editors. *Anaerobic fungi: biology, ecology and function.* New York, USA: Marcel Dekker Inc.; 1994; 169-190.

[37] Lee, S.S., Ha, J.K. and Cheng, K.J. (2000). Relative contributions of bacteria, protozoa, and fungi to in vitro degradation of orchard grass cell walls and their interaction. *Applied and Environmental Microbiology,* 66, 3807-3813.

[38] Stewart, C.S., Duncan, S.H., Richardson, A.J., Backwell, C. and Bebie, R. (1992). The inhibition of fungal cellulolysis by cell-free preparation from ruminococci. *FEMS Microbiology Letters,* 97, 83-88.

[39] Bernalier, A., Fonty, G., Bonnemoy, F. and Gouet, P. (1993). Inhibition of cellulolytic activity of *Neocallimastix frontalis* by *Ruminococcus flavefaciens. Journal of General Microbiology,* 139, 873-880.

[40] Roger, V., Bernalier, A., Grenet, E., Fonty, G., Jamot, J. and Gouet, P. (1993). Degradation of wheat straw and maize stem by a monocentric and polycentric rumen fungi, alone or in association with rumen cellulolytic bacteria. *Animal Feed Science and Technology,* 42, 69-82.

[41] Ha, J.K., Lee, S.S., Kim, S.W., Han, I.K., Ushida, K. and Cheng, K.J. (2001). Degradation of rice straw by rumen fungi and cellulolytic bateria through mono-, co- or sequential cultures. *Asian-Australasian Journal of Animal Sciences,* 14, 797-802.

[42] Joblin, KN. 1997. Interaction between ruminal fibrolytic bacteria and fungi. In: Onodera R, Itabashi H, Ushida K, Yano H, Sasaki Y editors. *Rumen microbes and digestive physiology in ruminants.* Basel: S. Karger; 1997; 3-10.

[43] Orpin, G.C; Joblin, K.N. The rumen anaerobic fungi. In: Hobson PN, Stewart editors. *The rumen microbial ecosystem.* London, United Kingdom: Elsevier; 1997; 140-195.

[44] Cheng, Y.F., Edwards, J.E., Allison, G.G., Zhu, W.Y. and Theodorou, M.K. (2009). Diversity and activity of enriched ruminal cultures of anaerobic fungi and methanogens grown together on lignocellulose in consecutive batch culture. *Bioresource Technology,* 100, 4821-4825.

[45] Marvin-Sikkema, F.D., Richardson, A.J., Stewart, C.S., Gottschal, J.C. and Prins, R.A. (1990). Influence of hydrogen-consuming bacteria on cellulose degradation by anaerobic fungi. *Applied and Environmental Microbiol*ogy, 56, 3793-3797.

[46] Müller, M. (1993). Hydrogenosome. *Journal of General Microbiology, 139*, 2879-2889.

[47] van der Giezen, M. (2009). Hydrogenosomes and mitosomes: conservation and evolution of functions. *Journal of Eukaryotic Microbiology,* 56, 221-231.

[48] Yarlett, N. and Hackstein, J.H.P. (2005). Hydrogenosomes: one organelle, multiple origins. *Bioscience,* 55, 657-668.

[49] Benchimol, M., Durand, R. and Almeida, J.C.D. (1997). A double membrane surrounds the hydrogenosomes of the anaerobic fungus *Neocallimastix frontalis. FEMS Microbiology Letters,* 154, 277-282.

[50] Benchimol, M. (2009). Hydrogenosomes under microscopy. *Tissue Cell,* 41, 151-168.

[51] James, T.Y., Kauff, F., Schoch, C.L. and 66 others. (2006). Reconstructing the early evolution of fungi using a six-gene phylogeny. *Nature,* 443, 818-822.

[52] Garcia-Vallvé, S., Romeu, A. and Palau, J. (2000). Horizontal gene transfer of glycosyl hydrolases of the rumen fungi. *Molecular Biology and Evolution,* 17, 352-361.

[53] Brookman, J.L., Mennim, G., Trinci, A.P.J., Theodorou, M.K. and Tuckwell, D.S. (2000). Identification and characterization of anaerobic gut fungi using molecular methodologies based on ribosomal ITS1 and 18S rRNA. *Microbiology,* 146, 393-403.

[54] Steenbakkers, P.J.M., Ubhayasekera, W., Goossen, H.J.A.M., van Lierop, E.M.H.M., van der Drift, C., Vogels, G.D., Mowbray, S.L. and Op Den Camp, H.J.M. (2002). An intron-containing glycoside hydrolase family 9 cellulase gene encodes the dominant 90 kDa component of the cellulosome of the anaerobic fungus *Piromyces* sp. strain E2. *Biochemical Journal,* 365, 193-204.

[55] Harhangi, H.R., Akhmanova, A.S., Steenbakkers, P.J.M., Jetten, M.S.M., van der Drift, C. and Op den Camp H.J.M. (2003). Genomic DNA analysis of genes encoding (hemi)cellulolytic enzymes of anaerobic fungus *Piromyces* sp. E2. *Gene,* 314, 73-80.

[56] Nicholson, M.J., Theodorou, M.K. and Brookman, J.L. (2005). Molecular analysis of the anaerobic rumen fungus *Orpinomyces*: insight into an AT-rich genome. *Microbiology-SGM*, 151, 121-133.

[57] Brownlee, A.G. (1989). What determines the amino acid composition of rumen microbes? *Asian-Australasian Journal of Animal Sciences,* 2, 441-443.

[58] Bayer, E.A., Belaich, J.P., Shoham, Y. and Lamed, R. (2004). The cellulosomes: multienzyme machines for degradation of plant cell wall polysaccharides. *Annual Review of Microbiology*, 58, 521-554.

[59] Doi, R.H. (2008). Cellulases of mesophilic microorganisms: cellulosome and noncellulosome producers. *Annals of the New York Academy of Sciences,* 1125, 267-279.

[60] Fontes, C.M.G.A. and Gilbert, H.J. (2010). Cellulosomes: highly efficient nanomachines designed to deconstruct plant cell wall complex carbohydrates. *Annual Review of Biochemistry,* 79, 655-681.

[61] Ljungdahl, L.G. (2008). The cellulase/hemicellulase system of the anaerobic fungus *Orpinomyces* PC-2 and aspects of its applied use. *Annals of the New York Academy of Sciences,* 1125, 308-321.

[62] Qiu, X., Selinger, B., Yanke, L.J. and Cheng, K.J. (2000). Isolation and analysis of two celulase cDNA from *Orpinomyces joyonii. Gene*, 245, 119-126.

[63] Aylward, J.H., Gobius, A.S., Xue, G.P., Simpson, G.D. and Dalrymple, B.P. (1999). The *Neocallimastix patriciarum* cellulase, CelD, contains three almost identical domains with high specific activities on Avicel. *Enzyme Microbiology and Technology,* 24, 609-614.

[64] Fujino, Y., Ogata, K., Nagamine, T. and Ushida, K. (1998). Cloning, sequencing, and expression of an endoglucanase gene from rumen anaerobic fungus *Neocallimastix frontalis* MCH3. *Bioscence Biotechnology and Biochemistry,* 62, 1795-1798.

[65] Li, X.L., Chen, H., Ljungdahl, L.G. (1997). Two cellulases, CelA and CelC, from the polycentric anaerobic fungus *Orpinomyces* strain PC-2 contain N-terminal docking domains for a cellulase-hemicellulase complex. *Applied and Environmental Microbiology,* 63, 4721-4728.

[66] Millward-Sadler, S.J., Hall, J., Black, G.W., Hazlewood, G.P. and Gilbert, H.J. (1996). Evidence that the *Piromyces* gene family encoding endo-1,4-mannases arose through gene duplication. *FEMS Microbiology Letters,* 141, 183-188.

[67] Eberhardt, R.Y., Gilbert, H.J. and Hazlewood, G.P. (2000). Primary sequence and enzymatic properties of two modular endoglucanases, Cel5A and Cel45A, from the anaerobic fungus *Piromyces equi. Microbiology,* 146, 1999-2008.

[68] Raghothama, S., Eberhardt, R.Y., Simpson, P., Wigelsworth, D., White, P., Hazlewood, G.P., Nagy, T., Gilbert, H.J. and Williamson, M.P. (2001). Characterization of a cellulosome dockerin domain from the anaerobic fungus *Piromyces equi. Nature Structural Biology,* 8, 775-778.

[69] Nagy, T., Tunnicliffe, R.B., Higgins, L.D., Walters, C., Gilbert, H.J. and Williamson, M.P. (2007). Characterization of double dockerin from the cellulosome of the anaerobic fungus *Piromyces equi. Journal of Molecular Biology*, 373, 612-622.

[70] Steenbakkers, P.J.M., Freelove, A., van Cranenbroek, B., Sweegers, B.M.C., Harhangi, H.R., Vogels, G.D., Hazlewood, G.P., Gilbert, H.L. and Op Den Camp, H.J.M. (2002). The major component of the cellulosomes of anaerobic fungi from the genus *Piromyces* is a family 48 glycoside hydrolase. *DNA Sequence,* 13, 313-320.

[71] Harhangi, H.R., Akhmanova, A.S., Emmens, R., van der Drift, C., de Laat, W.T.A.M., van Dijken, J.P., Jetten, M.S.M., Pronk, J.T. and Op den Camp, H.J.M. (2003). Xylose metabolism in the anaerobic fungus *Piromyces* sp. strain E2 follows the bacterial pathway. *Archives of Microbiology,* 180, 134-141.

[72] Jeffries, T.W. (1983). Utilization of xylose by bacteria, yeasts, and fungi. *Advances in Biochemical Engineering-Biotechnology,* 27, 1-32.

[73] Jin, Y.S. and Jeffries, T.W. (2003). Changing flux of xylose metabolites by altering expression of xylose reductese and xylitol dehydrogenase in recombinant *Saccharomaces cerevisiae. Applied Biochemistry and Biotechnology,* 105, 277-285.

[74] Thareja, A., Puniya, A.K., Goel, G., Nagpal, R., Sehgal, J.P., Singh, P.K. and Singh, K. (2006). *In vitro* degradation of wheat straw by anaerobic fungi from small ruminants. *Archives of Animal Nutrition,* 60, 412-417.

[75] Tripathi, V.K., Sehgal, J.P., Puniya, A.K. and Singh, K. (2007). Hydrolytic activities of anaerobic fungi from wild blue bull (Boselaphus tragocamelus). *Anaerobe,* 13, 36-39.

[76] Paul, S.S., Kamra, D.N. and Sastry, V.R.B. (2010). Fermentative characterictics and fibrolytic activities of anaerobic gut fungi isolated from wild and domestic ruminants. *Archives of Animal Nutrition,* 64, 279-292.

[77] Nielsen, B.B., Zhu, W.Y., Dhanoa, M.S., Trinci, A.P.J. and Theodorou M.K. (2002). Comparison of the growth kinetics of anaerobic gut fungi on wheat straw in batch culture. *Anaerobe,* 8, 216-222.

[78] Joblin, K.N., Matsui, H., Naylor, G.E. and Ushida, K. (2002). Degradation of fresh rey grass by methanogenic co-cultures of ruminal fungi grown in the presence or absence of *Fibrobacter succinogenes. Current Microbiology,* 45, 46-53.

[79] Williams, A.G., Withers, S.E. and Orpin, C.G. (1994). Effect of the carbohydrate growth substrate on polysaccharolytic enzyme formation by anaerobic fungi isolated from the foregut and hindgut of nonruminant herbivores and the forestomach of ruminants. *Letters in Applied Microbiology,* 18, 147-151.

[80] McCabe, B.K., Kuek, C., Gordon, G.L.R. and Phillips, M.W. (2001). Immobilization of monocentric and polycentric types of anaerobic chytrid fungi in Ca-alginate. *Enzyme and Microbial Technology*, 29, 144-149.

[81] McCabe, B.K., Kuek, C., Gordon, G.L.R. and Phillips, M.W. (2003). Production of β-glucosidase using immobilized *Piromyces* sp. KSX1 and *Orpinomyces* sp. 478P1 in repeat-batch culture. *Journal of Industrial Microbiology and Biotechnology,* 30, 205-209.

[82] Nagpal, R., Puniya, A.K. and Singh, K. (2010). In vitro activity of the anaerobic fungus *Ccaecomyces* sp. immobilized in alginate beads. *Journal of Animal and Feed Sciences,* 18, 758-768.

[83] Sridhar, M. and Kumar, D. (2010). Production of fibrolytic enzymes in repeat-batch culture using immobilized zoospores of anaerobic rumen fungi. *Indian Journal of Biotechnology,* 9, 87-95.

[84] Mesta, L., Heyraud, A., Joseleau, J.P. and Coulet, P.R. (2003). Catalytic properties of endoxylanase fusion proteins from *Neocallimastix frontalis* and effect of immobilization onto metal-chelate matrix. *Journal of Biotechnology,* 101, 253-265.

[85] Liu, J.H., Selinger, L.B., Cheng, K.J., Beauchemin, K.A. and Moloney, M.M. (1997). Plant seed oil-bodies as an immobilization matrix for a recombinant xylanase from the rumen fungus *Neocallimastix patriciarum. Molecular Breeding,* 3, 463-470.

[86] Mamen, D., Vadivel, V., Pugalenthi, M. and Parimelazhagan, T. (2010). Evaluation of fibrolytic activity of two different anaerobic rumen fungal isolates for their utilization as microbial feed additive. *Animal Nutrition and Feed Technology,* 10, 37-49.

[87] Sehgal, J.P., Jit, D., Puniya, A.K. and Singh, K. (2008). Influence of anaerobic fungal administration on growth, rumen fermentation and nutrient digestion in female buffalo calves. *Journal of Animal and Feed Sciences,* 17, 510-518.

[88] Dey, A., Sehgal, J.P., Puniya, A.K. and Singh, K. (2004). Influence of an anaerobic fungal culture (*Orpinomyces* sp.) administration on growth rate, ruminal fermentation

and nutrient digestion in calves. *Asian-Australasian Journal of Animal Sciences,* 17, 820-824.

[89] Samanta, A.K., Singh, K.K., Das, M.M. and Pailan, G.H. (2008). Effect of direct fed anaerobic fungal culture on rumen fermentation, nutrient utilization and live weight gain in crossbred heifers. *Indian Journal of Animal Sciences,* 78, 1134-1137.

[90] Bedfort, M.R. and Schulze, H. (1998). Exogenous enzymes for pigs and poultry. *Nutrition Research Reviews,* 11, 91-114.

[91] Azain, M.J., Li, X.L., Shah, A.K. and Davies, T.E. (2002). Separation of the effects of xylanase and β-glucanase addition on performance of broiler chicks fed barley based diets. *International Poultry Science Forum,* Abstract 46.

[92] Liu, J.R., Yu, B., Liu, F.H., Cheng, K.J. and Zhao, X. (2005). Expression of rumen microbial fibrolytic enzyme genes in probiotic *Lactobacillus reuteri. Applied and Environmental Microbiol*ogy, 71, 6769-6775.

[93] Liu, J.R., Yu, B., Zhao, X. and Cheng, K.J. (2007). Coexpression of rumen microbial beta-glucanase and xylanase genes in *Lactobacillus reuteri. Applied Microbiology and Biotechnology,* 77, 117-124.

[94] Shah, A.K., Cooper, A. and Adolphson, R.B. (2000). Use of an extremely high specific activity xylanase in ECF and TCF pulp bleaching. *Tappi Journal,* 83, 1-12.

[95] Geng, X.L. and Li, K.C. (2003). Deinking of recycled mixed office paper using two endoglucanases, CelB and CelE, from the anaerobic fungus *Orpinomyces* PC-2. *Tappi Journal,* 2, 29-32.

[96] Liu, J.H., Selinger, L.B., Tsai, C.F. and Cheng, K.J. (1999). Characterization of a *Neocallimastix patriciarum* xylanase gene and its product. *Canadian Journal of Microbiology,* 45, 970-974.

[97] Chen, Y.L., Tang, T.Y. and Cheng, K.J. (2001). Direct evolution to produce an alkalophilic variant from a *Neocallimastix patriciarum xylanase. Canadian Journal of Microbiology,* 47, 1088-1094.

[98] Tsai, C.T. and Huang, C.T. (2008). Overexpression of the *Neocallimastix frontalis* xylanase gene in the methylotrophic yeasts *Pichia pastoris* and *Pichia methanolica. Enzyme and Microbial Technology,* 42, 459-465.

[99] Li, X.L., Skory, C.D., Ximenes, E.A., Jordan, D.B., Dien, B.S., Hughes, S.R. and Cotta, M.A. (2007). Expression of an AT-rich xylanase gene from the anaerobic fungus *Orpinomyces* sp. strain PC-2 in and secretion of the heterologous enzyme by *Hypocrea jecorina. Applied Microbiology and Biotechnology,* 74, 1264-1275.

[100] van Wyk, N., den Haan, R. and van Zyl, W.H. (2010). Heterologous production of NpCel6A from *Neocallimastix patriciarum* in *Saccharomyces cerevisiae. Enzyme and Microbial Technology,* 46, 378-383.

[101] Kuyper, M., Hartog, M.M.P., Toirkens, M.J., Almering, M.J.H., Winkler, A.A., van Dijken, J.P. and Pronk, J.T. (2005). Metabolic engineering of a xylose-isomerase-expressing *Saccharomyces cerevisae* strain for rapid anaerobic xylose fermentation. *FEMS Yeast Reseasrch,* 5, 399-409.

[102] van Maris, A.J.A., Winkler, A.A., Kuyper, M., de Laat, W.T.A.M., van Dijken, J.P. and Pronk, J.T. (2007). Development of efficient xylose fermentation in *Saccharomyces cerevisiae*: xylose isomerase as a key component. *Advances in Biochemical Engineering-Biotechnology,* 108, 179-204.

[103] Madhavan, A., Tamalampudi, S., Ushida, K., Kanai, D., Katahira, S., Srivastava, A., Fukuda, H., Bisaria, V.S. and Kondo, A. (2009). Xylose isomerase from polycentric fungus Orpinomyces: gene sequencing, cloning, and expression in *Saccharomyces cerevisiae* for bioconversion of xylose to methanol. *Applied Microbiology and Biotechnology,* 82, 1067-1078.

[104] Bellissimi, E., van Dijken, J.P., Pronk, J.T. and van Marias, A.J.A. (2009). Effects of acetic acid on the kinetics of xylose fermentation by an engineered, xylose-isomerase-based *Saccharomyces cerevisiae* strain. *FEMS Yeast Research,* 9, 358-364.

[105] Fliegerova, K; Prochazka, J; Mrazek, J; Novotna, Z; Strosova, L; Dohanyos, M. Biogas and rumen fungi. In: Litonjua et al. editors. *Biogas: production, consumption and applications*. Hauppauge, New York, USA: Nova Science Publisher Inc; 2011; in press.

[106] Lockhart, R.J., Van Dyke, M.I., Beadle, I.R., Humphreys, P. and McCarthy, A.J. (2006). Molecular biological detection of anaerobic gut fungi (Neocallimastigales) from landfill sites. *Applied and Environmental Microbiol*ogy, 72, 5659-5661.

In: Fungi: Types, Environmental Impact and Role in Disease ISBN: 978-1-61942-671-9
Editors: Adolfo Paz Silva and María Sol

Chapter II

Carbohydrate Uptake and Catabolism in Fungi. Importance of Sugar Trafficking during Plant-Host Interactions

Walter A. Vargas[1,*] ***and Charles M. Kenerley***[2]
[1]Centro Hispano Luso de Investigaciones Agrarias, CIALE.
Universidad de Salamanca, Salamanca, Spain
[2]Department of Plant Pathology and Microbiology Texas A&M University, TX, US

Abstract

Fungal species have evolved different strategies to obtain and degrade resources available in their natural environments and/or their hosts. In general, these compounds can act as a carbon source or signaling molecules indicating the presence of the host in the surroundings. These molecules have been extensively studied as they contribute to host specificity and determine regulatory aspects of the host-fungus interaction. The knowledge and understanding of the metabolic traits and regulatory aspects of host-derived metabolites represent an interesting subject for research leading to the identification of novel targets for fungal disease control and management.

In this chapter, the most relevant findings on filamentous fungi carbohydrate metabolism are reviewed. This review also provides a comparative analysis of carbohydrate metabolism in symbiotic, pathogenic and saprophytic species including sugar uptake and further degradation through glycolysis and tricarboxylic acid pathway. Even though fungal species interacting with plants are the main subject of discussion, the human pathogens *Candida albicans* and *Aspergillus fumigatus* are also included in the analysis.

[*]E-mail: wvargas@usal.es.

1. Introduction

In order to grow and thrive in a wide range of hosts, pathogens and saprophytic fungal species have evolved an extremely flexible metabolism that enables them to assimilate diverse carbon sources. The ability to use a wide range of substrates increases the likelihood of successful colonization under varying conditions [1].

Fungal pathogenic species colonize a broad spectrum of hosts such as plants, insects and animals, and in some cases more than one host. The success of host colonization not only depends on the pathogenic intruder, but also on the metabolic status and the physiological conditions of the host. For instance, a biotrophic fungal species that infects a plant with greater levels of soluble sugars finds a valuable source of energy that can easily be acquired [2]. In contrast, a necrotrophic fungal species that infects decaying plant tissue mostly will find structural and complex polymers, which require an intricate degradation process [3]. The last example to mention, is the case of hemibiotrophic species where the same fungal hypha start with a biotrophic lifestyle (taking advantage mostly of plant metabolites) and develops later the necrotrophic stage to feed on dead and decaying tissue (using mostly structural components) [2]. These variations in lifestyle and cell metabolism demand a greater metabolic flexibility on the intruder and a high capacity to respond to changing nutritional conditions.

Strategies for successful fungal infections rely on the ability of fungi to respond quickly to environmental changes imposed by the host. These strategies include the establishment of different transport systems in the plasma membrane for the uptake of organic compounds (mainly carbohydrates), which is regulated by the composition and the concentration of these compounds in the host cells.

Most sugars can be metabolized by the glycolytic pathway into the key metabolite pyruvate. After that, pyruvate can enter fermentation or respiration pathways for energy production. Although both pathways produce NADH, respiration is energetically more efficient as it produces greater amounts of ATP after complete oxidation of organic compounds into CO_2 (through tricarboxilic acid cycle and oxidative phosphorylation).

Most of our understanding of carbohydrate degradation, respiration and fermentation in eukaryotes is based on the non-pathogenic ascomycete *Saccharomyces cerevisiae*. Fermentative yeasts, such as *S. cerevisiae*, usually inhabit glucose-rich environments, such as grapes and other fruits, and fermentation is stimulated by high glucose concentrations (even under aerobic conditions) while respiration is suppressed.

In contrast, filamentous fungi develop a preference for respiratory metabolism and a high-efficiency in ATP production. Moreover, different fungi will present varying metabolic rates and substrate preferences based on the their ecological niche and the strategies they employ for host infection.

Based on the host, and the colonizing strategies, fungi have developed alternative metabolic pathways and enzymatic machinery for the optimal use of resources. Plants represent an abundant source of carbohydrates and other compounds to support fungal growth. In the last decade, many reports have revealed metabolic aspects of fungi that support their success in plant colonization. Mycorrhizal species rely on enzymes from the plant cells to degrade sucrose and use the resulting monosaccharides as a source of energy and signals to regulate the symbiotic interaction [4-7].

A different scenario occurs in other symbiotic, pathogenic or saprophytic fungal species. In general, these species have evolved specific sucrose transporters and intracellular sucrose-hydrolyzing enzymes, or extracellular hydrolytic enzymes and hexose transporters to uptake the resulting monosaccharide.

These transporters also are available to take advantage of glucose released upon cell-wall degradation by the plant host, either as the result of the colonization strategy or as the primary source of carbon.

2. Carbohydrate Metabolism in Plants

In plants, carbohydrates play vital roles in cell physiology and molecular biology. Carbohydrates are synthesized *de novo* after the reductive CO_2 fixation through the Calvin cycle inside chloroplasts. After fixation, the anaplerotic pathways lead to the synthesis of more complex molecules in different compartments in the plant cell.

2.1. CO_2 Fixation and Carbohydrate Metabolism

An extensive and detailed literature on CO_2 fixation and carbohydrate biosynthesis is available. In this chapter, a summary of the most relevant aspects of plant carbohydrate fixation and carbon partitioning is presented.

CO_2 fixation through the Calvin cycle occurs in chloroplasts stroma resulting in triose phosphate molecules that are exported to the cytosol to enter further metabolic pathways.

The reductive reactions of photosynthesis are energized by the energy captured from light in the photosystems and converted into ATP and NADPH molecules. Thus, the chloroplasts provide all the basic machinery to support the autotrophic activities.

In plants, two of the most abundant forms of carbohydrates are sucrose and cellulose. These two compounds are key components for plant physiology and vital importance during plant-pathogen interactions.

2.2. Sucrose

Sucrose is the main soluble sugar involved in transport of carbon and energy from source leaves to heterotrophic tissues. Sucrose is a soluble disaccharide (α-D-glucopyranosyl-(1→2)-β-D-fructofuranoside) accumulated in the vacuole during the day and used during the night to support plant metabolism. This disaccharide plays important roles in plant cells; it is involved in gene expression regulation, stress tolerance mechanisms, and sugar-mediated signal transduction [8, 9].

For many decades sucrose metabolism has been widely studied in higher plants, and the roles sucrose plays in plants are tightly controlled by the synthesis and degradation rates [10] . Sucrose degradation can be initiated by the action of two types of enzymes, invertases that hydrolyze the molecule into glucose and fructose, or by sucrose synthases that cleave sucrose into fructose and UDP-glucose.

Sucrose utilization in plants is initiated in the cytosol, organelles and in the apoplast [11]. Its hydrolysis in the apoplast plays a vital role during plant-microbe interactions, being the main source of monosaccharides to support microbial growth.

Additional studies have revealed that the ratio of hexose to sucrose in the apoplast is an important cue in plant defense signaling [12].

2.3. Cellulose

Cellulose is one of the most abundant polymers in Nature as it is a key component of the cell walls of higher plants [13]. Plant cell walls are complex and elusive structures.

Their intricate composition has precluded biochemical studies on biosynthesis, assembly and even degradation. Plant cell walls are mainly composed of cellulose, a β-1, 4-linked glucose polymer that form microfibrils of both crystalline and non-crystalline structures.

In addition to cellulose, plant cell walls also contain other components such as glucans, lignin and glycoproteins arranged in a complex matrix that challenges biochemical and structural studies [13, 14].

However, it is well-known that β-1, 4 glucosidases are key components in cellulose degradation and important hydrolytic enzymes to reduce cellulose into glucose units [14, 15].

3. Carbohydrate Metabolism in Fungal Species

Above, the two most abundant carbon- and energy- rich components in plants were summarized in an introductory description for further discussion of their metabolism by fungal species.

When microorganisms encounter plants, they find an abundant source of nutrients to exploit. However, the ability to use these resources depends on metabolic components that allow the microorganisms to attack and degrade plant metabolites and/or structural components [14].

During the initial phase of fungal colonization, only small molecules can be translocated across the plasma membrane through specific transporters [1].

In the presence of more complex and high-molecular weight compounds, the fungal cells release hydrolytic enzymes to digest these plant molecules into their minimal units for incorporation into cell metabolism. In general, complex carbohydrates (such as cellulose) are reduced to their component monosaccharides, which are transported into fungal cells through hexose transporters located into the cell membrane [14].

In contrast, two alternative strategies have evolved to use sucrose. The disaccharide can be hydrolyzed by extracellular fungal invertases yielding fructose and glucose, which are transported by fungal hexoses carriers.

Alternatively, recent findings demonstrated that some fungi are able to uptake the disaccharide and hydrolyze the molecule by intracellular invertases [16, 17].

3.1. Degradation of Cellulose

Cellulose degradation has been studied in several microbes, including fungi, due to its industrial importance. However, in this chapter the degradation of cellulose is approached from the metabolic and physiological point of view for fungal cells.

Cellulose represents one important source of energy for saprophytic organisms. However, the use of this polysaccharide as a nutrient is limited for heterotrophic organisms due to the intricate structure of plant cell-walls. Microbes, specially a wide diversity of fungi, are considered the primary agents involved in the full-degradation of plant components.

To take full advantage of carbon and energy content of cellulose, fungi first break down the complex matrix of the plant cell-walls. Then, cellulolytic enzymes reduce cellulose into smaller oligosaccharides and glucose, which are further metabolized through glycolysis and respiration.

Several enzymes participate in cellulose degradation. The concomitant action of two different cellulose-degrading enzymes, an endoglucanase or cellulase (E.C.3.2.1.4) and cellobiohydrolases (EC3.2.1.91), is essential for cellulose breakdown and further metabolism. In the fungus *Trichoderma reesei* the expression of an extracellular cellulase system is induced upon culture in the presence of cellulose, but repressed when easily metabolizable sugars are available [18]. Species of *Trichoderma* colonize intercellular spaces of the host, and the initiation of host colonization includes the secretion of hydrolytic enzymes for degradation of plant cell-walls [19].

Genomic surveys, to identify homologs of cellulases and cellobiohydrolases in several fungi illustrated a correlation between the presence of cellulose-hydrolyzing enzymes, host-type, and colonization strategies (Table 1). In general, species capable of developing a necrotrophic lifestyle (*Colletotrichum graminicola* and *Botritis cinerea*) present both cellulase and cellobiohydrolase homologs in their genomes.

This pattern of cellulolytic enzymes is also evident in saprophytic strains able to grow on plant species (*Aspergillus fumigatus* and *Fusarium verticilliodes*). The presence of cellulose degrading enzymes in all these species, reflect the ability of these fungi to colonize and scavenge cellulose as a source of energy. However, *Ustilago maydis* and *Laccaria bicolor*, two species that develop intracellular biotrophic life styles in plants lack homologs to cellulose-degrading enzymes. This appears to be correlated with the concept that during biotrophic development metabolically active plant cells provide the nutrients for fungal growth and development [1, 2].

Also, the absence of these genes was apparent in the genome sequence of *S. cerevisiae* or *Candida albicans*, two ascomycete strains that ferment soluble sugars or infect animals, respectively. In contrast, *Trichoderma* species that could be characterized as biotrophic-like stains contain the hydrolytic machinery to degrade cellulose (Table 1) [19, 20]. Conservation of cellulases in *Trichoderma* species may be attributed to the strategy for host root colonization and the ability of these fungal species to colonize dead tissue and scavenge on plant litter [21].

The common aspect of all cellulose-degrading species is that these organisms are able to live at expenses of plant-derived cellulose, which provides an enormous source of carbon and energy to support fungal growth. However, in the case of *Trichoderma* the metabolic scenario seems more complex, as cellulose degradation is also part of the colonization mechanism.

Table 1. Genomic survey to identify putative invertases, cellulases, sucrose transporters and glucose transporters in fungal strains. *S. cerevisiae* invertase SUC2, *T. reesie* Cel2, *T. virens* TvSut and *T. harzianum* Gtt1 served as query for, invertases, sucrose transporters, glucose transporters and cellulases, respectively

	Invertases		Cellulases		Suc transporters	Glucose trasnporters	
	Gene ID	SS	Gene ID	SS	Gene ID	Gene ID**	
U. maydis	UM01945	19	none		UM02869	>20 homologs	UM03895
	UM03605	none			UM01281		
					UmSrt		
T. virens	Trive21923	none	Trive44205	24	Trive02154	>70 homologs	Trive76958
			Trive90504	18	Trive43166		
S. cerevisiae	Sacce1604	19	none		none	>30 homologs	Scce0873
C. graminicola	GLRG06686	none	6 homologs	21*	GLRG00135	>70 homologs	GLRG10740
	GLRG10490	none			GLRG04447		
	GLRG11332	none			GLRG08173		
F. verticillioides	FVEG12976	none	FVEG01770	18	FVEG01171	>100 homologs	FVEG09829
	FVEG08584	none	FVEG01037	17	FVEG14050		
	FVEG02083	15	FVEG10897	17			
B. cinerea	BC1T10247	18	BC1T06035	17	BC1T03637	>50 homologs	BC1T08389
	BC1T16010	24	BC1T14702	18			
			BC1T10880	16			

	Invertases		Cellulases		Suc transporters	Glucose trasnporters	
	Gene ID	SS	Gene ID	SS	Gene ID	Gene ID**	
L. bicolor	none		none		none	20 homologs	Lacbi304755
A. fumigatus	Afu2g01240	17	Afu6g01800	20	Afu5g06210	>75 homlogs	Afu3g14170
	Afu5g00480	22	Afu7g01540	18			
	Afu5g00530	21	Afu6g11610	26			
			Afu6g07070	17			
C. albicans	none		none		none	23 homologs	orf19.3668

SS, cleavage position of the putative secretion signal.

* Position of the secretion signal for the homolog with the highest similarity to the query.

* * Gene ID corresponding to the homolog with highest similarity to the query.

3.2. Degradation of Sucrose

One of the most abundant disaccharides in aquatic and terrestrial environments is sucrose, which is photosynthetically produced by plants, algae and cyanobacteria [22]. Due to the widespread production of sucrose, heterotrophic organisms have developed the ability to use sucrose as a carbon and energy source.

Fungal species colonizing plants, or environments with increased sucrose concentrations, may use sucrose not only to support growth but also to control host invasion. Sucrose degradation in fungal species may be achieved extracellularly or inside the cell, by extracellular or intracellular acid invertases, respectively. Acid invertases are β-fructofuranosidases belonging to the glycoside hydrolase family GH32 that catalyzes the hydrolysis of the disaccharide into glucose and fructose for further metabolism. In general, fungal invertases proved to be specific for β-fructofuranosides with the highest affinities for sucrose and raffinose. Both sugars act as transport forms of assimilates in plants, which should render primary targets for sucrose hydrolyzing enzymes of fungal origin during plant colonization [9, 23].

Invertases have been described in many fungal species including saprophytic, symbiotic, biotrophic and necrotrophic species. When a necrotroph attacks plants, dead plant cells are colonized as the hyphae scavenge for sources of carbon (including sucrose) to support their growth. However, for symbiotic species or those displaying biotrophic lifestyles, the ability to use sucrose may serve a dual function, as the disaccharide may be used as a nutrient or a signal to control the interaction [24, 25].

Invertases have been described in several pathogenic fungi such as *U. maydis*, *Uromyces fabae*, *B. cinera*, *C. graminicola*, *Aspergillus niger* and *Thermomyces lanuginosus* [26-30]. In the case of *U. fabae*, the expression of the fungal invertase was detected upon penetration of the fungus into the leaf with high expression levels in the haustoria. The study of this pathosystem suggested that the pathogenic infection established new sinks that competed with physiological sink organs in the host.

During biotrophic interactions of *U. maydis*, the secretion of a soluble invertase (UmSuc2) was expected to supply monosaccharides for hyphal growth favoring hexose uptake by the pathogen over hexose retrieval by host cells [29]. Genomic analysis revealed the presence of two genes encoding invertases in *U. maydis* genome Um01945 (UmSuc2) and Um03605. Sequence comparisons revealed that, in contrast to UmSuc2, Um03605 lacks any secretion signal and is expected to encode a putative intracellular invertase. Intracellular sucrose degradation in this fungal species has been demonstrated by the presence of invertase activity in cytoplasmic extracts of *U. maydis,* and the expression of a membrane-located sucrose carrier in the fungal cells during biotrophic development [17, 31]. However, the importance of Um03605 in *U. maydis* has not been tested. This redundancy of intracellular and extracellular sucrose metabolism points to an intricate regulatory network controlling the sucrose/hexose ratio in the host cells. A possible outcome of such metabolic network is a depletion of sucrose at the infection site, which increases the strength of the sink and the demand for photoassimilates to feed the fungal cells.

In the necrotrophic species *B. cinerea* three populations of acid invertases have been described in axenic cultures [26]. The authors reported the characterization of soluble, salt-soluble and tightly wall-bound fractions of acid invertases during saprophytic growth. The possibility of a precursor/product relationship between the soluble and tightly wall-bound compounds, with the salt-soluble fraction representing an intermediate of a vesicular transport form, was suggested. However, sequence analysis of *B. cinerea* genome revealed the presence of two genes encoding putative acid invertases (BC1T10247 and BC1T16010), with BC1T10247 displaying a putative signal peptide for secretion (Table 1). The expression of invertases and subcellular localization in this necrotroph is expected to be more complex than previously thought. The presence of invertase activity in *B. cinerea* cells proved to be constitutive, independent of the growth stage or cultural conditions, and the importance of this enzyme during plant pathogenesis has not been resolved.

The causal agent of anthracnose in maize plants, *C. graminicola*, also displays an array of complex carbohydrate metabolic activity that includes the expression and accumulation of invertases in the conidial spore matrix [32]. The presence of an invertase enzyme in spores was related to spore viability and protection. However, the recent release of the genome sequence of *C. graminicola* allows the identification of three genes encoding putative acid invertases lacking secretion signals (Table 1). Functional and biochemical studies will be necessary to understand the importance of invertases in *C. graminicola* and the role they play during anthracnose development in maize plants.

Acid invertases have been also identified and biochemically characterized in plant-symbiotic species. The soilborne and filamentous fungus *Trichoderma virens* accumulates a highly active intracellular invertase (TvInv) when colonizing plant roots or growing saprophytically in the presence of sucrose [24]. *Thrichoderma* species have been studied due to their biocontrol properties and importance for agriculture. Based on the results recently reported, sucrose may play two important functions in *T. virens*. First, as a signal molecule involved in the control plant root colonization, and secondly as a source of carbon and energy in the rhizosphere. The expression of *tvinv* not only contributes to the intracellular degradation of sucrose and further metabolism, but also grants the fungal cells with a signaling path likely to regulate host colonization.

The presence of invertases in plant-infecting fungal species provides fungal cells access to one of the most abundant compounds from plant cells. Moreover, the genomic complexity in various fungal species reflects the relevance that sucrose metabolism acquired in different fungal species, independently of their hosts or lifestyles. For instance, in *T. virens* the invertase enables the fungal cells to degrade sucrose exuded by plant roots. The capture of sucrose from the soil environment by fungal cells, suggests that the disaccharide is being depleted form the rhizosphere and a reduced source of carbon will be available for other soil microorganisms. This could also be interpreted as part of the strategy *T. virens* has evolved to compete with other microbes in the soil.

Carbon uptake by *T. virens* appears to be a complex mechanism and directly affects the outcome of root colonization and regulation of gene expression. When sucrose degradation is impaired in *T. virens* cells, greater levels of extracellular hydrolytic enzymes, including a β-1,4 glucanase, are produced. Interestingly, this deregulation of the hydrolytic enzymes is accompanied by an increased rate of host colonization. This phenomenon suggests that, as part of the mechanisms for the control of root colonization, the products of sucrose hydrolysis exert carbon catabolic repression on genes encoding cell-wall hydrolytic enzymes. The catabolic repression due to intracellular sucrose degradation establishes a metabolic network for a fine-tuning control of plant cell-wall degradation. Thus, when the hyphae of *T.* virens are growing inside the host, intracellular sucrose degradation contributes to regulate the production of cell-wall hydrolytic enzymes (Figure 1).

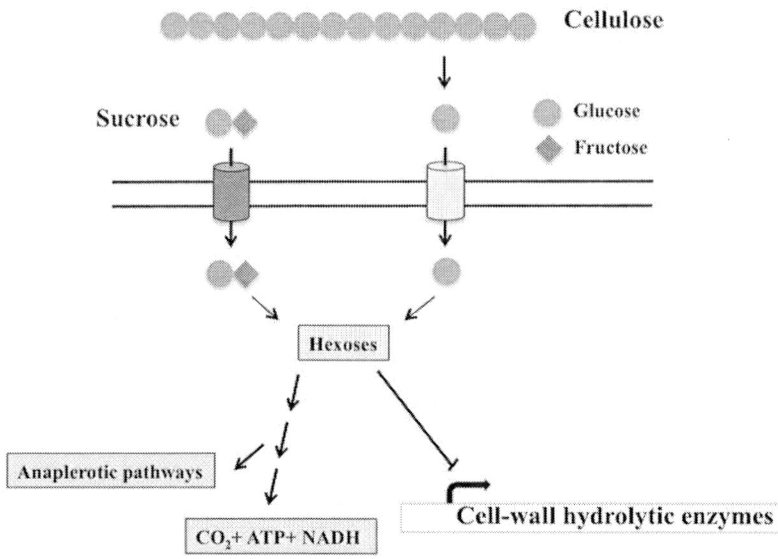

Figure 1. Schematic representation illustrates carbon flux from the plant towards *Trichoderma virens* cells. The figure describes glucose production form cellulose and the uptake of sucrose for further intracellular metabolism. The fate of hexoses is illustrated in the production of energy and reducing equivalents, or derived into the anaplerotic pathways. The scheme also exemplifies the carbon catabolic repression that glucose exerts on the expression of hydrolytic enzymes.

The global regulation of gene expression depending on sucrose metabolism would be of vital importance to control the interaction between both intruder and host cells. For instance, in the case of *T. virens* a feedback control exerted by sucrose metabolism on the production of hydrolytic enzymes may contribute to avoiding excessive host cell-walls degradation that could neglect the symbiotic interaction. Sequence searches of the genome database for the ectomycorrizal fungus *L. bicolor* did not retrieve homologs to SUC2 (the gene encoding an acid invertase in *S. cerevisiae*) or cellulolytic enzymes (Table 1). In general, mycorrhizal species can be considered as biotrophic fungi that rely on a metabolically active plant host to obtain monosaccharides and do not scavenge in plant litter [7, 33].

The ability to use sucrose was also present in animal-pathogenic fungal interactions. For instance, genes encoding invertases have been identified in several species belonging to the *Asperigillus* and other important animal pathogens. It is interesting to highlight that the genome sequence of *C. albicans* lacks any homolog to *S. cerevisiae* invertase. However, an intracellular maltase-like enzyme is important for sucrose hydrolysis, which is induced in *C. albicans* cell in the presence of the disaccharide [34].

4. Carbohydrate Transporters

4.1. Hexoses

Due to the importance of glucose metabolism in *S. cerevisiae*, hexose transporters have been extensively studied in this organism [35, 36]. Seventeen genes have been characterized as hexose transporters belonging to the major facilitators superfamily [37]. The glucose uptake system is established by a constitutive low-affinity system (*Km*, 15 to 20 mM) and a glucose-repressed high-affinity system (*Km*, 1 to 2 mM) [36, 38].

For the filamentous fungus *Neurospora crassa*, glucose uptake mechanisms involve a constitutive low-affinity system (*Km*, 8 mM) and an inducible high-affinity system (*Km*, 10 mM) [39-42]. In addition to glucose transporters, *N. crassa* contains specific carriers for fructose and galactose, which increases the complexity of the hexose transporter system [43, 44].

The physiological relevance and biochemical properties of hexose transporters from fermentative yeasts and filamentous fungi are determined by the environmental conditions of different substrates they colonize. For instance, fungal strains invading plants rely on an almost endless source of hexoses, either from cell wall degradation or from plant primary metabolism.

A few genes encoding glucose transporters in plant-colonizing filamentous fungi have been cloned and characterized to date; *hxtl* from *U. fabae* [45], *AmMstl* from *Amanita muscaria* [46], *gtt*1 from *Trichoderma harzianum* [47].

Recently, a comprehensive study describes the functional characterization of a group of five glucose transporters from the hemibiotrophic plant pathogenic *C. graminicola*, CgHXT1-CgHxT5 [48]. Biochemical studies of these five transporters defined, as in *N. crassa*, a complex hexose transporter system with both an inducible and a constitutively expressed system (during vegetative and pathogenic development) with varying affinities for glucose uptake. CgHXT4 presented low glucose transport activity and was assigned the function of a constitutively expressed glucose sensor. CgHXT1 and CgHXT3 are transiently expressed during the biotrophic stage of *C. graminicola* and display a high affinity for glucose, but also are able to transport other monosaccharides.

The properties of these two transporters provide *C. graminicola* an efficient system to derive monosaccharides produced by metabolically active maize cells [2]. Finally, CgHXT2 and CgHXT5 define two exclusive transporters of the necrotrophic stage of *C. graminicola* development on maize leaves enabling uptake of monosaccharides generated from dead plant tissues.

The hexose transporter system of *C. graminicola* represents a complex system differentially regulated at the transcriptional level during host colonization. These observations illustrate that the fungal cells are able to sense and specifically modify the metabolic scenario according to the nutrient availability and the presence of the host.

Only a few fungal genes have been functionally characterized and their biochemical activity confirmed. However, BLAST searches of genomes of several fungal species revealed the presence of a multycopy gene family in each of them (Table 1). There is a minimum of 20 putative members in *U. maydis* and *L. bicolor*, and at least 100 homologs are encoded in *F. verticilliodes*. The human pathogen *C. albicans* contains 23 putative homologs to Gtt1. Although, only five glucose transporters have been functionally characterized in *C. graminicola*, sequence analysis revealed the presence of more than 70 homologs to yeast transporters. Such a large gene family implies an intricate mechanism for hexose metabolism in this hemibiotrofic pathogenic fungus. However, only further functional studies will contribute to a better understanding of the physiological relevance of the whole gene family in every fungal species.

When colonizing a host, the fungal cells have to compete with host metabolism for the use of monosaccharides. This metabolic competition will demand rapid and efficient mechanisms to capture and degrade nutrients for successful colonization and pathogenic development. This observation suggests that hexose uptake may represent a key step on the regulatory networks controlled by monosaccharide metabolism.

The synthesis of different combinations of transporters and the catabolic enzymes offers a considerable number of metabolic scenarios for monosaccharide uptake and degradation in the fungal cell. This metabolic flexibility encountered in fungal species reflects the plasticity of fungal metabolism for host colonization.

4.2. Sucrose

Sucrose is one of the most abundant soluble metabolites produced by plants and transported towards sink tissues [49]. Due to the availability of considerable amounts of sucrose, in almost all plant tissues, this disaccharide represents a valuable source of carbon to pathogenic or saprophytic fungi.

Even though sucrose can be hydrolyzed by extracellular invertases and resulting monosaccharides further transported, fungi also have developed the ability to uptake sucrose from the medium for intracellular metabolism [16, 17, 50]. This phenomenon was recently identified and functional studies on sugar transporters demonstrated the existence of plant-like sucrose transporters in fungal species [16, 50].

Sucrose mobilization in plants is facilitated by well-characterized sucrose/H^+ symporters (Suc transporters) that move sucrose against concentration gradients [51-56]. Recent investigations have revealed that Suc transporters in fungal species display similar biochemical properties to plant counterparts. In agreement, the protein sequences identified in *T. virens* and *Schizosaccharomyces pombe* (TvSut and SpSUT1, respectively) share a common ancestor with plant Suc transporters as illustrated by phylogenetic reconstructions of the corresponding sequences [16, 50].

In contrast, the amino acid sequence of UmSrt (a sucrose carrier from *U. maydis*) is not related to plant sucrose symporters, but the protein displays similar biochemical properties to trasnporters from plants [17]. An interesting aspect of both *T. virens* and *U. maydis* is that their transporters are induced when the fungal cells are invading plant tissue or using sucrose as a sole carbon source.

In the case of *U. maydis,* UmSrt is a pathogenicity factor, as intracellular sucrose degradation is fundamental for hyphal development and host infection [17]. In contrast, in the symbiont *T. virens,* the activity of TvSut is involved in a metabolic network that controls the expression of genes important for interaction with host plant [16]. The occurrence of sucrose transporters in plant-colonizing fungi offers an increased metabolic plasticity to fungal cells and represents an additional checkpoint to regulate different physiological processes associated to the fungal lifestyle and host colonization strategies.

Sucrose uptake and intracellular metabolism enables *T. virens* to down-regulate hydrolytic activities, which probably avoids excessive tissue damage in the living host. In this way, the presence of sucrose allows the fungal cells to discriminate between living and dead plant tissue, which activates the specific pathways for the metabolism of the corresponding source of carbon, sucrose or cellulose, respectively.

The specificity, biochemical properties, and physiological role of sucrose transporters in every fungal species will presumably differ depending on the life style, evolutionary history of each species, and the natural environments they inhabit. In the case of the rhizosphere competent species *T. virens,* TvSut is related to the establishment of a beneficial association with plant roots, but in the case of a pathogenic fungus, sucrose uptake is associated with virulence factors [17].

5. Glycolysis and TCA

The nutritional state during host-pathogen interactions relies on a bidirectional flux and full degradation of nutrients. During the infection of plant tissues, fungal cells obtain simple carbohydrate molecules that are oxidized to generate energy and reducing equivalents to sustain the fungal life cycle.

In general, monosaccharides are incorporated into the glycolytic pathway to generate pyruvate, which later enters the tricarboxilic acid cycle (TCA), resulting in the net conversion of hexoses into CO_2, ATP and NADH. However, our knowledge of glycolysis, its regulation, and genes encoding glycolytic enzymes in filamentous fungal species is limited.

Most of what we know about glycolysis and the TCA cycle in eukaryotes is based on the non-pathogenic ascomycete *S. cerevisiae* [57]. To consider a more comprehensive view of glycolysis and the TCA cycle in filamentous fungi, the results of a genomic survey for fungal genes related to glycolysis and TCA cycle are summarized in Table 2.

This set of data was generated by a BLAST search in the genome sequences of the different fungal species using the corresponding genes from *S. cerevisiae.* All the species tested encode all the important genes for the glycolytic and TCA cycle for full degradation of carbohydrates. These genetic determinants, presented in Table 2, are in agreement with articles describing the major pathways of carbohydrate metabolism in the plant-interacting fungi *U. maydis* and *L. bicolor*, and the opportunistic human pathogen *C. albicans* [58-60].

Saccharomyces cerevisiae mainly metabolizes hexoses through fermentation instead of respiration, even under aerobic conditions [57].

Under high glucose conditions, glycolysis is induced and the TCA cycle suppressed in *S. cerevisiae* [61]. In contrast, transcriptional studies revealed that *C. albicans* prefers to oxidize carbohydrates through respiration under aerobic conditions and rely on fermentation only under anaerobic conditions [60] .

Similarly, species of filamentous fungi contain the whole gene set to successfully convert hexoses into pyruvate through glycolysis, which is further oxidized using the TCA cycle (Table 2). Contrasting to *S. cerevisiae* behavior, the aerobic fungi *T. reesie*, *N. crassa*, and *Aspergillus oryzae* show little or no repression of TCA when glucose concentration in the medium increases [62-64].

Although fungi generally display similar metabolic pathways, their transcriptional regulation can be quite different depending on the availability of resources or the presence of the host. Different combinations of transporters and/or the intracellular enzymatic machinery will contribute to the metabolic diversification of the fungal cell.

Despite the differences at the transcriptional level, contributing to metabolic diversity, differences in biochemical properties of the enzymes are also likely to determine the fitness of the pathogen during host infection.

Table 2. Comparative analysis of fungal genomes to identify glycolysis and TCA-related genes. Functionally characterized sequences from *S. cerevisiae* were used as query in BLAST analysis

GLYCOLYSIS	*U. maydis*	*T. virens*	*S. cerevisiae*	*C. graminicola*	*F. verticilliodes*	*B. cinerea*	*L. bicolor*	*C. albicans*
Hexokinase	++	++++	++++	++++	++++	+++++	++	++++
Phosphoglucoisomerase	++	+	+	+	+	+	+	+
Phosphofructokinase	+	+	+	+	+	++	+	++
Aldolase	+	+	+	+	+	+	++	+
Triose P isomerase	++	++	+	++	+	+	+	+
Glyceraldehyde 3P DH	+	++	+++	+	+	++	+++++	+
Phosphoglycerate Kinase	+	+	+	+	+	+	+	+
Phosphoglycerate Mutase	+	++	+	++++	+	++	+++	++
Enolase	+	++	+++++	+	+	+	+	+
Pyruvate Kinase	+	+	++	+	+	+	+	+
TCA								
Pyruvate DH A	++	+++	+	++	++	++	++	+
B	+	+++	+	++	++	++	++	+
Citrate syntase	+	++	+++	++++	+++	++	+++	+
Aconitase	++++	++++++	++++	+++++	+++++	++++	+++	++++
Isocitrate DH	+	+	+++	+	+	++	+	++
a Ketoglutarate DH	+	+	+	+	+	++	+++	+
Succinyl-CoA synthase	+	++	+	+	++	+	+	+
Succinate DH	++	++	+++	+++	++	++	+++++	+++
Fumarase	+	+	+	+	++	+	+	++
Malate DH	++	++	+++	++	++	++	+++	+++

plus symbols, represent the number of homolog sequences identified in the corresponding genome sequence.

Conclusions

In order to succeed in host colonization, fungal pathogens depend upon virulence factors, their metabolic flexibility and ability to adapt to changes in the environment, and the availability of resources provided by the host. The relevance of the multiplicity of genes encoding invertases and cellulases in fungi still remain enigmatic. The metabolic complexity of the carbon circuits in fungal cells lead to intricate regulatory pathway controlling carbon flux from the source (plant) to its final destination (fungal respiration), which is vital for host colonization and fungal life cycle.

Acknowledgments

This work was supported by Ministerio de Ciencia e Investigacion from Spain throughout the Sub-programa Juan de la Cierva (JCI-2009-05364).

References

[1] Divon H.H. & Fluhr R. (2007) Nutrition acquisition strategies during fungal infection of plants. *FEMS Microbiology Letters,* 266, 65-74.

[2] Voll L.M., Samans B., Doehlemann G., Münch S. & Sonnewald U. (2011) Common motifs in the response of cereal primary metabolism to fungal pathogens are not based on similar transcriptional reprogramming. *Frontiers in Plant Science,* 2:39.

[3] van Kan J.A.L. (2006) Licensed to kill: the lifestyle of a necrotrophic plant pathogen. *Trends in Plant Science,* 11, 247-253.

[4] Nehls U., Mikolajewski S., Magel E. & Hampp R. (2001) Carbohydrate metabolism in ectomycorrhizas: gene expression, monosaccharide transport and metabolic control. *New Phytologist,* 150, 533-541.

[5] Schaarschmidt S., Kopka J., Ludwig-Müller J. & Hause B. (2007) Regulation of arbuscular mycorrhization by apoplastic invertases: enhanced invertase activity in the leaf apoplast affects the symbiotic interaction. *The Plant Journal,* 51, 390-405.

[6] Schaarschmidt S., González M-C., Roitsch T, Strack D., Sonnewald U.& Hause B. (2007) Regulation of arbuscular mycorrhization by carbon. The symbiotic interaction cannot be improved by increased carbon availability accomplished by root-specifically enhanced invertase activity. *Plant Physiology,* 143,1827-1840.

[7] Nehls U. (2008) Mastering ectomycorrhizal symbiosis: the impact of carbohydrates. *Journal of Experimental Botany,* 59, 1097-1108.

[8] Hanson J. & Smeekens S. (2009) Sugar perception and signaling--an update. *Current Opinion in Plant Biology,* 12, 562-567.

[9] Wind J., Smeekens S. & Hanson J. (2010) Sucrose: metabolite and signaling molecule. *Phytochemistry,* 71, 1610-1614.

[10] Roby C., Cortès S., Gromova M., Le Bail J-L. & Roberts J.K.M. (2002) Sucrose cycling in heterotrophic plant cell metabolism: first step towards an experimental model. *Molecular Biology Reports,* 29, 145-149.

[11] Vargas W.A. & Salerno G.L. (2010) The Cinderella story of sucrose hydrolysis: Alkaline/neutral invertases, from cyanobacteria to unforeseen roles in plant cytosol and organelles. *Plant Science,* 178, 1-8.

[12] Bolton M.D. (2009) Primary metabolism and plant defense--fuel for the fire. *Molecular Plant Microbe Interactions,* 22, 487-497.

[13] Lodish H., Berk A., Zipursky S.L., Paul P., Baltimore D. & Darnell J. (2000) *Molecular Cell Biology,* Available from: http://www.ncbi.nlm.nih.gov/books/NBK21475/

[14] Wei H., Xu Q., Taylor L.E. 2nd, Baker J.O., Tucker M.P. & Ding S-Y. (2009) Natural paradigms of plant cell wall degradation. *Current Opinion in Biotechnology,* 20, 330-338.

[15] Lopez-Casado G., Urbanowicz B.R., Damasceno C.M.B. & Rose J.K.C. Plant glycosyl hydrolases and biofuels: a natural marriage. *Current Opinion in Plant Biology,* 11, 329-337.

[16] Vargas W.A., Crutcher F.K. & Kenerley C.M. Functional characterization of a plant-like sucrose transporter from the beneficial fungus *Trichoderma virens*. Regulation of the symbiotic association with plants by sucrose metabolism inside the fungal cells. *New Phytologist,* 189, 777-789.

[17] Wahl R., Wippel K., Goos S., Kämper J. & Sauer N. (2010) A novel high-affinity sucrose transporter is required for virulence of the plant pathogen *Ustilago maydis*. *PLoS Biology,* 8(2):e1000303.

[18] Ilmen M., Saloheimo A., Onnela M. & Penttila M. (1997) Regulation of cellulase gene expression in the filamentous fungus *Trichoderma reesei*. *Applied and Environmental Microbiology,* 63, 1298-1306.

[19] Harman G.E., Howell C.R., Viterbo A., Chet I. & Lorito M. (2004) *Trichoderma* species--opportunistic, avirulent plant symbionts. *Nature Reviews in Microbiology,* 2, 43-56.

[20] Martinez D., Berka R.M., Henrissat B., Saloheimo M., Arvas M., Baker S.E., et al. Genome sequencing and analysis of the biomass-degrading fungus *Trichoderma reesei* (syn. *Hypocrea jecorina*). *Nature Biotechnology,* 26, 553-560.

[21] Fisk M.C., Fahey T.J., Sobieraj J.H., Staniec A.C. & Crist T.O. Rhizosphere disturbance influences fungal colonization and community development on dead fine roots. *Plant and Soil,* 341, 279-293.

[22] Salerno G.L. & Curatti L. (2003) Origin of sucrose metabolism in higher plants: when, how and why? *Trends in Plant Science,* 8, 63-69.

[23] Farrar J. & Lewis D. (1987) Nutrient relations in biotrophic infections. In: Fungal Infection of Plants. Cambridge, UK: Cambridge University Press, p. 92-132.

[24] Vargas W.A., Mandawe J.C. & Kenerley C.M. (2009) Plant-derived sucrose is a key element in the symbiotic association between *Trichoderma virens* and maize plants. *Plant Physiology,* 151, 792 -808.

[25] Wind J., Smeekens S. & Hanson J. (2010) Sucrose: metabolite and signaling molecule. *Phytochemistry,* 71, 1610-1614.

[26] Geissmann M., Frey T. & Ruffner H.P. (1991) Occurrence and properties of acid invertase in cultures of *Botrytis cinerea*. *Mycological Research,* 95, 1321-1327.

[27] Chaudhuri A. (1994) Invertase in a thermophilic fungus, *Thermomyces lanuginosus*: its unusual behaviour and regulation. Available from: http://etd.ncsi.iisc. ernet.in /handle/2005/104

[28] Voegele R.T., Wirsel S., Möll U., Lechner M. & Mendgen K. (2006) Cloning and characterization of a novel invertase from the obligate biotroph *Uromyces fabae* and analysis of expression patterns of host and pathogen invertases in the course of infection. *Molecular Plant Microbe Interacteractions,* 19, 625-634.

[29] Horst R.J., Engelsdorf T., Sonnewald U. & Voll L.M. (2008) Infection of maize leaves with *Ustilago maydis* prevents establishment of C4 photosynthesis. *Journal of Plant Physiology,* 165, 19-28.

[30] Rubio M. & Navarro A. (2006) Regulation of invertase synthesis in *Aspergillus. Enzyme and Microbial Technolgy,* 3, 601-606.

[31] Callow J.A., Long D.E. & Lithgow E.D.(1980) Multiple molecular forms of invertase in maize smut infections. *Physiological Plant Pathology,* 16, 93-107.

[32] Bergstrom G.C. & Nicholson R.L. (1981) Invertase in the Spore Matrix of Colletotrichum graminicola. *Journal of Phytopathology,* 102, 139-147.

[33] Smith S.E. & Read D.J. Mycorrhizal symbiosis. Academic Press.

[34] Geber A., Williamson P.R., Rex J.H., Sweeney E.C. & Bennett J.E. (1992) Cloning and characterization of a *Candida albicans* maltase gene involved in sucrose utilization. *Journal of Bacteriology,* 174, 6992-6996.

[35] Boles E. & Hollenberg C.P. (1997) The molecular genetics of hexose transport in yeasts. *FEMS Microbiology Reviews,* 21, 85-111.

[36] Ozcan S. & Johnston M. (1999) Function and regulation of yeast hexose transporters. *Microbiology and Molecular Biology Reviews,* 63, 554-569.

[37] Marger M.D. & Saier M.H. Jr. (1993) A major superfamily of transmembrane facilitators that catalyse uniport, symport and antiport. *Trends in Biochemical Sciences,* 18, 13-20.

[38] Lagunas R. (1993) Sugar transport in *Saccharomyces cerevisiae. FEMS Microbiology Reviews,* 10, 229-242.

[39] Scarborough G.A. (1970) Sugar transport in *Neurospora crassa.* II. A second glucose transport system. *Journal of Biological Chemistry,* 245, 3985-3987.

[40] Scarborough G.A. (1971) Sugar transport in *Neurospora crassa.* III. An inositol requirement for the function of the glucose active transport system. *Biochemical and Biophysical Research Communications,* 43, 968-975.

[41] Schneider R.P. & Wiley W.R. (1971) Regulation of sugar transport in *Neurospora crassa. Journal of Bacteriology,* 106, 487-492.

[42] Schneider R.P. & Wiley W.R. (1971) Kinetic characteristics of the two glucose transport systems in *Neurospora crassa. Journal of Bacteriology,* 106, 479-486.

[43] Rand J.B. & Tatum E.L. (1980) Characterization and regulation of galactose transport in *Neurospora crassa. Journal of Bacteriology,* 141, 707-714.

[44] Rand J.B. & Tatum E.L. (1980) Fructose transport in *Neurospora crassa. Journal of Bacteriology,* 142, 763-767.

[45] Voegele R.T., Struck C., Hahn M. & Mendgen K. (2001) The role of haustoria in sugar supply during infection of broad bean by the rust fungus *Uromyces fabae. Proceedings of the National Academy of Sciences U.S.A.,* 98, 8133-8138.

[46] Nehls U., Wiese J., Guttenberger M. & Hampp R. (1998) Carbon allocation in ectomycorrhizas: identification and expression analysis of an *Amanita muscaria* monosaccharide transporter. *Molecular Plant Microbe Interactions,* 11, 167-176.

[47] Delgado-Jarana J., Moreno-Mateos M.A. & Benítez T. (2003) Glucose uptake in *Trichoderma harzianum*: role of *gtt1*. *Eukaryotic Cell,* 2, 708-717.

[48] Lingner U., Münch S., Deising H.B. & Sauer N. (2011) Hexose transporters of a hemibiotrophic plant pathogen: functional variations and regulatory differences at different stages of infection. *Journal of Biological Chemistry,* 286, 20913-20922.

[49] Koch K. (2004) Sucrose metabolism: regulatory mechanisms and pivotal roles in sugar sensing and plant development. *Current Opinion in Plant Biology,* 7, 235-246.

[50] Reinders A. & Ward J.M. (2001) Functional characterization of the alpha-glucoside transporter Sut1p from *Schizosaccharomyces pombe,* the first fungal homologue of plant sucrose transporters. *Molecular Microbiology,* 39, 445-454.

[51] Shakya R. & Sturm A. (1998) Characterization of source- and sink-specific sucrose/H^+ symporters from carrot. *Plant Physiology,* 118, 1473-1480.

[52] Noiraud N., Delrot S. & Lemoine R. (2000) The sucrose transporter of celery. Identification and expression during salt stress. *Plant Physiology,* 122, 1447-1455.

[53] Barth I., Meyer S. & Sauer N. (2003) PmSUC3: characterization of a SUT2/SUC3-type sucrose transporter from *Plantago major*. *The Plant Cell,* 15, 1375-1385.

[54] Yao Li C., Shi J.X., Weiss D. & Goldschmidt E.E. (2003) Sugars regulate sucrose transporter gene expression in citrus. *Biochemical and Biophysical Research Communications,* 306, 402-407.

[55] Flemetakis E., Dimou M., Cotzur D., Efrose R.C., Aivalakis G., Colebatch G., et al. (2003) A sucrose transporter, LjSUT4, is up-regulated during *Lotus japonicus* nodule development. *Journal of Experimental Botany,* 54, 1789-1791.

[56] Krügel U., Veenhoff L.M., Langbein J., Wiederhold E., Liesche J., Friedrich T., et al. (2008) Transport and sorting of the *Solanum tuberosum* sucrose transporter SUT1 is affected by posttranslational modification. *The Plant Cell,* 20, 2497-2513.

[57] Chambers A., Packham E.A. & Graham I.R. (1995) Control of glycolytic gene expression in the budding yeast (*Saccharomyces cerevisiae*). *Current Genetics,* 29, 1-9.

[58] Saavedra E., Ramos-Casillas L.E., Marín-Hernández A., Moreno-Sánchez R. & Guerra-Sánchez G. (2008) Glycolysis in *Ustilago maydis*. *FEMS Yeast Research,* 8, 1313-1323.

[59] Deveau A., Kohler A., Frey-Klett P. & Martin F. (2008) The major pathways of carbohydrate metabolism in the ectomycorrhizal basidiomycete *Laccaria bicolor* S238N. *New Phytologist,* 180, 379-390.

[60] Askew C., Sellam A., Epp E., Hogues H., Mullick A., Nantel A., et al. (2009) Transcriptional regulation of carbohydrate metabolism in the human pathogen *Candida albicans*. *PLoS Pathogens,* 5, e1000612.

[61] Johnston M. (1999) Feasting, fasting and fermenting. Glucose sensing in yeast and other cells. *Trends in Genetics,* 15, 29-33.

[62] Chambergo F.S., Bonaccorsi E.D., Ferreira A.J.S., Ramos A.S.P., Ferreira Júnior J.R., Abrahão-Neto J, et al. (2002) Elucidation of the metabolic fate of glucose in the filamentous fungus *Trichoderma reesei* using expressed sequence tag (EST) analysis and cDNA microarrays. *Journal of Biological Chemistry,* 277, 13983-13988.

[63] Xie X., Wilkinson H.H., Correa A., Lewis Z.A., Bell-Pedersen D. & Ebbole D.J. (2004)Transcriptional response to glucose starvation and functional analysis of a glucose transporter of *Neurospora crassa*. *Fungal Genetics and Biology,* 41, 1104-1119.

[64] Maeda H., Sano M., Maruyama Y., Tanno T., Akao T., Totsuka Y., et al. (2004) Transcriptional analysis of genes for energy catabolism and hydrolytic enzymes in the filamentous fungus *Aspergillus oryzae* using cDNA microarrays and expressed sequence tags. *Applied and Microbiology and Biotechnology,* 65, 74-83.

In: Fungi: Types, Environmental Impact and Role in Disease ISBN: 978-1-61942-671-9
Editors: Adolfo Paz Silva and María Sol

Chapter III

Fungi: Types, Environmental Impact and Role in Disease

***Gabriela Moreno-Coutiño*[1] *and Roberto Arenas*[2]**
[1] Attending physician, Mycology section
[2]Head Mycology section, Dermatology Department,
"Dr. Manuel Gea González" General Hospital, Mexico City, México

Abstract

Ecological changes, human demographic modifications, international travel and commerce as well as fungal adaptation to the environment and its conditions such as antifungal drugs by genetic recombination cause emergence or re-emergence of infectious diseases. Cryptococcus species complex is a large genus, where the pathogenic species C. neoformans (var. neoformans and var. grubii) and C. gatti are the agents of cryptococcosis. This is a ubiquitous fungus, responsible for disease in both immunocompetent and immunocompromised hosts and causes approximately 600,000 deaths/yr in North America.

C. neoformans is ubiquitous in the environment due to dissemination by pigeons following migratory and trade routes, and C. gattii is restricted to tropical/subtropical regions because it is not associated to pigeon guano, but to some tree species. Simultaneously, an increase of prevalence has been reported in patients living with HIV/AIDS, receiving aggressive cancer therapy or solid organ transplant recipients.

Coccidioides immitis and C. posadasii are responsible for the most common respiratory mycoses. C. immitis is found in California and C. posadasii elsewhere. Although this disease is considered endemic, epidemic outbreaks have been documented, mostly related to droughts, as well as immigration of susceptible individuals. Immunosuppressed patients are the most severely affected, with various clinical manifestations. Some associated factors have been identified regarding the epidemics, while new approaches to treatment and prevention are being applied.

Sporotrichosis is a subcutaneous mycosis caused by Sporothrix species complex fundamentally transmitted by traumatic inoculation with contaminated plants, wood, and other vegetable debris. Cases inoculated by animals, particularly cats, have been documented. This is sometimes considered an occupational disease because it commonly presents in gardeners and field workers, but epidemic outbreaks have occurred in several countries affecting people with different occupations, regardless of their immune status.

1. Introduction

Fungi have been classified since the 18th Century according to their morphology and physiology. However, new techniques like molecular genetics have enabled the corroboration or reclassification of fungal species.

So, although human pathogenic fungi remain basically the same, their classification and their spectrum of disease vary according to the climatic changes, human demographic alterations, travel, commerce and microbial adaptation that condition the emergence or reemergence of fungal infections. This is, new infectious diseases in a population or geographic range, or an increase of the number of cases in a determined region causing outbreaks.

The outbreaks can be understood based on the term "pathocenosis" as defined by Mirko Grmek in 1969: "The qualitatively and quantitatively defined group of pathological states present in a given population at a given time. The frequency and the distribution of each disease depend not only on endogenous (infectivity, virulence, route of infection, vector) and ecological factors (climate, urbanization, promiscuity) but also on frequency and distribution of all the other diseases within the same population".[1]

This term is originated with the understanding that most infections are caused by pathogens already present in the environment, and that, by different reasons as addressed by Grmek, gain a selective advantage over other microorganisms. This manifests as the capacity to infect new hosts, tolerate new environmental conditions, adaptation to nutrients and energy sources as well as predator evasion. This capacity will determine the abundance and distribution in nature. [2]

Fungi improve their survival rate by adapting to the new characteristics in the environment and other living organisms, microscopic or not such as other microbes, predators and symbiotic hosts.

With all these interacting factors, fungal infections vary their environmental impact and the degree of disease they cause, showing an important increase of prevalence in immunocompromised patients such as those that live with HIV/AIDS or receive treatment with immunosuppressive drugs.

This is particularly true for systemic fungal infections that are less common than superficial or subcutaneous mycosis, but are more severe. Traditionally they are divided into endemic and opportunistic infections. The former infect immunocompetent patients, and cause symptoms in a few, only under specific conditions. The latter infect immunocompromised patients, a group that is rapidly growing, and are associated to higher morbi-mortality.

For example, 5-45% of solid organ transplant recipients present fungal disease. Immunosuppressive drugs are an important risk factor, but also the donor of the organ may be infected. Transplanted lungs may act as a reservoir for pathogenic fungi, being subclinic in the donor and active in the recipient. [3]

All this constant evolving interactions determine the relation between the host, the fungus and their ecology, that ultimately influence the interaction with the habitat and the role in disease.

2. Cryptococcosis

Cryptococcus species complex is a large genus, where the pathogenic species C. neoformans (var. neoformans and var. grubii) and C. gatti are probably the only agents of human cryptococcosis.

C. neoformans is ubiquitous in the environment due to dissemination by pigeons following migratory and trade routes, while C. gattii is restricted to tropical/subtropical regions and is not associated to pigeon guano, but to some tree species.

Cryptococcus species complex are encapsulated yeasts that present multilateral budding. They are firm walled, spherical or ellipsoidal that characteristically show a thick capsule visible with Indian ink. Spore formation is the first step in pathogenesis, since these particles must be inhaled by the host directly from the environment, and are much smaller in size compared to yeasts. However, hyphae formation is essential for spore formation, and thus, fungal propagation. This occurs under specific conditions, such as in the presence of a mating partner, low nitrogen and water levels or darkness.

For investigative purposes, the yeast like strains are classified as BSL-2, but where basidiospores are generated, as in mating experiments, safety conditions at BSL-3 are recommended. [4]

The capsule is predominantly formed by the polysaccharides glucuronoxylomannan (GXM, 90-95%)) and galactoxylomannan (GalXM, 5-8%) and mannoproteins (MP, <1%). It is the main virulence factor. Five different serotypes have been identified (A,B,C,D and AD).[3]

Its size varies depending on environmental and host's characteristics, and the fungal species, with the intention to provide the best conditions for fungal survival. This capsule offers protection similar to the one provided by bacterial capsules and has an important role for virulence, immunomodulation and immune evasion from the host, as proved by a study where acapsular fungi were incapable of producing disease in murine models. Encapsulated fungi are able to withstand dehydration, rehydration and UV radiation facilitating long-distance dissemination more successfully than non-encapsulated organisms. But still, the final size of the spore and the capsule must be small enough to enable penetration through inhalation until it reaches the alveoli.

Other studies show that these fungi are less virulent in immunocompetent hosts. However, these same acapsular yeasts can cause disease in immunosuppressed patients, where low virulence is not a factor for disease prevention.

The particular characteristics and components of the capsule, distinguishes them from other fungi. The phenotypic variation allows rapid adaptation to the changes in the environment, with the intention to evade the host's immune system. The capsule is the first structure to contact the host, so its variability and versatility greatly influences fungal survival. For this reason, the induction and any changes present in the composition can be clinically relevant.

The capsule starts its adaptation process only hours after the infection is established, and continues during the course of the disease.

Some in vivo studies show that the lung environment acts as a capsule growth inducer. Other identified factors that alter the capsule's size are high CO2 concentration, iron deprivation and alkaline media.

Additional identified virulence factors are melanization and biofilm formation making the fungus less susceptible to the host's defenses and antifungal treatments.

In the case of C. gatti, its growth is related to many trees that given the exportation and cultivation to many "new" regions of the world, these species have provided a constantly growing niche for this particular species of Cryptococcus. Air and water currents, contaminated wood products, and animals including of course, migratory birds that conforms the ecologic environment collaborate to spore propagation. [5]

The environment influences greatly the spore distribution. The spore count is highest during the summer. Thunderstorms influence wind speeds, cold fronts and air pressure.

Parks, gardens, and buildings can be reservoirs, particularly for immunosuppressed patients.[6]

Environmental factors directly affect the number and distribution of fungal spores. Seasonal changes show that they influence spore concentrations secondary to temperature, precipitation and wind speed. For this reasons, C. gatti has recently emerged as pathogen in the northwestern region of North America. [7]

It emerged in British Columbia, Canada in 1999. Previously it had been identified in tropical and sub-tropical regions. In this area, C. gatti has been isolated from trees, soil, water and air. [8]

British Columbia has the largest documented population of C. gatti infected persons worldwide with an annual incidence of 25.1/million. The number of cases increased between the years of 1999-2001 probably due to the additive result of the awareness and efficiency of reporting this illness, greater fungal spore concentration and accumulation of exposed susceptible individuals. The incidence stabilized in 2002, only to increase between the years 2006 and 2007. C. gatti is considered a public health problem, and with the help of molecular biology it has been proven that the fungus has spread from Vancouver Island to the US Pacific Northwest, particularly to Washington and Oregon states where it is now considered an emerging infectious disease. As a result, there is an expansion of the native endemic zone and higher risk for infection in humans and animals.[9]

It infects predominantly immunocompetent patients, where it causes pneumonia or lung cryptococcomas. This infection can potentially disseminate to any other organ, but has predilection for the central nervous system (CNS) where it causes meningoencephalitis or brain cryptococcomas.

The incubation period is currently unknown, but estimated to be of 6 or 7 months. The estimates of incubation period for Cryptococcus neofornans var. grubii are much longer than those for C. gatti. [8] This is probably due to differences in host factors, exposure dose or variation in recognition of symptom onset. So, C. gatti has spread and expanded its ecological niche. Anthropogenic and natural dispersal of the spores occur and collaborate. However, it is still under study the influence of the reproductive cycle of the fungus in the virulence and resistance to environmental hazards. [10]

Besides C. gatti, the other human pathogenic species Cryptococcus neoformans var. neoformans (serotype D) and var. grubii (serotype A) grow in pigeon guano, as well as in bat guano in lesser degree.

C. neoformans var. grubii causes disease predominantly in HIV/AIDS patients. It has even been described as the only serotype capable of causing infection in this population. However, it has been found also in immunosuppressed patients by other reasons such under aggressive cancer therapy and solid organ transplantation.

Cryptococcus species complex causes 8% of invasive fungal infections in transplanted patients. Another study reported 0.7% of cryptococcosis in non-HIV infected patients and hematopoietic stem cell transplant (HSCT), 18% in solid organ transplant recipients (SOT), 9% in patients with hematologic malignancies, and 9% in patients with other malignancies.

The mortality for cryptococcosis in non-HIV infected patients is 30%. One study found that cardiac transplant patients are at higher risk of developing cryptococcosis and kindney recipients to have disseminated disease. [11]

Of 650 cases of renal transplant, 10 cases of cryptococcosis were identified. Eight cases developed meningitis, one a cerebral granuloma, and other pulmonary.

Prompt examination of CSF is warranted for an adequate diagnosis, with India ink preparation. [12]

This serotype is responsible for the majority of the infections worldwide.

Pigeons and other birds related to this fungus are cosmopolitan and common. This facilitates the worldwide dissemination of the fungi following the bird migration and trade routes, as well as the influence of the domestication of some bird species.

In Europe it is responsible for 20% of the infections in HIV positive patients. It is the most common cause of fatal nervous system (CNS) infection. Several reports mention this infection as the initial AIDS defining illness in 20-30% of AIDS patients in Africa, 4.3% in Singapore, and it is responsible for 20-40% of the AIDS attributable mortality.

Up to 70% of the recorded infections are estimated to occur in sub-Saharan Africa. Theses patients die within 6 months if left untreated. On the other hand, treatment can cause the paradoxical response seen in the immune reconstitution inflammatory syndrome in 10-42% of patients that initiate antiretroviral therapy for the first time with a high mortality rate between 33-66%.[13]

Diagnosis is based on clinical features, and/or presence of encapsulated yeasts at direct examination/histology and/or isolation of the fungus from any body site, and/or cryptococcal antigen detection and titration that are helpful for dissemination evaluation, severity of the infection and response to treatment.

For antigen detection, the threshold of 1:512 or higher is considered an independent factor for relapse in patients with AIDS, and despite the HIV status, the mortality is of 15-20% at three months, or even higher in some African and Asian countries.

Negative prognostic factors for mycological therapy failure are disseminated disease, high serum antigen titer at baseline and lack of fluocytosine during the induction therapy.[14]

After analyzing the ecological niches and reproductive forms, it is clear that C. grubii and C. neoformans are two varieties with different epidemiology. C. neoformans has a wider distribution, considered cosmopolitan because it is capable of reproducing sexually and asexually in pigeon guano that is found everywhere, contrary to C. grubii that can only perform the asexual part, limiting the ecological niche of this variety to the areas with the adequate tree species for its reproduction. Cryptococcus species reproduce clonally, but may also complete a sexual cycle of opposite or same-sex mating. It is probably that same sex mating originates the hypervirulent infectious propagules, with recombinant progeny with altered virulence. [4]

Cryptococcosis can be an asymptomatic, chronic or acute infection. After the initial pulmonary infection, the fungus can have systemic spread to any organ, but with particular predilection for the central nervous system. When symptomatic, the pulmonary infection causes coughing, pleuritic chest pain, fever, dyspnea, weight loss and malaise. Besides the meningitis, other common extrapulmonary sites of infection are the skin, prostate and eye. In HIV patients, it is estimated of 4.3%.[15]

Mortality is of 30-60% in HIV/AIDS infected patients, and higher in solid organ transplant recipients. This only supports the fact that an intact immune response to this fungus is essential for good prognosis. [16]

We can conclude that the actual most important factor in the impact of this disease is the constantly growing immunosuppressed population, particularly patients living with HIV/AIDS and transplant recipients. As the human to human transmission is extremely rare, it is always considered as acquired from the environment. And, as the environmental niches are expanding, this can result in pandemias. Cryptococcosis outbreaks occur due to climatic changes, rather than to the emergence of pathogenic strains, (although hybrid strains have been isolated). So the focus should be on the prevention of infection, particularly in patients at high risk. Physical barriers such as air filters and sealed windows in immunosuppressed wards are recommended, as well as exhaustive search for antigens or the fungus in tissues and serum of organ donors. Meticulous screening should also be performed in organ recipient patients to avoid reactivation after the initiation of immunosuppressive therapy, because, as reported in some studies, up to 82% of patients with cryptococcal pneumonia can have a negative serum cryptococcal antigen.[11, 17-20] C. neoformans has shown decreased susceptibility to fluconazole , itraconazole, voriconazole and amphotericin B. Even so, fluconazole remains being the drug of choice for empirical therapy of cryptococcosis.

Amphotericin B has good cure rates. In immunosuppressed patients the association with fluocytosine has better results, particularly reducing the risk of reactivation during maintenance therapy, which is recommended with fluconazole. Despite the administration of adequate treatment, mortality remains high, particularly in immunosuppressed patients. In patients living with HIV/AIDS, the paradoxical response of immune restoration inflammatory syndrome (IRIS) can present in those that are initiating highly active anti-retroviral therapy.

3. Coccidioidomycosis

This systemic infection, also known as “Valley Fever”, was described in 1892 in Argentina by Alexandro Posadas. He concluded that the etiologic agent was infectious, but it was until 1900 that Ophüls and Moffitt were able to classify it as a fungus.[21]

The genus Coccidioides includes only two species that were described in 2002, C. immitis and C. posadasii, and so far, no sexual phase has been identified. Both are thermally dimorphic fungi and morphologically identical. However, the differences are found genetically and epidemiologically. C. immitis is restricted to the San Joaquin Valley in the California area, and C. posadasii is found elsewhere, particularly in México and South

America inhabiting semiarid areas that characteristically have dry and alkaline soil with sand and silt. So for easier understanding they have been classified as follows:

- Group I: isolates not from California (C. posadasii)
- Group II: isolates from California (C. immitis)

In the saprophytic phase, the fungus is found 10-30 cm. below the surface of the soil. It is formed by septate hyphae and thick wall arthroconidia, which are separated by empty thin walled brittle cells formed by autolysis of alternate conidia along the hyphae. Thus, hyphae are dissecated by environmental temperature and arthroconidia are easily released into the air by soil disruption and wind. Host invasion is based upon inhalation of the arthrospores of the mycelial form that reach the lungs, or on rare occasions, after traumatic inoculation into tissue. The disease is not transmitted form person to person. [21]

Once in the host, in 48-72 hours, the arthroconidia transforms into a spherule with a thick double wall that measures 15-75 µm, filled with 800-1,000 endospores. After rupture of the spherule, each endospore turns into a new spherule that allows the infection to persist. This phase is influenced by CO2 increases, rather than by temperature although the dimorphism is temperature dependant, as in other dimorphic fungi such as Blastomyces sp., Histoplasma sp., Paracoccidioides sp., Penicillium marneffei and Spotothrix sp.

It is considered the most common respiratory mycosis in endemic areas because it is highly virulent. During the last 15 years the number of cases has increased, and now it is one of the most frequent causes of community acquired pneumonia. At the moment, apparently the main factors related to the increase in the incidence are environmental rather than genetic. However, susceptibility genes in humans are under study in an attempt to understand why in African-Americans and Filipinos the infection is more prevalent and have higher risk of presenting disseminated disease. Also at higher risk are pregnant women, patients living with HIV/AIDS or otherwise with cell mediated immunosuppression. [22]

It is important to point out that there is no evidence that supports that any race is more susceptible to present primary pulmonary disease, however, data confirm that African Americans are 14 times more likely to develop severe or disseminated disease, in phillipines the disease is 175 times more likely to disseminate, and 3 times more likely for Mexican Americans.[23]

Epidemic outbreaks have been widely documented before relating climate and disease in 1957. The main findings are a complex relation between drought indices, lagged precipitation, temperature, wind, seed and dust during the preceding 1-2 or years.

Research points to the conclusion that during drought years the number of organisms competing with C. immitis decreases. The spores in dry soil germinate after the rain season, reaching a higher density than usual because of the lack of competing organisms. In late summer and fall, the soil dries again, and the arthroconidia form as a consequence of hyphae dissecation, and thus, become airborne and potentially infectious. After this, any disturbance of the soil, commonly by wind, but other factors can participate, such as dust storms, disruption and aerosolization of the desert surface by construction, wildfires and earthquakes, can disperse the spores that may be then inhaled by the host.[24]

As a consequence of the natural cycle of the fungus, the highest number of exposures is during the fall, corresponding to dry windy weather, in patients that by occupational or recreational reasons spend time outdoors in endemic areas, and exposure decays in the rain season.

Some of the areas mostly studied are Arizona and California. In Arizona, a bimodal seasonality of the disease has been documented, with peak exposures during June and July, and then again in October and November, which are the driest and dustier months of the year.

Several of the incidence reports include the one in 1930-1940 where numerous cases were noted in the San Joaquin Valley in California. And then, in this same valley, during World War II (1939-1945) several training airfields were built, where a rate of new infections in military personnel was 8-25%/yr. The mortality was low, but morbidity high, interrupting military training.

Since then, other outbreaks have been reported in military members which often train in open spaces such as desserts and, because of their activities they create dusty conditions around them, aerosolizing and inhaling the arthroconidia from the environment. A cohort reported 17.5% of disseminated cases compared to the 1% reported in the literature.[25]

Other cases were reported during dust storms in 1977. During the 1980´s the cases of coccidioidomycosis was lower than previous years. This was explained by the increased amount of land employed for agriculture, and thus, the increase in irrigation of fields and by the use of air conditioning that notably decreases the amount of indoor dust.[22]

A decade later, during the years 1991-92 a dramatic increase was reported in California, probably associated to drought and peaked in January 1994 where an outbreak occurred after the Southern California earthquake that caused landslides and airborne dust containing infectious spores. It was not until 1996 that the epidemic seemed to be reducing.[26]

This major epidemic represented a 11-fold increase above the mean number of cases reported between 1955-90. Molecular analysis demonstrated that the responsible for the epidemic were the environmental factors and not a new pathogenic fungal strain.[27]

In the USA, the estimated incidence is 150,000 infections/yr., where Arizona and California, USA account for 95% of all cases reported. The current Arizona epidemic is linked to climate conditions. Arizona reported 2,695 cases in 2003 and 3,515 in 2005, in a wide age range between 30-75 years old.[21]

Although a previous study mentions that coccidioidomycosis was commonly found among adults over 65 years of age. This was at least, partially explained by the increased influx of older population into the Sun Belt in the South of California and Arizona. Many are primo infections, but reactivations were also reported. This highlights the importance of travel history of patients to endemic areas.[28]

Mexico has the largest number of reported cases in Latin America with a prevalence of 10-40% in the Northern states.[29]

Of all the reported cases, 40% will be symptomatic and it is estimated that 1-2% will develop disseminated disease, frequently in skin, bones, joints and meninges. This is particularly true for immunosuppressed patients, in which, the treatment is challenging and requires an aggressive approach.[25]

With the medical advances in organ transplantation, this disease may become an important diagnosis to discard prior to lung donation. In North Carolina, a fatal coccidioidomycosis was reported after bilateral lung transplantation, where the donor had previously traveled to Mexico and the recipient had no travel history to endemic areas. The immunosuppresive therapy in the recipient reactivated latent Coccidioides infection in the donor lungs.[30]

Diagnosis is frequently delayed, particularly in patients that live elsewhere and travel to endemic areas. Symptoms begin 1-3 weeks after exposure. Those with primary pulmonary coccidioidomycosis may present acute or subacute features of a flulike illness or progressive pneumonia, and in cases of disseminated disease the symptoms will depend on the affected organ. Dissemination by hematogenous spread occurs early although it may manifest months or years after the initial respiratory illness.

Controversy exists regarding treatment administration in self-limited pulmonary infection. Many experts recommend close follow up without specific antifungal therapy. Current guidelines recommend administration of treatment to patients at risk of developing severe or disseminated disease. The first line of treatment for primary coccidioidomycosis are azoles, especially fluconazole and itraconazole. For severe and disseminated disease, initial administration of amphotericin B followed by azole therapy is recommended.[23]

Serologic tests aid in diagnosis and follow up of patients. Some of the more commonly used tests are immunodiffusion, enzyme immunoassay and complement fixation. Combined results of these tests provide a better approach to the individual evolution, increasing the sensibility of the tests, particularly in immunocompromised patients.[31]

The coccidioidin skin test is useful for diagnosis and epidemiologic studies in immunocompetent population. For environmental detection, laboratory mice are inoculated with the fungus, which is then further studied. This is expensive and time consuming, so new techniques are under evaluation such as polymerase chain reaction (PCR).

Recovery confers lifelong immunity to reinfection and manifests as a positive coccidioidin skin test. This supports the development of a vaccine. The Valley Fever Vaccine Project has identified and cloned immunogenic proteins that have proven effective in the prevention of deaths and fungal burdens in murine models, and studies are being done now in humans.[29]

Prevention is of paramount importance. Simple measures as the use of a face mask while being outdoors exposed to the fungus will reduce the risk of infection. Dust reduction indoors is also important, particularly for people with high risk of infection. Prophylactic treatment is currently not recommended, and the hope for an efficient vaccine is still the best chance to diminish incidence.[25]

Public health plays a very important role in the control and prevention of this fungal infection. A better understanding of the climatic factors that influence outbreaks will allow a more precise timing for environmental sampling of soil, the adequate implementation of prevention campaigns and a better understanding of other risk factors.

As is expected, periodic epidemics of coccidioidomycosis will continue to present. With the current information, public health has a 18 month lead time to study the development of the fungal life cycle from drought to rain to windy seasons.[24]

4. Sporotrichosis

Sporotrichosis was described in 1898 in the American continent where it is a frequent disease, as well as in other countries with tropical and subtropical climates such as Asia and Australia.

The genus is conformed by ubiquitous dimorphic fungi that include saprophitic, insect, plant and human parasites. So far, only the asexual form has been identified. The Sporothrix complex includes several species of medical interest such as S. brasiliensis, S. globosa, S. mexicana and Sporotrhrix schenckii sensu stricto. Initially it was considered to be caused only by the fungus Sporotrhrix schenckii. Now we know that other molecular species are be involved in the disease so the fungi are now named S. schenckii species compelx. This is the etiological agent of human sporotrichosis.

In humans, the disease is acquired through traumatic inoculation with contaminated material. For this reason, sporotrichosis is traditionally considered an occupational disease because it causes infection in workers exposed to vegetation where the fungus is ubiquitous, such as those with gardening or agricultural activities.

Sporothrix species complex has been isolated mainly form cats and armadillos, but also from straw, wheat grains, fruits, tree bark, wood, shrub thorns, rose shrubs, ploughed land, and animals such as horses, dogs, snakes, rats, boars, camels, cattle, chimpanzees, dolphins, donkeys, fowl, mules, spiders, flies, dead insects and larvae, dust, animal droppings, algae and sea animals, caused either by bites or scratches.

Isolated cases report history of injury with fishbone of dry fish called ¨Ngari¨, a knife while cleaning fish, indoor swimming pool complex, steel wool and even in a container of intravenous fluid at a medical center.

The clinical forms are lymphatic, fixed cutaneous, disseminated, pulmonary and osteoarticular. Rare forms such as invasion of the central nervous system, ocular, sinuses, pericardium, larynx, mammary gland, spleen, liver and bone marrow can be seen occasionally, particularly in immunosuppressed patients.

In adults, lesions predominate in hands and arms, which are normally exposed to traumatism, especially in those that work with vegetables and plants. The majority of the children present the lesions in the face. This is explained because children frequently expose their faces and the epidermis is thinner than in adults.[32, 33]

Rarely the fungus can enter the host through the respiratory tract. The mycelial form is the one inoculated, and transforms into yeast once inside the host. Only one report of transmission of person-to-person has been documented. [34, 35]

Culture of the yeast is made in Sabouraud agar with or without antibiotics, incubated at 25 or 37 °C, and between 25-28 °C for the mycelial form in the same agar.

Melanin production can be seen in either morphological presentation. Its capacity to produce pigment at a very broad pH range has advantages for greater pathogenicity and survival. This has been proven with melanized strains that are more capable of causing infection than non-melanized ones.

The cell wall is composed of polysaccharides such as mannose, glucose and rhamnose. Peptido-rhamnomannans are used for immunological diagnosis as intradermal tests, and are extremely useful for epidemiological research.

S. schenckii species complex strains show a great difference in virulence and antifungal susceptibility. S. brasiliensis is one of the most virulent species, followed by S. schenckii, S. globosa, while S. mexicana and S. albicans had low or no virulence in murine models.[36]

Some epidemics have been recorded. So far, the largest one reported occurred in South Africa between 1941-1944, where 3000 gold miners in the Witwatersrand coal mines where infected from contaminated timber inside the mines. At this moment, South Africa is known to have special ecologic areas as well as India and Japan with annual incidence of 15-20 cases per year.[37]

In the 1980's, the cat was recognized as potential source of disease. Since then other animals have shown capacity to infect humans, although cats are the only species with zoonotic potential, capable of transmitting the disease by scratch or bite.[38]

In 1988, the USA reported an outbreak with 84 cases affecting 15 states.[35]

Between 1992-1993 five patients were reported in Oklahoma with sporotrichosis. This outbreak was associated with stored hay or hay bales harvested in the US plains states (Iowa, Kansas, Minnesota, Missouri, Nebraska, North and South Dakota).[39]

An outbreak in 9 of 65 tree nursery workers in Florida developed lymphocutaneous Sporotrichosis in 1994.[40]

Few studies have estimated the annual incidence of 0.2-2.4 cases per million persons nationwide in the USA.[37]

Sporotrichosis is endemic in South America, but the exact incidence in unknown. It has been estimated one case per 1000 persons in Peru. The Southern Andean mountains are considered an hypendemic area with an incidence of 48-60 cases/100,000.[35, 41-43]

Children have an incidence 3 times higher than adults and are more likely to have lesions on the face and neck. The main risk factors identified are owing a cat because up to 2.4% are colonized with the fungus, and thus, believed to act as reservoirs, playing in crop fields, having a dirt floor in the house, working mainly outdoors and having a ceiling made of raw wood or conditions associated with lower socioeconomic status.[37, 44]

Cats, dogs and humans were involved in a large epidemic initiated by infected cats in Brazil, between 1998-2004. This emergent infectious disease proved expansion in host and vector ranges, dominated by zoonosis. At this moment the main risk factor identified for this mycosis is taking care of sick cats, and the most frequent occupations reported are domestic activities (30%) and students (18%).[45]

After identifying cats as reservoirs, epidemiological studies have shown that over 70% of these pets do not receive adequate treatment or burial. This is determinant for the epidemic by sustained dissemination of the disease [42] making sporotrichosis the most common subcutaneous mycosis in this country [46]. In Colombia, 23% of the patients recall the inoculating traumatism, while the majority fails to remember "minor" trauma. Uruguay reported 136 cases in 16 years, of which the main risk factor was contact with armadillos in 81% of cases.[17]

A Mexican report describes an epidemic familial outbreak of four cases in two families that were infected after sleeping in an old rust-stained camping tent, that was shown to be the source of infection. Three of them presented multiple sites of inoculation.[48]

In Australia in 2000, sporotrichosis was diagnosed as an emerging disease as part of an epidemic reported in an area where previously no cases had been diagnosed. Hay manipulation was found to be the source of infection.[47, 49]

India and the Asia-Pacific region are also endemic. The latter is considered to have different environmental reservoirs from those in the Western world.[50, 51]

In Sikkim, India autochthonous cases of sporotrichosis were reported, along with the geographic characteristics of the Himalayas where it is located. The annual temperature ranges from 11.3-19.8°C with an average humidity of 79-95% and average rainfall of 3,581 mm.

Pathogenicity tests were performed in mice, by inoculation with posterior development of the infection in the animal with a positive culture for S. shenckii compex.[52]

Disseminated disease is frequently seen in immunocompromised patients. The areas involved are commonly skin, lungs, bones, joints, and central nervous system.

Predisposing factors other than HIV/AIDS are alcoholism, diabetes mellitus, hematological malignancies, chronic or obstructive pulmonary disease, long term treatment with steroids, chemotherapy, and transplant recipients.[33]

Disseminated sporotrichosis can be a manifestation of the immune reconstitution inflammatory syndrome (IRIS).[53]

The first case associated to the immune reconstitution inflammatory syndrome (IRIS) of sporotrichosis meningitis was reported in 2010. This is a very unusual presentation, either in immunosuppressed or immunocompetent population.[54]

This fungus has caused numerous cases of local skin or eye infections in laboratory personnel, mostly associated to accidents involving splashing culture material into the eye, scratching or injecting infected material, or being bitten by experimentally infected animals. No pulmonary infections have been reported as a result from laboratory exposure. BSL-2 is recommended for handling laboratory specimens and infected animals.

With global warming, exposure to natural allergens could increase, as most of allergens increase their number in warmer climates. Expectation exists regarding fungi that are able enough to adapt to any climate. Fungal growth is enhanced by increased temperatures and/or humidity, and could have an impact on the clinical presentation of sporotrichosis.

Examples of this are the two Japanese cases of sporotrichosis that occurred in an area with annual average temperature of <10°C, and the usual annual temperature of endemic areas for sporotrichosis are around 22°C. Other areas in Japan, with warmer climate have reported large case series with 1.5-1.7 per 1000 outpatient visits to the dermatology department in a University Hospital.[37, 55, 56]

Several studies have tried to conclude the most important environmental factor involved in the propagation of the disease. Some reports mention that climate might have a determinant role, since some countries have higher prevalence than others. Others mention that the temperate seasons influence the spore amount in the environment, being highest during autumn and winter. A report from Brazil supports this fact. They found that high humidity and rain, temperature of 15-20°C seen during autumn and winter are favorable for the fungal development.

However, nowadays more factors are involved. Global warming is altering ecosystems worldwide, animal commerce, human migration, human immunosuppressive disease and fungal adaptation to this changes have undeniable influence in the prevalence of the disease. Apparently, at the moment, the highest risk factor identified for the increasing number of cases, are the immunosuppressed patients that live with HIV.[40]

After the war of 1914-18, France showed a reduction in the number of cases of sporotrichosis, despite the fact that the fungus was still isolated form environmental sources. This could be explained, at least partially, by the improvement in living standards and nutrition after that date.[58].Basic prevention of the disease includes protection such as decreased environmental exposure particularly in endemic areas. These measures include wearing protective clothing during outdoors activities, limiting contact with cats and soil, and improvements in living spaces among others.[37]

The gold standard for diagnosis is culture. However, frequently other tests are performed prior to this, such as a biopsy. The standard serologic test for Sporotrichosis (slide latex agglutination) has low sensitivity (56%), so other techniques are usually combined to complement the patients´study.[40]

In biopsy specimens, the cigar-shaped form is usually present and aids greatly for diagnosis. However, a report of two patients with AIDS found that these yeasts were large (more than 8μ) and round, similar to Cryptococcus neoformans.[58]

Immunostaining of biopsy specimens is an easy and rapid method that increases the accuracy of diagnosis.[59].

The study of Ghosh et al., as mentioned by Lyon et al showed that rates of sporotrichin reactivity increase as age increases, indicating that immunity is acquired with aging in patients that not necessarily will manifest the disease.[37].

Approximately 10% of diagnosed patients do no require treatment because of spontaneous cure, and around 2% of clinically cured patients can present clinical relapse. Only few patients require hospitalization, with good cure rates, and only occasionally death has been reported. The prognosis is generally good, particularly for cutaneous forms and immunocompetent patients.[41]

In limited resource settings, the first line of treatment for cutaneous and lymphocutaneous disease is potassium iodide (PI) which is both economic and effective in 60-100% of the cases. In patients where PI can't be prescribed, itraconazole is a good option.[47]

In older adults the response rate to azoles ranges between 90-100% in lymphocutaneous disease. Osteoarticular disease has lower response rates. Amphotericin B is the first line treatment for disseminated disease. Nephrotoxicity and drug interactions are the two most common adverse effects, so close monitoring is recommended.[28]

Patients living with HIV/AIDS and disseminated disease should receive initial therapy with amphotericin B followed by prolonged systemic azole therapy.[60]

Conclusion

Changes in epidemiology and incidence of endemic diseases sometimes is presumed to be altered by catastrophic events such as floods, earthquakes, volcanic eruptions, and even war. Although these events certainly may alter the incidence of communicable diseases, the changes that influence epidemics are usually far more subtle, such normal air and water currants, bird migration and commercial interchange of prime material or animals. About the host's susceptibility, of increasing importance are medical advances and challenges such as acquired immunosuppression, either by steroid or chemotherapy, or HIV infection.

So the imbalance either in the environment, in the host or in both will manifest as an increase in the incidence of infectious diseases that have health, social and economic consequences of varying degrees, depending on severity and extension.

Public health is in charge of controlling and improving every possible factor that may favorably influence the diminishment of the number of cases. Sanitation facilities, the degree of crowding, underlying health status of the population are always determinant factors, the better the general status of the population, the better response they will have when confronted to an infectious disease.

Endemic mycoses and opportunistic fungal infections are related with environmental fungi. Climatic and demographic changes, as well as travel, commerce and fungal adaptation, play an important role in the emergence or reemergence of fungal diseases. This so called "pathocenosis" can be shown with three important mycoses: Cryptococcosis, coccidioidomycosis and sporotrichosis.

C. neoformans is ubiquitous and dissemination is due to pigeon guano, following migratory and trade routes, while C. gattii is restricted to tropical/subtropical regions and associated to some trees.

Coccidioides includes two species, C. immitis and C. posadasii, inhabiting semiarid areas that characteristically have dry and alkaline soil. Both fungi are morphologically identical, but with genetic and epidemiological differences. The former is restricted to the Valley of California, and the later is found elsewhere, particularly in México and South America.

Sporotrichosis is an occupational mycosis caused by Sporothrix complex, it includes 5 molecular species: S. brasiliensis, S. globosa, S. albicans, S. mexicana and Sporotrhrix schenckii sensu stricto. These species are a related with hyperendemic areas in South America and also with zoonotic infections in cats and armadillos.

In patients living with HIV/AIDS all these fungi usually cause disseminated severe disease and should be treated with amphotericin B followed by prolonged systemic azole therapy.

These three examples of fungal disease are constantly adapting to the changes in the environment and the host's characteristics.

References

[1] Gonzalez JP, Guiserix M, Sauvage F, Guitton JS, Vidal P, Bahi-Jaber N, Louzir H & Pontier D. (2010) Pathocenosis: a holistic approach to disease ecology. *EcoHealth*; 7: 237-241

[2] Nielsen K, De Obaldia AL & Heitman J. (2007). *Cryptococcus neoformans* mates on pigeon guano: implications for the realizad ecological niche and globalization. *Eukaryotic cell*, 6: 949-959

[3] Zaragoza O, Rodrigues ML, De Jesus M, Frases S, Dadachova E & Casadevall A. (2009). The capsule of the fungal pathogen *Cryptococcus neoformans*. *Advances in Applied Microbiology*, 68: 133-216

[4] Gupta G & Fries B. Variability of phenotypic traits in *Cryptococcus* varieties and species and the resulting implications for pathogenesis. (2010). *Future Microbiology,* 5:775-787

[5] Springer DJ & Chaturvedi V. (2010). Projecting global occurrence of *Cryptococcus gattii*. *Emerging Infectious Diseases,* 16:14-20

[6] Tang JW. (2009). The effect of environmental parameters on the survival of airborne infectious agents. *Journal of the Royal Society Interface*, 6: S737-S746.

[7] Byrnes III EJ, Li W, Lewit Y, Ma H, Voelz K, Ren P, Carter DA, Chaturvedi V, Bildfell RJ, May RC & Heitman J. (2010). Emergence and pathogenicity of highly virulent *Cryptococcus gattii* genotypes in the Northwest United States. *PLoS Pathogens*, 6: 1-16

[8] Mac Dougall L & Fyfe M. (2006). Emergence of *Cryptococcus gattii* in a novel environment provides clues to its incubation period. *Journal of Clinical Microbiology*, 44: 1851-1852.

[9] Galanis E & Mac Dougall L. (2010). Epidemiology of *Cryptococcus gattii*, British Columbia, Canada 1999-2007. *Emerging Infectious Diseases,* 16: 251-257

[10] Byrnes III EJ, Bildfell R, Frank SA, Mitchell TG, Marr K &Heitman J. (2009). Molecular Evidence that the Vancouver Island *Cryptococcus gattii* outbreak has expanded into the United States Pacific Northwest. *Journal of Infectious Diseases*, 199:1081-1086.

[11] Person AK, Kontoyiannis DP & Alexander BD. (2010). Fungal infections in transplant and oncology patients. *Infectious Diseases Clinics of North America,* 24:439-459

[12] Schröter GPJ, Temple DR, Husberg BS, Weil III R & Starzl TE. (1976). Cryptococcosis after renal transplantation: report of ten cases. *Surgery*, 79: 268-277

[13] Boulware DR, Meya DB, Bergemann TL, Wiesner DL & Rhein J. (2010) Clinical Features and Serum Biomarkers in HIV Immune Reconstitution. Inflammatory Syndrome after Cryptococcal Meningitis: A Prospective Cohort Study. *PLoS Medicine* 7(12): e1000384. doi:10.1371/journal.pmed.1000384

[14] Dromer F, Mathoulin-Pelissier S, Launay O, Lortholary O & The French Cryptococcosis Study. (2007). Determinants of disease presentation and outcome during Cryptococcosis: the crypto A/D study. *PLoS Medicine*, 4:21

[15] Bellamy R., Sangeetha S & Paton NI. (2004). AIDS –defining illnesses among patients with HIV in Singapore, 1985 to 2001: results form the Singapore HIV observational cohort study (SHOCS). *BioMed Central Infectious Diseases*, 4: 47

[16] Voels K & May RC. (2010). Minireviews: Cryptococcal interactions with the host immune system. *Eukaryotic Cell*, 9: 835-846.

[17] Nielsen K, De Obaldia AL & Heitman J. (2007). *Cryptococcus neoformans* mates on pigeon guano: implications for the realizad ecological niche and globalization. *Eukaryotic Cell*, 6: 949-959.

[18] Bovers M, Hagen F, Kuramae EE, Hoogveld HL, Dromer F, St-Germain G & Boekhout T. (2008). AIDS patient death caused by novel Cryptoccocus neoformans x C. gatti hybrid. *Emerging Infectious Diseases*, 14: 1105-1108.

[19] Goldman JD, Vollmer ME & Luks AM. (2010). Cryptococcosis in the immunocompetent patient. *Respiratory Care*, 55:1499-1503

[20] Hsueh PR, Lau YJ, Chuang YC, Wan JH, Huang WK, Shyr JM, Yan JJ, Yu KW, Wu JJ, Ko WC, Yang YC, Liu YC, Teng LJ, Liu CY & Luh KT. (2005). Antifungal susceptibilities of clinical isolates of *Candida* species, *Cryptococcus neoformans*, and *Aspergillus* species from Taiwan: Surveillance of multicenter antimicrobial resistance in Taiwan program data form 2003. *Antimicrobial Agents and Chemotherapy*, 49: 512-517

[21] Saubolle MA, McKellar PP & Sussland D. (2007). Epidemiologic, clinical and diagnostic aspects of coccidioidomycosis. *Journal of Clinical Microbiology*, 45: 26-30

[22] Kirkland TN & Fierer J. (1996). Coccidioidomycosis: a reemerging infectious disease. *Emerging Infectious Diseases*, 3: 191-199.

[23] Ruddy BE, Mayer AP, Ko MG, Labonte HR, Borovansky JA, Boroff ES & Blair JE. (2011). Coccidioidomycosis in African Americans. *Mayo Clinic Proceedings*, 86:63-69

[24] Comrie AC. (2005). Climate factors influencing Coccidioidomycosis seasonality and outbreaks. *Environmental Health Perspectives*, 113: 688-692.

[25] Lee RU & Crum-Cianflone NF. (2008). Increasing incidence and severity of coccidioidomycosis at a naval air station. *Military Medicine*, 173: 769-775.

[26] Watson JT, Gayer M & Connolly MA. (2007). Epidemics after natural disasters. *Emerging Infectious Diseases*, 13: 1-5.

[27] Fisher MC, Koenig GL, White TJ & Taylor JW. (2000). Pathogenic clones versus environmenetally driven population increase: analysis of an epidemic of the human fungal pathogen *Coccidioides immitis*. *Journal of Clinical Microbiology* , 38: 807-813.

[28] Kauffman CA. (2001). Fungal infections in older adults. *Clinical Infectious Diseases*, 33: 550-555

[29] Hector RF & Laniado-Laborin R. (2005). Coccidioidomycosis: a fungal disease of the Americas. *PLoS Medicine*, 2: e2

[30] Miller MB, Hendren R & Gilligan PH. (2004). Posttransplantation disseminated Coccidioidomycosis acquired from donor lungs. *Journal of Clinical Microbiology*, 42: 2347-2349.

[31] Blair JE, Coakley B, Santelli AC, Hentz JG & Wengenack NL. (2006). Serologic testing for symptomatic coccidioidomycosis in immunocompetent and immunosuppressed hosts. *Mycopathologia*, 162: 317-324.

[32] Rubio G, Sánchez G, Porras L & Alvarado Z. (2010). Esporotricosis: prevalencia, perfil clínico y epidemiológico en un centro de referencia en Colombia. *Revista Iberoamericana de Micología*, 27: 75-79

[33] Devi KR, Devi MU, Singh TN, Devi KS, Sharma SS, Singh LR, Singh HL & Singh NB. (2006). Emergence of Sporotrichosis in Manipur. *Indian Journal of Medical Microbiology*, 24: 216-219.

[34] Rees RK & Swartzberg JE. (2011). Feline-transmited Sporotrichosis: a case study from California. *Dermatology Online Journal* 17(6):2

[35] De Araujo T, Marques AC & Kerdel F. (2001). Sporotrichosis. *International Journal of Dermatology*, 40: 737-742.

[36] López-Romero E, Reyes-Montes MR, Pérez-Torres A, Ruíz-Baca E, Villagómez-Castro JC, Mora-Montes HM, Flores-Carreón A & Toriello C.(2011). *Sporothrix schenckii complex* and sporotrichosis, an emerging health problem. *Future Microbiology*, 6: 85-102.

[37] Lyon GM, Zurita S, Casquero J, Holgadoo W, Guevara J, Brandt ME, Douglas S, Shutt K, Warnock DW & Hajjeh RA. (2003). Population based surveillance and a case control study of risk factors for endemic lymphocutaneous Sporotrichosis in Peru. *Clinical Infectious Diseases*, 36: 34-39.

[38] Barros MBL, Schubach AO, Schubach TMP, Wanke B & Lambert-Passos SR. (2008). An epidemic of Sporotrichosis in Rio de Janeiro, Brazil: epidemiological aspects of a series of cases. *Epidemiology and Infection*, 136: 1192-1196.

[39] Dooley DP, Bostic PS & Beckius ML. (1997). Spook house sporotrichosis. A point-source outbreak of sporotrichosis associated with hay bale props in a Halloween haunted-house. *Archives of Internal Medicine*, 157: 1885-1887

[40] Hajjeh R, McDonnell S, Reef S, Licitra C, Hankins M, Toth B, Padhye A, Kaufman L, Pasarell L, Cooper C, Hutwagner L, Hopkins R & McNeil M. (1997). Outbreak of

Sporotrichosis among tree nursery workers. *Journal of Infectious Diseases*, 176: 499-504

[41] Neyra E, Fonteyne PA, Swinne D, Fauche F, Bustamante B & Nolard N. (2005). Epidemiology of human Sporotrichosis investigated by amplified fragment length polymorphism. *Journal of Clinical Microbiology*, 43: 1348-1352

[42] Alves SH, Boettcher CS, Oliveira DC, Tronco-Alves GR, Sgaria MA, Thadeu P, Oliveira LT & Santurio JM.(2010). *Sporothrix schenckii* associated with armadillo hunting in Southern Brazil: epidemiological and antifungal susceptibility profiles. *Revista Da Sociedade Brasileira de Medicina Tropical*, 43: 523-525

[43] Pappas PG, Tellez I, Deep AE, Nolasco D, Holgado W & Bustamante B. (2000). Sporotrichosis in Peru: Description of an area of hypendemicity. *Clinical Infectious Diseases*, 30: 65-70.

[44] Kovarik CL, Neyra E & Bustamante B. (2008). Evaluation of cats as the source of endemic sporotrichosis in Peru. *Medical Mycology*, 46: 53-56.

[45] Schechtman RC. (2010). Sporotrichosis: part II. *Skin Medicine*, 8: 275-280.

[46] Saraiva Freitas DF, do Valle AC, de Almeida Paes R, Bastos FI & Galhardo MCG. (2010). Zoonotic sporotrichosis in Rio de Janeiro, Brazil: a protracted epidemic yet to be curbed. *Clinical Infectious Diseases*, 50: 453

[47] Rubio G, Sánchez G, Porras L & Alvarado Z.(2010). Esporotricosis: prevalencia, perfil clínico y epidemiológico en un centro de referencia en Colombia. *Revista Iberoamericana de Micologìa,* 2;507: 75-79

[48] Campos P, Arenas R & Coronado H. (1994). Epidemic cutaneous sporotrichosis. *International Journal of Dermatology,* 33: 38-41.

[49] Feeney KT, Arthur IH, Whittle AJ, Altman SA & Speers DJ. (2007). Outbreak of Sporotrichosis, Western Australia. *Emerging Infectious Diseases* , 13: 1228-1231.

[50] Chakrabarti, A & Slavin, M.A. (2011). Endemic fungal infections in the Asia-Pacific region. *Medical Mycology*, 49: 337-344

[51] Mehta KI, Sharma NL, Kanga AK, Mahajan VK & Ranjan N. (2007). Isolation of *Sporothrix schenckii* from the environmental sources of cutaneous sporotrichosis patients in Himachal Pradesh, India: results of a pilot study. *Mycoses*, 50: 496-501

[52] Bhutia PY, Gurung S, Yegneswaran PP, Pradhan J, Pradhan U, Peggy T, Pradhan PK & Bhutia CD. (2001). A case series and review of sporotrichosis in Sikkim. *Journal of Infection in Developing Countries*, 5: 603 608.

[53] Gutierrez-Galhardo MC, do Valle AC, Fraga BL, Schubach AO, Hoagland BR, Monteiro PC & Barros MB. (2010). Disseminated sporotrichosis as a manifestation of immune reconstitution inflammatory syndrome. *Mycoses*, 53, 78-80.

[54] Galhardo MCG, Silva MTT, Lima MA, Nunes EP, Schettini LEC, de Freitas RF, Paes RD & Neves ED. (2010). *Sporothrix schenckii* meningitis in AIDS during immune reconstitution syndrome. *Journal of Neurology, Neurosurgery and Psychiatry*, 81: 696-699

[55] Bernard SM, Samet JM, Grambsch A, Ebi KL & Romeiu I. (2001). The potential impacts of climate variability and change on air pollution-related health effects in the United States. *Environmental Health Perspectives*, 109: 199-209.

[56] Inokuma D, Shibaki A & Shimizu H. (2010). Two cases of cutaneous sporotrichosis in continental/microthermal climate zone: global warming alert? *Clinical and Experimental Dermatology*, 35: 668-669

[57] Symmers WSC. (1968). Sporotrichosis in Ireland. *Ulster Medical Journal*, 37: 85-101.

[58] Rocha MM, Dassin T, Lira R, Lima EL, Severo LC & Londero AT. (2001). Sporotrichosis in patients with AIDS: report of a case and review. *Revista Iberoamericana de Micología*. 18: 133-136

[59] Marques MEA, Coelho KIR, Sotto MN & Bacchi CE. (1992). Comparison between histochemical and immunohistochemical mehtods for diagnosis of sporotrichosis. *Journal of Clinical Pathology, 45: 1089-1093*

[60] Vilela R, Souza GF, Fernandes-Cota G & Mendoza L. (2007). Cutaneous and meningeal sporotrichosis in a HIV patient. *Revista Iberoamericana de Micología*, 24: 161-163.

In: Fungi: Types, Environmental Impact and Role in Disease ISBN: 978-1-61942-671-9
Editors: Adolfo Paz Silva and María Sol

Chapter IV

Candida Species as Causative Agents in Infections with Special Emphasis on Candida Albicans

Jack Ho Wong and Tzi Bun Ng
School of Biomedical Sciences, Faculty of Medicine,
The Chinese University of Hong Kong, Shatin,
New Territories, Hong Kong,
China and Shenzhen Research Institute,
Shenzhen, China

Abstract

The intent of this article is to review recent literature on the role of *Candida albicans* in infection. Candidemia, invasive candidiasis in neonates, disseminated candidiasis, invasive candidiasis in critically ill patients, Candida cerebrospinal fluid infection following neurosurgery, Candida cerebral abscess in premature infants.

Hematogenous dissemination of fungus of the eyes manifested as chorioretinitis or endophthalmitis, oropharyngeal candidiasis, Candida endocarditis, Candida pneumonia in immunocompromised patients, intra-abdominal pancreatic fungal infection, recurrent vulvovaginal candidasis, Candida interdigital ulcers, Candida interdigital ulcers, congenital cutaneous candidiasis, and burn wound infection are discussed.

Introduction

Candida albicans is an important fungal pathogen that is associated with many infections. A wealth of information has accumulated on this fungus. This article serves to provide a bird eyes view of the diversity of diseases that *Candida albicans* is implicated in. A mention of the other *Candida* species is made where appropriate.

a) Candidemia

Das et al. [1] reviewed the epidemiology of candidemia in a UK tertiary referral center. The predisposing factors comprised the use of broad-spectrum antibiotics or intravascular device, admission to an intensive care unit, and recent surgery. Non-*Candida albicans* species accounted for about 60% of the incidents, higher than what had been reported from other UK centers. *C. albicans* was the most common, followed by *C. glabrata* and *C. parapsilosis*. Overall *C. tropicalis, C. krusei, C. norvegensis,* and *C. lusitaniae* accounted for 7% of the incidents. The 30-day mortality rate was 37%. Advanced age and the presence of septic shock were associated with mortality [1].

b) Non-albicans candidemia

Candidemia due to non-albicans *Candida* spp. is of special concern because of its high drug resistance and increase in prevalence. Early identification of non-albicans candidemia is important. Among patients with non-albicans candidemia, more patients had neutropenia and less patients had candiduria and intensive care unit stay compared with those with *C. albicans* candidemia. Poor renal function and shock are independent risk factors for fatal candidemia. The outcome of candidemia was poor [2].

c) Invasive candidiasis in neonates

Premature infants and neonates with severe digestive diseases are vulnerable. *C. albicans* is responsible. Morbidity primarily concerns brain and lungs. Diagnosis is often delayed due to polymorphic clinical expression. Treatment is based on amphotericin B and prophylactic treatment with fluconazole [3].

d) Disseminated candidiasis

Disseminated candidiasis is the third leading and often fatal nosocomial blood stream infection in the USA. Immunization with either a glycopeptide or a peptide synthetic vaccine, both derived from *C. albicans,* prevented disseminated candidiasis in normal mice by cell wall. A combined protracted GI-tract colonization with *Candida* and the possibility of serum antibody responses to the presence of the fungus makes this a good animal model for testing the effectiveness of vaccines designed to prevent human disseminated candidiasis [4].

Tobudic et al. [5] described two patients with acute myelogenous leukemia who developed cervical lymphadenitis and chronic disseminated infection due to *C. albicans.* Thus *C. albicans* infection may be present in leukemic patients with acute lymphadenitis. Evaluation for visceral dissemination should be conducted even when fungemia is absent.

e) Invasive candidiasis in critically ill patients

The incidence of invasive candidiasis in critically ill patients has increased over the past decade and is associated with considerable morbidity and mortality. *C. albicans* accounts for approximately half of the invasive candidiasis cases, with non- *albicans* species, such as *C. glabrata* rising in frequency. Since early, appropriate therapy has brought about better outcomes, antifungal therapy is implemented in critically ill patients with risk factors for candidemia (prophylaxis). Systemic antifungal therapy is also employed in patients at high risk for invasive candidiasis based on surrogate markers of infection such as colonization (preemptive therapy), or in patients with unresolving sepsis in spite of management (empirical therapy) [6].

Mortality due to invasive infections by *Candida* strains in critically ill patients remains high in spite of introduction of new drugs in the last decade. To improve clinical outcomes, use of early effective therapies against different strains of *Candida* is important. Empiric therapy with echinocandins in suspected invasive candidiasis in these patients has been suggested [7].

f) Intravenous catheterization

Intravenous catheterization can lead to colonization as well as a broad spectrum of infections ranging from catheter site infections to catheter-related bloodstream infections. Aktaş et al. [8] emphasized that it is critical for any individual hospital to assess periodically the distribution and susceptibility profiles of isolates obtained from catheter-related infections to set out rational empirical treatment strategies.

g) Biofilms –related infections

C. albicans is the most common fungus associated with biofilm-related infections. Biofilms are microbial communities encased in a matrix of extracellular polymeric substances. Biofilm growth is highly resistant to antimicrobial agents. The review by Tobudic et al. [5] discusses the factors affecting antifungal resistance and activity of mono- and combination therapy of different antifungal classes and antifungal activity *in vitro* and *in vivo* against *C. albicans* biofilms.

h) *Candida* cerebrospinal fluid infection following neurosurgery

Candida infections of the central nervous system after neurosurgery are uncommon but complex and associated with significant morbidity and mortality. Prompt institution of antifungal therapy and removal of infected devices are required [9].

i) *Candida pelliculosa* meningitis in HIV infection

C. pelliculosa is a non-albicans species rarely associated with disease. Ratcliffe et al. [10] described the first reported case of *C. pelliculosa* meningitis in HIV infection. This case highlights the risk of uncommon opportunistic infections in association with the immunocompromised and the importance of autopsy as a diagnostic tool in HIV infection.

j) *Candida* cerebral abscess in premature infants.

Mao et al. [11] found that serial magnetic resonance imaging is useful in the early diagnosis of cerebral abscess caused by *C. albicans* in premature infants.

k) Hematogenous dissemination of fungus of the eyes manifested as chorioretinitis or endophthalmitis

Hematogenous dissemination of fungus of the eyes can manifest as chorioretinitis or endophthalmitis. In decreasing order of frequency, the organisms identified were *C. albicans, C. parapsilosis, C. glabrata, C. tropicalis,* and others. Disseminated ocular fungal infection is rare in the present era of widespread prophylactic antifungal therapy [12].

l) Oropharyngeal candidiasis

Oropharyngeal candidiasis, a common fungal infection in HIV-positive patients and patients with hematological cancer, can be treated with either systemic or topical antifungal agents. Brito et al. [13] determined the antifungal susceptibility profile of oral *Candida* isolates from HIV-positive patients and control individuals. Amphotericin B, fluconazole, flucytosine, ketoconazole and nystatin were effective on most isolates from both groups. Predisposing factors of oral candidiasis could be local and/orsystemic. Local factors include wearing dentures, impaired salivary gland function and poor oral health. Systemic factors include antibiotics and some other drugs, malnutrition, diabetes, immunosuppression and malignancies. Management involves an appropriate antifungal treatment and oral hygiene. Predisposing factors should be treated or eliminated where feasible. Oral hygiene involves cleaning the teeth and dentures. Dentures should be disinfected daily and left out overnight Miconazole buccal tablets exhibit few drug interactions because of low systemic absorption and are generally well tolerated [14].

Yan et al. [15] determined the prevalence of oral candidiasis and multiple oral *Candida* infections in patients with primary Sjögren's syndrome (SS), and the clinical and drug susceptibility profile. 87% (26/30) of subjects had oral candidiasis, in which 42% (11/26) had multiple Candida spp. infection. Although *C. albicans* remains the predominant isolate, *C. tropicalis*, *C. glabrata*, *C. parapsilosis*, and *C. krusei* were present, alone or in combination. Chronic atrophic candidiasis is the most common clinical type of oral candidiasis in patients with SS. The resistance to azoles was remarkable, and the phenomenon of cross-resistance between itraconazole and fluconazole was observed.

Patients with primary SS carry a high risk of oral candidiasis and a high frequency of multiple *Candida* infections. The azole resistance patterns of *Candida* spp. support the necessity for drug susceptibility testing as a routine procedure for patients with oral *Candida* infections [15].

Richardson et al. [16] reviewed the impact of biofilms in oral candidosis with special reference to medically compromised patients. The concept of oral candidosis as a mixed candidal-bacterial biofilm infection has changed the understanding of its epidemiology, diagnosis, and the strategy of treatment. *C. albicans* is the most common cause of oral candidosis though other *Candida* species are often seen in medically compromised patients exposed to multiple courses of antifungal azoles. Treatment comprises mechanical destruction of the biofilm in conjunction with topical drugs.

Azole antifungals should be avoided for patients suffering from recurrent oral yeast infections because of a risk of selection and enrichment of resistant strains within the biofilm. Oral candidosis can also signify undiagnosed or poorly controlled disease like HIV infection or diabetes.

m) *Candida* endocarditis

Use of cardiac rhythm management devices (CRMD) has brought about an increased incidence of pacemaker and cardioverter-defibrillator infections. CRMD Candida endocarditis is rare and serious. Isolates include *C. albicans* and other *Candida* sp. Both targeted antifungal therapy and device removal are recommended for management [17].

n) *Candida* pneumonia in immunocompromised patients

Early detection and treatment of pulmonary fungal infection in immunocompromised patients has a profound impact on mortality. However, information available about the thin-section computed tomography findings of *Candida* pneumonia has largely been limited to isolated cases seen in reviews of various infections and to other acute lung diseases seen in immunocompromised patients [18].

o) Cystic fibrosis

The importance of filamentous fungi and/or yeasts in relation to underlying inflammation is poorly defined. Reihill et al. [19] found that treatment of human bronchial epithelial cells or cystic fibrosis bronchial epithelial cells with *C. albicans* whole cells did not alter cytokine secretion.

On the contrary, treatment of cystic fibrosis bronchial epithelial cells with *A. fumigatus* conidia ensued in a 1.45-fold increase in interleukin-6 and a 1.65-fold increase in interleukin - 8 secretion; in contrast there was much less secretion from human bronchial epithelial cells. The data indicate that infection with *A. fumigatus* but not *C. albicans* evokes a pro-inflammatory response in cystic fibrosis epithelial cells.

p) Intra-abdominal pancreatic fungal infection

C. albicans is frequently isolated in necrotizing pancreatitis. Prolonged use of prophylactic antibiotics, chronic indwelling devices, and minimally invasive or surgical interventions for pancreatic fluid collections further elevate the risk of pancreatic fungal infection.

Amphotericin B is the most effective treatment. No randomized controlled trials have examined the efficacy of antifungal prophylaxis using broad-spectrum antibiotics for prevention [20].

q) Recurrent vulvovaginal candidasis

Vulvovaginal candidiasis is the second most common vaginal infection. HIV-infection is a risk factor [21]. Recurrences of vulvovaginal candidosis are caused by identical *Candida* strains suggesting *C. albicans* persistence in the female anogenital area.

The external vulva is a site of *C. albicans* persistence and source of endogenous re-infection. Combined therapy with oral fluconazole and ciclopiroxolamin cream topically compared favorably with fluconazole maintenance regimen [22].

r) *Candida* interdigital ulcers

Interdigital ulcer is a rare condition while erosio interdigitalisblastomycetica is common for candidiasis. Luo et al. [23] reported four Chinese patients with Candida interdigital ulcers. *C. albicans* was isolated from a man and two women, and *C. tropicalis* from another woman. Biopsy specimens revealed yeast and mycelium as well as inflammatory infiltrate in necrotic tissue in two patients; only inflammatory cells were noted in the other two. Candida species may cause interdigital ulcer on hand or foot. Oral itraconazole and topical bifonazole may be an optional therapy

s) Congenital cutaneous candidiasis

Congenital cutaneous candidiasis, a very rare disorder with good prognosis that is manifest within the first few days of life, have symptoms ranging from diffuse skin eruption devoid of systemic symptoms to respiratory distress, hepatosplenomegaly, sepsis, and death. *C. albicans* was seen on skin biopsy. The disease implies a congenital intrauterine infection distinct from neonatal candidiasis, which manifests as thrush or diaper dermatitis. The infection is acquired from the maternal genital tract. Topical antifungal therapy is adequate if there is no systemic candidiasis [24].

t) Burn wound infection

Fungi bring about burn wound infection, the most devastating form of trauma, as part of microbial infection, fungemia, rare aggressive soft tissue infection and as opportunistic infections. The risk factors for fungal infection in burns comprise late surgical excision, open dressing, artificial dermis, prolonged hospital stay, age of burns, total burn size, body surface area, full thickness burns, central venous catheters, inhalational injury, antibiotics, steroid treatment, long-term artificial ventilation, fungal wound colonisation, hyperglycaemic episodes and immunosuppressive disorders. Rapid diagnosis and treatment are critical since mortality is otherwise extremely high. Emergence of resistance in non-albicans *Candida* spp., unusual yeasts and moulds in fungal burn wound infection, leaves very few fungi susceptible to antifungal drugs, leaving many patients susceptible [25].

Capoor et al. [26] conducted a study to find the incidence, predisposing factors, spectrum, clinical profile and antifungal susceptibility of fungal wound infection in burn patients. Fungal pathogens were *C. tropicalis* (14%), *C. parapsilosis* (5.6%), *Aspergillus niger* (2.8%) and one each of *C. albicans* (1.4%), *C. glabrata* (1.4%), *Syncephalestrum* (1.4%) and *Fusarium solani* (1.4%). All of the fungi were susceptible to amphotericin B.

Discussion

From the foregoing account it can be seen that *Candida* spp., especially *C. albicans*, are important causative agents of diseases that affect various parts of the human body. Some of these diseases have a high mortality. Expedient diagnosis and therapy are pivotal to reduce the mortality. Search for new and efficacious drugs are also of paramount importance.

Acknowledgments

The award of research grants (Research Funds for the Control of Infectious Diseases project reference no. 10090812 and 10090882) by Food and Health Bureau, the Government of Hong Kong Special Administrative Region, is gratefully acknowledged.

References

[1] Das, I., Nightingale, P., Patel, M. & Jumaa, P. (2011) Epidemiology, clinical characteristics, and outcome of candidemia: experience in a tertiary referral center in the UK. *International Infectious Diseases,* [Epub ahead of print]

[2] Chi, H.W., Yang, Y.S., Shang, S.T., Chen, K.H., Yeh, K.M., Chang, F.Y. & Lin, J.C. (2011) *Candida albicans* versus non-albicans bloodstream infections: The comparison of risk factors and outcome. *Journal of Microbiology, Immunology and Infection,* [Epub ahead of print]

[3] Brissaud, O., Tandonnet, O. & Guichoux, J. (2011) Invasive candidiasis in Neonatal Intensive Care Unit. *Archives of Pediatrics,* 18 Suppl 1, S22-32.

[4] Cutler, J.E., Corti, M., Lambert, P., Ferris, M. & Xin, H. (2011) Horizontal Transmission of *Candida albicans* and Evidence of a Vaccine Response in Mice Colonized with the Fungus. *PLoS One,* [Epub ahead of print]

[5] Tobudic, S., Kratzer, C., Lassnigg, A. & Presterl, E. (2011) Antifungal susceptibility of *Candida albicans* in biofilms. *Mycoses,* [Epub ahead of print]

[6] Echeverria, P.M., Kett, D.H. & Azoulay, E. (2011) *Candida* prophylaxis and therapy in the ICU. *Seminars in Respiratory and Critical Care Medicine,* 32, 159-73.

[7] Tobar, A.E., Silva, O.F., Olivares, C.R., Gaete, G.P. & Luppi, N.M. (2011) Invasive candidiasis in critically ill adult patient. *Revista Chilena de Infectologia*, 28, 41-9.

[8] Aktaş, E., Sarı, E.N., Seremet Keskin, A., Pişkin, N., Külah, C. & Cömert, F. (2011) Causative agents of intravenous catheter-related infections and their antibiotic susceptibilities. *Mikrobiyoloji Bulteni*, 45, 86-92.

[9] O'Brien, D., Stevens, N.T., Lim, C.H., O'Brien, D.F., Smyth, E., Fitzpatrick, F. & Humphreys, H. (2011) *Candida* infection of the central nervous system following neurosurgery: a 12-year review. *Acta Neurochirurgica (Wien),* 153, 1347-50.

[10] Ratcliffe, L., Davies, J., Anson, J., Hales, S., Beeching, N.J. & Beadsworth, M.B. (2011) *Candida pelliculosa* meningitis as an opportunistic infection in HIV: the first reported case. *International Journal of STD and AIDS,* 22, 54-6.

[11] Mao, J., Li, J., Chen, D., Zhang, J., Du, Y.N., Wang, Y.J., Li, X., Wang, R., Chen, L.Y. & Wang, X.M. (2011) Value of MRI in the diagnosis of cerebral abscess caused by *Candida albicans* in premature infants. *Zhongguo Dang Dai Er Ke Za Zhi,* 13, 621-6.

[12] Dozier, C.C., Tarantola, R.M., Jiramongkolchai, K. & Donahue, S.P. (2011) Fungal eye disease at a tertiary care center: the utility of routine inpatient consultation. *Ophthalmology*, 118, 1671-6.

[13] Brito, G.N., Inocêncio, A.C., Querido, S.M., Jorge, A.O. & Koga-Ito, C.Y. (2011) In vitro antifungal susceptibility of *Candida* spp. oral isolates from HIV-positive patients and control individuals. *Brazilian Oral Research,* 25, 28-33.

[14] Collins, C.D., Cookinham, S. & Smith, J. (2011) Management of oropharyngeal candidiasis with localized oral miconazole therapy: efficacy, safety, and patient acceptability. *Patient Preference and Adherence,* 5, 369-74.

[15] Yan, Z., Young, A.L., Hua, H. & Xu, Y. (2011) Multiple Oral Candida Infections in Patients with Sjogren's Syndrome -- Prevalence and Clinical and Drug Susceptibility Profiles. *Journal of Rheumatology,* [Epub ahead of print]

[16] Richardson, R. (2011) Oral candidosis- Clinical challenges of a biofilm disease. *Critical Reviews of Microbiology,* [Epub ahead of print]

[17] Halawa, A., Henry, P.D. & Sarubbi, F.A. (2011) Candida endocarditis associated with cardiac rhythm management devices: review with current treatment guidelines. *Mycoses,* 54, e168-74.

[18] Xie, L.X., Liu, S.Y., Chen, Y.S., Liu, K. & Xue, F. (2001) An animal experiment about early pulmonary candidiasis in immunosuppressive rabbits: thin-section CT images dynamically observed and proved by histopathological results. *Acta Radiologica,* 52, 743-9.

[19] Reihill, J.A., Moore, J.E., Elborn, J.S. & Ennis, M. (2011) Effect of *Aspergillus fumigatus* and *Candida albicans* on pro-inflammatory response in cystic fibrosis epithelium. *Journal of Cystic Fibrosis,* [Epub ahead of print]

[20] Trikudanathan, G., Navaneethan, U. & Vege, S.S. (2011) Intra-abdominal fungal infections complicating acute pancreatitis: a review. *American Journal of Gastroenterology,* 106, 1188-92.

[21] Oliveira, P.M., Mascarenhas, R.E., Lacroix, C., Ferrer, S.R., Oliveira, R.P., Cravo, E.A., Alves, A.P. & Grassi, M.F. (2011) Candida species isolated from the vaginal mucosa of HIV-infected women in Salvador, Bahia, Brazil. *Brazilian Journal of Infectious Diseases,* 15, 239-44.

[22] Beikert, F.C., Le, M.T., Koeninger, A., Technau, K. & Clad, A. (2011) Recurrent vulvovaginal candidosis: focus on the vulva. *Mycoses,* doi: 10.1111/j.1439-0507.2011.02030.x. [Epub ahead of print]

[23] Luo, D.Q., Yang, W., Wu, L.C., Liu, J.H. & Chen, W.N. (2011) Interdigital ulcer: an unusual presentation of Candida infection. *Mycoses,* doi: 10.1111/j.1439-0507.2011.02023.x. [Epub ahead ofprint]

[24] Jagtap, S.A., Saple, P.P. & Dhaliat, S.B. (2011) Congenital cutaneous candidiasis: a rare and unpredictable disease. *Indian Journal of Dermatology,* 56, 92-3.

[25] Capoor, M.R., Sarabahi, S., Tiwari, V.K. & Narayanan, R.P. (2010) Fungal infections in burns: Diagnosis and management. *Indian Journal of Plastic Surgery,* 43(Suppl), S37-42.

[26] Capoor, M.R., Gupta, S., Sarabahi, S., Mishra, A., Tiwari, V.K. & Aggarwal, P. (2011) Epidemiological and clinico-mycological profile of fungal wound infection from largest burn centre in Asia. *Mycoses,* [Epub ahead of print]

In: Fungi: Types, Environmental Impact and Role in Disease
Editors: Adolfo Paz Silva and María Sol
ISBN: 978-1-61942-671-9

Chapter V

Effect of Polyene Antibiotics on the Potential Pathogenicity Factors in Saprophytic Micromycetes

*I. L. Kuzikova**, *N. G. Medvedeva*†,
V. I. Sukharevich and T. D. Trosheva
Institution of Russian Academy of Sciences,
Saint Petersburg Scientific Research Center for Ecological Safety RAS,
St. Petersburg, Russia

Abstract

The influence of polyene antibiotics on formation by saprophytic micromycetes of such pathogenicity factors as pigments, exopolysaccharides, phospholipases, and catalase was examined. The tested polyenes included medicinal antibiotics amphotericin B (heptaene) and nystatin (tetraene), as well as hexaene antibiotic 30 which was specially isolated with a view to its use for protection against biodeterioration. It was shown that that all the polyene antibiotics tested promote the formation of exopolysaccharides and phospholipases. The effects exerted on the catalase activity range from promotive to inhibitory, depending on the specific test culture.

However, by contrast to chemical fungicides, polyenes inhibit synthesis of pigments in all the saprophytic micromycetes tested. It was demonstrated that, when the polyene antibiotics are combined with surfactants, the fungicidal concentrations of the polyenes decrease by nearly an order of magnitude. Combined agents comprised of the polyenes and surfactants also suppress formation of pathogenicity factors (pigments, exopolysaccharides, phospholipases, and catalase) in saprophytic fungi. These findings suggest that non-medicinal antifungal polyene antibiotics, taken both alone and in combination with surfactants, offer promise for protection of industrial materials against mycodeterioration.

---

* e-mail: ilkuzikova@ya.ru.
† ngmedvedeva@gmail.com.

Keywords: Polyene antibiotics, nystatin, amphotericin B, antibiotic 30, saprophytic micromycetes, pathogenicity factors, pigments, polysaccharides, phospholipases, catalase

Introduction

The contamination of human habitats with potentially hazardous opportunistic fungi is currently subject of intense research worldwide. Relevant studies undertaken in a range of cities and towns in England, Canada, the United States, Germany, Spain, Argentina, Turkey, Japan, Israel, Denmark, etc. demonstrated a considerable taxonomic diversity of the micromycete species occurring in those urban areas and revealed several regionally specific features of anthropogenic mycocenoses at both the genus and species levels [1-7]. Those studies showed that, depending on the country, the number of fungi-contaminated buildings varied from 20 to 50%. Up to 37 and over taxa were identified in the dwellings examined, among which Aspergillus, Penicillium, Cladosporium, Alternaria, Ulacladium, Chaetomium, and Stachybotris fungal genera were of the most frequent occurrence, as indicated by virtually all the studies. The fungi exhibit only certain variability in their qualitative and quantitative relationship depending on the urban climate and other conditions.

In Russia, studies on fungal contamination of human habitats have been undertaken primarily in Moscow [8], St. Petersburg and Leningrad Oblast [9-11], and Vladivostok [12].

High fungal content of living areas is largely associated with increased humidity due to roof, plumbing, and other leaks. The indoor air and interior wall and ceiling surfaces are most often contaminated by fungi belonging to Penicillium and Aspergillus genera, while Cladosporium, Stemphylium, Rhizopus, and Alternaria fungi are less common; the wallpaper in five apartments was contaminated with Stachybotris chartarum [13]. Importantly, out of the 37 micromycete species of 14 genera, isolated from housing and occupational air in St. Petersburg, 24% cause allergic responses. The presence of Aspergillus flavus, A. fumigatus, A. niger, A. ustus, A. versicolor, Cladosporium cladosporoides, Penicillium brevicompactum, P. cyclopium, P. chrysogenum, and P. funiculosum micromycetes in urban and indoor air represents a risk factor for human health [11]. The resultant anthropogenic mycocenoses that occur in living spaces of urban buildings are organized in a manner different from that characteristic for naturally occurring mycocenoses.

A convincing proof of an adverse impact exerted by anthropogenic mycocenoses on human health may be found in the fact that the incidence of mycoses caused by opportunistic fungi tends to increase year after year [14]. This issue is closely related to that of mycodeterioration in view of the increasing trends in the concentrations, exhibited by the destructor fungi occurring in human habitats, most of which are opportunistic pathogens [15].

It is recognized now that, out of ca. 100 000 micromycete species described to date, from 100 to 500 may cause diseases [15]. This is particularly relevant to molds, above all to those from Aspergillus, Penicillium, Fusarium, Acremonium, and Alternaria genera.

Among them, the greatest risk is posed by Aspergillus niger, Aspergillus flavus, Aspergillus glaucus, Aspergillus nidulans, and Aspergillus fumigatus species which are known to be source of infectious diseases, mycoses of different organs in human body and, under indoor conditions, of lung diseases in humans [16]. Instances are known of mycetomes produced by Aspergillus amstelodami, Aspergillus nidulans, Aspergillus glaucus; Aspergillus terreus fungi may cause bronchopneumonia [17]. In cases of immunodeficiency, the mortality among persons affected with A. flavus reaches 50% [18].

Allergenic properties are most frequently reported for Alternaria alternate, Cladosporium sp., Aspergillus flavus, A. fumigatus, A. niger, Penicillium chrysogenum, P. tardum, and Rhizopus stolonifer species [19, 20].

Important findings about fungi associations, especially those under damp indoor conditions, as being responsible for a variety of health effects were reported by Gray et al. [21] who proposed the "mixed mold mycotoxicosis" term for multisystem illnesses.

A severe risk posed by opportunistic fungi also comes from low-level exposures to toxins (mycotoxins) which are formed by many of them and are responsible for numerous acute and chronic systemic pathologies in humans and animals.

Internal toxic contamination occurs through inhalation of air contaminated with toxins, dust, and fungal components, as well as through intake of food contaminated with fungi. People living in damp indoor spaces, inside which the building materials and air are contaminated with toxins, display nonspecific symptoms such as cough, eye, skin, and respiratory tract irritations, headache, fatigue [22], and in some cases, lung bleeding [23].

Of much importance for toxin formation in indoor air are Aspergillus and Penicillium genera fungi [24, 25]. Numerous studies were dedicated to A. fumigatus which is characterized by secretion into the environment of various toxins: aflatoxins, gliotoxin, fumagillin, fumagatin, fumitremorgin, etc. [26, 27]. This fact is of particular ecotoxicological significance, since the above-mentioned fungi, widespread in nature, commonly cause biodeteriorations in different chemical materials and are responsible for contamination of dwellings, offices, and other spaces.

Along with toxins, a negative impact on human health may come from allergenic proteins, as well as from β(1→3)-glucan components which cause airway inflammations [28], volatile organics produced by fungal growth, and other secondary metabolites contained in fungal spores and mycelia [29, 30].

Pathogenicity of micromycetes depends on a complex of adaptive features through which they resist host defenses and invade into host organism.

These features of fungi, the pathogenicity factors, include: the ability to grow at 37°C, enhanced adherence ability, melanin formation in cells [31, 32], and encapsulation [33], as well as the extracellular secretion of enzymes, aspartic proteases, phospholipases [33, 34], and oxidases, dimorphism, etc. The degree of pathogenicity (virulence) is determined by the intensity level of these characteristics [26, 35].

Major pigments produced by fungi include 3,4-dihydroxyphenylalanine melanin and 1,8-dihydroxynaphthalene melanin. Melaninogenesis is one of the main adaptive and nonspecific responses through which micromycetes become resistant to adverse impacts. Eighty-five per cent of all the pigment compounds investigated to date were attributed to Aspergillus and Penicillium genera; the most abundant colored metabolites are those detected in A. glaucas, A. flavus, and A. terreus fungi.

However, it was not until recently that the both types of melanins were associated with fungal virulence toward humans, animals, and plants [31, 36].

The most intense research efforts aiming to elucidate the relationship between fungal virulence and melaninogenesis were dedicated to clinical isolates of Cryptococcus neoformans which cause infections that pose health hazards, particularly to individuals with weakened immune systems [37]. It was demonstrated [38] that the melanin-forming C. neoformans species are superior in virulence over albino mutants.

Melanin production inhibits phagocytosis; this pigment neutralizes the activity of other cationic antimicrobial peptides [39] and binds avidly to antifungal antibiotic amphotericin B and caspofungin [40].

There are only very limited number of studies concerned with melaninogenesis in mycelial fungi, although many of them form pigments that display broadly varying structures, properties, and functions [41].

Some researchers [42] treat melanin as an antioxidant factor which is comparable in importance to superoxide dismutase. It was demonstrated [43] that melanin-lacking mutants are much less virulent. For example, nonpigmented A. fumigatus strains are less pathogenic than are pigmented ones and more avidly bind to complement. They show enhanced susceptibility to antifungal agents, which finding was associated by Latgé with higher permeability of the walls due to lack of pigments [43].

Another well-known virulence factor in fungi is represented by polysaccharides (PSs) [44, 45].

The importance of polysaccharide capsule in pathogenesis was convincingly demonstrated by the example of C. neoformans fungi [46], whose non-encapsulated mutants are poorly virulent or avirulent for mice [47]. The cells isolated from an infected animal have much larger polysaccharide capsules than do control culture cells.

The major function of PSs as a virulence factor is to contribute to resistance of a pathogen to host immune responses [48].

During infection, a thick polysaccharide capsule is formed in C. neoformans, which may suppress the leukocyte migration from the blood stream into an inflamed area [49].

Kozel et al. [44, 50] reported that neutrophils and macrophages rapidly phagocytized non-encapsulated isolates in vitro, by contrast to the encapsulated isolates which were resistant to phagocytosis.

Extracellular PSs are essential for adherence of fungi to the host cells, e.g., for C. albicans fungi cell adherence to epithelial or endothelial cells [51].

The virulence factors of pathogenic fungi also include the production and secretion of some hydrolytic enzymes.

To penetrate into the host tissue, the microbial cells form constitutive and inducible hydrolytic enzymes that damage host cell membranes or impair their functions and cause destruction of the immune system cells and molecules in the host. For example, phospholipases are deemed to be important virulence factors for various microorganisms: bacteria [52], yeasts [26], and mycelial fungi [27, 43].

High phospholipases activity revealed in C. albicans strains correlates with high virulence [53] both in vitro and in vivo (in mice); phospholipases formation is regarded as an important, universal virulence factor in pathogenic fungi [54].

Ganendren et al. [55] demonstrated the capability of phospholipases to enhance the virulence of yeasts in experiments with the use of an inhibitor of phospholipases, alexidine dihydrochloride. Under the action of inhibitor, fungal infection displayed slow development.

The phospholipases activity of opportunistic mycelial fungi was studied as well [27, 43, 56, 57]. The ability to produce extracellular phospholipases was detected in A. fumigatus, which is the most common invasive pathogen of humans; its phospolipases activity tends to enhance with increasing culture growth temperature to 37°C [56].

Medical microbiologists treat catalase-peroxidase oxyreductase as a virulence-associated factor as well [58, 59].

In Mycobacterium tuberculosum, Vibrio cholerae, and Staphylococcus aureus the virulence is directly related to the catalase activity [58]. In Mycobacterium avium, a causative agent of tuberculosis, a decrease in catalase activity is accompanied by a noticeable decrease in its virulence [60]. At the same time, a low-temperature induction of catalase activity leads to enhanced virulence [61]. It was found [62] that, in cases of staphylococcal infection, catalase is directly involved in phagocytosis suppression via affecting the lysosomal apparatus of neutrophils and macrophages, thereby enhancing the pathogen resistance to antifungal agents.

Studies dedicated to mycelial fungi are scarce; they are mostly concerned with A. fumigatus, which is the most common and dangerous airborne fungal pathogen of humans.

Like with other microorganisms, A. fumigatus is attacked by the reactive oxygen species produced by phagocytic cells upon infection of a host; detoxification of H2O2 is effected by fungal catalases as well [63].

The formation of pathogenicity factors in mycelial fungi can be promoted by fungicides. For example, catamine AB, metacide, metatin, and trilan stimulate the production by saprophytic fungi of such pathogenicity factors as pigments, polysaccharides, and phospholipases [64]. This may lead to a greater abundance in the anthropogenic community of fungi that are potentially hazardous to human health, which necessitates the constant search for new environmentally friendly protective agents.

Many researchers believe that the adverse impact of synthetic fungicides can be limited by changing, either totally or partially, from synthetic to natural biocides, in particular, to microbial metabolites possessing antifungal activity. Antibiotics, both with medicinal and non-medicinal functions, have long been successfully applied in crop production [65, 66], cattle breeding [67], food and canning industry [68], and (though to a much lesser extent so far) for protection against biodeterioration.

Nystatin and griseofulvin proved to be efficient in protection of paint and varnish coatings against microphytes, destructive soil fungi [69]. Beneficial effects were demonstrated for calcium-alginate bioresistant fiber which was developed on the basis of antibiotic kanamycin [70].

There exist instances of successful use of antibiotics for protection of cultural monuments against mycodeterioration. For example, nystatin with its fungicidal properties found application in restoration of church frescoes and paintings damaged by fungi before and after the catastrophic flood in Florence [71]. Isothiazolinone chloride, a derivative of thiazole (thiazolidine ring is part of penicillin series antibiotic molecules) was recommended for ancient monument protection against destructive microorganisms [72, 73].

Two antibiotics, phytobacteriomicin produced by Actinomyces lavendulae and grisin synthesized by Act. griseus [74], hold promise as antifungal agents for protection of film materials.

Antibiotic imbricin is efficient against fungi that cause deterioration to library collections [75].

However, widespread use of antibiotics for non-medicinal purposes is the subject of continuing heated debates among users of antibiotics and medical experts. The latter, considering ever increasing number of antibiotic-resistant microorganisms, insist upon more severe restrictions regarding the use of antibiotics for these purposes. With both parties having valid reasons to persist in their respective opinions, this long-standing discussion has not culminated in finding solution to this problem as yet.

One such solution may be found in the search for new antibiotics suitable for non-medicinal purposes either via rediscovery of existing known compounds or via chemical synthesis [76]. Also, particular interest in this context is attracted by antibiotics for which the emergence of resistant microorganisms occurs rarely or very slowly. Such antibiotics include, e.g., macrolide polyene antifungal antibiotics [77]. Polyene antibiotics are characterized by high antifungal activity, and not least importantly, delayed emergence of resistant microorganisms [77, 78].

However, the specific action of polyene antibiotics on the physiological and biochemical properties (in particular, emergence of potential pathogenicity factors) of saprophytic micromycetes that cause mycodeterioration is studied poorly. This complicates the assessment of the potential of antifungal polyene antibiotics as protective agents against mycodeterioration.

Here, we examined how polyene antibiotics affect the potential pathogenicity factors in fungi: pigment formation, production of exopolysaccharides, and phospholipase and catalase synthesis. We used both medicinal antibiotics (nystatin, amphotericin B) and new hexaene antibiotic 30 of the mediocidin type, which was specially isolated by us with a view to its use for protection against mycodeteriotation [79].

Materials and Methods

We used Aspergillus niger Van Thieghem 1199, A. terreus Thom 1025, Penicillium funiculosum Thom 1115, and P. ochrochloron Biourge 1702 mycelial fungi (available from the All-Russian Collection of Microorganisms) which belong to the most widespread microorganisms able to induce degradation of various materials.

The cultures were supported on agarized Czapek medium with 2% starch. The fungi were cultivated by the surface method on the same agarized medium in Petri dishes, as well as in 250-ml flasks containing 50 ml of liquid medium, under static conditions or on a Certomat BS-1 shaker at 24–26°C and 240 rpm. The cultivation time was 5–8 days, depending on the cultivation technique and the purpose of the experiment.

The fungal biomass was determined by weighing, and the intracellular colored substances, by aqueous-alkaline extraction followed by precipitation upon acidification to pH 2.0, drying, and weighing [80].

The fungal exopolysaccharides contained in the native solution were precipitated with ethanol and separated from the solution by centrifugation at 5000 rpm for 15–20 min, after which the resulting precipitate was dried at 120ºC and weighed. The amount of the polysaccharides was expressed in milligrams per gram of absolutely dry biomass (a.d.b.).

We used medicinal polyene antibiotics nystatin (tetraene) and amphotericin B (heptaene), as well as new hexaene antibiotic 30 of the mediocidin type, which was specially isolated by us with a view to its use for protection against mycodeteriotation. The antibiotic concentration in the medium was varied depending on the fungal sensitivity and to the intent and effect that the degree of inhibition of fungal biomass growth did not exceed 60%. As surfactants we used ethylenediaminetetraacetate and chlorhexidine bigluconate which were selected by us in preliminary experiments on the basis of the following criteria: effect on fungal pathogenicity factors (formation of pigments, polysaccharides, and phospholipases) and the lowest concentration level at which synergistic effect was observed [81]. The phospholipases activity of the fungi was determined by the plate method [82]. After 5 days of incubation at 25°C the diameter of the colony (Dc) and the total diameter of the colony plus the precipitation zone (Dc+p.z.) were measured. The phospholipases activity was determined as the ratio of the diameter of the colony plus the precipitation zone to that of the colony and expressed as percentage of the control. The catalase activity was measured by following the decomposition of hydrogen peroxide, which was monitored titrimetrically. A unit of catalase activity was defined as the amount of the enzyme catalyzing the decomposition of 1 µmole of H2O2 in 1 min and expressed in units/a.d.b. g [83]. The experimental data were processed with the use of Statistica 5.5 software package.

Results and Discussion

For two medicinal polyenes, amphotericin B and nystatin, it was shown that these antibiotics are highly efficient inhibitors of the growth of saprophytic micromycetes of various genera.

The minimal fungicidal concentration (MFC) of amphotericin B and nystatin, determined with the use of Czapek liquid medium, was 0.5–50 and 1.0–20 µg/ml, respectively, depending on the test culture (Table 1).

Table 1. Minimal fungicidal concentration (MFC) of amphotericin B and nystatin

Fungus	MFC, µg/ml	
	amphotericin B	nystatin
Aspergillus niger	0.5±0.03	1.0±0.05
A. flavus	30.0±1.2	10.0±0.6
A. terreus	35.0± 1.7	10.0±0.5
Penicillium funiculosum	50.0±2.0	20.0±1.2
P. ochrochloron	20.0±1.2	20.0±1.0
Scopulariopsis brevicaulis	15.0±0.7	5.0±0.3
Trichoderma viride	15.0±0.7	5.0±0.3

Along with the growth inhibition, the examined polyenes in subbiocidal concentrations impede the production of pigments, multifunctional fungal metabolites. For example, pigment production in the case of Aspergillus niger, A. terreus, Penicillium funiculosum, and P. ochrochloron fungi decreased, on the average, by 46 and 59%, in the medium with amphotericin B and nystatin respectively (Fig. 1). These decreases were paralleled by dramatic decreases in the color intensity of the culture liquid.

This feature, revealed by us for the first time, represents one of the advantages offered by the polyenes over many synthetic fungicides (catamine AB, metacide, metatin, copper 8-oxyquinolate, salicylanilide, and trilan) which promote the formation of colored metabolites [64, 84] and is important in several aspects:

- Pigments belong to potential pathogenicity factors of opportunistic fungi;
- Pigments rank among biodeterioration factors in fungi, and a decrease in their production intensity also contributes to reduction of the damage caused by fungi responsible for biodeterioration;
- Pigments play an important protective role in fungi, and suppression of their biosynthesis leads to enhanced sensitivity of fungi to antifungal agents.

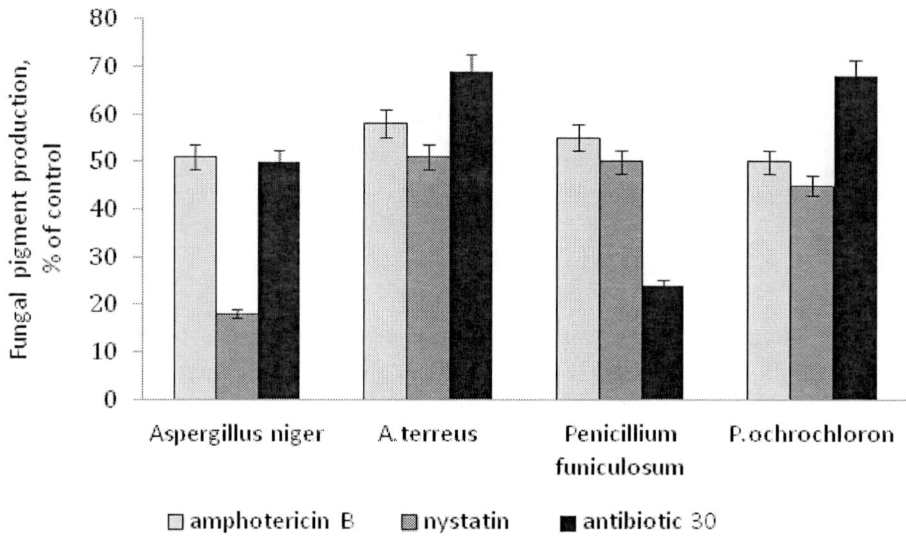

Figure 1. Effect of the polyene antibiotics on fungal pigment production. Control (without antibiotic) is taken as 100%.

The antifungal activity of the new polyene antibiotic 30 specially obtained by us with a view to its use for protection against mycodeteriotation was compared with those of highly purified medicinal polyene antibiotics amphotericin B and nystatin. Our experiments showed that the antifungal performance of antibiotic 30 is comparable with that of amphotericin B and is slightly poorer than that of nystatin (Table 2).

Table 2. Antifungal activity of antibiotic 30 and well-known medicinal polyene antibiotics

Antibiotic	Antibiotic, μg/disc	Fungal growth suppression zone, mm		
		Aspergillus niger	Penicillium ochrochloron	Candida utilis
Antibiotic 30	40	18.5±0.9	17.5±0.9	18.0±1.1
Antibiotic 30	80	19.5±1.0	18.0±1.1	15.0±0.9
Amphotericin B	40	16.5±1.0	9.0±0.5	15.0±0.8
Nystatin	80	25.0±1.3	29.5±1.8	23.0±0.9

Also, we examined how antibiotic 30 compares with the well-known antifungal polyenes in terms of the effect produced on the formation of pigments and exopolysaccharides (metabolites belonging to pathogenicity factors). Figure 1 shows that antibiotic 30 and the well-known polyenes amphotericin B and nystatin affect the fungal pigment formation along similar lines. For all the test cultures examined the pigments production decreased by 47% on the average. Studies of the biosynthesis of exopolysaccharides by fungi showed that it was promoted by both amphotericin B and nystatin in all the fungal test cultures examined by us: The amount of extracellular polysaccharides per gram of absolutely dry biomass increased by 23–48%, depending on the culture (Table 3).

A somewhat different effect was observed in the case of antibiotic 30 (Table 4). For example, the amount of polysaccharides released into the medium by A. niger and P. funiculosum cultures decreased by 13 and 51%, respectively. At the same time, antibiotic 30 promotes formation of exopolysaccharides in A. terreus and P. ochrochloron by 141 and 21%, respectively.

Enhanced synthesis of polysaccharides is undesirable, considering the protective function performed by polysaccharides which rank among the main pathogenicity factors in fungi.

As regards the activity of phospholipases (which belong to virulence factors of various microorganisms, including micromycetes), polyene antibiotic 30 caused the amount of phospholipases produced to increase slightly, by 10–25%, for all the test cultures examined (Table 5).

Table 3. Effect of amphotericin B, nystatin, and surfactant on polysaccharides production by fungi

Fungus	Antibiotic, μg/ml		Surfactant, %	Extracellular polysaccharides	
	ampho-tericin B	nystatin		mg/a.d.b g	% of control
Aspergillus niger	0.05	-	-	176±9	148
	-	-	0.1	82±4	69
	0.05	-	0.1	110±6	92
	-	0.07	-	107±5	142
	-	-	0.1	48±2	64
	-	0.07	0.1	69±4	92
A. terreus	1.0	-	-	239±12	123
	-	-	0.02	150±7	77
	1.0	-	0.02	170±8	87
	-	1.3	-	218±11	130
	-	-	0.02	129±6	77
	-	0.7	0.02	150±7	89
Penicillium funiculosum	5.0	-	-	260±12	118
	-	-	0.03	120±6	55
	5.0	-	0.03	150±7	68
	-	4.0	-	194±10	124
	-	-	0.02	79±4	50
	-	4.0	0.02	93±5	60

Table 4. Combined effect of antibiotic 30 and surfactant on fungal growth and exopolysaccharides production

Fungus	Antibiotic 30 μg/ml	Surfactant %	Biomass		Exopolysaccharides	
			g/l, a.d.b	% of control	mg/a.d.b. g.	% of control
Aspergillus niger	-	-	6.8±0.3	100	146.0±7.3	100
	1.0	-	4.4±0.2	65	127.0±7.6	87
	1.0	$5 \cdot 10^{-4}$	2.7±0.2	40	204.0±16.0	140
A. terreus	-	-	4.4±0.2	100	139.0±7.0	100
	5.0	-	4.0±0.2	62	297.0±21.0	241
	5.0	$1 \cdot 10^{-3}$	2.2±0.1	50	526±32.0	378
Penicillium funiculosum	-	-	5.2±0.2	100	691.0±35.0	100
	2.0	-	4.2±0.3	80	338.0±24.0	49
	2.0	$1 \cdot 10^{-4}$	2.5±0.3	48	288.0±17.0	42
P. ochrochloron	-	-	7.5±0.3	100	76.0±3.8	100
	2.0	-	6.7±0.3	89	92.0±5.5	121
	2.0	$5 \cdot 10^{-5}$	3.9±0.2	52	74.0±3.7	97

Table 5. Effect of antibiotic 30, taken alone and in combination with surfactant, on fungal phospholipase activity

Fungus	Concentration		Colony diameter Dc		Phospholipase activity	
	Antibiotic 30, μg/ml	Surfactant, %	mm	% of control	Dc+p.z. /Dc	% of control
Aspergillus terreus	-	-	53	100	1.32	100
	5.0	-	42	79	1.45	125
	5.0	$2 \cdot 10^{-3}$	37	70	1.35	102
Penicillium funiculosum	-	-	40	100	1.5	100
	5.0	-	30	75	1.87	125
	5.0	$2 \cdot 10^{-3}$	32	80	1.0	67
P. ochrochloron	-	-	52	100	1.25	100
	5.0	-	40	77	1.38	110
	5.0	$2 \cdot 10^{-3}$	38	73	1.0	80

Similar trends were observed in the variants with medicinal antibiotics nystatin and amphotericin B. It should be noted that chemical fungicides catamine AB, metacide, and trilan cause the phospholipase activity of Aspergillus and Penicillium genus fungi to increase much more significantly, by up to 200% [64].

In studies of the catalase activity as influenced by antibiotic 30 the concentration of the latter in the medium was varied depending on the fungal sensitivity as follows: 1.0–6.0, 5.0–8.0, 1.5–8.0, 0.2–0.4, and 0.2–1.0 μg/ml for Aspergillus niger, A. terreus, P. funiculosum, P. chrochloron, and A. fumigatus, respectively.

Figure 2 shows that, for three out of the five test cultures examined, the catalase activity increased by 19–137%, which increase was the larger the higher the antibiotic concentration. In the case of A. terreus and P. ochrochloron fungi we observed the inverse trend: With increasing antibiotic 30 concentration the catalase activity decreased to 64 and 86% of the control level, respectively. A similar trend was observed in the experiments with nystatin and amphotericin B.

To suppress the promotive effect of the polyene antibiotics on the synthesis of exopolysaccharides, phospholipases, and catalase (which represent pathogenicity factors for saprophytic fungi) we took advantage of the approach that we developed earlier [64]. It consists in the use of polyene antibiotics in combination with surfactants for the purpose of suppressing the protective functions of fungi.

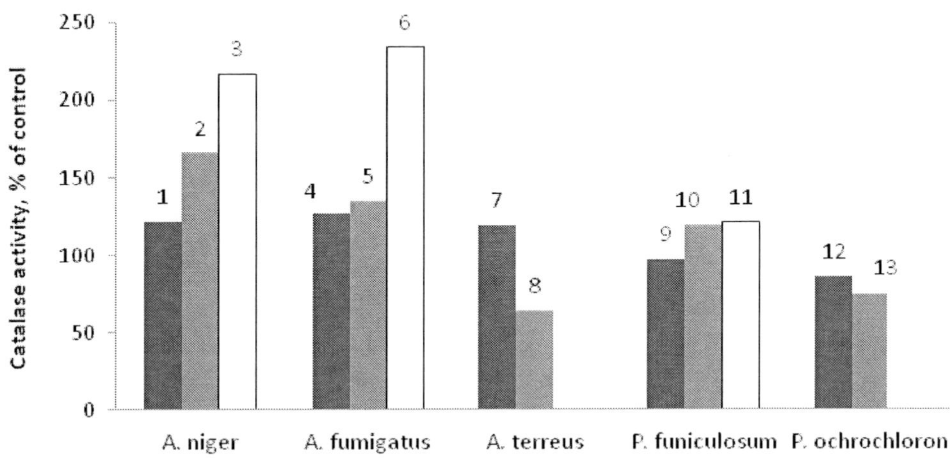

Figure 2. Effect of different concentrations of antibiotic 30 on the catalase activity in fungi. Antibiotic 30 concentration (μg/ml): 1 – 1.0; 2 – 2.0; 3 – 6.0; 4 – 0.2; 5 – 0.5; 6 – 1.0; 7 – 5.0; 8 – 8.0; 9 – 1.5; 10 – 4.0; 11 – 8.0; 12 – 0.2; 13 – 0.4. Control (without antibiotic) is taken as 100%.

Surfactants cause the cell membrane permeability in microorganisms to increase, thereby promoting the penetration of biocides into cells; they also can inhibit some protective, in particular, pathogenicity factors in fungi [81]. Preparation of complex agents based on polyenes and surfactants is of particular interest, since polyenes are also able to increase the membrane permeability and inhibit pigment formation in fungi.

We found that combined agents show a better performance than do individual antibiotics with respect to pigment production (Table 6).

Depending on the test culture, antibiotics 30 combined with surfactant causes the pigment production to decrease by 25–67% against 6–32% in the case of antibiotic 30 used alone under identical conditions. At the same time, for nystatin combined with surfactant such an effect was not observed: Under the action of nystatin used both alone and in combination with surfactant the pigment production by the test cultures decreases by 43 and 49%, on the average, respectively. Similar results were obtained for amphotericin B.

As regards polysaccharide formation by fungi as affected by combinations of amphotericin B and nystatin with surfactant, we found that, by contrast to individual antibiotics, the combined agents on their basis cause the extracellular PSs production to decrease to or even beyond the control level (Table 3). This phenomenon is most probably due specifically to the ability of surfactant to strongly suppress the polysaccharide production (Table 3).

Table 6. Combined effect of the polyene antibiotics and surfactant on fungal pigment production

Fungus	Antibiotic, μg/ml		Surfactant, %	Pigment, % of control
	nystatin	antibiotic 30		
Aspergillus niger	0.06	-	-	32
	0.06	-	$5 \cdot 10^{-2}$	30
	-	1.0	-	68
	-	1.0	$5 \cdot 10^{-4}$	33
A. terreus	0.5	-	-	62
	0.5	-	$5 \cdot 10^{-2}$	51
	-	5.0	-	102
	-	5.0	$5 \cdot 10^{-4}$	46
Penicillium funiculosum	5.0	-	-	75
	5.0	-	$2.5 \cdot 10^{-2}$	58
	-	2.0	-	80
	-	2.0	$1 \cdot 10^{-4}$	48
P. ochrochloron	-	2.0	-	94
	-	2.0	$5 \cdot 10^{-5}$	75

The effect exerted by hexaene antibiotic 30 in combination with surfactant on the extracellular PSs production depends on the properties of fungi: In Aspergillus genus it increases under the action of the combined agent, and in Penicillium genus, decreases relative to the level achieved with individual antibiotic 30 (Table 4).

As to fungal phospholipase activity, the combined agent comprised of antibiotic 30 and surfactant caused it to decrease by 23–58% relative to the level achieved with individual antibiotic 30. In the case of P. funiculosum and P. ochrochloron fungi the phospholipase activity even drops below the control level (Table 5). Similar results were obtained for combined agents based on nystatin and amphotericin B.

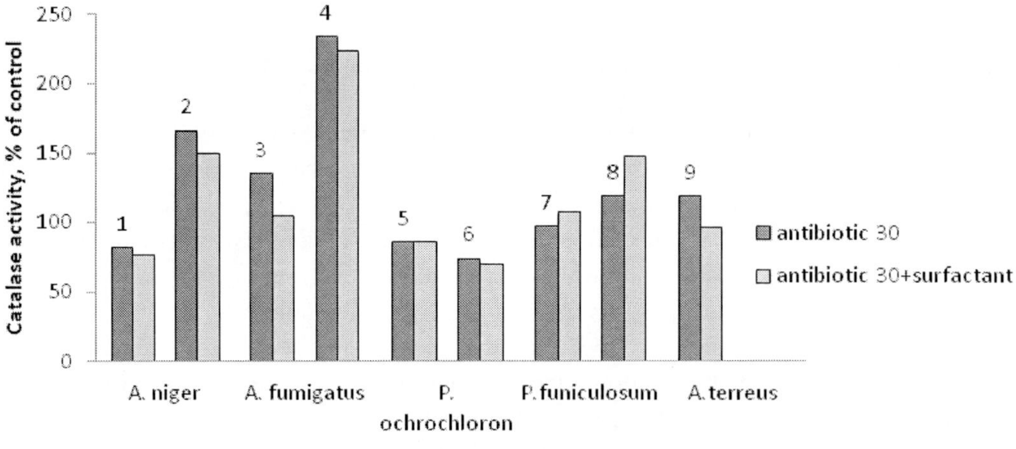

Figure 3. Catalase activity of fungi cultivated in a liquid nutritive medium containing antibiotic 30 taken alone or in combination with surfactant. Antibiotic 30 concentration (μg/ml): 1 – 1.5; 2 – 4.0; 3 – 0.5; 4 – 1.0; 5 – 0.2; 6 – 0.4; 7 – 1.5; 8 – 4.0; 9 – 5.0. Surfactant concentration $2 \cdot 10^{-5}$%.

Studies of the catalase activity of Aspergillus genus fungi showed that it decreases under the action of antibiotic 30 combined with surfactant (this decrease is the larger the higher the concentration of the antibiotic in its composition), though remains fairly high on the whole and exceeds the control level (Fig. 3).

A different response to the combined agent was exhibited by Penicillium genus fungi, for which no decrease in the catalase activity was observed, and by P. funiculosum fungi, for which it even slightly increased.

Table 7. Effect of the polyene antibiotics and surfactant on micromycete growth

Fungus	Antibiotic, μg/ml		Surfactant, %	Biomass, % of control
	amphotericin	nystatin		
Aspergillus niger	0.7*			0
	0.3			53
	0.3		0.1	0
		2.0*		0
		0.3		83
		0.3	0.1	0
A. terreus	45*			0
	10			75
	10		0.05	0
		15*		0
		3		50
		3	0.05	0
A. flavus	30*			0
	10			46
	10		0.1	0
		10*		0
		2.5	0.1	0
Penicillium funiculosum	48*			0
	15			63
	15		0.05	0
		25*		0
		10		68
		10	0.05	0

*– minimal fungicidal concentration

Similar trends in the fungal catalase activity were revealed for the polyene antibiotics amphotericin B and nystatin.

Studies of the combined effect of the polyene antibiotics and surfactants on the growth of saprophytic micromycetes showed that, depending on the test culture, the biocidal concentration of the antibiotics in the combined agents may decrease by a factor of 3.2, 4.5, and 4–10, on the average, for amphotericin B, nystatin, and antibiotic 30, respectively (Tables 7, 8).

Table 8. Combined effect of antibiotic 30 and surfactant on fungal growth

Fungus	Antibiotic 30, mg/ml	Surfactant, %	Biomass, % of control
Aspergillus niger	8.0*	0	0
	1.0	0	67
	1.0	$1 \cdot 10^{-3}$	0
A. terreus	8.0*	0	0
	2.0	0	50
	2.0	$1 \cdot 10^{-3}$	0
Penicillium ochrochloron	2.0*	0	0
	0.2	0	61
	0.2	$6 \cdot 10^{-4}$	0
P. funiculosum	6.0*	0	0
	1.0	0	47
	1.0	$2 \cdot 10^{-5}$	0

*– minimal fungicidal concentration.

Conclusion

Thus, the macrolide polyene antibiotics, both medicinal ones and that isolated by us with a view to its use for protection against microbiodeteriorations, are highly active fungicides possessing a broad antifungal action spectrum. Though structurally different in terms of the number of double conjugated bonds they contain, the polyenes amphotericin B (heptaene), nystatin (tetraene), and antibiotic 30 (hexaene) used by us exert a uniform effect on micromycetes belonging to different taxonomic groups, including those having dominance in biodeterioration. This uniformity is manifested in how these polyenes affect the physiological and biochemical properties of fungi. By contrast to chemical fungicides, all the examined polyenes strongly inhibit the pigment formation. In this feature, revealed by us for the first time, polyene antibiotics as potential inhibitors of microorganisms that cause biodeterioration have a doubtless advantage over other agents. This is due to the fact that pigments represent a pathogenicity factor in fungi, as well as a major biodeterioration factor, and enhance the resistance of destructor fungi to fungicides.

Another feature in which polyene antibiotics compare favorably with synthetic fungicides consists in rare occurrence and delayed development of fungal resistance.

It should be noted, however, that, similar to chemical fungicides, the polyene antibiotics examined by us promote formation of other pathogenicity factors in fungi, specifically, phospholipases, polysaccharides, and catalase, which at the same time belong to adaptability factors. The extent to which a particular factor will be affected varies depending on the properties of fungi and the antibiotic concentration.

The promotive effect exerted by the polyenes on the above-mentioned pathogenicity factors in saprophytic fungi can be reduced through combining polyenes with surfactants. Also, such combinations are beneficial in that they lead to increased fungal sensitivities to antibiotics, thereby decreasing the fungicidal concentrations of the polyenes by nearly an order of magnitude.

The above-discussed properties of the antifungal polyenes make them promising as environmentally friendly agents to be used for protection against mycodeterioration and also for preparation on their basis of even more efficient combined agents.

References

[1] Hunter, CA; Grant, C; Flannigan, B; Bravery, AF. Mould in buildings: the air spora of domestic dwellings. *International Biodeterioration and Biodegradation*, 1988, 24(2), 81-101.

[2] Koch, A; Heilemann, KY; Bischof, W; Heinrich, Y; Wickmann, HE. Indoor viable mold spores – a comparison between two cities (eastern Germany) and Hamburg (western Germany). *Allergy*, 2000, 55, 176-179.

[3] Shelton, BG; Kirkland, KH; Flanders, WD; Morris, GK. Profiles of airborne fungi in building outdoor environments in the United States. *Applied and Environmental Microbiology*, 2002, 68(4), 1743-1753.

[4] Gunnbjörnsdottir, MI; Norbäck, D; Plaschke, P; et al. The relationship between indicators of building dampness and respiratory health in young Swedish adults. *Respiratory Medicine*, 2003, 7, 302-307.

[5] Aira, MJ; Jato, V; Stchigel, AM; Rodriguez-Rajo, FJ; Piontelli, E; Aeromycological study in the Cathedral of Santiago de Compostela (Spain). *International Biodeterioration and Biodegradation*, 2007, 60, 231-237.

[6] Cül, H; Issever, H; Ayraz, Ő; Cügöк, G. Occupational and environmental risk factors for the sick building syndrome in modern offices in Istanbul: A cross sectional study. *Indoor and Built Environment*. 2007. V. 16, N1. P. 47-54.

[7] Aringoli, EE; Basilico Maria de la Luz; Altahus, RL; Basilico, JC. Multivariate analysis of fungal association in the indoor air of Argentinean houses. *International. Biodeterioration and Biodegradation*, 2008, 62, 281-286.

[8] Antropova, AB; Mokeeva, VL; Bilanenko, EN; Chekunova, LN; Zheltikova, TM; Petrova-Nikitina, AD. Aeromikota of dwellings in the city of Moscow. *Mikol. Fitopatol.,* 2003, 37, 1–11. (in Russian)

[9] Smolyanitskaya, OL. Species composition of micromycetes in some exhibition halls and depositories of Hermitage. *Mikol. Fitopatol.*, 2004, 38, 51–58. (in Russian)

[10] Sergeeva, LE; Vasil'ev, OD. Mycobiota of the Russian National Library depositories. *Mikol. Fitopatol.,* 2005, 39, 69–80. (in Russian)

[11] Ivanova, AM; Kirtsideli, IYu. Complexes of microscopic fungi in the air of St. Petersburg. *Mikol. Fitopatol.*, 2007, 41, 40–47. (in Russian)

[12] Egorova, LN; Klimova, YuA. Mycobiota of air in different premises in the city of Vladivostok. *Mikol. Fitopatol.,* 40,487–493, 2006. (in Russian)

[13] Kozlova, YaI; Vasil'eva, NV; Chilina, GA; Bogomolova, TS; et al. Mycogenic allergy in residents of premises contaminated with micromycetes. *Problems in Medical Mycology*, 2008, 10(2), 17-22. (in Russian)

[14] Van Burik, JA; Magee, PT. Aspects of fungal pathogenesis in humans. *Annu Rev Microbiol*, 2001, 55, 743-772.
[15] Antonov, VB. Influence of Biodeterioration on Human Health. In: Mater. Int. Conf. "Problems of durability of buildings and structures", October 10–12, 2007. St. Petersburg, 137–142. (in Russian)
[16] Landau, MA. *Molecular Mechanisms of Action of Physiologically Active Compounds*. Nauka. Moscow; 1981. (in Russian)
[17] Krylenkov, VA; Vlasov, DYu. Role of biodeterioration of building materials in the formation of human epidemiological environment. In: Mater. Int. Conf. "Problems of durability of buildings and structures", October 10–12, 2007. St. Petersburg, 171–175. (in Russian)
[18] Sheppard, DC; Rieg, G; Chiang, LY; Filler, SG; Edwards, JE; and Ibrahim, AS. Novel inhalational murine model of invasive pulmonary aspergillosis. *Antimicrob. Agents Chemother*, 2004, 48, 1908-1911.
[19] Mirakyan, ME. *Essays on Professional Fungal Diseases*. Erevan: Aiastan; 1981. (in Russian)
[20] Marfenina, OE; Kul'ko, AB; Ivanova, AE. Microscopic fungi in outdoor urban environment. *Mikol. Fitopatol.,* 2002, 36, 22–31. (in Russian)
[21] Gray, MR; Thrasher, ID; Crago, R; Madison, RA; Arnold L et al. Mixed Mold Mycotoxicosis: immunological changes in humans following exposure in Water-Damaged Building. *Archives of Environmental Health*, 2004, 58(7), 410-420.
[22] Hodgson, MJ; Morey, P; Leung, WY; Morrow, L; Miller, D; Jarvis, BB et al. Building-associated pulmonary disease from exposure to Stachybotrys chartarum and Aspergillus versicolor . *J. Occup. Environ. Med*, 1998, 40, 241-249.
[23] Gottschalk, C; Bauer, J; Meyer, K. Detection of satratoxin G and H in indoor air from a water-damaged building. *Mycopathologia*, 2008, 166 (2), 103-107.
[24] Varga, J; Kevei, F; Rinyu, E; Teren, J; Kozakiewicz. Ochratoxin production by Aspergillus species. *Appl. Environ. Microbiol*, 1996, 62(12), 4461-4464.
[25] Samson, RA; Frisvad, JC; Penicillium subgenus Penicillium: new taxonomic schemes mycotoxins and other extrolites. *Studies in Mycology*, 2004, 49, 257.
[26] Karkowska-Kuleta, J; Rapala-Kozik, M; Kozik A. Fungi phatogenic to humans: molecular bases of virulence of Candida albicans, Cryptococcus neoformans and Aspergillus fumigatus. *Acta Biochimica Polonica*, 2009, 56(2), 211-224.
[27] Tomme, JFC; Kauffman, HF. Putative virulence factors of Aspergillus fumigatus. *Clin. Exp. Allergy*, 2000, 30, 476-484.
[28] Chew, GL; Douwes, J; Doekes, G; Higgins, KM; van Strien, R; Spithoven, J; et al. Fungal extracellular polysaccharides, $\beta(1\rightarrow3)$-glucans and culturable fungi in repeated sampling of house dust. *Indoor Air*. 2001, 11, 171–178.
[29] Pasanen, AL; Korpi, A; Kasanen, JP; Pasanen, P. Critical aspects on the significance of microbial volatile metabolites as indoor air pollutants. *Environ. Int.,* 1998, 24, 703-712.
[30] Nielsen, KF. Mycotoxin production by indoor molds. *Fungal Genetics Biology*, 2003, 39, 103-117.
[31] Polak, A. Melanin as a virulence factor in pathogenic fungi. *Mycoses*, 1999, 33(5), 215-224.
[32] Butler, MI; Day, AW; Henson, JM. Pathogenic properties of fungal melanins. *Mycologia*, 2001, 93(1), 1-8.

[33] Schaller, M; Borelli, C; Korting, HC; Hube, B. Hydrolytic enzymes as virulence factors of Candida albicans. *Mycoses*, 2005, 48(6), 365-377.

[34] Cox, GM; McDade, HC; Chen, SCA; Tucker, SC; et al. Extracellular phospholipases activity is a virulence factor for Cryptococcus neoformans. *Molecular Microbiology*, 2001, 39(1), 166-175.

[35] Latge, JP, The pathobiology of Aspergillus fumigatus. *Trends in Microbiology*, 2001, 9(8), 382-389.

[36] Langfelder, K; Streibel, M; Jahn, B; et al. Biosynthesis of fungal melanins and their importance for human pathogenic fungi. *Fungal Genetics and Biology*, 2003, 38, 143-158.

[37] Currie, BP; Casadevall, A. Estimation of the prevalence of cryptococcal infection among HIV infected individuals in New York City. *Clin. Infect. Dis*, 1994, 19, 1029-1033.

[38] Kwong-Chung, KJ; Rhodes, JC. Encapsulation and melanin formation as indicators of virulence in Cryptococcus neoformans. *Infect. Immun*, 1986, 51, 218-223.

[39] Doering, TL; Nosanchuk, JD; Roberts, WK; and Casadevall, A. Melanin as a potential cryptococcal defence against microbicidal proteins. *Med. Mycol*, 1999, 37, 175-181

[40] Van Duin, D; Casadevall, A; and Nosanchuk, JD. Melanization of Cryptococcus neoformans and Histoplasma capsulatum reduces their susceptibility to amphotericin B and caspofungin. *Antimicrob. Agents Chemother*, 2002, 46, 3394-3400.

[41] Laykh, SP. *Microbial Melaninogenesis and Its Functions*. Moscow, Nauka; 1981. (in Russian)

[42] Jacobson, ES; Tinnell, SB. Antioxidant function of fungal melanin. *J. Bacteriol*, 1993,175, 7102-7104.

[43] Latge, JP. Aspergillus fumigatus and aspergillosis. *Clin. Microbiol. Rew*, 1999, 12(2), 310-350.

[44] Kozel, TR; Cuerlain, AS; Highison, BA; Highison, GJ. Role of the capsule in phagocytosis of Cryptococcus neoformans. *Rev. Infect. Dis*, 1988, 10, 8436-8439.

[45] Buchanan, KL; Murphy, JW. What makes Cryptococcus neoformans a pathogen? *Emerg. Infect. Dis*, 1998, 4, 71-83.

[46] Chang, YC; Kwon-Chung, RJ. Complementation of capsule-deficient mutation of Cryptococcus neoformans restores its virulence. *Molecular and Cellular Biology*, 1994, 14(7), 4912-4919.

[47] Fromptling, RA; Shadomy, HK; Jacobson, ES. Decreased virulence in stable acapsular mutants of Cryptococcus neoformans. *Mycopathologia*, 1982, 79, 23-29.

[48] Vecchiarelli, A. The cellular responses induced by the capsular polysaccharide of Cryptococcus neoformans differ depending on the presence or absence of specific protective antibodies. *Curr. Mol. Med*, 2005, 5, 413–420.

[49] Perfect, JR. Cryptococcus neofromans: a sugar-coated killer. In: Heitman J, Filler SG, Edwards JE Jr, Mitchell AP, editors. *Molecular principles of fungal pathogenesis*. Washington DC: ASM Press; 2006; PP. 281–303.

[50] Kozel, TR; and Cazin, JJr. Noncapsulated variant of Cryptococcus neoformans. I. Virulence studies and characterization of soluble polysaccharide. *Infect. Immun,* 1971, 3, 287-294.

[51] Masuoka, J. Surface glycans of Candida albicans and other pathogenic fungi: physiological roles, clinical uses, and experimental challenges. *Clin. Microbiol. Rev.*, 2004, 17(2), 281-310.

[52] Salyers, A; Witt, D. Virulence factors that damage the host. In: Salyers A, Witt D, editors. *Bacterial pathogenesis: a molecular approach*. Washington, DC: ASM Press; 1994; PP. 47-62.

[53] Barrett-Bee, K; Hayes, Y; Wilson, RG; Ryley, JF. A comparison of phospholipases activity, cellular adherence and pathogenicity of yeasts. *J. Gen. Microbiol,* 1985, 131, 1217-1221.

[54] Ghannoum, MA. Extracellular phospholipases as universal virulence factor in pathogenic fungi. *J. Med. Mycol*, 1998, 39, 55-59.

[55] Ganendren, R; Widmer, F; Singhal, V; Wilson, C; Sorrell, T; Wright, L. In Vitro Antifungal Activities of Inhibitors of Phospholipases from the Fungal Pathogen Cryptococcus neoformans. *Antimicrob. Agents. Chemother*, 2004, 48, 1561-1569.

[56] Birch, M; Robson, G; Law, D; Denning, D. Evidence of multiple extracellular phospholipase activities of Aspergillus fumigatus. *Infection and Immunity*, 1996, 64(3), 751-755.

[57] Shen, DK; Noodeh, AD; Kazemi, A; Grillot, R; Robson, G; Brugère, JF. Characterisation and expression of phospholipases B from the opportunistic fungus Aspergillus fumigatus. *FEMS microbiology letters*, 2004, 239(1), 87–93.

[58] Somov, GP; Litvin, VYu. *Saprophytism and Parasitism of Pathogenic Bacteria: Environmental Aspects*. Novosibirsk: Nauka, 1988. (in Russian)

[59] Chromy, BA; Choi, MW; Murphy, GA; Gonzales, AD et al. Proteomic characterization of Yersinia pestis virulence. *J. of Bacteriology*, 2005, 187, 8172-8180.

[60] Rotov, VI; Kozlov, VS. Catalase activity and virulence of Mycobacterium avium during long-term persistence in the environment. *Mikrobiol. Zh.*, 1976, 38, 640–641. (in Russian)

[61] Varvashevich, TN; Nikiforova, LS; Dratsenko, AF. Mechanisms of low–temperature regulation of microorganisms. In: Somov GP. editor. *Psychrophilic Pathogens*. Novosibirsk: Inst. Epidemiol. Mikrobiol Sib. Otd. Akad. Med. Nauk SSSR; 1986; 7–14. (in Russian)

[62] Mandell, GL. Catalase, superoxid dismutase and virulence of Staphylococcus aureus. In vitro and in vivo studies with emphasis on staphylococcal leukocyte interaction. *J. of Clinical Investigation*, 1975, 55(3), 561-566.

[63] Paris, S; Wysong, D; Debeaupuis, J; Shibuya, K; Philippe, B; Diamond, RD; and Latge, J. Catalases of Aspergillus fumigatus. *Infection and Immunity*, 2003, 71(6), 3551–3562.

[64] Kuzikova IL, Medvedeva NG, and Sukharevich VI. Complex fungicidal agents for protection against biodeterioration. In:Varfolomeev SD, Zaikov GE, Krylova LP, editors. *Biotechnology and the Ecology of Big Cities*. New York: Nova Science Publishers Inc.; 2011; 71–79.

[65] Yamaguchi H et al. In: Yamaguchi H., Kobayashi GS, Takahashi H. editors. *Recent progress in antifungal chemotherapy*. New York, Basel, Hong Kong: Marcel Dekker Inc.; 1992; 427–432.

[66] Titova, YuA; Novikova, II; Khlopunova, LB. Trichodermin based on secondary bioconversion of waste and its efficiency against cucumber diseases. *Mikol. Fitopatol.,* 2002, 36, 76–79. (in Russian)

[67] Egorov, NS. *Fundamentals of the Theory of Antibiotics*. Moscow: Vysshaya Shkola; 1986. (in Russian)

[68] Shenin, YuD; Belakhov,VV. Fungicidal activity of imbricin and its salts. *Antibiot. Khimioter.,* 1996, 41, 18–22. (in Russian)

[69] Lisina-Kulik, ES; Tyurin, BF; Makarova, LV; et al. Use of some antibiotics for protection of paint and varnish coatings against destruction by soil fungi, macrophytes. *Lakokras. Mater. Ikh Primen.* 1976, 4, 47–49. (in Russian)

[70] Ermilova, IA. *Theoretical and Practical Principles of Microbial Degradation of Chemical Fibers.* Moscow: Nauka; 1991. (in Russian)

[71] Gale, E; Candliff, E; Reinolds, P; et al. *The Molecular Basis of Antibiotic Action.* London: Wiley; 1972.

[72] Curri, SB. Biocide testing and enzymological studies on damaged stone and frescoes surfaces preparation of antibiograms. *Biochem. Exp. Biol*, 1979a, 15(1), 97–104.

[73] Curri, SB. The microbiological attack on the surface of Fidia's stabue ecrops and Pandrossos: II Evaluation of the activity of some biocides on the polluted marble. *Biochem. Exp. Biol*, 1979b, 15(3), 299–304.

[74] Vorob'eva, LI; Naumova, ES; Altukhova, LS; et al. Use of streptomycin antibiotics for treatment of film materials. *Microbiol. Zh.*, 1989, 51, 83–87. (in Russian)

[75] Sukharevich, VI; Dobrusina, SA; Medvedeva, NG. Imbricin: biocide for protection of paper against fungal biodeterioration. *Mikol. Fitopatol.,* 1997, 31, 36–41. (in Russian)

[76] Kim, BS; Moon, SS; Hwang, BK. Structure elucidation and antifungal activity of an anthracycline antibiotic, daunomycin, produced by Actinomadura roseola. *J. Agric. Food Chem*, 2000, 48, 1875–1881.

[77] Navashin, SM; Fomina, IP. *Rational Antibiotic Therapy*. Moscow: Meditsina; 1982. (in Russian)

[78] Patel, PV; Johnson, IR. Dominant mutation for nystatin resistance in yeast. *Appl. Microbiol.*, 1968, 16, 1654.

[79] Kuzikova, IL; Sukharevich, VI; Shenin, YuD; Medvedeva, NG. Biological abilities and identification of the polyene antifungal antibiotic perspective for protection from fungi biodeterioration. *Biology Bulletin,* 2010a, 37, 193–202. Original Russian Text: Kuzikova, IL; Sukharevich, VI; Shenin, YuD; Medvedeva, NG, 2010, published in *Izvestiya Akademii Nauk, Seriya Biologicheskaya*, 2010, 2. 238-247.

[80] Lysenko, SV; Lyakh, SP. Protective role of pigments in fungi isolated from lisosphere under UV exposure. *Mikrobiologiya*, 1977, 46, 867–877. (in Russian)

[81] Kuzikova, IL; Medvedeva, NG; Sukharevich, VI. Effect of fungicides and surfactants on the potential pathogenicity factors in fungi which cause mycodeterioration. *Mikol. Fitopatol.,* 2010b, 44, 441–447. (in Russian)

[82] Price, MF; Wilkinson, ID; Gentry, LO. Plate method for detection of phospholipase activity in Candida albicans. *Sabouraudia*, 1982, 20, 7-14.

[83] Borisova, VN. Enzyme Activators of Oxygen (Terminal Oxidases). In: Bilai VI, editor. *Methods of Experimental Mycology.* Kiev: Naukova Dumka; 1973; 95–102. (in Russian)

[84] Sukharevich, VI; Zaitseva, TB; Medvedeva, NG; Sukharevich, ME; Kuzikova, IL. Influence of biocides of different chemical nature on synthesis of pigments in cellulose-degrading fungi. *Mikol. Fitopatol.,* 2000, 34, 39–42. (in Russian)

In: Fungi: Types, Environmental Impact and Role in Disease ISBN: 978-1-61942-671-9
Editors: Adolfo Paz Silva and María Sol

Chapter VI

Ectomycorrhizal Fungi and their Role in Symbiosis

Minna J. Kemppainen and Alejandro G. Pardo*

Laboratorio de Micología Molecular, Departamento de Ciencia y Tecnología, Universidad Nacional de Quilmes (UNQ), and Consejo Nacional de Investigaciones Científicas y Técnicas (CONICET). Roque Sáenz Peña 352, (B1876BXD) Bernal, Provincia de Buenos Aires, Argentina

Abstract

Mycorrhiza is the most ancient and widespread mutualistic symbiotic association between plants and soil borne fungi. Mycorrhizae *sensu lato* are dual organs of absortion formed by mutualistic symbiotic fungi inhabiting healthy absorbing organs such as roots, rhizomes or thalli of most terrestrial plants and also many aquatics and epiphytes.

In nature the soil nutrient acquisition of plants occurs to great extent via these mycorrhizal organs. It is now widely accepted that mycorrhizae would have played a fundamental role in the colonization of the land by plants functioning as nutrient absorbing structures even before the appearance of roots.

Mycorrhizae are traditionally classified into categories on the basis of their fungal partners and macro- and microscopical anatomical structures of the dual organs.

Among them ectomycorrhiza (ECM) is typically formed between the fine roots of trees of boreal and temperate forest and fungal species usually belonging to Basidiomycota and Ascomycota.

In this chapter we discuss the evolutionary origin of ECM, some of its morphological and anatomical characters, the ecological significance of ECM and some of the molecular traits related to establishment and functioning of ECM structures between fungi and their plant hosts.

*Corresponding author: Alejandro G. Pardo, Phone: 54-11-4365-7100 (ext 4205), Fax: 54-11-4365-7132, E-mail: apardo@unq.edu.ar.

1. Mycorrhizal Symbiosis

1.1. An Ancient Interaction between Plants and Fungi

Mycorrhizae are symbiotic organs formed between over 92 % of terrestrial plant families (80 % of plant species) and fungi [1]. The existence of this specific interaction between plant roots and fungi was described already in 1842 by Nägeli and the term mycorrhiza (fungal root) was first mentioned in reference to this peculiar association between tree roots and fungi by Frank in 1885 [2]. Today it is known that mycorrhiza is the most prevalent symbiotic system on earth and it can be found in all terrestrial ecosystems. The definition for the actual use of the term mycorrhiza was set by Trappe as "dual organs of absorption formed when symbiotic fungi inhabit healthy absorbing organs (roots, rhizomes or thalli) of most terrestrial plants and many aquatics and epiphytes" [3]. In nature the soil nutrient acquisition of plants occurs to great extent via these mycorrhized organs, a process which is dependent on the activity of the symbiotic fungal partners [4]. Fossil data show the presence of mycorrhizal structures in the stems of *Aglaophyton major* in the Rhynie chert from 400 million years ago indicating that this type of symbiotic interaction is very ancient [5]. Moreover, it is now widely accepted that mycorrhizae would have played a fundamental role in the colonization of the land by plants functioning as nutrient absorbing structures prior to the development of roots [6-9].

1.2. Classification of Mycorrhizae

Mycorrhizae are traditionally classified into categories on the basis of their fungal associates and distinguishable macro– and microscopical anatomical structures of the formed dual-organs [4, 10]. As mutualistic interactions, plant plasmamembrane is never penetrated by the symbiotic fungi but the host and the fungal cells co-exist in a close physical contact in symbiotic structures characteristic for each mycorrhizal type. At the moment seven categories are recognized: endomycorrhiza or arbuscular mycorrhiza (AM), ectomycorrhiza (ECM), ectendomycorrhiza, ericoid mycorrhiza, arbutoid mycorrhiza, monotropoid mycorrhiza and orchid mycorrhiza. However, more detailed anatomical studies propose that this classification is to some extent too artificial and fixed as mycorrhizal interactions presenting transitive characteristics of several categories exist in nature [4]. Arbuscular mycorrhiza (formerly referred to as vesicular-arbuscular mycorrhiza) is in number the most abundant type of mycorrhizal interaction. Over 80 % of all mycorrhizal plant species are endomycorrhizal these including the majority of herbaceous plants and tropical trees. Arbuscular mycorrhiza is also believed to be the ancestral type of mycorrhiza of land plants [1, 11].

Mycorrhizae are considered mutualistic associations in which the fungus delivers mineral nutrients to the plant and the fungal symbiont receives carbon compounds from the host, but some exceptions to this rule are also known to exist. In monotropoid mycorrhiza, an interaction formed between achlorophyllous and heterotrophic plant species belonging to Ericales [12] and basidiomycete fungi such as *Tricholoma* and *Russula* [13, 14], it is the plant which depends on the fungus for its carbon acquisition and survival. The presence of the symbiotic monotropoid fungus has also been shown to be fundamental for seed germination of their myco-heterotrophic hosts [15]. Another exception to this mutualistic rule is presented by the orchid mycorrhiza, interaction formed by members of the plant family Orchidaceae, which consists of over 17000 species. This fungal-plant interaction might not be mutualistic because there is no evidence that the fungus gains any benefit from it. The fungus is however essential for the plant host and required for seed germination and establishment of seedlings [16]. The monotropoid and orchid mycorrhizae highlight the high functional and evolutionary flexibility of the mycorrhizae found in nature. They demonstrate how these plant-fungus interactions have specialized beyond their role in simply improving the mineral nutrition of the host. However, this appears to be the case for other types of mycorrhizae as well. Mycorrhizae can improve the host plant fitness via variable mechanisms many of which are not directly linked to the nutrient status of the host. Moreover, the activity of mycorrhizal fungi changes physiochemical characteristics of soils and mycorrhizal interactions are proposed to be one of the fundamental driving forces modeling plant communities and terrestrial ecosystems [17].

2. Ectomycorrhizal Symbiosis

2.1. Ectomycorrhiza

Ectomycorrhizae (ECM) are formed typically between the fine roots of around 10 % of plant families and this mycorrhizal type dominates in forest ecosystems of boreal and temperate regions as well as in some tropical ecosystems. Although not so abundant as AM, ectomycorrhizal host plants include woody species belonging to over thirty families. Many of these are trees of great ecological and economical interests such as pines, beeches, birches, poplars, oaks, dipterocarps and eucalypts. ECM is the primary mycorrhiza in Pinaceae, Araucariaceae, Cupressaceae, Gnetaceae, Polygonaceae, Nyctaginaceae, Myrtaceae, Salicaceae, Fabaceae, Fagales and Malvales [1]. In addition to trees also some shrubs and herbaceous species can have ECM associations.

Fungal species involved in ECM belong predominantly to the fungal phyla Basidiomycota (from here referred as basidiomycetes) and Ascomycota (from here ascomycetes). Also one fungal genus, *Endogone*, in Zygomycota has been described as ectomycorrhizal [4]. Almost 6000 basidiomycetes are identified as EMC fungi [18], and many of these are species forming macroscopic epigeous sexual structures, fruiting bodies. The number of ascomycete species involved in ECM was traditionally believed to be of minor importance and include species such as *Genococcum geophilum*, and *Tuber spp.* (over 100 species). This underestimation of ascomycetes as ECM fungi has been partly due to the fact that many of them do not form easily detectable or epigeneous fruiting bodies. The use of molecular techniques for species recognition directly on mycorrhizal organs is now however indicating that this group of fungal symbionts is more numerous than traditionally believed. They are found especially in alkaline soils and lower soil horizons and proposed to act as pioneer species during early forest successional stages [18]. Currently the total species richness of ECM fungi in nature is estimated to be between 20000 and 25000 [19]. Besides forming ECM many ECM fungi can also be engaged in other mycorrhizal types such as ectendo- monotropoid- and arbutoid mycorrhiza as well as in orchid mycorrhiza with myco-heterotrophic orchid genera [13, 14, 20, 21].

Even though described already by Frank in the 19th century [22] the understanding of the basic functions of ECM is still limited. The increased plant access to soil nutrients via the fungal mycelium and the flow of photosynthates from the plant host to the fungal symbiont have been broadly accepted for decades but the exact identity of the transferred compounds and the mechanisms controlling the establishment and function of the symbiotic interaction are still poorly understood.

2.2. Evolutionary Origin of ECM

The first fossil records of ECM are from *Pinus* roots from 50 million years ago [23]. The origin of this type of symbiosis is however most likely more ancient. It could date back to 130 million years ago, to the approximated period of appearance of ECM host plants [24], which also coincides with the currently proposed origin of basidiomycete fungi [25]. The topic of origin and age of ECM symbiosis is however under unresolved discussion [26, 27]. It is clear from phylogenic plant studies that the ECM symbiosis has several independent origins within existing angiosperms [1, 28].

This finding could reflect a possible evolutionary plant adaptation process to less nutrient rich environments by associating with symbiotic fungi. The given co-evolution could have consequently resulted in increased speciation of the plant and fungal symbionts and explain the current high number of ECM basidiomycetes [18]. Furthermore, the symbiotic lifestyle seems to have evolved several times among saprotrophic homobasiomycetes. Also multiple reversals to the free-living condition within this group of fungi have been detected by molecular phylogenetic analysis [29]. These findings may indicate that the ECM symbiosis is an evolutionary unstable interaction. However, the conclusions of Hibbett on basidiomycete reversals from ECM to the saprotrophic lifestyle have received strong criticism by Bruns and Shefferson [28].

2.3. ECM Morphology and Anatomy

Ectomycorrhiza is formed predominantly by tree lateral roots. Fungal colonization results in micro-and macroscopical modification of these roots making the ECM organs detectable with the naked eye. These visual features are dependent on the plant species and the fungal partner involved. For instance, the branching pattern, texture and color have led to the identification of different morphotypes which have been used as keys for identification of fungal symbionts [30]. Even though considerable variation exists between different morphological classes, ECM is defined by three fundamental structures:

1) A fungal mantle (or sheath) that covers the root to different extent.
2) Hyphal penetration between, but not within root cells forming a branched structure called Hartig net.
3) An extraradical mycelium that extends into the soil from the symbiotic organ.

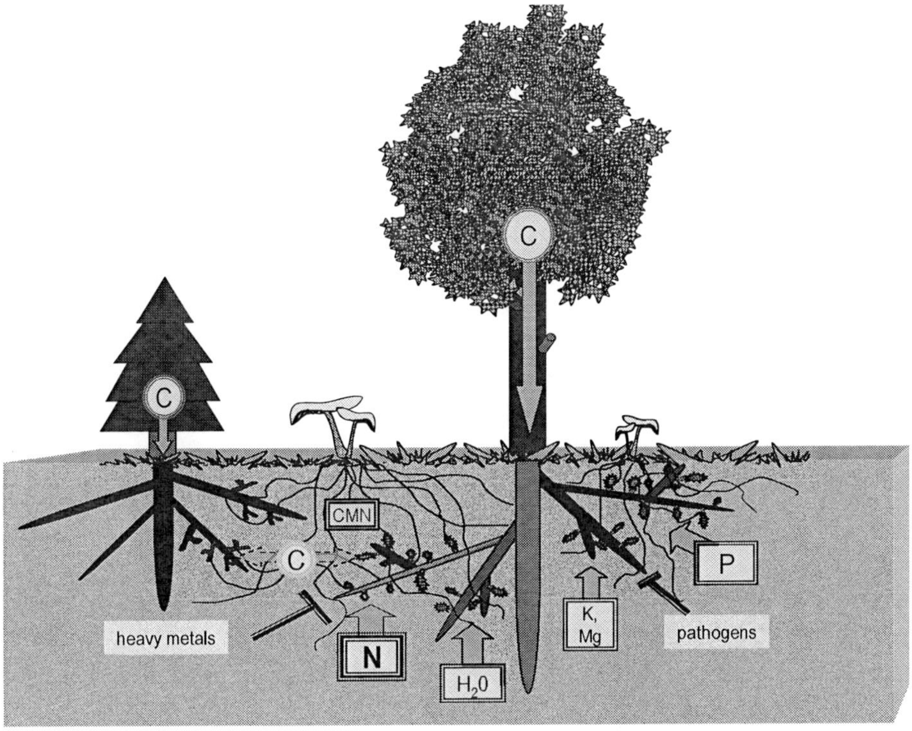

Figure 1. ECM improves plant nutrient and water uptake, and protects them from pathogens and heavy metals. The extraradical mycelium increases the root nutrient absorption area in soil. ECM fungi have also access to nutrient resources not directly available to host plants (organic N, P and rock-minerals). In this symbiotic interaction the fungus is supplied with plant hexoses derived from photosynthesis. Due to the low host specificity of ECM fungi different plant individuals and species are connected by common mycelial networks (CMNs) in forest ecosystems. Horizontal carbon and nitrogen flow between plants, via CMNs, has been demonstrated. C, carbon; N, nitrogen; P, phosphorus; K, potassium; Mg, magnesium.

The Hartig net creates the wide contact surface between the intraradical mycelium and plant root cells and is the physical interface where the exchange of nutrients takes place between the two symbionts. Both the plant and the fungal cell wall become modified in the symbiotic organ. Thc structure of the Hartig net shows significant variation between angiosperm and gymnosperm hosts. In conifers it involves both epidermal and cortical cell layers while in angiosperms usually only epidermal cells are surrounded by fungal mycelium [4, 10]. No AM-like penetrations of root cells are observed in a functional ECM. Plant cell penetrations and pathogenic-like behavior of ECM fungi have however been reported in senescent symbiotic organs [4]. The anatomical structures characteristic of an ECM root are presented in figure 1.

2.4. Ecological Significance of ECM

ECM has a direct improving effect on host plant growth and absorption of nutrients and water. This type of symbiosis is especially important for plants growing in ecosystems where nitrogen and phosphorus are the growth limiting nutrients [31]. The enhancement of plant nitrogen, phosphorus, and potassium uptake by mycorrhiza has been clearly demonstrated and sub-optimal nutritional conditions increase mycorrhization [4]. In template and boreal forest over 95 % of tree fine roots are ectomycorrhized and this symbiosis has a beneficial impact on plant growth both in natural and timber-production forests. All conifers are ectomycorrhizal and trees belonging to Pinaceae are never found in nature without mycorrhizal colonization [32].

The improved nutrient acquisition of the host in ECM depends on the vast extraradical mycelial networks which connects the symbiotic organs to soil. Ectomycorrhiza allows the plant to access a remarkable higher volume of soil than by the root system alone. The absorption surface area increase of ECM has been estimated to be 60-fold [33] and ECM mycelium can comprise up to 80 % of the total fungal biomass and 30 % of the total microbial biomass in boreal forest soil [34-36]. Besides increasing the surface area hyphae also have smaller diameter which allows them to exploit nutrients and water from smaller soil pores than plant roots [37].

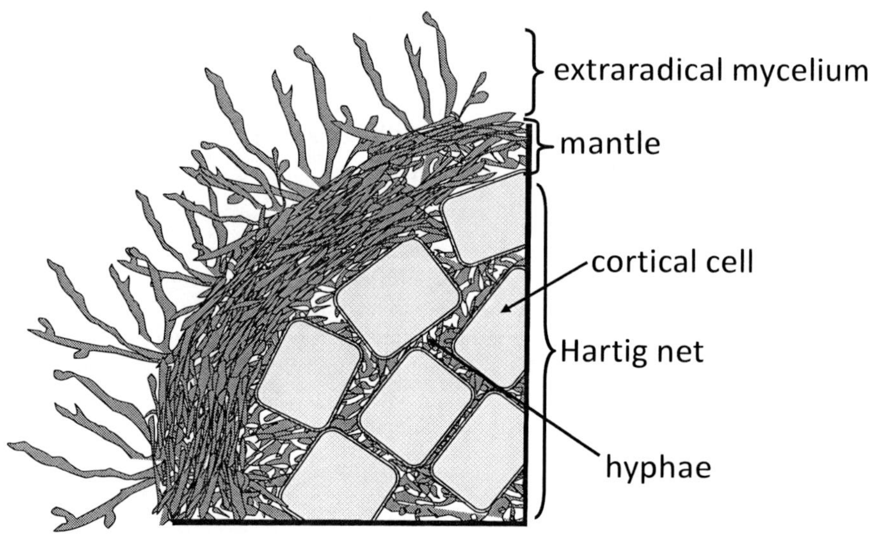

Figure 2. Cross section of an ECM root.

Moreover, it is the fungal capacity to utilize a wider range and often complex nutrient resources, not directly available to plants in soil, which increases host nutrition. In boreal and temperate forests ecosystems a significant pool of soil nutrients, especially nitrogen, is bound to organic resources not accessible to plants without fungal activities.

In addition to the nutritional effects ECMs are shown to protect plants against root pathogens [38]. This protective effect can be caused by a direct physical barrier created by the fungal mantle and/or via antagonistic effects of the symbiont against plant pathogens. Also general increased fitness of mycorrhizal plants which originates from their improved nutritional status offers an increased resistance to disease. Ectomycorrhiza can also protect the host plant roots against toxic concentrations of heavy metals [39]. The basic ECM functions which affect plant nutrient acquisition and improve host fitness are presented in figure 2.

2.5. Specificity of ECM Symbioses

One host plant can form ECM simultaneously with several different fungal species and the same fungal species can be engaged in ECM interaction with several taxonomically different hosts [40-42]. A minimum of 1000 fungal species are estimated to colonize trees in Scandinavian forest. This number vastly exceeds the number of tree species in the region [43]. Therefore, ECM is generally considered a relatively unspecific interaction even though a higher degree of host specificity, or preference, can exist in nature due to competition. This relative low host specificity of ECM fungi results in hyphal connections between different tree individuals and species in forest ecosystems. These connections are referred as common mycelial networks (CMNs) and they can function as channels for nutrient-flow between different plants (Fig. 2). Mobility of carbon has been demonstrated between birch and Douglas fir trees [44, 45], as well as transfer of nitrogen from N_2-fixing trees to non-N_2 fixing ones via CMNs [46, 47]. Moreover, besides ectomycorrhiza, ECM fungi can be involved in other types of mycorrhizae. This creates even greater potential complexity for CMN-links which can connect not only trees but also trees and surrounding shrubs and herbaceous plants species with each other. The ecological significance of these hyphal connections and the scale of horizontal nutrient flow via them are not clear yet but CMNs may play a role in promoting natural succession events in forest ecosystems. Figure 2 shows the outcome of the relative low host specificity of ECM fungi, the interconnections of different plant species via CMNs and the potential horizontal carbon flow between plant individuals in forest soil.

2.6. Basidiomycetes as ECM Fungi

The EMC symbiosis is formed by a vast number (more than 5500) of filamentous basidiomycetes [4]. These species typically form macroscopic fruiting bodies and several are also edible, of economical value, and collected and commercialized by man (Boletales, *Cantharellus spp.*, *Tricholoma matsutake*, ext.).Therefore, this group of mycorrhizal fungi has gained general public and scientific interest for centuries. Ectomycorrhizal basidiomycetes are clearly a polyphyletic group and many of them belong to large basidiomycete families like Amanitaceae, Boletaceae and Russulaceae [6]. They dominate in acidic, nitrogen limited forest soils with accumulation of complex nitrogen in organic humus layer. Ectomycorrhizal species are predominantly homobasidiomycetes with 5-6 μm wide hyphae and they inhabit upper organic soil horizons. They, unlike AM fungi, are not biotrophic and dependent on host plant for their growth capacity and hence can be cultivated as axenic cultures in laboratory. Despite the growth as saprotrophs *in vitro* the possible role of ECM fungi as facultative soil saprotrophs in nature is however under scientific debate.

The ligno- and cellulolytic capacity of these fungi is low when compared to saprotrophic basidiomycetes. They also occupy soil horizons where the availability of carbon compounds of high energy value is low.

These facts highlight the limited capacity of ECM fungi to acquire carbon directly from the soil matrix and support the idea of their dependency on carbon supply from the host trees in nature [48]. Most basidiomycetes are heterothallic with a mating system controlled by one till four multiallelic mating type loci (Aα, Aβ, Bα, Bβ) [49].

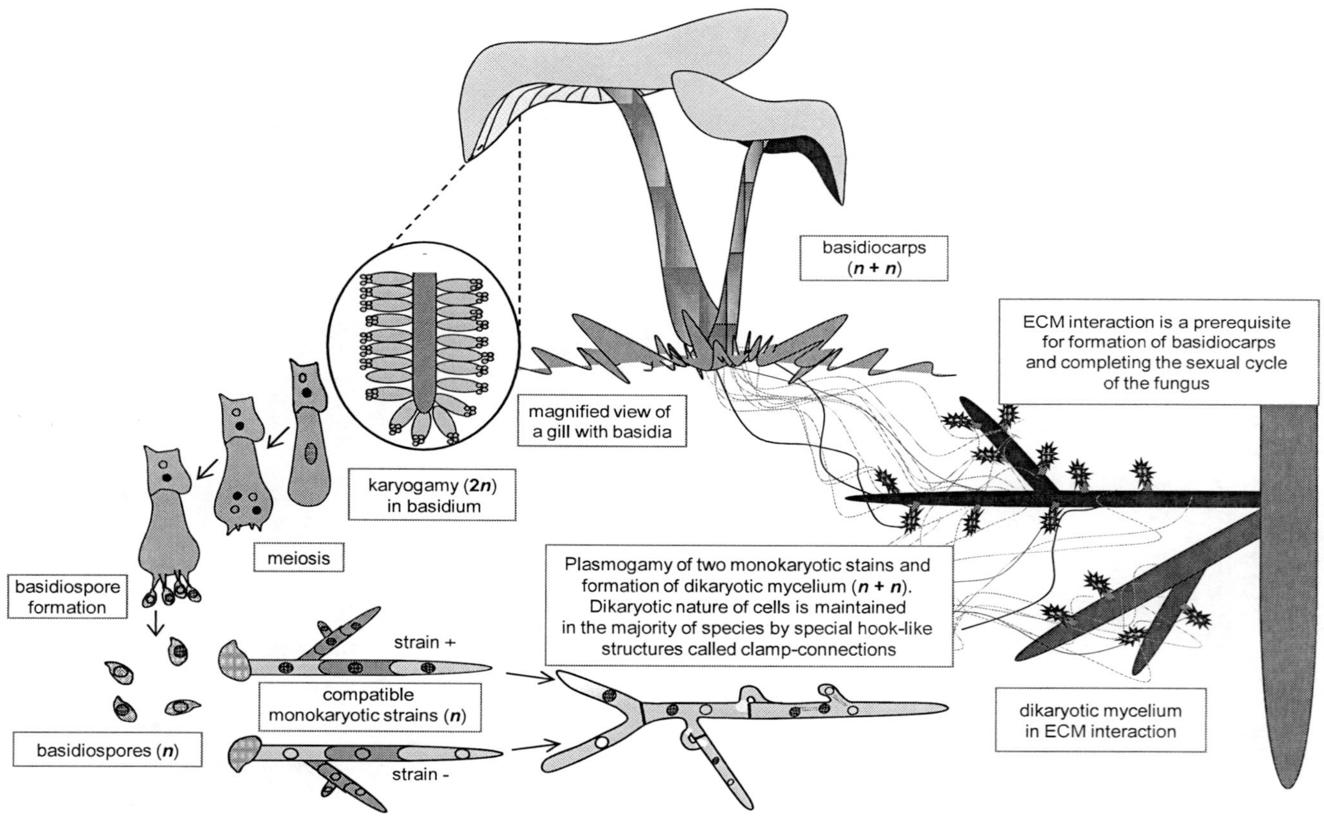

Figure 3. Life-cycle of an ECM basidiomycete.

The sexual structures, often macroscopic fruiting bodies, are formed for production of basidiospores via meiosis. These basidiospores germinate into monokaryotic mycelium which needs to anastomose with another compatible monokaryotic strain in order to produce a dikaryotic mycelium. In the dikaryotic mycelium karyogamy is postponed and the two compatible nuclei co-exist separately. This dikaryotic condition is maintained in the mycelium usually by special structures, the so-called clamp connections, which assure the sorting of two separate nuclei in each hyphal compartment separated by septa. It is the dikaryotic mycelium that can further complete the sexual cycle of the fungus. This generally requires the formation of ECM with the host plant for which reason the production of basidiocarps and basidiospores is not possible in axenic fungal cultures. On the other hand, monokaryotic strains are usually unable to form ECM though some exceptions to this rule are known to exist [50, 51]. Also intraspecific variation in mycorrhization capacity, host specificity, as well as in metabolic profiles is well reported for different ECM basidiomycetes [4]. A typical life-cycle of an ECM basidiomycete is presented in figure 3.

2.7. Establishment of ECM Symbiotic Structures

The successful establishment of an ECM symbiosis involves several steps. The interaction is initiated by a pre-colonization phase when the two symbionts are not yet in physical contact. This is followed by the cell-to-cell contact and finally by establishment of functional symbiotic structures between the plant and the fungus. All these events are controlled by a molecular cross-talk between the two symbiotic partners. This communication is reflected by mycorrhization which is accompanied by morphological changes and differential gene expression in both the ECM fungus and the plant host.

The precise molecular mechanisms responsible for successful ECM formation are still poorly understood. Akiyama et al. [52] isolated the strigolactone 5-deoxystrigol from the root exudates of *Lotus japonicus* and identified it as the compound responsible for the induction of hyphal branching, a critical step in host recognition by AM fungi [53]. The possible role of strigolactones, or related molecules, in ECM development has not yet been shown but significant amount of information has been gathered by researchers over the last 50 years covering different steps of ECM formation and the chemistry involved in them.

2.7.1. Plant-Fungus Cross-Talk during the Pre-Colonization Phase

Plant hosts and compatible ECM fungi are demonstrated to interchange chemical information already before the cell-to-cell contact takes place. During the pre-colonization phase compatible strains are attracted to the host while strains incapable of forming mycorrhiza with a given host are not [54]. The identity of the specific compounds, produced by the symbionts, and responsible for the ECM compatibility response have been hindered for a long time.

Ectomycorrhizal fungi and mycorrhization can also stimulate formation of plant lateral roots, cause dichotomization of conifer short roots and affect differentiation of host cells such as inhibition of root hair development. Also the hyphal morphology of the fungal symbiont is affected by the presence of the host plant. Due to these well documented morphological effects during the ECM development hormones, such as auxins and cytokinins, produced by ECM fungi and/or plant hosts, have gained great research interest [55]. Vicinity of plant roots induces hyphal branching. For instance eucalypt produces the flavonol rutin and the cytokinin zeatin which have been identified as branching factors [56]. Zeatin also stimulates the production of a tryptophane derivative, hypaphorine, in the ECM fungus *Pisolithus tinctorius*. This compound is highly produced by the fungus during mycorrhiza formation and it has a controlling effect on host root hair development [57, 58]. A competitive antagonism of plant auxins and fungal hypaphorine has been proposed to control ECM organogenesis [59]. In concordance with previous results auxin-induced genes have been shown to be up-regulated in *Pinus pinaster* mycorrhiza [60]. Also inoculation of *P. pinaster* with a *Hebeloma cylindrosporum* auxin overproducer strain results in increased mycorrhization rate and in some specific anomalic features in Hartig net anatomy. These data strongly suggest that fungal auxins are one of the compounds involved in control of ECM formation [61, 62]. However, not all ECM fungi are auxin producers, or they produce quantities not high enough to explain the plant response. Therefore, the role of fungal-origin ethylene, instead of auxin, as a controller of host root morphology in ECM has been proposed [63, 64].

The mycorrhizal induced small secreted protein 7 (*MiSSP7*) is the most highly symbiosis-upregulated gene from the ectomycorrhizal fungus *Laccaria bicolor* [65]. The *MiSSP7* gene encodes an effector protein which is indispensable for the establishment of mutualism [66]. MiSSP7 is secreted by the fungus upon receipt of diffusible signals from plant roots, imported into the plant cell via phosphatidylinositol 3-phosphate-mediated endocytosis, and targeted to the plant nucleus where it alters the transcriptome of the plant cell. A large fraction of the plant genes most highly upregulated by MiSSP7 are involved in modification of root architecture. Among them, auxin-responsive, cell wall remodeling and reactive oxygen species production genes are positively modulated. *Laccaria bicolor* transgenic strains with reduced expression of *MiSSP7*, launched by RNAi, do not enter in symbiosis with poplar [66]. MiSSP7 resembles effector proteins of pathogenic fungi, nematodes and bacteria, that are similarly targeted to the plant nucleus to promote colonization of the plant tissues. Whereas several effectors released by pathogens have been characterized to date, MiSSP7 is the first effector discovered in a mutualistic system.

2.7.2. Defense Reactions Involved in ECM Formation

In mycorrhized roots fungal hyphae and plant cells co-exist in a very intimate manner. The colonization of roots occurs typically without a strong defense reaction by the host plant. This shows that the symbiotic partners pass through a recognition process that leads to establishment of functional symbiotic organs. However, there is evidence of a weak plant defense reaction in the initial phase of the symbiotic interaction.

In the plant-pathogenic fungal interaction plant defense reactions are known to be stimulated by compounds called elicitors. Fungal cell wall fragments are shown to have the capacity to act as such elicitors. These are recognized by the plant, probably by their affinity to receptors in plant plasma membrane, and they initiate an intracellular signal cascade resulting in activation of host defense responses. These include flux of ions, protein phosphorylation, activation of heterotrimeric G-proteins, expression of defense genes, synthesis of reactive oxygen species such as H_2O_2, and production of phytoalexins [67].

Also several ECM fungi are demonstrated to liberate compounds that are recognized as elicitors and they result in activation of plant defenses in a highly similar manner than in pathogenic interactions [68-71, 74, 75]. A fast initial response of ion fluxes (K^+, Cl^-, Ca^{2+}, H^+) is followed by protein phosphorylations and production of H_2O_2. Later on production of enzymes such as peroxidases, β-1,3-glucanases and chitinases is induced in plant cells. Even though highly similar to the plant response against fungal pathogens, the defense reactions triggered in the early ECM interaction phase are significantly weaker. The eliciting activity of the fungus depends on the extent of compatibility between the symbiotic partners [76, 77]. Interestingly, a more pronounced plant defense reaction, expressed by chitinase production, has been detected when a good colonizer interacts with the host [78]. A role of host plant chitinases in inactivation of elicitors originating from ECM fungi and thus attenuating plant defense reaction in symbiotic interaction has been proposed [67]. Also compounds with auxin activity, produced by some ECM fungi might be involved in suppressing plant defense reactions. The demonstration by Plett et al. [66] that the most highly upregulated *L. bicolor* transcript in symbiotic tissues, MiSSP7, is an effector protein necessary for the penetration of fungal hyphae into plant tissues and for maintenance of symbiosis in the plant apoplast, would mean that the steps leading to the colonization of roots by ECM symbionts, including the use of effectors to subjugate plant defenses or alter plant cell wall characteristics, may be closer in mechanism to their pathogenic relatives than originally thought. This hypothesis is further supported by the finding that MiSSP7 binds to plasmamembrane phospholipids and enters the plant cell via endocytosis in a manner similar to pathogenic effectors containing RXLR or RLXR-like motifs.

2.7.3. Formation of Functional ECM Organs

Formation of functional ECM structures is initiated by hyphal contact with host roots which initiates mantle formation. The factors influencing physical contact between the cells of the two symbionts are not yet clear. The adhesion of hyphae to the root surface may involve interactions between the plant and fungal polysaccharides and glycoproteins [79-81]. A layer of extracellular fibrillar polymers, present on the hyphal surface of several ECM fungi before attachment seems to play a role in this process [79, 82]. Also interaction-specific lectins can be involved [83]. Once the contact is made with plant cells it is accompanied with lose of hyphal apical dominance, increased hyphal branching, swelling and cell wall modifications [76, 84-86]. In *Pisolithus tinctorius-Eucalyptus globulus* ECM the two principal fungal cell wall glycoproteins disappear during the symbiotic interaction. This process is accompanied by strong expression of symbiosis-regulated acid polypeptides (SRAPs) and hydrophobic polypeptides (hydrophobins) localized in the fungal cell wall [77, 86, 87]. Lateral root growth is slowed down as a reaction to fungal colonization [88]. Specific details of the colonization process show symbiotic partner-specific variations [4]. The Hartig net formation is generally initiated by inner-mantle hyphae which penetrate between epidermal cells. In nature newly formed roots can be colonized by the Hartig net hyphae of the mother root [89]. Penetration of the host root induces important morphological changes in hyphae and result in highly lobed structures which multiply the surface area of the fungal partner. These Hartig net hyphae are dense in cytoplasm and rich in ER (endoplasmic reticulum) and mitochondria indicating high metabolic activity. The hyphal penetration process is accompanied by breakdown of the middle lamella between the cortex cells [90, 91]. This indicates that enzymatic activities must be involved in the colonization process. Cell wall modifications affecting both symbionts are evident and plant and fungal cell walls became fused into a shared interfacial matrix. If fungal-born enzymes are involved in this process is not known. A tight physical contact is evident between the plant and the fungus forming the space where nutrient exchange takes place in functional ECM organs.

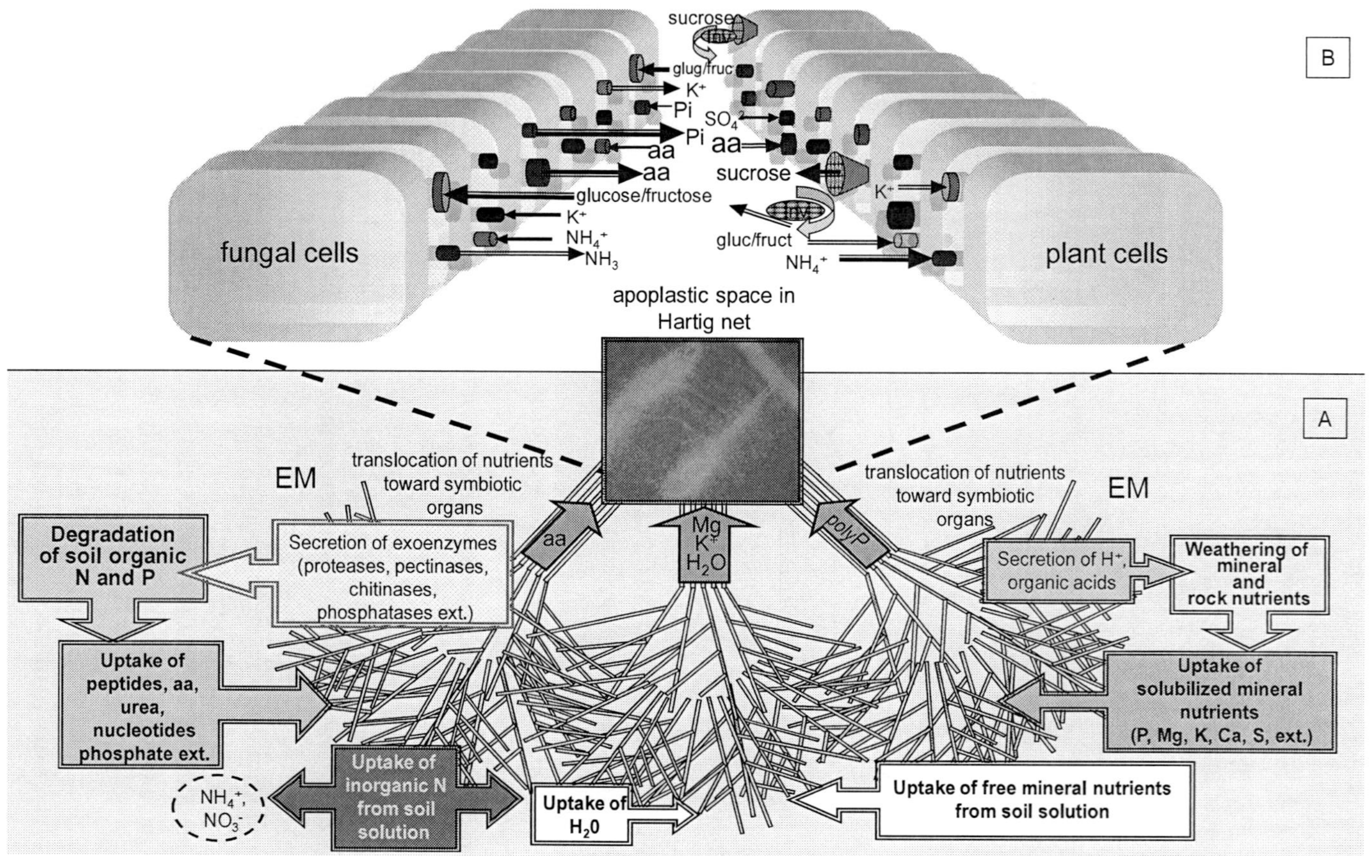

Figure 4. The functions of ECM. The uptake of nutrients and water from soil and translocation towards symbiotic organs by the extra-radical mycelium (EM) is presented in the lower half of the figure (A). The nutrient transfer events between fungal and plant cells at the apoplastic space of the Hartig net are presented in the upper half of the figure (B). (A) The EM multiplies the tree root nutrient and water absorption area. More importantly, EM has access to organic nutrient sources in soil which are not directly available to plants. Secretion of a wide range of fungal hydrolytic enzymes leads to degradation of soil biopolymers and liberation of nutrient compounds which can be taken up by the fungus. Also free inorganic N (ammonium and nitrate), P and micronutrients in the soil solution, as well as water are efficiently taken up by the EM. Moreover, ECM fungi are demonstrated to liberate nutrients from soil minerals and rocks by weathering. This is accomplished either by hydrogen or organic acid secretion which leads to lowering soil pH. Different micro- and macronutrients, and water are further translocated in the EM towards the symbiotic organs. The N translocation forms are proposed to be predominantly free amino acids while P is most probably translocated by the vacuolar system as polyphosphate.

2.8. Nutrient Transfer in ECM

It is generally accepted that in ECM a mutual benefit exist for both partners due to nutrient exchange in symbiotic organs.

The fungus receives carbon (C) as photosynthates and the plant receives nutrients, mainly nitrogen (N) and phosphorus (P). ECM also improves plant access to soil water resources and increases uptake of other marco- and micronutrients such as potassium (K) and magnesium (Mg). The current knowledge of nutrient uptake by the ECM extraradical mycelium, transport towards the symbiotic organs and exhange of nutrients and C compounds at the symbiotic interface is summarized in figure 4 and discussed below.

2.8.1. Nitrogen Transfer in ECM

In contrast to agricultural soils, which usually contain comparably high amounts of inorganic N, mainly in the form of nitrate [92], in ECM forests the amount of mineral N in soil is plant growth limiting being the nitrification rate generally low [93, 94]. The involvement of ECM in N uptake of host plants was already proposed by Frank [22] and during the following 100 years several studies have demonstrated that ECM indeed improves plant N nutrition [95, 96]. In ECM symbioses variable soil N compounds are taken up by the extraradical mycelium, metabolized by the ECM fungus, transported via mycelial networks towards symbiotic organs and subsequently transferred to the plant host.

Ectomycorrhization increases host plant N status. This is due to the vast extraradical fungal network which, firstly, multiplies the absorbing surface area of roots [97] and, secondly, creates access to N resources not directly utilized by plants. The latter depends on the fungal capacity to mobilize N from complex organic sources which typically accumulate in acid forest soils [98]. The destruction of the extraradical connections with symbiotic organs has shown to cause strong reduction in host plant ammonium absorption. This demonstrates that extraradical mycelium is the main structure responsible for N uptake in ECM plants. While soil inorganic N can be taken up by plants, the ammonium absorption capacity of fungal mycelium is also demonstrated to be higher than the one of plant roots. It is this higher affinity of fungal nutrient uptake systems which further contributes to improved host N nutrition when engaged in ECM symbioses [99]. Moreover, mycorrhization has been shown to cause down-regulation of host plant N uptake-systems. As a result ECM roots may become completely dependent on the fungal N acquisition machinery [100-102]. Hence by colonizing the majority of tree fine roots EMC fungi control the nutrient uptake of their hosts in nature.

Ectomycorrhizal fungi have evolved in N limiting ecosystems and they are clearly well adapted to low free N concentrations. Ammonium is the dominant inorganic N form in ECM forest soils. It is also the preferred N source for many ECM fungi even though the role of nitrate as N source may be more important also in these ecosystems than traditionally believed [95, 103, 104]. However, it is the fungal capacity to metabolize soil complex organic N resources which is likely to be fundamental for ECM function in nature. Ectomycorrhizal fungi can take up a range of organic N compounds such as amino acids, di-tripeptides, proteins and N containing secondary metabolites. The capacity of different ECM fungi to use different N sources can however vary between species and strains [105-112]. Fungi can also solubilize complex soil nutrients by secreting proteases, acid phosphatases and oxidases [113, 114].

The main core of molecular information on basic N metabolism of ammonium, nitrate, amino acid and peptide uptake of ECM fungi comes from studies in the basidiomycetes *Hebeloma spp.*, (for a review see Müller et al. [115], *Amanita muscaria* [116] and the ascomycete *Tuber borchii* [117-122]. These studies have led to cloning and characterization of several high and low affinity ammonium transporters and genes responsible for nitrate utilization and growth on organic N sources. The basic regulation and the expression profiles of these genes have been studied and their relevance both in symbiotic- and non-symbiotic mycelia has been assessed. The assimilation of ammonium into amino acids in ECM fungi takes place via the GS-GOGAT pathway (glutamine synthetase-glutamate synthetase) or via the GDH pathway (glutamate dehydrogenase). The relative importance of each pathway is however shown to vary between different ECM species [95].

Various studies indicate that the amino acid L-glutamine is most probably the major N translocation form in fungal mycelium in soil [123-125]. However, despite the growing knowledge on basic N metabolism of ECM fungi, the precise nature of nitrogenous compounds which are transferred from the fungus to the plant at the symbiotic interface is not clear. A traditional view proposes that L-amino acids (predominantly glutamine, glutamate) are the mobile molecules in ECM. This is based on the indirect evidence of ^{15}N-labelling experiments and on the fact that these compounds represent the major pool of free N in mycorrhizal organs [95]. Direct molecular proofs for this widely accepted hypothesis are however still missing. On the other hand, experiments of Simard et al., [33, 45, 126] demonstrated that C compounds can move via CMNs from one mycorrhized individual into another ECM host. This indicates that C skeletons can be transferred in ECM organs between the fungal and plant symbionts. These C compounds could easily be also nitrogenated molecules giving further support to amino acids as mobile molecules in the ECM interface.

Ammonium, liberated from L-arginine by the fungal urea-cycle, was demonstrated to be the transferred N form in the AM symbiosis [127]. This scientific breakthrough has initiated increased attention for the direct role of ammonium in ECM as well [128]. The experimental data supporting the ammonium hypothesis comes from gene expression profiles of *Populus tremula* ECM roots. Mycorrhization results in strong up-regulation of a plant high affinity ammonium importer in symbiotic organs indicating that direct ammonium transfer could take place in ECM [129]. Also recently the *Amanita muscaria* ammonium importer AmAMT2 has been shown to be down-regulated both in presence of elevated concentrations of ammonium and in ECM formed with *Populus*. This indicates that the fungal symbiont has, in contrast to the host plant, a reduced ammonium uptake capacity in symbiotic organs. This adds new evidence for the direct role of ammonium as possible transferred N form in ECM [130]. However, the fact that *Populus* species belong to the minority group of trees which can form both ECM and AM raises the question of the universal role of ammonium transfer in the ECM symbiosis. Is it possible that inorganic and organic N transfer can co-exist in ECM? A proportion of each could be regulated by the general availability of C substrates and/or the species engaged in the interaction [128].

2.8.2. Carbon Transfer in ECM

Compared with saprotrophic fungi, ECM symbionts have only a limited capacity or potential to degrade plant cell wall compounds such as cellulose and lignin [44, 65, 131, 132]. The true capacity of ECM fungi to compete for soil C resources is therefore highly questionable. In nature EMC fungi however exploit a special symbiotic niche and depend and feed on plant photosynthetic C products delivered at the Hartig net. This C flow between the plant host leaves, fed with radioactive CO_2, and the mantle of the ECM symbiont was demonstrated in 1957 by Melin & Nilsson [133].

Up to 30 % of plant net production can be consumed for the maintenance of the EMC interaction [107]. The photosynthetic activity of the host plant has been shown to affect mycorrhization rate and fruiting body production of the fungal symbiont [16, 134, 135]. Mycorrhization also increases the level of photosynthetic activity of the host plant probably by generating a strong C sink in symbiotic organs [136-138]. In *Picea abies-Amanita muscaria* mycorrhiza enhanced sucrose consumption and concentration gradient have been co-localized in ECM roots and within zones with major symbiotic interaction [139, 140].

As with N transfer the precise nature of the C compound transferred at the ECM interface has not been directly demonstrated yet. Sucrose is the major transport form of photosynthetic products in plants and sugars have thus been proposed as logical mobile C compound in ECM. However, also other C compounds such as carboxylic acids and amino acids could function as C transfer molecules. Ectomycorrhizal fungi prefer hexose monosaccharides and they cannot grow on sucrose. This is due to the lack of invertase activity in this group of fungi. The utilization of sucrose in symbiotic organs has been shown to depend on plant cell wall bound invertase [141, 142]. Consequently, sucrose must be exported from host cells into the apoplastic space, broken down into monosaccharides by plant invertase and these monosaccharides would be taken up by fungal cells. Two hexose transporters (AmMst1/2) with higher affinity for glucose and up-regulation in ECM have been characterized in *A. muscaria* [143, 144]. Also several *Laccaria bicolor* putative monosaccharide transporters have been detected as symbiosis upregulated [145]. These data of increased fungal hexose uptake capacity strongly point towards monosaccharides as major mobile C compounds in ECM.

The maintenance of a steep C gradient at the symbiotic interface can be considered fundamental for massive C transfer to take place in ECM. Therefore, once the C compounds are taken up by the ECM fungus from the apoplastic space these must be quickly metabolized via glycolysis, pentose phosphate pathway and metabolic pathways resulting in synthesis of the fungal storage compounds trehalose and glycogen. In concordance with the expected high carbohydrate metabolism in ECM fungi, increased fungal glycolysis rates have been detected in symbiotic mycelia of *Pisolithus microcarpus* and *Amanita muscaria* [146, 147].

On the other hand, the maintenance of uni-directional hexose uptake in symbiotic organs must be accompanied by reduced plant sugar re-uptake from the apoplastic space. This seems to happen as majority of plant hexose transporters are shown to be down-regulated in symbiosis [148, 149]. Interestingly, some poplar monosaccharide transporters are also demonstrated up-regulated in symbiotic organs [150, 151]. These transporters are postulated to be regulated at the protein level by phosphorylation and they could play a fundamental role in establishment and function of ECM symbiosis. An insufficient supply of nutrients from the fungal symbiont could result in activation of these transporters and retrieval of hexoses from the apoplastic space back to plant host. This could consequently shut down the symbiotic interaction [148].

2.8.3. Phosphorus Transfer in ECM

The role of mycorrhiza in improving plant P nutrition has gained great research interest for a long time. Arbuscular mycorrhiza is the predominant form of mycorrhiza in environments where P is the main plant growth limiting nutrient. These include agricultural plant communities as well as tropical forest ecosystems where AM is fundamental for plant P nutrition [152, 153]. Even though undeniably important for plant N nutrition, ECM also play an important role in improving host plant P acquisition [98, 154-156]. The movement of ^{32}P, fed to extraradical mycelium of ECM fungi, to the roots and finally to the needles of the symbiotic host *Pinus sylvestris* was demonstrated by Melin and Nilsson in 1950 [157].

Phosphorus can be present in soil as soluble forms readily available for plants. The concentration of this free P in soil is however demonstrated to be very low. This is mostly due to the negative charges of phosphates which cause their rapid sequestration by positive soil cations [158]. Therefore, the major P pool in soil is present as insoluble inorganic or organic compounds not directly available for plants. Ectomycorrhizal fungi are able to attack these organic P forms with extracellular enzymes such as phosphomonoesterases [4]. Also direct mineralization of rock phosphates by some ECM fungi has been demonstrated [159, 160]. The mechanisms of this so called "rock-eating" are not yet fully understood but the substrate acidification by ammonium uptake and organic acid production seem to be involved [161, 162]. Also a synergistic interaction by plant roots and other soil-organisms can contribute to this mineralization process [163].

Host plant P content has been demonstrated to be in correlation with the extent of extraradical mycelium of ECM suggesting that plant P originates dominantly from the symbiotic interaction [164]. Once taken up from soil mycorrhizal fungi accumulate P in their vacuoles as polyphosphates [165, 166] and these seem to be the main transport form of P towards the host plant. If these polyphosphates are present in cells as insoluble long chain granules or are those sample preparation artifacts is an area of ongoing scientific debate [4]. The P uptake of host plant roots is also increased by mycorrhization under low and high P conditions indicating that the fungal P uptake mechanism dominates in symbiosis [167]. Actually, in AM plants high affinity P transporters are down-regulated in a whole plant root system resulting in dependency of the fungal uptake mechanism [168, 169]. If similar down-regulation of plant P acquisition systems occurs in ECM, making the host totally dependent on the fungal uptake system is not known yet.

The most probable transfer form of P at the ECM interface is orthophosphate. The mechanisms involved in P efflux and influx in the Hartig net are however not yet described. Two *H. cylindrosporum* P transporters have been recently characterized. These were shown to be mainly expressed in extraradical mycelium and would be involved in P uptake from soil substrates [167].

2.8.4. Other Nutrients Transferred in ECM

While N and P metabolisms have been targets of vigorous research and debate, the involvement of ECM in transfer of other macro- and micronutrients, such as potassium (K), magnesium (Mg) and calcium (Ca), has gained less scientific attention. These are not considered growth limiting nutrients in ECM ecosystems but the involvement of symbiosis in plant uptake of these compounds as well is undeniable. In mycorrhization a thick mantle formation reduces the access of soil non-limiting nutrients to the roots changing the true availability of these compounds to the plant. This can consequently circuit the uptake of K and Mg, instead of direct plant uptake, via the symbiotic fungus in some symbiotic couples [4]. The ECM fungi have also been shown to weather rocks liberating not only P (see above) but also K and Mg [170-172].

Hatch in 1937 demonstrated that ECM could improve host plant K uptake [4]. Potassium is highly mobile in soil and its transport in the fungal hypha could also be linked to P transport as K^+-polyP [173]. Magnesium is also a highly mobile element in ECM forest soils but it is not considered growth limiting to plants. Transfer of this nutrient via ECM to plant host has been demonstrated by isotope labeling and shown to contribute up to 4 % of plant total Mg uptake [174]. Calcium uptake by ECM has equally been demonstrated [174, 175] but the significance of it for plant fitness in nature is not clear.

Concluding Remarks

The ECM symbiosis is a complex interaction between two eukaryotic organisms. Even though past and recent insights at the physiological, cellular and genetic level in this mutualistic symbiosis have been achieved our knowledge of this system is still limited. A more detailed comprehension of how the partners interact to establish and maintain a functional symbiosis will require the discovery of all the genes involved in this process. As genome sequencing, whole-genome expression profiles and the ability to genetically modify ECM fungi have became a reality (65, 66, 176-180) we envisage a promising future in the better understanding of this amazing relationship between fungi and trees.

Acknowledgments

We thank UNQ, CONICET and ANPCyT for financial support of our laboratory.

References

[1] Wang, B. & Qiu, Y.L. (2006). Phylogenetic distribution and evolution of mycorrhizas in land plants. *Mycorrhiza*, 16, 299-363.

[2] Frank, A.B. (1885). Ueber die auf Wurzelsymbiose beruhende Ernährung gewisser Baume durch unterirdische Pilze. *Berichte der Deutschen Botanischen Gesellschaft*, 3, 128-145. (Translated by Trappe JM; Proc 6th N Amer Conf Myc 18-25.)

[3] Trappe, J.M. (1996). What is a mycorrhiza? In: Azcon-Aguilar C, Barea J-M (eds) Mycorrhiza in integrated systems –from genes to plant development. Proceedings of the 4th European Symposium on Mycorrhizas, EC Report EUR 16728, Luxembourg, pp 3-6.

[4] Smith, S.E. & Read D.J. (2008). Mycorrhizal Symbiosis. 3rd edition. Academic Press, London.

[5] Remy, W., Taylor, T.N., Hass, H. & Kerp, H. (1994). Four hundred-million-year-old vesicular-arbuscular mycorrhizae. *Proceedings of the National Academy of Sciences of USA*, 91, 11841-11843.

[6] Brundrett, M.C. (2002). Coevolution of roots and mycorrhizas of land plants. *New Phytologist*, 154, 275-304.

[7] Malloch, D.W., Pirozynski, K.A. & Raven, P.H. (1980). Ecological and evolutionary significance of mycorrhizal symbiosis in vascular plants. *Proceedings of the National Academy of* Sciences *of USA*, 77, 2113-2118.

[8] Pirozynski, K.A. & Malloch, D.W. (1975). The origin of land plants: A matter of mycotrophism. *Biosystems*, 6, 153-164.

[9] Selosse, M.A. & Le Tacon, F. (1998). The land flora: a phototroph-fungus partnership? *Trends in Ecology and Evolution*, 13, 15-20.

[10] Peterson, R.L., Massicotte, H.B. & Melville, L.H. (2004). *Mycorrhizas: Anatomy and cell biology*. NRC Research Press, Ottawa, Canada.

[11] Cronquist, A,. (1981). An integrated system of classification of flowering plants. Columbia University Press, New York, USA.

[12] Leake, J.R. (1994). The biology of myco-heterotrophic (saprophytic) plants. Tansley Review N° 69. *New Phytologist*, 127, 171-216.

[13] Bidartondo, M.I. & Bruns, T.D. (2001). Extreme specificity in epiparasitic Monotropoideae (Ericaceae): widespread phylogenetic and geographical structure. *Molecular Ecology*, 10, 2285-2295.

[14] Young, B.W., Massicotte, H.B., Tackaberry, L.E., Baldwin, Q.F. & Egger, K.N. (2002). *Monotropa uniflora:* morphological and molecular assessment of mycorrhizae retrieved from sites in the sub-Boreal Spruce biogeoclimatic zone in central British Columbia. *Mycorrhiza*, 12, 75-82.

[15] Bruns, T.D. & Read, D.J. (2000). *In vitro* germination of non-photosynthetic, myco-heterotrophic plants stimulated by fungi isolated from the adult plants. *New Phytologist*, 148, 335-342.

[16] Peterson, R.L., Uetake, Y. & Zelmer, C. (1998). Fungal symbioses with orchid protocorms. *Symbiosis*, 25, 29-55.

[17] Johnson, N.C. & Gehring, C.A. (2007). Mycorrhizas: Symbiotic mediators of rhizosphere and ecosystem processes. In: Cardon ZG, Whitbeck JL, (eds). The Rhizosphere: An ecological perspective. Elsevier Academic Press, Amsterdam, The Netherlands, pp. 73-100.

[18] Molina, R., Massicotte, H. & Trappe, J.M. (1992). Specificity phenomena in mycorrhizal symbiosis: community-ecological consequences and practical implications. In Allen MF, (ed). "Mycorrhizal functioning" An Integrative Plant-Fungal Process. Chapman & Hall, London, UK, pp. 357-423.

[19] Rinaldi, A.C., Comandini, O. & Kuyper, T.W. (2008). Ectomycorrhizal fungal diversity: separating the wheat from the chaff. *Fungal diversity*, 33, 1-45.

[20] Molina, R & Trappe, J.M. (1982). Lack of mycorrhizal specificity by the ericaceous hosts *Arbutus menziesii* and *Arctostaphylos uva-ursi. New Phytologist*, 90, 495-509.

[21] Yu, T.E., Egger, K.N. & Peterson, R.L. (2001). Ectendomycorrhizal associations – characteristics and functions. *Mycorrhiza*, 11, 167-177.

[22] Frank, A.B. (1894). Die Bedeutung der Mykorrhiza-pilze für die gemeine Kiefer. *Forstwissenschaftliches Centralblatt*, 16, 185-190.

[23] LePage, B.A., Currach, R.S., Stockey, R.A., & Rothwell, G.W. (1997). Fossil ectomycorrhiza from the Middle Eocene. *American Journal of Botany*, 84, 410-412.

[24] Axelrod, D.I. (1986). Cenozoic history of some western American pines. *Annals of the Missouri Botanical Garden*, 73, 565-641.

[25] Berbee, M.L & Taylor, J.W. (2001). Fungal molecular evolution; gene tree and geologic time. In McLaughin DJ, McLaughin EG, Lemke PA, (eds). The Mycota: a comprehensive treatise on fungi as experimental systems for basic and applied research. Systematics and evolution, Part B. Springer, Berlin, Germany, pp. 229-245.

[26] Givnish, T.J. & Renner, S.S. (2004). Tropical intercontinental disjunctions: Gondwana breakup, immigration from the boreotropics, and transoceanic dispersal. *International Journal of Plant Sciences*, 2004 165:S1-S6.

[27] Pennington, R.T., Cronk, Q.C.B. & Richardson, JA. I(2004). ntroduction and synthesis: plant phylogeny and the origin of major biomes. *Philosophical Transactions of the Royal Society of London, Series B, Biological Sciences*, 359, 1455-1464.

[28] Bruns, T.D. & Shefferson, R.P. (2004). Evolutionary studies of ectomycorrhizal fungi: recent advances and future directions. *Canadian Journal of Botany*, 82, 1122-1132.

[29] Hibbett, D.S., Gilbert, L-B. & Donoghue, M.J. (2000). Evolutionary instability of ectomycorrhizal symbiosis in basidiomycetes. *Nature*, 407, 506-508.

[30] Ingleby, K., Mason, P.A., Last, F.T. & Fleming, L.V. (1990). Identification of ectomycorrhizas. *Institute of Terrestrial Ecology, Research Publication* N°. 5. HMSO, London, UK.

[31] MacFall, J.S., Slack, S.A. & Wehrli, S. (1992). Phosphorus distribution in the red pine roots and the ectomycorrhizal fungus *Hebeloma arenosa*. *Plant Physiology*, 100, 713-717.

[32] Kendrick, B. (1991). Fungal symbiosis and evolutionary innovations. In Margulis L, Fester R, (eds). Symbiosis as a source of evolutionary innovations: Speciation and morphogenesis. The MIT Press, Cambridge, USA, pp. 249-261.

[33] Simard, S.W., Jones, M.D., Durall, D.M., Perry, D.A., Myrold, D.D. & Molina, R. (1997b). Resiprocal transfer of carbon isotopes between ectomycorrhizal *Betula papyrifera* and *Pseudotsuga menziesii*. *New Phytologist*, 137, 529-542.

[34] Högberg, M.N. & Högberg, P. (2002). Extramatrical ectomycorrhizal mycelium contributes one-third of microbial biomass and produces, together with associated roots, half the dissolved organic carbon in a forest soil. *New Phytologist*, 154, 791- 795.

[35] Wallander, H. (2006). External mycorrhizal mycelia - the importance of quantification in natural ecosystems. *New Phytologist*, 171, 240-242.

[36] Wallander, H., Nilsson, L.O., Bååth, E. & Hagerberg, D. (2001). Estimation of the biomass and seasonal growth of external mycelium of ectomycorrhizal fungi in the field. *New Phytologist*, 151, 753-760.

[37] van Breemen, N., Finlay, R., Lundström, U., Jongmans, A.G., Giesler, R. & Olsson, M. (2000). Mycorrhizal weathering: a true case of mineral plant nutrition? *Biogeochemistry*, 49, 53-67.

[38] Chakravarty, C., Peterson, R.L. & Ellis, B.E. (1991). Interaction between the ectomycorrhizal fungus *Paxillus involutus*, dampingoff fungi and *Pinus resinosa* seedlings. *Journal of Phytopathology*, 132, 207-218.

[39] Colpaert, J.V. & van Assche, J.A. (1993). The effects of cadmium on ectomycorrhizal *Pinus sylvestris L. New Phytologist*, 123, 325–333.

[40] Bruns, T.D. (1995). Thoughts on the process that maintain local species-diversity of ectomycorrhizal fungi. *Plant and Soil*, 170, 63-73.

[41] Trappe, J.M. (1967). Pure culture synthesis of Douglas-fir mycorrhizae with species of *Hebeloma*, *Suillus*, *Rhizopogon* and *Astraeus*. *Forest Science*, 13, 121-130.

[42] Trappe, J.M. (1977). Selection of fungi for ectomycorrhizal inoculation in nurseries. *Annual Review of Phytopathology*, 15, 203-222.
[43] Knudsen, H., Hansen, L. (1991). New taxa and combinations in the Agaricales, Boletales and *Pisolithus arhizus*. *Nordic Journal of Botany*, 11, 477-481.
[44] Colpaert, J.V. & van Tichelen, K.K. (1996). Decomposition, nitrogen and phosphorus mineralization from beech leaf litter colonized by ectomycorrhizal or litter-decomposing basidiomycetes. *New Phytologist*, 134, 123-132.
[45] Simard, S.W., Jones, M.D., Durall, D.M., Perry, D.A., Myrold, D.D. & Molina, R. (1997a). Net transfer of carbon between ectomycorrhizal tree species in the field. *Nature*, 388, 579-582.
[46] He, X., Critchley, C., Ng, H. & Bledsoe, C. (2004). Reciprocal N ($^{15}NH_4^+$ or $^{15}NO_3^-$) transfer between nonN_2-fixing *Eucalyptus maculata* and N_2-fixing *Casuarina cunninghamiana* linked by the ectomycorrhizal fungus *Pisolithus sp. New Phytologist*, 163, 629–40.
[47] He, X.H., Critchley, C., Ng, H. & Bledsoe, C. (2005). Nodulated N_2-fixing *Casuarina cunninghamiana* is the sink for net N transfer from non-N_2-fixing *Eucalyptus macu*lata via an ectomy corrhizal fungus *Pisolithus* sp. supplied as ammonium nitrate. *New Phytologist*, 167, 897–912.
[48] Baldrian, P. (2009). Ectomycorrhizal fungi and their enzymes in soil: is there enough evidence for their role as facultative soil saprotrophs? *Oecologia*, 161, 657-660.
[49] Kües, U. & Casselton, L. (1992). Fungal mating type genes - regulators of sexual development. *Mycological Research*, 96, 993-1006.
[50] Kropp, B. & Fortin, J.A. (1988). The incompatibility system and relative ectomycorrhizal performance of monokaryons and reconstituted dikaryons of *Laccaria bicolor*. *Canadian Journal of Botany*, 66, 289-294.
[51] Lamhamedi, M.S., Fortin, J.A., Kope, H.H. & Kropp, B.R. (1990). Genetic variation in ectomycorrhiza formation by *Pisolithus arhizus* on *Pinus pinaster* and *Pinus banksiana*. *New Phytologist*, 115, 689-697.
[52] Akiyama, K., Matsuzaki, K. & Hayashi, H. (2005). Plant sesquiterpenes induce hyphal branching in arbuscular mycorrhizal fungi. *Nature,* 435, 824–827.
[53] Buée, M, Rossignal, M., Jauneau, A., Ranjeva, R. & Bécard, G. (2000). The pre-symbiotic growth of arbuscular mycorrhizal fungi is induced by a branching factor partially purified from plant root exudates. *Molecular Plant-Microbe Interactions*, 13, 693-698.
[54] Horan, D.P. & Chilvers, G.A. (1990). Chemotropism - the key to ectomycorrhizal formation. *New Phytologist*, 116, 297-301.
[55] Gogala, N. (1991). Regulation of mycorrhizal infection by hormonal factors produced by host and fungi. *Experientia*, 47, 331- 340
[56] Lagrange, H., Jay-Allemand, C. & Lapeyrie, F. (2001). Rutin, the phenolglycoside from eucalyptus root exudates, stimulates *Pisolithus* hyphal growth at picomolar concentrations. *New Phytologist*, 150, 349-355.
[57] Béguiristain, T.R. & Lapeyrie, F. (1997). Host plant stimulates hypaphorine accumulation in *Pisolithus tinctorius* hyphae during ectomycorrhizal infection while excreted fungal hypaphorine controls root hair development. *New Phytologist*, 136, 525-532.

[58] Ditengou, F.A., Béguiristain, T. & Lapeyrie, F. (2000). Root hair elongation is inhibited by hypaphorine, the indole alkaloid form the ectomycorrhizal fungus *Pisolithus tinctorius*, and restored by indole-3-acetic acid. *Planta*, 211, 722-728.

[59] Jambois, A., Dauphin, A., Kawano, T., Ditengou, F.A., Bouteau, F., Legué, V. & Lapeyrie, F. (2005). Compative antagonism between IAA and indole alkaloid hypaphorine must contribute to regulate ontogenesis. *Physiologia Plantarum*, 123, 120-129.

[60] Reddy, S.M., Pandey, A.K., Melayah, D., Marmeisse, R. & Gay, G. (2003). The auxin responsive gene Pp-C61 is up-regulated in *Pinus pinaster* roots following inoculation with ectomycorrhizal fungi. *Plant Cell and Environment*, 26, 681-691.

[61] Gay, G., Normand, L., Marmeisse, R., Sotta, B. & Debaud, J.C. (1994). Auxin overproducer mutants *of Hebeloma cylindrosporum* Romagnesi have increased mycorrhizal activity,. *New Phytologist*, 128, 645-657.

[62] Gea, L., Normand, L., Vian, B & Gay, G. (1994). Structural aspects of ectomycorrhiza of *Pinus pinaster* (Ait.) Sol. formed by an IAA-overproducer mutant *of Hebeloma cylindrosporum* Romagnesi. *New Phytologist*, 128, 659-670.

[63] Kaska, D.D., Myllylä, R. & Cooper, J.B. (1999). Auxin transport inhibitors act through ethylene to regulate dichotomous branching of lateral root meristems in pine. *New Phytologist*, 142, 49-58.

[64] Wallander, H., Nylund, J.E. & Sundberg, B. (1992). Ectomycorrhiza and nitrogen effects on root IAA: results contrary to current theory. *Mycorrhiza*, 1, 91-92.

[65] Martin, F., Aerts, A., Ahren, D., Brun, A., Danchin, E.G.J., Duchaussoy, F., et al. (2008). The genome of *Laccaria bicolor* provides insights into mycorrhizal symbiosis. *Nature*, 452, 88-92.

[66] Plett, J.M., Kemppainen, M., Kale, S.D., Kohler, A., Legué, V., Brun, A., Tyler, B.M., Pardo, A.G. & Martin, F. (2011). A secreted effector protein of *Laccaria bicolor* is required for symbiosis development. *Current Biology* 21, 1197-203.

[67] Salzer, P & Boller, T. (2000). Elicitor-induced reactions in mycorrhizae and their suppression. In: Podila GK, Douds DD, (eds).Current Advances in Mycorrhizal Research. APS Press, St. Paul, Minnesota, USA, pp. 1-10.

[68] Campbell, M.M. & Ellis, B.E. (1992). Fungal elicitor-mediated responses in pine cell cultures. 1. Induction of phenylpropanoid metabolism. *Planta*, 186, 409-417.

[69] Hebe, G., Hager, A. & Salzer, P. I(1999). nitial signaling processes induced by elicitors of ectomycorrhiza-forming fungi in spruce cells can also be triggered by G-protein–activating mastoparan and protein phosphatase inhibiting cantharidin. *Planta*, 207, 418-425.

[70] Mensen, R., Hager, A. & Salzer, P. (1998). Elicitor induced changes of wall-bound and secreted peroxidase activities in suspension-cultured spruce (*Picea abies*) cells are attenuated by auxins. *Physiologia Plantarum*, 102, 539-546.

[71] Salzer, P & Hager, A. (1993b). Effect of auxins and ectomycorrhizal elicitors on wall-bound proteins and enzymes of spruce [*Picea abies* (L.) Karst.] cells. *Trees*, 8, 49-55.

[72] Salzer, P., Hebe, G., Reith, A., Zitterell-Haid, B., Stransky, H., Gashchler, K. & Hager, A. (1996). Rapid reactions of spruce cells to elicitors released from the ectomycorrhizal fungus *Hebeloma crustiliniforme*, and inactivation of these elicitors by extracellular spruce cell enzymes. *Planta*, 198, 118-126.

[73] Salzer, P., Hübner, B., Sirrenberg, A. & Hager, A. (1997). Differential effect of purified spruce chitinases and β-1,3-glucanases on the activity of elicitors from ectomycorrhizal fungi. *Plant Physiology*, 114, 957-968.

[74] Sauter, M. & Hager, A. (1989). The mycorrhizal fungus *Amanita muscaria* induces chitinase activity in roots and in suspensioncultured cells of its host *Picea abies*. *Planta*, 179, 61-66.

[75] Schwacke, R. & Hager, A. (1992). Fungal elicitors induce a transient release of active oxygen species from cultured spruce cells depending on Ca^{2+} and protein kinase activity. *Planta*, 187, 136-141.

[76] Bonfante, P., Balestrini, R., Martino, E., Perotto, S., Plassard, C. & Mousain, D. (1998). Morphological analysis of early contacts between pine roots and two ectomycorrhizal *Suillus* strains. *Mycorrhiza* , 8, 1-10.

[77] Burgess, T., Laurent, P., Dell, B., Malajczuk, N. & Martin, F. (1995). Effect of fungal-isolate aggressivity on the biosynthesis of symbiosis-related polypeptides in differentiating eucalypt ectomycorrhiza. *Planta*, 195, 408-417.

[78] Albrecht, C., Burgess, T., Dell, B. & Lapeyrie, F. (1994). Chitinase and peroxidase activities are induced in eucalyptus roots according to aggressiveness of Australian ectomycorrhizal strains of *Pisolithus sp. New Phytologistogist*, 127, 217-222.

[79] Lei, J., Wong, K.K.Y. & Piché, Y. (1991). Extracellular concanavalin A-binding sites during early interactions between *Pinus banksiana* and two closely related genotypes of the ectomycorrhizal fungus *Laccaria bicolor*. *Mycological Research*, 95, 357- 363.

[80] Martin, F., Laurent, P., de Carvalho, D., Voiblet, C., Balestrini, R., Bonfante, P., Tagu, D. (1999). Cell wall proteins of the ectomycorrhizal basidiomycete *Pisolithus tinctorius* : identification, function, and expression in symbiosis. *Fungal Genetics and Biology*, 27, 161-174.

[81] Martin, F. & Tagu, D. (1995). Ectomycorrhiza development: A molecular perspective. In: Varma AK, Hock B, (eds). Mycorrhiza: Structure, function, molecular biology and biotechnology. Springer-Verlag, Berlin, Heidelberg, Germany, pp. 29-58.

[82] Lei, J., Ding, H., Lapeyrie, F., Piché, Y., Malajczuk, N. & Dexheimer, J. (1990). Ectomycorrhizal formation on the roots of *Eucalyptus globulus* and *Pinus caribaea* with two isolates of *Pisolithus tinctorius*: structural and cytochemical observations. In Nardon P, Gianinazzi-Pearson V, Grenier AM, Margulis L, Smith DC, (eds). Endocytobiology vol IV, INRA, Paris, France, pp. 123-126.

[83] Giollant, M., Guillot, J., Damez, M., Dusser, M., Didier, P. & Didier, E. (1993). Characterization of a lectin from *Lactarius deterrimus*. *Plant Physiology*, 101, 513-522.

[84] Brunner, I. & Scheidegger, C. (1992). Ontogeny of synthetized *Picea abies* (L.) Karst. - *Hebeloma crustuliniforme* (Bull. Ex St Amans) Quél. ectomycorrhizas. *New Phytologist*, 120, 359-369.

[85] Jacobs, P.F., Peterson, R.L., & Massicotte, H.B. (1989). Altered fungal morphogenesis during early stages of ectomycorrhiza formation in *Eucalyptus pilularis*. *Scanning Electron Microscopy*, 3, 249-255.

[86] Tagu, D. & Martin, F. (1996). Molecular analysis of cell wall proteins expressed during early steps of ectomycorrhizal development. *New Phytologist*, 133, 73-85.

[87] Tagu, D., Nasse, B. & Martin, F. (1996). Cloning and characterization of hydrophobins-encoding cDNAs from the ectomycorrhizal Basidiomycete *Pisolithus tinctorius*. *Gene*, 168, 93-97.

[88] Peterson, R.L. (1992). Adaptations of root structure in relation to biotic and abiotic factors. *Canadian Journal of Botany*, 70, 661-675.
[89] Wilcox, H.E. (1968). Morphological studies of the roots of Red pine, *Pinus resinosa*. II. Fungal colonization of roots and the development of mycorrhizae. *American Journal of Botany*, 55, 686-700.
[90] Massicotte, H.B., Peterson, R.L., Ackerley, C.A. & Ashford, A.E. (1987). Ontogeny of *Eucalyptus pilularis-Pisolithus tinctorius* ectomycorrhizae. II. Transmission electron microscopy. *Canadian Journal of Botany*, 65, 1940-1947.
[91] Massicotte, H.B., Peterson, R.L., Ackerley, C.A. & Piché, Y. (1986). Structure and ontogeny of *Alnus crispa-Alpova diplophloeus* ectomycorrhizae. *Canadian Journal of Botany*, 64, 177-192.
[92] Gorfer, M., Blumhoff, M., Klaubauf, S., Urban, A., Inselsbacher, E., Bandian, D., Mitter, B., Sessitsch, A., Wanek, W. & Strauss, J. (2011). Community profiling and gene expression of fungal assimilatory nitrate reductases in agricultural soil. *ISME Journal*, advance online publication doi: 10.1038/ismej.2011.53
[93] Keeney, D.R. (1980). Prediction of soil nitrogen availability in forest ecosystems: a literature review. *Forest Science*, 26, 159–171.
[94] Read, D.J. (1991). Mycorrhizas in ecosystems. *Experientia*, 47, 376–391.
[95] Alexander, I.J. (1983). The significance of ectomycorrhizas in the nitrogen cycle. In: Lee JA, McNeill D, Rorison IH, (eds). Nitrogen as and ecological factor. Blackwell Scientific Publications, Oxford, UK, pp. 69-93.
[96] Trappe, J.M. (2005). A.B. Frank and mycorrhizae: the challenge to evolutionary and ecologic theory. *Mycorrhiza*, 15, 277-281.
[97] Rousseau, J.V.D., Sylvia, D.M. & Fox, A.J. (1994). Contribution of ectomycorrhiza to the potential nutrient-absorbing surface of pine. *New Phytologist*, 128, 639-644.
[98] Pérez-Moreno, J. & Read, D.J. (2000). Mobilization and transfer of nutrients from litter to tree seedlings via the vegetative mycelium of ectomycorrhizal plants. *New Phytologist*, 145, 301-309.
[99] Javelle, A., Chalot, M., Söderström, B. & Botton, B. (1999). Ammonium and methylamine transport by the ectomycorrhizal fungus *Paxillus involutus* and ectomycorrhizas. *FEMS Microbiology Ecology*, 30, 355–366.
[100] Bailly, J., Debaud, J.C., Verner, M.C., Plassard, C., Chalot, M., Marmeisse, R. & Fraissinet-Tachet, L. (2007). How does a symbiotic fungus modulate expression of its host-plant nitrite reductase? *New Phytologist*, 175, 155-165.
[101] Gobert, A. & Plassard, C. (2002). Differential NO_3^- dependent patterns of NO_3^- uptake in *Pinus pinaster*, *Rhizopogon roseolus* and their ectomycorrhizal association. *New Phytologist*, 154, 509-516.
[102] Gobert, A. & Plassard, C. (2007). Kinetics of NO_3^- net fluxes in *Pinus pinaster*, *Rhizopogon roseolus* and their ectomycorrhizal association, as affected by the presence of NO_3^- and NH_4^+. *Plant Cell and Environment*, 30, 1309-1319.
[103] Laverman, A.M., Zoomer, H.R., van Verseveld, H.W. & Verhoef, H.A. (2000). Temporal and spatial variation of nitrogen transformations in a coniferous forest soil. *Soil Biology and Biochemistry*, 32, 1661-1670.
[104] Stark, J.M. & Hart, S.C. (1997). High rates of nitrification and nitrate turnover in undisturbed coniferous forests. *Nature*, 385, 61-64.

[105] Anderson, I.C., Chambers, S.M. & Cairney, J.W.G. (1999). Intra- and interspecific variation in patterns of organic and inorganic nitrogen utilization by three Australian *Pisolithus* species. *Mycological Research*, 103, 1579–1587.
[106] Chalot, M. & Brun, A. (1998). Physiology of organic nitrogen acquisition by ectomycorrhizal fungi and ectomycorrhizas. *FEMS Microbiology Reviews*, 22, 21–44.
[107] Finlay, R.D. & Söderström, B. (1992). Mycorrhiza and carbon flow to the soil. In: Allen MF, (ed). Mycorrhiza functioning. Chapman & Hall, London, UK, pp. 134-160.
[108] Guidot, A., Verner, M.C., Debaud, J.C. & Marmeisse, R. (2005). Intraspecific variation in use of different organic nitrogen sources by the ectomycorrhizal fungus *Hebeloma cylindrosporum*. *Mycorrhiza*, 15, 167–177.
[109] Keller, G. (1996). Utilization of inorganic and organic nitrogen sources by high-subalpine ectomycorrhizal fungi of *Pinus cembra* in pure culture. *Mycological Research*, 100, 989–998.
[110] Nygren, C.M.R., Eberhardt, U., Karlsson, M., Parret, J.L., Lindahl, B.D. & Taylor, A.F.S. (2008). Growth on nitrate and occurrence of nitrate reductase-encoding genes in a phylogenetically diverse range of ectomycorrhizal fungi. *New Phytologist*, 180, 875-889.
[111] Rangel-Castro, J.I., Danell E., & Taylor, A.F.S. (2002). Use of different nitrogen sources by the edible ectomycorrhizal mushroom *Cantharellus cibarius*. *Mycorrhiza*, 12, 131–137.
[112] Wallenda, T. & Read, D.J. (1999). Kinetics of amino acid uptake by ectomycorrhizal roots. *Plant Cell and Environment*, 22, 179-187.
[113] Bending, G.D. & Read, D.J. (1995). The structure and function of the vegetative mycelium of ectomycorrhizal plants. VI. Activities of nutrient mobilizing enzymes in birch litter colonized by *Paxillus involutus* (Fr.) Fr. *New Phytologist*, 130, 411-417.
[114] Nygren, C.M.R., Edqvist, J., Elfstrand, M., Heller, G. & Taylor, A.F.S. (2007). Detection of extracellular protease activity in different species and genera of ectomycorrhizal fungi. *Mycorrhiza*, 17, 241-248.
[115] Müller, T., Avolio, M., Olivi, M., Benjdia, M., Rikirsch, E., Kasaras, A., Fitz, M., Chalot, M. & Wipf, D. (2007). Nitrogen transport in the ectomycorrhiza association: the *Hebeloma cylindrosporum–Pinus pinaster* model. *Phytochemistry*, 68, 41-51.
[116] Nehls, U., Kleber, R., Wiese, J. & Hampp, R. (1999). Isolation and characterization of a general amino acid permease from the ectomycorrhizal fungus *Amanita muscaria*. *New Phytologist*, 144, 343–349.
[117] Guescini, M., Pierloeni, R., Palma, F., Zeppa, S., Vallorani, L., Potenza, L., Sacconi, G., Giomaro, G. & Stocchi, V. (2003). Characterization of the *Tuber borchii* nitrate reductase gene and its role in ectomycorrhizae. *Molecular Genetics and Genomics*, 269, 807-816.
[118] Guescini, M., Stocchi, L., Sisti, D., Zeppa, S., Polidori, E., Ceccaroli, P., Satarelli, R. & Stocchi, V. (2009). Characterization and mRNA expression profile of the *TbNre1* gene of the ectomycorrhizal fungus *Tuber borchii*. *Current Genetics*, 55, 59-68.
[119] Guescini, M., Zeppa, S., Pierleoni, R., Sisti, D., Stocchi, L. & Stocchi, V. (2007). The expression profile of the *Tuber borchii* nitrite reductase suggests its positive contribution to the host plant nitrogen nutrition. *Current Genetics*, 51, 31-41.

[120] Montanini, B., Betti, M., Marquez, A.J., Balestrini, R., Bonfante, P. & Ottonello, S. (2003). Distinctive properties and expression profiles of glutamine synthetase from plant symbiotic fungus. *Biochemistry Journal*, 373, 357-368.

[121] Montanini, B, Moretto, N., Soragni, E., Percudani, R. & Ottonello, S. (2002). A high-affinity ammonium transporter from the mycorrhizal ascomycete *Tuber borchii. Fungal Genetics and Biology*, 36, 22-34.

[122] Montanini, B., Viscomi, A.R., Bolchi, A., Martin, Y., Siverio, J.M., Balestrini, R., Bonfante, P. & Ottonello, S. (2006). Functional properties and differential mode of regulation of the nitrate trasporter from a plant symbiotic ascomycete. *Biochemistry Journal*, 394, 125-134

[123] Ek, H., Andersson, S., Arnebrant, K. & Söderström, B. (1994). Growth and assimilation of NH_4^+ and NO_3^- by *Paxillus involutus* in association with *Betula pendula* and *Picea abies* as affected by substrate pH. *New Phytologist*, 128, 629-637.

[124] Finlay, R.D., Ek, H., Odham, G. & Söderström, B. (1988). Mycelial uptake, translocation and assimilation of nitrogen from 15Nlabelled ammonium by *Pinus sylvestris* plants infected with four different ectomycorrhizal fungi. *New Phytologist*, 110, 59-66.

[125] Finlay, R.D., Ek, H., Odham, G., & Söderström, B. (1989). Uptake, translocation and assimilation of nitrogen from 15N-labelled ammonium and nitrate sources by intact ectomycorrhizal systems of *Fagus sylvatica* infected with *Paxillus involutus*. *New Phytologist*, 113, 47-55.

[126] Simard, S.W., Jones, D.M. & Durall, D.M. (2002). Carbon and nutrient fluxes within and between mycorrhizal plants. In: van der Heijden MGA, Sanders I, (eds). Mycorrhizal Ecology. Ecological Studies. Vol 157. Springer-Verlag, Heidelberg, Germany, pp. 33-74.

[127] Govindarajulu, M., Pfeffer, PE; Jin, HR; Abubaker, J; Douds, DD; Allen, JW; Bucking, H; Lammers, PJ; Shachar-Hill, Y. (2005). Nitrogen transfer in the arbuscular mycorrhizal symbiosis. *Nature*, 435, 819-823.

[128] Chalot, M., Blaudez, D. & Brun, A. (2006). Ammonia: a candidate for nitrogen transfer at the mycorrhizal interface. *Trends in Plant Science*, 11, 263–266.

[129] Selle, A., Willmann, M., Grunze, N., Gessler, A., Weiss, M. & Nehls, U. (2005). The high-affinity poplar ammonium importer PttAMT1.2 and its role in ectomycorrhizal symbiosis. *New Phytologist*, 168, 697-706.

[130] Willmann, A., Weiß, M. & Nehls, U. (2007). Ectomycorrhiza-mediated repression of the high-affinity ammonium importer gene *AmAMT2* in *Amanita muscaria*. *Current Genetics*, 51, 71-78.

[131] Nagendran, S., Hallen-Adams, H.E., Paper, J.M., Aslam, N. & Walton, J.D. (2009). Reduced genomic potential for secreted plant cell-wall-degrading enzymes in the ectomycorrhizal fungus *Amanita bisporigera*, based on the secretome of *Trichoderma reesei*. *Fungal Genetics and Biology*, 46, 427-435.

[132] Read, D.J. & Pérez-Moreno, J. (2003). Mycorrhizas and nutrient cycling in ecosystems – a journey towards relevance? *New Phytologist*, 157, 475-492.

[133] Melin, E & Nilsson, H. (1957). Transport of C^{14}-labelled photosynthate to the fungal associate of pine mycorrhiza. *Svensk Botanisk Tidskrift*, 51, 166-186.

[134] Högberg, P., Nordgren, A. & Ågren, G. (2002). Carbon allocation between tree root growth and root respiration in boreal pine forest. *Oecologia*, 132, 579-581.

[135] Lamhamedi, M.S., Godbout, C. & Fortin, J.A. (1994). Dependence of *Laccaria bicolor* basidiome development on current photosynthesis of *Pinus strobus* seedlings. *Canadian Journal of Forest Research*, 14, 412-415.

[136] Dosskey, M.G., Linderman, R.G. & Boersma, L. (1990). Carbon-sink stimulation of photosynthesis in Douglas fir seedlings by some ectomycorrhizas. *New Phytologist*, 115, 269-274.

[137] Loewe, A., Einig, W., Shi, L., Dizengremel, P. & Hampp, R. (2000). Mycorrhization and elevated CO_2 both increase the capacity for sucrose synthesis in source leaves of spruce and aspen. *New Phytologist*, 145, 565-574.

[138] Nylund, J.E. & Wallender, H. (1989). Effects of ectomycorrhiza on host growth and carbon balance in a semi-hydroponic cultivation system. *New Phytologist*, 112, 389-398.

[139] Hampp, R., Schaffer, C., Wallenda, T., Stülten, T., Johann, T. & Einig, W. (1995). Changes in carbon partitioning of allocation due to ectomycorrhiza formation: biochemical evidence. *Canadian Journal of Botany*, 73, 548-556.

[140] Rieger, A., Guttenberger, M. & Hampp, R. (1992). Soluble carbohydrates in mycorrhized and non-mycorrhized fine roots of spruce seedlings. *Zeitschrift für Naturforschung*, 47, 201-204.

[141] Salzer, P & Hager, A. (1991). Sucrose utilization of the ectomycorrhizal fungi *Amanita muscaria* and *Hebeloma crustuliniforme* depends on the cell wall-bound invertase activity of their host *Picea abies*. *Botanica Acta*, 104, 439-445.

[142] Salzer, P & Hager, A. (1993a). Characterization of wall-bound invertase isoforms of *Picea abies* cells and regulation by ectomycorrhizal fungi. *Physiologia Plantarum*, 88, 52-59.

[143] Nehls, U. (2004). Carbohydrates and nitrogen: nutrients and signals in ectomycorrhizas: In: Varma A, Abbot L, Werner D, Hampp R, (eds). Plant surface microbiology. Springer-Verlag, Berlin, Germany, pp. 373-392.

[144] Nehls, U., Wiese, J., Guttenberger, M. & Hampp, R. (1998). Carbon allocation in ectomycorrhizas: identification and expression analysis of an *Amanita muscaria* monosaccharide transporter. *Molecular Plant-Microbe Interactions*, 11, 167-176

[145] López M.F., Dietz, S., Grunze, N., Bloschies, J., Weiss, M. & Nehls, U. (2008). The sugar porter gene family of *Laccaria bicolor*: function in ectomycorrhizal symbiosis and soil-growing hyphae. *New Phytologist*, 180, 365-78.

[146] Duplessis, S., Courty, P.E., Tagu, D. & Martin F. (2005). Transcript patterns associated with ectomycorrhiza development in *Eucalyptus globulus* and *Pisolithus microcarpus*. *New Phytologist*, 165, 599-611.

[147] Schaeffer, C., Johann, P., Nehls, U. & Hampp, R. (1996). Evidence for an up-regulation of the host and a down-regulation of the fungal phosphofructokinase activity in ectomycorrhizas of Norway spruce and fly agaric. *New Phytologist*, 134, 697–702.

[148] Nehls, U. (2008). Mastering ectomycorrhizal symbiosis: the impact of carbohydrates. *Journal of Experimental Botany* 59, 1097-1108.

[149] Wright, D.P., Scholes, J.D., Read, D.J. & Rolfe, S.A. (2000). Changes in carbon allocation and expression of carbon transporter genes in *Betula pendula* Roth. colonized by the ectomycorrhizal fungus *Paxillus involutus* (Batsch) Fr. *Plant Cell and Environment*, 23, 39-49.

[150] Grunze, N., Willmann, M. & Nehls, U. (2004). The impact of ectomycorrhiza formation on monosaccharide transporter gene expression in poplar roots. *New Phytologist*, 164, 147-155.

[151] Nehls, U., Grunze, N., Willmann, M., Reich, M. & Küster, H. (2007). Sugar for my honey: Carbohydrate partitioning in ectomycorrhizal symbiosis. *Phytochemistry*, 68, 82-91.

[152] Bucher, M. (2007). Functional biology of plant phosphate uptake at root and mycorrhiza interfaces. *New Phytologist*, 1, 11-26.

[153] Smith, S.E., Smith, F.A. & Jakobsen, I. (2003). Mycorrhizal fungi can dominate phosphate supply to plants irrespective of growth responses. *Plant Physiology*, 133, 16-20.

[154] Finlay, R.D. & Read, D.J. (1986). The structure and function of the vegetative mycelium of ectomycorrhizal plants. II. The uptake and distribution of phosphorus by mycelial strands interconnecting host plants. *New Phytologist*, 103, 157-165.

[155] McComb, A.L. & Griffith, J.E. (1946). Growth stimulation and phosphorus absorption of mycorrhizal and non-mycorrhizal northern White Pine and Douglas Fir seedlings in relation to fertilizer treatment. *Plant Physiology*, 21, 11-17.

[156] van Tichelen, K.K. & Colpaert, J.V. (2000). Kinetics of phosphate absorption by mycorrhizal and non-mycorrhizal Scots pine seedlings. *Physiologia Plantarum*, 110, 96-103.

[157] Melin, E & Nilsson, H. (1950) Transfer of radioactive phosphorus to pine seedlings by means of mycorrhizal hyphae. *Physiologia Plantarum*, 3, 88-92.

[158] Vance, C.P., Uhde-Stone, C., & Allan, D.L. (2003). Phosphorus acquisition and use: critical adaptations by plants for securing a nonrenewable resource. *New Phytologist*, 157:423-447.

[159] Bowen, G.D & Theodorou, C. (1967). Studies on phosphate uptake by mycorrhizas. 14th IUFRO Congress, Munich, Germany, 5, 116-138.

[160] Stone, E.L. (1950). Some effects of mycorrhizae on the phosphorus nutrition of Monterey pine seedlings. *Proceedings - Soil Science Society of America*, 14, 340-345.

[161] Lapeyrie, F., Ranger, J. & Vairelles, D. (1991). Phosphate-solubilizing activity of ectomycorrhizal fungi in vitro. *Canadian Journal of Botany*, 69, 342-346.

[162] Mahmood, S., Finlay, R.D., Wallander, H. & Erland, S. (2002). Ectomycorrhizal colonisation of roots and ash granules in spruce forest treated with granulated wood ash. *Forest Ecology and Management*, 160, 65-74.

[163] Olsson, P.A. & Wallander, H. (1998)Interactions between ectomycorrhizal fungi and the bacterial community in soils amended with various primary minerals. *FEMS Microbiology Ecology*, 27, 195-205.

[164] Torres-Aquino, M. &; Plassard, C. (2004). Dynamics of ectomycorrhizal growth and P transfer to the host plant in response to low and high soil P availability. *FEMS Microbiology Ecology*, 48, 149-156.

[165] Ashford, A.E., Ling Lee, M., & Chilvers, G.A. (1975). Polyphosphate in eucalypt mycorrhizas: a cytochemical demonstration. *New Phytologistogist*, 74, 447-453.

[166] Ashford, A.E., Vesk, P.A., Orlovich, D.A., Markovina, A.L. & Allaway, W.G. (1999). Dispersed polyphosphate in fungal vacuoles in *Eucalyptus pilularis/Pisolithus tinctorius* ectomycorrhizas. *Fungal Genetics and Biology*, 28, 21-33.

[167] Tatry, M.V., Kassis, E.E., Lambilliotte, R., Corratgé, C., van Aarle, I., Amenc, L.K., Alary, R., Zimmermann, S., Sentenac, H. & Plassard, C. (2009). The differentially regulated phosphate transporters from the symbiotic fungus *Hebeloma cylindrosporum* and phophorus acquisition by ectomycorrhizal *Pinus pinaster*. *Plant Journal*, 57, 1092-1102.

[168] Burleigh, S.H. & Harrison, M.J. (1999). The down-regulation of *Mt4*-like genes by phosphate fertilization occurs systematically and involves phosphate translocation to the shoot. *Plant Physiology*, 119, 241-248.

[169] Liu, H., Trieu, A.T., Blaylock, L.A. & Harrison, M.J. (1998). Cloning and characterization of two phosphate transporters from *Medicago truncatula* roots: regulation in response to phosphates and to colonization by arbuscular mycorrhizal (AM) fungi. *Molecular Plant-Microbe Interactions*, 11, 14-22.

[170] Arocena, J.M., Glowa, K.R., Massicotte, H.B. & Lavkulich, L. (1999). Chemical and mineral composition of ectomycorrhizosphere soils of subalpine fir (*Abies lasiocarpa* (Hook.) Nutt.) in the Ae horizon of a Luvisol. *Canadian Journal of Soil Science*, 79, 25-35.

[171] Glowa, K.R,. Arocena, J.M. & Massicotte, H.B. (2003). Extraction of potassium and and/or magnesium from selected soil minerals by *Piloderma*. *Geomicrobiology Journal*, 20, 99-111.

[172] Paris, F., Bonnaud, P., Ranger, J. & Lapeyrie, F. (1995). *In vitro* weathering of phlogopite by ectomycorrhizal fungi. I. Effect of $K^+$and Mg^{2+} deficiency on phyllosilicate evolution. *Plant and Soil*, 177, 191-201.

[173] Jentschke, G., Brandes, B., Kuhn, A.J., Schröder, W.H. & Godbold, D.L. (2001). Interdependence of phosphorus, nitrogen, potassium and magnesium translocation by the ectomycorrhizal fungus *Paxillus involutus*. *New Phytologist*, 149, 327-337.

[174] Wiese, J., Kleber, R., Hampp, R. & Nehls, U. (2000). Functional characterization of the *Amanita muscaria* monosaccharide transporter AmMst1. *Plant Biology*, 2, 1-5.

[175] Melin, E. & Nilsson, H. (1955). ^{45}Ca used as indicator of transport of cations to pine seedlings by means of mycorrhizal mycelium. *Svensk Botanisk Tidskrift*, 49, 119-121.

[176] Kemppainen, M., Circosta, A., Tagu, D., Martin, F. & Pardo A.G. (2005). *Agrobacterium*–mediated transformation of the ectomycorrhizal symbiont *Laccaria bicolor* S238N. *Mycorrhiza* 16, 19-22.

[177] Kemppainen, M., Duplessin, S., Martin, F. & Pardo, A.G. (2008). T-DNA insertion, plasmid rescue and integration analysis in the model mycorrhizal fungus *Laccaria bicolor*. *Microbial Biotechnology*, 1, 259-269.

[178] Kemppainen, M., Duplessin, S., Martin, F. & Pardo, A.G. (2009). RNA silencing in the model mycorrhizal fungus *Laccaria bicolor*: gene knock-down of nitrate reductase results in inhibition of symbiosis with *Populus*. *Environmental Microbiology*, 11, 1878-1896.

[179] Kemppainen, M., Pardo, A.G. (2010). pHg/pSILBAγ vector system for efficient gene silencing in homobasidiomycetes: optimization of ihpRNA-triggering in the mycorrhizal fungus *Laccaria bicolor*. *Microbial Biotechnology*, 3, 178–200.

[180] Martin, F., Kohler, A., Murat, C., Balestrini, R., Coutinho, P., Jaillon, O., et al. (2010). Périgord black truffle genome uncovers evolutionary origins and mechanisms of symbiosis. *Nature*, 464, 1033-1038.

In: Fungi: Types, Environmental Impact and Role in Disease ISBN: 978-1-61942-671-9
Editors: Adolfo Paz Silva and María Sol

Chapter VII

Effects of Organic P and Soil Disturbance on Symbiotic Efficiency of Arbuscular Mycorrhizal Fungi, and Morphological Characteristics of the Regeneration Mycelia after Abscission

Wen Ke Liu[*]
Institute of Environment and Sustainable Development in Agriculture, Key Lab. for Agro-Environment & Climate Change, Ministry of Agriculture, Chinese Academy of Agricultural sciences, Beijing, China

Abstract

Two experiments were designed to investigate the effects of soil disturbance and organic P (Na-phytate and lecithin) on mycorrhiza formation and symbiotic efficiency of AMF associated with maize in greenhouse. The results showed that P forms did not affect the root colonization of BEG167 and *Glomus versifome* (*G.v*) with maize. *G.v* produced more extraradical fungal biomass than BEG167 when KH_2PO_4 and lecithin were applied. The treatments with lecithin and KH_2PO_4 had higher maize biomass than those supplied with Na-phytate. BEG167 and *G.v* differed in maize biomass contribution when Na-phytate was applied. Soil disturbance reduced significantly root colonization, and significantly or slightly decreased shoot biomass and P uptake of maize colonized by BEG141. However, mycorrhiza development and symbiotic efficiency of *G.v* were not influenced by soil disturbance. On cellulose acetate membrane, numerous regeneration hyphae with anastomosis structure, cluster structure and dichotomous structure were observed. The results suggested that rapid recovery of extraradical hyphae of AMF might compensate the P contribution loss caused by soil disturbance and buffer the decrease of propagule reserve in soil ecosystem.

[*] Email: liuwke@163.com.

1. Introduction

Arbuscular mycorrhizal fungi (AMF) are ubiquitous in natural and agricultural soils. They play important roles in plant nutrition and soil conservation[1]. Therefore, they are regarded as important beneficial microorganisms in low input agro-ecosystems and organic farming [2-3]. The persistence of AMF in ecosystems depends on the formation and survival of propagules (e.g. spore, hypae and infected roots) [4].

Therefore, abundance and diversity of AMF in agro-ecosystems are lower than natural ecosystems for low diversity of host plant and frequent soil disturbance caused by activities such as tillage [5].

Earlier results showed that soil disturbance reduced root colonization by AMF [6-7] and early-season P uptake [8-9] by crops. However, other experiments found similar or relatively high colonization and P uptake [10-12]. Until now, no uniform conclusions have been drawn for inconsistent data represented [13]. Factually, soil disturbance damages the extraradical hyphae web of AMF vertically, but the responses of AMF to disturbance have been poorly understood. Some reports indicated that AMF differed in their response to soil disturbance at genus level [14-15].

In order to investigate the effects of soil disturbance on mycorrhizal development and symbiotic efficiency of AMF, a greenhouse experiment was conducted with factorial design of two AMF species and three soil disturbance treatments.

Organic P accounts for 30-80% of total soil P [16] and is an important potential source of P for plants in most agricultural and natural soils. Generally, organic P contents in agricultural soils are low due to intensive cropping and high removal ratio of crop biomass, and also appears highly heterogeneous in the soil profile. Previous data showed that organic P (RNA, phytate and lecithin) utilization could be significantly improved with the contribution of AMF at low P level [17] and high P level [18].

In addition, organic P could improve hyphal density of AMF in soil in pot culture or long-term field experiment [18-19]. However, no investigations have been published about inter-specific variations in responses to organic P of AMF when organic P is applied solely in rhizosphere to simulate the heterogeneous P status in agricultural soil.

Redecker *et al.* (1995) [20] developed a cultivation system to recover extraradical fungal biomass of AMF, called a glass-bead cultivation system, which made it possible to explore directly the fungal responses to different soil conditions. Chen *et al.* (2001) [21] modified the method for studies on nutrient and trace metal uptake by arbuscular mycorrhiza, and Liu (2008) [22] used the glass-bead cultivation system to evaluate the soil adaptability of AMF. In this paper, the glass-bead cultivation system was used to collect fungal biomass to assess the response differences of AMF to maize supplied with various P forms in rhizosphere at 50 mg kg^{-1}.

2. Materials and Methods

The glass-bead cultivation system used in experiment 1 is a three compartmental rhizobox separated using 1 mm (both root and AMF hyphae can pass through) and 30 μm nylon nets (only AMF hyphae can pass through). Its configurations are length × width ×

height = (4 + 2 + 5) × 10 × 12 cm. The left compartment was used for maize growth are filled with 0.5 kg pot^{-1} soil, the middle one used as buffer zone contained quartz sand (0.3 kg pot^{-1}, diameter 0.5-2.0 mm), and the right one was designed for mycelium growth of AMF filled with glass beads (0.8 kg pot^{-1}, diameter 1.5-2.0 mm). Quartz sand and glass beads were purified and sterilized at 121℃ for 2 h.

Experiment 1 was conducted as a complete factorial design with two factors: AMF (BEG167 and *Glomus versifome*, *G.v*) and three phosphorus forms (Na-phytate, KH_2PO_4 and lecithin). In the plant growth compartment, 50 mg kg^{-1} P was applied as KH_2PO_4, Na-phytate and lecithin. There were six treatments in total with each treatment replicated four times.

The rhizobox used in experiment 2, length × width × height = (3 +10) × 9.5 × 8 cm, was separated into two compartments (plant compartment and hyphal compartment) by 30 μm nylon mesh to allow hyphae to pass through but not roots. Two AMF, *Glomus intraradices* (BEG141) and *G.v* were tested. Two inoculation treatments were designed. Each inoculation treatment had 15 replicates.

In experiment 1, 50 g inoculant of BEG167 and *G.v* for each pot was inoculated into the plant compartment using the layer inoculation method, while 20 g inoculant of BEG141 and *G.v* was used in experiment 2. All inoculants consisted of culture substrate, infected root segments, spores, and hyphae. In both experiments, maize (*Zea mays* L.) Nongda 108 was used as host plant. Seeds were surface sterilized in 10% (v/v) hydrogen peroxide for 10 min and rinsed with de-ionized water.

Three seeds were planted in each rhizobox and two seedlings were maintained after germination. When the maize was grown for 42 d, soil disturbance treatments were applied in the soil compartment. First, the soil in the soil compartment 0-2, 2-4 and 4-6 cm from the nylon net was excavated vertically with a sharp knife. Then it was replaced with the initial sterilized soil. Three replicates of *G.v* and BEG167 inoculation were used as controls. Each soil disturbance treatment was replicated three times.

The spare three inoculation pots for each AMF species were treated similarly as above but with disturbance 0-3, 2-5 and 4-7 cm respectively, with one pot for each disturbance treatment, but quartz sand was used as replacement. In addition, half a cellulose acetate membrane (0.45 μm, diameter 6 cm) split equally along the centre was inserted vertically with quartz sand, one point adhering to the nylon net vertically, then the membrane was buried with quartz sand. The method was modified from the inserted membrane technology reported by Baláz & Vosátka (2001)[23].

A black plastic membrane was laid over the quartz sand to retain water. Soil used in two experiments was collected from the plot with low phosphorus input, at the Changping Long-term Experimental Station, Beijing. Soil properties were: Olsen-P 5.21 mg kg^{-1}, pH 7.9 ($CaCl_2$), O.M. 12.1 g kg^{-1}. Soil was seized through a 2 mm sieve and sterilized at 121°C for 2 h.. The two experiments were harvested 56 d after maize planting in a greenhouse at temperature 20-30°C. Shoots and roots of maize were harvested separately, then carefully washed under a stream of water to remove adhering soil, blotted dry, weighted and cut into segments about 1 cm long and mixed thoroughly. Weighed samples were taken randomly for determination of root colonization (RC) and dry weight (dried in oven at 70°C for 48 h).

Approximate 0.5 g root sample was used to determine root colonization. The root sample was cleared with 10% KOH, and then stained with trypan blue in lactoglycerol. Percentage root length colonized was assessed under a dissecting microscope using a gridline intersect method [24]. The shoot samples were ground and dry-ashed. P concentrations of shoot were determined by the phosphovanado-molydbdate colorimetric method [25].

The extraradical fungal biomass in the glass-bead compartment was recovered by rinsing with a stream of water to remove adhering soil and glass beads with 0.025 mm mesh, then impurities were removed under microscope with forceps.

Then biomass was rinsed with deionized water carefully, dried in oven at 70°C for 48 h and weighed. The cellulose acetate membranes were recovered, stained with trypan blue and observed under microscope.

3. Results

3.1 Results of Experiment 1

RC and Fungal Biomass

Maize RC by the isolate BEG167 (28.3-31.8%) was significantly lower than the colonization percentage produced by *G.v* (60.2-61.3%), which was independent of P forms applied (Table 1). *G.v* produced more extraradical fungal biomass than BEG167 when KH_2PO_4 and Na-Phytate were applied. Three P forms made slightly differences in fungal biomass of two tested fungi (Table 1).

Shoot Dry Weight (SDW), P Concentration (SPC) and P Uptake (SPU)

The SDW of maize applied with KH_2PO_4 and lecithin was higher than that treated with Na-phytate.

No differences were found in SDW between two fungi inoculation treatments with KH_2PO_4 and lecithin, but BEG167 contributed less to SDW than *G.v*. Inoculation with *G.v* gave higher SPC and SPU than BEG167 when KH_2PO_4 was applied.

Table 1. Root colonization, fungal biomass, and shoot biomass and P nutrition of maize inoculated with *G.v* or BEG167 combined with various forms of P

AMF	P forms	RC (%)	Fungal biomass (mg·pot^{-1})	SDW (g·pot^{-1})	SPC (%)	SPU (mg·pot^{-1})
G..v	KH_2PO_4	60.3 a	3.27 a	14.47 a	0.17 a	24.18 a
	Na-phytate	61.3 a	2.71 ab	8.85 b	0.14 b	12.54 c
	Lecithin	60.2 a	2.30 abc	14.52 a	0.14 b	20.63 b
BEG167	KH_2PO_4	28.3 b	1.28 cd	14.13 a	0.14 b	19.34 b
	Na-phytate	31.8 b	0.88 d	7.38 c	0.13 b	9.81 d
	Lecithin	29.0 b	1.73 bcd	14.03 a	0.14 b	20.12 b

Notes: Values in table are means of four replicates. Different letters the means indicate significant difference at 5% level. NT represented no treatment.

3.2 Result of Experiment 2

Rc

Maize RC by BEG141 was higher than by *G.v* (Table 2). Soil disturbance significantly influenced the RC by BEG141 but not by *G.v*.

The RC by BEG141 decreased significantly from 85.6 to below 60% with disturbance.

Table 2. Root colonization, shoot biomass and P nutrition of maize inoculated with *G.v* or BEG141 combined with different disturbance patterns

AMF	Disturbance pattern (cm)	RC (%)	SDW ($g \cdot pot^{-1}$)	SPC (%)	SPU ($mg \cdot pot^{-1}$)
G.v	NT	20.5 c	5.19 ab	0.11 bcd	5.75 bcd
	0-2	24.6 c	5.25 ab	0.10 d	5.32 cd
	2-4	24.1 c	5.00 ab	0.11 d	5.27 d
	4-6	27.8 c	4.97 ab	0.11 cd	5.30 cd
BEG141	NT	85.6 a	5.58 a	0.13 a	7.19 a
	0-2	48.2 b	5.02 ab	0.12 abc	6.04 bc
	2-4	55.8 b	4.93 ab	0.12 a	6.11 abc
	4-6	54.2 b	5.35 a	0.12 ab	6.54 ab

Notes: Values in table are means of four replicates. Different letters the means indicate significant difference at 5% level. NT represents no treatment.

Biomass, SPC and SPU of Maize

Soil disturbance did not decrease SDW (Table 2). In addition, soil disturbance decreased the P contribution to maize of BEG141 inoculation.

For *G.v* inoculation, only 2-4 cm disturbance treatment decreased P contribution to maize.

Morphological Observation of Regenerated Hyphae

Numerous regenerated hyphae were found on the recovered cellulose acetate membranes no matter in which sites they were inserted, or which AMF was inoculated.

Apart from dichotomous structures, numerous anastomosis structures were frequently observed on both membranes inserted in the rhizobox inoculated with both BEG141 and *G..v* (Figure 1a-h).

Cluster Structure and Dichotomous Structure

On the cellulose acetate membrane, a lot of cluster structures of fungal regeneration hyphae were observed (Figure 2a-b).

In addition, dichotomous structures of AMF were found on the cellulose acetate membrane (Figure 3a-b).

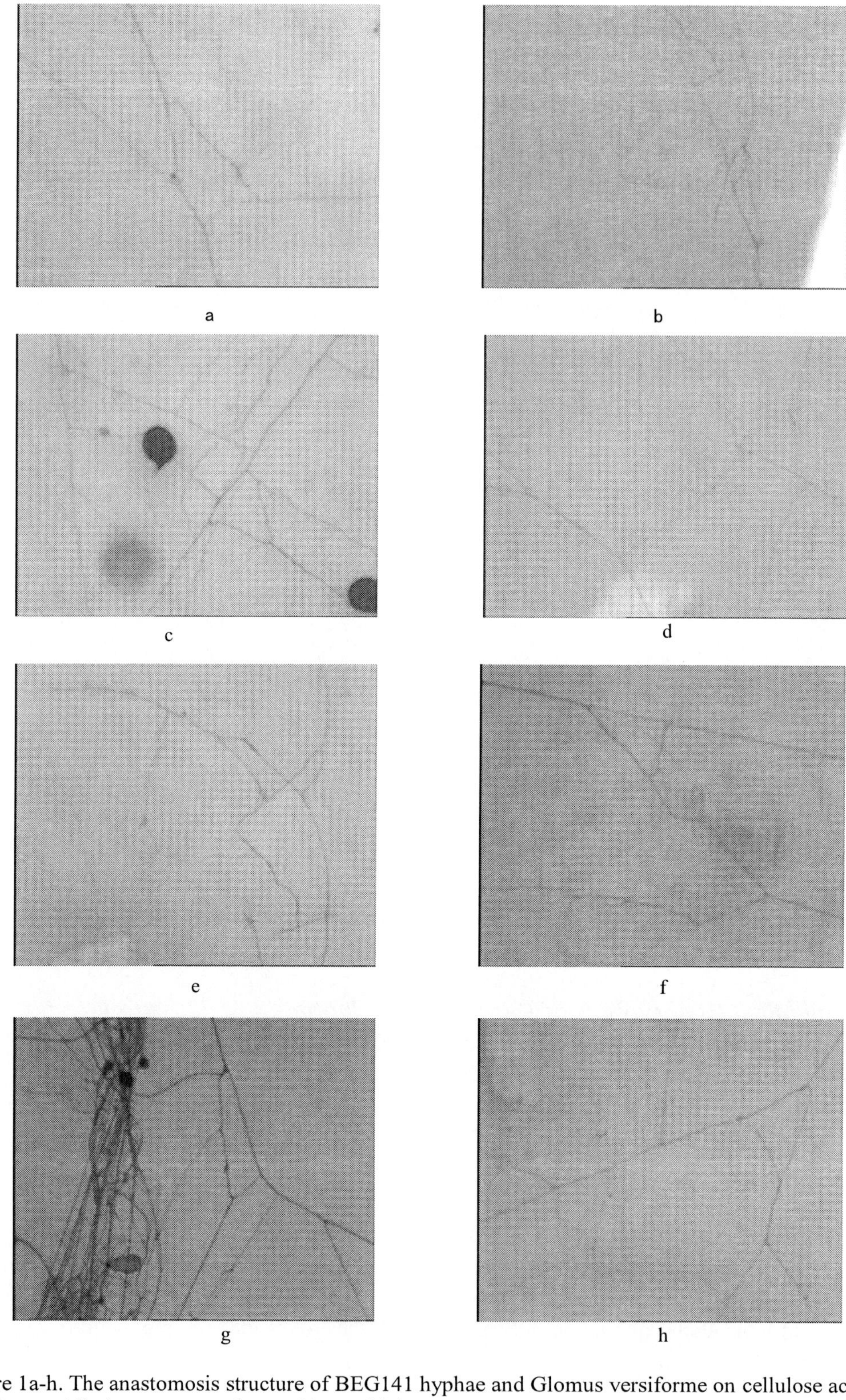

Figure 1a-h. The anastomosis structure of BEG141 hyphae and Glomus versiforme on cellulose acetate membrane. All pictures were taken at 20× multiple.

a b

Figure 2a-b. The cluster structures of BEG141 hyphae on cellulose acetate membrane.

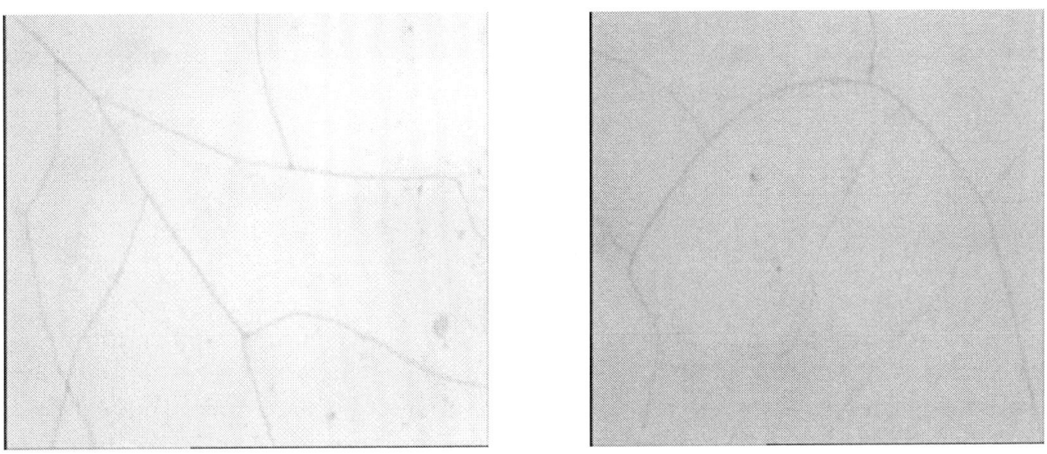

a b

Figure 3a-b. Dichotomous structures of BEG141 hyphae on cellulose acetate membrane (20x).

4. Discussions

Soil P content and disturbance by tillage are important factors that influence arbuscular mycorrhizal development in fields. Three fungi, BEG141, BEG167 and *G.v* are dominant or common species of the AMF community in agricultural soils of North China. The present data showed that P forms made no differences in RC, which was in agreement with the results of Feng *et al.* (2003)[17]. Previous studies had indicated that organic P could increase hyphal density of AMF in soil [18-19]. Fungal biomass and hyphal density in soil are two effective indices for AMF to evaluate the quantity of extraradical biomass. However, fungal biomass, including spores and mycelium, is more effective than hyphal density, mycelium only, for indicating the overall biomass of AMF. Therefore, two indices are essentially different when author's different species are compared arising to their biological characteristics (hyphal thickness, sporulation timing and quantity). To our knowledge, this is the first report of indirectly assessing the effects of organic P application on fungal biomass using the glass-bead cultivation method.

In the second experiment, BEG141 and *G.v* root colonization responded differently to hyphal severance caused by soil disturbance, which may explain the reasons why RC varies on a case-by-case basis in the field after tillage. However, the present data confirmed that extraradical hyphae could regenerate from original hyphae after being excised since no regenerated hyphae could grow 7 cm in two weeks (4-6 disturbance treatment). Therefore, the hyphal structures near the disturbance sites might serve as the inoculants.

BEG141 and *G.v* had different strategies to supply carbon for extraradical hyphae reestablishment. The carbon demand of extraradical hyphae re-establishment was large and urgent for AMF after being severed. A hypothesis of two kinds of strategies for carbon gain can be suggested. One way is to demand more carbon from the host plant, as with *G.v*, while the other is to discount part of the carbon that should be used in establishing intraradical fungal structure, as with BEG141. The data showed that RC of BEG141 decreased significantly from 85.6 to below 60%. McGonigle & Miller (2000) [13] suggested that the inconsistent effect of soil disturbance might result from specific component differences of the AMF community in field. Also, these authors indicated that the effect of soil disturbance on RC depends on the inoculum density of soil ecosystems. The present data showed that BEG141 and *G.v* responded differently in terms of root colonization, which indicates that the composition of the AMF community will determine the actual effect of soil disturbance on mycorrhiza formation. The data might partly explain the inconsistent results of soil disturbance on RC in past studies at AMF species level [14-15]. To the author's knowledge, this is the first evidence that there are interspecific differences in response of AMF to soil disturbance.

Rapid recovery of extraradical hyphae of AMF after disturbance might play great role in maintenance survival and biodiversity of the AMF community in intensively cultivation field. In addition, the results suggested that rapid recovery from disturbance damage of AMF might compensate for the P contribution loss caused by soil disturbance and buffer the decrease of propagule reserve in soil ecosystems. The ecological and genetic significance of anastomosis structures of AMF had been reviewed by Giovannetti et al. (2001) [26]. In conclusion, components of organic P might affect AMF biomass and their function differently for their heterogeneous distribution in soil profile, and also there are interspecific differences between AMF in the same community with respective response to soil disturbance in agro-ecosystems.

References

[1] Smith, S.E & Read, D.J. (1997). *Mycorrhizal Symbiosis*, 2nd edn..Academic Press; London, UK.

[2] Gosling, P., Hodge, A., Goodlass, G. & Bending G.D. (2006). Arbuscular mycorrhizal fungi and organic farming. *Agriculture, Ecosystems and Environment*, 113, 17–35.

[3] Douds, D., Nagahashi, Jr. G., Reider, C. & Hepperly, P.R. (2007). Inoculation with arbuscular mycorrhizal fungi increases the yield of potatoes in a high P soil. *Biological Agriculture and Horticulture*, 25, 67-78.

[4] Kabir Z. (2005). Tillage or no-tillage: Impact on mycorrhizae. *Canadian Journal of Plant Science*, 85, 23-29.

[5] Helgason, T., Daniell, T. J., Husband, R., Fitter, A.H. & Young, J. P. W. (1998). Ploughing up the wood-wide web. *Nature*, 394, 431..

[6] Mulligan, M.F., Smucker, A.J.M. & Safir, G.F. (1985). Tillage modifications of dry edible bean root colonization by VAM fungi. Agronomy Journal, 77, 140-144.

[7] McGonigle, T. & Miller, M. (1996). Development of fungal below ground in association with plants growing in disturbed and undisturbed soils. *Soil Biology and Biochemistry*, 28, 263-269.

[8] O'Halloran, I.P., Miller, M.H. & Arnold, G. (1986). Absorption of P by corn (*Zea mays* L.) as influenced by soil disturebance. *Canadian Journal of Soil Science*, 66, 287-302.

[9] Gavito, M.E. & Miller, M.H. (1998). Changes in mycorrhiza development in maize indeced by crop management practices. *Plant and Soil*, 198, 185-192.

[10] McGonigle, T., Evans, D.G. & Miller, M. (1990). Effect of degree of soil disturbance on mycorrhizal colonization and phosphorus absorption by maize in growth chamber and field experiment. *New Phytologist*, 116, 629-636.

[11] McGonigle, T.P., Miller, M.H. & Young, D. (1999). Mycorrhizae, crop growth, and crop phosphorus nutrition in maize-soybean rotations given various tillage treatments. *Plant and Soil*, 210, 33-42.

[12] Koske, R.E. & Germma, J.N. (1997). Mycorrhizae and succession in plantings of beachgrass in sand dunes. *American. Journal of Botany*, 84, 118-130.

[13] McGonigle, T. & Miller, M. (2000). The inconsistent effect of soil disturbance on colonization of roots by arbuscular mycorrhizal fungi: a test of the inoculum density hypothesis. *Applied Soil Ecology*, 14, 147-155.

[14] Reader, R.J. & Hart, M.M. (2004). Do arbuscular mycorrhizal fungi recover from soil disturbance differently? *Tropical Ecology*, 1, 97-111.

[15] Merrywheather, J. & Fitter, A.H. (1998). Patterns of arbuscular mycorrhiza colonization of the roots of *Hyacinthoides non-scripta* after disruption of soil mycelium. *Mycorrhiza*, 8, 87-91.

[16] Dalal, R.C. (1977). Soil organic phosphorus. *Advances in Agronomy*, 29, 83–117.

[17] Feng, G., Song, Y.C., .Li, X.L. & Christie, P. (2003). Contribition of arbuscular mycorrhizal fungi to utilization of organic sources of phosphorus by red clover in a calcareous soil. Applied Soil. *Applied Soil Ecology*, 22, 139-148.

[18] Tarafdar, J. C. & Marschner H. (1994). Phosphatase activity in the rhizosphere and hyphosphere of VA mycorrhizal wheat supplied with inorganic and organic phosphorus. *Soil Biology and Biochemistry*, 3, 387-395.

[19] Gryndler, M., Larsen, J., HrŠelová, H., Řezáčová, V., Gryndlerová, H. & Kubát, J. (2006). Organic and mineral fertilization, respectively, increase and decrease the development of external mycelium of arbuscular mycorrhizal fungi in a long-term field experiment. *Mycorrhiza*, 16, 159-166.

[20] Redecker, D., Thierfelder, H. & Werner, D. (1995). A new cultivation system for arbuscular mycorrhizal fungi on glass beads. *Angewandte Botanik*, 69, 189-191.

[21] Chen, B.D., Christie, P. & Li, X.L. (2001). A modified glass bead compartment cultivation system for studies on nutrient and trace metal uptake by arbuscular mycorrhiza. *Chemosphere*, 42, 185-142.

[22] Liu, W. K. (2008). N, P contribution and soil adaptability of four arbuscular mycorrhizal fungi. *Acta Agriculturae Scandinavica Section B-Soil and Plant Science*, 58(3):285 – 288.

[23] Baláz, M. & Vosátka, M. (2001). A novel insert membrane technique for studies of mycorrhizal extraradical mycelium. *Mycorrhiza*, 11, 291-296.

[24] Giovannetti, M.B. & Mosse, B. (1980). An evaluation of techniques for measuring vesicular arbuscular mycorrhizal infection in roots. *New Phytologist*, 84, 489-500.

[25] Kitson, R.E., & Mellon, M.G. (1944). Colorimetric determination of phosphorus as molybdovanado phosphoric acid. *Industrial & Engineering Chemistry Analytical Edition*, 16, 554–557.

[26] Giovannetti, M., Fortuna, P., Citernesi, A.S., Morini, S. & Nuti, M.P. (2001). The occurrence of anastomosis formation and nuclear exchange in intact arbuscular mycorrhizal networks. *New Phytologist*, 151, 717-724.

Section 2: Fungal Applications to Forest Ecosystems

In: Fungi: Types, Environmental Impact and Role in Disease ISBN: 978-1-61942-671-9
Editors: Adolfo Paz Silva and María Sol

Chapter VIII

The Value of Mycorrhizal Fungi for Sustainable and Durable Soils

Jacqueline Baar
NCOI, Postbus 447, 1200 AK Hilversum, The Netherlands

Abstract

The interest in maintaining and increasing biodiversity has increased worldwide since The Treaty on Biological Diversity of Rio de Janeiro in 1992. This century, awareness of sustainability has grown too. The awareness of biodiversity and sustainability is not only focused on the above-ground environment, also the below-ground soil environment is considered. The soil environment occurring globally provides services that humans draw from nature. The so-called ecosystem services include products like food and processes like nutrient transfer. Numerous people consider the ecosystem services as public benefits contributing to welfare for free and infinitely available. The ecosystem services have considerable economic value and the consideration on this value is growing. However, the services provided by the soil can be used within certain conditions for sustainability reducing the risks of irreversible changes. Durable and sustainable soil ecosystem services have more economic value, certainly on the long term, than those services used for welfare without any consideration for the environment and biodiversity. Sustainable soil ecosystem services require a good soil quality integrating the physical, chemical, and biological components of soils and their interactions. This implies elimination of soil degradation, reduction of input of chemical fertilizers and crop protection agents as well as decrease of dehydration of soil and management of beneficial soil organisms. A good soil quality determines agricultural sustainability, environmental quality and, consequently, plant, animal and human health. For such good soil quality, knowledge is needed on the contribution of soil-organisms to durable and sustainable soil. Mycorrhizal fungi are an important group of the soil-organisms being beneficial to sustainable soils. These soil fungi live in symbiosis with the great majority of plants, trees and crops. Evidence is growing that mycorrhizal fungi enhance the plant health and crop yield. The efficacy of mycorrhizal fungi vary with abiotic factors including soil type and structure. In this paper, the efficacy of mycorrhizal fungi in relation to abiotic factors will be presented based on recent data. Preconditions on the soil characteristics for sustainable soils of good quality and for optimal development of mycorrhizal fungi are provided. The economic value of optimal mycorrhizal fungal development is shown and related to soil ecosystem services. European policy makers are working on legislation for soil protection and management. The European Commission has proposed a Soil Framework Directive to address soil degradation across the European Union, and to repair the damage that has already been done. Indicators for soil quality could contribute to monitoring the condition of soils when reflected in standards, like soil quality metrics. Mycorrhizal fungi are indicative for changes in soils being sensitive to changes in the physical-chemical conditions of the soil. In this paper, it is shown that mycorrhizal fungi are reliable and sensitive indicators for disturbances in the soils to be used for the development of soil quality metrics. Different soil quality metrics with different categories ranging from very bad to very good are described based on a considerable number of root samples. The applicability of the soil quality metrics is discussed and their potential usefulness for the legislations for protection of soil quality is shown.

1. Ecosystem Services

The Treaty on Biological Diversity of Rio de Janeiro has contributed to the interest worldwide for maintaining and increasing biodiversity. The growing interest is based on the awareness that biodiversity is declining a thousand times faster nowadays than million years ago [1]. The accelerated rate of declining biodiversity raises concerns about possible consequences of such loss for ecosystem functioning, and moreover for the provision of ecosystem services and human well being [2-8]. Growing human needs and demands have led to an increase in resource demands imposed on ecosystems and greater global consumption of natural resources. This has resulted in a decline of ecosystem services, because the majority of the people are under the consideration that the ecosystem services are free, invulnerable and infinitely available. Most people take the ecosystems services as public benefits and as a consequence the ecosystems services are absent at any society's balance sheet [9].

However, ecosystem services have an economic value. Minimum estimates equal or exceed global gross national product, although most services lie outside the market and are difficult to calculate [10]. An approximate price tag of US $33 trillion a year was calculated on these fundamental ecosystem services, nearly twice the value of the world's gross national product. The two major ecosystem services, 'soil formation' and 'nutrient cycling', were respectively estimated to represent US $17.1 and 2.3 trillion [11].

2. Soil Ecosystem Services

Human society benefits from a multitude of resources and processes from natural and managed ecosystems [9]. The ecosystems do not comprise the most obvious above-ground environment only, also the below-ground soil environment should be taken into account [9, 12].

Soils are present worldwide in terrestrial and aquatic ecosystems forming the basis for human activities. In fact, the soil provides many of the services that humans draw from nature. The so-called ecosystem services provided by the soil can be used within certain conditions that reduce the risks of irreversible changes [13], and are linked to the concept of sustainability of the system. In fact, there is clear evidence that biodiversity has positive effects on the provision of the ecosystem services and that further biodiversity loss can only be expected to compromise service delivery [1].

In this context, four categories of ecosystem services can be distinguished: 1. the provisioning services; 2. the regulating services; 3. the supporting services; 4. the cultural services (see Table 1, [14]). The ecosystem services comprise a) biomass production within the framework of agriculture and nature; b) flexibility of soil uses; c) provision of an endurable environment and d) protection of the below- and above ground biodiversity.

Durable and sustainable soil ecosystem services require a good soil quality integrating the physical, chemical, and biological soil components and their interactions. A good soil quality determines agricultural sustainability, environmental quality and, consequently, plant, animal and human health [15]. This will need the reduction of inputs of chemical fertilizers and crop protection agents, the decrease of soil dehydration, the management of beneficial soil organisms, factors leading to the elimination of soil degradation. In fact, the absence of such threats means that soil biodiversity and ecosystem service performance can arbitrarily be set at 100% [16].

Table 1. The four categories of ecosystem services

Provisiong services
• food (including seafood and game), crops, wild foods, and spices • water • minerals (including diatomite) • pharmaceuticals, biochemicals, and industrial products • energy (hydropower, biomass fuels)
Regulating services
• carbon sequestration and climate regulation • waste decomposition and detoxification • purification of water and air • crop pollination • pest and disease control
Supporting services
• nutrient dispersal and cycling • seed dispersal • Primary production
Cultural services
• cultural, intellectual and spiritual inspiration • recreational experiences (including ecotourism) • scientific discovery

3. Mycorrhizal Fungi Part of Soil Biota

Soils contain abundant biological components including bacteria, protozoa, fungi, nematodes and earthworms. Generally, about 3,000 to 5,000 kg of soil living organisms occur in one hectare of soil. The upper soil layers generally contain the highest densities of soil organisms including earth worms, pot worms, nematodes, fungi and bacteria [15]. The abundance and diversity of micro-organisms in the soil provide information about the conditions of the surrounding soil. The soil micro-organisms, including bacteria, fungi, nematodes and earthworms, indicate whether physico-chemical conditions are suitable for their development on one hand, and on the other hand they fulfil key functions in the soil. Soil micro-organisms contribute to decomposition processes, nutrient cycling and improvement of soil structure.

Fungi are present in the soil communities for about 40%, and mycorrhizal fungi are part of the communities establishing a symbiotic relationship with the great majority of plants. Investigations showed occurrence of mycorrhizal fungi on 83% of dicotyledonous and 79% of monocotyledonous plants [17].

The mycorrhizal fungal hyphae associate with plant roots. Exchange of nutrients derived from the soil by the hyphae for carbon originating from the photosynthesis takes place in the mycorrhizal roots. Mycorrhizal fungi are of major importance for the protection of plants against drought. Another role of the fungi is increase of plant health and reduction of sensitivity to above- and below ground pathogens. Mycorrhizal fungi also contribute to the formation of soil aggregates [18].

The group of fungi that occurs on 80% of the plant families, including trees and crops, are arbuscular mycorrhizal fungi (AMF). These fungi constitute a key functional group of soil biota with the ability to contribute to crop productivity and ecosystem sustainability in new plant production strategies [9]. The AMF also improve the germination and growth of plants through increased uptake of available soil phosphorus (P) and other non-labile mineral nutrients essential for plant growth.

The AMF play an important role in enhancing the resistance against diseases and improving plant health. Moreover, these fungi have also 'non-nutritional' effects in amelioration of soil structure and prevention of erosion [18]. The beneficial effects of AMF on plant performance and soil health are essential for the sustainable soil management [19-20].

The ectomycorrhizal fungi (ECM) are ubiquitous symbiotic fungi occurring all over the world. The host range of these fungi includes most gymnosperm and angiosperm trees as well as economically important timber-producing tree species [21]. Trees in urban areas are also dependent in their growth on mycorrhizal fungi [22]. Thousands of ECM species occur worldwide. The majority of ECM belongs to the Basidiomycetes forming epigeous sporocarps [23]. Well-known examples are the Fly agaric (*Amanita muscaria*), King Bolete (*Boletus edulis*) and Chanterelle (*Cantharellus cibarius*). The ECM play an important role in the health and survival of many tree species. These fungi provide the trees several benefits, which include the enhanced ability to absorb water and important ions such as phosphorus and nitrogen, and protection from soil borne root pathogens such as *Fusarium oxysporum* and heavy metals [18].

4. Economic Value of Mycorrhizal Fungi

The application of mycorrhizal fungi attributes to improving soil quality, and a good soil quality contributes to healthy plant growth with low incidence of pathogens. Mycorrhizal fungi are available on the market in a variety of products. These products contain either *in vivo* or *in vitro* produced mycorrhizal fungi by different companies worldwide. This means that the producers of mycorrhizal fungi see market opportunities for their products. Users show growing interest applying effective mycorrhizal fungi [24]. In 2010, a selection of twenty four (potential) users in The Netherlands were questioned systematically. The users comprised companies in landscaping, advising companies in rural development or tree caring, tree specialists and plant growers indicating their preferences for purchase and utilization of mycorrhizal fungi. The majority of the interrogated (potential) users, 88%, prefers good quality mycorrhizal fungi that are effective. The users indicated that efficacy of mycorrhizal fungi means for them productive plant growth and healthy plants with no incidence of diseases. The users believe that it is essential that products appear on the market being appropriate for the plant species and the soil conditions. The general opinion was that products with a general mix of mycorrhizal fungi was too generalistic and too often showing no effectivity. Miller [25] had a similar opinion stating that promotion of plant growth by mycorrhizal fungi can be enhanced by increasing the effectiveness of the indigenous fungi as well as by introducing more effective species. Various studies showed the efficacy of plant and soil specific mycorrhizal fungi. For instance, growth promotion of leek plants was shown by a mycorrhizal fungus appropriate to leek and clay soils in a collaborative study by Orfanoudakis & Baar (oral comm.). The leek plants with AMF were healthy and showed no incidence of diseases. The increase in length of the leek was about 10% which could be equivalent for at least 5% yield increase (fig. 1). The yield of leeks is about 30 ton/ha and the trading price of leeks was about € 100,-/kg in 2011. An increase of yield with 5% results in 1500 kg extra leeks. The positive effect of the AMF *Glomus intraradices* strain in the clay soil in this experiment is interesting for application in commercial food production systems. Before application, however, an economic analysis should be undertaken evaluating the response to increasing the effectiveness of the mycorrhizal symbiosis relative to increasing yield with addition of phosphorus fertilizer [25].

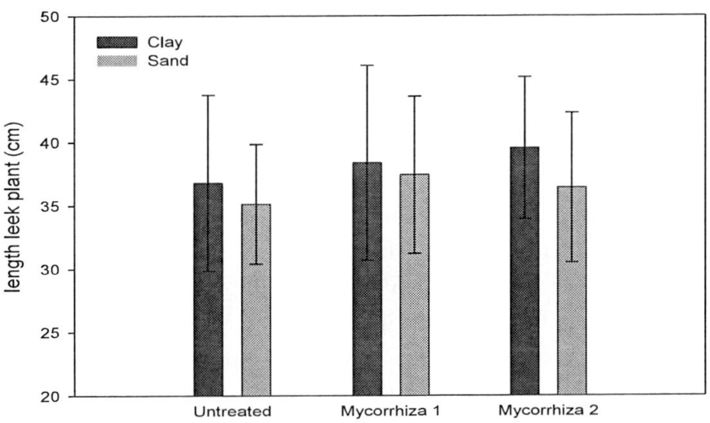

Figure 1. Enhanced growth of leek plants by AMF in clay and sandy soils.

In India, such an economic analysis was made when the fertilizer prices were doubling within a few years and biological solutions started to receive attention. Application of mycorrhizal fungi as a biological solution in combination with reduced fertilizer treatment resulted in considerable savings for the growers [26]. Such positive results show the promise of mycorrhizal technology for one of the most important ecosystem services: food production.

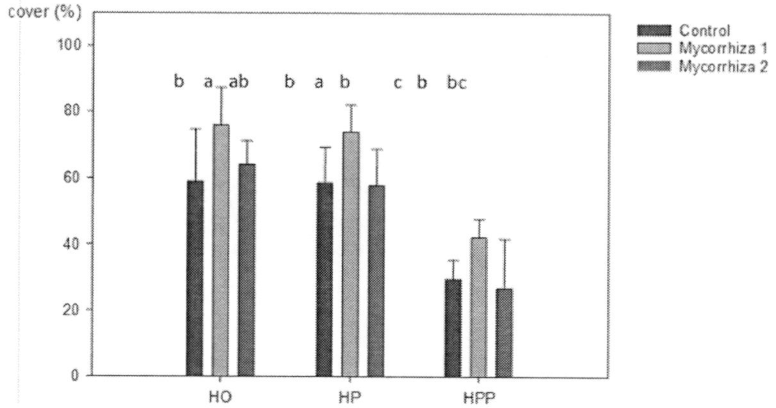

Figure 2. Degree of cover (%) by the plant species in pots per treatment. HO = undisturbed soil, HP = top soil removed and HPP = deep soil layer removed. Mycorrhiza 1 = mycorrhizal mixture 1 and mycorrhiza 2 = mycorrhizal mixture 2. Different letters = significant difference ($p<0.05$), error bars show standard deviations.

An economic analysis can also be made for the recovery of natural plant communities as part of restoration management. A restoration project in The Netherlands showed that stimulation of original natural plant species varied between the two mycorrhizal mixtures and the soil types [27]. Mycorrhizal mixture 1 was the most effective to a variety of natural grasses in the three sandy soil types that differed in ammonium, phosphate en potassium concentrations (fig. 2).

The efficacy of the AMF mixture 1 was highest in the most nutrient poor soil type, HPP. The AMF mixture 2 did not promote plant growth in HPP, but did in the most nutrient rich soil, HO (Fig. 2). This suggests that the application of mycorrhizal mixtures is only effective when the mycorrhizal fungi fit to the plant species and the soil conditions [27]. This is valid for different regions in the world. It only makes sense to use appropiate mycorrhizal products consisting of fungi that form symbiosis relationships with the target plant species. This means that a prerequisite for effective mycorrhizal products is that they contain indigenous species. Another prerequisite is sufficient colonizing fungal units for successful development of the mycorrhizal fungi. The efficacy and success of the application of mycorrhizal fungi in restoration practices is enhanced when information is obtained on the chemical properties of the soil as well as on the abundance and vitality of the present mycorrhizal fungi. Based on such information, the additive indigenous mycorrhizal fungi for the target plant species can be added to the soil in the appropriate composition and concentrations.

An economic analysis showed that the benefits of applying mycorrhizal fungi exceeded the costs making the application of mycorrhizal fungi attractive for the recovery of natural plant communities. The costs of the indigenous mycorrhizal fungi to be applied in an area of 10.000 m^2 (ha) was calculated at € 2500,-. The application of the mycorrhizal fungi can be calculated at €1000,- per ha. The total costs are € 3500,- per ha for improving the development of target pant communities. The effects of the application of mycorrhizal fungi are sustainable and have not to be repeated each year. Assuming that the effects of mycorrhizal fungi last for at least ten years, than the costs of the application of mycorrhizal fungi are € 350,- per ha per year [27].

Subsequently, the monetary values of the benefits were calculated. Therefore, the following benefits were identified: the non-user value and the recreational value. The non-use value is based on the people who attach importance and value to the existence of a specific nature area although they do not utilize that nature area. The non-use value of a nature area is dependant of the current and future use. Based on a study in Belgium to this value in nature forest reserve, the monetary non-use value is estimated at € 3500,- per ha per year [28]. The recreational value is difficult to estimate when there are no data on the number of visitors present. Therefore, the data from a Belgium study on the surplus of visitors by increasing the biodiversity in the target plant communities due the application of mycorrhizal fungi were used. Based on the Belgium study, the additional recreational value was estimated at € 1000,- per ha per year. In total the monetary value of the non-user value and the recreational value of a nature area in The Netherlands is estimated at € 4500,- per ha per year [27].

For forestry, two of the major factors for the use of ECM are the economic efficacy of inocula and the availability of practical and efficacious inocula [29]. Thus, it may not be surprising that quantitative field evidence of the use of mycorrhizal inoculations in wood production has been slow to emerge. A small-scale field trial in the Congo showed a volume increase of 30% at 50 months after out planting of eucalypts inoculated with *Pisolithus tinctorius* [30]. Similarly, in south-western Australia, growth increases of up to 80% were found 7 years after inoculation with *Amanita xanthocephala* [29]. Inoculation as a management option increases operational costs, and the advisability of adopting ECM technology is a measure of costs (inoculation) against benefits (increased productivity). The rationale for incurring the cost of inoculation is that the gain in improved biomass productivity will be greater than the cost of inoculation. The economic analysis was based on the planting of *Eucalyptus globulus* on 40 ha leased, ex-pasture sites [29]. Estimation was made of the monetary effect of the increase in mean annual increment (MAI) of the biomass of plantation trees in the order of 10, 20 and 30% over a base MAI of 22 m^3 ha^{-1} $year^{-1}$. At the base MAI rate, it was assumed that the stands would be harvestable in ten years. The inoculation with ECM would be of economic value when the growth improvements are attributable to the inoculated fungi [29]. Coppiced rotations would best reflect the average scenario in plantation operations in Western Australia. With such rotations, an increase in MAI of 30% and a stumpage price of \$ 20 m^3, would result in an extra \$ 952 ha^{-1} (net present value) return from the plantation at harvest. Therefore, ECM in Western Australia could be capable of improving MAI by 30% and then significant gains in revenue can be realized [29].

These examples showed economic profitable plant growth by the application of AMF and ECM. The economic benefits are highest when plant and soil specific mycorrhizal fungi are applied. Still, mycorrhizal fungi do not always contribute to plant growth for reasons not yet understood. The activity of AMF may be greatly limited by soil fumigation, non-responsive plant varieties, or rotations based primarily on non-mycorrhizal crops, or crops of low AM dependency [31]. Manipulation of agricultural systems to favor development of AMF is one way to achieving efficacy of mycorrhizal fungi. Another way is to breed plants for more optimal colonization by AMF as suggested by several authors [32-35]. Other ways are now explored in the technology of AMF allowing application of these fungi in high production agriculture with economic profitable yields.

5. Indicators for Soil Quality

The efficacy of AMF and ECM is generally influenced by the soil conditions. Optimal performance of AMF and ECM is usually found in good quality soils with the appropriate physical, chemical and biological characteristics resulting in productive and healthy plant growth. The good quality soils contribute to the ecosystem services of which agricultural food production is a major one. Thus far, however, good soil quality has not been defined by standardized measurements. Policy makers in Europe are working on legislation for protection and management of good quality soils. The European Commission has proposed a Soil Framework Directive to address soil degradation across the European Union, and to repair the damage that has already been done [16]. Indicators for soil quality could contribute to monitoring the condition of soils. The development of indicators would be the most valuable when the monitoring results are reflected in standards, which could be like soil quality metrics. Such soil quality metrics could be understandable for policy makers because they are comparable to the water quality metrics in use under the Water Directive Framework [12]. The interest in the development of standards for soil quality is increasing [15].

The interest in soil quality metrics is mostly based on indicators that can be determined easy and rapidly for relatively low costs. How to asses soil quality is an intriguing and challenging question [15]. The choice of indicative parameters for soil quality has not been resolved thus far. Moreover, standards for soil quality for reflecting the results on monitoring the soil quality status by the indicative parameters, like soil quality metrics, have not been proposed.

During the last decades, attempts have been made to develop tools for determining soil quality. Physical-chemical parameters were considered at first. Subsequently, chemical parameters were determined in combination with physical parameters. Concentrations of nutrients, including nitrogen, phosphate and potassium, were used for determining whether a soil has a good quality for planting crops or trees. These inferences are based mostly on the dead part of the soil and less on the living parts. For capturing the complete soil quality, the biological soil components have to be taken into consideration. Assessing soil quality with biological approaches was initiated in different countries in the world in 1992, but little information has been exchanged or published in the international literature [15].

For the Netherlands, attempts have been made by the use of a mixture of soil organisms, soil processes and soil functions for the development of a soil quality system [15]. However, thus far this has not resulted in any applicable soil quality metric because such a combination of soil parameters is complicated to interpret. Also, the analysis costs of the combination of biological indicative parameters are currently relative high being unattractive for users.

Mycorrhizal fungi, both AMF and ECM, are influenced by different soil conditions, indicating that these fungi are very responsive to the physical and chemical characteristics of soils. In addition, mycorrhizal fungi are highly sensitive to disturbances, like eutrophication by means of fertilisation, acidification, deep ploughing and movements of soil. The fungi respond rapidly to changes in the nutrient concentrations in the soil, such as nitrogen and phosphate. Mycorrhizal fungi react to disturbances, while other organisms still occur [36, 37].

Arbuscules are structures of AMF being physiological active in the exchange of nutrients, water and carbon. Arbuscules respond highly to changes in the chemical composition of the soil and they are extremely sensitive to high phosphate concentrations in soils [18, 38, 39]. The AMF structures disappear as soon as a soil contains high levels of plant available phosphate. Therefore, arbuscules are good indicators for changes in the physical-chemical soil conditions. The diversity of the AMF also reflects the situation of a soil. There are AMF occurring under high nitrogen levels while others occur under low nitrogen conditions [40]. These investigators observed that an increasing input of nitrogen was associated with the displacement of the larger-spored AMF species of *Scutellospora* and *Gigaspora* species with a concomitant proliferation of small-spored *Glomus* species. Nitrogen enrichment also significantly reduced spore abundance and modified the timing of AMF spore production in the most eutrophied site, and reduced hyphal root infection [40]. This implies that soil organisms like AMF respond to the physical-chemical soil conditions and are indicative for the conditions in the soil. The indicative value of soil organisms could be suitable for the development of soil quality standards, such as soil quality metrics.

Root tips associated with ECM hyphae are actively involved in the exchange of nutrients, water and carbon. The ECM hyphae on the roots tips and in the soil respond to changes in the chemical composition of the soil [36, 41]. The ECM fungi respond to nitrogen enhancement of the soil by a decrease in mycorrhizal root tip numbers and a change in species composition resulting in lower diversity. Another factor of influence on ECM is pH. Lowering of the pH results in reduced number of ECM and hyphae formation [36]. A change in the below-ground ECM community with increasing nitrogen levels of the soil and a declining pH in soils in Scots pine (*Pinus sylvestris*) forests was observed. Mycelia of *Inocybe lacera* and *Rhizopogon luteolus* were found at locations with relatively low nitrogen levels and high pH while hyphae of *Lactarius hepaticus* and *Russula* sp. were detected in soils with relatively high nitrogen levels and low pH [41].

6. Standards for Soil Quality

Recently, standards for soil quality based on mycorrhizal fungi were proposed [12]. The methodology for the development of the soil quality metrics is related to the methodology for the development of the water quality metrics within the Water Directive Framework. The water quality metrics are also based on biological parameters enabling distinction between bad, good and maximum ecological quality ratio's. The soil quality metrics for AMF in clay and sandy soils developed in 2010 are relatively simple [12]. Since 2010, novel data and insights have been obtained resulting in adjusted standards for soil quality (Tabel 1 and 2).

A soil quality metric based on observations on ECM is presented. Observations on ECM in The Netherlands from different locations ranging in nutrient levels, organic matter content and pH were used to define the soil quality metrics. The data originated from studies in The Netherlands [42].

Table 1. The soil quality metric for clay soil based at 310 samples of arbuscular mycorrhizal fungi. EQR = Ecological Quality Ratio

Soil quality metrics based on AMF in clay soils						
EQR	0-0,16	0,16-0,33	0,33-0,5	0,5-0,66	0,66-0,83	0,83-1,0
Category	Very bad	Bad	Inadequate	Moderate	Good	Very good
Arbuscules	0-1%	1-6%	6-15%	15-35%	35-70%	70-100%
Arbuscule/ hyphae	0-0,05	0,05-0,15	0,15-0,3	0,3-0,6	0,6-0,8	0,8-1,0
Number of taxa	0-1	1-2	2-4	4-6	6-8	>8

Table 2. The soil quality metric for clay soil based at 310 samples of AMF. EQR = Ecological Quality Ratio

Soil quality metrics based on AMF in sandy soil						
EQR	0-0,16	0,16-0,33	0,33-0,5	0,5-0,66	0,66-0,83	0,83-1,0
Category	Very bad	Bad	Inadequate	Moderate	Good	Very good
Arbuscules	0-5%	5-10%	10-25%	25-50%	50-70%	70-100%
Arbuscule/ hyphae	0-0,1	0,1-0,15	0,15-0,3	0,3-0,6	0,6-0,8	0,8-1,0
Number of taxa	0-1	1-2	2-4	4-6	6-8	>8

Tabel 3. The soil quality metric based at 500 samples of ECM. EQR = Ecological Quality Ratio

Soil quality metrics based on ECM					
EQR	0-0,2	0,2-0,4	0,4-0,6	0,6-0,8	0,8-1
Category	Bad	Inadequate	Moderate	Good	Very good
Fine roots with ectomycorrhizal fungi (ECM)	0-25%	25-50%	50-70%	70-90%	>90%
Number of taxa ectomycorrhizal fungi (ECM)	0-1	1-2	2-3	3-5	>5

7. Value of Soil Standards

Considering the requirements for monitoring soil conditions and for a soil quality metric, mycorrhizal fungi seem to be good indicators of soil conditions. The soil quality metrics are applicable in soil management measurements and provide repetitive results. Moreover, the assessment of soil quality by the biological indicators could fit in the policy making at the national and international level [12, 15].

The soil quality metrics presented are based on observations on AMF and ECM in a variety of locations ranging from low to high nutrient inputs. The soil quality metrics are based on more than three hundred root samples for AMF and five hunderd root samples for ECM in The Netherlands. This meets the requirement that there are sufficient data available for a first attempt to determine differences between bad and good locations. The reliability of the soil quality metrics for AMF and ECM can be raised by including a higher number of observations. For the Netherlands, sampling from different locations in The Netherlands is an ongoing process and will continue. Moreover, the soil quality metrics developed so far can lay a basis for the development of soil standards in other countries.

The soil standards lay a basis for a systematic relationship between the occurrence of AMF and soil conditions. The number of systematic studies on the relationship between abiotic soil variables and the development of mycorrhizal fungi is still low [43]. This is surprising because optimal performance of mycorrhizal fungi can only be obtained when soil conditions are known resulting in optimal plant growth and plant health. Therefore, the number of systematic studies on the relationship between physical-chemical composition of soils and the occurrence of mycorrhizal fungi at a larger scale, such as throughout Europe, have to be extended. The data derived from these studies will be indicative for soil quality throughout Europe. This can result in the development of more soil quality measures for more soil type's throughput Europe.

References

[1] Balvanera, P. Pfisterer, A. Buchmann, N. He, J-S. Nakashizuka, T. Raffaelli, D. & Schmid, B. (2006). Quantifying the evidence for biodiversity effects on ecosystem functioning and services. *Ecology letters*, 9, 1146-1156.

[2] Schläpfer, F. & Schmid, B. (1999). Ecosystem effects of biodiversity – a classification of hypotheses and exploration of empirical results. *Ecological Applications*, 9, 893–912.

[3] Chapin, F.S. Zavaleta, E.S. Eviner, V.T. Naylor, R.L. Vitousek, P.M. Reynolds, H.L. et al. (2000). Consequences of changing biodiversity. *Nature*, 405, 234–242.

[4] Loreau, M. Naeem, S. Inchausti, P. Bengtsson, J. Grime, J.P. Hector, A. et al. (2001). Biodiversity and ecosystem functioning: current knowledge and future challenges. *Science*, 294, 804–808.

[5] Kinzig, A. Pacala, S.W. & Tilman, D. (2002). *Functional Consequences of Biodiversity: Empirical Progress and Theoretical Extensions*. Princeton University Press, Princeton, NJ.

[6] Díaz, S. & Cabido, M. (2001). Vive la differénce: plant functional diversity matters to ecosystem functioning (review article). *Trends in Ecology and Evolution*, 16, 646–655.

[7] Hooper, D.U. Chapin, F.S. Ewell, J.J. Hector, A. Inchausti, P. Lavorel, S. et al. (2005). Effects of biodiversity on ecosystem functioning: a consensus of current knowledge. *Ecolological Monographs*, 75, 3–35.

[8] Srivastava, D.S. & Vellend, M. (2005). Biodiversity–ecosystem function research: is it relevant to conservation? *Annual Review of Ecology Evolution and Systematics*, *36*, 267–294.

[9] Gianinazzi, S. Gollotte, A. Binet, M-N. Van Tuinen, D. Redecker, D. & Wipf, D. (2010). Agroecology: the key role of arbuscular mycorrhizas in ecosystem services. *Mycorrhiza, 20*, 519–530.

[10] Pimm, L.S. (1997) The value of everything. *Nature, 387*, 231–232.

[11] Costanza, R. d'Arge, R. De Groot, R. Farber, S. Grasso, M. Hannon, B. Limburg, K. Naeem, S. O'Neill, R.V. Paruelo, J. Raskin, R.G. Sutton, P. & Van den Belt, M. (1997). The value of the world's ecosystem services and natural capital. *Nature, 387*, 253–260

[12] Baar, J. 2010. Development of soil quality metrics using mycorrhizal fungi. *Spanish Journal of Agricultural Research,* No1 Extra, S137-S143.

[13] Hanson, C. Ranganathan, J. Iceland, C. & Finisdore, J. (2008). *The Corporate Ecosystem Services Review.* World Resources Institute, Washington, USA.

[14] Millennium Ecosystem Assessment (MEA). 2005. *Ecosystems and Human Well-Being: Synthesis*. Island Press, Washington.

[15] Bloem, J. Hopkins, D.W. & Benedetti, A. (2008). *Microbiological methods for assessing soil quality*. CAB International, Wallingford, UK.

[16] Rutgers, M. Jagers op Akkerhuis, G.A.J.M. Bloem, J. Schouten, A.J. & Breure, A.M. (2010). *Priority areas in the Soil Framework Directive. The significance of soil biodiversity and ecosystem services*. RIVM RIVM Report 607370002/2010

[17] Wilcox, H.E. (1996). Mycorrhizae. In Y. Waisel, A. Eshel, & U. Kafkafi (Eds.) *Plant Roots: the hidden half* - second edition. Marcel Decker, Inc.

[18] Smith, S.E. & Read, D.J. (2008). *Mycorrhizal Symbiosis*. Academic Press, London, UK. 3rd Edition.

[19] Jeffries, P. Gianinazzi, S. Peretto, S. Turnau, K. & Barea, J.M. (2003). The contribution of arbuscular mycorrhizal fungi in sustainable maintenance of plant health and soil fertility. *Biololgy and Fertilility of Soils, 37*, 1–16.

[20] Barrios, E. (2007). Soil biota, ecosystem services and land productivity. *Ecological Economics, 64*, 269–285

[21] Peterson, R. L. Piche, Y. & Plenchette, C. (1984). Mycorrhizae and their potential use in the agricultural and forestry industries. *Biotechnological Advances, 2*, 101–120.

[22] Bainard, L.D. Klironomos, J.N. & Gordon, A.M. (2011). The mycorrhizal status and colonization of 26 tree species growing in urban and rural environments. *Mycorrhiza, 21*, 91-96.

[23] Harley, J. L. (1989). The significance of mycorrhiza. *Mycological Research, 92*, 129-139.

[24] Baar, J. & Josten, E. (2004). *Programme and abstracts of the COST-meeting: Role of Mycorrhiza in sustainable land management*. Applied Plant Research, WUR, The Netherlands.

[25] Miller, M.H. McGonigle, T.P. & Addy, H.D. (1995). An economic approach to evaluate the role of mycorrhizas in managed ecosystems. *Plant and Soil, 159*, 27-35.

[26] Gulati, A. & Cummings Jr. R. (2008). Mycorrhiza, a fungal solution for the farm economy. *The economic Times.*

[27] Baar, J. (2010). Restoration of plant communities in The Netherlands through the application of arbuscular mycorrhizal fungi. Symbiosis 52: 87-94.

[28] Rousseau S, Moons E (2007) Manual for an economic evaluation of the environmental policy: application for the Flander forest policy (In Dutch*). Journal for Enonomy and Managment*, LII, 1.

[29] Kuek, C. (1994). Issues concerning the production and use of inocula of ectomycorrhizal fungi. In A. D. Robson, L. K. Abbott, & N. Malajczuk (Eds.), *Management of Mycorrhizas in Agriculture, Horticulture and Forestry* (pp. 221-230). Kluwer Academic Publishers, Dordrecht.

[30] Garbaye, J. Delwaulle, J. C. & Diangana, D. (1988). Growth response of eucalypts in the Congo toectomycorrhizal inoculation. *Forest Ecology and Management*, 24,151-157.

[31] Ryan, M. & Graham, J.H. (2002). Is there a role for arbuscular mycorrhizal fungi in production agriculture? *Plant and Soil,* 244, 263-271.

[32] Parke, J.L. & Kaeppler, S.W. (2000). Effects of genetics among crop species and cultivars upon the arbuscular mycorrhizal symbiosis. In Y. Kapulnik, & D.D, Douds (Eds.), *Arbuscular mycorrhizas: physiology and function* (pp. 131–146). Kluwer, Dordrecht.

[33] Plenchette, C. Fortin, J.A. & Furlan, V. (1983). Growth responses of several plant species to mycorrhizae in a soil of moderate P-fertility I. Mycorrhizal dependency under field conditions. *Plant Soil*, 70,199–209.

[34] Gosling, P. Hodge, A. Goodlass, G. & Bending, G.D. (2006). Arbuscular mycorrhizal fungi and organic farming. *Agriculture Ecosystems & Environment*, 113, 17–35.

[35] Sawers, R.J.H. Gutjahr, C. Paszkowski, U. (2008). Cereal mycorrhiza: an ancient symbiosis in modern agriculture. *Trends in Plant Science*; 13, 93–97.

[36] Erland, S. Taylor, A.F.S (2002). Diversity of Ecto-mycorrhizal Fungal Communities in Relation to the Abiotic environment. In M.G.A. van der Heijden, & I.R. Sanders (Eds.), *Mycorrhizal Ecology* (pp. 163-200). Berlin, Springer.

[37] Baar, J. & Ozinga W.A. (2007). *Mycorrhizal fungi, key factor for sustainbale agriculture and nature*. KNNV-uitgeverij, Zeist, The Netherlands. 104 pp [In Dutch].

[38] Baar, J. Bergsma, H. & Steffen. F. (2008). The potential role of arbuscular mycorrhizal fungi for transition of highly fertilized grasslands into natural high biodiversity fields. In F. Feldman, & Y. Kapulnik, & J. Baar (Eds.), *Mycorrhiza Works* (pp. pp 217-228). Spectrum Phytomedizin, DPG-Publisher Braunschweig, Germany.

[39] Kahiluoto, H. Ketoja, E. & Vestberg, M. (2009). Contribution of arbuscular mycorrhiza to soil quality in contrasting cropping systems. *Agriculture, Ecosystems & Environment*, 134, 36-45.

[40] Egerton-Warburton, L.M. & Allen, E.B. (2000). Shifts in arbuscular mycorrhizal communities along an anthropogenic nitrogen deposition gradient. *Ecological Applications*, 10, 484-496.

[41] Smit, E. Veenman, C. & Baar, J. (2003). Molecular analysis of ectomycorrhizal basidiomycete communities in a *Pinus sylvestris* L. stand reveals long-term increased diversity after removal of litter and humus layers. *FEMS Microbiology Ecology*, 45, 49-57.

[42] Baar, J. (1995). *Ectomycorrhizal fungi of Scots pine as affected by litter and humus*. PhD thesis. Wageningen University, The Netherlands.

[43] Estaun, V. Camprubi, A. Joner, E.J. (2002). Selecting arbuscular mycorrhizal fungi for field application. In S. Gianinazzi, H. Schüepp, J.M. Barea, & K. Haselwandter (Eds.). *Mycorrhizal Technology in Agriculture* (pp 249-259). Birkhäuser Verlag, Basel, Switzerland.

In: Fungi: Types, Environmental Impact and Role in Disease ISBN: 978-1-61942-671-9
Editors: Adolfo Paz Silva and María Sol

Chapter IX

Fungi as Contributors to Cycling of Radionuclides in Forest Ecosystems

*J. Guillén** *and A. Baeza*
LARUEX (Environtmental Radioactivity Laboratory), Dpt. Applied Physics, Faclty of Veterinay Sciences, University of Extremadura, Cáceres, Spain

Abstract

Understanding the environmental behaviour of radionuclides in forest ecosystems is of great importance to ensure an adequate level of radioprotection. The accumulation of radionuclides by fungi and the process of their transfer from the soil play a significant part in this behaviour because of the ability of these organisms to accumulate heavy metals, including most radionuclides. As a consequence of radioactive fallout and nuclear accidents, anthropogenic radionuclides have entered forest ecosystems, with radiocæsium being among the most hazardous in the medium-to-long term. As a general rule, fungi are able to accumulate radiocæsium very efficiently. However, there is a large range of variation, about seven orders of magnitude, in its content as reported in the literature. In areas heavily affected by a radioactive deposition, the consumption of mushrooms by man or animals contributes to increasing their body's radiocæsium content. There have been fewer studies of radionuclides other than radiocæsium, but on occasions they can also be of importance. In areas with low or almost negligible levels of anthropogenic radioactive contamination, the content of radionuclides belonging to the natural radioactive decay series, such as ^{210}Po and ^{210}Pb, contributes significantly to the internal dose of the general population. The uptake of naturally occurring radionuclides also shows a great range of variation, although not as wide as that observed for radiocæsium. Not all species of fungi accumulate radionuclides at the same level. There is also a difference based on the fungus's nutritional mechanism, with mycorrhizae generally having a greater radiocæsium content than saprophytes. But there are also other factors that influence the accumulation and the transfer of radionuclides from the soil, such as the quantity of radionuclides deposited as a consequence of the fallout event, the chemical form of those radionuclides, and the soil's properties (pH, and organic matter and clay contents), among others.

*E-mail address: fguillen@unex.es.

1. Introduction

Fungi play an important role in the cycling of radionuclides within forest ecosystems, because they are able to accumulate heavy metals and radionuclides. These radionuclides are first accumulated by the fungus's mycelium and then translocated to the fruit bodies. The content of radionuclides, mainly radiocæsium, in fungi from heavily radioactively contaminated locations (e.g., areas affected by the Chernobyl accident) is greater than in other foodstuffs [1,2], in particular than in other forest products such as bilberries [3,4]. Although the fruit bodies' lifespan is very short and highly seasonal, with maximum fructification in autumn and a secondary fructification in spring, they have great significance for health and the environment. The consumption of fruit bodies collected in radioactively contaminated areas can even be a radiological hazard for the human population. An increase in the body's content of radiocæsium was observed among the population which often consumed wild mushrooms in such areas [2], and a seasonal increase was also observed in autumn linked to the time of maximal appearance of mushroom fruit bodies [5]. There is another indirect pathway which can also affect the human population, especially critical groups in the Arctic – the consumption of game. Reindeer and caribou eat mushrooms whenever available, and a rumen content of up to 20% of mushrooms has been reported in an individual moose [6]. Abundant spores of the fungi *Xerocomus* spp. and *Hypholoma capnoides* and an increased ^{137}Cs content have been observed in the fæces of roe deer [7]. In some countries, this consumption implies a significant, highly seasonal, enhancement of the radiocæsium content in the meat, with a maximum in the second half of the year [8,9]. Although radiocæsium is the most studied, fungi take up other anthropogenic and naturally occurring radionuclides. Their contribution to the internal dose can be in some cases, even higher than that of radiocæsium, as in the case of ^{210}Pb and ^{210}Po [10-12]. The accumulation and transfer of all these radionuclides depend on many variables, such as the quantity of radionuclides deposited as a consequence of the fallout event, the chemical form of those radionuclides, and the soil's properties (pH, and organic matter and clay contents), among others. Other variables depend on the characteristics of the considered species of fungus and its nutritional mechanism.

2. Determination of Radionuclides

Radionuclides are unstable isotopes that usually lose their excess energy by the emission of ionizing radiation. The emission probability is different for each radionuclide and it is expressed in terms of its half-life. The half-life of a radionuclide, $T_{½}$, is the time it takes for the activity to decrease by a half. There are three main types of ionizing radiation: α, β, and γ, whose main properties are the following:

1) α-particles are helium nuclei (2 protons + 2 neutrons). They have high values of mass and kinetic energy, the latter within the range 3.95–8.78 MeV[1]. Their interaction with matter is so strong that even a sheet of paper is enough to stop them, with all their

[1] An electron volt (eV) is a unit of energy equal to the amount of kinetic energy gained by a single unbound electron when it accelerates through an electric potential difference of 1 V. 1 eV = $1.609 \cdot 10^{-9}$ J.

energy being lost in crossing its thickness. Most of α-emitting radionuclides are naturally occurring: ^{238}U, ^{226}Ra, ^{210}Po, ^{232}Th, etc.

2) β-particles have various origins depending on the characteristics of the nucleus (β^-, β^+, EC or electron conversion, …), but the final result is the emission of electrons or positrons within a wide range of energies (0.01859–3.54 MeV). Due to their low mass and charge compared with α-particles, they can penetrate more deeply into matter, requiring a few centimetres of aluminium to stop them. Some β-radionuclides are: ^{3}H, ^{210}Pb, ^{90}Sr, ^{40}K, ^{137}Cs etc.
3) γ-radiation consists of high-energy photons, from a few keV to less than 10 MeV. Their penetrating power is greater than the previous particles, and they require several centimetres of lead to be stopped. This type radiation is frequently emitted immediately after α- or β-radiation. ^{40}K, ^{137}Cs, and ^{226}Ra, among others, are γ-emitting radionuclides.

Systems to detect radionuclides are different for each type of radiation, because they have different characteristics. Due to the low penetrating power of α and β particles, it is necessary to perform radiochemical separation procedures for their proper assay. In the case of γ-radiation, however, such radiochemical separations are usually unnecessary. The determination of α-emitting radionuclides is usually by means of silicon detectors within a vacuum chamber. Gas-flow proportional counters and LSC (liquid scintillation counting) are used to determine both β-particles and α-particles. The determination of γ-emitting radionuclides is by means of NaI(Tl) and HP(Ge) crystals. Sample preparation for γ-spectrometry is easier than for α- and β-spectrometry because usually no radiochemical procedures are involved. As an example, the preparation of the sample for the determination of ^{137}Cs in mushroom samples can simply consist of reducing their volume by drying and then ashing at 400 °C [13]. For the determination of ^{90}Sr, samples are usually ashed at 600 ºC to remove organic matter [14].

3. Sources of Radionuclides in the Environment

The radionuclides occurring in the environment can be classified into two main groups: natural and anthropogenic (or man-made). The naturally occurring radionuclides comprise those belonging to the natural radioactive decay series, ^{40}K, and cosmogenic radionuclides. There are three natural radioactive series present in nature: thorium, uranium, and actinium, also denominated A = 4n, A = 4n+2, and A = 4n+3 respectively, where A is the mass number of the radionuclide. These natural series begin with ^{232}Th (thorium series), ^{238}U (uranium series), and ^{235}U (actinium series), and finish with ^{208}Pb, ^{206}Pb, and ^{207}Pb, respectively. The isotope ^{40}K ($T_{½}$ = $1.277 \cdot 10^9$ yr) is also a naturally occurring radionuclide with a long half-life, and constitutes 0.012% of potassium. Cosmogenic radionuclides are formed by the interaction of high-energy cosmic rays with atomic nuclei in the atmosphere. Some examples are ^{7}Be, ^{10}Be, and ^{22}Na. The distribution of these radionuclides in the geosphere depends on the distribution of the geological media from which they derive and the processes which concentrate them at a specific location in specific media [15]. They can also be enhanced locally by human activities, and may be a cause for significant radiological concern because of NORM (Naturally Occurring Radioactive Material) industries. These industries use raw materials with high contents of natural radionuclides, and their bioavailability in their products, by-products, and residues may be enhanced due to physicochemical changes or to the form in which they are managed. Mining industries (uranium, iron, copper, ...), the production of fertilizers from phosphate rocks, and burning non-nuclear fuels (coal, oil, gas) are some examples of NORM activities [15].

Anthropogenic radionuclides were first introduced into the environment in 1945 as a consequence of the atom bomb blasts during World War II. During the 1950s and 1960s, there were a great number of atmospheric nuclear weapons tests, which released huge amounts of a multitude of radionuclides into the atmosphere. Due to atmospheric circulation, they became distributed worldwide and ultimately were deposited onto the soil [16]. This phenomenon is usually known as global fallout. Most of these radionuclides were short-lived and have already decayed, as is the case with ^{131}I ($T_{½}$ = 8.02 d). However, ^{137}Cs and ^{90}Sr have longer half-lives (30.07 and 28.79 yr, respectively), and are actually the most important man-made radionuclides environmentally. The estimated quantities of activity released were 948 and 622 PBq for ^{137}Cs and ^{90}Sr, respectively. Some radioisotopes of plutonium were also released but in lower amounts – 10.87 PBq for $^{239+240}Pu$. The annual deposition of ^{137}Cs and ^{90}Sr due to global fallout is shown in Fig. 1.

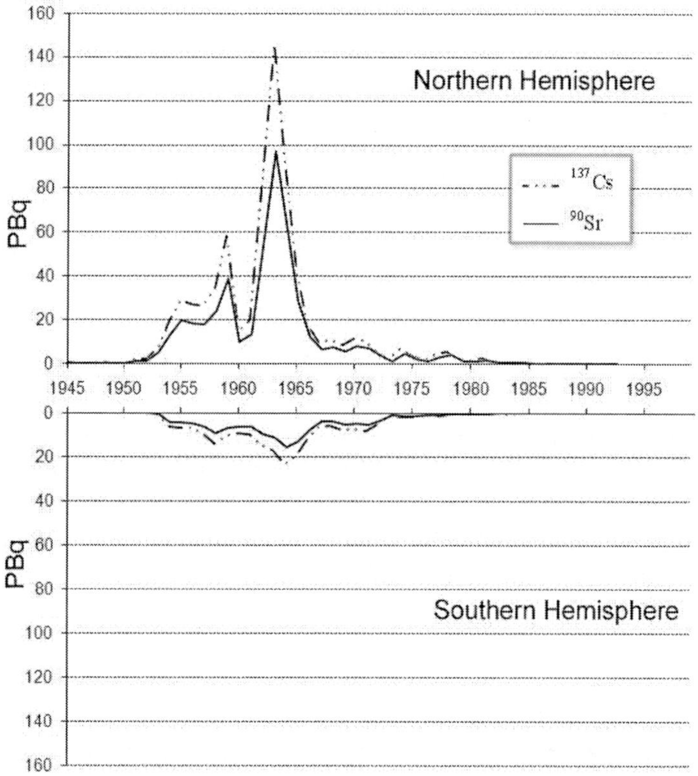

Figure 1. Annual deposition of ^{137}Cs and ^{90}Sr, expressed in petabecquerels (PBq), in the northern and southern hemispheres produced by atmospheric nuclear tests. Data from UNSCEAR [16].

There was a maximum around 1962, in which year the greatest number of tests were carried out. As the majority of tests took place in the northern hemisphere, the deposition there was greater than in the southern hemisphere. The deposition was inhomogeneous in latitude because of circulation patterns and airflows in the stratosphere and troposphere. It was greater in the temperate regions, with a maximum deposition of ^{90}Sr in the 40–50° latitude bands, and lower in the equatorial and polar regions [16].

There is another natural series named neptunium (A = 4n+1), which was extinct because the half-life of ^{237}Np ($T_{½}$ = $2.14 \cdot 10^6$ yr) is shorter than the Earth's age. However, it was re-introduced into nature by the atmospheric release of anthropogenic ^{241}Pu ($T_{½}$ = 14.4 yr) and ^{241}Am ($T_{½}$ = 432.2 yr), predecessors of ^{237}Np.

Other sources of radionuclides in the environment are related to releases from nuclear fuel reprocessing factories, plutonium fabrication plants, nuclear waste storage sites, and accidents in nuclear reactors such as Windscale, Three Mile Island, Chernobyl, and, more recently, Fukushima. The most serious accident was Chernobyl (26 April 1986), which marked a point of inflexion for radioecology and radioprotection. Large quantities of radionuclides were released, about 54, 85, and 10 PBq for ^{134}Cs, ^{137}Cs, and ^{90}Sr, respectively [16]. The most heavily contaminated areas (>3700 kBq/m^2 of ^{137}Cs) were located mostly within the 30 km radius around Chernobyl. However, it also seriously affected other countries such as Finland, Sweden, Norway, Germany, and Austria, where there were areas in which about 40-185 kBq/m^2 of ^{137}Cs were deposited [16]. The importance of the Fukushima deposition in Europe was lower than the Chernobyl one [17].

4. Radionuclide Content in Fruit Bodies

Fruit bodies of fungi are known to be capable of accumulating naturally occurring and anthropogenic radionuclides. Table 1 shows the range of activity levels reported in the literature worldwide. Radiocæsium is the most extensively analyzed radionuclide, due to its potential to represent a radiological hazard and easy measurement procedure. More than 90% of the literature consulted reports some radiocæsium content in fruit bodies. Other γ-emitting radionuclides, such as ^{40}K, can be determined simultaneously in the same γ-spectrometric assay. The determination of α- and β-emitters has been less frequent, which might be partially attributed to the need for radiochemical separation processes in their determination.

4.1. Antrhopogenic Radionuclides

The values of the radiocæsium content in fruit bodies collected worldwide after 1986 present a wide range of variation – about nine orders of magnitude (see Table 1). Firstly, this huge range is primarily a consequence of the different levels of radiocæsium deposition onto the soil. This can be clearly observed in Table 2, where the range of ^{137}Cs in fruit bodies is listed by country. Those countries more seriously affected by the Chernobyl accident presented higher ^{137}Cs contents, while more distant, and therefore less affected, countries presented lower contents. The range of variation reduces to one to three orders of magnitude when considering mushrooms collected in the same area or forest [32, 47, 84]. The ranges listed in Table 2 were reported after the Chernobyl accident in 1986. However, the data on the accumulation of ^{137}Cs before that date are much lower in value due to the differences between global and Chernobyl fallout. They range within 37–17 800 Bq/kg d.w. [34, 72, 110, 117, 118]. *Cortinarius armillatus, Paxillus, involutus, Xerocamus badius*, and *Rozites caperata* presented high pre-Chernobyl radiocæsium contents. The impact of Chernobyl is clearly detected in the ^{137}Cs content over the period 1984–1988, with there being a marked increase by up to ten times in samples of the same species collected before and after 1986 [34]. Because ^{134}Cs was also released in Chernobyl fallout, it was also detected in mushrooms. These data are not presented in table 2 because the Chernobyl fallout is characterized by the ^{134}Cs/^{137}Cs ratio, reflecting the situation of these isotopes within the reactor. Indeed, this ratio was used to distinguish between the ^{137}Cs content due to global fallout from that of Chernobyl origin [84, 91].

Secondly, the accumulation of ^{137}Cs depends on the species of mushroom [13]. Several species of fungi that presented higher radiocæsium contents than most are also listed in Table 2 (*Xerocumus badius*, *Paxillus involutus*, *Hebeloma sp.*, *Rozites caperata*, among others). The accumulation of radiocæsium was observed to present a log-normal frequency histogram in *X. badius* collected in Poland [84]. A similar distribution was observed in the content of different species of mushroom collected in Spain and Japan [13, 112], countries in which the contribution of Chernobyl fallout was negligible [119]. A single given species of mushroom can be used as an environmental biomonitor to map radiocæsium contamination, although it can present the shortcoming of it being difficult to find fruit bodies of the same species in different locations. *X. Badius* has been used for this purpose [73, 84]. The range of ^{137}Cs content for a single species has been found to vary by one to three orders of magnitude [84, 94]. According to analyses of variance, the variation among populations of *Suillus variegatus* has been estimated to be about 40%, with no significant differences found within the same population [94]. Some pigments present in the cap of the fruit bodies – derivatives of pulvinic acid, badione A1, and norbadione A2 – can bind potassium and cæsium efficiently [120]. These compounds are present in the cap of *X. badius* and *Boletus erythropus*. The ^{137}Cs content presents a positive correlation with the stable cæsium (^{133}Cs) content in mushrooms, the ratio ^{137}Cs/^{133}Cs being almost constant for a given forest ecosystem [100, 101, 115]. Both stable cæsium and radiocæsium were found to be positively correlated with rubidium [100, 112].

The accumulation of radiocæsium is not constant in time [67, 82, 98]. It may decrease (*Rozites caperatus, Lactarius torminosus*) [67, 82], remain constant (*Suillus variegatus*) [98], or even increase (*Cantharellus* spp.) [98]. This apparently strange behaviour is related to the layer of soil in which the mycelium is located and the time it takes for the radiocæsium deposited onto the surface soil to migrate downwards. Thus, species whose mycelium is located near the surface showed a decrease with time, while those with deeper mycelium may show an increase in radiocæsium content with time [121].

Radiostrontium is another long-lived radionuclide that has been released into the environment, although not so widely analyzed in mushrooms as radiocæsium. The range of variation worldwide is lower than that observed for radiocæsium – about four orders of magnitude. Although there is little data, differences of the ^{90}Sr content depending on the origin of the fallout can be observed. In locations close to Chernobyl the range was 7.8–1500 Bq/kg d.w. [23, 24], while in more distant countries – Poland, Germany, and Spain – the range was 0.25–6.3 Bq/kg d.w. [14, 19, 22, 25]. The differences between the locations affected by radiocæsium and ^{90}Sr released from Chernobyl can be attributed to the fact that radiocæsium is volatile while radiostrontium is an intermediate radionuclide, and therefore unable to be be transported over such long distances as radiocæsium [16]. *Paxillus involutus*, *Boletus edulis*, *Clytocibe sp.*, and *Lactarius deliciosus* presented the highest contents of ^{90}Sr. The range of the $^{90}Sr/^{137}Cs$ ratios for the same species but in different ecosystems was 0.001–0.70 [18, 25]. This indicated that the fruit bodies were able to accumulate radiocæsium preferentially over radiostrontium.

Plutonium was released from Chernobyl attached to refractory elements [16], so that it was deposited mainly in the vicinity. At first glance, the range of activity levels reported in the literature would imply that the mushroom uptake of plutonium is highly variable. However, when considering only samples from Ukraine, the range was 2.96–82.8 Bq/kg d.w. [24, 27], while the range for samples from other countries (Poland, Spain, Finland, and Japan) was 0.0009–0.164 Bq/kg d.w [14, 25-27]. *Paxillus involutus* and *Cantharellus cibarius* presented the highest content in Ukraine, and *Hebeloma cylindrosporum* and *Russula cessans* in Spain which had been affected mainly by global fallout [119]. In a given ecosystem, the uptake of plutonium was lower than of radiostrontium, and much lower than of radiocæsium. The uptake of americium was the lowest of all the long-lived artificial radionuclides considered, with the only data reported limited to Finland and Spain. Its origin may be considered to be global fallout, as the presence of ^{241}Am in the Chernobyl Reactor-4 core was minimal, and its parent ^{241}Pu was attached to refractory elements [16]. *Clitocybe* sp., *Hebeloma cylindrosporum*, and *Lycoperdon perlatum* presented the highest contents of ^{241}Am [14].

The content of several short-lived radionuclides (^{110m}Ag, ^{125}Sb, ^{103}Ru, and ^{131}I) has also been reported. These radionuclides were detected in mushrooms collected just after the Chernobyl accident. Their range of variation may be attributed to differences in the deposition of those radionuclides over several countries (Ukraine, Austria, Italy, Slovakia, and former Yugoslavia) [23, 24, 30-33].

4.2. Naturally Occurring Radionuclides

The most extensively studied naturally occurring radionuclide is ^{40}K. Its range of variation is about three orders of magnitude, considering the minimum and maximum values reported in the literature (see Table 1), although the usual values detected are within 1000–2000 Bq/kg d.w. independently of the species or location. For a single given species, the frequency distribution is symmetrical and almost Gaussian, differing from the log-normal distribution observed for radiocæsium [25, 84]. This implies that the uptake of ^{40}K (0.012% of natural potassium) is self-regulated by homœostatic processes. Although ^{40}K is an alkaline element, as also is ^{137}Cs, no correlations were detected between these radionuclides [13, 84]. This lack of correlation was also observed for the case of stable potassium with either radiocæsium or stable cæsium [100, 115].

Because of their long half-lives, the presence of uranium and thorium isotopes in mushrooms can be important. They can be measured directly by means of inductively coupled mass spectrometry (ICP-MS), X-ray fluorescence, and epithermal neutron activation analysis (ENAA) [52, 53]. The uranium and thorium contents determined by these techniques can be related to the activity levels of total uranium and total thorium, which consist of ^{238}U

(99.27%) and ^{232}Th (100%), respectively, taking into account the different half-lives of the isotopes. The relationship between mass and activity level is given in Eq. 1:

$$A(Bq) = \frac{\ln 2 \cdot N_A}{T_{1/2}(s) \cdot M} \cdot m(g) \qquad (1)$$

where A is the activity level in Bq, $T_{½}$ the half-life of the radionuclides in s, M the atomic mass, and N_A Avogadro's number.

Since uranium and thorium are not essential to the development of fungi, their content in mushrooms is lower than that of ^{40}K. Some differences have been observed between the uranium and thorium content in different countries. While in Poland and Czech Republic the range was (0.0011–0.270) and (0.0008–0.496) Bq/kg d.w. for ^{238}U and ^{232}Th, respectively [27, 53], the range observed in Spain was (0.120–47.4) and (0.061–10.7) Bq/kg d.w. for the same radionuclides [25, 27, 50, 52]. Although this was initially attributed to differences in the analytical techniques, ICP-MS vs X-ray fluorescence [53], no data were provided about the uranium and thorium content in the soils corresponding to Poland or the Czech Republic [27, 53]. Mushrooms collected from rubble piles in uranium mining areas in Germany had higher ^{238}U contents (range 0.1–259 Bq/kg d.w.) than those collected in unaffected areas (range 0.9–3.5 Bq/kg d.w.) [48]. The ratio ^{234}U/^{238}U in mushroom samples was close to unity, indicating that the two uranium isotopes were close to secular equilibrium[2] [27, 51]. In Kosovo and Serbia, mushrooms have been used as biomonitors for depleted uranium (DU)[3]. The ^{238}U content in those areas was within the range (0.7–11.3) Bq/kg d.w. [49, 51], similar to that observed in Spain and Germany. Evidence for DU was found in only one sample in which ^{236}U was detected [49]. *Tricholoma equestre*, *Macrolepiota procera*, and *Cantharellus cibarius* presented high uranium contents [52]. *Lepista nuda*, *Cantharellus cibarius*, and *Clitocybe gibba* presented high thorium contents [52].

Other radionuclides belonging to the natural uranium series are ^{226}Ra, ^{210}Pb, and ^{210}Po. The ^{226}Ra content in mushrooms was higher than that of the head of the series, ^{238}U (see Table 1). This may be attributed to their different chemical properties. Uranium is a heavy metal and considered to be toxic, while radium is an alkaline earth element similar to calcium. The frequency distribution for ^{226}Ra was observed to be symmetrical and of Gaussian type [13]. As had been the case with uranium, the ^{226}Ra content was observed to be greater in mushrooms collected in rubble piles from uranium mining zones [48]. *Lactarius deliciosus*, *Lycoperdon* sp., and *Terfezia arenaria* showed high ^{226}Ra contents [13]. The isotope ^{210}Pb occurs as a descendant of the decay of ^{226}Ra, but radon occurs between these two radionuclides in the decay chain. Because radon is a gas, it emanates from the soil into the atmosphere. There, it decays to ^{210}Pb which is deposited onto the surface layers of soil. Thus, its origin might be direct deposition of ^{210}Pb onto the fruit bodies. However, comparisons with the uptake of stable lead showed that mushrooms mainly take ^{210}Pb up directly from the soil [54]. Correlations between ^{210}Pb and stable lead have been reported, and the frequency distribution of the radioisotope was found to be asymmetric [10]. For some species such as

---

[2] Secular equilibrium is a situation in which the quantity of a radioactive isotope remains constant because its production rate, due to decay of a parent isotope, is equal to its decay rate.

[3] Depleted uranium is uranium with a lower ^{235}U content than naturally occurring uranium, which has a ^{235}U/^{238}U activity ratio of about 0.4.

Lactarius deliciosus and *Lycoperdon perlatum*, the ^{210}Pb content was greater than that of its predecessor, ^{226}Ra [10]. *Fomes fomentarius*, *Sarcasphaera crassa*, and *Lycoperdon perlatum* presemted high ^{210}Pb contents. The ^{210}Pb ($T_{½}$ = 22.3 yr) decays to ^{210}Bi ($T_{½}$ = 5.013 d) and then to ^{210}Po ($T_{½}$ = 138.376 d). Because of this decay chain and the short half-lives of ^{210}Bi and ^{210}Po, the data reporting ^{210}Po content in mushrooms should be accompanied by the corresponding ^{210}Pb content [11, 12]. Most mushrooms accumulate ^{210}Po preferentially over ^{210}Pb, with ^{210}Po/^{210}Pb ratios within the range 0.87–320 [11, 12]. Therefore, ^{210}Pb and ^{210}Po were usually not in secular equilibrium within the fruit bodies. This might be a consequence of the greater ^{210}Po content in the litter layer of the soil [12]. *Leccinum vulpilum*, *L. versipelle*, and *Suillus luteus* presented high ^{210}Po contents [12].

Table 1. Range of the content of anthropogenic and naturally occurring radionuclides in fruit bodies of fungi collected in different countries. The data, which are reported as Bq/kg f.w., were corrected by assuming a dry/wet ratio of 0.1. * The uranium and thorium contents determined by conventional methods were converted to activity levels using the factors $9.28 \cdot 10^{-5}$ and $2.46 \cdot 10^{-4}$ g/Bq for total uranium and thorium, respectively

Anthropogenic radionuclides		
Radionuclide	Range (Bq/kg d.w.)	References
^{137}Cs	0.4 – 50700000	See table 2
^{90}Sr	0.0002 – 1500	[14, 18-25]
$^{238,239+240}$Pu	0.0002 – 82.8	[14, 22, 24-29]
^{241}Am	0.0054 – 0.067	[14, 25, 28]
^{110m}Ag	2.1 – 1872	[30-32]
^{125}Sb	130 – 10310	[24]
^{103}Ru	10 - 1480	[23, 31, 33]
^{131}I	11 - 48	[23]
Naturally Occurring Radionuclides		
^{40}K	70 – 3520	[20, 23, 32, 34-47]
^{234,235,236,238}U	0.0011 – 47.4	[25-29, 48-53]*
^{228,230,232}Th	0.0008 – 13	[21, 25, 27, 50, 52, 53]*
^{226}Ra	0.021 – 512	[10, 13, 20, 26]
^{210}Pb	0.75 – 289	[10-12, 48, 54, 55]
^{210}Po	2.22 – 1174	[11, 12, 48, 56]
^{7}Be	1.5 - 133	[13, 20]

The ^{210}Pb and ^{210}Po content in mushrooms can contribute significantly to the internal dose by ingestion, being even higher than the radiocæsium contribution [11, 12].

The content of the cosmogenic radionuclide ^{7}Be in mushrooms is associated with its deposition from the atmosphere with rainfall [122]. Due to its short half-live, $T_{½}$ = 53.4 d, its presence in mushroom samples implies its recent deposition shortly before their fructification. The frequency distribution was of log-normal type [13]. *Cantharellus cibarius*, *Pleurotus eryngii*, and *Lactarius deliciosus* presented the highest values [13, 20].

Table 2. Range of ^{137}Cs content in mushrooms after 1986, expressed in Bq/kg d.w., in different countries worldwide, together with the species that showed the greatest accumulations of radiocæsium and the corresponding literature references. The data, which are reported as Bq/kg f.w., were corrected by assuming a dry/wet ratio of 0.1

EUROPE			
Country	Range (Bq/kg d.w.)	Species	References
Austria	148 -37370	Xerocomus badius, Hydnum repandum, Rozites caperata	[33, 57, 58]
Belgium	160 – 102000	Cortinarius brunneus, Cortinarius armillatus, Laccaria amethysana, Paxillus involutus	[46, 59-62]
Bulgaria	41 – 990	Craterellus cornucopioïde, Lactarius deliciosus	[40, 46]
Czech Republic	50 - 150700	Laccaria amethysana, Cortinarius armillatus, Xerocomus badius	[40, 46, 63-65]
Denmark	212 -13343	Rozites caperata, Cortinarius alboviolaceus	[7, 66, 67]
Finland	10 - 121400	Hydnum sp., Rozites caperata	[68-70]
France	2.5 – 5595	Cantharellus sp., Rozites caperata, Xerocomus badius, Hydnum repandum	[40, 54]
Germany	90 - 11290	Hebeloma sp., Hydnum repandum, Paxillus involutus, Rozites caperata	[9, 19, 71-73]
Great Britain	0.4 - 30500	Hydnum repandum, Boletus badius, Cantarellus cibarius	[74-76]
Hungary	0.6 -714	Tricholoma terreum, Suillus granulatus	[46, 77]
Italy	95 - 135575	Cantharellus lutescens, Clytocibe infandibuliformis	[31, 32, 35, 78-80]
Lithuania	42 – 3400	Tricholoma equestre, Cantharellus cibarius	[40]
Norway	500 - 445000	Lactarius sp., Amanita fulva, Amanita vaginata, Cortinarius armillatus, Rozites caperata	[8, 81, 82]
Poland	40 – 156700	Xerocomus badius, Sarcodon imbricatum	[3, 22, 27, 34, 40, 46, 47, 55, 83-86]
Romania	70 – 360	Hydnum repandum, Lactarius deterrimus, Boletus sp.	[40, 46, 80, 87]
Slovakia	323 – 966	Suillus luteus, Russula aeruginea	[65, 88]
Spain	0.5 – 647	Hebeloma cylindrosporum, Lactarius deliciosus	[13, 20, 89]
Switzerland	3 – 2000	Boletus edulis, Rozites caperata, Xerocamus badius	[90]
Sweden	20 -950000	Dermocybe cinnamomea, Hygrophorus sp., Rozites caperata, Russula decolorans, Suillus variegates, Cortinarius collinitus	[18, 91-100]
Turkey	9.84 - 401	Sarcodon scabrosus, Lepista nuda, Suillus bovinus	[101]
Belarus/Ukraine/Russia	9.6 – 50700000	Paxillus involutus, Xerocomus badius, Lactarius torminosus	[24, 27, 46, 91, 93, 102-108]
Former Yugoslavia	1.5 – 117000	Cortinarius armillatus, Laccaria amethysana	[30, 91, 109]
AMERICA			
Country	Range (Bq/kg d.w.)	Species	References
Canada	2 – 560	Leucopaxillus giganteus, Laccaria laccata	[91, 93, 110]
USA	30 – 50	Armillaria gallica, Boletus edulis	[91]
Mexico	2 - 1521	Clavariadelphus truncatus, Gomphus floccosus, Cortinarius caerukescens	[39, 111, 112]
ASIA			
Country	Range (Bq/kg d.w.)	Species	References
Japan	0.92 -16300	Hebeloma sp.,Tricholoma flavovirens	[36, 38, 43, 44, 113-116]
Taiwan	1.1 – 7.3	Ganoderma tsuga, Lentinula edodes	[41]
China	2.1	Boletus sp.	[80]

5. Nutritional Mechanism of Fungi

The accumulation of radionuclides depends on the nutritional mechanism of the fungi. In fact, mycorrhizal fungi accumulate more radiocæsium than saprophytes, and these have similar values of accumulation to those of parasites [13, 73, 91, 92, 109]. Only in Northern Italy has the radiocæsium accumulation by saprophytic mushrooms been reported to be greater than by mycorrhizae [78]. However, those authors attributed this to the fact that the samples were collected after the recent deposition of Chernobyl fallout, when it was still mostly in the surface soil [78]. There are not enough data reported for other anthropogenic radionuclides to observe any dependence on the nutritional mechanism.

With respect to naturally occurring radionuclides, mycorrhizal and saprophytic mushrooms have similar contents of ^{40}K and ^{226}Ra [13, 38]. The cosmogenic radionuclide ^{7}Be content was higher in saprophytic than in mycorrhizal mushrooms, because it is associated with rainfall deposition onto the surface soil [13]. Mycorrhizal and saprophytic mushrooms presented similar ^{210}Pb content, although the median value for saprophytes was slightly higher [10].

6. Distribution within the Fruit Bodies

Radionuclides are heterogeneously distributed within the fruit bodies of fungi. In most species, the cap has a greater radiocæsium content than the stipe [36, 37, 95, 118]. A similar distribution has been observed for ^{40}K [29, 36]. The distribution pattern of ^{210}Pb and ^{210}Po seems to be species dependent. In some species of the Boletaceae family, the cap content was higher than that of the stipe [12]. The ratio cap/stipe was (0.55–2.01) and (0.64–712) for ^{210}Pb and ^{210}Po, respectively [12, 56]. In some cases, the distribution for the same species varied with forest type for a single species [12]. That variation may be associated with the development stage of the fruit bodies. In young fruit bodies, the stipe presented higher activity than the cap (with gills) for ^{134}Cs, ^{85}Sr, and ^{60}Co. The pattern was reversed when mature fruit bodies were considered [29]. In that study under controlled laboratory conditions, it was also observed that the flesh of the cap and the gills presented about the same ^{134}Cs, ^{85}Sr, and ^{60}Co content [29]. A high radiocæsium content of the gills was also detected in wild mushrooms [37, 81, 118]. The distribution of other radionuclides has been less studied. The content of ^{226}Ra was found preferentially in the cap, but only two species were reported, *Pleurotus eryngii* and *Tricholoma equestre* [29]. The distribution of uranium, thorium, and plutonium within fruit bodies is not yet fully documented. Higher cap contents of these radionuclides were reported for *Pleurotus eryngii*, while the reverse was found in *Tricholoma equestre* [29].

7. Soil-to-Mushroom Transfer

The soil-to-mushroom transfer processes of radionuclides are usually quantified in terms of the so-called transfer factors. The classic definition is simply termed the transfer factor, TF, or content ratio, CR, being the ratio between the content in the mushroom and in the soil (see Eq. 2). In the case of radionuclides deposited onto the soil as a consequence of radioactive fallout, the aggregate transfer factor, T_{agg}, has been used, defined as the ratio between the content in the mushroom and the deposition of the radionuclide (see Eq. 3):

$$TF,CR = \frac{C_{mushroom}(Bq/kg\ d.w.)}{C_{soil}(Bq/kg)} \tag{2}$$

$$T_{agg} = \frac{C_{mushroom}(Bq/kg\ d.w.)}{Soil\ Deposition(Bq/m^2)} \tag{3}$$

where $C_{mushroom}$ is the radionuclide content in the mushroom, C_{soil} the radionuclide content in the soil, and Soil Deposition the amount of the radionuclide deposited per unit area.

These transfer factor definitions imply that the radionuclides have reached an equilibrium state in the ecosystem being considered. There is no temporal dependence in Eqs. 2 and 3. Variations with time are to be expected, especially for the T_{agg} values due to the downwards migration of radionuclides once deposited onto the soil [123]. According to its definition, the use of T_{agg} is unsuitable for naturally occurring radionuclides because they are already present in the soil, except in cases in which an anthropogenic deposition occurred, e.g., as a consequence of a NORM activity. The T_{agg} data represent a useful tool for a rough but rapid estimation of the transfer, especially immediately after an accidental release [124].

The definition of TF also presents some conceptual difficulties. The radioactive content of mushroom is clearly delimited, but what is the radionuclide content in the soil? There is no unified definition for which soil depth to use, this being left to the criterion of each researcher. However, there is a growing trend to use the content of the layer of soil in which the mycelium is located, and from which radionuclides will be transferred to the fruit bodies [113, 114, 121, 124]. Fungal mycelium mobilizes nutrients and metals, radionuclides among them, directly from minerals through the excretion of such organic acids as citric and oxalic acids [125, 126]. Thus, the location of the mycelium in the soil, and therefore the zone from which radionuclides are taken up, is highly important. Another difficulty that arises is whether all the radionuclide content is available for transfer to the fruit bodies, or whether part of it is strongly bound to soil particles. Such attachment will depend on various factors, among which are the quantity and chemical form of the radionuclides released, the time elapsed between deposition of the radionuclides and the determination of the speciation, the physicochemical characteristics of the soil, and the climatic conditions.

7.1. Location of Mycelium in Soil

There are different methods to locate the mycelium in soil: mechanical isolation of fungal mycelium directly from the soil; the determination in the soil of chemical compounds present in the cell walls of the mycelium; and the comparison of the ratio $^{137}Cs/^{134}Cs$ in different layers of soil and in the fruit bodies. In the first, fungal structures are extracted from soil samples, and the species are identified by DNA analyses or by recognition of those structures as being characteristic of a certain species [99, 127]. In the second, two of the chemical compounds present in the mycelium cell walls that can be sought in the soil layers are chitin and ergosterol. The former does not distinguish between living and dead fungal biomass, and is also present in insects in the soil. Ergosterol, however, is not only the main sterol present in most fungi, being a principal component of their cell walls, but is also absent from practically all other living matter, including the roots of plants [128]. Another advantage is that ergosterol degrades faster than chitin, so that it can be used as an indicator of the amount of functional mycelium present in a soil. But it also has the shortcoming that it is not fungal species specific. It can be used to estimate in which layer of soil most of the mycelium is located, but is unable to identify the species.

The application of these location methods can generate depth profiles of the distribution of mycelium in soil [105, 129, 130]. The vertical distribution was found to be fairly homogeneous for *Tylospora* spp. and *Piloderma fallax*, and strongly superficial for *Hydnellum peckii* [106]. In a pine forest, the ergosterol content was maximal in the surface layer of soil (0–5 cm), generally decreasing with depth [130].

These methods can also be used to estimate the radiocæsium associated with the mycelium in the soil, given determinations of the mycelium content within the range (12–138) mg fungal d.w./ g soil d.w. using ergosterol [129], and a mean value of 13.8 mg fungal mycelium/ g soil by mechanical extraction [105]. The first approach was to assume that the radiocæsium content was the same in the fruit bodies and in the mycelium, resulting in about 32% of the radiocæsium in the soil being associated with the mycelium [129]. This, however, was only a very rough estimate because the accumulation of ^{137}Cs in mycelium is not the same as in fruit bodies [105, 106]. Taking this into account, it was estimated that a mean value of 15% of the total inventory of ^{137}Cs in the soil was in the mycelium [105]. The ^{137}Cs associated with mycelium is not constant with time, but varies depending on the development stage of the fungus. Mycorrhizae of *Suillus variegatus* and *Lactarius rufus* were reported as having greater radiocæsium contents before the production of fruit bodies than during that process [95].

Finally, the ^{137}Cs/^{134}Cs method can be used to locate most of the mycelium in forest soil [60, 121]. The main shortcoming of this method is the necessity for ^{134}Cs to be present in the soil, which implies that recent radioactive fallout has occurred, as was the case after Chernobyl. It is based on the hypothesis that ^{134}Cs and ^{137}Cs (pre- and post-Chernobyl) have the same bioavailability, and therefore that the ^{137}Cs/^{134}Cs ratio detected in the fruit bodies must be the same as in the soil horizon in which the mycelium is located. By way of example using this procedure, *Xerocomus badius* was located in the L- or O_f- horizon, and *Russula cyanoxantha* in the A_h- or B-Horizon [121].

7.2. Bioavailability of Radionuclides in Soil

Soil characteristics such as pH, the clay and organic matter contents, and the climatic conditions to which the ecosystem is subject influence the bioavailability of radionuclides, and hence their transfer to fruit bodies. Radiocæsium, for instance, is more mobile in acidic soils [131, 132], while clay content can reduce its availability since the element tends to be irreversibly bonded to clay particles at the frayed edge sites (FES) [133]. The presence of humic and fulvic acids in a soil are also important: they can form complexes with plutonium and americium [134], and can reduce the adsorption of ^{137}Cs by clays [135].

To identify and estimate the association of radionuclides with soil particles, sequential extraction procedures are particularly useful. Several approaches have been described in the literature. They can be classified into methods based on a sequential extraction of the soil with selective reagents, and those in which the soil is sequentially treated with chemical solutions of increasing replacement/dissolving power [136]. The first group of procedures is mainly based on that developed by Tessier [137]. There is some discussion regarding the meaning of each fraction, because there is no unequivocal definition of the speciation scheme in the literature, and there is also the problem that the fractions are mainly operationally defined. The fractions usually considered in the literature are: water soluble (extracted with H_2O), exchangeable (extracted with NH_4OAc), easily reduced ($NH_2OH \cdot HCl$), oxidizable (H_2O_2 adjusted to pH 2, and NH_4OAc in HNO_3), acid-digestible (HNO_3 or HCl), and residual [138-140]. The second group is based on techniques usually used for soil testing in agriculture [141, 142]. The fractions considered are: water soluble (extracted with H_2O), exchangeable (extracted with NH_4OAc), dilute-acid extractable (extracted with HCl 1M), concentrated-acid extractable (extracted with HCl 6M), and residual. The radionuclides associated with the dilute-acid extractable fraction are considered to be potentially available, being remobilizable by a change in the climatic conditions affecting the soil. The concentrated-acid extractable fraction is considered to be persistently bound to soil particles. According to this, the available fraction can be assumed to comprise the exchangeable plus dilute-acid fractions [50, 123]. Radiostrontium is generally the most available radionuclide in a soil [130, 142]. The availability of radiocæsium is less, depending strongly on the soil's clay content [133]. Among the naturally occurring radionuclides, ^{210}Pb presents the greatest availability [10], followed by ^{226}Ra and uranium [50, 130]. Thorium and ^{40}K present very low availability [50, 130].

The bioavailability of radionuclides in soil is not constant with time. It changes as part of the radionuclides associated with the more labile fractions moves to deeper layers of soil or is taken up by plants and fungi. The temporal evolution has been studied in soils affected by Chernobyl fallout, observing different trends. The exchangeable fraction of radiocæsium decreased while increasing the lapsus of time since the deposition occurred [139, 142], which is usually known as ageing effect. The ^{90}Sr increased in the exchangeable and dilute-acid fractions (extracted with NH_4OAc and HCl 1M respectively), and a decrease in the concentrated-acid fraction (extracted with HCl 6M) over a time interval of 8 years [142]. This was due to the kinetics of the dissolution of condensed fuel particles ("hot particles"), which in various soils typical of the Chernobyl exclusion zone was dependent on the soil's acidity [143], the solution rate being greater at lower pH. This dependence was also observed in experiments under laboratory conditions with ^{134}Cs and ^{85}Sr as tracers in a slightly basic loam soil [144]. It was observed that the exchangeable fraction (extracted with NH_4OAc) was related to time by the expression: At^{-B}, with A and B being fitting parameters [144].

7.3. Transfer Factors

Table 3 gives the ranges of the TF values reported for anthropogenic and naturally occurring radionuclides. Values of T_{agg} are not listed because they cannot be defined for naturally occurring radionuclides. For radiocæsium, as was the case in the analysis of its accumulation, it presented the highest TF values and a great variability. The TF and T_{agg}

values are also species dependent [37, 130, 145]. Correlations between the transfer of ^{137}Cs and stable cæsium have been reported [113]. The TF values for mycorrhizal mushrooms are also higher than for saprophytes [146]. The TF values for ^{40}K are generally higher than those of ^{90}Sr. The TF values for the other radionuclides are usually less than unity, indicating that there is no significant accumulation.

In the values listed in Table 3, no correction has been made to take account of the availability of radionuclides. A modification of the TF, termed the available transfer factor (ATF), can be used to estimate this effect (see Eq. 4):

$$ATF = \frac{C_{mushroom}(Bq/kg\ d.w.)}{C_{soil}(Bq/kg)\cdot AF(\%)} \quad (4)$$

where AF is the percentage of available radionuclides in the soil. With ATF, there is an increase in the values of the transfer factors, these being greater for those radionuclides with lower availability [10, 14, 50, 130].

Table 3. Range of the TF values for anthropogenic and naturally occurring radionuclides in fruit bodies of fungi collected in different countries

	TF Anthropogenic radionuclides	
Radionuclide	Range	References
^{137}Cs	0.001 – 2941	[21, 37, 61, 73, 86, 101, 103, 111, 113, 130, 148, 149]
^{90}Sr	0.001 – 3.6	[14, 21, 130]
$^{238,239+240}Pu$	0.105 – 0.59	[14]
^{241}Am	0.13 – 0.60	[14]
	TF Naturally Occurring Radionuclides	
^{40}K	0.17 – 5.68	[37, 101, 130]
$^{234,235,236,238}U$	0.043 – 0.49	[50]
$^{228,230,232}Th$	0.030 – 0.47	[50]
^{226}Ra	0.05 – 1.5	[10, 130]
^{210}Pb	0.053 – 0.51	[10]

Conclusion

Fungi play an important role in cycling radionuclides in forest ecosystems. They can act as sinks of radionuclides, because they can take up and accumulate a great variety of them, both anthropogenic and naturally occurring. The main factor that influences the accumulation of anthropogenic radionuclides is the quantity and chemical form of the fallout. Radiocæsium is the most extensively studied radionuclide because it represented a radiological hazard for humans after the Chernobyl accident. In countries where heavy radioactive fallout occurred, the radiocæsium content was found to be higher than in other, less affected countries. Similar differences, although less marked, have been observed for radiostrontium and plutonium. However, in areas with lower levels of anthropogenic radioactive contamination, naturally occurring radionuclides, such as ^{210}Pb and ^{210}Po, should also be monitored because of their great radiotoxicity and potentially hazardous nature. An enhancement of uranium series radionuclides has also been reported in mushrooms influenced by NORM activities. Thus, the fruit bodies of fungi can be used as biomonitors for the radioactive pollution of a given ecosystem.

Several factors can also influence the uptake of radionuclides. Not all species of fungi accumulate them at the same level. In particular, the nutritional mechanism of the fungus can be important, and can have different effects for different radionuclides. Mycorrhizal fungi accumulate more radiocæsium than saprophytes, but for ^{40}K and radium both nutritional types show similar contents.

Accumulation in fungal fruit bodies is the consequence of radionuclide transfer processes, mainly from the soil. These processes depend on the layer of soil in which the mycelium is located, because it takes up radionuclides along with nutrients from that layer. The mycelium can also accumulate a significant percentage of the total inventory of radiocæsium in soils. The bioavailability of radionuclides in the soil also plays an important role in the transfer, because not all radionuclides contained in the soil are available for transfer to the fruit bodies.

Acknowledgments

We are grateful to the Autonomous Government of Extremadura (Junta de Extremadura) for the financial support to the LARUEX research group (FQM001).

References

[1] Horyna, J. (1991) Wild mushrooms – the most significant source of internal contamination. *Isotopenpraxis*, 27(1), 23-24.

[2] Skuterud, L., Travnikova, I.G., Balonov, M.I., Strand, P. & Howard, B.J. (1997) Contribution of fungi to radiocaesium intake by rural populations in Russia. *Science of the Total Environment* 193, 237-242.

[3] Mietelski, J.W. & Jasińska, M. (1996) Radiocesium in billberries from Poland: comparison with data for mushroom samples. *Journal of Radioecology,* 4, 15-25.

[4] IAEA. *Environmental consequences of the Chernobyl accident and their remediation: twenty years of experience. Report of the Chernobyl Forum Expert Group 'Environment'.* Radiological Assessment Reports Series, IAEA, Vienna, 2006.

[5] Shutov, V.N., Bruk, G.Ya., Basalaeva, L.N., Vasilevitsky, V.A., Ivanova, N.P. & Kaplun, I.S. (1996) The role of mushroom and berries in the formation of internal exposure doses to the population of Russia after the Chernobyl accident. *Radiation Protection Dosimetry*, 67(1), 55-64.

[6] Johanson, K.J., Bergström, R., Eriksson, O. & Erixon A. (1994) Activity concentrations of ^{137}Cs in moose and their forage plants in Mid-Sweden. *Journal of Environmental Radioactivity*, 22, 251-267.

[7] Strandberg, M. (1994) Mushroom spores and ^{137}Cs in faeces of the roe deer. *Journal of Environmental Radioactivity*, 23, 189-203.

[8] Hove, K., Pedersen, Ø., Garmo, T.H., Solheim, H. & Staaland, H. (1990) Fungi: a major source of radiocesium contamination of grazing rumiants in Norway. *Health Physics,* 59(2), 189-192.

[9] Zibold, G., Drissner, J., Kaminski, S., Klemt, E. & Miller, R. (2001) Time-dependence of the radiocaesium contamination to roe deer: measurement and modelling. *Journal of Environmental Radioactivity,* 55, 5-27.

[10] Guillén, J., Baeza, A., Ontalba, M.A. & Míguez, M.P. (2009) ^{210}Pb and stable lead content in fungi: its transfer from soil. *Science of the Total Environment*, 407, 4320-4326.

[11] Guillén, J., Baeza, A. & García, E. (2009) ^{210}Po and ^{210}Pb in mushrooms and their contribution to the internal dose in Spain. *Proceedings of International Topical Conference on Po and radioactive Pb isotopes*, Seville, Spain.

[12] Vaarama, K., Solatie, D. & Aro, L. (2009) Distribution of ^{210}Pb and ^{210}Po concentrations in wild berries and mushrooms in boreal forest ecosystems. Science of the Total Environment, 408, 84-91.

[13] Baeza, A., Hernández, S., Guillén, F.J., Moreno, G., Manjón, J.L. & Pascual, R. (2004) Radiocaesium and natural gamma emitters in mushrooms collected in Spain. *Science of the Total Environment*, 318, 59–71

[14] Baeza, A., Guillén, J., Mietelski, J.W. & Gaca, P. (2006) Soil-to-fungi transfer of ^{90}Sr, $^{239+240}Pu$, and ^{241}Am. *Radiochimica Acta*, 94, 75-80.

[15] IAEA (2003) Extent of Environmental Contamination by Naturally Occurring Radioactive Material (NORM) and Technological Options for Mitigation. *IAEA Technical reports series nº 419*, Vienna.

[16] UNSCEAR (2000) *United Nations Scientific Committee on the Effects of Atomic Radiation.UNSCEAR 2000 Report to the General Assembly, with scientific annexes*, New York.

[17] Masson, O., Baeza, A., Bieringer, J., Brudecki, K., Bucci, S., Capai, M., Calvalho, F.P., Commam, O., Cosma, C., Dalheimer, A., Didier, D., Depuydt, G., De Geer, L.E., De Vismes, A., Gini, L., Groppi, F., Gudnason, K., Gurriaran, R., Hainz, D., Halldórsson, Ó., Hammond, D., Hanley, O., Holeý, K., Homoki, Zs., Ioannidou, A., Isajenko, K., Jankovic, M., Katzlberger, C., Kettunen, M., Kierepko, R., Komtro, R., Kwakman, P.J.M., Lecomte, M., Leon Vintro, L., Leppänen, A.-P., Lind, B., Lujaniene, G., Mc Ginnity, P., Mc Mahon, C., Malá, II., Manenti, S., Manolopoulou, M., Mattila, A., Mauring, A., Mietelski, J.W., Møller, B., Nielsen, S.P., Nikolic, J., Overwater, R.M.W., Pálsson, S.E., Papastefanou, C., Penev, I., Pham, M.K., Povinec, P.P., Ramebäck, H., Reis, M.C., Ringer, W., Rodríguez, A., Rulík, P., Saey, P.R.J., Samsonov, V., Schlosser, C., Sgorbati, G., Silobritiene, B.V., Söderström, C., Sogni, R., Solier, L., Sonck, M., Steinhauser, G., Steinkopff, T., Steinmann, P., Stoulos, S., Sýkora, I., Todorovic, D., Tooloutalaie, N., Tositti, L., Tschiersch, J., Ugron, A., Vagena, E., Vargas, A., Wershofen, H. & Zhukova, O. (2001) Tracking of airborne radionuclides from the damaged Fukushima Dai-Ichi nuclear reactors by European networks. *Environmental Science & Technology*, dx.doi.org/10.1021/es2017158, (in press).

[18] Mascanzoni, D. (1990) Uptake of ^{90}Sr and ^{137}Cs by mushrooms following the Chernobyl accident. In: Desmet, G., Nassimbeni, P. & Belli, M. (Eds), *Proc. CEC Workshop Transfer of radionuclides in natural and semi-natural environments*, Elsevier Applied Science, London, 459-467.

[19] Römmelt, R., Hiersche, L., Schaller G. & Wirth, E. (1990) Influence of soil fungi (basidiomycetes on the migration of Cs-134+137 and Sr-90 in coniferous forest soils. In: Desmet, G., Nassimbeni, P.& Belli, M. (Ed.), *Proc. CEC Workshop Transfer of radionuclides in natural and semi-natural environments*, Elsevier Applied Science, London, 152-160.

[20] Paniagua, J.M. (1991) Concentración radiactiva en suelos de la provincia de Cáceres. Modelización dosimétrica por irradiación externa. Ph Thesis, Univertisy of Extremadura, Cáceres, Spain (in Spanish).

[21] Kozhevnikova, T.L., Mishenkov, N.N., Martyushova, L.N. & Krivolutskii, D.A. (1993) Accumulation of ^{90}Sr and ^{137}Cs by fruit bodies of mushrooms. *Russian Journal of Ecology*, 24(6), 391-394.

[22] Mietelski, J.W., LaRosa, J. & Ghods, A. (1993) ^{90}Sr and $^{239+240}$Pu, ^{238}Pu, ^{241}Am in some samples of mushrooms and forest soil from Poland. *Journal of Radioanalytical and Nuclear Chemistry*, 170(1), 243-258.

[23] Romeo, F. & Signore, A. (1994) La presenza di micro e macroelementi nei fungí: una rassegna. *Rivista di Merceologia*, 33(IV), 255-308 (in Italian).

[24] Lux, D., Kammerer, L., Rühm, W. & Wirth, E. (1995) Cycling of Pu, Sr, Cs, and other longliving radionuclides in forest ecoystems of the 30-km zone around Chernobyl. *Science of the Total Environment*, 173/174, 375-384.

[25] Baeza, A., Guillén, J. & Mietelski, J.W. (2004) Uptake of alpha and beta emitters by mushrooms collected and cultured in Spain *Journal of Radioanalytical and Nuclear Chemistry*, 261(2), 375-380

[26] Yamamoto, M., Shiraishi, K., Los, I.P., Kamarikov, I.Y. & Buzinny, M.G. (1995) Alpha-emitting radionuclide content in food samples as related to the Chernobyl accident. *Journal of Radioanaytical and. Nuclear Chemistry Letters*, 201(6), 459-468.

[27] Mietelski, J.W., Baeza, A.S., Guillen, J., Buzinny, M., Tsigankov, N., Gaca, P., Jasińska, M. & Tomankiewicz, E. (2002) Plutonium and other alpha emitters in mushrooms from Poland, Spain and Ukraine. *Applied Radiation and Isotopes*, 56, 717–729.

[28] Outola, I. (2003) Effect of industrial pollution on the distribution of Pu and Am in soil and on soil-to-plant transfer of Pu and Am in a pine forest in SW Finland. *Journal of Radioanalytical and Nuclear Chemistry*, 257(2), 267-274.

[29] Baeza, A., Guillén, F.J., Salas, A. & Manjón, J.L. (2006) Distribution of radionuclides in different parts of a mushroom: Influence of the degree of maturity. *Science of the Total Environment*, 359, 255-266.

[30] Byrne, A.R. (1998) Radioactivity in fungi in Slovenia, Yugoslavia, following the Chernobyl Accident. *Journal of Environmental Radioactivity* 6, 177-183.

[31] Battiston, G.A., Degetto, S., Gerbasi, R. & Sbrignadello, G. (1989) Radioactivity in mushrooms in northeast Italy following the Chernobyl accident. *Journal of Environmental Radioactivity*, 9, 53-60

[32] Gentili, A., Grenigini, G. & Sabbatini, V. (1991) Ag-110m in fungi in central Italy after the Chernobyl accident. *Journal of Environmental Radioactivity*, 13, 75-78.

[33] Teherani, D.K. (1987) Accumulation of ^{103}Ru, ^{137}Cs and ^{134}Cs in fruitbodies of various mushrooms form Austria after the Chernobyl accident. *Journal of Radioanalytical and Nuclear Chemistry*, 117(2), 69-74.

[34] Bem, H., Lasota, W., Kuśmierek, E. & Witusik, M. (1990) Accumulation of ^{137}Cs by mushrooms from Rogozno area of Poland over the period 1984-1988. *Journal of Radioanalycal and Nuclear Chemistry*, 145(1), 39-46.

[35] Borio, R., Chicchini, S., Cicioni, R., Degli Esposti, P., Rongoni, A., Sabatini, P., Scampoli, P., Antonini, A. & Salvadori, P. (1991) Uptake of radiocesium by mushrooms. *Science of the Total Environment* 106, 183-190.

[36] Muramatsu, Y., Yoshida, S. & Sumiya, M. (1991) Concentrations of radiocesium and potassium in basidiomycetes collected in Japan. *Science of the Total Environment*, 105, 29-39.

[37] Heinrich, G. (1993) Distribution of radiocesium in the different parts of mushrooms. *Journal of Environmental Radioactivity*, 18, 229-245.

[38] Yoshida, S. & Muramatsu, Y. (1994) Acumulation of radocesium in basidiomycetes collected from Japanese forests. *Science of the Total Environment*, 157, 197-205.

[39] Gaso, M.I., Cervantes, M.L., Segovia, N. & Salazar, S. (1996) Soil-fungi radiocesium transfer in forest ecosystems in Mexico. *Environment International*, 22(1), S365-S368.

[40] Calmet, D., Boursier, B., Bouisset, P., Guiard, A. & Barker, E. (1998) Mushroom as a reference material for intercomparison exercises and as bioindicator of radiocesium deposition in soil (France and Central European countries). *Applied Radiation and Isotopes*, 49(1-2), 19-28.

[41] Wang, J., Wang, C., Lai, S. & Lin, Y. (1998) Radioactivity concentrations of ^{137}Cs and ^{40}K in basidiomycetes collected in Taiwan. *Applied Radiation and Isotopes*, 49(1-2), 29-34.

[42] Kalač, P. (2001) A review of edible mushroom radioactivity. *Food Chemistry*, 75, 29-35.

[43] Ban-Nai, T., Muramatsu, Y. & Yoshida, S.(2004) Concentration of ^{137}Cs and ^{40}K in mushrooms consumed in Japan and radiation dose as a result of their dietary intake. *Journal of Radiation Research*, 45, 325-332.

[44] Kuwahara, C., Fukumoto, A., Ohsone, A., Furuya, N., Shibata, H., Sugiyama, H. & Kato, F. (2005) Accumulation of radiocesium in wild mushrooms collected from a Japanese forest and cesium uptake by microorganisms isolated from the mushroom-growing soils. *Science of the Total Environment*, 345, 165-173.

[45] Gaso, M.I., Segovia, N., Morton, O., Lopez, J.L., Machuca, A. & Hernandez, E. (2007) Radioactive and stable metal bioaccumulation, crystalline compound and siderophore detection in *Clavariadelphus truncatus*. *Journal of Environmental Radioactivity*, 97, 57-89.

[46] Szántó, Z., Hult, M., Wäjten, U. & Altzitzoglou, T. (2007) Current radioactivity content of wild edible mushrooms: a candidate for an environmental reference material. Journal of Radioanalytical and Nuclear Chemistry, 273(1), 167-170.

[47] Mietelski, J.W., Dubchak, S., Błażec, S., Anielska, T. & Turnau, K. (2010) ^{137}Cs and ^{40}K in fruiting bodies of different fungal species collected in a single in southern Poland. *Journal of Environmental Radioactivity*, 101, 706-711.

[48] Wichterey, K. & Sawallisch, S. (2002) Naturally occurring radionuclides in mushroom from uranium mining regions in Germany. *Radioprotection*, 37, 353-358.

[49] Jia, G., Belli, M., Sansone, U., Rosamilia, S. & Gaudino, S. (2004) Concentration, distribution and characteristics of depleted uranium (DU) in the Kosovo ecosystem: A

comparison with the uranium behaviour in the environment uncontaminated by DU. *Journal of Radioanalytical and Nuclear Chemistry*, 260(3), 481-494.

[50] Baeza, A. & Guillén, J. (2006) Influence of the soil bioavailability of radionuclides on the transfer of uranium and thorium to mushrooms. *Applied Radiation and Isotopes*, 64(9), 1020-1026.

[51] Žunić, Z.S., Mietelski, J.W., Błażec, S., Gaca, P., Tomankiewicz, E., Ujić, P., Čeliković, I., Čuknić, O. & Demajo, M. (2008) Tracer of DU in samples of environmental bio-monitors (non-flowering plants, fungi) and soil from target sites of the Western Balkan region. *Journal of Environmental Radioactivity*, 99, 1324-1328.

[52] Campos, J.A. & Tejera, N.A. (2010) Bioconcentration factors and trace elements bioaccumulation in scorocaprs of fungi collected from quartzite acidic soils. *Biological Trace Elemement Research*, DOI 10.1007/sl2011-010-8853-4.

[53] Borovička, J., Kubrová, J., Rohovec, J., Řanda, Z. & Dunn, C.E. (2011) Uranium, thorium and rare earth elements in macrofungi: what are the genuine concentrations? *Biometals*, DOI 10.1007/sl10534-011-9435-4.

[54] Kirchner, G. & Daillant, O. (1998) Accumulation of ^{210}Pb, ^{226}Ra and radioactive cesium by fungi. *Science of the Tota. Environment*, 222, 63-70.

[55] Malinowska, E., Szefer, P. & Bojanowski, R. (2006) Radionuclides content in Xerocomus badius and other commercial mushrooms from several regions of Poland. *Food Chemistry*, 97, 19–24.

[56] Skwarzec, B. & Jakusik, A. (2003) ^{210}Po bioaccumulation by mushrooms from Poland. *Journal of Environmental Monitoring*, 5, 791–794.

[57] Teherani, D.K. (1988) Determination of ^{137}Cs and ^{134}Cs radioisotopes in various mushrooms from Austria one year after the Chernobyl incident. *Journal of Radioanalytical and Nuclear Chemistry*, 126(6), 401-406.

[58] Ismail, S.S., Dolezel, P. & Xarg, V. (1995) ^{137}Cs, trace and toxic elements distribution in Austrian mushrooms. *Journal of Radioanalytical and Nuclear Chemistry, Letters*, 200(4), 315-336

[59] Andolina, J. & Guillite, O. (1990) Radiocesium availability and retention sites in forest humus. In: Desmet, G., Nassimbeni, P. & Belli, M. (Eds), *Proc. CEC Workshop Transfer of radionuclides in natural and semi-natural environments*, Elsevier Applied Science, London, 657-663.

[60] Fraiture, A., Guillitte, O. & Lambinon, J. (1990) Interest of fungi as bioindicators of the radiocontamination in forest ecosystems.. In: Desmet, G., Nassimbeni, P. & Belli, M. (Eds), *Proc. CEC Workshop Transfer of radionuclides in natural and semi-natural environments*, Elsevier Applied Science, London, 477-484.

[61] Guillitte, O., Fraiture, A. & Lambinon, J. (1990) Soil-fungi radiocesium transfer in forest ecosystems. In: Desmet, G., Nassimbeni, P. & Belli, M. (Eds), *Proc. CEC Workshop Transfer of radionuclides in natural and semi-natural environments*, Elsevier Applied Science, London, 468-476.

[62] Lambinon, J., Fraiture, A., Gasia, M.C. & Guillitte, O. (1998) La radiocontamination des champignons sauvages en Wallonie (Belgique) suite a l'accident de Tchernobyil. In.: *IVe symposium international de radioecology de Cadarache. Impact des accidents d'origine nucléaire sur l'environment, Tome 2.* Centre d'etudes nucléaires de Cadarache, E-37 – E44 (in French).

[63] Řanda, Z. & Benada, J. (1990) Mushrooms – significant source of internal contamination by radiocaesium. In: Desmet, G., Nassimbeni, P. & Belli, M. (Eds), *Proc. CEC Workshop Transfer of radionuclides in natural and semi-natural environments*, Elsevier Applied Science, London, 169-178.
[64] Cibulka, J., Šišák, L., Pulkrab, K., Miholová, D., Száková, J., Fučíková, A., Slámová, A., Stěhulová, I. & Barláková, S. (1996) Cadmium, lead, mercury and caesium levels in wild mushrooms and forest berries from different localities of the Czech Republic.*Scientia Agriculturae Bohemica*, 17(2), 113-129.
[65] Dvořák, P., Kunová, V., Beňová, K., Ohera & M. (2006) Radiocesium in mushrooms from selected locations in the Czech Republic and the Slovak Republic. *Radiation and Environmental Biophysics*, 45(2), 145-151
[66] Strandberg, M. (1994) Radiocesium in a Danish pine forest ecosystem. *Science of the Total Environment*, 157, 125-132.
[67] Strandberg, M. (2004) Long-term trends in the uptake of radiocesium in *Rozites caperatus*. *Science of the Total Environment*, 327, 315-321.
[68] Ikäheimonen, T.K., Klemola, S. & Ilus, E. (2003) Behaviour of gamma emitters, Sr-90 and transuranic elements in forest environment. *Proceedings of the International Conference on Isotopic and Nuclear Analytical Techniques for Health and Environment*, IAEA, Vienna.
[69] Kostiainen, E.T. (2005) Variation in ^{137}Cs uptake of edible mushrooms. In: Strand, P., Børretzen, P. & Jølle, T. (Eds) *Proceedings from the 2nd International Conference on Radioactivity in the Environment,* 270–273.
[70] Kostiainen, E.T. (2007) 137Cs in Finnish wild berries, mushrooms and game meat in 2000-2005. *Boreal Environment Research*, 12, 23-28.
[71] Rückert, G., Diehl, J.F. & Heilgeist, M. (1990) Radioactivity levels in mushrooms collected in the area of Karlsruhe during 1987 and 1988. *Zeitschrift fur Lebensmittel-Untersuchung und –Forschung* 1990, 190(6), 496-500.
[72] Kammerer, L., Hiersche, L. & Wirth, E. (1994) Uptake of radiocaesium by different species of mushrooms. *Journal of Environmental Radioactivity*, 23, 135-150.
[73] Heinrich, G. (1992) Uptake and transfer factors of ^{137}Cs by mushrooms. *Radiation Environmental Biophysics*, 31, 39-49.
[74] Barnett, C.L., Beresford, N.A., Self, P.L., Howard, B.J., Frankland, J.C., Fulker, M.J., Dodd, B.A. & Marriott, J.V.R. (1999) Radiocaesium activity concentrations in the fruit bodies of macrofungi in Great Britain and an assessment of dietary intake habits. *Science of the Total Environment*, 231, 67-83.
[75] Barnett, C.L., Beresford, N.A., Frankland, J.C., Self, P.L., Howard, B.J. & Marriott, J.V.R. (2001) Radiocaesium intake in Great Britain as a consequence of the consumption of wild fungi. *Mycologist*, 15(3), 98-104.
[76] Toal, M.E., Copplestone, D., Johnson, M.S., Jackson, D. & Jones, S.R. (2002) Quantifying ^{137}Cs aggragated transfer coefficients in a semi-natural woodland ecosystem adjacent to a nuclear reprocessing facility. *Journal of Environmental Radioactivity*, 63, 85-103.
[77] Vaszari, E.E., Tóth, V. & Tarján, S. (1992) Determination of radioactivities of some species of higher fungi. *Journal of Radioanalytical and Nuclear Chemistry*, 165(6), 345-350.

[78] Giovanni, C., Nimis, P.L. & Padovani, R. (1990) Investigations of the performance of macromycetes as bioindicators of radioactive contamination. In: Desmet, G., Nassimbeni, P. & Belli, M. (Eds), *Proc. CEC Workshop Transfer of radionuclides in natural and semi-natural environments*, Elsevier Applied Science, London, 485-491.

[79] Ingrao, G., Belloni, P. & Santatoni, G.P. (1992) Mushrooms as biological monitors of trace elements in the environment. *Journal of Radioanalytical and Nuclear Chemistry*, 161(1), 113-120.

[80] Marzano, F.N., Bracchi, P.G. & Pizzetti, P. (2001) Radioactive and conventional pollulants accumulated by edible mushrooms (*Boletus sp.*) are useful indicators of species origin. Environmental Research Section A, 85, 260-264.

[81] Bakken, L.R. & Olsen, R.A. (1990) Accumulation of radiocaesium in fruit bodies of fungi. In: Desmet, G., Nassimbeni, P. & Belli, M. (Eds), *Proc. CEC Workshop Transfer of radionuclides in natural and semi-natural environments*, Elsevier Applied Science, London, 664-668.

[82] Amundsen, I., Gulden, G. & Strand, P. (1996) Accumulation and long term behaviour of radiocaesium in Norwegian fungi. *Science of the Total Environment*, 184, 163-171.

[83] Grabowski, D., Muszynski, W., Petrykowska, M., Rubel, B., Smagala, G. & Lada, W. (1994) Activity of cesium-134 and cesium-137 in game and mushrooms in Poland. *Science of the Total Environment*, 157, 227-229.

[84] Mietelski, J.W., Jasińska, M., Kubica, B., Kozak, K. & Macharski, P. (1994) Radioactive contamination of Poslish mushrooms. *Science of the Total Environment*, 157, 217-226.

[85] Zagrodzki, P., Mietelski, J.W., Krośniak, M. & Petelenz, B. (1994) Accumulation of radiocesium in forest litter and selected regions of Poland and its influence on litter-to-mushroom transfer factor. In: Kučera, J., Obrusník, I. & Sabbioni, E. (Eds) *Nuclear analytical methods in the life sciences*, 273-277.

[86] Pietrzak-Flis, Z., Radwan, I., Rosiak, L. & Wirth, E. (1996) Migration of ^{137}Cs in soils and its transfer to mushrooms and vascular plants. *Science of the Total Environment*, 186, 243-250.

[87] Tanase, C., Pui, A., Oprea, A. & Popa, K. (2009) Translocation of radioactivity from substrate to macromycetes in the Crucea (Romania) unramiun mining area. *Journal of Radioanalytical and Nuclear Chemistry* 2009, 281, 563-567.

[88] Čipáková, A. (2004) ^{137}Cs content in mushrooms from localities in eastern Slovakia. *Nukleonika*, 49(1), S25-S29 .

[89] Arrond,. E. (1988) Mediciones de radiactividad en setas de la región Vasconavarra. *Revista Ibérica de Micología*, 5, 45-49 (in Spanish).

[90] Foidevaux, P., Dell, T. & Tosell, P. (2006) Radionuclides in foodstuffs and food raw material. In: Pöschl, M. & Nollet, L.M.L. (Eds.) *Radionuclide concentrations in food and the environment*, CRC Taylor & Fancis Group, USA

[91] Smith, M.L., Taylor, H.W. & Sharma, H.G. (1993) Comparison of the post-Chernobyl ^{137}Cs contamination of mushrooms from Eastern Europe, Sweden, and North America. *Applied and Environmental Microbiology*,, 59(1), 134-139.

[92] Guillitte, O., Melin, J. & Wallberg, L. (1994) Biological pathways of radionuclides originating from the Chernobyl fallout in a boreal forest ecosystem. *Science of the Total Environment*, 157, 207-215.

[93] Smith, M.L., Taylor, H.W. & Sharma, H.D. (1994) Using γ-ray spectroscopy for the study of Chernobyl fallout on edible mushrooms. *Journal of Radioanalytical and Nuclear Chemistry*, 180(1), 109-113.

[94] Dahlberg, A., Nikolova, I. & Johanson, K.J. (1997) Intraspecific variation in ^{137}Cs in sporocarps of *Suillus variegatus* in seven Swedish populations. *Mycological Research*, 101(5), 545-551.

[95] Nikolova, I., Johanson, K.J. & Dahlberg, A. (1997) Radiocaesium in fuitbodies and mycorrhizae in ectomycorrhizal fungi. *Journal of Environmental Radioactivity*, 37(1), 115-125.

[96] McGee, E.J., Synnott, H.J., Johanson, K.J., Fawaris, B.H., Nielsen, S.P., Horrill, A.D., Kennedy, V.H., Barbayiannis, N., Veresoglou, D.S., Dawson, D.E., Colgan, P.A. & McGarry, A.T. (2000) Chernobyl fallout in a Swedish spruce forest ecosystem *Journal of Environmental Radioactivity*, 48, 59-78.

[97] Andersson, I., Lönsjö, H. & Rosén, K. (2001) Long-term studies on the transfer of ^{137}Cs from soil to vegetation and to grazing lambs in a mountain area in northern Sweden. *Journal of Environmental Radioactivity*, 52, 45-66.

[98] Mascanzoni, D. (2001) Long-term ^{137}Cs contamination of mushrooms following the Chernobyl fallout. *Journal of Radioanalytical and Nuclear Chemistry*, 249(1), 245-249.

[99] Rosén, K., Vinichuk, M., Nikolova, I. & Johanson, K. (2011) Long-term effects of single potassium fertilization on ^{137}Cs levels in plants and fungi in a boreal forest ecosystem. *Journal of Environmental Radioactivity*, 102, 178-184.

[100] Vinichuk, M., Rosén, K., Johanson, K.J. & Dahlberg, A. (2011) Correlations between potassium, rubidium and cesium (^{133}Cs and ^{137}Cs) in sporocarps of *Suillus variegatus* in a Swedish boreal forest. Journal of Environmental Radioactivity, 102, 386-392.

[101] Karadeniz, Ö. & Yaprak, G. (2007) Dynamic equilibrium of radiocesium with stable cesium within the soil-mushroom system in Turkish pine forest. *Environmental Pollution*, 148, 316-324.

[102] Hoshi, M., Yamamoto, M., Kawamura, H., Shinohara, K., Shibata, Y., Kozlenko, M.T., Takatsuji, T., Yamashita, S., Namba, H., Yokoyama, N., Izumi, M., Fujimura, K., Danilyuk, V.V., Nagataki, S., Kuramoto, A., Okajima, S., Kiikuni, K. & Shigematsu, I. (1994) Fallout radioactivity in soil and food samples in the Ukraine: Measurements of iodine, plutonium, cesium, and strontium isotopes. *Health Physics*, 67(2), 187-191

[103] Tsvetnova, O.B. & Shcheglov, A.I. (1994) Cs-137 content in the mushrooms of radioactive contaminated zones of the European part of the CIS. *Science of the Total Environment*, 155, 25-29.

[104] Travnikova, I.G., Shutov, V.N., Bruk, G.Ya., Balonov, M.I., Skuterud, L., Strand, P., Pogorely, Ju.A. & Burkova, T.F. (2002) Assessment of current exposure levels in different populations groups of the Kola Peninsula. *Journal of Environmental Radioactivity*, 60, 235-248.

[105] Vinichuk, M.M. & Johansson, K.J. (2003) Accumulation of ^{137}Cs by fungal mycelium in forest ecosystems of Ukraine, *Journal of Environmental Radioactivity*, 64, 27-43.

[106] Vinichuk, M.M., Johansson, K.J., Rosén, K. & Nilsson, I. (2005) Role of the fungal mycelium in the retention of radiocaesium in forest soils. *Journal of Environmental Radioactivity*, 78, 77–92.

[107] Yoshida, S., Muramatsu, Y., Dvornik, A.M., Zhuchenko, T.A. & Linkov, I. (2004) Equilibrium of radiocesium with stable cesium within the biological cycle of

contaminated forest ecosystems, *Journal of Environmental Radioactivity*, 75(3), 301-313.

[108] Ramzaev, V., Andersson, K.G., Barkovski, A., Fogh, C.L., Mishine, A. & Roed, J.(2006) Long-term stability of decontamination effect in recreational areas near the town Novozybkov, Bryansk Region, Russia. *Journal of Environmental Radioactivity*, 83, 280-298.

[109] Vilic, M., Barisic, D., Kraljevic, P. & Lulic, S. (2005) concentration in meat of wild boars (*Sus scrofa*) in Croatia a decade and a half after the Chernobyl accident. *Journal of Environmental Radioactivity*, 81, 55-62.

[110] Mihok, S., Schwartz, M. & Wiewel, A.M. (1989) Bioconcentration of fallout ^{137}Cs by fungi and read-backed voles (*Clethrionomys gapperi*). *Health Physics*, 57(6), 959-966.

[111] Gaso, M.I., Segovia, N., Herrera, T., Perez-Silva, E., Cervantes, M.L., Quintero, E., Palacios, J. & Acosta, E. (1998) Radiocesium accumulation in edible wild mushrooms from coniferous forests around the Nuclear Centre of Mexico. *Science of the Total Environment*, 223, 119-129.

[112] Gaso, M.I., Segovia, N., Morton, O., Cervantes, M.L., Godinez, L., Pea, P. & Acosta, E. (2000) ^{137}Cs and relationships with major trace elements in edible mushrooms from México. *Science of the Total Environment*, 262, 73-89.

[113] Tsukada, H., Shibata, H. & Sugiyama, H. (1998) Transfer of radiocaesium and stable caesium from substrata to mushrooms in a pine forest in Rokkasho-mura, Aomori, Japan. *Journal of Environmental Radioactivity*, 39(2), 149-160.

[114] Yoshida, S. & Muramatsu, Y. (1994) Radiocesium concentrations in mushrooms collected in Japan. *Journal of Environmental Radioactivity*, 22, 141-154.

[115] Ban-Nai, T., Muramatsu, Y. & Yoshida, S. (1997) Concentrations of ^{137}Cs and ^{40}K in edible mushrooms collected in Japan and radiation dose due to their consumption. *Health Physics*, 72(3), 384-389.

[116] Yoshida, S. & Muramatsu, Y. (1998) Concentrations of alkali and alkaline earth elements in mushrooms and plants collected in a Japanese pine forest, and their relationship with ^{137}Cs. *Journal of Environmental Radioactivity*, 41(2),183-205.

[117] Grüter, H. (1966) Verhalten einheimischer Pilzarten gegenüber dem Spaltproduckt Caesium-137. *Zeitschrift fur Lebensmittel-Untersuchung und –Forschung*, 134, 173-179 (in German).

[118] Haselwandter, K. (1978) Accumulation of radioactive nuclide ^{137}Cs in fruitbodies of basidiomycetes. *Health Physics*, 34, 713-715.

[119] Ferrero, J.L., Jordá, M.L., Milió, J., Monforte, L., Moreno, A., Navarro, E., Senent, F., Soriano, A., Baeza, A., del Río, M. & Miró, C. (1987) Atmospheric radioactivity in Valencia, Spain, due to the Chernobyl reactor accident. *Health Physics*, 53 (5), 519-524.

[120] Aumann, D.C., Clooth, G., Steffan, B. & Steglich W. (1989) Complexation of cesium 137 by the cap pigments of the Bay Boletus (*Xerocomus badius*). *Angew. Chem. Int. Ed. Engl.*, 28(4), 453-454.

[121] Rühm, W., Kammerer, L., Hiersche, L. & Wirth, E. (1997) The ^{137}Cs/^{134}Cs ratio in fungi as indicator of the major mycelium location in forest soil. *Journal of Enrivonmental Radioactivity*, 35(2), 129-148.

[122] Baeza, A., Del Río, L.M., Jiménez, A., Miró, C., Paniagua, J.M. & Rufo, M. (1996) Analysis of the temporal evolution of atmospheric ^{7}Be as a vector of the behavior of

other radionuclides in the atmosphere. *Journal of Radioanalytical and Nuclear Chemistry*, 207(2):331 –344.

[123] Fesenko, S.V., Soukhova, N.V., Sanzharova, N.I, Avila, R., Spiridonov, S.I., Klein, D., & Badot, P.M. (2001) ^{137}Cs availability for soil to understorey transfer in different types of forest ecosystems. *The Science of the Total Environment*, 269, 87-103.

[124] Steiner, M., Linkov, I. & Yoshida, S. (2002) The role of fungi in the transfer and cycling of radionuclides in forest ecosystems. *Journal of Environmental Radioactivity*, 58, 217-241.

[125] Gadd, G.M. (1999) Fungal production of citric and oxalic acid: importance in metal speciation, physiology and biogeochemical processes. *Advances in Micriobial Physiology*, 41, 47-92.

[126] Landeweert, R., Hoffland, E., Finlay, R.D., Kuyper, T.W. & Van Breemen, N. (2001) Linking plants to rocks: ectomycorrhyzal fungi mobilize nutrients from minerals. *Trends in Ecology & Evolution*, 16(5), 248-254.

[127] Nikolova, I., Johanson, K.J. & Clegg, S. (2000) The accumulation of ^{137}Cs in the biological compartment of forest soils. *Journal of Environmental Radioactivity*, 47, 319-326.

[128] Weete, J.D. (1974) *Fungal lipid biochemistry. Distribution and Metabolism*. Plentum Press, New York.

[129] Olsen, R.A., Joner, E. & Bakken, L.R. (1990) In: Desmet, G., Nassimbeni, P. & Belli, M. (Eds) *Proc. CEC Workshop Transfer of radionuclides in natural and semi-natural environments*, Elsevier Applied Science, London, 657-663.

[130] Baeza, A., Guillén, J. & Bernedo, J.M. (2005) Soil-fungi transfer coefficients: importance of the location of mycelium in the soil and of the differential availability of radionuclides in soil fractions. *Journal of Environmental Radioactivity*, 81, 89-106

[131] Schell, W.R. & Tobin, M.J. (1990) Deposition and mobility of chemical elements in forest and wetland environment. In: Desmet, G., Nassimbeni, P. & Belli, M. (Eds) *Proc. CEC Workshop Transfer of radionuclides in natural and semi-natural environments*, Elsevier Applied Science, London, 118-128.

[132] Baeza, A., Del Río, M., Jiménez, A., Miró, C. & Paniagua, J. (1995) Relative sorption of ^{137}Cs and ^{90}Sr in soil: influence of particle size, organic matter content and pH.*Radiochimica Acta*, 68, 135-140.

[133] Ohnuki, T. (1994) Sorption characteristics of cesium on sandy soils and their components. *Radiochimica Acta*, 65, 75-80.

[134] Lee, M.H. & Lee, C.W. (2000) Association of fallout-derived ^{137}Cs, ^{90}Sr and ^{239,240}Pu with natural organic substances in soils. *Journal of Environmental Radioactivity*, 47, 253-262.

[135] Dumat, C. & Stauton, S. (1999) Reduced adsorption of caesium on clay minerals caused by various humic substances. *Journal of Environmental Radioactivity*, 46, 187-200.

[136] Alexakhin, R.M. & Krouglov, S.V. (2001) Soil as the main compartment for radioactive substances in terrestrial ecosystems. In: Bréchignac, F., & Howard, B.J. (Eds.) *Radioactive pollutants. Impact on the environment*, EDP Sciences, Les Ulis, France, 149-174.

[137] Tessier, A., Cambell, P.G.C. & Bisson, M. (1979) Sequential extraction procedure for speciation of particulate trace metals. *Analytical Chemistry*, 7, 201-215.

[138] Riise, G., Bjørnstad, H.E., Lien, H.N., Oghton, D.H. & Salbu, B. (1990) A study of the radionuclide association with soil components using a sequential extraction procedure. *Journal of Radioanalytical and Nuclear Chemistry Articles*, 142(2), 531-538.

[139] Forsberg, S., Rosén, K., & Bréchignac, F. (2001) Chemical availability of ^{137}Cs and ^{90}Sr in undisturbed lysimeter soils maintained under controlled and lose-to-real conditions. *Journal of Environmental Radioactivity*, 54, 253-265.

[140] Puhakainen, M., Riekkinen, I., Heikkinen, T., Jaakkola, T., Steinnes, E., Rissanen, K., Suomela, M. & Thørring, H. (2001) Effect of chemical pollution on forms of ^{137}Cs, ^{90}Sr and $^{239,240}Pu$ in Artic soil studied by sequential extraction. *Journal of Environmental Radioactivity*, 52, 17-29.

[141] Pavlotskaya, F.I. (1974) *Migration of radioactive products from global fallout in soils.* Moskow, Atomizdat (in Russian).

[142] Krouglov, S.V., Kurinov, A.D. & Alexakhin, R.M. (1998) Chemical fractionation of ^{90}Sr, ^{106}Ru, ^{137}Cs and ^{144}Ce in Chernobyl-contaminates soils: an evolution in the course of time. *Journal of Environmental Radioactivity*, 38 (1), 59-76.

[143] Kashparov, V.A., Ahamdach, N., Zvarich, S.I., Yoschenko, V.I., Maloshtan, I.M. & Dewiere, L. (2004) Kinetics of dissolution of Chernobyl fuel particles in soil in natural conditions. *Journal of Environmental Radioactivity*, 72, 335-353.

[144] Baeza, A., Paniagua, J.M., Rufo, M., Sterling, A. & Barandica, J. (1999) Radiocaesium and radiostrontium uptake by turnips and broad beans via leaf and root absorption. *Applied Radiation and Isotopes*, 50, 467-474.

[145] Kaduka, M.V., Shutov, V.N., Bruk, G.Ya., Balonov, M.I., Brown, J.E. & Strand P. (2006) Soil-dependent uptake of ^{137}Cs by mushrooms: experimental study in the Chernobyl accident areas. *Journal of Environmental Radioactivity*, 89, 199-211.

[146] Gillett, A.G. & Crout, N.M.J. (2000) A review of ^{137}Cs transfer to fungi and consequences for modelling environmental transfer. *Journal of Environmental Radioactivity*, 48, 95-121.

[147] Rühm, W., Steiner, M., Kammerer, L., Hiersche, L. & Wirth, E. (1998) Estimating future radiocaesium contamination in fungi on the basis of behaviour patterns derived from past instances of contamination. *Journal of Environmental Radioactivity*, 39(2), 129-147.

[148] Karadeniz, Ö. & Yaprak, G. (2011) Soil-to-mushroom transfer of ^{137}Cs, ^{40}K, alkali-alkaline earth element and heavy metals in forest sites of Izmir, Turkey. *Journal of Radioanalytical and Nuclear Chemistry*, 288, 261-270.

In: Fungi: Types, Environmental Impact and Role in Disease ISBN: 978-1-61942-671-9
Editors: Adolfo Paz Silva and María Sol

Chapter X

Fundamental Fungal Strategies in Restoration of Natural Environment

M. A. Karaman**, *M. S. Novaković and M. N. Matavuly
Microbiology Study Group, Department of Biology and Ecology, Faculty of Sciences, University of Novi Sad, Novi Sad, Serbia

Abstract

Fungi play vital role in nutrient cycling by achieving process of decomposition of organic matter and breakdown of complex compounds in nature. This role of fungi coupled with their fundamental strategy of adaptation to various environmental factors may be used for designing systems to enable the elimination of biopolymers and man-made xenobiotics from the environment via biosorption. Soil contaminants like heavy metals, radionuclides, polyaromatic hydrocarbons-PAHs and chemicals used in agriculture that are toxic and carcinogenic agents could be diminished or removed by activity of fungal exoenzymes. In this review, special attention was paid to lignolityc enzyme system expressed by white-rot fungi recognized as successful agent in bioremediation of a large variety of chemicals that are, like lignin, relatively long lived in the environment. Furthermore, fungi are very important in natural cycling of metal ions due to their great accumulation potential for heavy metals (Pb, Hg, Cd) and radionuclides (^{137}Cs), implicating them as good bioindicators of the pollution in urban and industrial areas and in contaminated forest ecosystems. These processes in macrofungi are influenced by environmental factors like metal concentrations of soil and substrate, pH, organic matter and contamination by atmospheric deposition as well as fungal factors like fungal structure, biochemical composition, decomposition activity, development of mycelium and sporocarps or portion of fruiting body. Concentration of radionuclides in fungi is determined by the amount of radioactivity precipitation, concentration of stable - non radioactive or analogous element in soil, soil characteristics (mineral composition, pH) and its taxonomic and nutritional identity.

Macrofungi maintain ecological balances used as bioindicators or as remediation agents of contaminated environment. Also, edibility and medicinal properties are of a

*E-mail address: maja.karaman@dbe.uns.ac.rs.

great importance for humans while majority of edible and medicinal mushrooms can accumulate high amounts of heavy metals and other environmental pollutants. This chapter summarizes relevant biological features of fungi (position in tree of life, nutritional strategy, enzyme systems), especially lignicolous macrofungi (mostly white-rot), as a basic tool for resolving physiological, ecological and biotechnological potential of fungi in changing polluted environment thus restoring the natural environment.

1. Introduction

Fungi are ubiquitous in natural environments representing one of the most important organisms in the world. This is not only due to their huge diversity and abundance, which are connected to their vital roles in ecosystem function, but also because of their influence on humans or human related activities. We recommend that potential application of fungi by humans should be preceded with the following actions: 1) taxonomical examinations, 2) impacts of pollutants contained in fruiting bodies (mycelium) of mushrooms on the environment, 3) heavy metal and radio-ecological analysis of terrain, microhabitats and fruiting bodies (mycelia) of fungal species.

1.1. Where do Fungi Belong?

It is supposed that the number of fungal species has been approaching almost 1,500.000, today, although only 5% of fungi is described [1]. Fungi represent a large, mysterious group of organisms for a long time while scientists are still working hard to find a proper place for them among the other groups of living organisms. Their specific characteristics and huge diversity confused scientists worldwide, often leading them in a wrong direction in classification. In the Linnaean (Carl von Linné) two Kingdom system of classification, fungi were included in the Plant Kingdom [2] according to their immobility and mode of nutrient absorption ability. This traditional idea of classification prevailed almost until the middle of the XIX century when new approaches (three Kingdom system by Haeckel [3], two Empire system by Chatton [4], four Kingdom system by Copeland [5], were established, but non of them have recognized fungi as a separate group.

Progress in electron microscopy and biochemical techniques has highlighted important differences between living organisms, distinguishing fungi as organisms substancially different from others in nutrition, by apsorption, cell organization and structure, storage compounds, haploid nuclei, photomorphogenesis, hormonal system, etc. In classification proposed by Whittaker in 1969, fungal organisms have finally gained the position they deserved – the position of separate Kingdom Fungi in Five Kingdom system including Monera, Protista, Planta, Fungi and Animalia according to their multicellular cell organisation and living style [6].

New revolution in biological classification began with the use of molecular phylogenetic analyses in the 1970s, based primarily on the ribosomal RNA (rRNA) genes which are highly preserved and present in all organisms containing enough information. Comparison of small rRNA subunit, done by Carl Woese, demonstrated that there are three evolutionary diverse groups of organisms, named domains, two prokaryotic – Bacteria and Archaea, and one eukaryotic – Eucarya [7, 8]. In this phylogenetic tree, plant, fungal and animal kingdoms form a cluster at the top of the Eucarya Domain and are often termed as «crown» eukaryotes [9]. Analyses of rDNA and protein-coding genes proved that fungi are more closely related to animals, being their closest relatives, than to plants ([10], [11], [12]). In the last three decades, molecular phylogenetic studies provided a better understanding of fungal diversity and caused constant changes inside the fungal tree of life. It has been demonstrated that several groups of organisms traditionaly classified and studied as fungi are acctualy outside of this group [13]. Slime molds, previously classified as a phylum Myxomycota inside the Kingdom Fungi, were proved to belong to the kingdom Protozoa/Amebozoa [14], devided into four phyla [9]. Oomycota, Hyphochytridiomycota and Labyrinthulomycota have been moved to the kingdom Straminipila [15] / Chromista ([16], [17], [18]). In contrast to these findings, certain organisms which have earlier been placed in other eukaryotic groups have, for the first time, found themselves inside the Kingdom Fungi. Some examples are: *Pneumocystis* – pathogen, once classified as a protozoan, now the member of the Taphrinomycotina in the Ascomycota ([19], [20], [21]), *Hyaloraphidium* - thought to be algal genus, now in the fungal phylum Chytridiomycota [21], Microsporidia – for a long time considered as a most primitive, early divergent eukaryotic group classified as a specific protozoan phylum, now recognized as a highly specialized and reduced fungi, included in the kingdom Fungi or at least concerned as a sister group ([22], [23], [24], [25]).

Four large groups which have been traditionally recognized as the «true fungi»: Chytridiomycota, Zygomycota, Ascomycota and Basidiomycota, also suffered dramatic changes regarding molecular phylogeny. In the 2001, arbuscular mycorrhizal fungi previously known as the order Glomerales/Glomales, which are primarly filamentous and lack flagella, were excluded from the Zygomycota and recognised as a unique phylum Glomeromycota [26]. According to phylogenetic studies previously reported, Ascomycota and Basidiomycota have been proven to represent a monophyletic clade and sister taxa. The clade containing these two taxa is now classified as a subkingdom Dikarya [27], although earlier was recognised as Dicaryomycota [28]. As it was demonstrated by the analysis of rDNA, Glomeromycota is a sister clade with the Dikarya ([13], [26], [29], [30]) forming a clade labeled as «Symbiomycota» (since most of the members form symbioses) [30]. Traditional Chytridiomycota and Zygomycota represent a basal fungal phyla, with earliest divergence, that have been long known as polyphyletic and paraphyletic ([13], [31]).

Molecular phylogenetic studies have brougth us priceless opportunity to take a deeper look into the tree of life and better understanding of relations among living beings. When fungi are concerned, these techniques showed us that we can not rely completelly on morphological traits and that many of an undiscovered species and phylogenetic relations lie hidden in the world of genes. In the recent classification of Hibbett *et al.*, based on the monophyly supported by the number of published molecular phylogenetic studies, new changes have been proposed for many of the basal fungal lineages [32]. The Chytridiomycota is retained as the phylum, containing two classes: Chytridiomycetes and Monoblepharidomycetes. The orders Blastocladiales and Neocallimastigales have been raised to the level of phylum: Blastocladiomycota (already in James *et al.* [33]) and Neocallimastigomycota. The Zygomycota is not accepted as the phylum and it′s former members are distributed among the phylum Glomeromycota and four subphyla *incertae sedis*: Mucoromycotina, Kickxellomycotina, Zoopagomycotina and Entomophtoromycotina [32].

The contemporary researches based on molecular phylogenetic analyses recognized fungal kingdom as one of five eukaryotic kingdoms containing seven phyla: Chytridiomycota, Neocallimastigomycota, Blastocladiomycota, Microsporidia, Glomeromycota, Ascomycota, Basidiomycota (Figure 1). Subkingdom Dikarya contains only two phyla: Ascomycota and Basidiomycota, representing monophyletic clade of "crown fungi" which are recognized as macro-fungi due to clearly visible fruiting-bodies. As opposed to these, micro-fungi comprise the microscopic organisms, yeasts and molds, that are commonly recognized as producers of toxic substances.

Organisms that we used to know as Fungi Imperfecti – Deuteromycetes – Mitosporic fungi, are now also being classified thanks to the molecular studies. Many of them already found their place inside diferent phyla while the rest of them are still waiting to be classified.

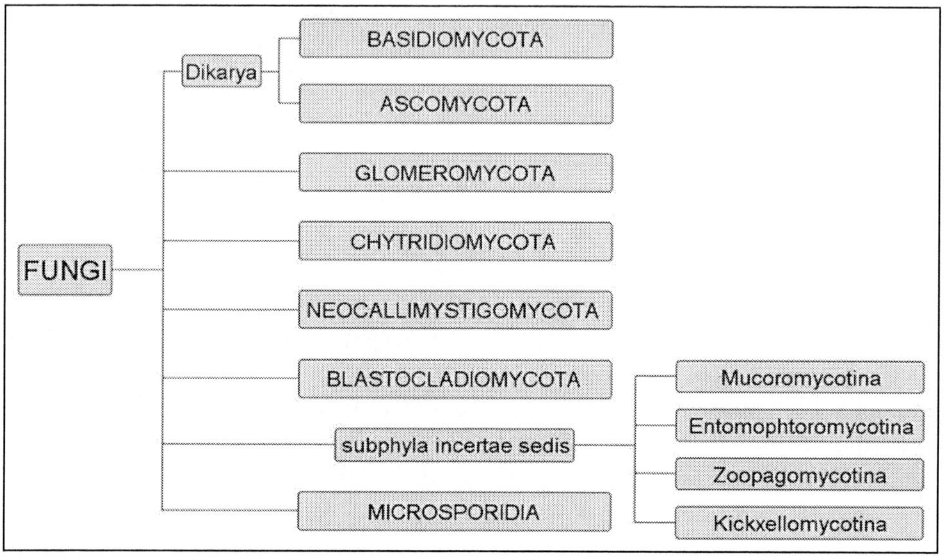

Figure 1. Seven phyla of Fungal Kingdom based on Hibbet et al., 2001[32].

1.2. Mycological Terms

Here are briefly explained some of mycological terms frequently used in this chapter. Macrofungi is the term usualy used for a macroscopical fungi, visible to the naked eye, which form a large fruiting bodies (members of the phyla Ascomycota and Basidiomycota, although a few are Zygomycota). Macrofungi with a morphologically different kind of the fruiting bodies are called pufballs, stinkhornes, bird's nests, morels, earth stars, truffles, coral fungi, etc. Macrofungi can be terrestrial (saprobes or mycorrhizal symbonts) or pathogens of plants, animals and fungi. Mushroom is a fleshy fruiting body of a different Basidiomycota, which is produced above ground (epigeous) and consists of a stem and a cap, with gills or pores underneath the cap. Fruiting body (also known as sporocarp, carpophore or fruit body) is a spore-bearing structure, rising from a substratum and representing part of the sexual phase of a fungal life cycle. Sporocarp of the Basidiomycota is also named basidiocarp and of the Ascomycota – ascocarp. On the other hand, mycelium (plural: mycelia) is a vegetative part of the fungus, growing inside the substratum and consisting of hyphae. Hypha is the basic structural unit of filamantous fungi, in the form of cylindrical, thread-like, apical growing structure which can be divided into compartments containing septum (pl. septa) or coenocytical (without septa). Terricolous fungi are the fungi growing on, or under the soil surface. Saproxylic fungi are the one growing on the wood substratum, while the term lignicolous (or sometimes xylophagous) is refered particulary to the fungi which have ability to decompose wood (wood-decaying). Fungi fruiting on woody substrata are usually either saprobes (degrading dead organic matter) or plant pathogens (using live organic matter).

1.3. Nutritional Groups

Generally, fungi belong to a group of *heterotrophic organisms*, with few exceptions of their ability to supply themselves chemotrophically by inorganic carbon, which resulted in their adaptation to the use of different organic substrates. There are three general nutritional groups of fungi: saprotrophic or saprobic, which grow on a substrate formed after death of organisms (term saprophytic has been replaced recently by terms saprotrophic or saprobic since they do not belong to plants), parasites, which attack living organisms and mutualistic fungi, which form associations of a mutual benefit with a variety of organisms.

Most fungi live as saprotrophs obtaining their nutrients from dead organic matter such as wood, leaf litter, soil, dung, dead animal and fungi causing catabolic dissimilation processes, thanks to extracellular digestion enzymes which are secreted by mycelia. In that way nutrients that are locked-up in the form of complex organic compounds release thus making them useable by other living organisms. This is making fungi as vital components of healthy natural ecosystems, especially forests. Deacon (2006), [9] emphasizes that saprotrophic fungi are distinguished in their behavior and capability to degrade certain types of substrates, as a consequence of colonising the same resource of nutrients in a different time and sometimes are also overlapped. Pioneer species use simple soluble substrates and usually cannot degrade the complex structural polymers. They are good competitors, characterized by rapid growth and short life cycle. Polymer-degrading fungi, which are substrate-specific, colonize and decompose the major structural polymers of hosts (such as cellulose, hemicellulose, chitin). They have an extended growth phase and are able to defend a food source via antibiosis or by taking away mineral nutrients from the substrate. Some fungi are specialized for the degradation of resistant organic materials (lignin, suberin, keratin, etc.) thereby achieving access to polymers. They are capable of defending the substrate by antagonism or inhibition, and have a mechanism to obtain mineral nutrients that were utilized by previous colonizers.

Among parasitic fungi, three major groups are recognized: 1 - obligate parasitic fungi, 2 - parasites of weakness and 3 - facultative parasitic fungi. Some parasites have very specific host requirements and may only attack a single species, while others may parasitize a variety of hosts genera [34]. Obligate parasitic fungi, also known as plant, animal and human pathogens, attack living organisms causing tissue damage and diseases that can lead to death of the host organism. Opposite to them, parasites of weakness are not able to attack healthy organism but require host weakness for its development. They colonize initially damaged, weakened or immuno-compromised host organisms. Those fungi that can infect host only on the places of damaged tissue, such as the surface of a wound, are often reffered to as a wound parasites. Facultative parasitic fungi are the ones that after death of the host keep on using the same tissues but now as a saprotrophs. It means that they are also capable of using dead organic substrate as the carbon source. In the nature, it can often be difficult to determine clear boundaries among these sub-categories [35].

Third nutritional group comprises fungi that form mutualistic associations with plants, animals and prokaryotes. Some of the best known beneficial associations in wich fungi are involved are lichens, with algae as a partners, and mycorrhizae, essential relationships with the roots of almost all living plants. In lichens, fungi protect algae from external effects and provide with water and mineral nutrients while algae supply fungi by organic nutrients. In mycorrhizal association, the plant provides the fungus with water and organic compounds, while the fungus provides the plant with the scarce minerals such as phosphorous [36]. Mycorrhizal fungi can protect plants against pathogenic fungi and microorganisms, as well as from the harmful effects of heavy metals and radionuclids. They do not grow without their host and many of them are host-specific, growing with one kind of plant.

Interesting mutualistic relationships occure between fungi and animals. Some ants and termites cultivate fungi in a special gardens“ (preparing a substratum for fungal growth and introducing them to it) which serve as a food source for their larvae. Certain wood beatles and wasps also inoculate tunnels created in the wood with fungi wich they farm for food. Gut sistem of ruminants (such as moose, cow, etc.) contains community of microorganisms, including fungi, that help these animals to digest cellulose [37]. Fungivorous animals, especially those that feed on a hypogeous fungi, have a significant role in a long distance spore dispersal and there we have another example of the animal/fungus mutualism (for many species of a small mammals fungi are a dominant food) [38]. All of these exemples proved the ubiquity of fungi in nature as well as their importance for the functioning of other organisms.

2. Wood-Decaying Fungi

Doubtless, fungi represent an exceptionally significant component in nature since they are the main agents that cause the decomposition of organic matter in terrestrial and aquatic ecosystems. One of their most important roles is degradation of wood, since they are the only organisms that can completely degrade all of its components [39]. Lignicolous fungi inhabit substrata that differ in size (tiny twigs, huge trunks), state of decay and moisture content [40]. Their decay activities are conected to the recycling of lignocellulosic and mineral nutrients back into the ecosystem but also on maintenance of mycorrhizal fungi in seasonnally dry forests since woody debris acts as a moisture sink. Moreover, many ectomycorrhizal fungi are climbing up on wood when fruiting and simultanously may obtain nutrients from the wood by degrading organic compounds.

In addition, wood presents the major storehouse of the carbon fixed by photosynthesis in the Biosphere, making wood-decaying fungi essential agents in the biogeochemical cycling of carbon in nature. Wood material is mostly made of fibrous and soft, amorphous cellulose and hemicellulose imbedded in glue-like, hard-to-degrade lignin (it is chemically complex, variable, nonhydrolysable and water-insoluble), building complex lignocellulose substrates which are unavailable to most saprotrophic organisms. Another obstacles for the wood-decomposers are very low content of available nitrogen (ratio C : N = 500 : 1), low content of phosphorus and toxic compounds concentrated in the heartwood [9]. Most efficient wood-decaying fungi are found in the phylla Basidiomycota and Ascomycota. Many of them are highly specialized in the sense of tree species or decomposition stage. There are a number of classifications of wood-degrading fungi according to different criteria, and categorization given here is based on a way of attacks on lignocellulose complex, i.e. according to the primary enzyme activity: 1 - white-rot fungi, 2 - brown-rot fungi, and 3 - soft-rot fungi.

2.1. White-Rot Fungi and Lignolitic Enzymes (Ligninases)

White-rot basidiomycetous fungi are the only organisms known to mineralise both lignin and carbohydrate components of wood to carbon dioxide and water due to posessing the specific enzymes for extracellular oxidation and depolimerisation of lignin [41].

The term white-rot fungi is derived from bleached residue of cellulose after complete lignin degradation and to a half extent the degradation of hemicellulose and cellulose. Contrary, the brown-rot fungi merely modify lignin while removing carbohydrates in wood [39]. Enzymatic „combustion", a process wherein enzymes generate reactive intemediates, whithout directly control the reactions of lignin breakdown has been proposed as the mechanism of lignin biodegradation [42]. Wood decay and the biogeochemical cycling are the final consequences of lignin biodegradation.

Lignin is an aromatic polymer forming up to 30% of woody plant tissues providing rigidity and resistance to biological attack [42]. It is the most abundant renewable aromatic polymer on the Earth, composed of non-phenolic (80-90%) and phenolic structures [43]. It has been shown that fungi degrade lignin by secreting enzymes collectively termed "ligninases". Its structure is complex and the process of its biodegradation is based on specific machanism of few lignolytic (ligninolytic) enzymes: laccases, manganese dependent peroxidases – MnP, lignin peroxidases-LiP, and non-specific strong oxidants able to destabilise and fragment lignin [41]. These include two lignolytic families: 1) phenol oxidase (laccase) and 2) peroxidases (lignin peroxidase (LiP) and manganese-dependent peroxidase (MnP)) [44]. This enzyme system is extracellular, relatively non-specific and cause generation of enzymatic and non enzymatic oxidants. The activity of these enzymes are influeneced by pH, temperature, substrate concentration, the pesence of mediators and vertryl alcohol, cofactors (Cu^{2+}, Mn^{2+}) inhibitors, and organic acids (e.g. citric, oxalic, and tartaric). White-rot basidiomycetes such as *P. chrysosporium* and *T. versicolor* ([45], [43]) have been found to be the most efficient lignin-degrading microorganisms studied.

Lignin peroxidase (LiP) called ligninase is the enzyme, heme-dependent-peroxidase with an unusually high redox potential and low optimum pH. It is able to oxidize metoxylated aromatic rings without a free phenolic group, generating cation radicals that can react further by variety of pathways. Hence, LiP is able to oxidize the non-phenolic part of lignin, but it was not detected in many lignin-degrading fungi. In addition, it has been widely accepted that the oxidative lignolytic enzymes are not able to penetrate the cell walls due to their size. Thus, it has been suggested that prior to the enzymatic attack, low-molecular weight diffusible reactive oxidative compounds have to initiate changes to the lignin structure ([46], [47]).

Manganese peroxidase (MnP) is also a heme-peroxidase that shows stronger preference for Mn(II) as its reducing substrate. It oxidise only phenolic substrates and the product Mn(III) forms a complex with organic acids and diffuses away to oxidize other materials, such as lignin. It has lower redox potential then LiP and it does not oxidize nonphenolic lignin models [42].

Laccase is a copper-containing oxidase and it does not require peroxide. Similarly to Mn-peroxidase it oxydizes only those lignin model compounds with a free phenolic group, forming phenoxy radicals.

Cellobiose is the heme-flavin enzyme that oxidizes cellobiose and some other carbohydrates and reduces quinones and the radicals produced by the action of lignin peroxidase, Mn-peroxidase and lacase on lignin model compounds.

The activity of these enzymes depends on a supply of H_2O_2 from glucose-1-oxidase, glucose-2-oxidase, glyoxyl oxidase, aryl alcohol oxidase, and methanol oxidase. Intracellular metabolism also supplies oxalate and other organic acids which can chelate Mn(III). It is clear that the lignin myco-degradation is extremely dependent on the supply of oxygen and therefore does not take place in a water-saturated conditions. Also degradation of lignin can be potentially dangerous for the fungus because some intermediates of this process can be mycotoxic. White-rot fungi produce such compounds but they can detoxify them by polymerizing it into pigments similar to melanin [41].

The most important white-rot fungi are the two main tree pathogens: *Armillaria mellea* and *Heterobasidion annosum*, and many saprotrophic mushrooms, including stump colonizers such as *Coriolus versicolor*, *Xylaria hypoxylon* and *Xylaria polymorpha* which degrade all components of wood including lignin. However, they have distinctive and very strong ability of nitrogen supply and also have been shown to recycle effectively nitrogen within their mycelium [48].

2.2. Brown-Rot Fungi

Only aproximatelly 6% of the wood-rotting fungi are brown-rots [49]. They are capable of cellulose and hemicellolose degradation but leave the lignin more or less intact in the form of brown structure on the wood surface, often cracked and appearing as a stacked bricks. Mechanism of cellulose and hemicellulose degradation is based on extracellular production of hyghly reactive hydroxyl radical (OH·radicals) that are produced through Fenton reactions to start degradation. Many of fungi such are *Laetiporus sulphureus* are excreting high levels of oxalic acid, hydroquinones and glycopeptides that would serve for binding and directly reducing Fe^{3+} to Fe^{2+} to provide reactants for subsequent hydroxyl radical formation [50]. Brown-rot decay almost exclusively occurs in terrestrial systems, with the greatest diversity and impact taking place in temperate and tropical zones [51].

Brown-rot fungi are predominantly Basidimycota, mostly confined to nine families (Polyporaceae, Auriculariaceae, Coniophoraceae, Sparassidaceae, Corticiaceae, Stereaceae, Tricholomataceae, Coprinaceae and Paxilaceae) and more commonly associated with conifer forests. Representatives such as mushroom *Schizophyllum commune, Fomes fomentarius, Daedalea quercina* and *Piptoporus betulinus* form a macroscopic, sometimes rather large console-shaped fruit bodies on a dead tree trunks ([52], [53], [54], [55]). Interesting member of the „brown-rot fungi" group is a so-called „dry-rot" fungus, e.g. *Serpula lacrymans.* It causes the same type of wood decay as the other members of the „brown-rot" group, but the term „dry-rot" derives from the fact that this cellulose degrading fungus, during the process, can produce a sufficient amount of moisture that allows unimpeded growth even in extremely dry conditions [9].

2.3. Soft-Rot Fungi

Soft rot causing fungi belong to the phylum Ascomycota and the group of mitosporic fungi. Comparing to white-rot and brown-rot fungi, they are not agressive decay organisms and may not be good competitors in normal conditions. Thus, they are usually found in damp environments and woods with limited access to oxygen [51]. Soft-rot fungi decompose cellulose and hemicellolose, with little or no effect on lignin, producing typical chains of cavities inside the cell wall. Most of them require high levels of nitrogen, which they draw from the wood, if available, or the environment such as soil and water [9].

3. Enzymology of Wood Degradation

Cellulolytic complex of deuteromycetes species of the genera *Trichoderma* (*T. harzianum*, *T. viride, T. reesei* and *T. koningii*) ([56], [57], [58], [59], [60], [61]) and *Gliocladium* (*G. virens, G. roseum*) ([61], [62], [63]) and anamorphs of Ascomycotina (*Chaetomium erraticum*, *Penicillium funiculosum* and *Fusarium solani*) ([64], [65], [66]), hydrolyze crystalline cellulose applying joint action of enzymes (endo-β-1,4-glucanase, exo-β-1,4-glucanase or celobiohydrolase (celobiase) and β-glucosidase).

Endoglucanase activity produces new shorter chains with accessible chain ends which then are attacked by celobiase. However, since the hydrolysis of glycosidic bonds is potentially reversible reaction, the separation of crystalline regions from cellulose chains will be prevented or slowed down due to the limitations caused by intramolecular and intermolecular hydrogen bonds. Therefore, these two enzymes must act sequentionally as well as successively, quickly one after the other, to prevent re-formation of glycosidic bonds.

Lignocellulose degradation by fungi is performed by complex mixtures of cellulases [67], hemicellulases [68] and ligninases ([69], [70]), reflecting the complexity of the materials. Cellulases and most hemicellulases belong to a group of enzymes known as glycoside hydrolases. Currently more than 2500 glycoside hydrolases have been identified and classified into 115 families [71]. Fungal cellulases (hydrolysis of β-1,4-glycosidic bonds) have been mostly found within a 11 glycoside hydrolases families ([71], [72]).

3.1. Fungal Extracellular Cellulases

Hydrolysis of the β-1,4-glycosidic bonds in cellulose can be achieved by many different enzymes known as cellulases which use two different catalytic mechanisms, the retaining and the inverting mechanisms ([67], [72]). Many different fungal species have the ability to degrade cellulose by producing extracellular fungal cellulose-degrading enzymes including endo-cleaving (endoglucanases) and exo-cleaving (cellobiohydrolases). Endoglucanases can hydrolyze glycosidic bonds internally, inside the cellulose chains, whereas cellobiohydrolases act preferentially on chain ends. The products of these enzymatic reactions are mostly a disaccharide cellobiose and, to a lesser extent, cello-oligosaccharides, which will be further hydrolyzed by the third group of enzymes - β-glucosidases [73]. Cellulases mostly have a small independently folded carbohydrate binding module responsible for binding the enzyme to the crystalline cellulose and thus enhance the enzyme activity [67]. Currently, many carbohydrate binding modules have been identified and classified into 54 families, however only 20 families have been found in fungi [74]. Endoglucanases (EGs), endo-1,4-β-glucanases (EC 3.2.1.4, endocellulase) also referred to as carboxymethylcellulases (CMCase), are named after the artificial substrate used to measure the enzyme activity. EGs initiate cellulose breakdown by attacking the amorphous regions of the cellulose, making it more accessible for cellobiohydrolases by providing new free chain ends. This has been shown by the effect of the enzyme on carboxymethyl-cellulose and amorphous cellulose [75]. For the fungal EGs, optimal pH ranges mostly between 4.0 and 5.0 and optimal temperature is from 50 to 70 °C. Many fungi produce multiple EGs, for example, *Trichoderma reesei* produces at least 5 EGs whereas three EGs were isolated from white-rot fungus *Phanerochaete chrysosporium* ([76], [77]). In addition, some EGs lack a carbohydrate binding module while some other EGs with carbohydrate binding module have been described. Cellobiohydrolases (CBHs) EC 3.3.1.91, exocellulase) preferentially hydrolyze β-1,4-glycosidic bonds from chain ends, producing cellobiose as the main product. CBHs have been shown to create a substrate-binding tunnel with their extended loops which surround the cellulose ([78], [79]). Similar to EGs, CBHs are monomers with no or low glycosylation with optimal pH between 4.0 and 5.0, but the optimal temperatures are wider, from 37 to 60 °C. Some CBHs can act from the non-reducing ends and others from the reducing ends of the cellulosic chains, which increases the synergy between opposite-acting enzymes. Cellobiose, the end product of CBHs, acts as a competitive inhibitor, which can limit the ability of the enzymes to degrade all of cellulose molecules in a system ([76], [78]). From many filamentous species such as *T. reesei, T. harzianum, G. virens* ([57] [63]) and basidiomycetes such as white-rot and brown-rot fungi β-glucosidases (BGLs) have been isolated (EC 3.2.1.21). By using the retaining mechanism, β-glucosidases hydrolyze soluble cellobiose and cellodextrins (attacking β-1,4-glycosidic bonds) to glucose, and are thus competitively inhibited by glucose [80]. BGLs show the highest variability among the cellulolytic enzymes due to their structure and localization. While some BGLs have a simple monomeric structure with around 35 kDa molecular mass (e.g. *Pleurotus ostreatus*) [81], some others have dimeric (e.g. *Sporobolomyces singularis* with 146 kDa) [82] or even trimeric structures with over 450 kDa, e.g. *Pisolithus tinctorius* [83]. In addition, most of BGLs are glycosylated and in some cases, such as the 300 kDa monomeric BGL from *Trametes versicolor*, the glycosylation degree is up to 90% [84]. Regarding localization, BGLs can be grouped into three different types including intracellular, cell wall-associated and extracellular [85]. Optimum pH for the enzymes relies on enzyme localization while the optimum temperature ranges from 45 to 75 °C.

3.2. Fungal Hemicellulases

Several different enzymes are needed to hydrolyze hemicellulose, due to its heterogeneity [86]. Xylan is the most abundant component of hemicellulose contributing over 70% of its structure. Xylanases are able to hydrolyze β-1,4 linkages in xylan and produce oligomers which can be further hydrolyzed into xylose by β-xylosidase ([57], [58]). Not surprisingly, additional enzymes such as β-mannanases, arabino-furanosidases or α-L-arabinanases are needed, depending on the hemicellulose composition which can be mannan-based or arabinofuranosyl-containing [87]. Similar to cellulases, hemicellulases are usually modular proteins and have other functional modules, such as carbohydrate binding modules, in addition to their catalytic domains. Also similarly to cellulases, most of the hemicellulases are glycoside hydrolases, although some hemicellulases belong to carbohydrate esterases which hydrolyze ester linkages of acetate or ferulic acid side groups ([87], [88]). Hemicellulases belong to 20 different glycoside hydrolases families and all of them, except 4, 8, 52 and 57 families, have been found in fungi. Similarly to cellulases, aerobic fungi such as *Trichoderma* and *Aspergillus* secrete a wide variety of hemicellulases in high concentrations and these work in a synergistic manner ([87], [74]).

3.3. Fungal Mechanisms of Oxidative (Non-Enzymatic) Lignocellulose Degradation

A few decades ago, non-enzymatic mechanisms for plant cell wall polysaccharide degradation were also considered and over the time more evidence for these was found. The non-enzymatic degradation mechanism is mostly assisted by oxidation through the production of free hydroxyl radicals (OH·). In fact, many white-rot and brown-rot fungi have been shown to produce hydrogen peroxide (H_2O_2) which enters the Fenton reaction, resulting in release of OH· ([89], [90]). These free radicals attack polysaccharides as well as lignin in plant cell walls in a nonspecific manner providing some cleavages which make it easier for the lignocellulolytic enzymes to penetrate ([91], [43]). Three different pathways have been found for the generation of free radicals including cellobiose dehydrogenase (CDH) catalyzed reactions, low molecular weight peptides/quinone redox cycling and glycopeptide-catalyzed Fenton reactions [76]. CDH, an extracellular monomeric protein with some glycosylation has been identified in a number of wood-degrading and cellulose-decomposing fungi including basidiomycetes (mostly white-rot fungi) and ascomycetes growing on cellulose. The enzyme is able to oxidize cellobiose, higher cellodextrins and other disaccharides or oligosaccharides with β 1,4 linkages. In addition, CDH with (in Ascomycetes) or without (in Basidiomycetes) carbohydrate binding modules has been identified since, even in the absence of carbohydrate binding modules, they are able to bind to cellulose through hydrophobic interactions [92]. CDH production is higher due to cellulases and hemicellulases activity ([93], [94]). It is now widely accepted that CDH is able to degrade and modify all three major components of the lignocellulosic residues (cellulose, hemicelluloses and lignin) by producing free OH· in a Fenton-type reaction [76]. It was also found that white-rot and brown-rot fungi produce low molecular weight chelators which are able to penetrate into the cell wall. For example, *Gloeophyllum trabeum* produces a low molecular weight peptide (known as short fiber generating factor, SFGF) which can degrade cellulose into short fibers by an oxidative reaction ([43], [95]). It has also been reported that some of these low molecular weight compounds are quinones which have to be converted to hydroquinones by some fungal enzymes and then through Fenton reaction, free hydroxyl radicals will be produced ([43], [74]). Different glycopeptides with different molecular weight (ranging from 1.5 to 12 kDa) have been found in many brown-rot fungi such as *G. trabeum* [96] and white-rot fungi such as *P. chrysosporium* ([47], [97]). Similar to the other mechanisms, glycopeptides are able to catalyze redox reactions and thus produce free hydroxyl radicals.

4. Biotechnological Application of Lignolytic Enzymes

There are many possible applications based on specific enzyme system of white-rot fungi to serve humans and it is resonable to expect that their range will continue to expand. Environmental conditions have an important influence on the synthesis and activities of lignolytic enzymes in conversion of natural and agricultural biomass or enzymatic conversion of pollutants and aroma-chemical precursors. In addition, enzymology and physiology of lignin catabolism of white-rot fungi are also a suprime tool in controlling synthesis and degradation of structurally similar organic compounds. The ability to produce various types of peroxidases and laccases, in addition to H_2O_2 and hydroxyl radicals is considered the initiating key to the degradation of many types of complex compounds by the white-rot fungi [50]. Considerable potential of lignolytic enzymes could be employed in a number of biotechnological applications.

4.1. Biopulping, Biobleaching and Decolorisation of Industrail Efluents

Biopulping is the directed treatment of plant biomass with lygnolytic microorganisms and enzymes prior to chemical and mechanical pulping to obtain a product enriched in polysaccharides and cellulose in particular. This process involving white-rot fungi is dependent on different factors such as biochemical caracteristics, fungal strain and substrate, culture conditions and incubation time [41]. Some efforts has been made toward developing cell free biopulping process that use isolated lignolytic enzymes which was overlapped because of the difficulties in costs and difficulties of generating and maintaining optimum conditons for enzyme activity. Biobleaching is the biological removal or destruction of smaller quantities of residual lignin and other coloring matter that remain after process of pulping. MnP and laccase have been shown to possess these abilities. White-rot fungi may have some applications in remediation of industrial wastewaters by decolorising the water and degrading toxic compounds. Effluent water may contain underivatised lignin, lignin derivatives, lignosulphate, tannins, phenolics and other colored and toxic compounds. Strain of *Lentinus edodes* was proved to remove 73% of the color in 5 days and olive waste water was reduced in color for 45%, total organic carbon by 75% and total phenols by 60% within 4 days [98]

4.2. Biosorption

Process when microbial biomass (dead or living) is used to remove metals from solutions is denoted as biosorption. In the past few years macrofungi have appeared as potential agents for the remediation of wastewater containing toxic metal ions. Fungal filamentous or hyphen structure features are very important in the effective substrate colonization, with hyphen that have apical growth and production of high amount of lateral branches and complex fruiting bodies. Living cells accumulate metals in two phases: physical–chemical reaction e.g. adsorption, intracellular uptake through the plasmalemma or simply membrane adsorption and extracellular precipitation in or around the cell wall [99]. Dead cells can only adsorb metals on the cell walls. Accumulation of some radionuclides (actinides) is prevalently a consequence of adsorption while for caesium it is a characteristic process of intracellular uptake.

In a recent study, adsorption potential of *P. ostreatus* showed that this species is an efficient biosorbent because of fast metal removal rate, remarkable biosorption capacity for Ni, Cr, Cu and Zn and high regeneration ability [100]. Wood-inhabiting basidomycets are an useful source of mycelial biomass for biosorption of metal ions due to ease of cultivation, high yield and non-hazardous nature. The most important role in white-rot fungi is dedicated to polysaccharides, proteins or pigments that have a good capacity for heavy metals binding [101]. The adsorption of heavy metals to the mycelia of white-rot fungi fits the Langmuir and Freundlich adsoption isoterm ([101], [100]).

Mycelia of four white-rot fungi, *D. quercina*, *G. applanatum*, *Stereum hirsutum* and *Schizophyllum commune* cultured in a liquid media showed that the preference for specific heavy metal is species-specific and differrent when a mixture of metals is offered. Pb content was maximal in *S. hirstum*, while *G. applanataum* contained maximal values of Cd, Al and Ca [102].

4.3. Xenobiotic Compound Degradation (Mycoremediation)

Major classes of resistant and environmentally toxic or hazardous substances are chlorinated organics, polycyclic aromatic hydrocarbons (PAHs), nitrated organics and textile dyes [41]. White-rot fungi have attracted the highest interest due to their ability to degrade simultaneously large variety of chemicals. Polychlorinated biphenyls (PCBs) are degraded by *Pleurotus ostreatus* and *Trametes versicolor* strains ([103], [104]). Recent study reporeted that *G. lucidum* is a promising white-rot fungus to degrade PAHs such as phenantrene and pyrene in the environment [105].

Lignolytic enzymes can transform chemical pollutants that are relatively long-lived in the environment due to its high molecular weight and low (aqueous) insolubility. They can cause biotransformation of a group of lipophilic pollutants that are generated during incomplete combustion of organic carbons such are fossil fuels, wood and municipal solid waste ([106], [107], [108]) or industrial effluents into benign or less harmfull products. This is the most attractive property that can be used in processess of biodegradation and biodeterioration of soils, sediments or water where PAHs are accumulated. PAHs are also componenets of wood preservatives. Species of brown-rot fungi such are *Fomitopsis*, *Lentinus* and *Laetiporus sulphureus* are the most studied fungi able to degrade xenobiotic compounds, although specific mechanisms are still not well characterized [50].

In comparative studies on lignin and PAHs degradation using 130 wild basidiomycetous fungi, *Phellinus* species showed better degradation in solid state fermentation as the most efficient strain. According to different predominant enzyme activity of manganese-dependant peroxidase (MnP) in *Phellinus* sp., contrary to *Laetiporus sulphureus* where lignin peroxidase was predominant, the route of PAHs degradation was species specific [108]. The species *Lentinus (Panus) tigrinus* showed high PAHs degradation rate being more pronounced than in *Irpex lacteus* [107]. Moreover, a preliminary assessment of strain/carrier combination is fundamental prior to field-scale mycoremediation.

4.4. Enhacement of Digestibility of Ruminant Feed and Production of Edible Fungi

Animal feed can be upgraded with white-rot fungi by enhancing digestibility of polysaccahrides from agricultural or forest residues containing lignin. Biological delignification of animal feed will be predominately established in the future in concern with the production of edible fungi [42]. White-rot basidiomycets, e.g. *Pleurotus ostreatus*, are actively involved in the re-circulation of carbon at a global level, since being lignin-degrading ones. Ability of white-rot fungi to use waste lignocellulosic material, e.g. agricultural waste (wheat straw, bagasse, coffee pulps, etc.) and brown coal that are abundant, cheap and non utilitarian, can be used for production of food for humans.

Some lignin degrading fungi produce edible sporocarps and can directly convert lignocellulose into food for humans. Fruiting bodies of *Lentinula edodes*, *Auricularia polytricha*, *Pleurotus ostreatus* and *Flammulina velutipes* are functional food, comprising both food and medicine. Edible mushrooms represent high value food, which are used for human nutrition because of their excellent flavour, texture, and can be eaten fresh or used in dry form for additives or for making healthy beverages. They are mostly used in a form of dietary supplements from Asian to Europe and North America.

At the same time macrofungi give the opportunity to the mankind in resolving stress problems. Recently, different genera of basidiomycetous fungi have been used as sources of natural bioactive metabolites of pharmacological interest, with various medical effects: antitumor, immunomodulating, cardiovascular, antimicrobial, antiparasitic, antidiabetic, hepatoprotective, etc. Biological activity and chemical composition of fungal species varies broadly and depends on strain and substrate and mode of growing. With regard to the fact that production of oxygen radicals provokes pathological changes in organisms, antioxidants derived from these edible and medicinal fungi could diminish pathological disorders in humans [109], [110], [111], [112].

Many fungal bioactive compounds (extracellular and intracellular), as products of primary or secondary metabolism, can be brought in human use both as novel medicines e.g. antibiotics [113], cholesterol-lowering agents [114], immunosuppressive drugs [115], agents against insects [116] and microorganisms [117], as well as for improvement of the antioxidant status during aging ([118], [119]). It is assumed that natural compounds, being part of the ecosystem, might be more compatible and less toxic to humans and the environment. Basidiomycetous macrofungi presently produce about 5% of all antibiotics, but there are studies showing that up to 40% of them are able to produce such medically important substances. According to these investigations their interaction with environmental components, especially with pollutants, attracted the attention of scientific public worldwide [120].

World market have been increasing over past decades ([121], [122]). *Ganoderma* species are the most investigated white-rot mushrooms for its pharamaceutical applications [123]. It can be harvested on timber logs, tree stumps, expanding cultivation of these edible and medicinal species in the future. Cultivation of *L. edodes* is usually on chestnut and oak, while *P. ostreatus* is grown on poplar originating substrate. Inoculation of logs is done by drilling holes in the timber and insertation dowels infected with fungal mycelium or insertation of cultures grown on sawdust. Fruiting depends on appropriate reduced temperature and carbon dioxide concentration. It should be noted that sterilised substrate is necessary tool for the cultivation, as well as reliable control of insect pests and parasitic fungi without resorting to pesticides.

5. Relations between Polluted Environment and Fungi

5.1. Interactions of Heavy Metals with Fungi, Especially White-Rot Macrofungi

Like some other microbial groups (e.g. lichens) macrofungi can accumulate metals from their environment by means of the physico-chemical and biological mechanisms [99]. Unlike lichens that are good bio-accumulators (bio-indicators) of atmospheric pollution ([124], [125], [126]), filamentous fungi are one of best accumulators of metal ions from soil, due to the biological properties of their vegetative mycelia, which are closly associated with the host roots and soil. Mycelium, living in the soil for several months or even years, is a potent absorptive biomass for accumulation of elements, but fungal fruiting bodies show even higher concentrations of these elements.

Mineral nutrients have many roles in fungal life taking part in composition of organic compounds, in activation of specific biochemical paths, etc. Metals are directly or indirectly incorporated in: growth, differentiation, reproduction and the overall metabolic (enzymatic) activities. According to Gadd [99], there is an ultimate chain that connects artificial or natural environment with fungi and metals (Figure 2). Environment affects all life stages of fungi as well as the availability and type of metal. Metal may has positive or negative impact on all life stages of fungi, being dependent on the concentration and type of metal, but also can change physico-chemical characteristics of the environment. Fungi can eliminate metals and organo-metaloids from solutions both by physical and biological processes and by transforming them into other forms. Fungi affect the environment by their metabolic activities and growth.

First studies of the interaction between heavy metals and fungi were organized along with the phytopathological examinations on fungicide preparations, which are based on toxicitie of metals to wood-decaying fungi ([127]). The early preparations of biocides were based on mercury but due to its high toxicity it was superseded by cooper-based word preservatives. Copper-chrom-arsenate (CCA) preservatives have been held in high repute since 1933 but recently their harmful effect was disclosed, giving priority to chromium and arsenic-free wood preservatives. Organotin compounds such as bis-(tri-n-butyl-tin)-oxide (TBTO) and n-butyl-tin-naphthenate have been used in a wood protection against wood-destroying molds and lower fungi but also against brown-rot fungi [101].

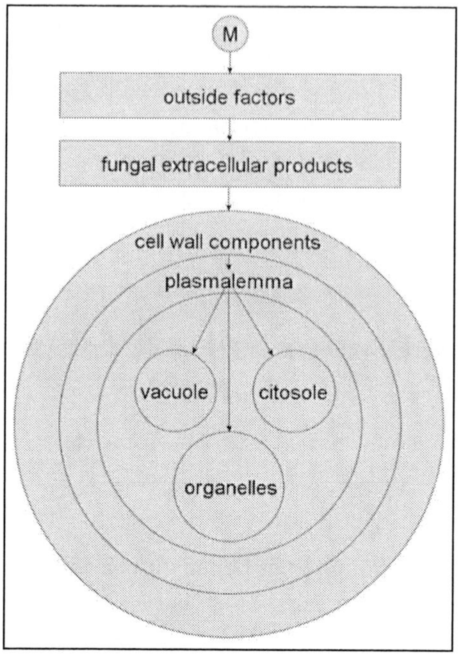

Figure 2. Interactions of heavy metals with fungi.

The environmental problems as well as toxicity tests showing harm of toxic metals, metalloids, radionuclides and organo-metaloids to living resources simulate new phase of investigations involving intake and translocation of metals and radionuclides through the sporocarp of edible fungal species ([128], [129]).

Essential microelements mostly belong to a group of heavy metals which activity is strongly specific, prevalently catalytic when present in very low concentrations. Essential elements for fungi are potassium (K), sodium (Na), magnesium (Mg), calcium (Ca), manganese (Mn), iron, (Fe), copper (Cu), zinc (Zn), cobalt (Co) and nickel (Ni), while unessential are rubidium (Rb), cesium (Cs, aluminium (Al), cadmium (Cd), silver (Ag), aurum (Au), mercury (Hg), lead (Pb), chromium (Cr) etc. Their toxicity relies on species, growth stage, physico-chemical characteristics of metals and environmental factors.

5.2. Environmental Factors and uptake of Metals

The impact of heavy metals on microorganisms depends on the following factors: 1) physico-chemical factors (pH, Eh, anion content, moisture, aeration, content of clay and organic matter), 2) chemical properties and concentration of heavy metal (Hg>Ag>Cu>Cd>Zn>Pb>Cr>Ni>Co); 3) species specific caracteristics and age of microorganism (e.g. age of mycelia and intervals between fructification [130]; 4) temeperature and timing of exposure; 5) chemical composition of microorganisms in a given habitat [131] (Figure 3).

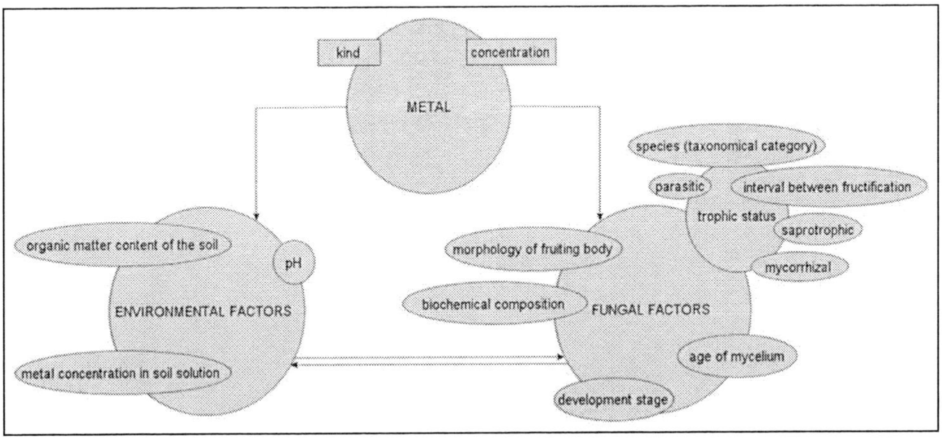

Figure 3. Main factors affecting the accumulation of heavy metals.

The uptake of metals is influenced by metal origin, transport mode (atmospheric deposition or sewage sludge) and biochemical and chemical factors (pH, carrier molecules, etc.) [132].

5.3. Toxicity Mechanisms

Although biotoxicity and bioaccumulation is considered as related to nonessential elements, toxicity of higher concentrations of essential elements is also established (e.g. higher concentrations of Ca^{2+} precipitate phosphate and reduce sporocarp formation) [133]. On the contrary, copper is toxic to the most of fungi even at very low concentration [101]. In a study which analysed the effect of Cu and Zn pollution on the terrestrial fungi in Sweden, most species were affected at intermediate Cu level (600 - 4000μg) while species from *Amanita* genus appeared tolerant to Cu and species belonging to genus *Cortinarius* decreased in abundance near the mill [134].

Wood–rotting fungi are mostly supplied with metals via wood that contains lower concentration and availability of heavy metal ions then soil [55]. Soil is highly influenced by industrial pollution and near motorways or gas-work sites this contamination is often accompanied by the presence of high levels of polycyclic aromatic hydrocarbons PAHs ([107], [55]). Atmospere is the most important transport medium for metals that derived from various sources and due to gravitation aerosol precipitate on vegetation, soil and waters [135]. It has been noticed recently that fruit bodies of white-rot fungi receive significant amounts of heavy metals from the atmosphere ([55], [101]).

Toxic activity of heavy metals is a result of one or more irregular metabolic processes or chemical reactions. Toxicity of heavy metals comprise the following: 1) blocking of functional groups of enzymes and transport systems (e.g. Cd, Pb, Ag, have great affinity towards sulfhydryl groups), 2) removing (by precipitation or chelating) or substitution of essential metal ions from biomolecules or organelles, 3) conformational changes of organic molecules, 4) denaturation and inactivation of enzymes (Ag and Cu have great affinity towards active parts of enzymes, and going through the cell membranes with organic molecules in a form of chelate); 5) degradation of integrity of cell membrane and membrane of organelles (Ag, U, Au, Cd and Cu, changes membrane permeability). Disruptions of cell membrane is associated with loss of K^+ ions and increase of cell permeability [99] caused by depolarization of electrochemical gradient.

Indirect mechanism of toxicity is connected with induction of free radicals generated in oxydoreductive reactions. In aerobic organisms, the processes of lipid peroxidation is provoked by metals [136] resulting in changing of lipids to peroxy-alkyl radicals and hydroperoxides of fat acids. Complexes with metals that are solubile in lipids can go through the Fenton reaction with hydroperoxides and accelerate this process, what is happening in aqueous solutions of free ions and complexes:

$$Fe^{2+} + H_2O_2 \longrightarrow Fe^{3+} + OH^- + OH^{\bullet}, O_2^- + Fe^{3+} \longrightarrow O_2 + Fe^{2+}$$

5.4. Effects of Specific Heavy Metals on Fungi

5.4.1. Lead

The lead shows silmilar reactions as Cu and Hg, forming insoluble sulfids, and has affinity for sulfhydryl groups [99]. It causes the damage of plasma membrane and forms a very stable organic chelates. Moreover, the presence of phosphates increases the tolerance of fungi to lead by formation of salts wich are hard to dissolve. Thus, the effects of lead on fungi are not a direct result of toxicity, but the consequence of soluble phosphates lack [137].

5.4.2. Copper

Although the Cu and Ni in nature often occure together and have similar toxic properties, they do not have joint effect on fungi [138]. Copper also forms insoluble sulphids. It's toxicity depends on the affinity for sulfhydryl groups and other tiolate groups that represent reactive centers of many enzymes [99], but also on binding to amino and imino groups [139]. Metallothioneins, γ-glutamyl peptides and other tiolates are involved in the detoxification process in fungi.

5.4.3. Chrome

Cr^{3+} and Al have many common features. By formation of $[Cr(H_2O)_5OH]^{2+}$, chrome lowers the pH of the solution [140]. Cr-sulphids are being hydrolyzed by water, which is why metalothionins and γ-glutamyl peptides can not bind it. Potential mechanisms of detoxification are: 1 - binding to hydroxy-, carboxyl- and methoxy- ligand groups in the form of chelates or organic polymers; or 2 – binding to polyphosphates in the vacuole [99].

5.4.4. Zinc

Zinc belongs to the moderately toxic elements for fungi. According to electronegativity it is between Cu and Al. At low pH it forms acid compounds $[Zn(H_2O)_5OH]^+$ [140]. The most efficient mechanisms of detoxification are binding to γ-glutamyl peptides and polyphosphates of the vacuole, as well as to metallothioneins to a lesser degree [99]. Zn serves as a nutrient in small amounts but is toxic in high doses. Ectomycorrhizal fungi exhibit completely different mechanisms of resistance to Zn and Cu.

5.5. Defence Mechanisms (Resistance/Tolerancy)

Toxicity of any metal depends on adaptability of fungi to environmental factors. Two groups of defence mechanisms are distinguished: resistant/tolerant and sensitive. Some authors recognized two basic defence mechanisms: 1) strategy of exclusion when concentration of heavy metal is on constantly low concentration until the critical one when metal is urgently transported in cell and performs its toxicity, 2) accumulative strategy when metal is actively concentrated in organism [141]. Resistance of fungi towards toxic metals can be defined as the ability of organisms to survive in the presence of high concentrations of metals by applying different extracellular mechanisms of defence as a direct answer to the damaging impact of metals. The defence is usually based on immobilisation of heavy metals using extracellular and intracellular chelating compounds e.g. synthesis of metallothioneins (copper or silver binding in *Agaricus bisporus* [142] or γ-glutamil peptides, mycophosphatin and phosphoserines (cadmium binding in *A. macrosporus* [143]. On the contrary, tolerance is a passive mechanism towards toxic metals that can be defined as the ability of organism to survive in a presence of toxic metals by applying intracellular abilities such as insolubility of cell-wall, production of extracellular polysaccharides and metabolic excretes which role is to detoxify metals using linkage or precipitation [144].

Moreover, existing tolerance/resistance on specific metals thus not include the analogues ability towards the other metals, implying these properties of fungi as strongly metalospecific. If multiple tolerance exists, then it reflects the toxic level of many specific metals in substrates ([145], [146]).

Biological mechanisms that are involved in these processes are: 1) extracellular precipitation, 2) formation of complexes and crystallization, 3) transformation of metals by oxidation, reduction, mutilation and dealkylation, 4) biosorption in cell wall, on pigments or extracellular polysaccharides, 5) in capacity for transport or completely membrane insolubility for specific element, 6) active transport of metals from the cell and 7) intracellular compartmentation and precipitation. Different organisms use directly or indirectly many different defence strategies against toxicities. Many fungal species among all taxonomic groups can be found in heavy metal polluted areas and soils contaminated with Cu, Cd, Pb, As, Zn. This pollution has impact on fungi by reducing the diversity and density of fungal populations or by strong selective pressure affecting the population of resistant/tolerant speccis [99]. Somc spccics bccomc tolcrant to hcavy mctals, prcvcaling in such arcas. Thc species *Laccaria laccata* and *Schizophyllum commune* are most often found in areas with high metal contents (([55] [134], [147]). In unpolluted areas, examples of resistant strains indicate that survival of these organisms is more dependent on intracellular fungal features then on adaptations on physico-chemical environmental factors. Humus or lignin degrading fungi in polluted areas are able to select ecotypes with tolerance to Co, Cu, Hg, Ni, Zn if produce tirosinase, whereas polyphenol degrading fungi and terrestrial fungi (e.g. *M. oreades*) develop tolerance towards metals [148].

5.5.1. Extracellular Mechanisms (Precipitation and Complex Forming)

Many extracellular products (e.g. organic acids: lemon acid or oxalic acid) of fungi can form chelate complexes with heavy metals or precipitate it by forming insoluble crystals of oxalate around cell walls or in extra cellular medium. White-rot fungi produce extracellular hyphal sheat, composed mainly of polysaccharides, including β-1-3 with β-1-6-linkages which binds calcium oxalate crystals. In white-rot and brown-rot fungi, extracellular and cell-wall associated binding is more important [101]. Oxalate is the most typical metal chelator which is mostly produced in brown-rot fungi (e.g. *D. quercina*), but also in some white-rot fungi (e.g. *P. ostreatus, P. chrysosporium* and *T. versicolor*) [149]. Many filamentous fungi excrete molecules of high affinity towards Fe, called syderofors, in order to complex Fe in a form of helates from the extracellular matrix ([150], [151]).

Cell wall is the first barrier of fungi towards heavy metals since it represents a place of controlling uptake of both soluble substances and water in the cell. Physico-chemical interactions comprise the following processes: ion exchange, adsorption, complex formation, precipitation and crystallization [152]. A first step in the interaction between heavy metals and a cell is a connection to the groups that contain oxygen (carboxylic, phosphate, phenolic, alcoholic, hydroxyl, carbonyl and metoxyl). Chitin and chitosan are the most important polysaccharide constituents of the cell wall and the major donors of functional –NH groups that are dedicated to the fixing metals Cu^{2+}, Co^{2+}, Cd^{2+}, Mn^{2+}, Zn^{2+}, Mg^{2+}, Ni^{2+}, Ca^{2+} ([153], [100]). Another group of heavy metal binding compounds produced by fungi are melanin and phenol polymers associated with the cell wall which have also ability to compex metal ions that decrease in the following order Cu>Ca>Mg>Zn ([154], [155]). Some fungal melanins are efficient bioabsorbers of copper. Melanin is a pigment that is important in the cell answer to the stress conditions and is located in the cell wall or on the extracellular surface in a granular form.

5.5.2. Intracellular Mechanisms

Intracellular mechanism of tolerance comprise processes of detoxification based on the: 1) chelation of metals in the cytosol with organic acids (malonic and citric), 2) forming salts between inorganic acids and heavy metals, and 3) accumulation of heavy metals in the cell organelles, especially in vacuole.

Metalollothioneins are citoplasmatic proteins, 10kD, which synthesis is induced by high concentrations of metals Zn, Cd, Cu, Hg, and Ag [156]. They are small polypeptides wealth in cysteine being able to chelate metals by sulphydril group SH-. Short γ-glutamil peptides are synthesized in cytoplasm by enzyme phytochelatin-synthase, containing three amino acids - glutamine (Glu), cystine (Cys) and glycine (Gly), and further transported to vacuole [156]. They are known as phytochelatines and represent the main detoxifying elements in filamentous fungi [157].

In white-rot fungi neider phytochelatins nor metallothioneins are registred, but they might be replaced with onother types of peptide or protein molecules found in certain species (e.g copper and cadmium binding peptide was found in *P. ostreatus* or copper binding peptide from *Grifola frondosa* with MW of 2240 Da ([101], [132]).

Fungal vacuoles have the most important role in macromolecules degradation, storage of metabolites and ions of cytoplasm and in regulation of pH homeostasis since they are capable of regulation of both essential and nonessential metals concentration in cytosole. Polyphosphates in vacuole are dedicated for binding two-valent cations Mg^{2+} and Mn^{2+} [158] by localization (compartmentation) of these elements Ca^{2+}, Mg^{2+}, Mn^{2+}, Zn^{2+}, Co^{2+}, Fe^{2+} inside the cell.

Fungi are capable for chemical transformation of metals and metalloids by the processes of oxidation, reduction, metilation and dealcilation, thanks to the enzymes they produce, making them less toxic or vaporizing them from the environment ([159]). Moreover, fungi are able to change local microhabitats causing changings necessary for efficient degradation of pH or metabolic excretion. *Phaeolus schweinitzii* can degrade trimethyl lead [160].

Heavy metals, in general, are potent inhibitors of enzymatic reactions, hence in white-rot fungi these metals influence extracellular enzymes involved in wood degrading process. Low concentrations of essential heavy metals are necessary for the development of these enzymes since they are a constitutional part of its structure (e.g. Mn is incorporated in a structure of mangan peroxidase - MnP, Cu is a cofactor of the enzyme laccase). The positive effect of Cu addition on the production of laccase was observed in various fungi but was proven to be strongly regulated at the level of transcription in *Trametes versicolor* [161].

5.6. Accumulation of Metals in Macrofungi

Although macrofungi have been traditionally used in human nutrition for centuries, in the last decades fungi were pronounced to be especially good sources of healthy food [162], natural antioxidants ([119], [163], [164]) or other biologically active compounds, including antimicrobials, cytotoxic and immunomodulative substances ([118], [165]). Hence, its consumption is lately increasing in many countries. Since fungal species could be an important portion of the human diet in the future, it is necessary to investigate more the chemical constituent and nutritional quality of both wild and cultivated mushrooms.

Screening of occurrence of trace elements in fungal sporocarps has been carried out due to the two main reasons: they can be used as bioindicators of environmental pollution, especially soil contamination [166], and some edible or medicinal species can accumulate high levels of heavy metals making its consumptions as detrimental to human's health [167]. Furthermore, the knowledge on heavy metal levels in mushrooms is alo very important for assessing their transfer along the food chains.

Some authors discuss macrofungi only as rough bioindicators of soil pollution with heavy metals [168], because of their tremendous bioaccumulative ability. This makes them unapropriate for usage for an exact assessment of soil pollution at the given place; but more represent an useful tool for distinguishing polluted versus unpolluted areas. By comparing the two Slovene areas (surroundings of thermal power and lead smelter) with high emissions of heavy metals with respect to levels of As, Cd, Pb and Hg in 16 fungal species, it was found that the average heavy metal levels (mg/kg d.w.) coincides fairly well with data on heavy metal emissions as well as their burdens in soil and other biota (e.g. vegetables, domestic animals, roe deer), confirming macrofungi as a good biomonitoring tool. Although physiological mechanisms of uptake and accumulation of heavy metals and radionuclides in fruiting bodies of macrofungi are rather speculative, the fact that macrofungi can be used in human nutrition makes this problem very important for the future.

Excessive levels of heavy metals (Pb, Zn, Cd, Hg) in sporocarps of macrofungi, mostly Basidiomycota, can be frequently noticed in polluted urban or industrial environment but also in forest ecosystems that are polluted ([169], [170]). Mostly investigated macrofungi are ectomycorhizal species [171], that are in tight connection with wood, thus making a great impact on surviving and productivity of forests in polluted areas influenced by acid rain, chemicals and minerals that are applying in agriculture. On the contrary, lignicolous fungi were far less analyzed due to the fact that their dominant role in forest ecosystems is in maintaining balance since they are wood parasites holding attention of phytopathologists.

During the last decade of 20th century, important changes in populations of Basidiomycota fungi was noticed, mostly a decrease and loss of many species in central Europe [171], and in south Scandinavia [169]. By comparing polluted and unpolluted areas it is concluded that the main cause of this situation is a devastating consequences of air pollution. It influences dramatically changes of forest soil, such as acidification, decline in pH, exhaustion of buffer capacity, and increased mobilization of some ions (eg. Al becomes soluble), thus increasing their concentration in soil solution and making them available for the fungi [148]. These changes inactivate normal metabolism of mycorrhizal species, leading to inabilities of making positive symbiosis with secondary roots of wood leading to lethal consequences [171].

High concentrations and bioaccumulation of different major and trace elements were reported in the European forests ([172], [99], [173], [128], [174], [175], [176]) and also in Japanese forests ([177], [178]) and in Turkey [179]. Many of them demonstrated that mushrooms tended to accumulate Cu, Zn, Rb, Cd and Cs. Kalac *et al.*, [128], observed that Hg, Pb, and Cu were accumulated by tericolous *Lepista nuda* and *Lepiota rhacodes*. Higher concentrations of Pb, Cd, Zn and Hg are found in macrofungi (prevalently ectomycorhizal that are in close contact with roots of wood) from urban or industrial areas ([169], [170], [171]) but also in wood ecosystems that are influenced by contaminators. Data on the general patterns of macro- and microelement accumulations in fruiting bodies of wild-growing lignicolous fungi are scarce ([55], [175]).

According to data based on metal content profiles of 92 mushrooms species [132], different mechanisms are involved in uptake of heavy metals by fungi. It is assumed that uptake ability is genetically coded at a genus- or species- levels beside the physicochemical influence only if the standard procedures are established, e.g. evaluation of contamination, substrate pH and mushroom growth. Saprotrophic terrestrial species, especially the genus *Agaricus* exhibits a strong affinity to Cu, Ag and Cd. Also wood-decomposing species, including mycorrhizal, or those frequently growing in forest area (Boletales, Aphylophoralles, Auriculariales, Lycoperdales, Sclerodermatales, Tremellales) have a tendency to accumulate Cr, Mn, Se particularly *Boletus*, *Suillus* and *Xerocomus* and Pb [132].

Atmospheric deposition of heavy metals from air (fumes, dust and aerosols), especially in the case of Cd, Pb and Hg, is evident in wood-decaying fungi (mercury in *Pleurotus eryngii*, lead in *Fistulina hepatica*) that possess perennial fruiting bodies of huge size. This indicates the importance of wood decaying fungi as bioindicators of air pollution. On the other hand, there are comprehensions that none of the fungi can be considered as effective bioindicator of heavy metal contamination, although the fruiting bodies may serve in determining level of pollution on field survey ([180] [181] [182]). In a recent study that analyzed concentrations of neodymium, lead, thorium and uranium in wild-growing macrofungi, the species specificity was determined as dominant factor of accumulation [166]. Only thorium and uranium were highly incorporated in wood-decaying fungi *Hypholoma fasciculare*, pointing to the substrate composition as an important factor to be considered as it was indicated by several authors [182]. According to previous study which dealt with macro and microelements concentration in tericolous and lignicolous fungi from 5 locations in National park Frushka Gora in Serbia [176], cluster analysis classified fungi mostly by location. The same species from different locations contained different mineral contents, indicating that the accumulation ability is not only genetically coded but also influenced by environmental factors.

Table 1. Data on mean content of nine trace elements (mg/kg d.m.) in wood-decaying mushrooms from unpolluted areas published from 2000-2011

Species	Ca	Mg	Fe	Zn	Cu	Cr	Pb	Mn	Cd	References	Ref.	Region
P. squamosus			139.00	71.00	23.00	9.50	0.80	19.00	0.85	Sarikurkcu, 2011	167	
F. hepatica		898.3	38.90	34.43	7.38	4.79	0.14	7.19	0.07	Ouzouni, 2009	184	Greece
	1640.0	969.0	41.1	50.2	30.9	2.41	42.7	26.8	2.52	Michelot et al., 1998	132	France
M. gigamteus	792.0	1510.0	267	48.7	21.9	0.9	20.8	7.07	3.62	Michelot et al., 1998	132	France
	440,59	2045,45	2504,36	44,21	9,52	7,75	3,25			Karaman , 2002	223	Serbia
Agrocybe aegerita							8.2		27.8	Cocchi et al., 2006	227	Italy
Agrocybe praecox	168C	1250	179	66.6	19.8	3.67	32.4	65.9	3.05	Michelot et al., 1998	132	France
Armillaria mellea							11.5		17.6	Cocchi et al., 2006	227	Italy
			480.9	90.3	30.8		1.0	31.3	0.3	Ita et al., 2006	185	Nigeria
		1063.1	499.0	54.12	17.38	4.20	0.49	55.59	1.67	Ouzouni, 2009	184	Greece
Armillaria tabescens		1150.7	60.40	64.45	17.47	4.37	0.79	11.18	1.80	Ouzouni, 2009	184	Greece
A. polymyces	564,05	1175,70	931,24	52,79	29,51	1,94	3,56			Karaman , 2002	223	Serbia
C. cibarius							11.3		4.2	Cocchi et al., 2006	227	Italy
							4.86			Campos et al., 2009	166	Spain
C. cornucopioides	194C	1000	426	165	49.3	1.94	30.7	223	2.06	Michelot et al., 1998	132	France
		866.3	118.2	54.29	32.49	1.57	nd	22.09	0.38	Ouzouni, 2009	184	Greece
Hirneola auricular-judae							31.3		1.6	Cocchi et al., 2006	227	Italy
	5051	979	58	26.6	1.8	2.3	29.5	23.6	2.1	Ouzouni, 2009	184	Greece
Auricularia mesenterica	545C	980	187	76.5	12.3	3.74	47.4	21.3	3.99	Ouzouni, 2009	184	Greece
Polyporus frondosis			731.6	120.1	34.4		0.4	37.3	0.2	Ita et al., 2006	185	Nigeria
G. applanatum			560.7	137.4	60.8		0.7	25.8	0.3	Ita et al., 2006	185	Nigeria
	4135,5	946,4	446,32	25,14	15,90	4,42	3,53			Karaman , 2002	223	Serbia
G. lucidum			604.8	60.1	43.8		0.7	30.4	0.3	Ita et al., 2006	185	Nigeria
	564,05	1175,7	2290,9	34,87	7,24	3,98	3,72			Karaman , 2002	223	Serbia
Pleurotus sapidus			473.5	98.4	39.2		0.8	28.4	0.1	Ita et al., 2006	185	Nigeria
Pleurotus ostreatus			407.7	90.6	45.9		0.4	39.8	0.3	Ita et al., 2006	185	Nigeria

Table 1. (Continued)

Species	Ca	Mg	Fe	Zn	Cu	Cr	Pb	Mn	Cd	References	Ref.	Region
	625,80	1114,39	148,10	111,07	5,20	3,52	<4,17			Karaman , 2002	223	Serbia
L. sulphureus			337.5	95.1	18.8		1.1	19.5	0.2	Ita et al., 2006	185	Nigeria
	242,07	1020,40	553,01	57,93	9,71	2,74	<6,39			Karaman , 2002	223	Serbia
Nectria innabarina			277.2	30.1	29.3		1.9	19.3	0.2	Ita et al., 2006	185	Nigeria
Polyporus brumalis	2700	677	81.6	72.9	17	3.8	45.2	63.4	3.23	Michelot et al., 1998	132	France
H. fasciculare	2240	728	160	82.4	28.4	3.3	33.6	16	2.7	Michelot et al., 1998	132	France
							3.50			Campos et al., 2009	166	Spain
Xylaria polymorpha	1750	699	55.2	71.8	12.7	1.3	28.4	14.9	1.96	Michelot et al., 1998	132	France
O. olearius							3.60			Campos et al., 2009	166	Spain
	418,86	1544,52	603,82	47,01	24,57	2,22	4,53			Karaman , 2002	223	Serbia

Analysis of microelement content in 22 species was found to be specific and depended dominantly on availability of these elements from their substrates to fungi. It is revealed that specific accumulators (e.g. ecotypes) of particular element were created as a defense mechanisms in the course of the evolution, or as a consequence of stress adaptation by which they could exclude or amortize unfavorable effects of heavy metals presence in the environment [183].

The mycelium network provides extensive contact with substratum and optimum absorption of nutrients. It is designed to accumulate all kind of elements, including heavy metals in its sporocarps and reach much higher concentrations than those of the substrate [166]. In as much as a mycelium has the great surface of hyphae that could absorb and accumulate metals, the majority of them (Cu, Zn, Cd) are captured in fungal biomass in the layer of humus. Basidiomycetous fungi degrade the upper layer of humus which containes polyphenolic compounds (lignin, humic acid, fulvic acid, humin). Enzymes (phenoloxidases) are dedicated to effectiveness of ion binding and chelates forming via ion exchange. With regard to this the high tolerance against heavy metals can be realized among the fungi that have ability of polyphenol degradation [148].

Several studies have pointed out the importance of specific elements to fungal strains. The mean macroelement and microelement concentration across all terrestrial and lignicolous fungi tested by Rudawska and Lewski, [175], was in the following order: N>K>P>S>Ca>Mg>Al>Zn>Fe>Mn>Pb>Cd, or in the similar order in the study of Ouzouni *et al.*, [184]: Mg>Fe>Zn>Mn>Cu>Ni>Cr>Co>Pb>Cd. For ten wild edible mushrooms analyzed by Ita *et al.* [185], the heavy metal accumulating potential generally decreased Fe>Zn>Cu>Mn>Pb>Cd (Table 1).

In a recent work of Karaman and Matavulj [183], tericolous and lignicolous wild growing species were analyzed where the highest mean concentration of macroelements (d.w.) was found for N (3.08%) > K (1.83%) > P (0.3 %) > Na (15.09 mg %) > Ca (2226.85 μg/g) > Mg (1384.24 μg/g) > Fe (928.03 μg/g) > Zn (58.14 μg/g) > Cu (17.60 μg/g) > Cr (3.94 μg/g) and Pb (3.88 μg/g). When transfer factors (TF) for the lignicolous fungi are analyzed, some facts should be take into account. First of all the volume of analyzed material of mycelium is just a part of generally present content of biomass in substratum, and also it is present in longer period of time than it is concerned by this analysis. Also the mycelium is in a possibility to accumulate nutritive substances both from dead (saprothrophic) and live (parasitic) material. The content of mineral and trace elements in fruiting bodies, although they are not in direct contact with the soil, can be influenced by the content in soil via mycelial cords and rhizomorphes. This is the possible explanation of higher element concentrations in fungi in relation to the same in soil. Moreover, saprobic lignolitic fungi are not in direct contact with the soil in contrast to mycorhizal fungi

In previous study of Karaman and Matavulj, [183], the highest level of Fe was detected in the lignicolous species *M. giganteus* (2504,36 μg/g) while the highest level of Zn (139,35 μg/g) was found in the species *Schizophyllum commune*. The highest content of Pb and Cr was found in *C. atramentarius*_(9,72 μg/g, 13,36 μg/g, respectively) reaching toxic concentrations. Species which contained more than 1000 μg/g Fe acted as super-accumulators (lignicolous species *M. giganteus, G. lucidum* and *Sch. commune* and tericolous species *C. atramentarius, F. velutipes* and *P. vernalis*), whereas accumulators of all microelements were *M. giganteus* (exept Zn), *Sch. commune* (exept Pb), and *G. applanatum* (except Fe and Zn). The species that accumulate all of microelements were those posessing large surphace area of mycelium due to rhizomorphs - *A. polymyces*, and *O. olearius*, and one strictly lignicolous fungus – *S. hirsutum,* which could be pronounced as a super-accumulator species of microelements, especially Pb and Cr [183]. The best accumulator species were tericolous possibly bioindicator species *C. atramentarius* (especially for Fe, Pb and Cr) and *P. vernalis* (especially for Pb and Cu), due to the highest concentration of Pb in the soil of the urban sampling site. The highest Tf =7.5 for Pb indicates that it was not merely the result of the species accumulator ability, but also, the result of specific chemical composition of soil (at pH<7, lower concentration of Mn and Zn influenced on the higher accumulation of Fe). This is in accordance with the data quoting that saprotrophic species could change physico-chemical composition of environment, showing their influence on availability of other elements in substrate and their accumulation in metabolically active hyphe.

According to the significant variations in microelement concentration among the analyzed species, and between the same species from the different sites, especially Fe and Pb content (i.e. *C. versicolor, S. hirsutum*) we assume that genetically different ecotypes of a species can be distinguished according to the ability of metal uptake from substrata. This phenomenon of so called intraspecific antagonism was already reported for macrofungal population of *Coriolus versicolor* [186]. Similiar situation could be expected in populations dispersed on different habitats. This can partially explain the inhomogeneous chemical content of lignicolous fungi derived from different locations ([176], [183]).

5.6.1. Influence of Mycorrhiza on Increasing Tolerance of Plant to Heavy Metals

Mycorrhizal fungi are found associated with most higher plants in which host plant gives the products of photosynthesis to fungi (especially carbon) while the mycorrhizal fungus supplies the host plant with water, less mobile mineral nutrients (phosphorus), improving uptake of NH_4^+, Ca^{2+}, Zn^{2+} and SO_4^{2-} ions or supplying amino acids and growth substances by excreting or by hyphae degradation. Besides protection from fungal pathogens and soil toxins [99], fungi protect plants from negative effects of heavy metals. This is accomplished by formation of compact mycelia layers around the root, making barrier between the root and the toxic metals (physiologically) to evolve genetically metal tolerant populations. There are suggestions that major evolutionary role of mycorrhizae is to facilitate the acclimation of plants to local soil conditions [187]. Ectomycorrhizal species arrest heavy metals in cell walls or vacuole by lowering metal concentration in soil solution. In addition, complex of hyphae can protect translocation of metals in plants reducing their toxicities. The two species, *Amanita muscaria* and *Paxillus involutus*, were proved to increase tolerance of *Betula spp.* towards Zn [188]. Otherwise, some fungal species could increase an uptake of heavy metals by plants, suggesting that specific mechanisms are still not known enough.

5.7. Interactions between Radionuclides and Fungi

5.7.1. Radionuclide in the Global Environment and Forest Ecosystems

Artificial radionuclides were discharged into the global environment through nuclear weapons tested until 1963, reaching maximum in 1982 ($^{137}Cs \approx 9{,}6x10^{17}$ Bq) [189]. Soon after the Chernobyl accident in 1986, about $3.8x10^{16}$ Bq of ^{137}Cs decay was released into the environment [189]. After the radioactive deposition originating from atmosphere by precipitations, a high accumulation of radioactive Cs in living organisms was registered, at first in lichens and mosses and later in fungal species [190]. A high transfer factor from soil to fungal sporocarps was recorded after similar events in Japanese forests [191]. The ability of fungi to take up Cs from the substrate (soil and wood) points to these organisms as promising bioindicators of soil pollution. Dighton *et al.,* [192], suggested that much of the absorbed Cs is biologically bound within the fungal tissue and that they have the capacity of holding all the potentially "labile" soil Cs. The main source of radionuclide for lignicolous fungi is the atmosphere, especially for beryllium that is cosmogeny radionuclide ([193], [183]), confirming that these fungi are good bioindicators of air pollution.

Since most fungi occur in forest ecosystems, literature data dealing with wild-growing species may be found in plentiful supply ([191], [194]). Forests are complex ecosystems composed of diverse plant associations, various vegetative strata and multi-layered soil profiles determining radionuclide biochemistry (transfer of radionuclides) characterized by high variability in contaminated areas [194]. It seems that forests are more endangered than the agricultural areas, taking up the major amount of radionuclides via air [195].

Their nutrition via absorption makes them specific bioindicators of radionuclide contamination. Usually, Cs contamination of fruiting bodies is for a degree of magnitude higher than in vascular plants occurring in the same area ([196], [197]). The elementary composition of fruiting bodies distinguish from the same of plants due to high ^{137}Cs, Cs and Rb concentrations and low Ca and Sr concentrations [178].

5.7.2. Radionuclides in Macrofungi

Fungi represent the greatest living biomass and source of enzymes in the forest soil, in organic horizons in particular, where mostly saprotrophic fungi take part in the process of litter decomposition ([198], [199]). Furthermore, fungi take up both nutrients (stable elements) and radionuclides (^{137}Cs or ^{90}St) from the soil aqueous solution, using the same enzymes via specific carrier molecules as well as energy.

Saprotrophic and mycorrhizal Basidiomycota fungi are able to accumulate radioactive ^{137}Cs ([192], [200], [201], [202]) with low transport rate, forming the major pull of radiocesium in the soil [192]. It is quoted that Cs is dominantly present in the upper surface area of soil at a depth of 0-5 cm and deeper than 5cm ([203], [204]), being mostly found in fungal mycelia in amount of 30-50% [178]. There are estimations that soil fungi act as a sink of radiocesium [205]. This statement is a consequence of binding of Cs, together with complexes of organic matter and clay particles, to mineral surface area, resulting in its low concentration in soil solution and low migration withinin soil profiles [206]. According to Nikolova et al., [207], since the production of fruiting bodies was different in different years due to weather conditions, fungal fruiting bodies may accumulate only between 0.01% and 0.1% of the total Cs available in different years. The atmospheric precipitations are associated with the migration of Cs from the soil surface to deeper layers, provoking higher pollution of saprotrophic fungal species with surface mycelia after several years [208], whereas the symbiotic species with deeper mycelia occurring in deeper soil zones (>5 cm) showed higher amounts of radiocesium after an extend period of time ([178], [209]).

There are many dilemmas about the relationship between the contamination rate, pedological characteristics and different nutritional groups of fungi, suggesting symbiotic and lignicolous species as less contaminated than saprotrophic ones [210]. Besides the Cs distribution in soil zones, another important factor affecting the contamination of macrofungi by radionuclides or heavy metals is mycelia habitat, namely their localization in soil profiles ([178], [208]). Furthermore, a significant fungal redistribution of Cs in the forest soil is needed during the production of fruiting bodies ([211], [212]). Saprotrophic fungi occurring on decomposing material above or within the surface layers of soil are first to be contaminated following deposition. Mycorrhizal mushrooms living in close association with trees may be the most contaminated in the medium or long after deposition while the contamination of parasitic mushrooms relies upon a degree of host tree/plant contamination [194]. Besides, there might be expected an important fungal redistribution of Cs in the forest soil during the production of fruitbodies [198].

It is quoted that about 90% of fungal biomass is concentrated in soil in a form of mycorrhizas, whereas only 10% goes for fruiting bodies e.g. sporocarps ([213], [214]), implicating the occurrence of ^{137}Cs and other mineral nutrients redistribution [207]. It is documented that radioactive Cs can be completely immobilized by fungi in soils and can be further incorporated in nutritional chain by eating fungal fruiting bodies (insects, snails, deer, goats, cattle etc.).

The translocation of nutrients and radionuclides within a mycelium depends on the nutritional type and fungal species [198]. The ectomycorrhizal fungi form complex rhisomorphs which contain large vessel hyphae and radiocesium might be translocated with bulk flow of nutrients or water along the rhizomorphs. Fungi with rhisomorphs thus contain higher concentrations of radiocesium what is evidenced in *Armillaria* species [198].

The concept of transfer factors and concentration ratios are used to quantify the transfer of radionuclides from soil to fungal fruitbodies, usually expressed as the amount of radioactivity per unit weight, dry weight (Bq/kg d.w.) or on a fresh weight basis (Bq/kg f.w.). The commonly accepted concept of 10% of dry matter of mushroom fruitbody mostly used for calculation [189], seems to be discussible. Since activity levels of fungal sporocarps may considerably vary over a small distance as related to various factors [198], other types of transfer factors are suggested (e.g. aggregated transfer factors - Taq), representing the ratio of the activity of fungal sporocarp divided by the total deposition on soil (Bq/m^2) [198].

5.7.3. Health Risk

Mushrooms are of special interest as they represent naturally occurring foods that are collected in forest ecosystems. The Cs transfer factors to mushrooms are widely variable (3-4 degrees of magnitude) due to species characteristics, mycelia depth and nutritional type of mushroom. It is rather speculative in which part of fungi (mycelia network or fruiting body) is higher accumulation of radiocesium, although there are data pointing to the fruiting bodies that act as the final sink while mycelium only mediates its transport and facilitates dilution of the contaminanat [215]. Therefore the fruiting body and in particular its cap is the final place of radiocesium accumulation. This finding is very important as mushroom sporocarps are mostly used in culinary or as the source of bioactive substances. Fortunatelly, the radioactive contamination can be considerably decreased by soaking or cooking of dried or frozen mushroom slices [189].

Accepted statutory limit for foods is 600 Bq/kg of fresh weight i.e. 6kBq/kg of dry matter (d.m.) for mushrooms [189]. According to the European Communities published Council Regulation (CEC, 1987), the maximum permitted level of ^{137}Cs for mushrooms is 12,5 kBq/kg d.m. while 10 kBq/kg d.m.is recomended by International Atomic Energy Agnecy (IAEA, 1994).

5.7.4. Factors affecting the Radioactive Accumulation in Fungi

The concentration of radioactive Cs generally relies upon the forest type [216], soil pH [217], fungal species [192] and mycelium habitat, possibly representing the most relevant factor [178]. On the contrary, the concentration of radionuclides in fungi is determined by a great number of factors among which the amount of radioactive precipitation on the first week of accident, concentration of stable (non radioactive) or analogous element in soil, soil characteristics (mineral composition, pH) and taxonomic and ecological identity of fungus are emphasized [216] (Figure 4). Long-lived product of fission ^{137}Cs is the most frequently investigated radionuclide.

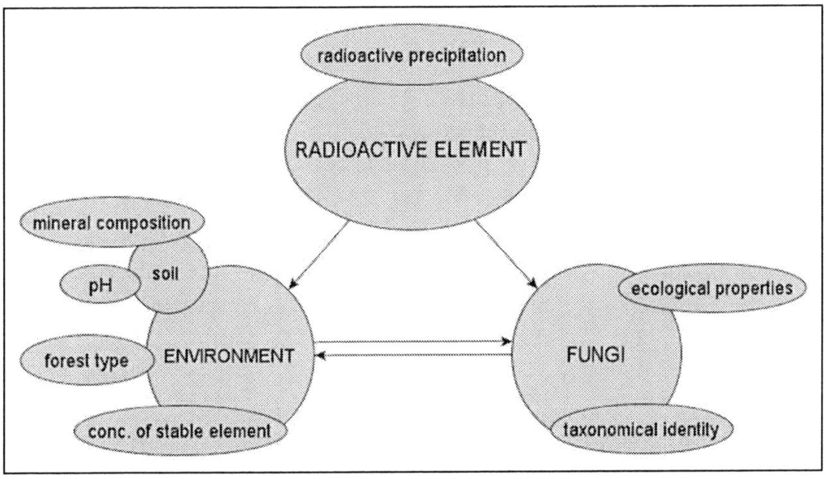

Figure 4. Main factors affecting the radioactive accumulation in fungi.

5.7.4.1. Species Factors (Bioindicators)

There are considerations that the species characteristics (taxonomic and nutritinal) are much more important than the soil characteristics or environmental conditions (e.g. pH, moisture content) ([192], [203], [216]). Some families are known as Cs-accumulators (e.g. Cortinariaceae, Clavariaceae, Entolomataceae and Strophariaceae) while Cs-discriminator are Helvellaceae and Lycoperdaceae [216]. Families, bioindicators of environmental contamination are Amanitaceae, Boletaceae and Russulaceae [210].

After the Chernobyl accident, heavily accumulating species were *Xerocomus badius, Xerocomus chrysenteron, Suillus variegatus, Rozites caperata* and *Hydnum repandum*. In former Yugoslavia, there were only few analyses dealing with the heavy metal concentration in macrofungi of Slovenia [172] and Montenegro [221]. The results presented show that the highest mean values of Cs were found in the species *Cortinarius armillatus* (44 kBq/kg d.w.), *C. traganus* (12,4 kBq/kg d.w), *Rozites caperata* (22 kBq/kg d.w), *Xerocomus badius* (9,3 kBq/kg d.w.), *Laccaria amethystina* (43 kBq/kg d.w.) and in *Boletus edulis* (300 Bq/kg d.w.) while the activity concentration was 110Ag was 500 Bq/kg.

In a recent work of Karaman and Matavulj, [183], the species *P. squarrosa* showed the highest activity concentration level of 55(4) Bq/kg (dry. wt.) and the highest TF values for essential naturally occurring radionuclide - potassium ^{40}K and the potassium analog - caesium ^{137}Cs, by accumulating the both elements in spite of the fact that the soil was alkaline (pH =6.01) that indicates stronger binding of Cs to soil complexes. Therefore, we imparted special importance to this species as a potential bioindicator species to be used in the radioecological studies of fungi. Since *P. squarrosa* belongs to the family *Strophariaceae* (Cs-accumulator family) ([210], [216]), the presented results confirmed the literature data published elsewhere. However, *P. squrrosa* is the only strictly parasitic species, so the possibility of using nutrients which flow through the host tree xylem is much realistic for real parasitic than for wood-decaying saprotrophes. All these facts recognize parasitic lignicolous species as promising bioindicators. In addition, these data has proved that both the systematic position and the ecological identity (Sap, P, Myc) play an important role in determining the Cs-content of different fungal species.

Furthermore, the activity concentrations of ^{137}Cs in the medicinal species *G. lucidum, C. versicolor, G. applanatum,* and *M. giganteus* from the Frushka Gora low mountain varied in a narrow range from <2.2 (*C. versicolor)* to 9.4±2.8 *(G. applanatum),* exhibiting values ten times lower than in *P. sqarrosa* (55.4 Bq/kg d.m.) [183]. Accordingly, significant differences within the same species from different localities were not recorded, indicating that the accumulation of radionuclides by fungi is species-specific ([192] [203], [216]). Moreover, the species *G. lucidum* was the only one species that showed high transfer factor for ^{40}K, but low influx of ^{137}Cs, what is possibly genetically determined feature commending this fungi to the attention if used in pharmacology. In addition, a high value of a cosmogeny radionuclide ^{7}Be was recorded in *S. hirsutum* species (140±30) while it was absent in wood bark due to a its fallout on the outside surface of the fruiting bodies. The active surface in this species is maximized because of its pilose upper surface ("Hairy Stereum") ([176], [183]). Various mushroom parts contain various Cs content. High concentrations of ^{137}Cs are recorded in fungal parts with high metabolism rate or abundant cytoplasm. Lamellas showed the highest values of ^{137}Cs, followed by cap and finally the stipe ([202], [219]). It was shown that Cs highly relies upon cap pigments [206].

5.7.4.1.1. ^{137}Cs uptake and Nutritional Fungal Groups

Different taxonomical and nutritional groups of fungi contain different concentrations of radioactive Cs. Great variations in Cs concentration were reported, indicating the following order: mycosymbionts > saprophytic > lignicolous fungi ([210], [216], [220]), although the influx of ^{137}Cs into hyphae of several basidiomycetes showed considerable variations in saprotrophic species exhibiting its highest rates, whereas in mycorrhizal species the lowest ([99], [192]). Lignicolous fungi (e.g. wood-decaying fungi) seem to have the lowest concentration of Cs due to their habitat (wood) showing much lower concentration of nuclides than the soil layers. This indicates that there are two transfer factors, namely the one is a transfer factor operant from soil to sporocarp and the other is a transfer of water and mineral elements through the floem via wood to fungi ([216], [219]). Some authors assume that fungi are accumulators of alkaline metals but not of other fission products derived from Chernobyl [220]. The essential element potassium is chemically very similar to radioactive Cs and therefore can be replaced by nonessential Cs. In addition, some authors think that the accumulation is species specific and that accumulator species do not possess mechanisms effective enough to distinguish these two elements [202]. There is no correlation between concentration of ^{40}K and ^{137}Cs ([203], [222]).

Influx of Cs in hyphae is very variable (85-275 nmol Cs/g d.w /h) and show higher values in saprotrophic fungi than in mycorrhizal fungi that do not belong to Basidiomycetes [192].

5.7.4.1.2. The Impact of Mycorrhizal Species on Plant Defense against Radionuclides

In mycorrhizal interactions fungi supply plants with carbon and energy, whereas plant supply fungi with water and inorganic substances. In spruce forests, it is noticed that Cs is captured by hyphae sheet and extracellular mycelia, indicating that the potency of ectomycorrhizal species to uptake heavy metals and radionuclides is much higher than that of saprotrophic fungi. Although mechanisms of uptake of K and Cs in cell roots is still not well undestood, it has been documented that mycorrhizoidal spruce seedlings contain lower amounts of Cs than the seedlings lacking fungi mycelia [206].

5.7.4.2. Environmental Factors

Some scientists consider that the Cs uptake is not highly species-dependent, and that it is more affected by location and time of sampling [220]. There are significant variations of ^{137}Cs and ^{134}Cs intake within the same genus as well as among individuals within a species. Since there are great variations of ^{134}Cs and ^{137}Cs intake within the same genus and among strains of the same species [203], some authors state that sites and temporal factor (time of collecting) have the priority [220]. The pH value of soil also have an effect on the availability of Cs, permitting that Cs is more dissolved in acid and more bound in alkaline soils [218]. It is also well documented that the highest availability of Cs to fungi is in positive correlation both with the highest Cs content of soil and the highest humus content (e.g. wither humus zone) as well as the lowest pH values or lower content of essential minerals, especially K. In contrast, lower sand content and highest clay content may cause lower fungal contamination. The highest ^{137}Cs concentration recorded for *G. applanatum* (188±13 Bq/kg d.w.) is explained by specific locality features [176] because maximum Cs values in soil were recorded at the slope margine. This result can be explained by the fixation of radiocesium at the lowest point of terrain profile where the abundance of both the organic matter and humus was reported. Also, the uncultivated forest soil provokes vertical migration of radiocesium, resulting in its binding to the humus layer which makes a large film of organic complexes while leaching to deeper layers is minimized even though it is a very slow process (a few mm per year) [218]. The activity of mycelium in the humus zone is more or less pronounced. Efficient binding of Cs to the clay minerals makes Cs inaccessible for plants, but some fungi can break down and absorb most of the components not available to plants.

6. Antioxidant Enzymes and Free Radical Scavengers as Biomarkers of Environmental Stress in Mushrooms

Biomarkers are defined as a biological response that can be related to exposure to an environmental contaminant. In a broad context they can include measuring such endpoints as reproduction and growth, or behavioral changes. Concerning aerobic organism defense systems on a cellular level, exposure to pollutants causes the production of potent oxidants and free radicals capable of damaging important cell components such as proteins and DNA. In response, the cell initiates antioxidant enzyme systems and produces free radical scavengers in order to prevent cellular injury and maintain cell homeostasis. The induced biomarker response can then be measured and related to measured concentrations of the contaminant affecting the fruiting body (mycelia) of mushroom.

Concerning our observations dealing with the biological activity (antioxidative, antimicrobial) of lignicolous fungal species collected from northern Serbian Voyvodina Province with respect to the total phenol content or phenolic acid concentration [119], we assumed that the specific bioactivity of fungal species is much dependent on the environmental and stress conditions that are the consequence of processes taking place in the strongly narrow surroundings of the growing sporocarp. These results pointed to the importance of conservation and characterization of fungal culture collection in order to sustain a specific genetical and physiological (metabolic) status that could be further used in mycotechnology or pharmacology.

On the other hand, wood as a substratum shows high potassium, except for *D. tricolor, G. lucidum* and, *C. versicolor* where the lowest concentrations even for ^{40}K were found [204]. Our results, ([223], [224], [225]), are in agreement with the results quoted above. In addition to the species registered above, we suggest the following species: *P. gibbosa*, *G. applanatum* and *D. Quercina,* whereas *A. polymyces, P. squarrosa, O. olearius* and *M. giganteus* should be emphasized as those exhibiting the highest ^{40}K content. The recorded differences might be the result of the morphological and functional differences between these groups of organisms. The last three species presented are mushroom-like fungi with high water content and the highest turnover rate enabling easy transfer of essential and natural radionuclids from soil. In samples of fungal fruiting bodies, ^{40}K concentrations ranged from 45±19 to 1710±120 Bq/kg (dry. wt.) while ^{137}Cs concentrations were between <2.2 and 36±4 Bq/kg. The presented values are in agreement with the literature data on wood-decaying fungi ([204], [226]).

Conclusion

Fungi are ubiquitous in natural environment, especially in forest ecosystems. Achieving processes of lignocellulose decaying and decomposing, wood inhabiting fungi play their vital role in nutrient cycling in the environment, transforming CO_2 and H_2O into the original form.

The occurrence and distribution of major and trace elements both in fruiting bodies of macrofungi (mushrooms) as well as in substrate (soil, wood) samples are inevitable, not only with regard to the basic biological knowledge about elementary composition of fungi in wood ecosystems (physiology and ecology), but, also, from practical aspects of toxicology and environmental protection (conservation of fungal species). Hence these data can be used in order to understand the long term behaviour of radionuclides and toxic elements in forest ecosystems and to speculate on the migration effects of chemical elements, especially the artificial elements in the future.

From the ecological point of view, examination of content of heavy metals and radionuclides in fruiting bodies might indicate the potential application of some macrofungal species as bioindicators of environmental contamination. Unlike lichens and mosses being mostly bioindicators of air pollution, lignicolous species might be interesting as indicators of substrate (wood) and soil contaminations.

In addition, it is important to ensure that the contamination of wood products is under control, safe for humans. Since the fungal species could be an important portion of the human diet in the future, chemical constituents and nutritional quality of both wild and cultivated mushrooms should be fully investigated. We will emphasize again that the contamination of selected fungal species with heavy metals should be continuously monitored at a local scale where polluted areas are situated as well as at a national scale.

For the potential application of wild growing lignicolous fungi as sources of food or pharmacologically active substances it is recommended to examine the following: 1) species genotype specificities, 2) impact of pollutant content of sporocarp (mycelium) on the environment 3) radioecological analysis of terrain, microhabitats and fruit bodies (mycelia) of fungal species.

Accordingly, potential application of fungi should be ensured by the accurate taxonomical determination of fungal species, analyses of impact of pollutants detected in fruiting bodies (mycelium) of mushrooms upon humans and heavy metal and radio-ecological monitoring of microhabitats. The best and the safest way of revealing all the fungal mysteries is certainly to bring into unity morphological, ecological, biochemical and molecular studies as well as to gather researchers from different fields to work together.

Acknowledgments

This work was supported in part by the Provincial Secretariat for Science and Technological Development, Vojvodina, Serbia ("Molecular and phenotypic diversity of taxa of economical and epidemiological importance, and endangered and endemic species in Europe") No 114-451-2139/2011-02 and by the Ministry of Education and Science of the Republic of Serbia within project No III43002.

This work is dedicated to the memory of our teacher and good friend, Prof. Dr. Ljiljana Čonkić, her kindness, generosity and humanity.

References

[1] Mueller, G.M. & Bills, G.F. (2004). Introduction. In: Mueller, G.M., Bills, G.F., Forster, M.S., editors. *Biodiversity of fungi: inventory and monitoring methods.* San Diego: Elsevier Academic Press; pp. 1-4.

[2] Linnaeus, C. (1735). *Systemae Naturae, sive regna tria naturae, systematics proposita per classes, ordines, genera & species.*

[3] Haeckel, E. (1866). *Generelle Morphologie der Organismen*, Berlin: Reimer; 1866.

[4] Chatton, E. (1925). *Pansporella perplexa.* Réflexions sur la biologie et la phylogénie des protozoaires. *Annales des Sciences Naurelles, comprenant la Zoologie,* 10, 7, 1–84.

[5] Copeland, H. (1938). The kingdoms of organisms, *The Quarterly Review of Biology,* 13, 383–420.

[6] Whittaker, RH. (1969). New concepts of kingdoms of organisms. *Science,* 163, (3863), 150-160.

[7] Woese, C.R., Kandler, O. & Wheelis, M. (1990). Towards a natural system of organisms: proposal for the domains Archaea, Bacteria, and Eucarya. *Proceedings of the National Academy of Sciences of the United States of America,* 87, 12, 4576–4579.

[8] Woese, C.R. (2000). Interpreting the universal phylogenetic tree. *Proceedings of the National Academy of Sciences of the United States of America,* 97, 15, 8392-8396.

[9] Deacon, J. (2006). *Fungal Biology*. 4th edition. Oxford: Blackwell Publishing; 2006.

[10] Baldauf, S.L. & Palmer, J.D. (1993). Animals and fungi are each others closest relatives-congruent evidence from multiple proteins. *Proceedings of the National Academy of Sciences of the United States of America,* 90, 24, 11558-11562.

[11] Berbee, M.L. & and Taylor J.W. (2000). Fungal molecular evolution: gene trees and geologic time. In: McLaughlin, DJ, McLaughlin, EG, Lemke, PA, editors. *The Mycota VII. Part B. Systematics and Evolution.* Berlin: Springer-Verlag; pp. 229-245.

[12] Richards, T.A. & Cavalier-Smith, T. (2005). Myosin domain evolution and the primary divergence of eukaryotes. *Nature*, 436, 7054, 1113-1118.

[13] Lutzoni, F., Kauff, F., Cox, C.J., McLaughlin, D., Celio, G., Dentinger, B., Padamsee, M., Hibbett, D., James, T.Y., Baloch, E., Grube, M., Reeb, V., Hofstetter, V., Schoch, C., Arnold, A.E. Miadlikowska, J., Spatafora, J., Johnson, D., Hambleton, S., Crockett, M., Shoemaker, R., Sung, G.-H., Lücking, R., Lumbsch, T., O'Donnell, K., Binder, M., Diederich, P., Ertz, D., Gueidan, C., Hansen, K., Harris, R.S., Hosaka, K., Lim, Y.-W., Matheny, B., Nishida, H., Pfister, D., Rogers, J., Rossman, A., Schmitt, I., Sipman, H., Stone, S., Sugiyama, J., Yahr, R. & Vilgalys, R. (2004). Assembling the fungal tree of life: progress, classification, and evolution of subcellular traits. *American Journal of Botany,* 91, 10, 1446-1480.

[14] Cavalier-Smith, T. (1993). Kindom Protozoa and its 18 Phylla. *Microbiological Reviews,* 57, 4, 953-94.

[15] Cavalier-Smith, T., & Chao, E.E.Y. (2006). Phylogeny and Megasystematics of Phagotrophic Heterokonts (Kingdom Chromista). *Journal of Molecular Evolution,* 62, 4, 388–420.

[16] Dick, M.W. (2001). Straminipilous fungi, systematics of the peronosporomycetes including accounts of the marine straminipilous protists, the plasmodiophorids and similar organisms. Dordrecht sh NL: Kluwer Academic Publishers, 2001.

[17] Porter, D. (2005). Labyrinthulomycetes. [Internet, updated 2010]. Available from: http://tolweb.org/Labyrinthulomycetes/21791 in: The Tree of Life Web Project, http://tolweb.org/

[18] Silverside, A.J. (2011). Index to the kingdom Fungi [Internet, updated 2011]. Available from: http://bioref.lastdragon.org/index2.html

[19] Landvik, S., Eriksson, O.E. & Berbee, M.L. (2001). *Neolecta* - a fungal dinosaur? Evidence from beta-tubulin amino acid sequences. *Mycologia*. 2; 93, 6, 1151-63.

[20] Liu, Y.J. & Hall, B.D. (2004). Body plan evolution of ascomycetes, as inferred from an RNA polymerase II phylogeny. *Proceedings of the National Academy of Sciences of the United States of America,* 101, 13, 4507-4512.

[21] Moncalvo, J.M. (2005). Molecular systematics - major fungal phylogenetic groups and fungal species concepts. In: Xu, J.P., editor. *Evolutionary Genetics of Fungi*. Norfolk: Horizon Scientific Press, 2005; pp.1-33.

[22] Hirt, RP; Logsdon, JM jr; Healz, B.; Dorey, MW; Ford-Doolittle W; Embley, TM. Microsporidia are related to Fungi: evidence from the largest subunit of RNA polymerase II and other proteins. *Proceedings of the National Academy of Sciences of the United States of America,* 1999 Jan 19;96(2):580-5.

[23] *Keeling*, P.J. (2003). Congruent evidence from alpha-tubulin and beta-tubulin gene phylogenies for a zygomycete origin of *microsporidia. Fungal Genetics and Biology, 38, 3, 298-309.*

[24] Gill, E.E. & Fast, N.M. (2006). Assessing the microsporidia-fungi relationship: combined phylogenetic analysis of eight genes. *Gene*, 375,103-9.

[25] Liu, Y.J., Hodson, M.C. & Hall, B.D. (2006). Loss of the flagellum heppened only once in the fungal lineage: phylogenetic structure of kindom Fungi inferred from RNA polymerase II subunit genes. *BMC Evolutionary Biology,* 6, Article No. 74. Available from: www.biomedcentral.com/1471-2148/6/74.

[26] Schüßler, A., Schwarzott, D. & Walker, C. 82001). A new fungal phylum, the Glomeromycota: phylogeny and evolution. *Mycological Research,* 105, 12, 1413-1421.

[27] James, T.Y., Kauff, F., Schoch, C.L., Matheny, P.B., Hofstetter, V., Cox, C.J., Celio, G., Gueidan, C., Fraker , E, Miadlikowska, J., Lumbsch, H.Th., Rauhut, A., Reeb, V., Arnold, A.E., Amtoft , A., Stajich , J.E., Hosaka , K., Sung , G.-H., Johnson, D., O'Rourke, B., Crockett, M., Binder, M., Curtis, J.M., Slot, J.C., Wang, Zh., Wilson, A.W., Schüßler, A., Longcore, J.E., O'Donnell, K., Mozley-Standridge, Sh., Porter, D., Letcher , P.M., Powell , M.J., Taylor , J.W., White, M.M., Griffith, G.W., Davies, D.R., Humber, R.A., Morton, J.B., Sugiyama, J., Rossman, A.Y., Rogers, J.D., Pfister' D.H., Hewitt, D., Hansen, K., Hambleton, S, Shoemaker, R.A., Kohlmeyer, J., Volkmann-Kohlmeyer, B., Spotts, R.A., Serdani, M., Crous, P.W., Hughes, K.W., Matsuura, K., Langer, E., Langer, G., Untereiner, W.A., Lücking, L., Büdel, B., Geiser, D.M., Aptroot, A., Diederich, P., Schmitt, I., Schultz, M., Yahr, R., Hibbett, D.S., Lutzoni, F., McLaughlin, D.J., Spatafora, J.W. & Vilgalys, R. (2006). Reconstructing the early evolution of the fungi using a six gene phylogeny. *Nature*. 443, 7113, 818–822.

[28] Schaffer, R.L. (1975). The major groups of Basidiomycetes. *Mycologia*. 66, 1-18.

[29] Sugiyama, J. (1998). Relatedness, phylogeny and evolution of the fungi. *Mycoscience*, 39, 487-511.

[30] Tehler, A., Little, D.P. & Farris, J.S. (2003). The full-length phylogenetic tree from 1551 ribosomal sequences of chitinous fungi, Fungi. *Mycological Research,*;107, Pt 8, 901-916.

[31] Tanabe, Y., Watanabe, M.M. & Sugiyama, J. (2005). Evolutionary relationships among basal fungi (Chytridiomycota and Zygomycota): Insights from molecular phylogenetics. *The Journal of General and Applied Microbiology*, 51, 5, 267-276.

[32] Hibbett, D.S., Binder, M., Bischoff, J.F., Blackwell, M., Cannon, P.F., Eriksson, O.E., Huhndorf, S., James, T., Kirk, P.M., Lücking, R., Thorsten Lumbsch, H., Lutzoni, F., Matheny, P.B., McLaughlin, D.J., Powell, M.J., Redhead, S., Schoch, C.L., Spatafora, J.W., Stalpers, J.A., Vilgalys, R., Aime, M.C., Aptroot, A., Bauer, R., Begerow, D., Benny, G.L., Castlebury, L.A., Crous, P.W., Dai, Y.C., Gams, W., Geiser, D.M., Griffith, G.W., Gueidan, C., Hawksworth, D.L., Hestmark, G., Hosaka, K., Humber, R.A., Hyde, K.D., Ironside, J.E., Kõljalg, U., Kurtzman, C.P., Larsson, KH., Lichtwardt, R., Longcore, J., Miadlikowska, J., Miller, A., Moncalvo, J.M., Mozley-Standridge, S., Oberwinkler, F., Parmasto, E., Reeb, V., Rogers, J.D., Roux, C.,

Ryvarden, L., Sampaio, J.P., Schüssler, A., Sugiyama, J., Thorn, R.G., Tibell, L., Untereiner, W.A., Walker, C., Wang, Z., Weir, A., Weiss, M., White, M.M., Winka, K., Yao, Y.J., Zhang, N. (2007). A higher-level phylogenetic classification of the Fungi. *Mycological Research*, 111, Pt 5, 509-47.

[33] James, TY; Letcher, PM; Longcore, JE; Mozley-Standridge, SE; Porter, D; Powell, MJ; Griffith, G.W. & Vilgalys, R. (2007). A molecular phylogeny of the flagellated fungi (Chytridiomycota) and description of a new phylum (Blastocladiomycota). *Mycologia*, 98, 6, 860–871.

[34] Carlile, M.J., Watkinson, S.C. & Graham, W.G. (2006). *The Fungi*. Second edition. Amsterdam: Elsevier Academic Press; 2006.

[35] Harley, J.L. (1971). Fungi in ecosystems. *Journal of Applied Ecology*, 8, 3, 1-16.

[36] Kendrick, B. (2003). The Fifth Kingdom. Third edition [CD-ROM]. Sidney B.C.: Mycologue Publications; 2003.

[37] Malloch, D. (2009). The Mycology Web Pages. Saint John: New Brunswick Museum; Available from: http://website.nbm-mnb.ca/mycologywebpages/MycologyWebPages.html

[38] Werner, P. (2005). Mycodigest: Of Mammals and Mutualists. *The Mycena News* [internet]. 56, 12, 1-4. Available from: http://www.mssf.org/mycena-news/pdf/0512mn.pdf

[39] Schwarze, F. (2007). Wood decay under the microscope. *Fungal Biology Reviews*, 21, 4, 133–170.

[40] Huhndorf, S.M., Miller, A.N. & Fernández, F.A. (2004). Molecular systematics of the Sordariales: the order and the family Lasiosphaeriaceae redefined. *Mycologia*, 96, 368-387.

[41] Ralph, J.P. & Catcheside, D.E.A. (2002). Biodegradation by white-rot fungi. In: Osiewacz H.D., editor. *The Mycota, A Comprehensive Treatise on Fungi as Experimental Systems and Applied research,* Vol. X. Berlin: Springer-Verlag; 2002; pp. 303-326.

[42] Reid, I.D. (1995). Biodegradation of lignin. *Canadian Journal of Botany,* 78, S1011-S1018

[43] Wang, L., Yan, W., Chen, J., Huang, F. & Gao, P. (2008). Function of the iron-binding chelator produced by *Coriolus versicolor* in lignin biodegradation. *Science in China Series C: Life Sciences*, 51, 3, 214-221.

[44] Martínez, A.T., Speranza, M., Ruiz-Dueñas, F.J., Ferreira, P., Camarero, S., Guillén, F., Martínez, M.J., Gutiérrez, A. & del Río, J.C. (2005). Biodegradation of lignocellulosics: microbial, chemical, and enzymatic aspects of the fungal attack of lignin. *International Microbiology*, 8, 3, 195-204.

[45] Moredo, N., Lorenzo, M., Domínguez, A., Moldes, D., Cameselle, C. & Sanroman, A. (2003). Enhanced ligninolytic enzyme production and degrading capability of *Phanerochaete chrysosporium* and *Trametes versicolor*. *World Journal of Microbiology and Biotechnology*, 9, 665-669.

[46] Srebotnik, E., Messner, K. & Foisner, R. (1988). Penetrability of White Rot-Degraded Pine Wood by the Lignin Peroxidase of *Phanerochaete chrysosporium*. *Applied Environmental Microbiology*. 54, 11, 2608-2614.

[47] Tanaka, H., Itakura, S. & Enoki, A. (1999). Hydroxyl radical generation by an extracellular low-molecular-weight substance and phenol oxidase activity during wood

degradation by the white-rot basidiomycete *Phanerochaete chrysosporium. Holzforschung*, 53, 21-28.

[48] Levi, M.P. & Cowling, E.B. (1969). Role of nitrogen in wood deterioration. VII. Physiological adaptation of wood-destroying and other fungi to substrates deficient in nitrogen. *Phytopathology*, 59, 460-468.

[49] Gilbertson, RL. (1981). North American wood-rotting fungi that cause brown rots. *Mycotaxon*, 12, 2, 372-416.

[50] Bagley, S.T. & Richter, D.L. (2001). Biodegradation by brown rot fungi, Chapter 16. In: Osiewacz, H.D, editor. *The Mycota, A Comprehensive Treatise on Fungi as Experimental Systems and Applied research,* Vol. X. Berlin: Springer-Verlag; 2001; pp. 327-341.

[51] Jurgens, J.A. (2010). Fungal biodiversity in extreme environments and wood degradation potential [PhD thesis]. Hamilton: University of Waikato; Available from: University of Waikato Library.

[52] Matavulj, M. (1995). Lignicolous macrofungi of some plant associations of the Vrshachke planine mountains. *Matica Srpska Proceedings for Natural Sciences*, 88, 59-64.

[53] Matavuly, M., Karaman, M., Bokorov, M., Radnovicy, D. & Zeremski, Y. (1998). Lignicolous macrofungi of the Danube-Tisza-Danube Major Canal bank region of the Bezdan-Tisza-Klek section. *Tiszakutató Ankét; Mintázat és háttérmechanizmus a szünbiológiában; Szegedi ökológiai napok*, 98; 17.-19.11.1998, Kivonatok, Szeged, 1998. pp. 43-44.

[54] Matavulj, M., Karaman, M., Gojković, I. & Đurđević, S. (2005). Lignicolous macrofungi of the Bardača floodplain region. *Matica Srpska Proceedings for Natural Sciences*, 109, 161-167.

[55] Škrbić, B., Milovac, S. & Matavulj, M. (2011). Multielement profiles of soil, road dust, tree bark and wood-rotten fungi collected at various distances from high-frequency road in urban area. *Ecological Indicators*. In print: doi:10.1016/j.ecolind.2011.05.023. Available from: *www.sciencedirect.com/science/article/pii/S1470160X11001609*

[56] Matavulj, M., Erbežnik, M. & Bokorov, M. (1983). Poplar stump decaying fungi. In: Klug, M.J. & Reddy, C.A., editors. *Current perspectives in microbial ecology: proceedings of the Third International Symposium on Microbial Ecology; 1983 August 07-12; East Lansing, Michigan State University, U.S.A*. Publisher: Washington, D.C.: American Society for Microbiology; 1984. K.14, bnp. 56. *ISBN:* 0914826603: 9780914826606

[57] Todorović, R., Matavulj, M., Grujić, S. & Petrović, J. (1986). Some properties of xylanase and cellulolytic enzyme complex of *Trichoderma harzianum* C1-R1. *Mikrobiologija*, 23, 1, 33-38.

[58] Todorović, R., Grujić, S. & Matavulj, M. (1987). Effect of reaction end-products on the activity of cellulolytic enzymes and xylanase of *Trichoderma harzianum. Microbios Letters*, 36, 113-119.

[59] Gomes, I., Gomes, J., Steiner, W. & Esterbauer, H. (1992). Production of cellulase and xylanase by a wild strain of *Trichoderma viride. Applied Microbiology and Biotechnology,* 36, 5, 701-707.

[60] Nidetzky, B., Steiner, W., Hayn, M. & Claeyssens, M. (1994). Cellulose hydrolysis by the cellulases from *Trichoderma reesei*: a new model for synergistic interaction. *Biochemical Journal*, 298, 705-710.

[61] Kubicek, C.P. & Harman, G.E. (1998). *Trichoderma* and *Gliocladium,* basic biology, taxonomy and genetics. *Vol. 1 of Trichoderma* and *Gliocladium,* Gary E. Harman, *ISBN* 0748408061, 9780748408061; London: Taylor & Francis, 1998.

[62] Matavulj, M., Erbežnik, M., Bokorov, M., Todorović, R. & Gantar, M. (1984). Cellulase activity of *Gliocladium sp.* C2-R1 isolated from poplar stump. In: Klug, M.J. & Reddy, C.A., editors. *Current perspectives in microbial ecology: proceedings of the Third International Symposium on Microbial Ecology; 1983 August 07-12; East Lansing, Michigan State University, U.S.A.* Publisher: Washington, D.C.: American Society for Microbiology; 1984. K.15, bnp. 56. *ISBN:* 0914826603: 9780914826606

[63] Todorović, R., Grujić, S. & Matavulj, M. (1990). Purfication and characterization of ß-glucosidase from *Gliocladium virens*. *Jurnal of the Serbian Chemical Soc*iety, 55, 7, 389-393.

[64] Soni, R., Sandhu, D.K. & Soni, S.K. (1999). Localisation and optimisation of cellulase production in *Chaetomium erraticum. Journal of Biotechnol*ogy, 73, 1, 43-51.

[65] Wood, T.M. & McCrae, S.I. (1977). Cellulase from *Fusarium solani*: purification and properties of the C_1 component. *Carbohydrate Research*, 57, 117-133.

[66] Van Wyk, J.P.H. (1999). Saccharification of paper products by cellulase from *Penicillium funiculosum* and *Trichoderma reesei. Biomass Bioenergy*, 16, 3, 239-242.

[67] Bayer, E.A., Chanzy, H., Lamed, R. & Shoham, Y. (1998). Cellulose, cellulases and cellulosomes. *Current Opinion in Structural Biology*, 8, 5, 548-557.

[68] Ljungdahl, L.G. (2008). The cellulase/hemicellulase system of the anaerobic fungus *Orpinomyces* PC-2 and aspects of its applied use. *Annals of the New York Academy of Sciences*, 1125, 308-321.

[69] Sanchez, C. (2009). Lignocellulosic residues: biodegradation and bioconversion by fungi. *Biotechnology Advances,* 27, 2, 185-94.

[70] Weng, J.K., Li, X., Bonawitz, N.D., Chapple, C. (2008). Emerging strategies of lignin engineering and degradation for cellulosic biofuel production. *Current Opinion in Biotechnology*, 19, 2, 166-172.

[71] Cantarel, B.L., Coutinho, P.M., Rancurel, C., Bernard, T., Lombard, V. & Henrissat, B. (2009): The Carbohydrate-Active EnZymes database (CAZy): an expert resource for Glycogenomics. *Nucleic Acids Research,* 37 (Database issue suppl. 1): D233-D238.

[72] Sandgren, M., Stahlberg, J. & Mitchinson, C. (2005). Structural and biochemical studies of GH family 12 cellulases: improved thermal stability, and ligand complexes. *Progress in Biophysics and Molecular Biology*, 89, 3, 246-291.

[73] Kumar, R., Singh, S. & Singh, O.V. (2008). Bioconversion of lignocellulosic biomass: biochemical and molecular perspectives. *Journal of Industrial Microbiology and Biotechnology*, 35, 377-391.

[74] Dashtban, M., Schraft, H. & Qin, W. (2009). Fungal Bioconversion of Lignocellulosic Residues; Opportunities & Perspectives. *International Journal of Biological Sciences*, 5, 6, 578-595.

[75] Percival-Zhang, Y.H., Himmel, M.E. & Mielenz, J.R. (2006). Outlook for cellulase improvement: screening and selection strategies. *Biotechnology Advances*, 24, 5, 452-481.

[76] Baldrian, P. & Valaskova, V. (2008). Degradation of cellulose by basidiomycetous fungi. *FEMS Microbiology Reviews*, 32, 3, 501-521.

[77] Foreman, P.K., Brown, D., Dankmeyer, L., Dean, R., Diener, S., Dunn-Coleman, N.S., Goedegebuur, F., Houfek, T.D., England., Kelley, A.S., Meerman, H.J., Mitchell, T., Mitchinson, C., Olivares, H.A., Teunissen, P.J., Yao, J. & Ward, M. (2003). Transcriptional regulation of biomass-degrading enzymes in the filamentous fungus *Trichoderma reesei*. *Journal of Biological Chemistry*, 278, 34, 31988-31997.

[78] Divne, C., Stahlberg, J., Teeri, T.T. & Jones, T.A. (1998). High-resolution crystal structures reveal how a cellulose chain is bound in the 50 A long tunnel of cellobiohydrolase I from *Trichoderma reesei*. *Journal of Molecular Biology,* 275, 2, 309-325.

[79] Rouvinen, J., Bergfors, T., Teeri, T., Knowles, J.K. & Jones, T.A. (1990). Three-dimensional structure of cellobiohydrolase II from *Trichoderma reesei*. *Science*, 249, 4967, 380-386.

[80] Dan, S., Marton, I., Dekel, M., Bravdo, B.A., He, S., Withers, S.G., Shoseyov, O. (2000). Cloning, expression, characterization, and nucleophile identification of family 3, *Aspergillus niger* beta-glucosidase. *Journal of Biological Chemistry*, 275, 4973-4980.

[81] Morais, H., Ramos, C., Matos, N., Forgács, E., Cserháti, T., Almeida, V., Oliveira, J., Darwish, Y., Ilés, Z. (2002). Liquid chromatographic and electrophoretic characterisation of extracellular beta-glucosidase of *Pleurotus ostreatus* grown in organic waste. *Journal of chromatography. B, Analytical technologies in the biomedical and life sciences*, 770, 1-2, 111-119.

[82] Ishikawa, E., Sakai, T., Ikemura, H., Matsumoto, K. & Abe, H. (2005). Identification, cloning, and characterization of a *Sporobolomyces singularis* beta-galactosidase-like enzyme involved in galacto-oligosaccharide production. *Journal of Bioscience and Bioengineering,* 99, 4, 331-339.

[83] Cao, W.G. & Crawford, D.L. (1993). Purification and some properties of beta-glucosidase from the ectomycorrhizal fungus *Pisolithus tinctorius* Strain Smf. *Canadian Journal of Microbiology,* 39, 125-129.

[84] Evans, C.S. (1985). Properties of the beta-D-glucosidase (cellobiase) from the wood-rotting fungus, *Coriolus versicolor*. *Applied Microbiology and Biotechnology,* 22, 128-131.

[85] Cai, Y.J., Chapman, S.J., Buswell, J.A. & Chang, S.T. (1999). Production and distribution of endoglucanase, cellobiohydrolase, and beta-glucosidase components of the cellulolytic system of *Volvariella volvacea*, the edible straw mushroom. *Applied and Environmental Microbiology,* 65, 2, 553-559.

[86] Saha, B.C. (2003). Hemicellulose bioconversion. *Journal of Industrial Microbiology and Biotechnology,* 30, 5, 279-291.

[87] Shallom, D. & Shoham, Y. (2003). Microbial hemicellulases. *Current Opinions in Microbiology*, 6, 3, 219-228.

[88] Bourne, Y. & Henrissat, B. (2001). Glycoside hydrolases and glycosyltransferases: families and functional modules. *Current Opinion in Structural Biology*, 11, 5, 593-600.

[89] Guillén, F., Martínez, A.T. & Martínez, M.J. (1992). Substrate specificity and properties of the aryl-alcohol oxidase from the ligninolytic fungus *Pleurotus eryngii*. *European Journal of Biochemistry,* 209, 2, 603-611.

[90] Suzuki, M.R., Hunt, C.G., Houtman, C.J., Dalebroux, Z.D. & Hammel, K.E. (2006). Fungal hydroquinones contribute to brown rot of wood. *Environmental Microbiology,* 8, 12, 2214-2223.

[91] Call, H.P. & Mücke, I. (1997). History, overview and applications of mediated lignolytic systems, especially laccase-mediator-systems (Lignozyme®-process). *Journal of Biotechnology*, 53, 163-202.

[92] Renganathan, V., Usha, S.N. & Lindenburg, F. (1990). Cellobiose oxidizing enzymes from the lignocellulose-degrading basidiomycete *Phanerochaete chrysosporium* - interaction with microcrystalline cellulose. *Applied Microbiology and Biotechnology*, 32, 609-613.

[93] Henriksson, G., Johansson, G. & Pettersson, G. (2000). A critical review of cellobiose dehydrogenases. *Journal of Biotechnology,* 78, 93-113.

[94] Baminger, U., Subramaniam, S.S., Renganathan, V. & Haltrich, D. (2001). Purification and characterization of cellobiose dehydrogenase from the plant pathogen *Sclerotium (Athelia) rolfsii. Applied and Environmental Microbiology*, 67, 4, 1766-1774.

[95] Yang, W., Liu, J., Wang, W., Zhang, Y. & Gao, P. (2004). Function of a low molecular peptide generated by cellulolytic fungi for the degradation of native cellulose. *Biotechnology Letters,* 26, 23, 1799-1802.

[96] Enoki, A., Hirano, T. & Tanaka, H. (1992). Extracellular substance from the brown rot basidiomycete *Gloeophyllum trabeum* that produces and reduces hydrogen peroxide. *Material und Organismen, (Berlin),* 27, 247-61.

[97] Tanaka, H., Yoshida, G., Baba, Y., Matsumura, K., Wasada, H., Murata, J., Agawa, M., Itakura, Sh. & Enoki, A. (2007). Characterization of a hydroxyl-radical-producing glycoprotein and its presumptive genes from the white-rot basidiomycete *Phanerochaete chrysosporium. Journal of Biotechnology,* 128, 500-511.

[98] Esposito, E., Canhos, P.V. & Duran, N. (1999). Screening of lignin degrading fungi for removal of color from kraft mill waste water with no additional carbon source. *Biotechnology Letters,* 13, 8, 571–576.

[99] Gadd, G.M. (1993). Interactions of fungi with toxic metals. *New Phytologist*, 124, 25-60.

[100] Javaid, A., Bajwa, R., Shafique, U. & Anwar, J. (2011). Removal of heavy metal by adsorption on *Pleurotus ostreatus*. *Biomass and Bioenergy*, 35, 5, 1675-1682.

[101] Baldrian, P. (2003). Interactions of heavy metals with white-rot fungi. *Enzyme and Microbial Technology*, 32, 78-91.

[102] Gabriel, J., Rychlovsky, P. & Krenželok, M. (1995). Beryllium content in some wood-rotting fungi in Czech Republic. *Toxicological and Environmental Chemistry*, 50, 233-236.

[103] Maksimović, M., Vojinović-Miloradov, M., Matavulj, M., Mihajlović, B. & Buzarov, D. (1997). Biodegradation of polychlorobiphenyls (PCB) by microfungi in the water Danube. *Limnologische Berichte Donau 1997*, *32. Konferenz der IAD der SIL,* Wien, Österreich; 1997. pp. 47-50.

[104] Novotný, C., Vyas, B.R., Erbanová, P., Kubátová, A. & Sasek, V. (1997). Removal of PCBs by various white rot fungi in liquid cultures. *Folia Microbiologica*, 42, 2, 136–140.

[105] Novotný, C., Erbanová, P., Sasek, V., Kubátová, A., Cajthaml, T., Lang, E., Krahl, J., Zadrazil, F. (1999). Extracellular oxidative enzyme production and PAH removal in soil by exploratory mycelium of white rot fungi. *Biodegradation*, 10, 3, 159–168.

[106] Ting, W.T.E., Yuan, S.Y., Wu, S.D. & Chang, B.V. (2011). Biodegradation of phenanthrene and pyrene by *Ganoderma lucidum*. *International Biodeterioration and Biodegradation*, 4, 65, 238-242.

[107] Covino, S., Cvancarova, M., Muyikar, M., Svoboda, K., D`annibale, A., Petruccioli, M., Federici, F., Kresinová, Z. & Cajthaml, T. (2010). An efficient PAH-degarading *Lentinus (Panus) tigrinus* strain: Effect of inoculum formulation and polluant bioavailability is solid matrices. *J Hazard Mater*. 183, 1-3, 669-676.

[108] Arun, A. & Eyini, M. (2011). Comparative studies on lignin and polycyclic aromatic hydrocarbons degradation by basidiomycetes fungi. *Bioresource Technol*. 102, 17, 8063-8070.

[109] Mimica-Dukic, N., Karaman, M., Matavulj, M., Pavkov, R. & Popovic, M. (2002). Lignicolous fungi as potential natural antioxidants. *Proceedings of the 50th Annual Congress of the Society of Medicinal Plant Research*. Barcelona, Spain, 2002 Sept 8-12. p. 126.

[110] Karaman, A.M., Mimica-Dukic, M.N., Knežević, P.N. & Matavulj, N.M. (2007). Antioxidative and antibacterial activity of some lignicolous basidiomicetous fungi from Serbia. *International Journal of Medicinal Mushr*ooms, 9, 3-4, 330-331.

[111] Karaman, M., Mimica-Dukić, N. & Matavulj, M. (2009). Lignicolous fungi from northern Serbia as natural sources of antioxidants. *Central European Journal of Biology*, 4, 3, 387-396.

[112] Karaman, M., Matavuly, M., Sepcic, K., Canadanovic-Brunet, J. & Kaisarevic, S. (2010). Antioxidative, hemolytic, cytotoxic and neurotoxic activities of several lignicolous fungi from Serbia. *Proceedings of the 9th International Mycological Congress, IMC9 - The Biology of Fungi*, Edinburgh, 1-6. August 2010.- P4.227. p 47. Available from: www.imc9.info.

[113] Engler, M., Anke, T. & Sterner, O. (1998). Production of Antibiotics by *Collybia nivalis, Omphalotus olearius,* a *Favolashia* and a *Pterula* species on natural substrates. *Zeitschrift für Naturforschung C. A Journal of Biosciences,* 53, 5-6, 318-324.

[114] Gunde-Cimerman, N. & Cimerman, A. (1995). *Pleurotus* fruiting bodies contain the inhibitor of HMG CoA reductase-lovastatin. *Experimental Mycology,* 19, 1, 1-6.

[115] Liu, G.-T. (1999). Recent advances in research of pharmacology and clinical application of *Ganoderma (*P. Karst) species *(Aphyllophoromycetdeae*) in China. *International Journal Medicinal Mushrooms,* 1, 63-67.

[116] Stadtler, M. & Sterner, O. (1998). Production of bioactive secondary metabolites in the fruit bodies of macrofungi as a response to injury. *Phytochemistry*, 49, 4, 1013-1019.

[117] Smania, Jr.A., Monache, F.D., Loguericio-Leite, C., Smania, E.F.A. & Gerber, A.L. (2001). Antimicrobial activity of Basidiomycetes. *International Journal Medicinal Mushrooms,* 3, 87-93.

[118] Karaman, M., Kaišarević, S., Somborski, J., Kebert, M. & Matavuly, M. (2009). Biological activities of the lignicolous fungus *Meripilus giganteus* (Pers.:Pers.) Karst. *Archives of Biological Sciences*, 61, 4, 353-361.

[119] Karaman, M., Jovin, E., Malbaša, R., Matavuly, M. & Popović, M. (2010). Medicinal and edible lignicolous fungi as natural sources of antioxidative and antibacterial agents. *Phytotherapy Research,*24, 10, 1473–1481.

[120] Molitoris, H.P. (1994). Mushrooms in Medicine. *Folia Microbiologica,* 39, 2, 91-98.

[121] Chang, S.T. (1999). Global impact of edible and medicinal mushrooms on human welfare in the 21st century: Nongreen revolution. *International Journal of Medicinal Mushrooms,* 1, 1-8.

[122] Wasser, S.P. & Weis, A.L. (1999). Medicinal Properties of Substances Occuring in Higher Basidiomycetes Mushrooms: Current Perspectives (Rewiew). *International Journal of Medicinal Mushrooms,* 1, 31-62.

[123] Paterson, R.R.M. (1985). *Ganoderma* – a therapeutic fungal biofactory. *Phytochemistry*, 67 1985-2001.

[124] Bargagli, R. (1990). Assessment of metal air pollution by epiphytic lichens: the incidence of crustal materials and possible uptake from substrate barks. *Studia Geobotanica*, 10, 97-103.

[125] Gordon, C.A., Herrera, R. & Hutchinson, T.C. (1995). The use of a common epiphytic lichen as a bioindicator of atmospheric inputs to two Venezuelan cloud forests. *Journal of Tropical Ecology*, 11, 1-26.

[126] Bennett, J.P. (1995). Abnormal chemical-element concentrations in lichens of Isle-Royale National-Park. *Environmental and Experimental Botany*, 35, 3, 259-277.

[127] Somers, E. (1961). The fungitoxicity of metal ions. *Annals of Applied Biology*, 49, 246-253.

[128] Kalac, P. & Svoboda, L. (2000). A review of trace elements concentration in edible mushrooms. *Food Chemistry*, 69, 273–281.

[129] Kalac, P. (2001). A review of edible mushroom radioactivity. *Food Chemistry*, 75, 29-35.

[130] Svoboda, L., Zimmermannova, K. & Kalac, P. (2000). Concentrations of mercury, cadmium, lead and copper in fruiting bodies of edible mushrooms in an emission area of copper smelter and a mercury smelter. *Science of the Total Environment,* 246, 1, 61-67.

[131] Govedarica, M., Milošević, N., Jarak, M. (1997). Heavy metals and soil microorganisms. In book: Kastori. R., editor. *Heavy metals in the environment.* [Teški metali u životnoj sredini] Novi Sad: Institute of Field and Vegetable Crops; 1997. (In Serbian).

[132] Michelot, D., Siobud, E., Dore, J-C., Viel, C., Poirier, F. (1998). Update on metal content profile in mushrooms – toxicological implications and tentative approach to the mechanisms of bioaccumulation. *Toxicon*. 36, 12, 1997-2012.

[133] Chiu, S.W., Chan, Y.H., Law, S.C., Cheung, K.T. & Moore, D. (1998). Cadmium and manganese in contrast to calcium reduce yield and nutritional values of the edible mushroom *Pleurotus pulmonarius*. *Mycological Research,* 102, 4, 449-57.

[134] Rühling, Å., Bååth, E., Nordgren, A. & Söderström, B. (1984). Fungi in metal contaminated soil nera the Gusum Brass Mill, Sweden. *Ambio*. 13, 1, 34-36.

[135] Bogdanović, D., Ubavić, M. & Hadžić, V. (1997). Heavy metals in soil. In book: Kastori, R. Editor, *Heavy metals in the environment.* Novi Sad: Institute of field and vegetable crops; 1997. (In Serbian)

[136] Melhorn, H., Seufert, G., Schmidt, A. & Kunert, K.J. (1986). Effect of SO_2 and O_3 on production of antioxidants in conifers. *Plant Physiol*ogy, 82, 1, 336–338.

[137] Babich, H. & Stotzky, G. (1977). Effect of cadmium on fungi and on interactions between fungi and bacteria in soil: influence of clay minerals and pH. *Applied and Environmental Microbiology,* 33, 5, 1059–1066.

[138] Ruhling, A. & Tyler, G. (1973). Heavy metal deposition in Scandinavia. *Water Air & Soil Pollution*, 2, 445–455.

[139] Ross, I.S. (1975). Some effects of *heavy metals* on fungal cells. *Transactions of British Mycological Society*, 64, 175-193.

[140] Cotton, F.A. & Wilkinson, G. (1988). *Advanced Inorganic Chemistry*. Fifth edition. New York: Wiley-Interscience; 1988.

[141] Baker, A.J.M. (1981). Accumulators and excluders-strategies in the response of plants to heavy metals. *Journal of Plant Nutrition,* 3, 643–654.

[142] Munger, K., Lerch, K. & Tschierpe, H.J. (1982). Metal accumulation in *Agaricus bisporus*: influence of Cd and Cu on growth and tyrosinase activity. *Experientia*, 38, 1039-1041.

[143] Meisch, H.U., Beckmann, I. & Schmitt, J.A. (1983). A new cadmium-binding phosphoglycoprotein, cadmium-mycophosphatin, from the mushroom *Agaricus macrosporus*. *Biochimica and Biophysica Acta*, 745, 3, 259–266.

[144] Mehra, R.K. & Winge, D.R. (1991). Metal ion resistance in fungi: molecular mechanisms and their related expression. *Journal of Cellular Biochemistry*, 45, 1, 30-40.

[145] Antonovics, J., Bradshaw, A.D. & Turner, R.G. (1971). Heavy metal tolerance in plants. *Advances in Ecological Research,* 7, 1–85.

[146] Foy, C.D., Chaney, R.L. & White, M.C. (1978). The physiology of metal toxicity in plants. *Annual Review of Plant Biology*. 29, 511-566.

[147] Garbaye, J., Kabre, A., Le Tacon, F., Piou, D. & Mousain, D. (1979). Mineral fertilization and fructification of higher fungi in a beech forest. *Annales des Sciences Forestières*. 13, 1, 34-36.

[148] Hoiland, K. (1995). Reaction of some decomposer basidiomycetes to toxic elements. *Nordic Journal of Botany*. 15, 3, 305-318.

[149] Shimada, M., Akamatsu, Y., Tokimatsu, T., Mii, K. & Hattori, T. (1997). Possible biochemical roles of oxalic acids as a low molecular weight compound involved in brown-rot and white–rot wood decay. *Journal of Biotechnology*, 53, 103-113.

[150] Winkelmann, G., Van der Helm, D. & Neilands, J.B. (1987). *Irontransport in microbes, plants and animals*. Weinheim: VCRVerlagsgesellschaft; 1987.

[151] Winkelmann, G. (1992). Structures and functions of fungal siderophores containing hydroxamate and complexone type iron binding ligands. *Mycological Research*, 96, 529-534.

[152] Peberdy, J.F. (1990). Fungal cell walls - a review. In: Kuhn P.J., Trinci A.P.J., Jung M.J., Coosey M.W., Copping L.E., editors. *Biochemistry of cell walls and membranes in fungi*. Berlin: Springer-Verlag; 1990, pp, 5-30.

[153] Sakaguchi, T. & Nakajima, A. (1982). Recovery of uranium by chitin phosphate and chitosan phosphate. In: Mirano S., Tokura S., editors. *Chitin and chitosan*. Tottori: Japanese Society of Chitin and Chitosan, 1982, pp, 177-182.

[154] Zunino, H. & Martin, J.P. (1977). Metal binding organic macromolecules in soil. I. Hypothesis interpreting the role of soil organic matter in the translocation of metal ions from rocks to biological systems. *Soil Science*, 123, 65-76.

[155] Wang, J. & Chen, C. (2009). Biosorbents for heavy metals removal and their future. *Biotechnology Advances,* 27, 2, 195-226.

[156] Kastori, R., Petrović, N. & Arsenijević-Maksimović, I. (1997). Heavy metals and plants. In: Kastori, R. editor. *Heavy metals in the environment*. [Teški metali u životnoj sredini] Novi Sad: Institute of Field and Vegetable Crops; 1997. (In Serbian)

[157] Grill, E., Loffler, S., Winnacker, E.-L. & Zenk, M.H. (1989). Phytochelatins, the heavy metal-binding peptides of plants, are synthesized from glutathione by a specific y-glutamylcysteine dipeptidyl transpeptidase (phytochelatin synthase). *Proceedings of the National Academy of Sciences of the United States of America,* 86, 6838-6842.

[158] Kihn, J.C., Dassargues, C.M. & Mestdagh, M.M. (1988). Preliminary ESR study of Mn(II) retention by the yeast *Saccharomyces*. *Canadian Journal of Microbiology,* 34, 11, 1230-1234.

[159] Gadd, G.M. (1992). Microbial control of heavy metal pollution. In: Fry, J.C., Gadd, G.M., Herbert, R.A., Jones, C.W. & Watson-Craik, I., editors. *Microbial control of environmental pollution*. Cambridge: Cambridge University Press, pp, 59-88.

[160] Macaskie, L.E. & Dean, A.C.R. (1990). Trimethyl lead degradation by free and immobilized cells of an *Arthrobacter* sp. and by the wood decay fungus *Phaeolus schweinitzii*. *Applied Microbiology and Biotechnology*, 33, 1, 81-87.

[161] Collins, P.J. & Dobson, A.D.W. (1997). Regulation of Laccase Gene Transcription in *Trametes versicolor*. *Applied and Environmental Microbiology*, 63, 9, 3444-3450.

[162] Breene, W.M. (1990). Nutritional and medicinal value of specialty mushrooms. *Journal of Food Protection,* 53, 10, 883-894.

[163] Elmastas, M.. Isildak, O. Turkekul, I. & Temur, N. (2007). Determination of antioxidant activity and antioxidant compounds in wild edible mushrooms. *Journal of Food Composition and Analysis*, 20, 337-345.

[164] Barros, L., Calhelha, R.C., Vaz, J.A., Ferreira, I.C.F.R., Baptista, P. & Esteviinho, L.M. (2007). Antimicrobial activity and bioactive compounds of Portugese wild edible mushrooms methanolic extracts. *European Food Research and Technology*, 225, 151-156.

[165] Lau, C.B.S., Ho, C.Y., Kim, C.F., Leung, K.N., Fung, K.P., Tse, T.F., Chan, H.H.L. & Chow, M.S.S. (2004). Cytotoxic acyivities of *Coriolus versicolor* (Yunzhi) extract on human leukemia and lymphoma cells by induction of apoptosis. *Life Sciences*, 75, 7, 797-808.

[166] Campos, J.A., Tejera, N.A. & Sanchez, C.J. (2009). Substrate role in the accumulation of heavy metals in sporocarps of wild fungi. *BioMetals*, 22, 5, 835-841.

[167] Sarikurkcu, C., Copur, M., Yildiz, D. & Akata, I. (2011). Metal concentration of wild edible mushrooms in Soguksu National Park in Turkey. *Food Chemistry*, 128, 731-734.

[168] Al Sajegh Petkovšek, S., Poličnik, H., Ramšak, R., Mavec, M. & Pokorny, B. (2010). Ecological remediation of the Šoštanj thermal power plant with respect to sustainable development of the Šalek valley, Slovenia. *Thermal Science*, 4, 3, 773-782.

[169] Rühling, A. & Söderström, B. (1990). Changes in fruitbody of mycorrhizal and litter decomposing macromycetes in heavy metal polluted coniferous forests in North Sweden. *Water, Air, & Soil Pollution*, 49, 375-387.

[170] Melgar, M.I., Alonso, J., Pérez-López, M. & Garcia, M.A. (1998). Influence of some factors in toxicity and accumulation of cadmium from edible wild macrofungi in NW Spain. *Journal of Environmental Science and Health*, B33, 4, 439-455.

[171] Fellner, R. & Pešková, V. (1995). Effects of industrial pollutants on ectomycorrhizal relationships in temperate forests. *Canadian Journal of Botany*, 73, 1, 1310-1315.

[172] Byrne, A.R., Ravnik, V. & Kosta, L. (1976). Trace elements concentrations in higher fungi. *Science and the Total Environment*, 6, 1, 65-78.

[173] Vetter, J. (1994). Data on arsenic and cadmium contents of some common mushrooms. *Toxicon*, 32, 1, 11-15.

[174] Malinowska, E., Szefer, P. & Falandysz, J. (2004). Metals bioaccumulation by bay bolete, *Xerocomus badius*, from selected sites in Poland. *Food Chemistry*, 84, 3, 405-416.

[175] Rudawska, M. & Leski, T. (2005). Macro- and microelements contents in fruiting bodieis of wild mushrooms from the Notecka forest in west-central Poland. *Food Chemistry*, 92, 3, 499-506.

[176] Karaman, M. & Matavuly M. (2005). Heavy metals and macroelements in some macrofungi of National Park Frushka Gora (Serbia). *Annals of the Faculty of Engineering Hunedoara, Romania*, 1, 69-76.

[177] Yoshida, S. & Muramatsu, Y. (1997). Determination of major and trace elements in mushroom, plant and soil samples collected from Japanese forests. *International Journal of Environmental Analitical Chemistry*, 67, 49-58.

[178] Yoshida, S. & Muramatsu, Y. (1998). Concentrations of alkali and alkaline earth elements in mushrooms and plants collected in Japanese pine forest, and their relationship with ^{137}Cs. *Journal of Environmental Radioactivity*, 41, 2, 183-205.

[179] Turkekul, I., Elmastas, M. & Tuzen, M. (2004). Determination of iron, copper, manganese, zinc, lead and cadmium in mushroom samples from Tokat, Turkey. *Food Chemistry*, 84, 389-392.

[180] Kalač, P., Wittingerová, M. & Stašková, I. (1989). The contents of seven biogenic trace elements in edible mushrooms. *Potravinarske Vedy*, 7, 131–136. (in Czech).

[181] Kalač, P., Nižnanská, M, Bevilaqua, D. & Stašková, I. (1996). Concentrations of mercury, copper, cadmium and lead in fruiting bodies of edible mushrooms in the vicinity of a mercury smelter and a copper smelter. *Science and the Total Environment*. 177, 1-3, 251–258.

[182] Svoboda, L., Zimmermannová, K. & Kalac, P. (2000). Concentrations of mercury, cadmium, lead and copper in fruiting bodies of edible mushrooms in an emission area of a copper smelter and a mercury smelter. *Science and the Total Environment*, 246, 1, 61-66.

[183] Karaman, A.M. & Matavulj, N.M. (2005). Macroelements and heavy metals in some lignicolous and tericolous fungi. *Matica Srpska Proceedings for Natural Sciences*, 108, 255-267.

[184] Ouzouni, P.K., Petridis, D., Koller, W.D. & Rifanakos, KA. (2009). Nutritional value and metal content of wild edible mushrooms collected from West Macedonia and Epirus, Greece. *Food Chemistry*, 115, 4, 1575-1580.

[185] Ita, B.N., Essien, J.P. & Ebong, G.A. (2006). Heavy metals in fruiting bodies of edible and non-edible mushrooms from the Niger delta region of Nigeria. *Journal of Agriculture and Social Sciences*, 2, 2, 1813-2235.

[186] Dix, J.N. & Webster, J. (1995). *Fungal ecology*. London: Chapman & Hall; 1995.

[187] Wilkinson, D.M. (1998). The evolutionary ecology of mycorrhizal networks. *Oikos*, 82, 2, 407-410.

[188] Brown, M.T. & Wilkins, D.A. (1985). Zinc tolerance of mycorrhizal Betula. *New Phytologist*, 99, 101-106.

[189] Kalač, P. (2001). A review of edible mushroom radioactivity. *Food Chemistry*, 75:29-35.

[190] Mattson, S.; Liden, K. (1975). Cesium-137 in carpets of the forest moss *Pleurozium schreberi:* 1961–1973. *Oikos*, 26, 323–327.

[191] Sugiyama, H., Shibata, H., Isomura, K. & Iwashima, K. (1994). Concentration of Radiocesium in Mushrooms and Substrates in the Sub-Alpine Forest of Mt. Fuji Japan. *Journal of the Food Hygienic Society of Japan*, 35, 1, 13-22

[192] Clint, G.M., Dighton, J. & Rees, S. (1991). Influx of ^{137}Cs into hyphae of basidiomycete fungi. *Mycological Research*, 95, 9, 1047-1051.

[193] Gabriel, J., Rychlovsky, P. & Krenželok, M. (1995). Beryllium content in some wood-rotting fungi in Czech Republic. *Toxicological and Environmental Chemistry*, 50, 233-236.

[194] Calmon, P., Thiry, Y., Zibold, G., Rantavaara, A. & Fesenko, S. (2009). Transfer parameter values in temperate forest ecosystems: a review. *Journal of Environmental Radioactivity*, 100, 9, 757-766.

[195] Heinrich, G., Muller, H.J., Oswald, K. & Gries, A. (1989). Natural and artificial radionuclides in selected styrian soils and plants before and after the reactor accident in Chernobyl. *Biochemie und Physiologie der Pflanzen*, 185, 55-67.

[196] Johanson, K.J., Bergström, R., Eriksson, O. & Erixon, A. (1994). Activity concentrations of ^{137}Cs in moose and their forage plants in mid-Sweden. *Journal of Environmental Radioactivity*, 22, 3, 251–267.

[197] Nikolova, I., Johanson, K.J. & Clegg, S. (2000). The accumulation of ^{137}Cs in the biological compartment of forest soils. *Journal of Environmental Radioactivity*, 47, 319–326.

[198] Steiner, M., Linkov, I. & Yoshida, S. (2002). The role of fungi in the transfer and cycling of radionuclides in forest ecosystems. *Journal of Environmental Radioactivity*, 58, 217-241.

[199] Smith, S.E. & Read, D.J. (1997). *Mycorrhizal Symbiosis*. Second edition. London: Academic Press Inc., 1997.

[200] Haselwandter, K. (1978). Accumulation of the radioactive nuclide ^{137}Cs in fruitbodies of basidiomycetes. *Health Physics*, 34, 6, 713-715.

[201] Elstner, E.F., Fink, R., Holl, W., Lengfelder, E. & Ziegler, H. (1987). Natural and Chernobyl-caused radioactivity in mushrooms, mosses and soil-samples of defined biotops in SW Bavaria. *Oecologia*, 73, 4, 553-558.

[202] Byrne, A.R. (1988). Radioactivity in Fungi in Slovenia, Yugoslavia, Following the Chernobyl Accident. *Journal of Environmental Radioactivity*, 6, 2, 177-183.

[203] Elstner, E.F., Fink, R., Holl, W., Lengfelder, E. & Ziegler, H. (1989). Radioactivity in mushrooms, mosses and soil samples of defined biotops in SW Bavaria - two years after "Chernobyl". *Oecologia*, 80, 173-177.

[204] Yoshida, S. & Muramatsu, Y. (1994). Concentartions of Radiocesium and Potassium in Japanese Mushrooms, *Environmental Science*, 7, 1, 63-70.

[205] Olsen, R.A., Joner, E. & Bakken, L.R. (1990). Soil fungi and the fate of radiocaesium in the soil ecosystem - a discussion of possible mechanisms involved in the radiocaesium accumulation in fungi, and the role of fungi as a Cs-sink in the soil. In: Desmet, G., Nassimbeni, P., Belli, M., editors. *Transfer of radionuclides in natural and semi-natural environments*. London: Elsevier Applied Science; 1990; pp, 657–663.

[206] Brunner, I., Frey, B. & Reisen, T.K. (1996). Influence of ectomycorrhization and cesium/potassium ratio on uptake and localisation of cesium in Norway spruce seedlings. *Tree Physiology*, 16, 8, 705-711.

[207] Nikolova, I., Johanson K.J. & Dahlberg A. (1997). Radiocaesium in Fruitbodies and Mycorrhizae in Ectomycorrhizal Fungi. *Journal of Environmental Radioactivity*, 37, 1, 115-125.

[208] Giovani, C., Nimis, P.L. & Padovani, R. (1989). Radiocaesium in macromycetes of N.E. Italy in the first three years after the Chernobyl accident. In: *Radiation protection, Selected topics, International Symposium,* Dubrovnik, 2-6 October 1989, pp, 410-415.

[209] Shand, C.A., Cheshire, M.V., Smith, S., Vidal, M. & Rauret, G. (1994). Distribution of radiocaesium in organic soils. *Journal of Environmental Radioactivity*, 23, 3, 285-302.

[210] Wasser, S.P. & Grodzinskaya, A.A. (1993). Content of Radionucleides in Macromycetes of the Ukraine in 1990-1991. In: Pegler D.N., Boddy L., Ing B., Kirk P.M., editors. *Fungi of Europe: Investigation, Recording and Conservation*. Kew: RBG; 1993; pp, 189-210.

[211] Melin, J., Wallberg, L. & Suomela, J. (1994). Distribution and retention of cesium and strontium in Swedish boreal forest ecosystems. *Science of the Total Environment*, 157, 93-105.

[212] Johansen, M.P., Hakonson, T.E., Whicker, F.W. & Breshears, DD. (2003). Pulsed redistribution of a contaminant following forest fire: cesium-137 in runoff. *Journal of Environmental Quality*, 32, 6, 2150-2157.

[213] Read, D.J. (1991). Mycorrhizas in ecosystems. *Experientia*, 47, 376–391.

[214] Read, D.J. (1991). Mycorrhizas in ecosystems-nature's response to the Law of the Minimum. In: Hawksworth D.L., editor. *Frontiers in mycology*. Wallingford: CAB International; 1991; pp, 101–130.

[215] Bazala, M.A., Golda, K. & Bystrzejewska-Piotrowska, G. (2008). Transport of radiocesium in mycelium and its translocation to fruitbodies of a saprophytic macromycete. *Journal of Environmental Radioactivity*, 99, 7, 1200-1202.

[216] Heinrich, G. (1992). Uptake and transfer factors of ^{137}Cs by mushrooms. *Radiation and Environmental Biophysics*, 31, 1, 39-49.

[217] Eckl, P., Hofmann, W. & Turk, R. (1986). Uptake of natural and man-made radionuclides by lichens and mushrooms. *Radiation* and Environmental *Biophysics*, 25, 1, 43-54.

[218] Oolbekkink, G.T. & Kuyper, T.W. (1989). Radioactive caesium from Chernobyl in fungi. *The Mycologist*, 3, 3-6.

[219] Heinrich, G. (1993). Distribution of radiocesium in the different parts of mushrooms. *Journal of Environmental Radioactivity*, 18, 3, 229-245.

[220] De Meijer, R.J., Aldenkamp, F.J. & Jansen, A.E. (1988). Resorption of cesium radionuclides by various fungi. *Oecologia*, 77, 268-272.

[221] Stanković, S., Stanković, A. & Krainčanić, M. (1993). Radioactive contamination of biosphaere of National Park Durmitor. *Elaborate, INEP*. 1993.

[222] Ruckert, G. & Diehl, J.F. (1987). *Anreicherung von Cäsium-137 und Cäsium-134 in 34 Pilzarten nach dem Reaktorunglück von Tschernobyl. Zeitschrift fur Lebensmitteluntersuchung und Forschung*, 185, 91-97.

[223] Karaman, M. (2002). Content of Macroelements and Heavy Metals in sporocarps of dominantly present Basidiomycotina fungi from the Frushka Gora Mountain and their antioxidative activity [*Master Degree Thesis*]. Novi Sad, Serbia: University of Novi Sad; 2002 (In Serbian, English Summary).

[224] Karaman, M., Matavulj, M. & Čonkić, Lj. (2002-2003). Radionuclides in sporocarps of medicinaly important mushrooms of the Frushka Gora hill. *Zbornik radova Prirodno-matematičkog fakulteta. Serija za biologiju.- Review of Research of Faculty of Sciences of Novi Sad, Biology Series,* Vol. 31-32, pp, 33-42.

[225] Karaman M., Matavulj M., Petrović O., Knežević P. & Simeunović J. (2010). Fungal interactions with haevy metals. *Proceedings of the 3rd International Scientific Conerence „Remediation 2010",* Serbian Chamber of Commerce, Belgrade, May, 11-12. 2010, pp, 65-73. (In Serbian, English summary).

[226] Muramatsu, Y., Yoshida, S. & Ogawa, M. (1994). Radicesium concentrations in mushrooms collected in Japan. *Journal of Environmental Radioactivity*, 22, 141-154.

[227] Cocchi, L., Vescovi, L., Petrini, L.E. & Petrini, O. (2006). Heavy metals in edible mushrooms in Italy. *Food Chemistry*, 98, 277–284.

In: Fungi: Types, Environmental Impact and Role in Disease ISBN: 978-1-61942-671-9
Editors: Adolfo Paz Silva and María Sol

Chapter XI

Arbuscular and Ectomycorrhiza Fungi: Useful Biological Tool to Promote Establishment of Exotic Trees in Arid and Semi-Arid African Areas

***Nathalie Diagne*[*1, 2, 3], *Diegane Diouf*[1, 2], *Sergio Svistoonoff*[3], *Aboubacry Kane*[1, 2], *Kandioura Noba*[1], *Laurent Laplaze*[2,3], *Claudine Franche*[3], *Didier Bogusz*[3] and *Robin Duponnois*[4]**

[1] Département de Biologie Végétale,
Université Cheikh Anta Diop (UCAD), BP, Dakar, Sénégal
[2] Laboratoire Commun de Microbiologie IRD/ISRA/UCAD,
Centre de Recherche de Bel Air, BP, Dakar, Sénégal
[3] Equipe Rhizogenèse, Unité mixte de Recherche Diversité et Adaptation et Développement des Plantes (DIADE),
Institut de Recherche pour le Développement (IRD), BP, France
[4] Laboratoire des Symbioses Tropicales et Méditerranéennes (LSTM),
Campus International de Baillarguet, Montpellier, Cedex, France

Abstract

The reduction of vegetation cover has considerably increased. A lot of efforts are being undertaken to repair degraded lands. In African countries, exotic plants were established in this purpose to rehabilitate affected ecosystems. These exotic plantations play a key role in improving soils fertility partly because they are able to establish a symbiotic relationship with mycorrhizal fungi which promote plant development and allow better resistance of plant to stresses. Some of these reforestation programs were not successful in certain zones but this situation is overcome when these exotic trees are transferred together with appropriated mycorrhizal fungi from their zone of origin.

[*]Corresponding author: nathalie.diagne@ird.fr; Nathalie.diagne@gmail.com.

Nevertheless, furthers studies carried out with exotic trees also showed that inoculation with local or more specific mycorrhizal fungi strains can strongly stimulate plant development. Therefore, it is important to find mycorrhizal fungi strains that are more beneficial for the establishment of exotic trees in foreign areas.

In this chapter, we will expose i) the major reforestation programs carried out in arid and semi-arid African zones with exotic trees, discuss ii) the role of mycorrhizal symbiosis in their establishment and iii) in the mitigation of negative influence of these exotic trees in soil microbiota.

Keywords: Reforestation, Exotic trees, Arbuscular mycorrhizal fungi, Ectomycorrhizal fungi

1. Introduction

Desertification and degradation of ecosystem are major threats for arid and semi-arid zones. They induce the loss of vegetation cover and reduce ecosystems productivity. Land desertification provokes a disturbance of vegetation cover which increases soil erosion resulting to the loss of available nutrients and of microbial propagules. However, low density of mycorrhizal propagules in degraded soils may limit the success of revegetation programs [1] so, to rehabilitate degraded ecosystems and improve soil biological characteristic, establishment of suitable plant cover is necessary [2, 3]. Generally exotic trees are introduced to rehabilitate degraded areas. Among them fast growing trees such as Casuarina, Acacia, Pine and Eucalyptus should have preference. They increase ecosystems productivity and improve local population livelihood. The introduced species are well adapted to harsher environments. They are very resistant to stresses such as drought, salinity etc. This tolerance to stresses is particularly due to their capacity to form a relationship with ectomycorrhizal (ECM) and endomycorrhizal (AMF) fungi. These microorganisms are essential component for sustainable ecosystems stability ([4, 5, 6]. Inoculation with mycorrhizal fungi is one of the alternatives to improve forest products and environmental quality [7]. In this purpose, inoculation with appropriate fungi is a promising biological tool for improving restoration program in degraded zones [8]. In association with plants, these microorganisms play a great role in plant development [9, 10, 11] and improve plant nutrition by increasing nutrient uptake such as water, P and N_2 [6]. In return, they benefit from plant carbohydrates [6, 12].

Inoculation of exotic trees with mycorrhizal fungi improve significantly their growth. So to improve the early establishment of exotic trees, transferred plants are inoculated with ectomycorrhizal and/or endomycorrhizal fungi because they allow plant to resist to stresses such as drought [13-14], salinity [15], to improve water and nutrient uptake particularly phosphorus and N_2 [6] and they play a major role in heavy metal phytoremediation [16, 17]. Although mycorrhizal fungi differ in their ability to influence plant growth. Field experiments showed that establishment of exotic trees is successful when they are associated and transferred with appropriated mycorrhizal fungi from the plant's zone of origin [18, 19]. In contrast, numerous studies showed that plant development can be stimulated by local mycorrhizal or specific selected mycorrhizal fungi [20, 21]. These studies illustrate the importance of selecting suitable mycorrhizal fungi to associate to exotic trees for their successful establishment in arid and semi-arid zones.

In this chapter we will describe the major reforestation programs carried out in arid and semi-arid areas in Africa; highlight the role of mycorrhizal symbiosis in the establishment of exotic trees and their role in counterbalancing the negative influence of exotic trees on soil bacteria community.

2. The Exotic Plants Established in African Arid and Semi-Arid Zones

To select exotic trees for introducing in new areas, the commonly method use is environmental matching. In African arid and semi-arid zones, species which belong to the following families: Casuarinaceae; Leguminoseae; Pinaceae and Myrtaceae are introduced to rehabilitate degraded ecosystems [22].

2.1. Casuarinaceae

The Casuarinaceae family has a native area which extends from Australia to South East Asia.

Table 1. Exotic trees planted in certain African arid and semi-arid zones and their isolate fungal partners

Plant family	Zones of introduction	Type of mycorrhizal symbiosis
Casuarina	Benin, Burkina Faso, Cameroon, Central African Republic, Chad, Congo, Cote d'Ivoire, Democratic Republic of Congo, Djibouti, Egypt, Eritrea, Ethiopia, Gabon, Gambia, Ghana, Guinea, Guinea-Bissau, Kenya, Liberia, Madagascar, Mali, Mauritania, Niger, Nigeria, Senegal, Sierra Leone, Somalia, South Africa, Sudan, Tanzania, Togo, Uganda, Zimbabwe…	ECM AMF
Australian Acacia	Algeria, Cap Verde, Burkina Faso, Egypt, Guinea, Ivory coast, Libya, Madagascar, Mali, Mauritania, Morocco, Niger, Nigeria, Senegal, South Africa, Tunisia…	ECM AMF
Eucalyptus	Cogon, Egypt, Ethiopia, Kenya, Morocco, Nigeria, Tanzania, Senegal, South Africa, Zimbabwe…	ECM AMF
Pinus	Cameroon, Cogon, Ghana, Ivory coast, Kenya, Liberia, Malawi, Nigeria, Senegal, South Africa, Uganda, Zambi, Zimbabwe…	ECM

Sources: Bâ et al., (2010) [22]; NRC, (1984) [25] ; NAS, (1980) [107]
ECM: Ectomycorrhizal fungi
AMF: Arbuscular mycorrhizal fungi.

This family comprises 4 genera (*Allocasuarina*, *Casuarina*, *Ceuthostoma* and *Gymnostoma*) and approximately 97 species [23, 24]. Among these genera, Casuarina is the most planted in Africa where it is present in more than thirty countries (Diagne *et al.*, 2011 unpublished) (Table 1). *C. equisetifolia*, *C. glauca* and *C. cunninghamia* were widely established in several zones in Africa [25].

Casuarina species were established in the Niayes regions, in the Nile Delta and Valley, the Sémé zone and the coastal dunes of KwaZulu-Natal in Senegal, Egypt, Benin and South Africa respectively. In these zones, Casuarina have a great interests and are used in the stabilization of sand dunes, the protection against wind, the reclamation of salt-affected land and the production of firewood and timber. Casuarinaceae trees have the ability to form the both symbionts: ectomycorrhizal and endomycorrhizal [26, 27]. However, ectomycorrhiza was more common with the genus *Allocasuarina* than the genus *Casuarina* [28, 29].

2.2. Leguminoseae

The Leguminoseae subfamily is formed by three families: the Caesalpinioideae, the Papilionoideae and the Mimosoideae. This subfamily contains about 2800 species from Australia, America, tropical and subtropical areas. The widely spread family in the Leguminoseae is the Mimosoideae. In this family, Acacia is the largest genus with 1200 species [30]. Among Acacia species, Australian Acacia which are fast growing trees are largely planted in arid and semi-arid countries (Table 1) to promote rehabilitation on degraded lands [11].

Acacia play an important role in these zones; they involve in fuel wood production, act also as a windbreaker and provide fodder for browsing livestock [31, 32]. Due to the capacity to fix nitrogen [33] as Casuarina genera, Acacia species have an additional role by improving ecosystem fertility [32]. More ever, Australian Acacia are able to form a relationship with the ectomycorrhizal and/or arbuscular mycorrhizal fungi [34] which improve plant growth and development [9, 35, 36, 37, 38]. Acacia species have a medicinal importance; they are generally used to treat malaria and syphilis [39].

2.3. Myrtaceae

The Myrtaceae is an important family containing about130 genera and 3,800 to 5,600 species. Among these genera, Eucalyptus and Melaleuca native from Australia, New Guinea, Indonesia and some South Pacific islands are largely introduced in Africa. The *Eucalyptus* genus contains approximately 700 species [40] and is one of the most widely grown trees in exotic plantations worldwide. In the tropics, around 10.06 million ha of eucalypt were planting at the end of 1990 [41]. This specie is widely present in African countries and is most planted in South Africa, Ethiopia, Kenya, Tanzania and Zimbabwe (Table 1). The Myrtaceae is an ecologically and economically family because species such as Eucalypts are cultivated to provide timber, pulp industries, shade and is used in pharmacology industries. Eucalyptus plant forms both arbuscular and ectomycorrhizal fungi on the same root [42-43].

2.4. Pinaceae

The Pinaceae family includes eleven genera. The most spread in African countries is the genus *Pinus*. There are between 105 and 125 species of pine. *Pinus* are widely introduced in many zones in Africa such as South Africa, Kenya, Tanzania, Uganda, Zimbabwe, Zambia, Malawi, Cameroon, Congo, Nigeria, Liberia, Ivory Coast and Guinea (Table 1) [18, 19, 44]. Pine is an important tree; it's planted in these African countries for the reforestation of marginally degraded agricultural land. They are commercially important tree species due to their production of timber and wood pulp. This specie is associated only with ectomycorrhizal fungi [19].

3. Benefits of Mycorrhiza in the Establishment of Exotic Tree Species

Due to the absence or the presence of intracellular penetration of plant cells by the fungi, we can distinct different types of mycorrhizal symbioses: arbuscular mycorrhiza, ectomycorrhiza, ectendomycorrhiza, arbutoid mycorrhiza, monotropoid mycorrhiza and orchidaceous mycorrhiza [6]. Among the different types cited above, ectomycorrhiza and arbuscular mycorrhizal are the most represented in the world [6]. These microorganisms improve the development of exotic tree species [45, 46, 47].

3.1. Ectomycorrhizal Fungi

Ectomycorrhiza are found with 3 and 5% of Angiosperms and Gymnosperms families [48]. This symbiotic association occurred with more than 5000 fungal species belonging to the Ascomycetes and Basidiomycetes. The ectomycorrhiza fungi are characterized by the development of structures such as: the sheat, the Hartig net, the extra radical mycelium and the absence of intracellular penetration of plant cells by the fungi.

Since extra radical EM fungal hyphae hydrolyze soil organic P [49], ectomycorrhizal fungi play an importance role, by improving nutrient mineral uptake particularly Pi which contributing to plant growth enhancement [6, 50]. They are currently used in reforestation program and improve the growth and the early establishment of exotic trees. These mycorrhizal fungi have an additional great role; they can counterbalance stress conditions such as nutrient deficiency, drought, heavy metals, soil disturbances etc. [4, 6] and enhance plant health by improving protection again pathogens [51]. The sheat of EM fungi acts as a barrier avoiding pathogens penetration [52]. These fungi produce substances such as phenolic component and antibiotic which inhibits pathogens development [53].

The majority of exotic trees (Casuarina, Acacia and Eucalyptus) have the ability to form both symbioses.

3.1.1. Ectomycorrhizal Fungi and Plant of Casuarinacea Family

Species of Casuarinaceae family forms a symbiosis relationship with ECM [26, 27]. It has been demonstrated that inoculation with symbiotic microorganisms can improve

Casuarina survival and growth [29, 47, 54], especially on degraded lands. However, few data about *Casuarina* inoculation with ectomycorrhiza in African countries are available. It has been shown that *Casuarina glauca* colonization with *Pisolithus* and *Scleroderma* is not effective because fungi have failed to form a Hartig net [29]. Despite that, these same fungi have colonized *Allocasuarina verticillata* with the formation of the mantle and the Hartig net. Under axenic conditions, Duponnois *et al.,* [29] observed that inoculation with *Pisolithus* and *Scleroderma* improves *Allocasuarina verticillata* growth. In others introduced zones such as China and particularly the South China, inoculation with ectomycorrhizal fungi significantly improve *Casuarina* growth [55].

Ectomycorrhiza are more frequent with the genus *Allocasuarina* where as endomycorrhiza are found on *Casuarina* [26, 29, 56]. Growth response of *A. verticillata* to ectomycorrhizal inoculation was higher than to endomycorrhizal inoculation despite the positive role of AMF on plant growth [57]. These authors conclude that inoculation with compatible mycorrhiza is recommended when Casuarina trees are established at zones where they have not previously been planted. In African countries, little studies on Casuarina inoculation have been done.

Beside the mycorrhizal symbiosis, Casuarina forms symbiosis with the soil nitrogen fixing bacteria of *Frankia* genus. In certain plantations, Casuarinaceae are only inoculated with *Frankia*. We hypothesized that the early establishment of these trees could be improving by these bacteria. It is thus important to inoculate with ECM to strengthen the positive effect obtain when Casuarinaceae plants are inoculated with *Frankia*. This kind of study seem to be very interesting because it has been previously demonstrated that the positive effect of the dual inoculation (*Frankia* and ECM) on Casuarinaceae growth is stronger comparing to the growth obtaining when one of these microorganisms are used as inoculum [47]. These results highlight the positive role of the inoculation of Casuarinaceae with mycorrhiza which increases their development.

3.1.2. Ectomycorrhizal Fungi and Plant of Leguminoseae Family

Australian Acacia plant of the Leguminoseae is one of the exotic trees associated with ECM which are well known to promote plant growth. It has been found by Duponnois and Bâ, [58] that the ECM commonly associated to Acacia is the genus *Pisolithus*. These ectomycorrhiza improve significantly the growth of Australian acacias in arid and semiarid conditions [59]. Some studies showed that *Pisolithus albus* have a positive effect on *Acacia holosericeae* growth [11, 59]. Similar results were obtained in Nigeria when *A. auriculiformis* is inoculated with ECM [60]. In experimentations carried out in Senegal, Duponnois *et al.,* [9] demonstrated that COI 024 (*Pisolithus sp*), increased plant development under glasshouse conditions and decreased the multiplication of the root-knot nematode *Meloidogyne javanica.*

3.1.3. Ectomycorrhizal Fungi and Plant of Myrtaceea Family

As noted with Acacia, Eucalyptus an exotic tree widely spread in Sahelian zone forms symbiosis with both ECM and AMF. In nursery or field conditions, inoculation of Eucalyptus by ECM increases plant growth. The commonly mycorrhizal fungi associated with Eucalyptus in its zones of introduction are ECM of the *Pisolithus* and *Scleroderma* genera [61, 62, 63]. Eucalyptus growth increases in response to ectomycorrhizal inoculation [64]. After testing the effectiveness of 16 fungal isolates, it has been concluded that ECM enhances *Eucalyptus globulus* Labill. and *E. diversicolor* F. Muell. growth [65]. On both eucalypt species testing *E. globulus* and *E. diversicolor*, the largest growth was obtaining with *P. tinctorius*. An increasing phosphorus uptake by the plant is observed when eucalypt is inoculated with ECM [64, 65]. In nursery condition in Congo, inoculation with *P. tinctorius* (strain from north of America), show the positive effect of fungi on eucalypt development [61]. Despite the effectiveness of the North American *Pisolithus*, this strain was eliminated by a local *Scleroderma* just one year after inoculation. This study shows the importance of selecting compatible mycorrhizal fungi for developing successful inoculation programmes using eucalyptus.

Outside their native zones, *Pisolithus* and *Scleroderma* are found to be most efficient for Eucalyptus mycorrhization [62, 63].

3.1.4. Ectomycorrhizal Fungi and Plant of Pinaceae Family

However, among the most introduced exotic species, Pinus are the one genera forming only ectomycorrhiza. Studies carried out by Zhu *et al.,* [66] illustrate the role of ectomycorrhizal fungi in pine growth. These authors show that without ectomycorrhizal fungi, some pine species are unable to growth optimally. Moreover, Hacskaylo, [67] showed that for the establishment of pine in the tropics, compatible ectomycorrhiza fungi could be necessary introduced. The early establishment of *Pinus radiata* in Kenya does not success unless they are associated with specific ectomycorrhizal fungi from an ancient pines plantation [18]. This problem was overcome by introduced soil containing propagules (spore and mycorrhizal root) from old South African pines plantation. This type of inoculum is widely used in West and East Africa to improve pine establishment [19, 68]. Positive effect of ECM inoculation on *P. halepensis* was observed in containers filled with a natural forest soil (Morocco) [69]; who reported that inoculation with *Pisolithus* strain *PH4* stimulate significantly *P. halepensis* growth.

Mycorrhizal fungi are involved in pine resistance to stresses. An accumulation of iron in ectomycorrhizal fungi has been noted under alkaline soil conditions [70]. These results, illustrate the influence of ectomycorrhiza fungi in pine growth and the iron sequestering by ECM.

3.2. Arbuscular Mycorrhiza

Arbuscular mycorrhizal symbiosis is the most widely distributed association in plants. It occurs between more than 80% of land plants including Angiosperms, Gymnosperms, Pteridophytes and Bryophytes [27] and around 200 fungal belonging to the Glomeromycota order [71]. This symbiosis is the oldest dated approximately 400 million years (Devonian) [72]. Arbuscular mycorrhiza are obligate biotrophs; they can not be cultured in the absence of plant host and therefore can not complete their life cycle [6]. The symbiosis with arbuscular mycorrhiza is an ubiquitous association which occurs in all terrestrial ecosystems [4, 73]. The endomycorrhiza are characterized by the development of the fungi within plant roots system [6]. AMF are very beneficial for plants host. In association with these plants, they play a great role due to the fungi mycelium which is able to explore a huge volume of soil and improve nutrients uptake for plants while the plant supplies carbohydrates to the fungi [4, 6]. These AMF improve plant resistance to biotic and abiotic stresses such as drought stress [13], salt stress [14-15], cold stress [74], resistance to heavy metals and pollution [16, 17] and they protect plants from pathogens [75, 76, 77]. Mycorrhizal symbiosis reduces the negative effect of nematodes on root diseases [75, 78]. Inoculation with AMF limits pathogens growth and improves plant development [76, 79].

Mycorrhizal fungi reduce pathogens performance by stimulating defensive responses [80, 81] or by altering root exudates used by these pathogens [82].

3.2.1. Arbuscular Mycorrhiza and Plant of Casuarinaceae Family

Plants of Casuarinaceae family are able to form both symbioses. However, arbuscular mycorrhiza are more associated with species of *Casuarina* genera. Inoculation of Casuarina with AMF enhances plant growth [12, 47, 54, 55, 83]. Growth response to a mixture of indigenous AMF strains of *C. equisetifolia* in non sterile soil is stronger than growth obtained when plants are inoculated with a mixture of exotic AMF strains (Diagne *et al.* 2011 unpublished).

Casuarina tolerance to stress increases when plants are associated with AMF. It has been demonstrated that in association with Casuarina, AMF improve plant tolerance to drought [57] and to flooding [84] and directly increase their biomass. With AMF, *C. equisetifolia* is better adapted to flooding condition compare to non inoculated Casuarina [84].

Outside their native plantations, knowledge about arbuscular mycorrhizal association with Casuarina in plantation is still very limited. However, some AMF are found associated with Casuarina in Algeria, Libyan and Egypt deserts. These AMF may be involved in Casuarina drought tolerance [85].

3.2.2. Arbuscular Mycorrhiza and Plant of Leguminoseae Family

While some Australian Acacia are ectomycorrhizal plant hosts, arbuscular mycorrhizal fungi are also observed in certain Australian Acacia [33].

Inoculation with *G. intraradices* significantly increases *A. holosericea* growth in Senegal [59]. Similar results were obtained in field conditions in Burkina Faso [86]. Experimental carried out in greenhouse in saline conditions with *A. mangium* and *A. auriculiformis* showed that AMF have strongly stimulated plant growth [36]. A better plant tolerance to salt stress was reported by these authors when the two exotic Australian acacia were inoculated with AMF and *Rhizobia*.

3.2.3. Arbuscular Mycorrhiza and Plant of Myrtaceae Family

About the interaction between arbuscular mycorrhizal fungi and Eucalyptus species, a few data are available because ECM symbioses are more studied than AMF symbioses. However Asai, [87] and Maeda, [88] observed arbuscular mycorrhizal fungi structures on eucalypts. Moreover, arbuscular mycorrhizal have been reported on *E. Pilularis* Sm. and *E. paniculata* Sm. [89]. Recently, it has been observed that inoculation of *E. globulus* by *Glomus claroideum*, *G. viscosum*, *G. intraradices* and *G. constrictum* increases significantly plant shoot dry weights [90]. Arbuscular mycorrhizal fungi stimulate significantly eucalypts growth. Eucalypts stem dry weight has been increased 49% compared to the control twenty weeks after inoculation [91].

In study carried out in Algeria, Adjoud-Sadadou and Hallis-Hargas, [92] found all the structures of arbuscular mycorrhiza within roots of four Eucalyptus species. Generally, AMF are found on young Eucalyptus while ECM are often found on old Eucalypts [93]. Studies dealing with both mycorrhizal symbioses (ECM and AMF) show a temporal replacement of arbuscular mycorrhizal by ectomycorrhizal with host ageing [92].

However, Chilvers and Pryor, [94] demonstrated that eucalypts typically form ectomycorrhiza. In some introduced areas such as Senegal, Congo, China and India eucalypts are principally associated with *Pisolithus* and *Scleroderma* genera [62, 95, 96, 97].

4. Mycorrhizal Symbiosis can Mitigate the Negative effect of Exotic Tree Species on Soil Microbial Functioning

Exotic tree species are planted in arid and semi arid ecosystems to recover degraded lands. These trees which are able to growth in poor sites have a key role in sustainable ecosystems balance. This effect is particularly influencing by the relationship between plants, soil and microorganisms such as mycorrhizal fungi.

Despite this positive effect numerous studies have demonstrated that exotic trees alter the structure and the functionally of soil microbial community (Table 2) [86, 98, 99, 100]. Bilgo *et al.*, [86] reported that *A. holosericeae* can significantly alter soil bacterial community by affecting their function (Table 2). Studies carried out by Remigi *et al.*, [100] show that microbial activity is greater in the crop soil in compare to the activity observed in plot with *A. holosericeae* plants (Table 3).

Table 2. Catabolic evenness and average SIR responses (µg CO2 g^{-1} soil h^{-1}) with each substrate group (carboxylic acids, amino-acids, amides and carbohydrates) of the soils collected inside the *Acacia holosericea* plantation (uninoculated plots, *Glomus intraradices* inoculated plots) and outside the *A. holosericea* plantation (Grassland). From Bilgo et al., (2011)

Treatments	Grassland	Control	G. intraradices inoculation
Catabolic evenness	17.9 (0.37)[a] a[b]	20.4 (1.02) b	22.9 (0.19) c
Carboxylic acids	827.2 (4.02) c	188.5 (5.26) a	228.9 (4.81) b
Amino-acids	166.3 (9.83) c	132.5 (1.25) a	142.2 (4.09) ab
Amines	100.0 (17.4) a	136.7 (16.4) a	131.3 (10.1) a
Amides	197.5 (5.95) c	98.3 (14.53) a	165.0 (3.53) b
Carbohydrates	145.1 (6.16) c	83.3 (6.67) a	68.3 (9.18) a

[a] Standard error of the mean.
[b] Data in the same line followed by the same letter are not significantly different according to the Newman-Keul's test ($p<0.05$).

With *A. holosericeae* growing in field condition after fifty years of plantation in Burkina Faso, the higher catabolic evenness was observed in soil collected inside *A. holosericeae* inoculated with *G. intraradices* (Table 2). Similar effect has also been obtained with *E. camaldulensis* an exotic specie [101]. The evaluation of the average SIR (Substrate Induces Respirations) responses showed that average SIR of carboxylic acids, amino-acids and amides is greater in grassland followed by the SIR response in the soils collected inside the *Acacia holosericea* plantation inoculated with *G. intraradices* while the data recorded in plantation with *A. holosericeae* trees uninoculated are lower [86] (Table 2).

Assessment of the total microbial activity show that the FDA hydrolysis is lower in these soils collected under uninoculated *A. holosericea* compared with the soil under the grassland [86]. Similar results were obtained for the deshydrogenase activity, the microbial biomass and the alkaline phosphatase. The negative effect of exotic trees could also been counterbalanced by inoculating with ectomycorrhizal fungi [37, 100]. In a *A. holosericea* field experiment carried out in Senegal and inoculated with *P. albus* [11, 59]; it has been noted a great disturbance in soil community that affecting the genotypic and the symbiotic effectiveness of Bradyrhizobia community [37]. These authors show that, ectomycorrhizal inoculation of *A. holosericea* modified the Bradyrhizobia soil diversity by promoting the multiplication of bacteria belonging only to the IGS groups (A and D).

It has been demonstrated that by changing the functions and diversity of soil microbial communities, ectomycorrhizal fungi select microorganisms potentially beneficial to plant growth (i.e. fluorescent pesudomonads that solubilising phosphate) [69, 102]. Ouahmane *et al.,* [69] show a higher number of fluorescent pseudomonads under *Pisolithus* inoculated plants.

Table 3. Microbial biomass and chemical characteristics of the soils sampled inside the *A. holosericea* plantation from each treatment (uninoculated plots and *Pisolithus. albus*-inoculated plots) and outside the *A. holosericea* plantation (crop soil). From Remigi et al., (2008)

Soil origin	Total nitrogen (%)	Total carbon (%)	Total phosphorus (mg kg^{-1})	Soluble phosphorus (mg kg^{-1})	Microbial biomass (µg C g^{-1}soil)
Crop soil	0.032a	0.398a	67.0a	4.6a	278.7b
Plantation with uninoculated trees	0.024a	0.338a	73.2ab	4.9a	227.5a
Plantation with P. Albus IR100-inoculated trees	0.046a	0.600b	84.2b	7.1b	231.2a

[a] Data followed by the same letter in each column are not significantly different according to results of the Newman-Keuls test ($p<0.05$).

These results highlight the selective effect of mycorrhizal fungi on soil microbiota [103, 104] and show clearly that exotic trees may be responsible for strong changes in soil bacteria community.

Furthermore, introduction of exotic trees have negative effect on soil nutrient content such as total carbon and total nitrogen, (Table 3; 4) [86, 100]. This negative effect is mitigated when exotic trees are associated with mycorrhizal fungi (Table 3; 4). By increasing the microbial activity and the nutrient content inside soil under exotic trees, mycorrhiza fungi enhance plant development. After 7 years of plantation in field conditions in Senegal, Remigi *et al.,* [100] show that inoculation with *P. albus* IR100 significantly increases *A. holosericea* growth. A higher height, stem diameter, leaf biomass, wood biomass, small branch biomass, wood biomass, total above ground biomass and litter biomass were obtained with inoculated plants while these data are lower when *A. holosericea* is uninoculated (Table 5).

Table 4. Microbial biomass and chemical characteristics of the soils sampled inside the Acacia holosericea plantation (uninoculated plots, *Glomus intraradices*-inoculated plots) and outside the *A. holosericea* plantation (Grassland). From Bilgo et al., (2011)

Treatments	Grassland	Control	G. intraradices inoculation
Total nitrogen (%)	0.058 (0.006)[a] b[b]	0.041 (0.002) a	0.050 (0.002) b
Total carbon (%)	0.676 (0.064) b	0.577 (0.015) a	0.669 (0.034) b
Soluble phosphorus (mg kg^{-1})	6.9 (1.4) a	8.4 (1.0) a	17.4 (1.0) b
Hyphal length (mg−1 soil)	1.3 (0.5) a	1.9 (0.3) a	3.9 (0.4) b
qCO2 (μg C-CO2 g^{-1} MBC h^{-1})	0.10 (0.031) a	0.17 (0.014) a	0.33 (0.023) b

a Standard error of the mean.

[b] Data in the same line followed by the same letter are not significantly different according to the Newman-Keuls test ($p<0.05$).

Table 5. Effect of P. albus IR100 inoculation on the growth of *A. holosericea* after 7 years of plantation in the field. From Remigi et al., (2008)

Treatments	Treatment Height (m)	Stem diam (cm)	Leaf biomass (kg per tree)	Small branch biomass (kg per tree)	Wood biomass (kg per tree)	Total above ground biomass (kg per tree)	Litter biomass (kg m^{-2})
Control (no inoculation)	4.84[a] a	11.8a	19.1a	16.9a	27.1a	63.1a	1.98a
P. albus IR100 inoculation	6.43b	23.2b	82.6b	75.1b	153.2b	310.9b	6.19b

[a] Data followed by the same letter in each column are not significantly different according to results of the Newman-Keuls test ($p<0.05$).

However, it has been also demonstrated that, AMF symbiosis influence the structure and functioning of the mycorrhizosphere bacteria community [105, 106] so, negative effect of exotic tree are significantly moderate when established species are associated with efficient AMF [86, 101]. With the case of *E. camaldulensis,* a limit of the development of others herbaceous was observed and this effect is stronger when plants are uninoculated with AMF. These results illustrate the great role of AMF which mitigate the negative influence of exotic trees [101].

Conclusion

In conclusion, this chapter shows that exotic trees are widespread in arid and semi-arid African zones. It also highlights the positive role of mycorrhizal symbiosis in enhancing exotic plant growth and improving their resistance to stresses. These microorganisms promote also the early establishment of plant and can counterbalance the negative influence of exotic trees in soil microbiota. However, growth stimulation of exotic trees outside their native areas depends on the origin of fungal isolate (mycorrhiza fungi from local zone, indigenous mycorrhizal fungi or specific selected mycorrhiza). Taken together, this chapter suggests that for the well development in foreign countries, exotic trees may be associated with appropriate fungal isolate.

References

[1] Duponnois, R., Plenchette, C. & Bâ, A.M. (2001). Growth stimulation of seventeen fallow leguminous plants inoculated with *Glomus aggregatum* in Senegal. *European Journal of Soil Biology*, 37, 181-186.

[2] Skuji nš, J. & Allen. M. F. (1986). Use of mycorrhizae for land rehabilitation. *World Journal of Microbiology and Biotechnology*, 2, 161-176.

[3] Morgan, R. P. C., Rickson, R. J. & Wright, W. (1990). Regeneration of degraded soils, p. 69-85. In J. Albaladejo, M. A. Stocking, and E. Diaz (ed.), Soil degradation and rehabilitation in Mediterranean environmental conditions. *Consejo Superior de Investigaciones Cientificas, Murcia*, Spain.

[4] Smith, S.E. & Read, D.J. (1997). *Mycorrhizal symbiosis* (2nd eds.). Academic, San Diego, USA.

[5] van der Heijden, M.G.A., Klironomos, J.N., Ursic, M., Moutoglis, P., Streitwolf-Engel, R., Boller, T., Wiemken, A. & Sanders, I.R. (1998). Mycorrhizal fungal diversity determines plant biodiversity, ecosystem variability and productivity. *Nature*, 396, 69-72.

[6] Smith, S.E. & Read, D.J. (2008). *Mycorrhizal Symbiosis*, Third Edition, Academy Press.

[7] Marcela C.P. & Marta, N.C. (2011). Mycorrhizal Interactions for Reforestation: Constraints to Dryland Agroforest in Brazil. *International Scholarly Research Network. Ecology*, doi:10.5402/2011/890850.

[8] Pigott, C.D. (1982). Survival of mycorrhizas formed by *Centroccocum geophilum* in dry soils. *New Phytologist*, 92, 513-517.

[9] Duponnois, R., Founoune, H., Bâ, A.M., Plenchette, C., El jafaari, S., Neyra. & M, Ducousso, M. (2000). Ectomycorhization of *Acacia holosericea* A. Cunn. ex G. Don by *Pisolithus spp.* in Senegal : effect on plant growth and on the roots- knot nematode *Meloidogyne javanica*. *Annals of Forest Science*, 57, 345-350.

[10] Duponnois, R., Colombet, A., Hien, V. & Thioulouse, J. (2005a). The Mycorrhizal fungus *Glomus intraradices* and rock phosphate amendment influence plant growth and microbial activity in the rhizosphere of *Acacia holosericea*. *Soil Biology and Biochemistry*, 37, 1460-1468.

[11] Duponnois, R., Founoune, H., Masse, D. & Pontanier, R. (2005b). Inoculation of *Acacia holosericea* with ectomycorrhizal fungi in a semi-arid site in Senegal: growth response and influences on the mycorrhizal soil infectivity after 2 years plantation. *Forest Ecology and Management*, 207, 351-362.

[12] He, X.H. & Critchley, C. (2008). *Frankia* Nodulation, Mycorrhization and Interactions between *Frankia* and Mycorrhizal Fungi in *Casuarina* Plants. In: *Mycorrhiza 3; Ajit Varma (eds); Springer Berlin chap VI; (Biomedical and life Sciences)* 767-781.

[13] Ruiz-Lozano, J.M, Porcel, R. & Aroca, R. (2006). Does the enhanced tolerance of arbuscular mycorrhizal plants to water deficit involve modulation of drought-induced plant genes? *New Phytologist*, 171, 693-698.

[14] Zuccarini, P. & Okurowska, P. (2008) Effects of mycorrhizal colonization and fertilization on growth and photosynthesis of sweet basil under salt stress. *Journal of Plant Nutrition*, 31, 497-513.

[15] Evelin, H., Kapoor, R. & Giri, B. (2009). Arbuscular mycorrhizal fungi in alleviation of salt stress: a review. *Annals of Botany*, 104, 1263-1280.

[16] Leyval, C., Weissenhorn, I., Berthelin, J. & Dexheimer, J. (1994). Influence des métaux lourds sur la germination des spores de champignons endomycorhiziens à arbuscules dans les sols. *Acta Botanica Gallica*, 141, 523-528.

[17] Göhre, V. & Paszkowski, U. (2006). Contribution of the arbuscular mycorrhizal symbiosis to heavy metal phytoremediation. *Planta*, 223, 1115-1122.

[18] Mikola, P. (1970). Mycorrhizal inoculation in afforestation. *Internatinal rewiev of forestry research*, 3, 123-196.

[19] Delwaulle, J.C., Diangana, D. & Garbaye, J. (1987). Augmentation de la production du pin des Caraïbes dans la région côtière du Congo par l'introduction du champignon *Pisolithus tinctorius*. *Revue Forestière Française*, 39, 409-417.

[20] Caravaca, F., Barea, J.M., Palenzuela, J., Figueroa, D., Alguacil, M.M. & Roldán, A. (2003). Establishment of shrub species in a degraded semi arid site after inoculation with native or allochthonous arbuscular mycorrhizal fungi. *Applied Soil Ecology*, 22, 103-111.

[21] Ouahmane, L., Hafidi, M., Thioulouse, J., Ducousso, M., Kisa, M., Prin, Y., Galiana, A., Boumezzough, A. & Duponnois, R. (2007). Improvement of *Cupressus atlantica* Gaussen growth by inoculation with native arbuscular mycorrhizal fungi. *Journal of Applied Microbiology*, 103, 683-690.

[22] Bâ, A. M., Diédhiou, A. G, Prin, Y., Galiana, A. & Duponnois, R. (2010). Management of ectomycorrhizal symbionts associated to useful exotic tree species to improve reforestation performances in tropical Africa. *Annals of Forest Science*, 67 doi: 10.1051/forest/2009108.

[23] Wilson, K.L. & Johnson, L.A.S. (1989). *Casuarinaceae*. In: George AS (eds), Flora of Australia, vol 3 Hamamelidales to Casuarinales. Australian Government Publishing Service, Canberra.

[24] Bisby, F.A., Roskov, Y.R., Ruggiero, M.A., Orrell, T.M., Paglinawan, L.E., Brewer, P.W., Bailly, N. & van Hertum, J. (2007). Species 2000 and ITIS catalogue of life: 2007 annual checklist. Digital resource at www.catalogueoflife.org/annual-checklist/2007/. *Species* 2000, Reading, UK.

[25] National Research Council (1984). Casuarinas: nitrogen-fixing trees for adverse sites. *Office of International Affairs*, Washington DC, USA.

[26] Bâ, A. M., Sougoufara, B. & Thoen, D. (1987). The triple symbiosis of *Casuarina equisetifolia* in Senegal. In: Sylvia, D.M,, Hung, L.L., Graham, J.H. (Eds.), *Mycorrhizae in the next decade*, Gainesville, USA, University of Florida, p. 121.

[27] Wang, B. & Qiu, Y.L. (2006). Phylogenetic distribution and evolution of mycorrhizas in land plants. *Mycorrhiza*, 16, 299-363.

[28] Thoen, D., Sougoufara, B. & Dommergues, Y. (1990). In vitro mycorrhization of *Casuarina* and *Allocasuarina* species by *Pisolithus* isolates. *Canadian Journal of Botatny*, 68, 2537-2542.

[29] Duponnois, R., Diedhiou, S., Chotte, J.L. & Sy, M.O. (2003). Relative importance of the endomycorrhizal and/or ectomycorrhizal associations in *Allocasuarina* and *Casuarina* genera. *Canadian Journal of Microbiology*, 50, 691-696.

[30] Pedley, L. (1986). Australian acacia: taxonomy and phytogeography, p. 11–16. *In* J. W. Turnbull (ed.), *Australian acacia in developing countries.* ACIAR, Canberra, Australia.

[31] Cossalter, C. (1987). Introduction of Australian acacias into dry, tropical West Africa. *Forest Ecology and Management*, 16, 367-389.

[32] Cossalter, C. (1986). Introducing Australian Acacias in dry, tropical Africa. In: Turnbull J.W. (Ed.), Australian acacias in developping countries. Proceedings of an International Workshop at the Forestry Training Center, Gympie, Australia 1986. *ACIAR, Camberra*, 1986, pp. 118- 122.

[33] Ducousso, M. & Thoen, D. (1991). Les types mycorhiziens des *Acacieae*. in: Physiologie des Arbres et Arbustes en zones arides et semi-arides. Groupes d'Etude de l'Arbre. Paris, France. pp: 175-182.

[34] Founoune, H., Duponnois, R., Meyer, J.M., Thioulouse, J., Masse, D., Chotte, J.L. & Neyra, M. (2002). Interactions between ectomycorrhizal symbiosis and fluorescent *pseudomonads* on *Acacia holosericea*: isolation of Mycorrhiza Helper Bacteria (MHB) from a Soudano-Sahelian soil. *FEMS Microbiology Ecology*, 41: 37-46.

[35] André, S., Galiana, A., Le Roux, C., Prin, Y., Neyra, M. & Duponnois, R. (2005). Ectomycorrhizal symbiosis enhanced the efficiency of two Bradyrhizobium inoculated on *Acacia holosericea* plant growth. *Mycorrhiza*, 15, 357-364.

[36] Diouf, D., Duponnois, R., Ba, A.T., Neyra, M. & Lesueur, D. (2005). Symbiosis of *Acacia auriculiformis* and *Acacia mangium* with mycorrhizal fungi and *Bradyrhizobium* spp. improves salt tolerance in greenhouse conditions. *Functional plant biology*, 32, 1143- 1152.

[37] Faye, A., Krasova-Wade T., Thiao M., Thioulouse, JM., Neyra, M., Prin, Y., Galiana, A., Ndoye, I., Dreyfus, B. & Duponnois, R. (2009). Controlled ectomycorrhization of an exotic legume tree species *Acacia holosericea* affects the structure of root nodule bacteria community and their symbiotic effectiveness on *Faidherbia albida*, a native Sahelian Acacia. *Soil Biology and Biochemistry*, 41, 1245-1252.

[38] Aggangan, N.S., Moon, H.K. & Han, S.H. (2010). Growth response of *Acacia mangium* Willd. Seedlings to arbuscular mycorrhizal fungi and four isolates of the ectomycorrhizal fungus *Pisolithus tinctorius* (Pers.). *Coker and Couch New Forests*, 39, 215-230.

[39] Ene, A.C., Ameh, D.A., Kwanashie, H.O., Agomo, P.U. & Atawodi, S.E. (2008). Preliminary *in vivo* antimalarial screening of petroleum ether, chloroform and methanol extracts of fifteen plants grown in Nigeria. *Journal of Pharmacological and Toxicology*, 3, 254-260.

[40] Potts, B.M. & Pederick, L.A. (2000). Morphology, phylogeny, origin, distribution and genetic diversity of eucalypts. In: *Diseases and pathogens of eucalypts* (Keane PJ, Kile GA, Podger FD, Brown BN, eds). CSIRO Publishing, Australia: 11-34.

[41] FAO. (1993). Forest resources assessment 1990 – tropical countries. *Forestry Paper* 112. FAO, Rome, Italy.

[42] Chilvers, G.A., Lapeyrie, F. & Horan, D.P. (1987). Ectomycorrhizal vs. endomycorrhizal fungi within the same root system. *New Phytologist*, 107, 441-448.

[43] Boudarga, K., Lapeyrie, F. & Dexheimer, J. (1990). A technique for dual vesicular-arbuscular endomycorrhizal/ectomycorrhizal infection of *Eucalyptus* in vitro. *New Phytologist*, 114, 73-76.

[44] Momoh, Z.O. & Gbadegesin, R.A. (1980). Field performance of *Pisolithus tinctorius* as a mycorrhizal fungus of pines in Nigeria. In: Mikola P. (Ed.), *Tropical mycorrhiza research*, Oxford University Press, New York, USA, pp. 72-79.

[45] Garbaye, J. (1991). Utilisation des mycorhizes en sylviculture. In: Strullu D.G. (Ed.), *Les mycorhizes des arbres et des plantes cultivées, Technique et Documentation Lavoisier*, Paris, pp. 197-248.

[46] Rincón, A., de Felipe, M.R. & Fernández-Pascual, M. (2007). Inoculation of *Pinus halepensis* Mill. with selected ectomycorrhizal fungi improves seedling establishment 2 years after planting in a degraded gypsum soil. *Mycorrhiza*, 18, 23-32.

[47] Elumalai, S. & Raaman, N. (2009). *In vitro* synthesis of *Frankia* and mycorrhiza with *Casuarina equisetifolia* and ultrastructure of root system. *Indian Journal of Experimental Biology*, 47, 289-297.

[48] Fortin, J.A, Plenchette, C. & Piché, Y. (2008). *Les mycorhizes, la nouvelle révolution verte.* (Eds. Multimondes et Quae), 131p.

[49] Plassard, C. & Dell, B. (2010). Phosphorus nutrition of mycorrhizal trees. *Tree Physiology*, 30, 1129-1139.

[50] Martin, F. & Nehls, U. (2009). Harnessing ectomycorrhizal genomics for ecological insights. *Current Opinion in Plant Biology*, 12, 508-515.

[51] Marx, D.H. (1969). The influence of ectotropic mycorhizal fungi on the resistance of pine roots to pathogenic infections: I. Antagonism of mycorrhizal fungi to root pathogenic fungi and soil bacteria. *Phytopathology*, 59, 153-163.

[52] Marx, D.H. (1972). Ectomycorrhizae as Biological Deterrents to Pathogenic Root Infections. *Annual Review of Phytopathology*, 10, 429-454.

[53] Duchesne, L.C., Ellis, B.E. & Peterson, R.L. (1989). Disease suppression by the ectomycorrhizal fungus *Puxillus incolutus:* contribution of oxalic acid. *Canadian Journal of Botany*, 67: 2726-2730.

[54] Rajendran, K. & Devaraj, P. (2004). Biomass and nutrient distribution and their return of *Casuarina equisetifolia* inoculated with biofertilizers in farmland. *Biomass Bioenergy*, 26:235-249.

[55] Zhong, C., Zhang, Y., Chen, Y., Jiang, Q., Chen, Z., Liang, J., Pinyopusarerk, K., Franche, C. & Bogusz, D. (2010). *Casuarina* research and applications in China. *Symbiosis*, 50, 107-114.

[56] Dell, B., Malajczuk, N., Bougher, N.L. & Thomson, G. (1994). Development and function of *Pisolithus* and *Scleroderma* ectomycorrhizas formed in vivo with *Allocasuarina, Casuarina* and *Eucalyptus. Mycorrhiza*, 5, 129-138.

[57] Zhang, Y., Zhong, C. L., Chen, Y., Chen, Z., Jiang, Q.B., Wu, C. & Pinyopusarerk, K. (2010). Improving drought tolerance of *Casuarina equisetifolia* seedlings by arbuscular mycorrhizas under glasshouse conditions. *New Forests*, 40, 261-271.

[58] Duponnois, R. & Bâ, A.M. (1999). Growth stimulation of *Acacia mangium* Willd by *Pisolithus* sp. in some senegalese soils. *Forest Ecology and Management*, 119, 209-215.

[59] Duponnois, R., Plenchette, C., Prin, Y., Ducousso, M., Kisa, M., Bâ A.M. & Galiana A. (2007). Use of mycorrhizal inoculation to improve reafforestation process with Australian *Acacia* in Sahelian ecozones. *Ecological Engineering*, 29, 105-112.

[60] Osonubi, O., Mulongoy, K., Awotoye, O.O., Atayese, M.O. & Okali D.U.U. (1991). Effects of ectomycorrhizal and vesicular-arbuscular mycorrhizal fungi on drought tolerance of four leguminous woody seedlings. *Plant Soil*, 136, 131-143.

[61] Garbaye, J., Delwaulle, J.C. & Diangana, D. (1988). Growth response of eucalypts to mycorrhizal inoculation in the Congo. *Forest Ecology and Management*, 24, 151-157.

[62] Chen, Y.L., Liu S. & Dell, B. (2007). Mycorrhizal status of Eucalyptus plantations in South China and implications for management. *Mycorrhiza*, 17, 527-535.

[63] Chen, Y.L., Kang, L.H. & Dell, B. (2006). Inoculation of *Eucalyptus urophylla* with spores of Scleroderma in a nursery in south China: comparison of field soil and potting mix. *Forest Ecology and Management*, 222, 439-449.

[64] Thomson, B.D., Grove, T.S., Malajczuk, N. & Hardy, G.E. StJ. (1994). The effectiveness of ectomycorrhizal fungi in increasing the growth of *Eucalyptus globulus* Labill. in relation to root colonization and hyphal development in soil. *New Phytologist*, 126, 517-524.

[65] Burgess, T. I, Malajczuk, N. & Grove, T.S. (1993). The ability of 16 ectomycorrhizal fungi to increase growth and phosphorus uptake of *Eucalyptus globulus* Labill. and *E. diversicolor* F. Muell. *Plant Soil*, 153, 155-164.

[66] Zhu, J., Li, F., Xu, M., Kang, H. & Wu, X. (2008). The role of ectomycorrhizal fungi in alleviating pine decline in semiarid sandy soil of northern China: an experimental approach. *Annals of Forest Sciences*, 65, 304.

[67] Hacskaylo, E. (1971). *Mycorrhizae, US Government Printing Office,*Washington, DC, 255 p.

[68] Marx, D.H. (1980). Ectomycorrhizal fungus inoculations: a tool for improving forestation practices. In: Mikola P. (Ed.), *Tropical mycorrhiza research,* Oxford University Press, New York, USA, pp. 13-71.

[69] Ouahmane, L., Revel, J.C., Hafidi, M., Thioulouse, J., Prin, Y., Galiana, A., Dreyfus, B. & Duponnois, R. (2009). Responses of *Pinus halepensis* growth, soil microbial catabolic functions and phosphate-solubilizing bacteria after rock phosphate amendment and ectomycorrhizal inoculation. *Plant and soil*, 320, 169-179.

[70] Hauer, R.J. & Dawson, J.O. (1996). Growth and iron sequestering of pin oak (*Quercus palustris*) seedlings inoculated with soil containing ectomycorrhizal fungi. *Journal of Arboriculture*, 22, 122-130.

[71] Schüßler, A. (2002). Molecular phylogeny, taxonomy, and evolution of arbuscular mycorrhiza fungi and *Geosiphon pyriformis*. *Plant and Soil*, 244, 75-83.

[72] Remy, W., Taylor, T.N., Hass, H. & Kerp H, (1994). Four hundred-million-year-old vesicular arbuscular mycorrhizae. *Proceedings of the National Academy of Sciences of the United States of America*, 91, 11841-11843.

[73] Morton, J.B., Bentivenga, S.P. & Bever, J.D. (1995). Discovery, measurement, and interpretation of diversity in symbiotic endomycorrhizal fungi. *Canadian Journal of Botany*, 73S, 25-32.

[74] Paradis, R., Dalpé, Y. & Charest, C. (1995). The combined effect of arbuscular mycorrhizas and short-term cold exposure on wheat. *New Phytologist*, 129, 637-642.

[75] Dehne, H. W. (1982). Interaction between vesicular-arbuscular mycorrhizal fungi and plant pathogens. *Phytopathology*, 72, 1115-1982.

[76] Liu, J., Maldonado-Mendoza, I., Lopez-Meyer, M., Cheung, F., Town, C.D. & Harrison, M.J, (2007). Arbuscular mycorrhizal symbiosis is accompanied by local and systemic alterations in gene expression and an increase in disease resistance in the shoots. *The Plant Journal*, 50, 529-544.

[77] Sayeed, M. & Siddiqui, Z.A. (2008). Arbuscular Mycorrhizal Fungi as Potential Bioprotectants against Plant Pathogens. Mycorrhizae: Sustainable. *Agriculture and Forestry*, 61-97.

[78] St-Arnaud, M., Hamel, C., Vimard, B., Caron, M. & Fortin, J.A. (1995). Altered growth of *Fusarium oxysporum* f.sp. *Chrysanthemi* in an in vitro dual culture system with the vesicular arbuscular mycorrhizal fungus *Glomus intraradices* growing in *Daucus carota* transformed roots. *Mycorrhiza*, 5, 431-438.

[79] Borowicz,V. A. (2001). Do arbuscular mycorrhizal fungi alter plant-pathogen relations? *Ecology*, 82, 3057-3068.

[80] Volpin, H., Elkind, Y., Okon, Y. & Kapulnik, Y. (1994). A vesicular arbuscular mycorrhizal fungus (*Glomus intraradices*) induces a defense response in alfalfa roots. *Plant Physiology*, 104, 683-689.

[81] Morandi, D. (1996). Occurrence of phytoalexins and phenolic compounds in endomycorrhizal interactions and their potential role in biological control. *Plant and Soil*, 185, 241-251.

[82] Graham, J.H. & Menge, J.A. (1982). Influence of vesicular-arbuscular mycorrhizae and soil phosphorus on take-all disease of wheat. *Phytopathology*, 72, 95-98.

[83] Shah, S.K., Shah, R.P., Xu, H.L. & Aryal, U.K. (2006). Biofertilizers: An Alternative Source of Nutrients for Sustainable Production of Tree Crops. *Journal of Sustainable Agriculture*, 29, 85-95.

[84] Osundina, M, A. (1997). Nodulation and growth of mycorrhizal *Casurarina equisetifolia* J.R and G. First in response to flooding. *Biology and Fertility of Soils*, 26, 95-99.

[85] Mejstřík, V.K. & Cudlín, P. (1983). Mycorrhiza in some plant desert species in Algeria. *Plant Soil*, 71, 363-366.

[86] Bilgo, A., Sangare, S. K, Thioulouse, J., Prin, Y., Hien, V., Galiana, A., Baudoin, E., Hafidi, M., Bâ, A.M. & Duponnois, R. (2011). Response of native soil microbial functions to the controlled mycorrhization of an exotic tree legume, *Acacia holosericea* in a Sahelian ecosystem. *Mycorrhiza*, DOI 10.1007/s00572-011-0390-2.

[87] Asai, M.T. (1934). Uber das Vorkommen und die Bedeutung der Wurzelpilze in den Landpflanzen. Jpn. *Journal of Botany*, 1, 107-150.

[88] Maeda, M. (1954). The meaning of mycorrhiza in regard to systematic botany. *Kumamoto Journal of Sciences*, 3, 57-84.

[89] Khan, A. G. (1978). Vesicular-arbuscular mycorrhizas in plants colonizing black wastes from bituminous coal mining in the Illawarra region of New South Wales. *New Phytologist*, 81, 53-63.

[90] Arriagada, C., Pacheco, P., Pereira, G., Machuca, A., Alvear, M. & Juan Antonio Ocampo J. A. (2009). Effect on arbuscular mycorrhizal fungal inoculation on *Eucalyptus globulus* seedlings and some soil enzyme activities under application of sewage sludge amendment. *Journal of Soil Science and Plant Nutrition,* 9, 89-101.

[91] Adjoud, D., Plenchette, C. & Halli-Hargas, R. (1996). Response of 11 *Eucalyptus* species to inoculation with three arbuscular mycorrhizal fungi. *Mycorrhiza*, 6, 129-135.

[92] Adjoud-Sadadou D. & Halli-Hargas, R. (2000). Occurrence of arbuscular mycorrhiza on aged Eucalyptus. *Mycorrhiza*, 9, 287-290.

[93] Boudarga, K. & Dexheimer, J. (1988). Etude ultrastructurale des endomycorhizes a vesicules et arbuscules de jeunes plants *d'Eucalyptus camaldulensis* (Denhardt) Myrtacees. Bulletin de la Société Botanique de France 135, 111-121.

[94] Chilvers, G.A. & Pryor, L.D. (1965). The structure of eucalypt mycorrhizas. *Australian Journal of Botany*, 13, 245-259.

[95] Thapar, H.S., Singh, B. & Bakshi, B.K. (1967). Mycorrhizae in *Eucalyptus. Indian Forester*. 93, 756-759.

[96] Thoen, D. & Ducousso, M. (1989). Mycorrhizal habit and sclerogenesis of *Phlepobus sudanicus* (Gyrodontaceae) in Senegal. *Agriculture Ecosystems and Environment*, 28, 519-523.

[97] Dell, B., Malajczuk N. & Dunstan, W.A. (2002) Persistence of some Australian *Pisolithus* species introduced into eucalypt plantations in China. *Forest Ecology and Management,* 169, 271-281.

[98] Kourtev, P. S., Ehrenfeld, J. G. & Haggblom, M. (2003). Experimental analysis of the effect of exotic and native plant species on the structure and function of soil microbial communities. *Soil Biology and Biochemistry*, 35, 895-905.

[99] Kourtev, P. S., Ehrenfeld J. G. & Huang, W. (2002). Exotic plant species alter the microbial community structure and function in the soil. *Ecology*, 83, 3152-3166.

[100] Remigi, P., Faye, A., Kane, A., Deruaz, M., Thioulouse, J., Cissoko, M., Prin, Y., Galiana, A., Dreyfus B. & Duponnois, R. (2008). The exotic legume tree species *Acacia holosericea* alters microbial soil functionalities and the structure of the arbuscular mycorrhizal community. *Applied and Environmental Microbiology*, 74, 1485-1493.

[101] Kisa, M., Sanon, A., Thioulouse, J., Assigbetse, K., Sylla, S., Spichiger, R., Dieng, L., Berthelin, J., Prin, Y., Galiana, A., Lepage, M. & Duponnois, R. (2007). Arbuscular mycorrhizal symbiosis counterbalance the negative influence of the exotic tree species *Eucalyptus camaldulensis* on the structure and functioning of soil microbial communities in a sahelian soil. *FEMS Microbiology Ecology*, 62, 32-44.

[102] Frey-Klett, P., Chavatte, M., Clausse, M.L., Courrier, S., Le Roux, C., Raaijmakers, J., Martinotti, M.G., Pierrat, J.C. & Garbaye, J. (2005). Ectomycorrhizal symbiosis affects functional diversity of rhizosphere fluorescent pseudomonads. *New Phytologist*, 165, 317-328.

[103] Meyer, J.R & Linderman, R.G. (1986). Selective influence on populations of rhizosphere or rhizoplane bacteria and actinomycetes by mycorrhizas formed by *Glomus fasciculatum*. *Soil Biology and Biochemistry*, 18, 191-196.

[104] Andrade, G., Mihara, K. L. & Linderman, R. G. (1998). Soil aggregation status and rhizobacteria in the mycorrhizosphere. *Plant Soil*, 202, 89-96.

[105] Marschner, P. & Timonen, S. (2005). Interactions between plant species and mycorrhizal colonization on the bacterial community composition in the rhizosphere. *Applied Soil Ecology*, 28, 23-36.

[106] Marschner, P., Yang, C.H., Lieberei, R. & Crowley, D.E. (2001). Soil and plant specific effects on bacterial community composition in the rhizosphere. *Soil Biology and Biochemistry*, 33, 1437-1445.
[107] N.A.S. (1980) Firewood Crops. Shrub and Tree Species for Energy Production. *National Academy of Sciences*, Washington, DC.

Reviewed by:
Yves Prin, LSTM (Laboratoire des symbioses tropicales et méditerranéennes), Montpellier, Yves.Prin@cirad.fr, Tel: 33 4 67 59 37 86, Fax : +33 4 67 59 38
Silvio Gianinazzi, INRA (Institut National de la Recherche Agronomique), Dijon, silvio.gianinazzi@dijon.inra.fr, Téléphone +33 (0) 3 80 69 31 46, Fax 03 80 69 37 53
Thierry Beguiristain, LIMOS (Laboratoire des Interactions Microorganismes-Minéraux-Matière), UMR 7137) du CNRS et de l'Université Henri Poincaré de Nancy Université, Nancy, thierry.beguiristain@limos.uhp-nancy.fr, Tel : 03 83 68 42 96, Fax : 03 83 68 42 84

In: Fungi: Types, Environmental Impact and Role in Disease ISBN: 978-1-61942-671-9
Editors: Adolfo Paz Silva and María Sol

Chapter XII

Disentangling Functions of Fungal Endophytes in Forest Trees

B. R. Albrectsen[1] and J. Witzell[2]
[1] Umeå Plant Science Centre, Department of Plant Physiology,
Umeå University, SE, Umeå, Sweden
[2] Swedish University of Agricultural Sciences,
Faculty of Forest Sciences,
Southern Swedish Forest Research Centre,
SE, Alnarp, Sweden

Abstract

Endophytic fungi are known to be abundant colonizers of the internal tissues of forest trees, but their ecological functions are still largely unknown. Recent studies indicate that endophytes may associate with tree's resistance and tolerance properties, and they are thus potential bio-agents that could be utilized in sustainable forest protection and management. To gain a better understanding of the endophytes' potential role in shaping forest health we need more evidence in the form of ecological studies of endophyte communities, in various tissues, across space, and time.

The recent advances in molecular methods have given us new and effective tools to obtain such data. Studies of endophyte functions are further facilitated with the development of new high through-put screening methods for substrate use and competitive ability. Fungi are known as chemical factories of natural compounds with biological properties. Beside their potential as antagonists against pests and diseases, the tree-associated endophytic fungi therefore also appear as an emerging source of novel biomolecules for industrial or clinical applications outside forestry.

This chapter presents some of the current methodological approaches that are likely to be valuable in studies on endophyte diversity in forest trees, and discusses the goals and impacts of the studies that aim at disentangling the beneficial potential of fungal endophytes in trees. A new concept, bioactive symbiosis, is suggested as a general framework for these studies.

Introduction

The mutualistic potential of plant's endophytic fungi has received considerable attention after Bacon et al (1977) isolated the fungus *Neotyphodium* from pastures where cattles suffered from fescue toxicity syndrome [1]. The fungus' ability to produce ergot alkaloids provides its host with a defensive advantage against herbivores referred to as defensive mutualism [2]. *Neotyphodium* is tightly bound to its host, it gets its nutrients from the plant, and its hyphae grow systemic throughout the grass tissue and infect seeds so that germinating seedlings will inherit the fungal associate. Most research on endophytic fungi has been confined to studies of Clavicipitaceous grass endophytes like *Neotyphodium* (Table 1). This group of fungi consists of relatively few species which are more or less tightly associated with their grass host. Beside anti-herbivory advantages, these endophytes provide their host with other advantages like tolerance to abiotic stress and pathogens [3].

Table 1. The number of scientific papers about grass versus tree endophytes, envisioned through search in Web of Science, Timespan=All Years. Databases=SCI-EXPANDED, SSCI, A&HCI, CPCI-S, CPCI-SSH

Endoph*	Fung*	Herbivor*	Grass*	Tree*	Hits
X	X	X	X		179
X	X	X		X	39
X	X		X		950
X	X			X	480
X			X		1290
X				X	604

However, the many studies that confirm the mutualistic relationship with clavicipitaceous endophytes have been criticized for all being conducted in well fertilized pasture systems [4]. Under plentiful environmental conditions the costs of hosting the endophyte may not limit plant fitness, and mutual benefits may prevail, whereas the cost benefit balance may swift under less favourable conditions, leading to antagonistic relationship between the plant and the endophyte [5]. Such relational swifts have been described in natural populations of *Festuca rubra,* along gradients of increasing herbivory, nutrient availability, and soil moisture [6].

Collaborative associations between mycobionts and photobionts are known from mychorriza and lichens, and obligate endosymbiosis is also well described at the cell level (i.e. prokaryote origin of mitochondria and chloroplasts). Indeed, some scientists regard collaborative associations between organisms as the main driver of evolution [7]. The endophytic fungi, by definition, live invisibly and symptomless for at least part of their life time within their host [8], and no plants studied so far (including cold season grasses like fescue) have been found to be free of endophytic fungi. Screenings of endophytic communities often show that they exhibit high diversity, representing a wide range of taxonomic groups in varying abundances. There are some reported examples of non-clavicipitaceous endophytes that seem to have anti-herbivory effects, but so far the non-clavicipitaceous studies are too few to support the generality of defensive mutualistic associations [4] (Table 1). However there is increasing evidence of conditioned relationships between a plant's phenotype and its associated endophytic community [9-11].

Endophytic fungi provide examples of both symbiogenic and more loosely associated plant-fungal mutualistic relationships. This is also captured in the classification by Rodriguez et al. [12] who divide fungal endophytes after host range, colonization and transmission patterns, tissue specificity and symbiotically conferred fitness benefits. Class 1 sensu Rodriguez et al. [12] contains the Clavicipitaceous grass endophytes, which include both systemic, seed transferred, and asymptomatic fungi and non-systemic, ascospore transferred, symptomatic grass endophytes. The systemic endophytes are known to express low taxonomic diversity and high infection rates [3] suggesting that low diversification may be characteristic of endophytes that are more tightly connected with their host plants. If this is a general pattern it should also be relevant for non-clavicipitaceous endophytes, and the tightly associated fungi that Rodriquez et al. [12] group as class 2 should have the highest potential for mutualism. The class 2 endophytes include fungi that colonize below and above ground plant parts extensively and are transmitted vertically or horizontally. Many of them provide documented fitness benefits for the plant, including promoted habitat adaptation, increased biomass, and tolerance against abiotic and biotic stress. These fungi colonize the host plants in diverse ways and appear to be taxonomically less diverse compared to other non-clavicipitaleous groups. In the classification by Rodriguez et al. [12], the remaining non-clavicipitaleous fungi are divided after tissue affinity, and more precisely after whether they colonize above ground (class 3) or below ground (class 4) plant parts. These fungi are transmitted horizontally and are less directly coupled with the host plant, as compared to the class 2 endophytes. However little is still known about the specific life cycles of individual endophyte species, or biodiversity of entire host plant communities; and the taxonomic classification of endophytes is likely to change as our knowledge about these topics grows.

Endophytic Fungi in Trees

Forest trees harbor a plethora of endophytic fungi that live at least a part of their life within the tissues of their host plants without causing symptoms [13-15]. Due to the hidden life of endophytes and their apparently high diversity, as well as the large size and extended life span of forest trees, it has been particularly difficult to address questions related to tree endophyte composition and function. In trees, endophytes are likely to be among the first potential decomposers of dying and dead leaves and wood. Endophytes may therefore significantly affect the dynamics of the decomposition process by affecting the flow of carbon, nitrogen and other nutrients in forest ecosystems. In recent years, the interest for endophytic fungi in forest trees has increased (Fig 1), and while the main emphasize has been on phylogenetic and taxonomic questions, attention has also been given to anti-biotic effects and the production of bio-active compounds [5, 16-18].

Trees are generally rich in ascomycetes and deuteromycetes and the composition of tree's endophyte communities may be related to host species or host clonal properties [19-27], which may be over-layed with the impact of environmental factors [13, 19, 28-29]. Endophytes in trees include known and unknown species with expected functions ranging from mutualistic and antagonistic, to novel or unknown [22, 24]. Although their function, in most cases, is still unknown, the apparent diversity invites for intelligent screening of these fungi for novel natural products [17-18], and there is a need to develop new fast methods to recognise and classify fungal endophytes [23].

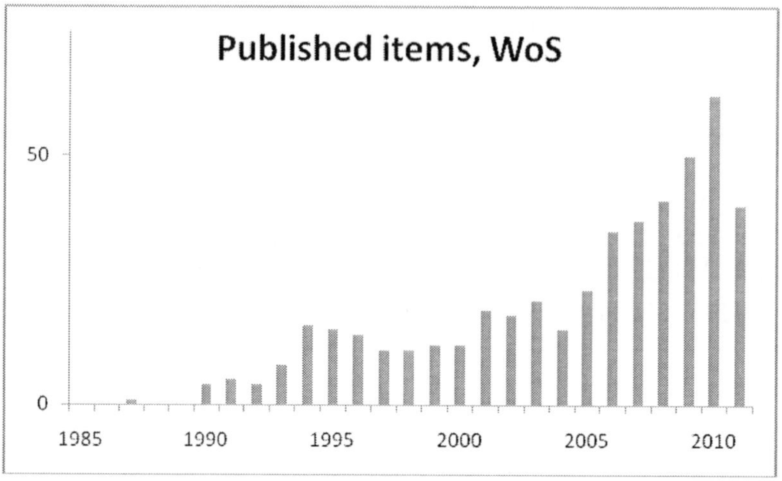

Figure 1. The publication history of 480 scientific papers that contain the words "endoph*", "fung*", and "tree". Search in Web of Science, Timespan=All Years. Databases=SCI-EXPANDED, SSCI, A&HCI, CPCI-S, CPCI-SSH.

No tight association with a systemic fungus similar to that of *Neotyphodium* in the cold season grasses has so far been reported for trees, and unequivocal examples of a shared defence strategy are sparse. However, several studies have suggested that endophytic fungi may negatively affect development and survival of insect herbivores, e.g. in Douglas fir (Pseudotsuga merziesii) [30], White elm (*Ulmus americana*) [31], Scots pine (*Pinus sylvestris*) [32], Garryara oak (*Quercus garryara)* [33-34], and Emory oak (*Quercus emoryi*) [35]. Furthermore the results of Arnold et al. [22] indicate that a limit in pathogen damage on *Theobroma cacao* was related with the presence of fungal endophytes, and Newcombe et al. [36] reported limited rust damage when poplar leaves had been pre-infected with endophytes. Other studies, however, suggest a lack of direct endophytic effects on tree's resistance (for example in *Betula spp.* [29, 37]), different effects to different herbivore species (*Q. garryara*, [35]), or negative effects on parasitoids implying that endophyte presence can associate with a positive feed-back on herbivore survival (in *Q. emoryi* [38]). In general, the literature seems to support a view that tree endophytes are indicators of host properties, either by association or possibly as part of shaping them.

Endophytes receive nutrients from their host and therefore basically have a parasitic life style, but at the same time the endophyte share fitness advantages with its host, and what strengthens the plant may consequently also benefit the endophyte. If an endophyte does not affect host interactions with other organisms it may be considered neutral (Fig 2a). If on the other hand the endophyte takes active part in plant defence this may happen in two distinctively different ways: either through direct interactions with the antagonist (Fig 2b), or indirectly through enhancement of the plant's tolerance or ability to defend itself (Fig 2c). In the case of simple, well described systems like *Neotyphodium* the endophyte component may be manipulated and studies can be performed to disentangle the nature of the interaction. In the case of diverse communities of fungi like those found in trees, studies may have to be performed at the community level before single effects may be recognized and studied specifically. There is some evidence that entire communities of endophytes may be associated with specific tree species, genotypes and environments including tree specific phenolic profiles (Bailey et al. 2005, Albrectsen et al. 2010) [25, 27].

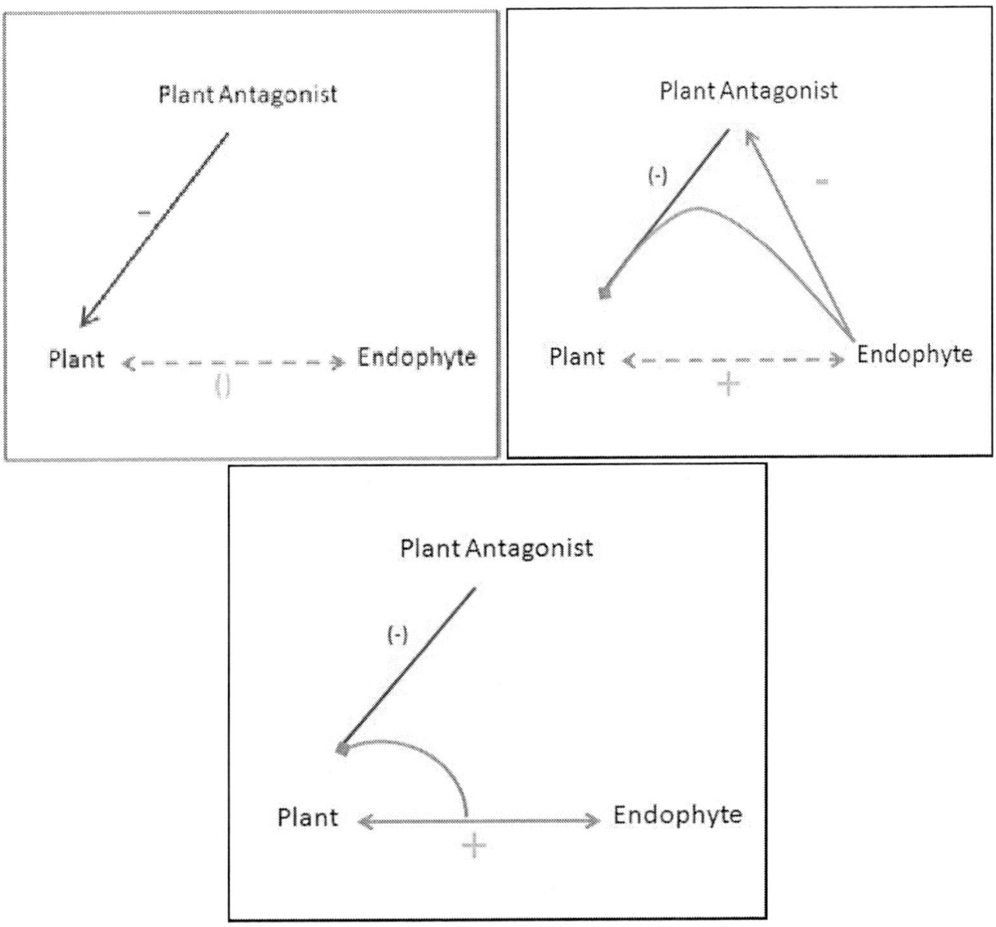

Figure 2. Defensive mutualism is an example of indirect plant defence, when the plant collaborates with another organism to obtain resistance benefits. The presence of endophytic fungi in plant tissues may be neutral when the endophyte does not have any impact on the plants interaction with other organisms (a). Defensive mutualism is when the endophyte causes resistance advantages for the host plant. This may happen if the endophyte directly limits the antagonist (b) or if it enhances the plants resistance properties and thus indirectly limits the negative effect of the antagonist (c).

Defining the Endophyte Communities with New Methods

Endophyte communities have conventionally been studied by culture-dependent methods where surface sterilization of tissues and subsequent growth of isolates on nutrient-rich media has been followed by morphotyping on the basis of colony characters and examination of spores (e.g., [22]), in some cases combined with Sanger-based DNA-sequencing [27]. Using this approach, large numbers of endophytes can be detected and the method provides a technically uncomplicated and cost-effective method to compare differences in the diversity of certain endophyte classes across plant materials, or to examine the presence of target fungi in a sample. Clearly, our current knowledge about the endophyte communities is greatly indebted to the skillful use of this approach in earlier studies (e.g. [22, 39]). However, researchers using the culture-dependent approach are also frequently pointing out the obvious limitations of the method. The basic limitation is that many endophytes are nonculturable on standard media or even on a selective media, remain sterile, or may escape the analysis in other ways, e.g. due to their extremely slow growth rate. Therefore, a comprehensive characterization of endophyte communities in trees may rarely be successful using this approach alone.

The field of characterizing entire microbial communities is, however, under rapid development, and next-generation sequencing (NGS) techniques are rapidly replacing traditional Sanger based techniques [40-42]. The instrument capacity has exploded over the past decade and what was a major sequencing task only a few years ago is now achievable for standard capacity facilities [41]. Metagenomics is a culture-independent analysis that uncovers sequence-based information of collective microbial genomes contained in an environmental sample [42]. The metagenomic analysis of microbial communities are based on DNA or RNA extraction which are used for different purposes: DNA libraries may be used for bioprospecting, and DNA sequences for assessment of taxonomic diversity and metabolic potential; RNA extracts may provide information about community functions and active community members [43].

Recent years have witnessed an explosive increase in metagenomic studies applied to fungal communities, based on the pyrosequencing technology. For example, Jumpponen and Jones [44] used 454-pyrosequenced [45] DNA-tagged [46, 47] PCR amplicons to study the fungal richness, diversity and community composition in the *Quercus macrocarpa* phyllosphere. A striking result of most of these analyses is the richness of uncultured taxa in the environmental samples, and methodology has been developed to set standards for both sampling and subsequent analyses [40, 48-49]. The progress in metagenomics is highly relevant for disentangling endophyte function in trees, and information on taxonomic diversity and associations are suddenly accessible through these high through-put methods. However, even the NGS approach is not free of errors due to artefacts and innate biological traits. In connection to validation of NGSs, and pyrosequencing in particular, concerns have been raised about problems associated with, e.g., nucleic acid extraction, primer selection and PCR, cell number and biomass in the samples as well as copy number instability within and between fungal species. All these factors may bias and skew the data and limit the utility and relevance of the information that can be extracted from these data [50]. Furthermore, the generally nonstandardized presentation of the NGS data is complicating the interpretation and comparison across studies [48]. It should also be kept in mind that targeting of the NSG analysis to endophyte communities is still dependent on selection of appropriate tissue samples. The use of DNA barcoding that employs short gene regions, e.g. internal transcribed spacer regions 1 and 2 (ITS) as the standard fungal DNA barcoding gene e.g. [51], or multilocus barcode approach [52], is currently broadly used as a sequence-based method to identify fungi. Yet, the identification of "unknown" fungi on basis of sequence data is strictly dependent on the availability of reliable voucher specimens [53] and quality of query tools and databases.

In addition to characterization of endophyte diversity, knowledge about functional properties is also highly needed, i.e. niche use strategies. The substrate-utilization profiles of isolated tree endophytes can be effectively recorded, e.g., using a phenotype microarray platform. This method is based on redox chemistry in a 96 well microtiter format, employing cell respiration as a universal reporter [54-55]. Briefly, a standardized cell suspension of each tested isolate is inoculated into the wells of microtiter plates, which have been loaded with pre-configured sets of organic substrates (specific arrays, i.e., plates, for carbon (C), nitrogen (N), phosphorus (P) or sulphur (S) compounds are commercially available). The development of color in wells correlates with the ability of cells to oxidize the compounds and is recorded over an incubation period with the help of a fully automatic system. The data is analyzed with a bioinformatics software. The method thus allows a high-throughput screening and comparison of phenotypes at cellular level. Using methods such as phenotype microarrays it is possible to gain detailed information about the substrate-use profiles of different fungi, which will help us to assess not only the competitive interactions between endophytes, but also determine which traits of forest trees may best provide specific qualitative or quantitative information about endophyte diversity.

Studies on Endophytes of Forest Trees: Knowledge Gaps and Links to Forest Management

As discussed above, endophytic fungi are likely to have diverse functions in trees, e.g., as enhancers of tree resistance to different stress factors or as a pool of potential opportunistic pathogens, respectively. These functions can be due to the general abundance and diversity of the endophyte community, or they may be attributed to the presence and frequency of specific endophytes. In either case, it is essential to understand how endophyte communities are formed, maintained and limited under different conditions and within different tissues of trees, and to what degree, and how, the different endophytes differentiate niches provided by the tree tissues. This understanding may significantly add to our ability to apply silvicultural practices (e.g. fertilization, intelligent choice of regeneration materials, maintenance of forest and stand structures) in a sustainable way, and where possible, to utilize endophytes to promote forest health and tree vitality.

In macro-ecology, the assembly, structure and dynamics of communities has been explained both by neutral theories and by tradeoff-based theories of interspecific competition as explained by Tilman [56] and refs. therein. The interspecific competition for resources is likely to be a central mechanism structuring the endophyte communities in trees, but also demographic stochasticity [56] may play a significant role in shaping of the prevailing endophyte assemblies in forest trees: The annual invasion of deciduous foliage [57] is likely to be dominated by the habitat-specific inoculum potential and therefore the availability of viable propagules in the environment is bound to have a marked impact on the endophyte communities in leaves. The endophyte communities in deciduous leaves should thus be more readily affected by environmental disturbances and resource pulses (e.g. fertilization and thinning operations), and more strictly recruitment limited [56], than the endophyte communities in the perennial tissues (bark and xylem) of the same trees. The colonization of the perennial tissues is likely to occur cumulatively over longer periods of time. In these tissues, the endophyte diversity may be mainly limited by the competition from resident fungi that occupy the relatively nutrient poor niche or micro-habitat. This effect may especially limit the invasion by fungi that represent the same or similar functional groups than the resident fungi [56].

Little is still known about the biochemical mechanisms that allow the maintenance of endophytes in planta. Once established, the endophytes have to cope with the chemical and physical environment inside the host, including constitutive and induced defensive mechanisms. The characteristic trait of the endophytic life style, i.e., the asymptomatic presence in the host plant, necessitates effective neutralization of plant defensive responses, either by avoidance or suppression. For instance, it is still unclear if the tree genotypes that show high resistance against aggressive fungal pathogens may also deter endophytic fungi. It is, however, apparent that the nature of the endophyte-host interaction can change from neutral to mutualistic, pathogenic or saprotrophic association [5, 14, 15, 37] and depending on the condition most fungi can adopt any one of these life-styles [58]. In a recent study, Alvarez-Loayza et al. [59] investigated the effects of light availability on disease expression by *Diplodia mutila*, a common endophyte/pathogen on the tropical palm (*Iriartea deltoidea*) that prefers shady conditions and successfully colonizes wet areas. They found that light conditions appear to influence where on the endosymbiont-pathogen continuum *D. mutila* is placed: while high light triggered the pathogenicity, apparently through light-induced production of H_2O_2 that triggers necrosis in the host tissues, low light promoted endosymbiotic development. This response constrained recruitment of endophyte-infested seedlings to shaded understory by reducing seedling survival in direct light [59]. It is conceivable that similar mechanisms, involving the interaction of environmental change and plant biochemistry, can trigger shifts from endophytic to pathogenic life style also in forest trees. This issue is of high contemporary interest for forestry, because the dual role of fungal communities as potential forest pathogens and as important part of the forest's biodiversity directly applies to the concept of sustainable forest management.

Conclusion

Clearly, the effect of fungal communities on stress tolerance and resistance of trees can be substantial. As a habitat-specific epigenetic mechanism [60] the fungal endophyte communities may profoundly challenge any attempts to improve the growth and resistance of trees by manipulation of plant genome or environment. Therefore, we propose that the assembly and dynamics of endophytic communities should not be ignored in silviculture e.g. forest health management [61], and in discussions of forest improvements for example when promoting genetically modified manipulations for desirable traits (GM trees). Through their ability to produce manifold of extracellular chemicals with various biological activities [16, 62] and manipulate host defensive chemistry [63], endophytes may contribute to at least part of the variations that have been recorded in different studies on plant defenses, such as phenolic compounds, and thereby push the plant quality beyond thresholds that matter e.g. to insect herbivores. Only through improved understanding of the ecological and biochemical mechanisms that regulate the recruitment and establishment of endophytic fungi in trees we can proceed to utilize the fungal communities as an integrated means to improve the health and vitality of forest trees. As a general guideline for future studies, we suggest that the apparently versatile (direct and indirect) functions of endophytes in forest trees could be fruitfully investigated under a concept of bioactive symbiosis, which we define as the ability of endophytes to shape plant resistance, either positively or negatively (depending on the conditions), against antagonists of any kind, including virus, bacteria, fungi, and herbivores.

References

[1] Bacon, CW; Porter JK, Robbins; JD Luttrell ES. (1977) Epichloë typhina from toxic tall fescue grasses. *Applied Environtal Microbiology*, 34, 576-581.

[2] Clay, K. (1988) Fungal endophytes of grasses - a defensive mutualism between plants and fungi. *Ecology*, 69, 10-16.

[3] Clay, K; Schardl, C. (2002) Evolutionary origins and ecological consequences of endophyte symbiosis with grasses. *American Naturalist*, 160, S99-S127.

[4] Saikkonen, K; Saari, S; Helander, M. (2010) Defensive mutualism between plants and endophytic fungi? *Fungal Diversity*, 41, 101-113.

[5] Schulz, B; Boyle, C. (2005) The endophytic continuum. *Mycological Research*, 109, 661-686.

[6] Saona, NM; Albrectsen, BR; Ericson, L; Bazely, DR. (2010) Environmental stresses mediate endophyte-grass interactions in a boreal archipelago. *Journal of Ecology*, 98, 470-479.

[7] Margulis, L. (1993) *Symbiosis in Cell Evolution* (2nd edn) W.H. Freeman & Co.

[8] Wilson, D. (1995) Endophyte - the evolution of a term, and clarification of its use and definition. *OIKOS*, 73, 274-276.

[9] Waller, F; Achatz, B; Baltruschat, H; Fodor, J; Becker, K, Fischer, M; Heier, T; Huckelhoven, R; Neumann, C; von Wettstein, D; Franken, P; Kogel, KH. (2005) The endophytic fungus Piriformospora indica reprograms barley to salt-stress tolerance, disease resistance, and higher yield. *Proceedings of the national academy of sciences of the United States of America*, 102, 13386-13391.

[10] van der Heijden, MGA; Bardgett, RD; van Straalen, NM. (2008) The unseen majority: soil microbes as drivers of plant diversity and productivity in terrestrial ecosystems. *Ecology Letters*, 11, 296-310.

[11] Aly, AH; Debbab, A; Proksch, P. (2011) Fungal endophytes: unique plant inhabitants with great promises. *Applied Microbiology and Biotechnology*, 90, 1829-1845.

[12] Rodrigues, RJ; White, JF; Arnold, AE; Redman, RS. (2009) Fungal diversity and functional roles. *New Phytologist*, 182, 214-220.

[13] Petrini, O. Fungal (1991). Endophytes of Tree Leaves. In Microbial Ecology of Leaves Andrews, J.H. and Hirano, S.S. (eds) Brock / Springer Series in Contemporary Bioscience p 179-196.

[14] Saikkonen, K. (2007) Forest structure and fungal endophytes. *Fungal Biological Reviews*, 21, 67–74.

[15] Sieber, TN. (2007) Endophytic fungi in forest trees: are they mutualists? *Fungal Biological Reviews*, 21, 75–89.

[16] Stierle, A; Strobel, G; Stierle, D. (1993) Taxol and taxane production by taxomyces-andreanae, an endopytic fungus of pacific yew. *Science*, 260, 214-216.

[17] Krohn, K; Biele, C; Drogies, KH; Steingröver, K; Aust, HJ; Draeger, S; Schulz, B. (2002) Biologically avtive secondary metabolites from fungi 18. Fusidilactones, a new group of polycyclic lactones from an endophyte, Fusidium sp. *European Journal of Organic Chemistry*, 2331-2336.

[18] Schulz, B; Boyle, C; Draeger, S; Römmert, AK; Krohn, K. (2002) Endophytic fungi: a source of novel biologically active secondary metabolites. *Mycological Research*, 106, 996-1004.

[19] Petrini, O; Fisher, PJ. (1990) Occurrence of fungal endophytes in twigs of Salix fragilis and Quercus robur. *Mycological Research*, 94, 1077-1080.

[20] Suryanarayanan, TS; Murali, TS; Venkatesan, G, (2002) Occurrence and distribution of fungal endophytes in tropical forests across a rainfall gradient. *Canadian Journal of Botany*, 80, 818-826.

[21] Suryanarayanan, TS; Venkatesan, G; Murali, TS. (2003). Endophytic fungal communities in leaves of tropical forest trees: Diversity and distribution patterns. *Current Science*, 85, 489-493.

[22] Arnold, AE;, Mejia, LC; Kyllo, D; Enith, I; Rojas, EI; Maynard, Z; Robbins, N; Herre, EA. (2003). Fungal endophytes limit pathogen damage in a tropical tree. *Proceedings of the national academy of sciences of the United States of America*, 100, 15649-15654.

[23] Arnold AE, Lutzoni F (2007) Diversity and host range of foliar fungal endophytes: Are tropical leaves biodiversity hotspots? *Ecology*, 88, 541-549.

[24] Ganley, RJ; Brunsfeld, SJ; Newcombe, G. (2004) A community of unknown, endophytic fungi in western white pine. *Proceedings of the national academy of sciences of the United States of America*, 101, 10107-10112.

[25] Bailey, JK; Deckert, R; Schweitzer, JA; Rehill, BJ; Lindroth, RL; Gehring, C; Whitham, TG. (2005). Host plant genetics affect hidden ecological players: links among Populus, condensed tannins, and fungal endophyte infection. *Canadian Journal of Botany*, 83, 356- 8137.2008.

[26] Manoharachary, C; Sridhar, K; Singh, R; Adholeya, A; Suryanarayanan, TS; Rawar, S; Johri, BN. (2005). Fungal biodiversity: Distribution, conservation and prospecting of fungi from India. *Current Science*, 89, 58-71.

[27] Albrectsen, BR; Björkén, L; Varad, A; Hagner, Å; Wedin, M; Karlsson, J; Jansson, S. (2010) Endophytic fungii in European aspen (populus tremula) leaves – diversity, detection, and a suggested correlation with herbivory resistance. *Fungal Diversity*, 41: 17-28.

[28] Elamo, P; Helander, ML; Saloniemi, I; Neuvonen, S. (1999) Birch family and environmental conditions affect endophytic fungi in leaves. *Oecologia*, 118, 151-156.

[29] Lappalainen, JH; J. Koricheva, J; Helander, ML; Haukioja, E. (1999). Densities of endophytic fungi and performance of leafminers (Lepidoptera: Eriocraniidae) on birch along a pollution gradient. *Environmental Pollution*, 104, 99-105.

[30] Carroll, G. (1995) Forest endophytes - pattern and process. *Canadian Journal of Botany*, 73, 1316-1324.

[31] Webber, JF. (1981) A natural biological control of Dutch elm disease. *Nature*, 292, 449-451.

[32] Saikkonen, K; Helander, ML; Ranta, H; Neuvonen, S; Virtanen, T; Suomela, J; Vuorinen, P. (1996) Endophyte-mediated interactions between woody plants and insect herbivores. *Entomological Experimentalis et Applicata*, 80, 269-271.

[33] Faeth, SH; Hammon KE. (1996) Fungal endophytes and phytochemistry of oak foliage: determinants of oviposition preference of leafminers? *Oecologia*, 108, 728-736.

[34] Faeth, SH; Hammon, KE. (1997) Fungal endophytes in oak trees; long term patterns of abundance and associations with leafminers. *Ecology*, 78, 810-819.

[35] Wilson, D; Carroll, GC. (1997) Avoidance of high-endophyte space by gall-forming insects. *Ecology*,78, 2153-2163.

[36] Newcombe, G; Martin, F; A, Kohler. (2010) Defense and nutrient mutualisms in Populus. In: *Genetics and Genomics of Populus*. S. Jansson et al. (eds.) Springer-Verlag, Berlin, Heidelberg, New York.

[37] Ahlholm, JU; Helander, M;, Lehtimäki, S; Wäli, P; Saikkonen, K. (2002) Vertically transmitted fungal endophytes: different responses of host-parasite systems to environmental conditions. *OIKOS*, 99, 173-183.

[38] Preszler, RW; Gaylord, ES; Boecklen, WJ. (1996) Reduced parasitism of a leaf mining moth on trees with high infection frequencies of an endophytic fungus. *Oecologia*, 108, 159-166.

[39] Fröhlich, J; Hyde, KD; Petrini, O. (2000). Endophyte fungi associated with palms. *Mycological Research*, 104, 1202-1212.

[40] Wooley, JC; Godzik, A; Friedberg, I. (2010) A Primer on Metagenomics. *PLoS Computunal Biology*, 6(2).

[41] Mardis, ER. (2011) A decade's perspective on DNA sequencing technology. *Nature*, 470, 198–203.

[42] Riesenfeld, CS; Schloss, PD; Handelsman, J. (2004) Metagenomics: Genomic analysis of Microbial Comunities *Annual Review Genetics*, 38, 525-52.

[43] Simon, C; Daniel, R. (2011) Metagenomic analyses: past and future trends. *Applied and Environmental Microbiology*, 77, 1153-1161.

[44] Jumpponen, A; Jones, KL. (2010) Seasonally dynamic fungal communities in the Quercus macrocarpa phyllosphere differ between urban and nonurban environments. *New Phytologist*, 186, 496–513.

[45] Margulies, M; Egholm, M; Altman, WE; Attiya, S; Bader, JS; Bemben, LA; Berka, J; Braverman, MS; Chen, YJ; Chen, ZT et al. (2005) Genome sequencing in microfabricated high-density picolitre reactors. *Nature*, 437, 376–380.

[46] Hamady, M; Walker, JJ; Harris, JK; Gold, NJ; Knight, R. (2008) Error-correcting barcoded primers for pyrosequencing hundreds of samples in multiplex. *Nature Methods*,5, 235–237.

[47] Meyer, M; Stenzel, U; Hofreiter, M. (2008) Parallel tagged sequencing on the 454 platform. *Nature Protocols*, 3, 267–278.

[48] Nilsson, HR; Tedersoo, L; Lindahl, BD; Kjøller, R; Carlsen, T; Quince, C; Abarenkov, K; Pennanen, T; Stenlid, J; Bruns, T; Larsson, K-H; Kõljalg, U; Kauserud, H. (2011) Towards standardization of the description and publication of next-generation sequencing datasets of fungal communities. *New Phytologist*, 191, 314-318.

[49] Yilmaz, P. and 98 authors. (2011) Minimum information about marker gene sequence (MIMMARKS) and minimum information about any (x) sequence (MIxS) specification. *Nature Biotechnology*, 29, 415-420.

[50] Amend, AS; Seifert, KA; Bruns, TD. (2010) Quantifying microbial communities with 454 pyrosequencing: does read abundance count? *Molecular Ecology*, 19(24), 5555-5565.

[51] Seifert, KA. (2009) Progress towards DNA barcoding of fungi. *Molecular Ecological Research*, 9, 83-89.

[52] Roe, AD; Rice, AV; Bromilow, SE; Cooke, JEK; Sperling, FA (2010) Multilocus species identification and fungal DNA barcoding: insights from blue stain fungal symbionts of the mountain pine beetle. *Molecular Ecological Research*, 10, 946–959.

[53] Brock, PM; Döring, H; Bidartondo, M I. (2009) How to know unknown fungi: the role of a herbarium. *New Phytologist*, 181: 719-724.

[54] Garland, JL; Mills, AL. (1991) Classification and characterization of heterotrophic microbial communities on the basis of patterns of community-level sole-carbon-source utilization. *Applied Environtal Microbiology*, 57, 2351-2359.

[55] Garland, JL. (1997) Analysis and interpretation of community-level physiological profiles in microbial ecology. *FEMS Microbiological Ecology*, 24, 289-300.

[56] Tilman, D. (2004) Niche tradeoffs, neutrality, and community structure: A stochastic theory of resource competition, invasion, and community assembly. *Proceedings of the national academy of sciences of the United States of America*,101,10854–10861.

[57] Helander, H; Ahlholm, J; Sieber, TN; Hinneri, S; Saikkonen, K. (2007) Fragmented environment affects birch leaf endophytes. *New Phytologist*,175, 547–553.

[58] Rodriguez, RJ; Redman, RS. (1997) Fungal lifestyles and ecosystem dynamics: biological aspects of plant pathogens, plant endophytes, and saprophytes. *Advances in Botanical Research*, 24, 169– 193.

[59] Alvarez-Loayza, P; White, JF; Torres, MS; Balslev, H; Kristiansen, T; Svenning, JC; Gil, N (2011) Light converts endosymbiotic fungus to pathogen, influencing seedling survival and niche-space filling of a common tropical tree, Iriartea deltoidea. *PLoS One*, 6(1): e16386.

[60] Rodriquez, R; Henson, J; Van Volkenburgh, E; Hoy, M; Wright, L; Beckwith, F; Kim, YO; Redman, R. (2008) Stress tolerance in plants via habitat-adapted symbiosis. *The ISME Journal*, 2, 404–416.

[61] Newcombe, G. (2011). Endophytes in forest management: four challenges. In: Pirttilä AM; Frank, AC (eds) Endophytes of forest trees. Biology and Applications. Springer, Dortrecht, pp 251-262

[62] Strobel, G; Daisy, B. (2003). Bioprospecting for microbial endophytes and their natural products. Microbiol. *Molecular Biology Review*, 67, 491-502.

[63] White, JF; Torres, MS. (2010). Is plant endophyte-mediated defensive mutualism the result of oxidative stress protection? *Physiologia Plantarum*,138, 440-446.

In: Fungi: Types, Environmental Impact and Role in Disease ISBN: 978-1-61942-671-9
Editors: Adolfo Paz Silva and María Sol

Chapter XIII

Fungal Disease and the Diversity and Long-Term Dynamics of Forest Communities

Takashi Osono*
Center for Ecological Research, Kyoto University

Abstract

Fungal diseases influence the consequences of competition between host and non-host plants and their co-existence, leading to changes in the species richness and diversity in forest ecosystems in which the hosts are embedded. Long-term studies of forest communities infested by serious fungal diseases have demonstrated the crucial importance of fungal pathogens in the maintenance of forest biodiversity. Here, I review long-term dynamics of plant populations and communities infested by fungal diseases and discuss the roles of fungal pathogens in the maintenance of the biodiversity of forest communities. Fungal diseases causing the death of host tissues at the shoot level, such as zonate leaf blight of dogwood and Swiss needle cast of Douglas-fir, have been shown to cause short-term changes in the shoot dynamics and primary productivity of individual host trees. However, the impacts of these diseases on long-term dynamics of host populations and forest communities remain to be investigated. In contrast, fungal diseases causing severe mortality of host populations, such as dogwood anthracnose, beech bark disease, Dutch elm disease, chestnut blight, phytophthora dieback, and laminated root rot, have been demonstrated to significantly impact the diversity and long-term dynamics of forest communities. Infestation by these fungal diseases leads to a decline of the host populations and replacement by other plant species, which can result in an increase, decrease, or undetectable changes in the number of plant species in forest communities. The response of forest diversity to diseases is variable and appears to depend on the virulence of fungal pathogens, relative abundance of hosts in forest communities, and environmental factors.

* E-mail address: tosono@ecology.kyoto-u.ac.jp.

1. Introduction

Fungal pathogens are responsible for the dynamics of host plant populations, the consequences of competition between hosts and non-host plants, and hence the diversity and dynamics of natural plant communities [13, 23, 47]. For example, experiments using a pair of competing plant species in pots clearly demonstrated that the presence of host-specific fungal disease alters competitive outcomes and co-existence between the plants [2, 4, 9]. These observations are consistent with mathematical models simulating the effects of fungal disease on plant competition [14]. Bradley et al. [7] showed in another pot experiment that pathogens maintained the diversity of a plant community consisting of multiple plant species across generations.

Difficulties are expected, however, in demonstrating the impacts of fungal diseases on the dynamics of host populations and on the diversity of natural forest communities consisting of diverse, long-lived tree species. To detect the fungal impacts, long-term censuses of forest trees are obviously needed in infested forest stands. Such census data have been scarce, but a growing number of publications have become available since the late 1980s, especially since 2000, reporting long-term changes in host populations and forest stands infested with several major fungal diseases. The purpose of this chapter is to synthesize these data on the dynamics of host populations and long-term changes in forest communities infested with fungal diseases and to discuss the effects of fungal pathogens on the diversity of forests.

This chapter mainly focuses on the interaction between fungi and forest canopy trees in natural conditions, because canopy trees define much of the structure of a forest community and loss of these trees has major consequences for forest dynamics. In addition, responses of understory and forest floor vegetations to fungal diseases are taken into account in some cases. Effects of fungal diseases on the diversity at multiple trophic levels, such as herbivores, carnivores, and decomposers, are out of the scope of this chapter because the space is too limited for a full explanation. This chapter accordingly emphasizes fungal pathogens attacking mature trees. A number of studies have already demonstrated that fungal diseases cause rapid loss of seeds, seedlings, and saplings on the forest floor [75, 77, 81, 96] and reduce the numbers of pollens and seeds produced in flowers [83]. Rapid turnover of these juvenile populations makes it easier to detect the effect of fungal pathogens on population dynamics of host plants. The roles of fungal diseases attacking juvenile plants in the regeneration of forest trees and the maintenance of forest diversity have been well documented [15, 44] and are not included in this chapter.

2. Linking Short-Term Effects of Pathogen Infection to Long-Term Dynamics of Forest Diversity

Fungal pathogens exert short-term, negative effects on growth and survival of their hosts, which can result in long-term changes in forest communities in which hosts are embedded (Figure 1). Pathogens attack individuals of susceptible host trees and cause such symptoms as defoliation, wilt, canker, and root rot, leading to reductions in primary production and fecundity, water defect, and occasionally the death of the individual trees that the pathogens infected. These are the short-term effects of pathogen on host occurring at timescales of months to years. As a result, fungal disease in host populations can cause changes of the population features of the host, for example, reduction of the population growth rate and increased mortality. Such changes in population features in turn affect interspecific interactions between host and non-host tree species comprising forest communities; that is, the competitive ability of host species relative to other neighbor non-host trees can be reduced. For example, if pathogens are specialists and attack their hosts in a density-dependent manner, then rare tree species should increase in relative abundance [15, 44]. Similarly, pathogens may promote coexistence of trees if there is a tradeoff between the competitive ability of a tree and its susceptibility to disease [4, 10, 14]. Hence, pathogens can influence long-term dynamics of forest diversity through changes in the community structure of trees on timescales of decades to centuries.

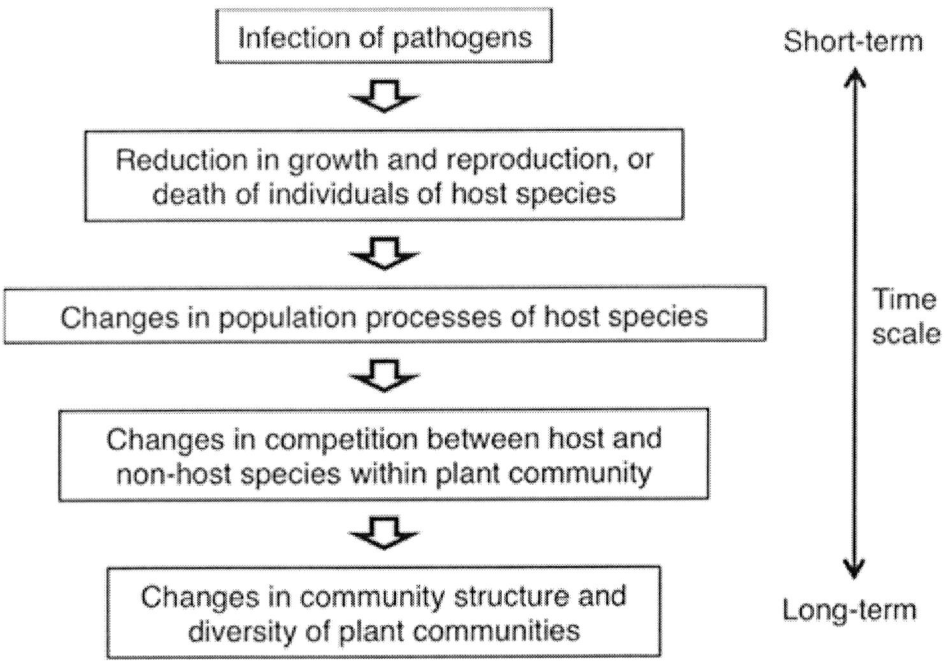

Figure 1. Effect of fungal disease on the diversity of plant community.

3. Fungal Disease and Mortality of Hosts at Shoot and Individual Levels

Individual adult trees generally harbor multiple infections of pathogenic fungal species, but not all of these pathogens are influential with respect to forest diversity. Pathogens usually evolve into less virulent ones because high virulence that can destroy host populations, such as that found in newly introduced, non-indigenous invasive pathogens, has negative outcomes for the survival of the pathogens and thus is less adaptive [28, 68]. A typical consequence of such evolution is biotrophic fungi such as rusts and powdery mildews that live obligatorily within plant tissues but are generally not so virulent as to kill shoots or individual trees (Figure 2).

Necrotrophic fungi of relatively low virulence can infect shoots of adult trees that are physiologically stressed by shading, desiccation, or attack by herbivores or other pathogens. These necrotrophic fungi can kill hosts at the shoot level and cause partial defoliation, reduced production of flowers, pollen, and seeds, and limited root death. The impacts of these necrotrophic pathogens at the individual tree and host population levels are expected to be relatively small (Figure 2), unless these pathogens attack hosts continuously and exert negative effects on host health cumulatively. These pathogens can persist as an endemic component of natural ecosystems [29]. Their impacts on short-term dynamics of shoot populations within individual trees have been investigated for such pathogens as zonate leaf blight on leaves of giant dogwood (Table 1). However, their impacts on long-term dynamics of forest communities and diversity are difficult to detect and yet to be demonstrated. Case studies of Swiss needle cast (Table 1) will have useful implications for scaling-up the mortality at the shoot level to the productivity and fitness of individual host trees.

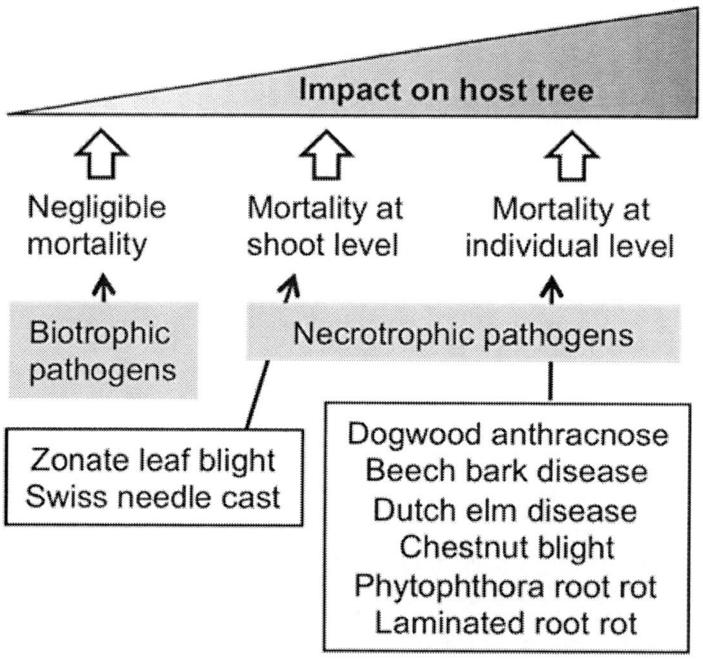

Figure 2. Fungal disease and mortality of hosts at shoot and individual levels.

In contrast, pathogens causing such diseases as butt rot and stem canker kill individual hosts relatively quickly and hence have greater impacts on forest dynamics and diversity. Those that spread rapidly and have high incidence are termed epidemic diseases. Epidemics, particularly those caused by exotic pathogens, may provoke dramatic, permanent changes in forest communities [18, 50, 52, 53]. Examples include beech bark disease, Dutch elm disease, chestnut blight, and phytophthora dieback (Table 1). Some native pathogens can also cause rapid mortality and stand-replacing effects in a forest, for example, *Phellinus weirii*, the fungus responsible for laminated root rot (Table 1). *Discula destructiva*, the causal agent of Dogwood anthracnose (Table 1) attacks leaves at the shoot level, but in North America (where it was introduced as an exotic pathogen) the fungus causes severe defoliation and kills mature individual trees quickly. Several studies have demonstrated that these fungal diseases have significant impacts on long-term dynamics of host and non-host trees and forest diversity.

Table 1. A summary of fungal diseases mentioned in this chapter

Disease	Pathogen	Major symptom	Host	Effects on diversity[a]
Zonate leaf blight [58]	*Haradamyces foliicola*	Defoliation	Host non-specific; dogwood is the most susceptible	Nt
Swiss needle cast [94]	*Phaeocryptopus gaeumannii*	Defoliation	Douglas-fir	Nt
Dogwood anthracnose [17, 71]	*Discula destructiva*	Defoliation	Dogwood	Nt
Beech bark disease [42]	*Nectria coccinea* var. *faginata*, *N. galligena*	Stem canker	Beech	Nt
Dutch elm disease [8]	*Ophiostoma ulmi*, *O. novo-ulmi*	Wilt	Elms	Pr/Nd/Re
Chestnut blight [3]	*Cryphonectria parasitica*	Stem canker	American chestnut	Pr
Phytophthora dieback [35]	*Phytopathora cinnamomi*	Root rot	Host non-specific	Nd/Re
Laminated root rot [32]	*Phellinus weirii*	Root and butt rot	Conifers; Douglas-fir is the most susceptible	Pr/Nd/Re

*Effects on forest tree diversity: Pr promoted, Nd not detected, Re reduced, Nt not tested.

4. Fungal Disease Alters the Dynamics of Shoot Populations: A Case Study of Zonate Leaf Blight

The effects of fungal disease on the defoliation and the dynamics of shoot populations were demonstrated for zonate leaf blight (ZLB). ZLB is one of the most destructive foliar diseases of woody plants in Japan, and causes severe defoliation of diseased leaves during the early growing season, occasionally leading to the death of trees. *Haradamyces foliicola*, the causal agent of ZLB, was recently described as a new genus and species [58]. Its pathogenicity has been confirmed for many tree species, with Cornaceae and Theaceae being highly susceptible to ZLB. Osono et al. [66] described the leaf phenology of giant dogwood (*Swida controversa,* Cornaceae) to follow the seasonal pattern and to quantify ZLB infection and defoliation. The study was conducted during the growing season in 2003, when a severe outbreak of ZLB was evident. The observation started in April, when the buds broke to expand new shoots. The proportion of shoots with blighted leaves increase from July, resulting in a 66% decrease of leaf number per shoot in August compared to the leaf number in July (Figure 3). The leaf number per shoot increased slightly again in September, due to compensatory expansion of apical buds that would have broken in the spring of the succeeding year. The leaf number then decreased thereafter due to ZLB and natural leaf senescence. Obviously, the number of leaves on shoots of giant dogwood was much lower during the growing season in the year of severe defoliation than in a normal year without a ZLB outbreak (Figure 3). Such severe defoliation can reduce the annual net primary production of individual dogwood trees. The study of Osono et al. [66] quantified the defoliation and the dynamics of shoot populations within individual trees, but the loss of primary production due to ZLB was not quantitatively evaluated at the shoot level or at the individual tree level, nor was its impact on the population dynamics of the host tree.

5. Scaling-up the Dynamics of Infested Shoots to the Loss of Productivity in the Canopy of Individual Trees: A Case Study of Swiss Needle Cast

The dynamics of diseased shoot populations was scaled up to the primary production of individual trees for a Douglas-fir (*Pseudotsuga menziesii*) foliar disease, Swiss needle cast (SNC). The frequency and severity of SNC have increased since the mid-1980s in the Pacific Northwest region [33], leading to intensive studies regarding the impact of this pathogen. *Phaeocryptopus gaeumannii*, the causal agent of SNC [94], infects needles to produce fruiting bodies that physically block stomata, leading to a reduction in CO_2 assimilation [54], photoinhibition [55], and a reduction in stomatal conductance in the remaining functional stomata [56]. These physiological alterations result in defoliation of infested needles. The defoliation due to SNC pathogen then changes the foliage age-class structure and vertical foliage distribution in tree canopies [89] and can drastically reduce the leaf area index [90]. According to these findings, Manter et al. [57] developed a mathematical model of needle-level stomatal conductance and net CO_2 assimilation (A_{net}) that includes the physiological impacts of SNC and can be scaled-up to whole-canopy estimates of stomatal conductance and A_{net}.

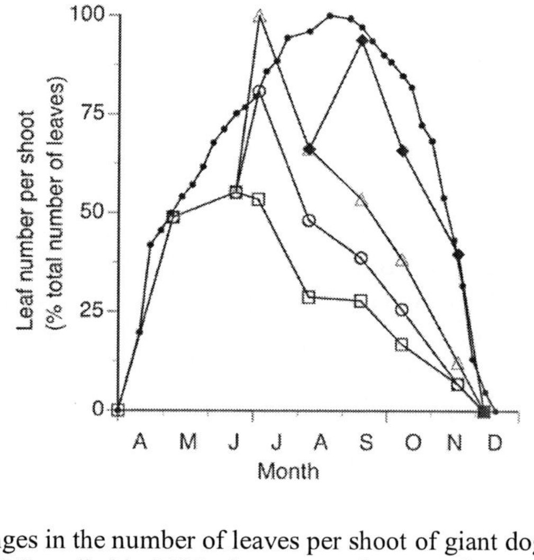

Figure 3. Phenological changes in the number of leaves per shoot of giant dogwood (*Swida controversa*). Osono et al. [66] studied in 2003 when severe damage of zonate leaf blight was observed during the growing season; each current-year shoot was divided into: first-order (□), second-order (○), third-order (△), and apical shoot (υ). The leaf phenology was compared with that in 1997 when the disease was not detected [36]; shoot-orders within current-year shoot were not distinguished (•). Leaf number per shoot was expressed as the proportion relative to the maximum number of leaves over the growing seasons.

The findings were that: i) for the needle age classes where colonization levels reached 25% pseudothecia density (as percent of stomata with visible fruiting bodies), estimated total carbon budgets were negative; ii) however, at the whole-canopy level, all trees maintained a positive carbon budget due to the large contribution from current-year needles that remained unaffected by the fungus for the first 6 months of development, or until the emergence of pseudothecia; and iii) the abscission of the older, more-heavily diseased foliage has a significant mitigating effect on whole-canopy A_{net}. In the latter case, SNC-associated reduction in A_{net} per unit leaf was estimated to reduce whole-canopy A_{net} more markedly without needle abscission than when older, more-heavily diseased needles were abscised.

The study of Manter et al. [57] shows the usefulness of mathematical approaches to scaling-up needle-level effects of fungal pathogens to canopy-level changes in primary productivity of individual host trees. In addition, further scaling-up to population and community levels is needed to examine the impacts of fungal infection at the shoot level on forest diversity. Long-term censuses of host tree populations and forest communities, as well as surveys of the infection rate at the shoot- and canopy-levels and changes in the primary production of individual trees are required for necrotrophic fungi of relatively low virulence. Practically, however, it would be difficult to link short-term effects of fungal disease at the shoot level, such as SNC, to long-term dynamics of forest communities, unless severe mortality of host trees had been observed.

6. High Mortality of Flowering Dogwood Caused by Dogwood Anthracnose Can Lead to Replacement by Non-Host Trees

Dogwood anthracnose causes severe defoliation and rapid mortality in trees of Cornaceae. Since the 1970s, this disease has greatly depressed populations of flowering dogwood (*Cornus florida*), a major component of understory trees throughout the eastern United States [84]. *Discula destructiva*, the causal agent of dogwood anthracnose [71], is considered to be exotic, possibly of Asian origin, but how the fungus reached North America remains unknown.

Rapid decreases in dogwood density have been reported in several populations [61, 78]. The mortality of flowering dogwood was the highest among co-occurring *Acer*, *Betula*, and *Quercus* species in the southern Appalachian Mountains [95]. The study of Williams and Moriarity [93] provided a striking example of fatal damage in a dogwood population in northwestern Pennsylvania: it reported that all living trees had symptoms of dogwood anthracnose, that non-diseased trees were not encountered, and that all dead trees had died due to the disease.

Flowering dogwood with smaller stem diameter, including saplings and seedlings, generally has higher levels of foliar infection and higher mortality than larger trees [12, 37, 45]. In trees of reproductive size, higher levels of anthracnose infection are associated with lower fruit production [12]. These disease-induced changes suppressed the regeneration of dogwood in these forests, leading to the establishment of non-host tree species. For example, Hires and Evans [37] reported an increase in two bird-dispersed tree species in a forest understory concurrent with declining dogwood populations, after 12-19 years of incidence of dogwood anthracnose. Jenkins and White [45] found that Eastern hemlock (*Tsuga canadensis*) has largely filled the compositional void created by the loss of dogwood in forest stands infested with dogwood anthracnose after 20 years of incidence. In these two studies, unfortunately, whether the replacements of tree species changed the overall species richness of the forest stands was not examined. Whether such a reduction in flowering dogwood altered the forest diversity has been yet to be demonstrated. To test the hypothesis that the loss of flowering dogwood in the forest understory would have an effect on forest diversity, long-term censuses are needed to follow the replacement of dogwood by saplings of dominant canopy tree species in the forest understory.

7. Minor Effect of Beech Decline due to Beech Bark Disease on Forest Communities

Beech bark disease (BBD) is caused by one of two ascomycete species, *Nectria coccinea* var. *faginata* or *N. galligena*, infecting minute feeding wounds made by the beech scale insect (*Cryptococcus fagisuga*) in the bark [42]. BBD was introduced to North America in the 1890s and then spread rapidly and became epidemic because American beech (*Fagus grandifolia*), a dominant tree species in the northeastern United States and southeastern Canada, was highly susceptible to the disease.

Shigo [79] described the spread of BBD in three stages. The first 'advancing front' stage consists of rising scale insect populations and low levels of *Nectria*. The 'killing front' stage follows, characterized by large populations of scale insects, severe outbreak of *Nectria*, and high beech mortality. The 'aftermath' stage consists of evidence of prior mortality, few older trees, and beech thickets made of small trees of sprout origin. In the 2000s, infested forests are entering the aftermath stage of the disease's progression, and BBD appears to have become an endemic component of forests [30].

Larger stems of American beech experienced higher mortality rates caused by BBD than smaller stems [25, 30, 74], in contrast with dogwood anthracnose. Sprouting from roots of dying larger stems also resulted in the establishment of smaller stems [24, 74]. Consequently, the distribution of beech diameter classes changed greatly in forest stands infested with BBD. Forrester et al. [25], for example, studied the changes in forest structure over a 15-year period to assess potential effects of BBD in an Adirondack hardwood forest. They reported an increase of beech density but no apparent change in its total basal area. This means that the increased number of small stems of beech compensated the decrease in basal area due to the death of large stems after BBD infestation.

Gaps were formed as a result of the death of large stems of beech after a BBD outbreak [48]. Observed in these gaps were replacement by eastern hemlock [74, 85], stimulated radial growth of sugar maple (*Acer saccharum*) [20], and regeneration of sprouted small stems of American beech [25]. Mathematical models to estimate long-term impacts of BBD on forest structure also indicated changes in the size structure of the beech population and an increase of eastern hemlock [49, 67], consistent with the field observations. BBD promotes seedling regeneration of hemlock because large logs of infested beech abundantly supplied to the forest floor serve as seedbeds [62, 67].

Few data have been available regarding the changes in the diversity of forest communities after BBD epidemics. Mortality rates of American beech reached 80-90% immediately after the introduction of BBD to North America [41], but long-term census data indicated that mortality rates were not so high after the 1970s [65]. Indeed, the increased number of small stems compensated the decrease in basal area of large stems caused by BBD infestation, as mentioned above, such that most beech populations have not become extinct. This was demonstrated by Twery and Patterson [85], who compared size structures of six beech populations between the 1930s to early 1960s and the 1980s, which corresponded to before and after the BBD epidemic, respectively, and found no noticeable changes in tree species composition. The authors speculated that BBD had caused minor compositional changes in most of the areas studied. Similarly, Forrester et al. [25] reported significant changes in the size class distribution of American beech and sugar maple but no consistent changes in tree species composition in the forest canopy. These results imply minor effects of BBD on the diversity of tree species in infested forests, but long-term censuses are needed to monitor the dynamics of forests infested with BBD.

8. Possible Impacts of Dutch Elm Disease on Forest Tree Diversity

A body of long-term census data are available for forest stands in North America and Europe that suffered epidemics of Dutch elm disease (DED) during the 20th century. These data indicate that number of tree species increased, was unchanged, or decreased over decades before and after the epidemics. It should be noted that these long-term changes in tree species richness are not simply attributable to DED because these studies lacked data of 'control' forest stands that escaped DED epidemics during the study periods. We can only estimate the possible contribution of DED to forest dynamics, relative to other factors affecting the long-term dynamics, such as senescence and death of trees, interspecific competitive interaction, windthrow, and anthropogenic disturbances such as harvesting and drainage.

Widespread mortality of elms (*Ulmus* spp.) due to DED epidemics has been reported since the 1910s in North America and Europe, and persists until today. An ascomycete fungus, *Ophiostoma ulmi*, was responsible for the first epidemic starting from the 1910s and peaking in the 1930s, whereas the second epidemic starting from 1960s was caused by a closely related species, *O. novo-ulmi* [8]. These fungi, spread by elm bark beetles, are considered to be exotic pathogens introduced to these continents.

A common pattern was noted regarding the population dynamics of elms infested with DED regardless of the study site [5, 69, 70, 72]. First, most of the canopy and subcanopy stems were killed due to DED infestation. Subsequent seedling regeneration and growth of sprouts from rootstocks of infected trees were substantial and vigorous under gaps formed by the death of canopy trees, leading to a temporary increase in the number of elm individuals. However, DED continued to kill elm trees that regenerated to reach the canopy and subcanopy layers. Consequently, young elms dominated the forest understory in forest stands infested with DED, but mature elms were excluded from the canopy [1, 59]. Huenneke [43] reported that shrub species typical of open habitats, such as *Alnus rugosa*, were abundant in multiple-tree gaps created by the death of two or more elms, whereas seedlings of canopy tree species were more abundant in single-tree gaps. The greater the number of dead elm stems in the canopy layer, the higher the density increase in the understory [21].

Hale et al. [31] described the changes in floodplain forests along the Lower Wisconsin River by comparing the data in the 1950s (before the DED outbreak) with those in 2001 (ca. 50 years after the DED outbreak). The number of large stems of American elm (*Ulmus americana*) declined during the 50-year period due to DED. In two study plots that had not been harvested, a minor change in tree species richness was found for one plot, whereas a decline in the number of tree species was found for the other plot. This decline was primarily associated with a shift of the forest to a previously unrecorded later successional stage dominated by maples and ashes, a possible result of river regulation to reduce floodplain disturbance. Possible effects of elm decline and concomitant formation of gaps could also be expected to affect the forest dynamics and tree species richness, but this could not be examined due to the lack of a 'control' plot that escaped DED infestation. Similarly, von Oheimb and Brunet [88] summarized secondary succession of deciduous forests in southern Sweden over a 77-year period between 1925 and 2002. The density of Wych elm (*U. glabra*) declined during that period due to the DED outbreak, as did the number of tree species in study plots (from 40 species in 1925 to 30 species in 2002). The decline in the number of tree species was due to the relative increase of a few dominant tree species in *Quercus*, *Fraxinus*, and *Fagus* and concomitant decrease in minor tree species due to competitive exclusion. Causal relationships between the elm decline and the decreased number of tree species remain unclear because of the lack of control plots, as in the case of Hale et al. [31]. In contrast to these two studies, the number of tall tree species increased from 5 species in 1972 before the DED outbreak to 10 species in 1996 after the DED outbreak in forest plots located in the southwestern part of the Czech Republic [76]. The density of Wych elm decreased in the canopy layer but increased in the understory during that period. The increased number of tall tree species was attributable not only to DED infestation but also to forest decline associated with the death of dominant *Picea* and *Abies* species due to acidic precipitation originating from anthropogenic activities. The three studies mentioned here documented the changes of diversity in forest stands infested with DED over a 24- to 77-year period. Elms were one of the predominant tree species in these study sites before the infestation, but were excluded from the canopy layer after the DED outbreak. No consistent pattern of changes of tree species richness was noted among these studies, or the impact of DED on the diversity was not evident due to the lack of 'control' treatments. Obviously, long-term observations of infested forest plots are not always enough to separate the effects of fungal disease from those of other factors affecting long-term dynamics and diversity of forests.

9. Increased Evenness and Species Richness due to the Decline of a Dominant Tree Species Caused by Chestnut Blight

American chestnut (*Castanea dentata*) was once predominant tree species in upland forest communities of eastern North America, accounting for 40 to 85% of the total basal area, until it was virtually eliminated by chestnut blight caused by a bark-inhabiting canker fungus, *Cryphonectria parasitica* (formerly *Endothia parasitica*), at the beginning of the 20th century [3]. *Cryphonectria parasitica* is an invasive pathogen from Asia [64], producing canker on twigs and stems that rapidly kills susceptible chestnuts. Studies of long-term vegetation changes indicated that the canopy dominated by chestnut was replaced by more than one species in infested forest stands, leading to an increase in evenness of species distribution and enhanced diversity [22, 80, 87]. The changes in forest diversity were not attributable solely to chestnut blight, because these studies had no 'control' plots that were not infested with chestnut blight, like the studies of Dutch elm disease. However, the data of the size distribution of regenerating tree species clearly support the notion that the change of diversity after the near-complete disappearance of the dominant chestnut trees was closely associated with chestnut blight infestation.

Elliott and Swank [22] used permanent forest plot inventories in the Coweeta Basin, North Carolina, sampled first in 1934 before chestnut blight introduction and twice afterward in 1969-72 and 1988-93 after the infestation. American chestnut was predominant tree species, contributing 36% of the total basal area in 1934. After chestnut blight infestation, however, it declined to less than 1% of the total basal area in 1988-1993 and was replaced by *Quercus* spp., *Acer rubrum*, and *Liriodendron tulipifera*, which had been relatively minor components in 1934. The number of tree species per plot declined slightly from 14.0 to 13.4 species during the period, but the diversity increased significantly over time, which was attributed to an increase in the evenness of species distribution. Diameter size class distribution of the tree species indicated that *Quercus* spp., *A. rubrum*, and *L. tulipifera* recruited in canopy gaps increased their relative abundance after the demise of American chestnut.

Stephenson [80] performed a similar investigation in upland forest communities in southwestern Virginia, in which American chestnut once made up 56% and 85% of the canopy in 1932, before chestnut blight introduction. In 1982, after chestnut blight infestation, chestnuts had completely disappeared from the canopy. This was followed by increases of *Quercus*, *Acer*, and *Betula* species, resulting in an increased number of canopy tree species. Similarly, Vandermast and van Lear [87] examined remaining stumps of chestnut in riparian forest stands and showed that forest stands co-dominated by the three *Quercus* species had been mono-dominated by chestnut. Unfortunately, Vandermast and van Lear [87] had no historical data on tree species composition and no description of forest diversity changes before and after chestnut blight infestation.

10. Comparison of Infested and Control Plots to Demonstrate the Effect of Phytophthora Dieback on Long-Term Forest Diversity

Phytophthora cinnamomi is a destructive and widespread soil-borne pathogen in Oomycetes causing damping-off in a wide range of vascular plant species [35]. This fungus causes the disease called phytophthora dieback on all continents except Antarctica and was introduced to Australia, possibly from Asia, before the first identification in 1930. More than 950 plant specie have been listed as susceptible and vulnerable hosts in Australia, including *Eucalyptus* and *Xanthorrhoea*, which are major components of Australian vegetation, and many genera of endemic plant taxa have a high proportion of susceptible species, a significant number of which are rare and threatened [35]. Consequently, *P. cinnamomi* has been identified as a key threatening agent of Australia's biodiversity [11]. Impacts of phytophthora dieback on forest diversity have been documented in southeastern and southwestern parts of Australia, where the destruction of susceptible vegetation has been observed. Species of *Phytophthora* also occur across extensive areas of tropical rainforest in northern Queensland, but most sites were reported to display no obvious signs of dieback or patch death [27].

Weste [91] described three phases of disease cycle in native vegetation infested with phytophthora dieback, on the basis of a 30-year observation of eucalyptus (*Eucalyptus* spp.) forest and woodland in Victoria, Australia. Epidemic disease was observed with death of all susceptible plant species in Phase 1, followed by the colonization of infested sites by field-resistant plant species in Phase 2. Phase 3 was characterized by the regeneration of susceptible plant species followed by a decline in pathogen populations. In the study of Weste [91], control plots were established adjacent to infested plots, making it possible to separate the effects of pathogen on plant diversity in each phase from the long-term changes in vegetation under a pathogen-free condition. Moreover, plant species were divided into susceptible, resistant, and unknown (in terms of disease susceptibility), so that long-term changes in species composition and species richness of plants of different susceptibility could be followed in relation to pathogen outbreak.

Weste et al. [92] established infested plots and pathogen-free (control) plots in three vegetation types of different moisture regimes in 1977 and followed the changes in the number of plant (trees + herbs) species until 2000 (Figure 4).
Eucalyptus and herbs susceptible to *P. cinnamomi* declined, and the relative abundances of resistant plants increased in the three vegetation types. In an open woodland site, the total number of plant species was relatively unchanged during the 24-year period in control plots, whereas the number of susceptible species decreased in infested plots until 1995. This indicated that the decrease in species richness in infested plots was attributable to phytophthora dieback. Similarly, the number of susceptible species decreased in infested plots of a low open forest and a tall open forest site. It should be noted, however, that the number of plant species also decreased in control plots of these vegetation types, suggesting that the decrease in susceptible species was not due only to the disease. Slight increases of susceptible species from 1995 to 2000 in infested plots of open woodland and tall open forest suggest that these plots proceeded to Phase 3 of the disease cycle described by Weste [91].

McDougall et al. [60] investigated the effects of phytophthora dieback on the number of plant (trees and herbs) species in jarrah forest of Western Australia. In the study area, populations of susceptible species including a dominant tall tree, *Eucalyptus marginata*, were eliminated due to the attack by *P. cinnamomi* during more than 70 years of impact. The overall numbers of plant species in 20- and 50-year-old infested sites were not significantly different from that in uninfested, healthy sites. Vegetation of infected sites had lower cover and fewer species of trees and shrubs, and higher cover and more species of annual plants, than vegetation of uninfested sites. Uninfested and infested sites were floristically similar, and 15 plant species that increased their cover at infested sites and hence were resistant to *P. cinnamomi*, were uncommon or absent at uninfested sites. Thus, the infestation of phytophthora dieback resulted in the decline of dominant trees and establishment of annual plants, but no apparent change in the total number of plant species.

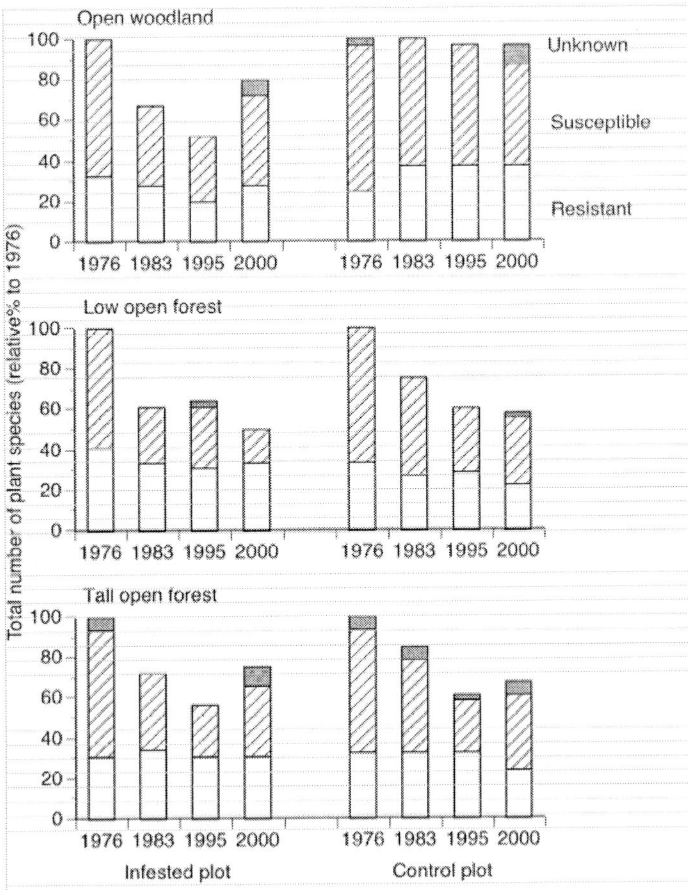

Figure 4. Changes in species richness of plants in plots infested and not infested with phytophthora dieback. Data after Weste et al. [92]. Infested and uninfested (control) plots were established in 1976 at three forest types dominated by Eucalyptus trees in Victoria, Australia and studied for vegetation changes for 24 years. Open woodland was located in the driest of the three forest types, whereas tall open forest was in the wettest, with low open forest in the intermediate moisture condition. No phytophthora dieback was observed in control plots during the study period. All plant species were divided into three groups (susceptible, resistant, unknown) according to their susceptibility to phytophthora dieback.

Phytophthora species other than *P. cinnamomi* are now also recognized as disease agents causing dramatic epidemics in agriculture and forestry around the world [34]. One current example is *Phytophthora ramorum*, the causal agent of sudden oak death in western North America [26]. This is an exotic pathogen introduced to that area in the mid 1990s. Establishment of permanent plots in infested sites and vegetation surveys started in 2000 [73]. It is noteworthy that high-resolution aircraft imagery has been introduced, along with field surveys, to estimate the rate and spatial-temporal patterns of tree mortality associated with *P. ramorum* [51, 63]. This new approach that incorporates landscape processes into plant pathology, referred to as 'landscape pathology' [40], will provide useful information regarding pathogen spread, geophysical features of disease expression, and impacts on forest diversity at the landscape scale.

11. Forest Diversity inside and outside of 'Infection Center' of Laminated Root Rot

Phellinus weirii, the causal agent of laminated root rot in Douglas-fir and other conifers, is a native root pathogen in evergreen coniferous forests of the Pacific Northwest [32]. The fungus spreads from tree to tree as ectotrophic mycelia or across root contacts and penetrates the root through intact bark, destroying phloem and cambium. The infection leads to the decomposition of woody roots and reductions in foliage moisture, chlorophyll, and nitrogen [82], and infected trees may live in a declining state for several years until death. Douglas-fir, mountain hemlock (*Tsuga mertensiana*), and firs (*Abies grandis* and *A. concolor*) are the most susceptible, whereas western hemlock (*Tsuga heterophylla*) and western redceder (*Thuja plicata*) are little affected by the fungus. Laminated root rot is among the most important disturbance agents in old-growth forests in the Pacific Northwest, and contributes to the maintenance of diversity in the natural ecosystems [32].

Infection by *P. weirii* advances about 30 cm per year and results in the formation of canopy gaps (called mortality centers or infection centers) that can reach one hectare in area [19]. Cromack et al. [16] reported that soil microbial biomass was lower, whereas needle decomposition and nutrient mineralization were faster, in regrowth forests inside infection centers than in adjacent old-growth forests that were not infected and thus can be regarded as 'controls'. Moreover, the regeneration of resistant tree species is observed at the center of infection centers, and hence a regrowth sequence of resistant tree species from the edge to the center of the infection center can be regarded as secondary succession. Boone et al. [6] showed successional changes in stem density, tree biomass, and soil organic matter in regrowth sequences within infection centers. Thus, infection centers of *P. weirii* are a useful system allowing one to test the hypothesis that this root pathogen can influence the forest diversity, by comparing forest vegetations inside versus outside of infection centers within old-growth forests and by examining regeneration processes of forest trees within infection centers.

Holah et al. [38] examined the species richness of tall trees, shrubs, and herbs inside and outside of *P. weirii* infection centers in low-elevation Cascade and Coast ranges of western Oregon. Plant species compositions were significantly different inside versus outside of infection centers at six study sites, with the cover of herbs increased at the expense of that of susceptible Douglas-fir in the infection centers. However, responses to laminated root rot varied among tree species and between study sites. For example, the cover of the shrub *Acer circinatum* in infection centers increased at one site, decreased at two sites, and was unchanged at the other three sites, compared to that in adjacent old-growth forests; the removal of Douglas-fir in the canopy due to *P. weirii* stimulated the growth of *A. circinatum* in some sites, whereas in other sites regeneration of other tall trees resulted in the suppression of *A. circinatum*. There was a tendency for shade-tolerant trees such as western hemlock to increase in infection centers at Cascade sites, whereas at Coast sites, tall trees were relatively minor and understory shrubs increased their covers [39]. Consequently, changes in plant diversity in infection centers were also variable among study sites [38]. Hill's diversity index in infection centers was higher at two sites and lower at one site than those at adjacent old-growth forests, and unchanged at three sites. Therefore, the response of forest diversity to the reduced number of Douglas-firs by *P. weirii* was not predictable and may depend on the ecological traits of plant species involved, species composition of the site, and site characteristics.

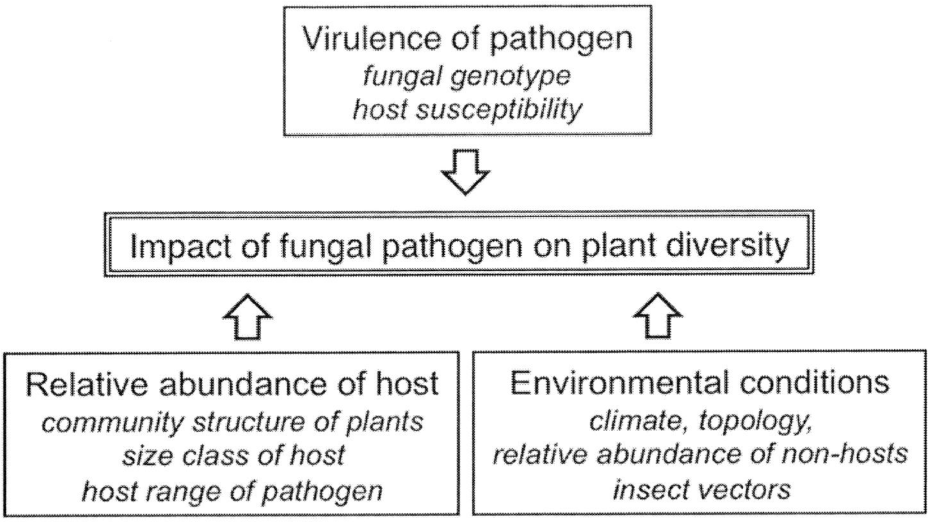

Figure 5. Factors affecting potential magnitude of impact of fungal pathogens on plant diversity.

Conclusion

Accumulating evidence demonstrates that pathogenic fungi serve as agents influencing forest diversity, especially those that are able to kill individual hosts relatively quickly (i.e., dogwood anthracnose, beech bark disease, Dutch elm disease, chestnut blight, phytophthora dieback, and laminated root rot). The potential magnitudes of the impact of fungal pathogen on forest diversity depend on the three-way interaction of virulence of pathogen × relative abundance of host × environmental factors (Figure 5). Virulence of pathogen is the product of the interaction between the genotype of the pathogen and the susceptibility of the host. Relative abundance of host in the forest community represents the ecological status of host populations, which is strongly related to the impact of the fungal pathogen. This is related to whether the host is dominant or subordinate, or rare, i.e. the community structure of trees, size class of the host (forest canopy, subcanopy, or understory), and host range of the pathogen in the particular forest stand. A similar discussion is applicable for the mycorrhizal dependence of dominant and/or subordinate plants, which determines the effect of mycorrhizal fungi on plant diversity [86]. Environmental conditions include physical and biological factors such as climate, topology, relative abundance of non-host plants competing with hosts, and insect vectors (if present) that influence not only host-pathogen interactions but also long-term dynamics of host and non-host populations. The responses of forest diversity to the infestation of fungal pathogens, which can be an increase or a decrease, or no change, of the number of plant species, depend on the interactions of these factors and hence are difficult to predict. It is thus difficult to draw general conclusions regarding the effects of fungal disease on the diversity and long-term dynamics of forest communities, because of the limited number of case studies.

Long-term observations of forest plots are necessary to demonstrate the effects of fungal pathogens on forest diversity and dynamics in the field. The comparison of case studies in this chapter addresses three recommendations that long-term studies of forest dynamics should incorporate. First is the establishment of 'control' plots that are not infested by pathogens. Effects of fungal pathogens were not clearly isolated from those of other factors affecting long-term dynamics of forest dynamics in studies of beech bark disease, Dutch elm disease, and chestnut blight, because of the lack of control plots. Conversely, they were successfully delimited in studies of phytophthora dieback and laminated root rot because of the establishment of control plots that were free of fungal infestation. It will be impossible, however, to establish control plots when fungal diseases of high virulence, such as exotic epidemics, devastate local host populations. Then, secondly, recording of 'mortality factor' for dead trees is useful to evaluate the relative importance of pathogens and other mortality factors such as drought, windthrow, and herbivory on long-term dynamics of host populations. A record of symptoms of dead plants is routinely made in studies of population dynamics of seedlings on the forest floor [77, 96]. This will enable not only the quantification of fungal disease as a mortality factor but also the evaluation of interactive effects of fungal diseases and other mortality factors. Thirdly, attention should be paid to vegetation dynamics in canopy gaps produced due to tree death caused by fungal diseases. Investigating the replacement of tree species and the changes in plant diversity in these gaps, like that described for beech bark disease, chestnut blight, and laminated root rot, will help to understand the ecological mechanisms underlying long-term changes in forest diversity.

Environmental factors influence the spatial patterns of fungal pathogens and host plants and their interactions and the severity of infestation, as summarized above. Studies at the landscape level that take the spatial heterogeneity of environmental conditions into account are thus important for predicting the spatial patterns of their impacts on forest diversity [40]. However, few studies of fungal diseases, including those mentioned in this chapter, take a landscape perspective into consideration. Stand-level investigations can bias the outcome if study plots are unrepresentative of disease on a broader scale. It should be borne in mind that local studies might tend to overemphasize disease because they are often performed in those stands with the highest disease incidence [40]. It is thus worth mentioning that spatial analyses using high-resolution aircraft imagery and Geographic Information System were successful in analyzing the spatial patterns of tree mortality caused by sudden oak death [51, 63]. This approach of 'landscape pathology' not only provides useful insights into variable responses of forest diversity to fungal pathogens but also helps in scaling-up stand-level measurements to landscape-level consequences in forest diversity [40].

Three further comments will address possible future research directions. First, mathematical approaches are useful to scale-up the impact of fungal diseases of low virulence causing mortality at the shoot level to forest diversity. These pathogens do not kill individual host trees, but reduce net primary production and plant vigor, which may alter intraspecific and interspecific interactions with adjacent trees. The study of Manter et al. [57] on Swiss needle cast is an excellent reference for scaling-up shoot-level events to tree-level consequences. Alternatively, partial defoliation of host trees in the canopy can ameliorate light conditions on the forest floor, which would affect the regeneration of host and non-host seedlings and saplings. Mathematical studies will again be useful to evaluate the relative importance of such a process that potentially affects forest diversity but is difficult to demonstrate by fieldwork. Secondly, the impacts of fungal pathogens are not necessarily direct; they can often be indirect. For example, Papaik et al. [67] demonstrated that the resistance of beech trees to windthrow is severely reduced by BBD. Forest disturbances of anthropogenic origin such as acid precipitation and clear-cutting can also affect the incidence of fungal pathogens [29, 76]. Interactive effects of fungal pathogens with other natural and anthropogenic disturbance regimes should be taken into consideration. Finally, the biodiversity of forests itself can affect disease transmission and prevalence [46]. Further studies are needed to elucidate potential connections between diversity and disease from the perspective of biodiversity conservation and management.

Acknowledgments

We thank Dr. Kenji Seiwa for useful discussions; and Dr. Elizabeth Nakajima for critical reading of the manuscript. This work was supported by Global COE Program A06 of Kyoto University.

References

[1] Abram, M.D. (1986) Historical development of gallery forests in northeast Kansas. *Vegetatio* 65, 29-37.

[2] Alexander, H. M. & Holt, R. D. (1998) The interaction between plant competition and disease. *Plant Ecol. Evol. Syst.* 1/2, 206-220.

[3] Anagnostakis, S.A. (1987) Chestnut blight: the classical problem of an introduced pathogen. *Mycologia* 79, 23-37.

[4] Ayers, P.G. & Paul, N.D. (1990) The effects of disease on interspecific plant competition. *Asp. Appl. Biol.* 24, 155-162.

[5] Boggess, W.R. & Bailey, L.W. (1964) Brownfield woods, Illinois: woody vegetation and changes since 1925. *Am. Midl. Nat.* 71, 392-401.

[6] Boone, R.D.; Sollins, P.; and Cromack, Jr K. (1988) Stand and soil changes along a mountain hemlock death and regrowth sequence. *Ecology* 69, 714-722.

[7] Bradley DJ, Gilbert GS, Martiny JBH (2008) Pathogens promote plant diversity through a compensatory response. *Ecol. Lett.* 11:461-469.

[8] Brasier, C.M. & Mehrotra, M.D. (1995) *Ophiostoma himal-ulmi* sp. nov., a new species of Dutch elm disease fungus endemic to the Himalayas. *Mycol. Res.* 99, 205-215.

[9] Burdon, J.J. & Chilvers, G.A. (1977) The effect of barley mildew on barley and wheat competition in mixtures. *Aust. J. Bot.* 25, 59-65.

[10] Burdon, J.J. & Chilvers, G.A. (1982) Host density as a factor in plant disease ecology. *Annu. Rev. Phytopathol.* 20, 143-166.

[11] Cahill, D.M.; Rookes, J.E.; Wilson, B.A.; Gibson, L.; and McDougall, K.L. (2008) *Phytophthora cinnamomi* and Australia's biodiversity: impacts, predictions and progress towards control. *Aust. J. Bot.* 56, 279-310

[12] Carr, D.E. & Banas, L.E. (2000) Dogwood anthracnose (*Discula destrictiva*): effects of and consequences for host (*Cornus florida*) demography. *Am. Midl. Nat.* 143, 169-177.

[13] Castello, J.D.; Leopold, D.J.; and Smallidge, P.J. (1995) Pathogens, patterns, and processes in forest ecosystems. *Bioscience* 45, 16-24.

[14] Chilvers, G.A. & Brittain, E.G. (1972) Plant competition mediated by host-specific parasites - a simple model. *Aust. J. Biol. Sci.* 25, 749-756.

[15] Connell, J.H. On the role of natural enemies in preventing competitive exclusion in some marine animals and in rainforest trees. In: den Boer PJ and Gradwell GR, editors. *Dynamics of Populations*. Wageningen: Center for Agricultural Publishing and Documentation; 1971; 298-312.

[16] Cromack, Jr. K.; Entry, J.A.; and Savage, T. (1991) The effect of disturbance by *Phellinus weirii* on decomposition and nutrient mineralization in a *Tsuga mertensiana* forest. *Biol. Fertil. Soils* 11, 245-249.

[17] Daughtrey, M.L.; Hibben, C.R.; Britton, K.O.; Windham, M.T.; and Redin, S.C. (1996) Dogwood anthracnose: understanding a disease new to North America. *Plant Dis.* 80, 349-357.

[18] Desprez-Loustau, M.L.; Robin, C.; Buee, M.; Courtecuisse, R.; Garbaye, J.; Suffert, F.; Sache, I.; and Rizzo, D.M. (2007) The fungal dimension of biological invasions. *Trends Ecol. Evol.* 22, 472-480.

[19] Dickman, A. & Cook, S. (1989) Fire and fungus in a mountain hemlock forest. *Can. J. Bot.* 67, 2005-2016.

[20] DiGregorio, L.M.; Krasny, M.E.; and Fahey, T.J. (1999) Radial growth trends of sugar maple (*Acer saccharum*) in an Allegheny northern hardwood forest affected by beech bark disease. *J. Torrey Bot. Soc.* 126, 245-254.

[21] Dunn, C.P. (1986) Shrub layer response to death of *Ulmus americana* in southeastern Wisconsin lowland forests. *Bull. Torrey Bot. Club* 113, 142-148.

[22] Elliott, K.J. & Swank, W.T. (2008) Long-term changes in forest composition and diversity following early logging (1919-1923) and the decline of American chestnut (*Castanea dentata*). *Plant Ecol.* 197, 155-172.

[23] Ellison, A.M.; Bank, M.S.; Clinton, B.D.; Colburn, E.A.; Elliott, K.; Ford, C.R.; Foster, D.R.; Kloeppel, B.D.; Knoepp, J.D.; Lovett, G.M.; Mohan, J.; Orwig, D.A.; Rodenhouse, N.L.; Sobczak, W.V.; Stinson, K.A.; Stone, J.K.; Swan, C.M.; Thompson, J.; von Holle, B.; and Webster, J.R. (2005) Loss of foundation species: consequences for the structure and dynamics of forested ecosystems. *Front. Ecol. Environ.* 3, 479-486.

[24] Fahey, T.J. (1998) Recent changes in an upland forest in south-central New York. *J. Torrey Bot. Soc.* 125, 51-59.

[25] Forrester, J.A.; McGee, G.G.; and Mitchell, M.J. (2003) Effects of beech bark disease on aboveground biomass and species composition in a mature northern hardwood forest, 1985 to 2000. *J. Torrey Bot. Soc.* 130, 70-78.

[26] Frankel, S.J. (2008) Sudden oak death and *Phytophthora ramorum* in the USA: a management challenge. *Aust. Plant Pathol.* 37, 19-25.

[27] Gadek, P. Introduction and overview. In: Gadek P, editor. Patch Deaths in Tropical Queensland Rainforests: association and impacts of *Phytophthora cinnamomi* and other soil borne organisms. Cairns: Cooperative Research Centre for Tropical Rainforest Ecology and Management; 1999; 1-7.

[28] Gilbert, G.S. (2002) Evolutionary ecology of plant diseases in natural ecosystems. *Annu. Rev. Phytopathol.* 40, 13-43.

[29] Gilbert, G.S. & Hubbell, S.P. (1996) Plant diseases and the conservation of tropical forests. *Bioscience* 46, 98-106.

[30] Griffin, J.M.; Lovett, G.M.; Arthur, M.A.; and Weathers, K.C. (2003) The distribution and severity of beech bark disease in the Catskill Mountains, N.Y. *Can. J. For. Res.* 33, 1754-1760.

[31] Hale, B.W.; Alsum, E.M.; and Adams, M.S. (2008) Changes in the floodplain forest vegetation of the lower Wisconsin river over the last fifty years. *Am. Midl. Nat.* 169, 454-476.

[32] Hansen, E.M. & Goheen, E.M. (2000) *Phellinus weirii* and other native root pathogens as determinants of forest structure and process in western North America. *Annu. Rev. Phytopathol.* 38, 515-539.

[33] Hansen, E.M.; Stone, J.K.; Capitano, B.R.; Rosso, P.; Sutton, W.; Winton, L.; Kanaskie, A.; and McWilliams, M. (2000). Incidence and impact of Swiss needle cast in forest plantations of Douglas-fir in coastal Oregon. *Plant. Dis.* 84, 773-778.

[34] Hansen, E.M. (2008) Alien forest pathogens: *Phytophthora* species are changing world forests. *Boreal Env. Res.* 13, 33-41.

[35] Hardham, A.R. (2005) *Phytophthora cinnamomi. Mol. Plant Pathol.* 6, 589-604.

[36] Hirayama, K. & Sakimoto, M. (1999) Classification of canopy and subcanopy tree species based on their leaf phenology in a cool temperate conifer-hardwood forest, Kyoto. *For. Res.* 71, 19-25.

[37] Hiers, J.K. & Evans, J.P. (1997) Effects of anthracnose on dogwood mortality and forest composition of the Cumberland Plateau (U.S.A.). *Conserv. Biol.* 11, 1430-1435.

[38] Holah, J.C.; Wilson, M.V.; and Hansen, E.M. (1993) Effects of a native forest pathogen, *Phellinus weirii*, on Douglas-fir forest composition in western Oregon. *Canadian. Journal for Forest Research* 23, 2473-2480.

[39] Holah, J.C.; Wilson, M.V.; and Hansen, E.M. (1997) Impacts of a native root-rotting pathogen on successional development of old-growth Douglas fir forests. *Oecologia* 111, 429-433.

[40] Holdenrieder, O.; Pautasso, M.; Weisberg, P.J.; and Lonsdale, D. (2004) Tree diseases and landscape processes: the challenge of landscape pathology. *Trends Ecol. Evol.* 19, 446-452.

[41] Houston, D.R. (1984) What is happening to the American beech? *The Conservationist* May-June, 22-25.

[42] Houston, D.R. (1994) Major new tree disease epidemics: beech bark disease. *Annu. Rev. Phytopathol.* 32, 75-87.

[43] Huenneke, L.F. (1983) Understory response to gaps caused by the death of *Ulmus americana* in central New York. *Bull. Torrey Bot. Club* 110, 170-175.

[44] Janzen, D.H. (1970) Herviores and the number of tree species in tropical forest. *Am. Nat.* 104, 501-528.

[45] Jenkins, M.A. & White, P.S. (2002) *Cornus florida* L. mortality and understory composition changes in western Great Smoky Mountains National Park. *J. Torrey Bot. Soc.* 129, 194-206.

[46] Keesing, F.; Belden, L.K.; Daszak, P.; Dobson, A.; Harveil, C.D.; Holt, R.D.; Hudson, P.; Jolles, A.; Jones, K.E.; Mitchell, C.E.; Myers, S.S.; Bogich T.; and Ostfeld, R.S. (2010) Impacts of biodiversity on the emergence and transmission of infectious diseases. *Nature* 468, 647-652.

[47] Kranz, J. (1990) Fungal diseases in multispecies plant communities. *New Phytol.* 116, 383-405.

[48] Kransy, M.E. & DiGregorio, L.M. (2001) Gap dynamics in Allegheny northern hardwood forests in the presence of beech bark disease and gypsy moth disturbances. *For. Ecol. Manag.* 144, 265-274.

[49] Le Guerrier, C.; Marceau, D.J.; Bouchard, A.; and Brisson, J. (2003) A modelling approach to assess the long-term impact of beech bark disease in northern hardwood forest. *Can. J. For. Res.* 33, 2416-2425.

[50] Leibhold, A.M.; MacDonald, W.L.; Bergdahl, D.; and Mastro, V.C. (1995) Invasion by exotic forest pests: a threat to forest ecosystems. *For. Sci. Monograph* 41, 1-49.

[51] Liu, D.; Kelly, M.; Gong, P.; and Guo, Q. (2007) Characterizing spatial-temporal tree mortality patterns associated with a new forest disease. *For. Ecol. Manag.* 253, 220-231.

[52] Loo, J.A. (2009) Ecological impacts of non-indigenous invasive fungi as forest pathogens. *Biol. Invasions* 11, 81-96.

[53] Lovett, G.M.; Canham, C.D.; Arthur, M.A.; Weathers, K.C.; and Fitzhugh, R.D. (2006) Forest ecosystem responses to exotic pests and pathogens in eastern North America. *Bioscience* 56, 395-405.

[54] Manter, D.K.; Bond, B.J.; Kavanagh, K.L.; Rosso, P.H.; and Filip, G.M. (2000) Pseudothecia of Swiss needle cast fungus, *Phaeocryptopus gaeumannii*, physically block stomata of Douglas fir, reducing CO_2 assimilation. *New Phytol.* 148, 481-491.

[55] Manter, D.K. (2002) Energy dissipation and photoinhibition in Douglas-fir needles with a fungal-mediated reduction in photosynthetic rates. *J. Phytopathol.* 150, 674-679.

[56] Manter, D.K. & Kavanagh, K.L. (2003) Stomatal regulation in Douglas fir following a fungal-mediated chronic reduction in leaf area. *Trees* 17, 485-491.

[57] Manter, D.K.; Bond, B.J.; Kavanagh, K.L.; Stone, J.K.; and Filip, G.M. (2003) Modelling the impacts of the foliar pathogen *Phaeocryptopus gaeumannii*, on Douglas-fir physiology: net canopy carbon assimilation, needle abscission and growth. *Ecol. Model.* 164, 211-226.

[58] Masuya, H.; Kusunoki, M.; Kosaka, H.; and Aikawa, T. (2009) *Haradamyces foliicola* anam. gen. et sp. nov., a cause of zonate leaf blight disease in *Cornus florida* in Japan. *Mycol. Res.* 113, 173-181.

[59] McCune, B.; Cloonan, C.L.; and Armentano, T.V. (1988) Tree mortality and vegetation dynamics in Hemmer Woods, Indiana. *Am. Midl. Nat.* 120, 416-431.

[60] McDougall, K.L.; Hobbs, R.J.; and Hardy, G.E.S.J. (2002) Vegetation of *Phytophthora cinnamomi*-infested and adjoining uninfested sites in the northern jarrah (*Eucalyptus marginata*) forest of Western Australia. *Aust. J. Bot.* 50, 277-288.

[61] McEwan, R.W.; Muller, R.N.; Arthur, M.A.; and Housman, H.H. (2000) Temporal and ecological patterns of flowering dogwood mortality in the mixed mesophytic forest of eastern Kentucky. *J. Torrey Bot. Soc.* 127, 221-229.

[62] McGee, G.G. (2000) The contribution of beech bark disease-induced mortality to coarse woody debris loads in northern hardwood stands of Adirondack Park, New York, U.S.A. *Can. J. For. Res.* 30, 1453-1462.

[63] Meentemeyer, R.K.; Rank, N.E.; Shoemaker, D.A.; Oneal, C.B.; Wickland, A.C.; Frangioso, K.M.; and Rizzo, D.M. (2008) Impact of sudden oak death on tree mortality in the Big Sur ecoregion of California. *Biol. Invasions* 10, 1243-1255.

[64] Milgroom, M.G.; Wang, K.; Zhou, Y.; Lapari, S.E.; and Kaneko, S. (1996) Intercontinental population structure of the chestnut blight fungus, *Cryphonectria parasitica*. *Mycologia* 88, 179-190.

[65] Mize, C.W. & Lea, R.V. (1979) The effect of the beech bark disease on the growth and survival of beech in northern hardwoods. *Eur. J. For. Path.* 9, 243-248.

[66] Osono, T.; Mori, A.; and Koide, K. (2004) Defoliation of giant dogwood (*Swida controversa*) caused by zonate leaf blight. *Ap. For. Sci.* 13, 161-164.

[67] Papaik, M.J.; Canham, C.D.; Latty, E.F.; and Woods, K.D. (2005) Effects of an introduced pathogen on resistance to natural disturbance: beech bark disease and windthrow. *Can. J. For. Res.* 35, 1832-1843.

[68] Parker, I.M. & Gilbert, G.S. (2004) The evolutionary ecology of novel plant–pathogen interactions. *Annu. Rev. Ecol. Evol. Syst.* 35, 675-700.

[69] Parker, G.R. & Leopold, D.J. (1983) Replacement of *Ulmus americana* L. in a mature east-central Indiana woods. *Bull. Torrey Bot. Club* 110, 482-488.

[70] Peterken, G.F. & Mountford, E.P. (1998) Long-term change in an unmanaged population of wych elm subjected to Dutch elm disease. *J. Ecol.* 86, 205-218.

[71] Redin, S.C. (1991) *Discula destructiva* sp. nov., cause of dogwood anthracnose. *Mycologia* 83, 633-642.

[72] Richardson, C.J. & Cares, C.W. (1976) An analysis of elm (*Ulmus americana*) mortality in a second-growth hardwood forest in southeastern Michigan. *Can. J. Bot.* 54, 1120-1125.

[73] Rizzo, D.M.; Garbelotto, M.; and Hansen, E.M. (2005) *Phytophthora ramorum*: integrative research and management of an emerging pathogen in California and Oregon forests. *Annu. Rev. Phytopathol.* 43, 309-335.

[74] Runkle, J.R. (1990) Eight years change in an old *Tsuga canadensis* woods affected by beech bark disease. *Bull. Torrey Bot. Club* 117, 409-419.

[75] Sahashi, N.; Kubono, T.; and Shoji, T. (1995) Pathogenicity of *Colletotrichum dematium* isolated from current-year beech seedlings exhibiting damping-off. *Eur. J. For. Path.* 25, 145-151.

[76] Samonil, P. & Vrska, T. (2008) Long-term vegetation dynamics in the Sumava Mts. natural spruce-fir-beech forests. *Plant Ecol.* 196, 197-214.

[77] Seiwa, K.; Miwa, Y.; Sahashi, N.; Kanno, H.; Tomita, M.; Ueno, N.; and Yamazaki, M. (2008) Pathogen attack and spatial patterns of juvenile mortality and growth in a temperate tree, *Prunus grayana*. *Can. J. For. Res.* 38, 2445-2454.

[78] Sherald, J.L.; Stidham, T.M.; Hadidian, J.M.; and Hoeldtke, J.E. (1996) Progression of the dogwood anthracnose epidemic and the status of flowering dogwood in Catoctin Moutain Park. *Plant Dis.* 80, 310-312.

[79] Shigo, A.L. (1972) The beech bark disease today in the Northeastern US. *J. For.* 70, 286-289.

[80] Stephenson, S.L. (1986) Changes in a former chestnut-dominated forest after a half century of succession. *Am. Midl. Nat.* 116, 173-179.

[81] Taniguchi, T.; Tanaka, C.; Tamai, S.; Yamanaka, N.; and Futai, K. (2008) Identification of *Cylindrocladium* sp. causing damping-off disease of Japanese black pine (*Pinus thunbergii*) and factors affecting the disease severity in a black locust (*Robinia pseudoacacia*)-dominated area. *J. For. Res.* 13, 233-240.

[82] Thomson, A.J.; Goodenough, D.G.; Barclay, H.J.; Lee, Y.J.; and Sturrock, R.N. (1996) Effects of laminated root rot (*Phellinus weirii*) on Douglas-fir foliar chemistry. *Can. J. For. Res.* 26, 1440-1445.

[83] Travers. S.E.; Gilbert, G.S.; and Perry, E.F. (1998) The effect of rust infection on reproduction in a tropical tree (*Faramea occidentalis*). *Biotropica* 30, 438-443.

[84] Trigiano, R.N.; Caetano-Anolles, G.; Bassam, B.J.; and Windham, M.T. (1995) DNA amplification fingerprinting provides evidence that *Discula destructiva*, the cause of dogwood anthracnose in North America, is an introduced pathogen. *Mycologia* 87, 490-500.

[85] Twery, M.J. & Patterson, W.A. (1984) Variations in beech bark disease and its effects on species composition and structure of northern hardwood stands in central New England. *Can. J. For. Res.* 14, 565-574.

[86] Urcelay, C. & Diaz, S. (2003) The mycorrhizal dependence of subordinates determines the effect of arbuscular mycorrhizal fungi on plant diversity. *Ecol. Lett.* 6, 388-391.

[87] Vandermast, D.B. & van Lear, D.H. (2002) Riparian vegetation in the southern Appalachian mountains (USA) following chestnut blight. *For. Ecol. Manag.* 155, 97-106.

[88] von Oheimb, G. & Brunet, J. (2007) Dalby Söderskog revisited: long-term vegetation changes in a south Swedish deciduous forest. *Acta Oecol.* 31, 229-242.

[89] Weiskittel, A.R.; Maguire, D.A.; Garber, S.M.; and Kanaskie, A. (2006) Influence of Swiss needle cast on foliage age-class structure and vertical foliage distribution in Douglas-fir plantations in north coastal Oregon. *Can. J. For. Res.* 36, 1497-1508.

[90] Weiskittel, A.R. & Maguire, D.A. (2007) Response of Douglas-fir leaf area index and litterfall dynamics to Swiss needle cast in north coastal Oregon, USA. *Ann. For. Sci.* 64, 121-132.

[91] Weste, G. (2003) The dieback cycle in Victorian forests: a 30-year study of changes caused by *Phytophthora cinnamomi* in Victorian open forests, woodlands and heathlands. *Aust. Plant Pathol.* 32, 247-256.

[92] Weste, G.; Brown, K.; Kennedy, J.; and Walshe, T. (2002) *Phytophthora cinnamomi* infestation - a 24-year study of vegetation change in forests and woodlands of the Grampians, Western Victoria. *Aust. J. Bot.* 50, 247-274.

[93] Williams, C.E. & Moriarity, W.J. (1999) Occurrence of flowering dogwood (*Cornus florida* L.), and mortality by dogwood anthracnose (*Discula destructive* Redin), on the northern Allegheny Plateau. *J. Torrey Bot. Soc.* 126, 313-319.

[94] Winton, L.M.; Stone, J.K.; Hansen, E.M.; and Shoemaker, R.A. (2007) The systematic position of *Phaeocryptopus gaeumannii*. *Mycologia* 99, 240-252.

[95] Wyckoff, P.H. & Clark, J.S. (2002) The relationship between growth and mortality for seven co-occurring tree species in the southern Appalachian Mountains. *J. Ecol.* 90, 604-615.

[96] Yamazaki, M.; Iwamoto, S.; and Seiwa, K. (2009) Distance- and density-dependent seedling mortality caused by several diseases in eight tree species co-occurring in a temperate forest. *Plant Ecol.* 201, 181-196.

Section 3: Biological Control of Livestock Parasites by Means of Fungi

In: Fungi: Types, Environmental Impact and Role in Disease ISBN: 978-1-61942-671-9
Editors: Adolfo Paz Silva and María Sol

Chapter XIV

The Role of Fungi in the Control of Animal Parasites – Classification, Mode of Action and Practical Applications

L. M. Madeira de Carvalho[*1], *F. A. Bernardo*[1] *and A. Paz-Silva*[2]
[1] Centro de Investigação Interdisciplinar em Saúde Animal, Parasitologia e Doenças Parasitárias, Faculdade de Medicina Veterinária, Universidade Técnica de Lisboa (CIISA/FMV/UTL), Pólo Universitário do Alto, Avenida da Universidade Técnica, Lisboa, Portugal
[2] Grupo para el Estudio de las Enfermedades de Équidos, Dpto. Patoloxía Animal, Facultade de Veterinaria de Lugo, Universidade de Santiago de Compostela, España

Abstract

Fungi are a wide group of organisms, being free-living or parasitic on both plants and animals. Some of them can infect insects and nematodes, being these last ones called nematophagous or nematode trapping fungi (NTF).This behaviour can have practical applications, namely in the field of biological control of gastrointestinal nematodes, since they kill their exogenous larval stages. There is still a gap in the knowledge of their mode of action and their use in regular deworming programs associated with anthelmintics. The purposes of this chapter are to enrol the most important groups of fungi used in parasite control (namely the NTF), the way they interact with parasites and how they can be used in parasite control programs along the time.

As an example of their practical application, data concerning the use of the nematophagous fungi *Duddingtonia flagrans* in the control of horse strongylosis in Portugal will be described. Field trials were performed in horse stud farms near Lisbon, evaluating the capacity of *D. flagrans* to reduce strongyle infection. Field work involved

[*]Email address: madeiradecarvalho@fmv.utl.pt.

always two groups of horses with mixed strongyle infection: one receiving a daily oral dose of *D. flagrans* (D), while the control one (C) received a similar amount of food without fungus. The L3 were reduced significantly in faecal cultures and pastures of D groups, being more effective in warmer seasons, in both pasture and stabled horses. Therefore, *D. flagrans* can be easily associated with the classic control methods of strongylosis in a horse management system involving pasture and stabled animals.

Keywords: Biological control –nematophagous fungi - *Duddingtonia flagrans* - Strongyles – Horse – Portugal

1. Introduction

Since the introduction of carbon disulfide in 1917, until the discovery of moxidectin in 1990, the anthelmintics have been used regularly in equine deworming, becoming increasingly affordable, diverse, with high margin of safety, efficacy and spectrum of action. On the other hand, the control of equine strongylosis has been based on intensive and almost exclusively of anthelmintics. This approach has overlooked a more rational management of anti-parasitic compounds and pasture hygiene measures, as well as the demand for alternative and / or complementary anti-parasitic measures regarding horse gastrointestinal parasites [1, 2].

Much attention of parasitological research of horses focus on cyathostomins since 1990s due to the lack of knowledge of their biology (namely on the hypobiosis intrinsic mechanism), their pathogenicity, their resistance to various groups of anthelmintics (piperazine, benzimidazoles and tetra-hydropirimidines) and difficulty to kill and/or remove their encysted L3 and L4 larval stages in the wall of large intestine with the average anthelmintic compounds [3-5].

Thus, strongyles remain an important cause of disease in horses, and their resistance to anthelmintics has been almost parallel to the development of new molecules, being problematic not only its occurrence, but early diagnosis and prevention of resistant populations [6].

In ruminants, the emergence of resistant Trichostrongyles since the mid-twentieth century, led to its sustainable and integrated control, also gradually adapted to the control of equine strongylosis. This new approach to "management" of the parasite populations, involves the improvement of pasture and its use, the breeding of animals more resistant to parasitic infections, the development of vaccines for the major helminthic diseases, and its control in quality and quantity by biological methods in a supplementary feeding system [7-9].

Among the various organisms studied as potential biological control agents there are dung arthropods, bacteria and fungi. These fungi that infect and destroy gastrointestinal nematodes act on their exogenous development stages in the pasture. These fungi have evolved independently and are distributed by various taxonomic groups, with ones attacking nematodes eggs, others are endoparasites and still others with predatory characteristics, depending on how they use nematodes as nutritional resource [3, 10, 11, 12].

These fungi belong to the Hyphomycetes group, also known as nematophagous fungi,and are able to trap and digest the free forms of the soil nematodes. Many of these fungi produce spores have resistance or saprophytic phases in the absence of their hosts. In addition, there are pathogenic for organisms that are not its target. There are over 200 described species of nematophagous fungi; most belong to the so-called imperfect fungi or Deuteromycetes. These fungi are a heterogeneous group and ubiquitous, living normally in the saprophytic form and occupying different niches in the soil, where they can also feed one wide range of free-living nematodes, either as a resource primary or secondary [12].

Predator nematophagous fungi or nematode trapping fungi (NTF) constitute the largest group, characterized by the production of a system of hyphae that form a "network" or "trap" (the original "trapping device"), where they capture infective larvae alive (3rd stage larvae or L3) of nematode parasites with an exogenous phase of development in the pasture. Later expansions of these "traps" or "networks" penetrate L3, grow and absorb the contents of infective larvae of several gastrointestinal nematode parasites of domestic animals (particularly Trichostrongylidae and Strongylidae of ruminants, horses and pigs), promoting its immobilization and his death. The nematode captured suffers piercing and fungus penetrates and grows inside its body and the content is quickly consumed (Figures 1.1 and 1.2) [3, 13].

The entire network is covered by a mucilaginous layer, where nematodes are attached and its production seems to be stimulated by the struggle of nematodes trapped. The capture of nematodes by nematophagous fungi is a combination of mechanical power (captured bodies) and the release of enzymes, hydrolytic extracellular serum proteases, such as chitinases and collagenases, which digest their cuticle (mainly composed of proteins). An hour after the penetration, the fungus forms a bulb inside the infective nematode and in a few hours completely occupies the body of this stage. The digestion process can last up to a week, after which trophic hyphae are smooth and the fungus grows again saprophytic, until the presence of nematodes in the middle encourages again its predatory activity [12].

There are different infective structures, but irrespective of the shape, physical or chemical mode of action, the result is always the same: the death of the nematode (Table 1.1) [14].

Figure 1.1. "The *Duddingtonia flagrans* are trapping us" [15].

Although their biology and ecology have been studied since the 1940s, its potential use in the control of herbivores gastrointestinal nematodes, found expression only from the 1980s, when he stepped up testing "in vitro" and "in vivo" with a few species of nematophagous fungi, in particular *Duddingtonia flagrans* and *Arthrobotrys oligospora*. More recently, *A. robusta*, *Monacrosporium thaumasium* and *Pochonia chlamydosporia* have been used in both *in vitro* and *in vivo* tests [8, 12, 15-17].

These fungi are cosmopolitan, being primarily in the microbial community of soil, pasture and organic debris in natural and agricultural soils, and in all areas where there is decaying organic material, locations also conducive to the development of diverse populations of parasitic nematodes [2, 3].

The biggest problem posed by the use of these fungi as biological control agents, is the need for their presence in faecal masses of the hosts for an effective reduction of infectious forms in the faeces micro-ecosystem and consequently in the pasture. However, it is known that after being on the soil, the fresh feces of ruminants and horses are quickly colonized by these fungi, and are also frequently ingested during grazing, with subsequent excretion in feces [18]. Several field works showed their ease of administration *per os* in ruminants and horses, the continuing viability and mode of action after passage through the digestive tracts of ungulates [19-22].

Table 1.1. Typical infection structures of some nematophagous fungi [14]

Infection structure	Species	Taxonomic classification
Adhesive nets	*Arthrobotrys oligospora*	Deuteromycetes
	A. conoides	
	A. musiformis	
	A. superba	
	Duddingtonia flagrans	
Adhesive branches	*Monacrosporium gephyropagum*	Deuteromycetes
Adhesive knobs	*M. ellipsosporum*	Deuteromycetes
	M. haptotylum	
Constricting rings	*A. dactyloides*	Deuteromycetes
	A. brochopaga	
Adhesive knobs and adhesive spores	*Nematoctonus concurrens*	Basidiomycetes
Adhesive spores	*N. leiosporus*	Basidiomycetes
	Drechmeria coniospora	Deuteromycetes
	Hirsutella rhossoliensis	
Ingested spores	*Harposporium anguillulae*	Deuteromycetes
Zoospores	*Catenaria anguillulae*	Chytridiomycetes
	Haptoglossa dickii	Oomycetes
Adhesive hyphae	*Stylopage hadra*	Zygomycetes
	Cystopage cladospora	
Toxic droplets	*Pleurotus ostreatus*	Basidiomycetes
Appressoria	*Pochonia chlamydosporia*	Deuteromycetes

This type of biological control has been studied in the control of intestinal strongyles (Nematoda: Strongyloidea) of the horse and is considered safe for these animals, causing a strong rate of capture and destruction of infective forms of those nematodes. The feeding of horses in the pasture with nematophagous fungi, may provide an alternative and / or complement to anthelmintics, proven in studies that achieved a significant reduction in the feces and of L3 on pasture, particularly cyathostomin L3, reducing substantially the risk of infection of the host (Fig. 1.2) [3, 22-24].

Based on these data, between 1997 and 2001 laboratory and field studies were conducted to evaluate the performance and effectiveness of the fungus *Duddingtonia flagrans*, variety Troll A, in the control of equine strongyles in Portugal (Table 1.2). These trials were conducted in the laboratories of the Faculty of Veterinary Medicine, Technical University of Lisbon (FVM/TUL) and on farms in Ribatejo, being performed in horses for the first time in the Iberian Peninsula. This study involved the FVM /TUL, the Horse Stud Farm "Companhia das Lezírias, SA" (CL), the National Stud Horse Farm, Fonte Boa, Santarém (CN) and the biotechnology company "CHR HANSEN - BioSystems" Denmark, the latter being the responsible entity for producing and distribution of the fungus *D. flagrans*.

Table 1.2. Types of tests carried out in Portugal with *Duddingtonia flagrans* (1997/2001)

Date	Type of test	Type of experience	Animals	Animal husbandry	Local
1998	Laboratory	Nematophagous activity in vitro	--	--	FVM/TUL
1997/1998	Field and Laboratory	Nematophagous activity in vitro and in vivo	Colts 1 year	Pasture	CL and FVM/TUL
1999	Field and Laboratory	Nematophagous activity in vitro and in vivo	Colts 1-3 years	Pasture	CL and FVM/TUL
999	Field and Laboratory	Nematophagous activity in vitro and in vivo	Yearlings 3 years	Pasture/Stable	CL and FVM/TUL
2000/2001	Field and Laboratory	Nematophagous activity in vitroandin vivo	Yearlings 3-4 years	Pasture/Stable	CN and FVM/TUL

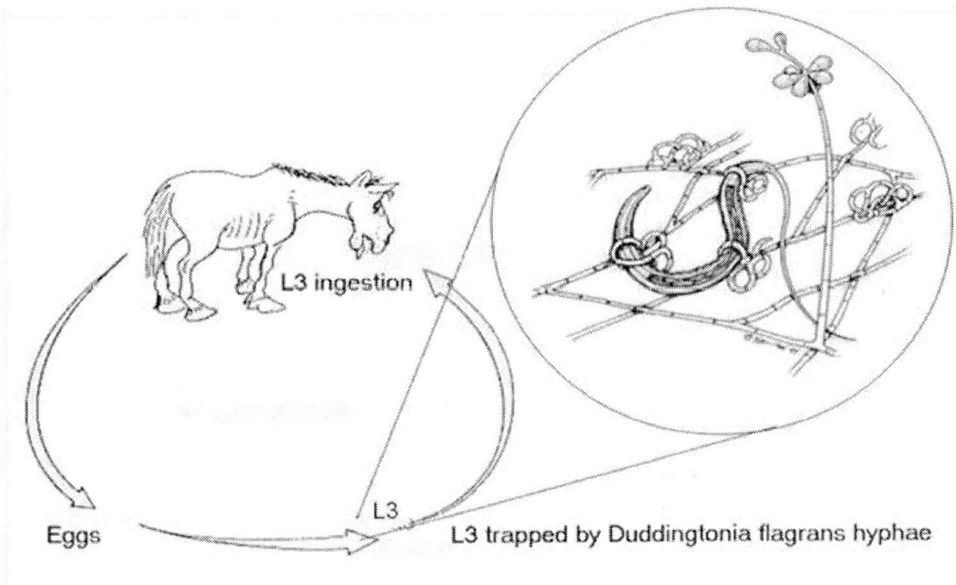

Figure 1.2. *Duddingtonia flagrans* nematophagous fungi act on existing L3 in faecal micro-environment and pasture, through a network or trap that retains or traps to destroy them completely [23].

The aim of this chapter is to describe the laboratory and field tests carried out in Portugal with the fungus *Duddingtonia flagrans,* var. Troll A, in the 1997-2001 period, which allowed:

a) Assessment of the nematocide activity in a Petri dish culture,to highlight some aspects of *in vitro* "nematophagous action" using light microscopy;
b) Assessment of the *in vivo* efficacy of *D. flagrans* var. Troll A in horses The reduction in the egg shedding levels, the number of L3 in faecal cultures and the contamination level of the pasture of horses fed fungi, compared with a control population, both in grazing animals or penned.

2. Experimental Work

2.1. In Vitro Trials

The work was done from a unique culture of *D. flagrans* var. Troll A, isolated in the laboratory environment of "CHR HANSEN - Biosystems, A / S" in Denmark. This culture was maintained in Petri dishes with Sabouraud Dextrose Agar medium (SDA), streptomycin sulphate and chloramphenicol, both at the concentration of 50 ppm at room temperature under anaerobic conditions in a sealed wrapper until the time of use (Adrian Gillespie, pers. com.).

It was made a subculture from solid medium to solid medium, with metal shaft and sterilization care. With an inoculum of *D. flagrans*, three cultures by surface seeding in Petri dishes with the same medium were performed and incubated at 25°C for 3-5 days under aerobic conditions in a humid chamber. (For observing the development of the fungus without stimulus to grow, three other Petri dishes were prepared with no infective larvae).

Each of the three initial dishes were added 500 μl of an aqueous suspension with 1000-2000 third stage larvae (L3), obtained by the Roberts & O'Sullivan faecal culture technique, from horses with a natural mixed strongyle infection [25]. After faecal culture, the larvae were washed with sterile water and centrifuged at 1500 rpm for 5 min and then stored at 4°C until the time of inoculation of plates. Observation was performed at 48 hours, 7 days, 15 days and 30 days post-inoculation. To this end, four subsamples were taken with a metal rod from different points of the surface of the culture medium and immediately observed at the microscope with slide and coverslip. During this procedure the total number of L3 were counted, as well as the number of living and dead L3, calculating the average percentage reduction of larval population for the four observations. At this stage, the characteristics of the development structures of this fungus and induced changes in nematodes were observed.

In another experiment, cultures were made in Petri dishes with Sabouraud agar, chloramphenicol at 25 μg / ml and tetracycline 30 μg / ml, using a cotton swab to make a seed in monolayer culture from the original *D. flagrans* var. Troll A (alternatively were used penicillin / streptomycin and kanamycin sulfate at a dilution of 250 μg / ml and fungisone dilution of 1 / 100 in sterile water to wash the larvae and prevent contamination with other fungi). With this procedure was intended to evaluate the effectiveness of trapping larvae in cultures with 24, 48 and 72 hours. L3 counts before infection and 24 hours post-infection were carried out (to calculate the rate of larval entrapment), proceeding to their isolation by the Baermann technique [26].

Table 2.1. Results of the total reduction in the average number of L3 larvae in cultures with *Duddingtonia flagrans* (observations at 48 h, 7, 15 and 30 days post-infection)

	23/01/98				28/01/98				6/02/1998				20/02/1998			
Cultures	tot L3.	dead L3	liv. L3	% red.	tot L3.	dead L3	liv. L3	% red.	tot L3.	dead L3	liv. L3	% red.	tot L3.	dead L3	liv. L3	% red.
1	133	100	33	75,2	86	84	2	97,7	471	435	36	92,4	150	144	6	96
2	170	156	14	91,8	368	360	8	97,8	500	470	30	94	83	82	82	98,8
3	151	132	19	87,4	72	64	8	88,9	200	193	7	96,5	35	34	34	97,1
Aver.	151	129	22	84,8	175	169	6	94,9	390	366	24	94,3	89	87	41	97,3

Tot L3. Total number of L3; dead L3 No. of dead L3; liv. L3 - No. of living L3, % red - % Reduction.

Table 2.2. Average percentage of L3 reduction

	48H	7 DAYS	15 DAYS	30 DAYS
Culture 1	75.2	97.7	92.4	96
Culture 2	91.8	97.8	94	98.8
Culture 3	87.4	88.9	96.5	97.1
Average	84.8	94.8	94.3	97.3

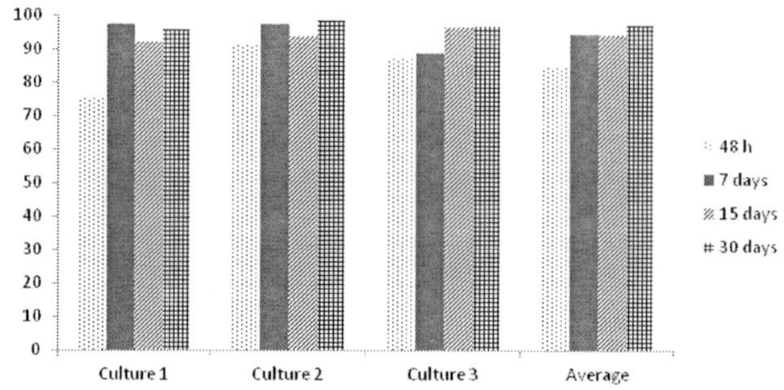

Figure 2.1. Percentages average reduction of L3 in a Petri dish at 48 h, 7, 15 and 30 days post inoculation.

To count larvae, 24 h later the first 10 ml of water taken from the Baermann apparatus were collected into tubes and submitted to centrifugation at 1500 rpm for 5 minutes. Then sediment was analysed between slide and coverslip and the existing larvae were counted.

The observed results of the nematophagous behavior of *D. flagrans* against a mixed population of strongyle infective larvae showed two distinct phenomena: in cultures with sequential observation at 48 hours, 7, 15 and 30 days, there were average rates of larval reduction of 84.8%, 94.9%, 94.3% and 97.3%, respectively, considered a very high level of destruction, even for this kind of control (Tables 2.1 and 2.2., Figure 2.1). Simultaneously, in cultures performed in order to watch the development of the fungus, we saw a rapid growth of the mycelium till the 7th day of observation. On day 15, that growth diminished a bit, and the number of chlamydospores increased, a situation that lasted until the 30th day of observation. After 3-5 days of culture, we observed a pronounced development of *D. flagrans,* an image similar to a tuft of cream color. When observed with slide and coverslip at the optical microscope the subsamples of culture had a wealth of vegetative elements, notably the mycelium, which formed a marked and three-dimensional network that holds the infective larvae. They also show frequent conidia and chlamydospores as the culture ages, the latter being especially important because they are the typical resistance form of this fungus in the presence of adverse conditions in the environment (Figure 2.14). Although it was not quantified, they increased after the introduction of nematode L3 in the plates. Even the larvae that remained alive and with some mobility, sometimes showed changes in the cuticle and others were completely involved by the fungal vegetative forms mentioned above (Figure 2.14). On the second phase of work with monolayer cultures of *D. flagrans* showed that after contact with the L3 for 24, 48 and 72 h there were high retention rates of larvae in Baermann apparatus, with average trapping percentages of 98.7%, 98.7 and 96.9%, respectively. In more recent trials rates of 100% trapping were also achieved for cultures with the same duration (Tables 2.3 and 2.4 and Figure 2.2).

Table 2.3.Total collected larvae and average percentage of trapped L3 in Baermann apparatus of *D. flagrans* cultures in July 1998

	24 H CULTURES				48 H CULTURES				72 H CULTURES		
	Nº tot.L3	L3 coll.	% trap.		Nº tot.L3	L3 coll.	% trap.		Nº tot.L3	L3 coll.	% trap.
Culture 1	120	1	99.2	Culture 4	123	0	100	Culture 7	125	7	94.4
Culture 2	130	1	99.2	Culture 5	103	4	96.1	Culture 8	128	1	99.2
Culture 3	138	3	97.8	Culture 6	115	0	100	Culture 9	180	5	97.2
Average	129	2	98.7	Average	114	1	98.7	Average	144	4	96.9

No. tot.L3 - Total number of L3; L3 coll. - No. of collected L3; % trap. -% of trapped L3.

Table 2.4.Total collected larvae and average percentage of trapped L3 in Baermann apparatus of *D. flagrans* cultures in September 1998

	24 H CULTURES				48 H CULTURES				72 H CULTURES		
	Nº tot.L3	L3 coll.	% trap.		Nº tot.L3	L3 coll.	% trap.		Nº tot.L3	L3 coll.	% trap.
Culture 1	91	0	100	Culture 3	98	0	100	Culture 5	108	0	100
Culture 2	100	0	100	Culture 4	82	0	100	Culture 6	94	0	100
Average	96	0	100	Average	90	0	100	Average	101	0	100

No. tot.L3 - Total number of L3; L3 coll. - No. of collected L3; % trap. - % of trapped L3.

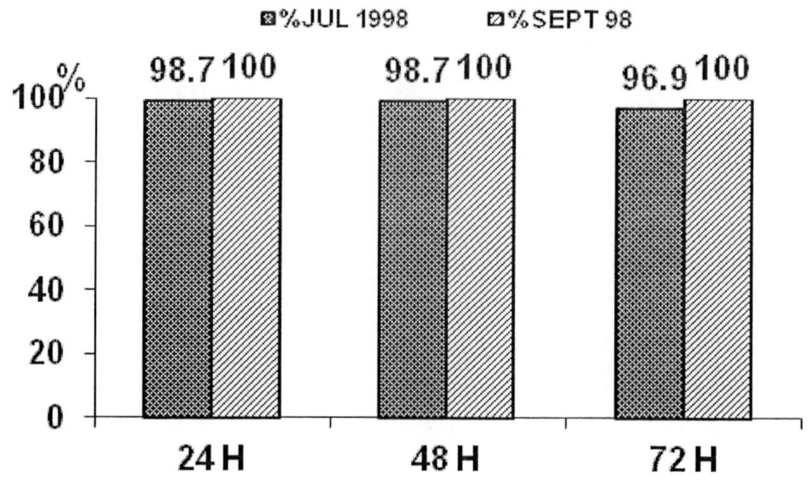

Figure 2.2. Percentage of horse strongyle L3 trapped in the Baermann apparatus at 24, 48 and 72 h.

Most of the *in vitro* work with the horse strongyles L3 was carried out with the faeces of horses naturally infected, which were added to the fungal spores to assess their nematophagous ability. These experiments evaluated the behavior of the fungus in a Petri dish by adding a mixed population of infective strongyle larvae.

First there was a chronological evolution of the fungus *D. flagrans* associated with a food substrate, namely a greater development of macroscopic colonies, which is a consequence of induction of its propagation in the presence of nematodes, namely after the larvae are added to the medium, as reported by other authors [3, 19].

This fungus has a high nematophagous action since at the end of 48 hours of incubation with L3 in the medium, about 85% of the larvae were already dead. The high percentages obtained in the following observations, 94.8, 94.3 and 97.3%, are consistent with the rate of production of mycelium and chlamydospores, which reach their maximum after 2-3 weeks of culture. There is no significant production of these structures beyond day 21 of culture, promoting also during this period the maximum trapping effect and larval destruction, as observed for this fungus when challenged by *Ostertagia ostertagi* larvae [27].

Thus, there is no great variation in the rate of mortality of free-living forms on days 7, 15 and 30 of culture, meaning that the fungus has reached its maximum production of vegetative forms and the maximum larval retention was achieved. In some samples, larval forms presented themselves still alive and with motility. But sometimes the cuticle showed changes and was surrounded by a thin mycelium or hyphae with chlamydospores, as for *O. ostertagi* L3 [19, 27].

Experiments carried out with monolayer culture of 24, 48 and 72 hours, also allowed us to assess, at the macroscopic level, the rapid development of colonies of *D. flagrans* because its observation was carried out in single layer colonies. Second, even in recent cultures can be achieved good levels of control *in vitro*, and this phenomenon is due to a more intimate contact of the fungus with the larvae and the rapid induction of nematode traps, preventing its isolation in the Baermann apparatus according to Gillespie (personal com.).

The results indicate that the nematophagous fungus *D. flagrans* var. Troll A has a pronounced action on the free life forms of equine strongyles, as described for other varieties of this fungus [3, 22, 23, 24].

While it is necessary to better understand the complex phenomena of inter-relationship between these fungi and gastrointestinal nematodes of domestic animals, we must consider that this is a promising and innovative pest control. This nematocide action may have a practical application in the control schemes, because *D. flagrans* var. Troll A can be used as an adjunct or complement to the classic anti-parasitic treatment or schedules.

The work now described is part of a broader investigation that spanned the study of the behavior of *D. flagrans* under field conditions, where the horses were submitted to regular treatment with this agent and that allowed us to supplement the data *in vitro* with the evaluation of its effectiveness *in vivo*. These tests are described in the next chapter.

2.2. Field Trials

2.2.1. Experimental Design

Tests on grazing animals took place in two periods, from November 1997 to March 1998 and February to July 1999, using about 10 foals -12 months of age, with a natural infection with intestinal strongyles. The initial set of animals was divided into two groups according to body condition and the eggs per gram (EPG) [27]. There has been a first individual fecal sample collection to determine the level of parasitism by EPG, ordering the horses into two groups so that the level of infection was similar in both.

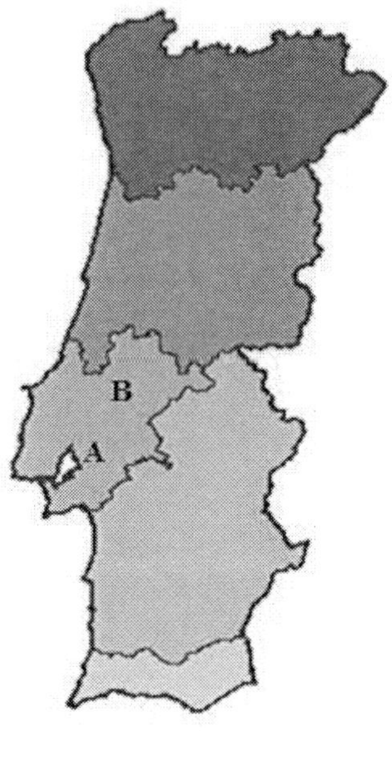

Figure 2.3. Location of field trials in Portugal: - A: Companhia das Lezírias horse farm, CL - 2 field trials with horses in a pasture and one trial with stabled horses; B – National Horse Stud farm, CN: 2 trials with stabled horses.

Before they start receiving periodic fungi feeding, animals from both groups were dewormed at the beginning of the test with a broad spectrum anthelmintic, so that the level of infection was not too high and to minimize the risk of a clinical strongylidosis. The group that received *D. flagrans* fungi was designated D and a daily dose of fungal material was supplied with the food, while the second group, control, called C, received the same amount of food without fungi.

Tests on penned animals in stables had a similar design, although one occurred in May-July 1999 at CL stud farm, and two other in February-August 2000 and June-September 2001 in the CN National Horse Stud Farm (Figure 2.3).

2.2.2. Animals and Management

In the first trial with horses grazing in 1997/98, 18 male foals were used, divided into two groups with 9 animals. In the second trial in 1999 three groups were used, one group with 9 male foals treated (D), and two control groups, the initial 12 female foals (C1) and a second, with 16 yearlings of 1-2 years (C2). During this last test and due to management imperatives of the farm, C1 was transferred in May to another pasture further away, so from there the group D was compared only with C2 group, which were located in a plot of land close to the initial test and included foals raised by the normal equine grazing production system of CL (Fig. 2.4). All animals belonged to Lusitano or Lusitano crossbreed of CL stud farm, with about 200-250 kg bodyweight, weaned in October at the beginning of the trial.

The animals were kept in a semi-extensive system from October 1997 to March 1998 (first trial) and February to July 1999 (second trial), staying in two adjacent paddocks with spontaneous pasture and similar dimensions. All horses grazed *ad libitum,* receiving a daily ration of grain, straw, hay and even a vitamin supplement (Neolait).

The horses were all vaccinated against tetanus and influenza (Tetagripiffa® vaccine) and dewormed with pyrantel pamoate (PYRP) (Strongid®) on 13 November 1997 in the 1st test and on 18 February 1999 the 2nd test, prior to the fungus administration. (Administration of an anthelmintic before the fungus is a condition for this type of control to the extent that the fungi have mainly a preventative action to eliminate infective larvae on pasture).

Fungi were administered to animals in group D, a daily dose of $5x10^5$ chlamydospores / kg bodyweight (equivalent to 10-20 g of fungal material per day, dry weight), mixed with the concentrate or silage. Group C received an equal amount of food but no fungi. The administration of *D. flagrans* began on 25thNovember 1997 and 19thFebruary 1999, in the 1st and 2nd trial, respectively. The level of parasitism detected in the 2nd test led to a new deworming in April 1999, groups D and C1 with doramectin (DRM) (Dectomax®) and C2 with ivermectin (IVM) (Eqvalan®).

The animals were grazing in the paddocks under regular observation during the study periods, to detect a reaction to taking the product and certifying that the animals ate the fungal material (Figure 2.4).

In tests with stabled animals the protocol was identical, having been used in the 1999 trial foals and adult horses 3-4 years old, distributed by groups C and D, with food (straw, oats and hay) and water *ad libitum.* These animals had regular access to the grazing paddocks located next to the barn. The animals of groups D and C were dewormed with IVM oral paste (Eqvalan®) and the amount of *D. flagrans* fungal material was individually administered the same dosage used for grazing animals on a daily basis *per os*, mixed with food from 20 May to 27 July 1999 (Figure 2.5).

In the second experiment with stabled animals, only 6 animals of 3-4 years were used in each group C and D, remaining stabled most of time, only leaving for riding. Deworming was carried out in both groups with subcutaneous IVM (Ivomec®) and the dose of *D. flagrans* given daily the group D was mixed with the granulated food, taking place between February and August 2000. The same animals were used on 2001for another stable trial, where 3 groups were tested: one fed fungi only (D), and two other that fungi was associated with a previous deworming with injectable ivermectin (D+IVM) and injectable moxidectin (D+MOX) (Figure 2.6).

All these experimental studies have been submitted to official review and approval by the General Directorate of Veterinary Medicine, Ministry of Agriculture, to safeguard the welfare and minimum level of animal suffering. In addition to these assumptions and if necessary, additional measures were agreed with the board of the farms and veterinary surgeons to minimize any discomfort and eliminate potential weaknesses inherent to experimentation, including the onset of diarrhea and spasms, by administering supplemental anti-parasitic, antibiotics, anti-inflammatory and anti-spasmodic drugs.

2.2.3. Fungal Material

In these trials we used the fungus *Duddingtonia flagrans* variant "Troll A", which normally occurs in Denmark, isolated and identified in Trollesminde, near Roskilde [11, 28].

Fungi were previously selected, cultivated and added to barley grains using industrial microbiological methods and packed in airtight plastic bags at the Laboratories "CHR HANSEN - BioSystems, A/S" (Fig. 2.4E and 2.4F). (The content of each small package was previously calculated for each day, for all the animals and overall weight of group D. In the case of animals kept in stable, this calculation was made directly to each individual). The packages were stored at room temperature and barley grains with fungi were mixed and homogenized with a food supplement in the feeder and administered orally at a daily dose of $5x10^5$ spores/kg body weight.

2.2.4. Faecal Tests for Egg Counts and L3 Culture

To diagnose strongyle infection level fecal sample collection was performed on weekly basis at the beginning and then every two weeks. The sampling were carried out individually (from the rectum) at the beginning, middle and end of each test, and from the pasture (fresh feces) on the remaining dates. EPG levels were calculated with McMaster technique, with detection threshold of 50 eggs per gram of feces. Results were expressed as the arithmetic average of total EPG per group for each sampling date [24, 29].

Faecal cultures were held individually at the beginning, middle and end of tests for grazing animals (parallel to the individual EPG counts) and the remaining sampling dates. Basically, from each sample taken from pasture a 4 g sub-sample was taken and gathered until reaching a total of 36 g for each group culture.

For each batch of animals were made 3-5 group cultures and whether they were individual or group based, were incubated for 11-14 days at a temperature of 26-28 ° C and relative humidity of 70-80%. The L3 were collected by the method of plastic cups and petri dishes [24, 30, 31]. After storage at 4°C, L3counts were done to calculate the average number of L3 / g.

Figure 2.4. A and B - Group C and D in the 1997/98 trial, C and D - Groups D and C1 in the 1999 test, E and F - Administration Duddingtonia flagrans in packages at the feeder to Group D and notice barley grains with the fungus mixed with silage.

Feces weighed prior to determination of the No. L3 / g feces and the percentage of development calculated with the following formula:

- % Larval development L3 = Average L3 g $^{-1}$ / average EPG for each sample date x100.

The average percentage of larval reduction induced by *D. flagrans* was determined by the formula [32]:

- % Reduction = (100 – L3 g $^{-1}$ in cultures D/L3g $^{-1}$ group in the cultures of group C) x100

Figure 2.5. Animals of the 1999 test. Horses in boxes and stalls (above) and in paddocks (below) used by groups C and D in this semi-housed horse production.

The L3 larvae of equine strongyles were observed in 100 μl of water suspension between slide and coverslip, proceeding to larval count and identification using differential staining with Lugol's solution. The total number of L3 was calculated for 10 ml of liquid and determined the number of L3/culture and L3/g feces, followed by the calculation of development percentage according to the average EPG for each sampling date. The results were expressed as the arithmetic mean of all counts made. The L3 identification was based on the keys of several authors [33-35].

Figure 2.6. Stabled animals used 2000 trials. We can see horses in stalls and the feeding system, used for both Groups C and D.

2.2.5. Observation of Fungal Development in Feces

During the first field trial in 1997/1998 with grazing animals, the fecal samples were collected individually and one gram of sample from each foal in each group was taken to another fecal culture with a total of 9 grams totaling 3 cultures for each group, for direct observation of *D. flagrans* mycelium [24]. This evaluation was performed by microscopic observation during four collection dates to check the growth and development of the fungus for 2-4 weeks after initiation of culture.

The development of micellar *D. flagrans* was classified as follows: 0, no development; +, small mycelium on the surface of the stool; + +, micellar network with more developed three-dimensional aspect; + + +, fully formed micellar network, tuft like with respect to the surface of the fecal sample.

2.2.6. Contamination Level of the Pasture

The larval counts were carried out in the pastures used by both groups weekly at the beginning of each test, and biweekly for the remainder of the study. The method used for herbage sampling was the double W with modifications and adaptations according to several authors, meaning that at least a total of 400 herbage samples were collected per each test group paddock [31, 36, 37].

The methods of extraction, storage and counting of infective larvae from pasture are in line with the mentioned above for fecal cultures in 2.2.4. The L3 were extracted by washing the herbage, centrifugation and filtration with decreasing mesh sieves of 1000, 500 and 20 μm, concentrated with vacuum pump and isolated through Baermann apparatus. The final sample of herbage was dried and the level of contamination was expressed by the formula: N º L3/kg dry herbage = Total Nº L3 (in 10 ml) x 1000/dry herbage weight (grams) [31]. The percentage reduction of larval population on the pasture was calculated using an adaptation of the formula in section 2.2.4[32].

2.2.7. Hematologic Parameters

During the first experiment with grazing animals conducted in 1997/1998, three blood samples were collected for assessment of hematic parameters in both groups of animals under study: in November 1997 and January and March 1998. The parameters analyzed included complete blood count, leukocyte formula, albumin, globulin, total protein, glucose and the enzymes ALT and AST.

For cell counts were used MS9™ and BPC BioSED™ PRIME Automatic Photometer counters and the remaining profiles were performed with the spectrophotometry technique (total protein, albumin, globulins) using "Photometer BPC BioSed srl – BPC Prime" and the technique of photometry reflectance (glucose) with the device "Reflotron®". Hematological reference values were considered from classical sources [38, 39].

2.2.8. Meteorological Data

Information on rainfall, relative humidity and average minimum and maximum temperatures, were obtained from the records of the weather station at Coruche location, the nearest to the place of experience. These climate variables were used for the field trials with animals on the pasture, from November 1997 to March 1998 and February to July 1999 (Figure 2.15).

2.2.9. Statistical Analysis

The descriptive statistics data were obtained with the "MICROSOFT EXCEL, v.2000®" and the differences between the treated (D) and control (C) groups were analyzed by the Mann-Whitney test program "GraphPad InStat for Windows Version 3.5, 2003." For this test, values of $p<0.1$were considered biologically significant and $p<0.05$ significant.

2.2.10. Results

a) EPG and L3 levels

EPG levels were similar for both groups at the beginning of the two trials of horses grazing. However, in the 1st test on 17/2/98 and at the end of the 2nd test (between groups C2 and D), significant differences were determined by the Mann-Whitney test ($p <0.05$). It was also observed in both experiments the reappearance of eggs was always first in controls and later in *Duddingtonia* groups (Figure 2.7A and B).

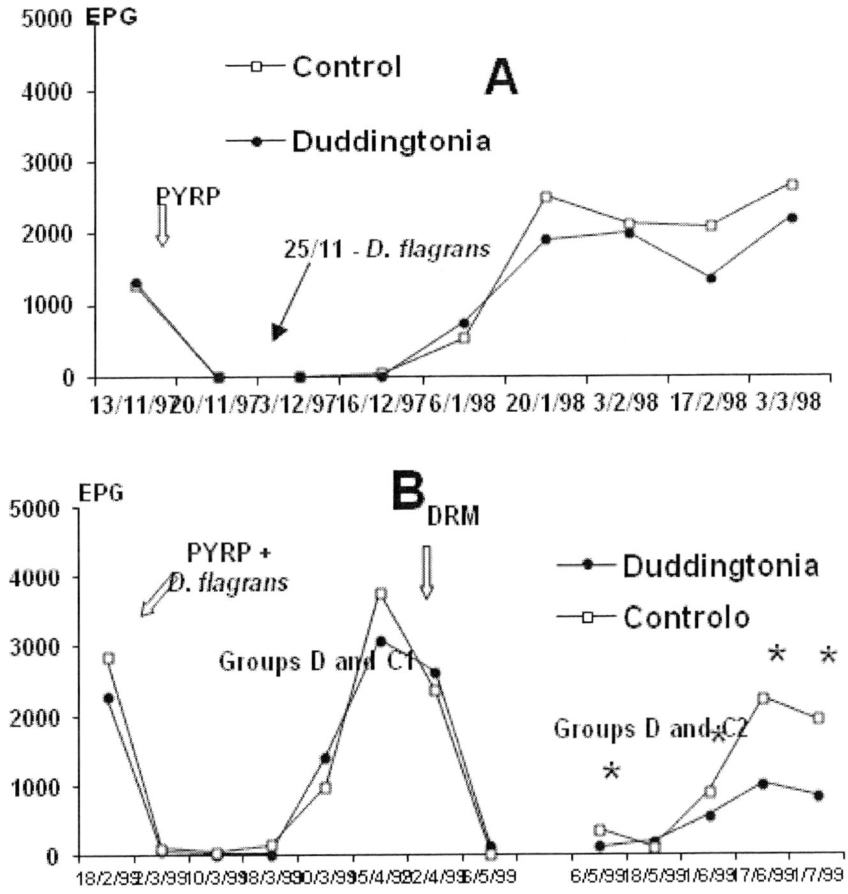

Figure 2.7. EPG values (arithmetic means) and control groups *Duddingtonia* in 1997/98 (A) and 1999 (B). PPIR - pyrantel pamoate; DRM - Doramectin * Significant differences by Mann-Whitney test between groups for values of $p <0.05$.

The fecal cultures showed a significant reduction in the number of L3/g in D groups when compared to C groups in both trials. In 1997/98, the differences were more evident from December to January, with a value of $p < 0.05$ at 6/1/1998 and $p < 0.1$ in other sampling dates with more pronounced differences (Figure 2.8A).

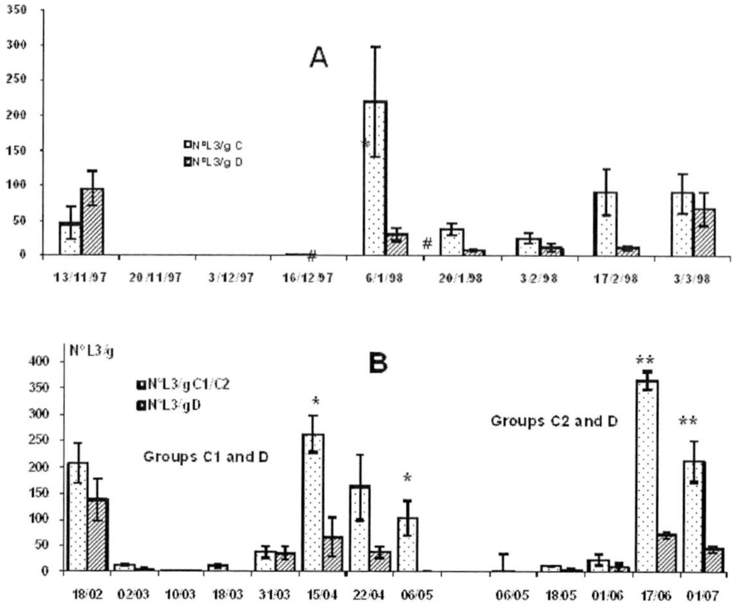

Figure 2.8. Mean values of L3/g of feces in fecal cultures of trials 1997/98 (A) and 1999 (B) in the control (C, C1 and C2) and *Duddingtonia* (D). Significant differences by Mann-Whitney test: # $p < 0.1$ * $p < 0.05$, ** $p < 0.01$. (The bars represent the standard error of mean).

There were also found significant reductions in the number of L3/g of feces in animals that received *D. flagrans* in 1999 and in the remaining, since group D values were always lower than those of groups C1 and C2 (Figure 2.8b). The percentage of larval development and larval reduction after the addition of fungi to animals from groups D, showed lower values of the animals in groups C, except the sampling carried out in March 1999. The average proportion of L3 developed in trial 1between group C / group D was 6:1 and in trial 2, the proportion was 22:1 (C1 / D) and 2.6:1 (C2 / D). Overall, the average of L3/g feces in group D was lower than that of group C in the two field studies and the percentage of larval reduction was 70% in 1997/98 and 62 and 72.2% in 1999 (Table 2.5 and Figure 2.9). The values of proportional abundance of control and *Duddingtonia* groups showed a predominance of L3 belonging to genus *Cyathostomum sl* (99.8 to 100%). In groups C and D were also identified L3 of genera *Poteriostomum* and *Triodontophorus,* while the L3 species *G. capitatus*, *O. robustus* and *T. axei* were identified only in animals of control groups (Table 2.6).

In these field experiments, the L3 *Cyathostomum sl* did not disappear completely in the cultures, and their number increased substantially to about 1.5 months after deworming with PYRP in both trials. A similar situation also occurred in deworming with DRM in 1999.

Table 2.5. Percentage of L3 development according to average EPG for each sample date and percentage of larval reduction (stars represent the significance analysis of larval development under nematophagous action of *Duddingtonia* by Mann-Whitney test)

Trial 1				Trial 2							
Date	Dev D %	Dev C %	% Reduction	Date	Dev D %	Dev C1 %	%Reduction	Date	Dev D %	Dev C2 %	%Reduction
3-12-97	0	11	100	2-3-99	8.79	13.47	59.61	6-5-99	0.43	0.72	79.53
16-12-97	0.25	1.87	86.63	10-3-99	9.28	4.57	18.74	18-5-99	2.54	13.28	62.77
6-1-98	4.03	41.14	90.20	18-3-99	0.82	8.24	98.63	1-6-99	1.8	2.75	59.71
20-1-98	0.42	1.51	72.19	31-3-99	2.57	3.84	3.97	17-6-99	7.19	16.38	80.22
3-2-98	0.60	1.18	49.15	15-4-99	2.17	6.99	74.60	1-7-99	5.3	10.96	78.72
17-2-98	0.83	4.35	80.88	22-4-99	1.46	6.86	76.33				
3-3-98	3.04	3.39	10.32	6-5-99	0.43	516.14	99.48				
Aritm. mean	1.53*	9.21	69.91	--	3.65**	80.01	61.62	--	3.45***	8.82	72.19
Min.	0.25	1.18	10.32	--	0.43	3.84	3.97	--	0.43	0.72	59.71
Max.	4.03	41.14	100	--	9.28	516.14	99.48	--	7.19	16.38	80.22
Stand. Dev.	1.60	15.84	30.83	--	3.28	208.23	40.86	--	2.50	5.84	10.07
Stand. Err.mean	0.65	5.98	11.63	--	1.24	78.58	15.42	--	1.12	2.60	3.80

*p=0.05; **p=0.073; ***p=0.22.

Table 2.6. Average proportional abundance of genera/species of horse strongyles (L3) in different test groups (Percentage of gen./spp. in 100 larvae counted)

	Proportional abundance (%)								
Grupo	*Cyathostomum*	*Gyaloc. capitatus*	*Oesoph. robustus*	*Poteriostomum*	*Stron. equinus*	*Stron. edentatus*	*Triodontophorus*	*Stron.vulgaris*	*Trich. axei*
Duddingtonia 97/98	100	0	0	0	0	0	0	0	0
Control 97/98	99.48	0	0	0.23	0	0	0.29	0	0
Duddingtonia 99	99.82	0	0	0.04	0	0	0.14	0	0
Control C1 99	99.40	0.34	0.01	0.22	0	0	0.03	0	0
Control C2 99	99.21	0	0	0.59	0	0	0	0	0.20

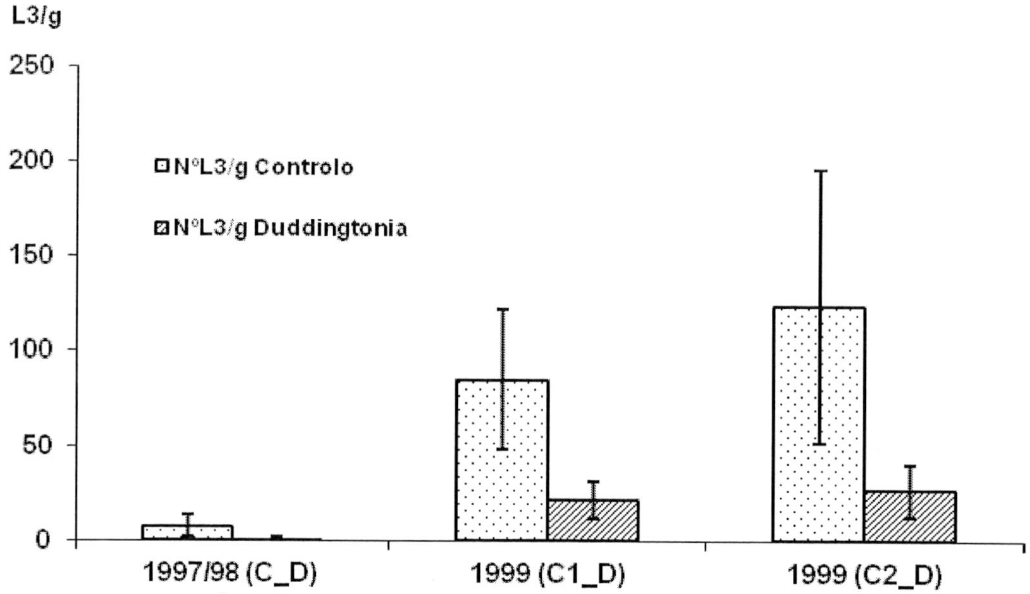

Figure 2.9. Global Values arithmetic means) of L3/g of feces in the control (C) and *Duddingtonia* (D) groups in trials conducted in the period 1997/99. (The bars represent the standard error of mean).

In the trial of horses stabled in 1999 at the CL horse farm, the group D showed a lower level of egg shedding than average EPG of group C, respectively 212 and 791, with a significant difference at $p < 0.05$. The average larval population of group D during the trial was 474 infective larvae in culture, while group C had 1538 larvae, which means more than threefold increase in infectious forms, a difference also significant at $p < 0.05$ (Figures 2.10 and 2.11).

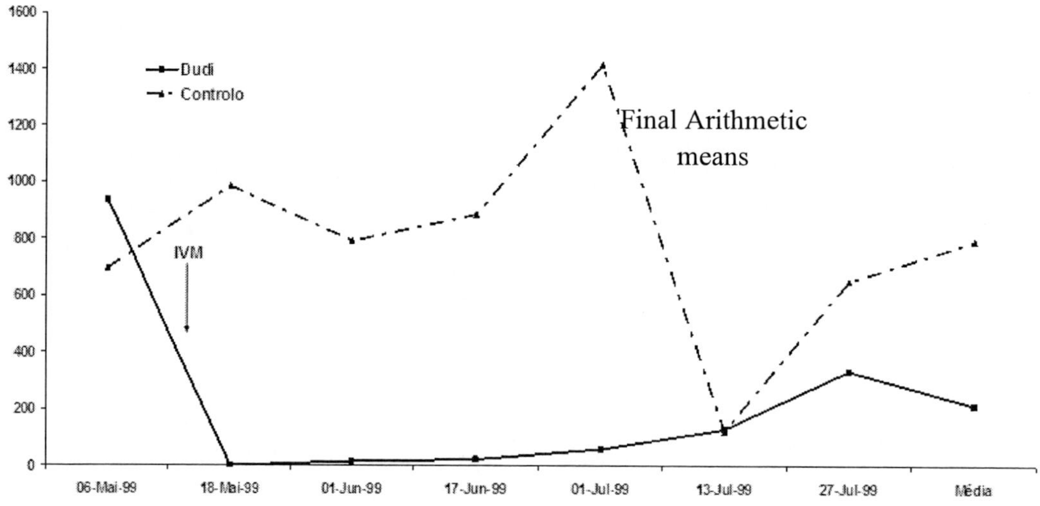

Figure 2.10. EPG values (arithmetic means) of control and *Duddingtonia* groups. Trial with stabled horses from May to July 1999. IVM – Ivermectin (Eqvalan®).

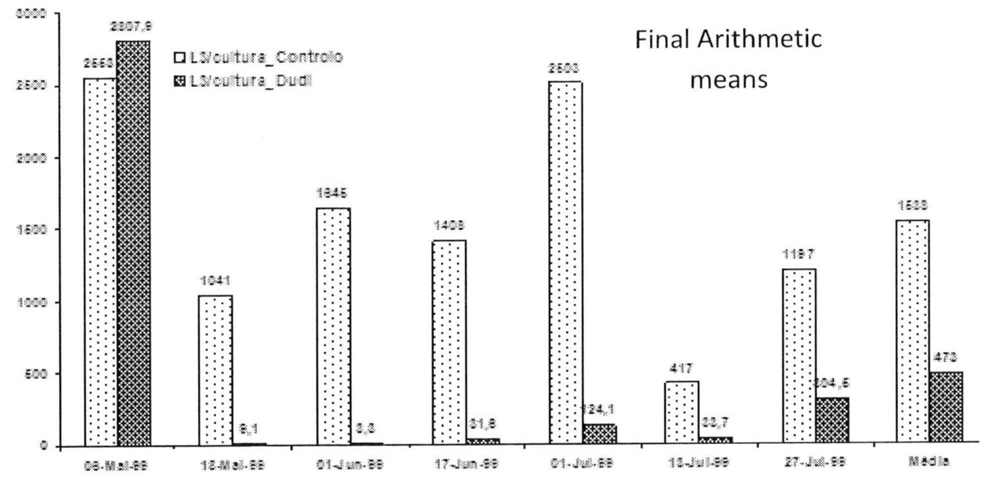

Figure 2.11.Mean values of N°L3/fecal culture of control group (C) and *Duddingtonia* (D).Trial with stabled horses from May to July 1999.

The EPG values of 2000 trial with stabled horses showed significant differences (p= 0.004) between groups C and D (keeping the latter with negative values for 3 months). The N° L3/culture in group D was significantly reduced (p = 0.01) with the average proportions ofL3 numbers between groups C and D of 95.1:1 (Figures 2.12 and 2.13).

Finally, in the trial with 3 groups of horses fed fungi in 2001, average percentage of larval development after administration was 18.5%, 9.3% and 12% respectively for groups *Duddingtonia* (A), *Duddingtonia* + Moxidectin (B) and *Duddingtonia* + Ivermectin (C).

These three groups had a lower larval development, particularly in groups B (moxidectin) and C (ivermectin). The average percentage of population reduction in larval treatment groups when compared with control group was 27.6%, -89% and 29.4%, respectively for groups A, B and C.

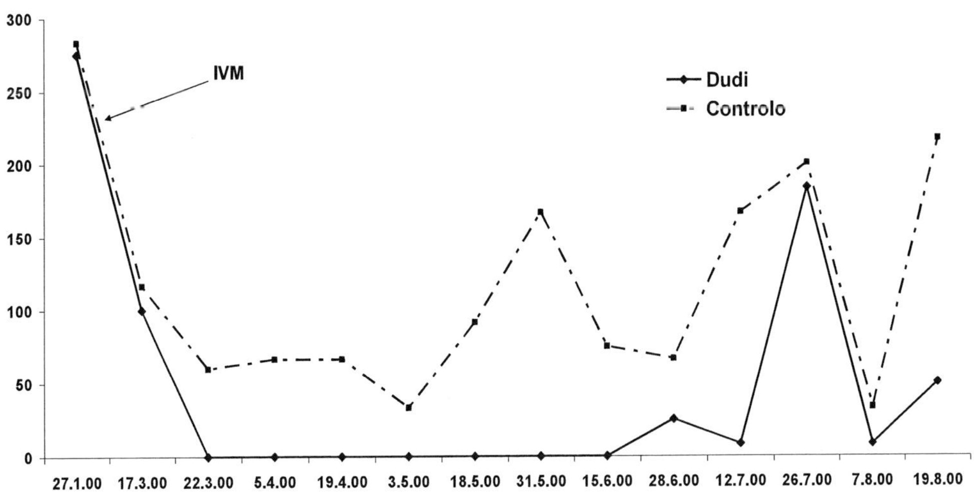

Figure 2.12. EPG values (arithmetic means) of control and *Duddingtonia* groups in the stable trial held in January- August 2000 at the CN national horse stud farm. IVM – Ivermectin (Ivomec®).

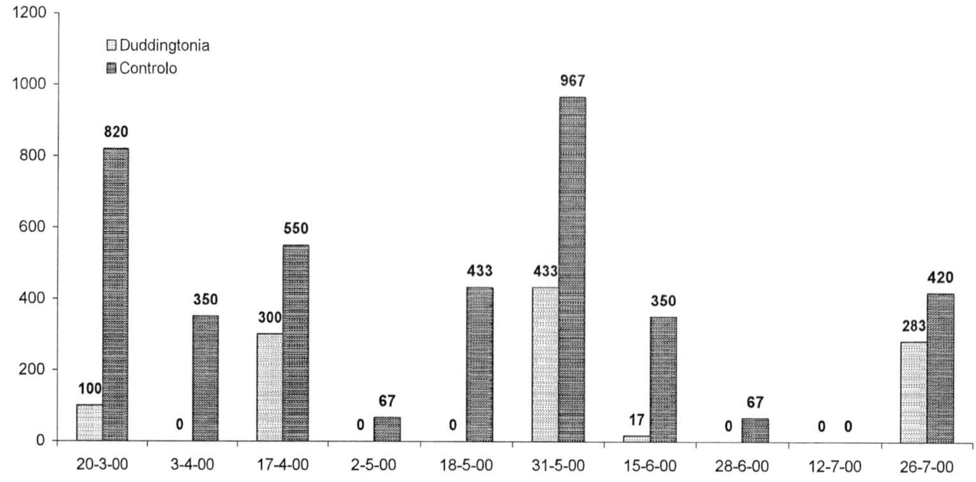

Figure 2.13. Mean values of N°L3/fecal of control and *Duddingtonia* groups in the stable trial held in January-August 2000 at the CN national horse stud farm.

b) Fungal development in the feces

The observation was made only in the testing of 1997/98 and in all cultures *Duddingtonia* group showed the development of the fungus, since the lower level to maximum, which demonstrated that the animals ate the fungus and this was able to form the typical three-dimensional micellar network with high larval destruction (Figure 2.14).

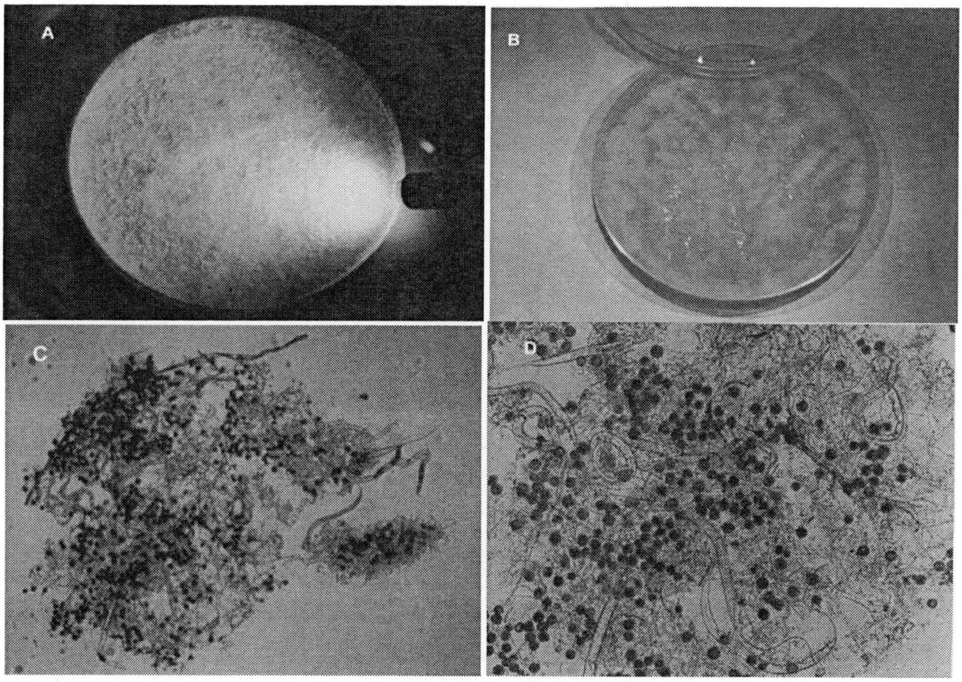

Figure 2.14. (Continued).

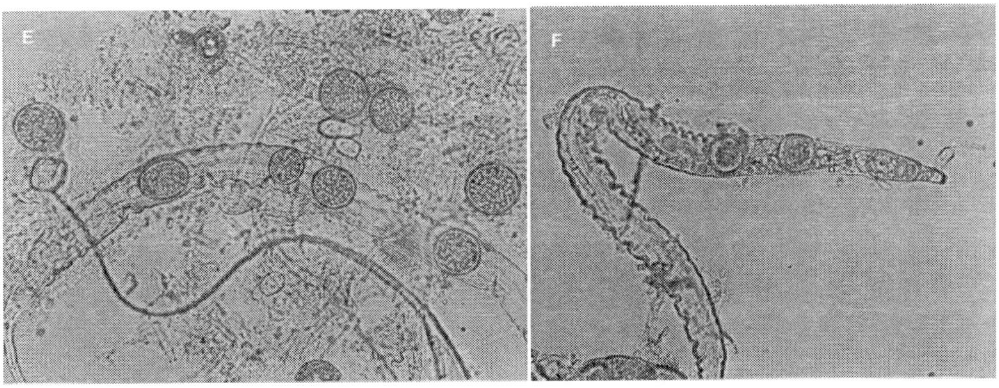

Figure 2.14. A - *D. flagrans* in a Petri dish with 5 days; B - Same, with 21 days, with a tuft aspect; C - *D. flagrans* network trapping L3 larvae of horse strongyles (40X magnification); D - As previous, note that chlamydospores are easily spotted (darker round spots) (100X magnification); E and F - Nematophagous activity of *D. flagrans* with chlamydospores and conidia in the environment surrounding the larva (E) and hyphae and chlamydospores inside the L3 (F) (200X magnification) (Original).

Samples of the control group also showed some degree of fungal development, but no typical forms of *D. flagrans* were observed, namely the network and chlamydospores (Table 2.7).

Table 2.7. Fecal cultures of *Duddingtonia* and Control samples for observation of fungal development

		Cultures in 4/12/1997				
		CONTROL			DUDDINGTONIA	
Observation	C1*	C2*	C3*	D1	D2	D3
11-12-97	+	+	+	++	++	++
19-12-97	+	+	+	+++	+++	+++
8-1-98	0	+	+	+++	++	+++
14-1-98	0	+	+	+++	++	+++
		Cultures in 17/12/1997				
		CONTROL			DUDDINGTONIA	
Observation	C1*	C2*	C3	D1	D2	D3
8-1-98	+	+	0	+	+	++
14-1-98	++	++	0	++	+	+++
5-2-98	++	++	0	++	++	++
		Cultures in 8/1/1998				
		CONTROL			DUDDINGTONIA	
Observation	C1*	C2*	C3*	D1	D2	D3
14-1-98	+++	+++	+++	+++	+++	+++
10-2-98	+++	+++	++	+++	+++	+++
2-3-98	+++	+++	++	+++	+++	+++
		Cultures in 5/2/1998				
		CONTROL			DUDDINGTONIA	
Observation	C1*	C2*	C3*	D1	D2	D3
10-2-98	++	+	++	+++	+++	+++
2-3-98	+++	++	++	+++	++	+++

*The control samples had cultures with other fungi, but were not detected forms of *D. flagrans*.

c) Pasture Contamination

The variation of L3 on pasture control group followed the fluctuations of precipitation and relative humidity during periods of study, while those grazing

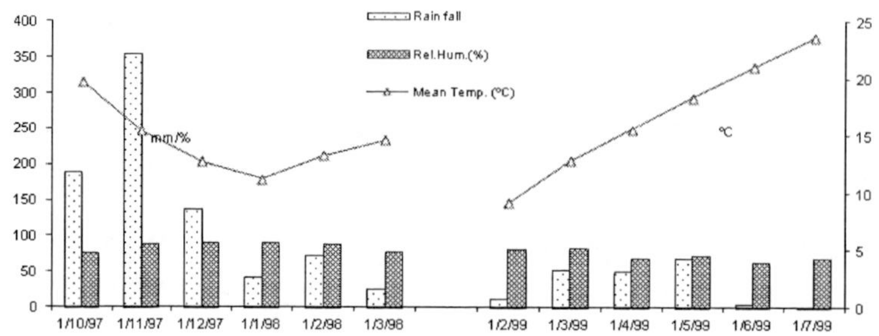

Figure 2.15. Values of total rainfall, relative humidity and mean temperature for the study area during the period of trials: October 1997 to March 1998 and February to July 1999.

At baseline of 1997/98, the number of L3/kg dry herbage raised progressively in the paddocks of both groups, though it was noted lower values for the pasture used by animals of group D.

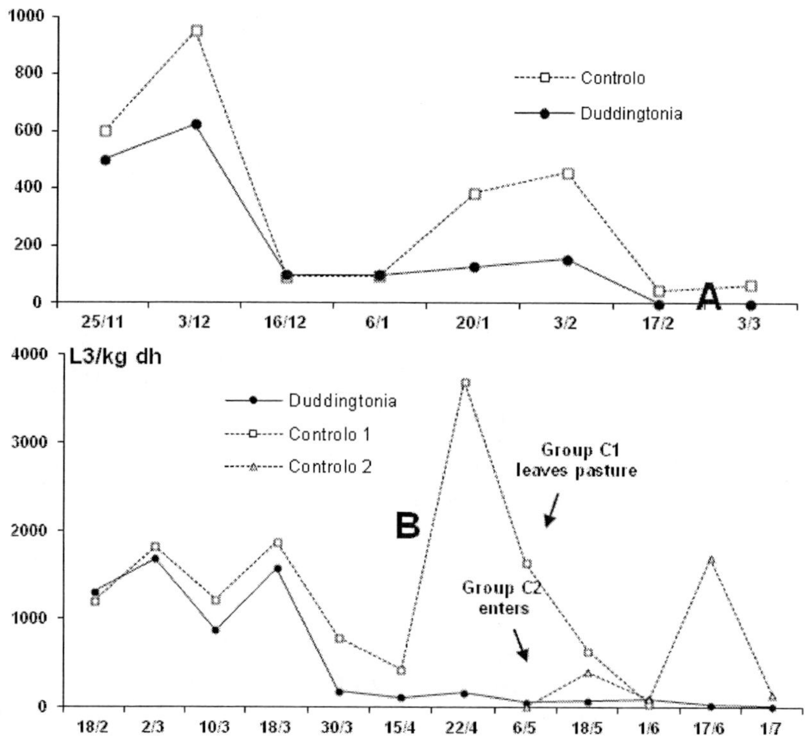

Figure 2.16. Variation in the number of L3/kg dry herbage (dh) in grazing control and *Duddingtonia* groups. October 1997 to March 1998 (A) and February to July 1999 (B).

Although we have not found significant differences ($p = 0.8$) between the two groups until the end of this trial, the contamination level of the paddock of the group fed fungi was reduced to zero in the last two samples taken in February/March, while the control group paddock showed infective larvae on the grass. All L3 collected in both pastures belonged to the genus *Cyathostomum sl* (Figure 2.16A). At the beginning of 1999 trial we recorded a similar variation in the levels of L3 on pasture, although from March the group D paddock showed a progressive decrease in the contamination, which lasted until July. This contamination compared to control groups showed a significant difference between groups C1 and D ($p= 0.038$), until C1 horses were transferred to another pasture. The difference between pasture contamination of groups C2 and D showed no significant differences ($p= 0.15$), but L3/kg dry herbage were always higher in control group compared to D pastures (Figure 2.16B). In trial 2, were isolated primarily infective larvae of the genus *Cyathostomum sl* (99.26% in group D and 99.67% in group C1), but were also isolated L3 *S. vulgaris* in group D (0.74% of total) and *Triodontophorus* spp. in group C1 (0.33% of total).

The global values of L3/kg dry herbage in both trials involving pasture always showed lower levels of contamination of the grass for animals receiving *D. flagrans*. The average reduction in pasture larval contamination of animals fed with the fungus was 50.5% in 1997/98, 60.3% in the first 1999 trial and 70.4% in the second one in 1999 (Figure 2.17).

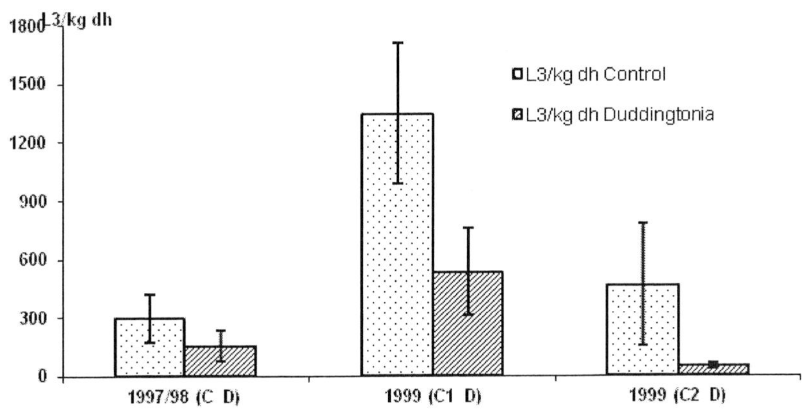

Figure 2.17.- Global values (arithmetic means) of N° L3/kg dh in pastures used by the control group (C) and *Duddingtonia* (D) in tests conducted from 1997 to 1999 (the bars represent the standard error of mean).

d) Blood parameters

The values of blood parameters between the two groups showed no significant difference ($p<0.05$), standing in the general ranges of default values. The group D values were closer to the standard ones when compared with the control group, although the differences were minimal (Table 8).

e) Discussion

The trials performed with the fungus *Duddingtonia flagrans,* variety Troll A, showed that administered *per os* to horses at a dose of $5x10^5$ was successfully eliminated in the feces and in sufficient quantity to a significant reduction of infective larvae in fecal cultures and pasture. Other field studies also demonstrated the action of nematophagous *D. flagrans* on equine strongyle L3 and ruminant Trichostrongylidae L3 in grazing systems. This is always important to confirm, because depending on the pasture and the ecosystem in which it is integrated, more factors may interfere with the effectiveness of this fungus [19, 22, 24, 40, 41].

The elimination of eggs showed no differences between the two groups in the two trials of grazing horses, as observed previously [24]. This was probably due to the fact that fungi do not act on adults, since they act only on the forms of pre-parasitic stages of strongyles in the environment. On the other, the colts ingested cyathostomin larvae during the spring and summer, which probably became hypobiotic during autumn and winter, just before the start of 1997/98 and 1999 trials, respectively.

Deworming with pyrantel pamoate may explain the short reappearance of eggs (about 1 month), because this anthelmintic is not effective hypobiotic stages [42]. Although the fungus *D. flagrans* has substantially reduced L3 in cultures and pasture, EPG peaks observed in January/February 1998 may also be related to the maturation of hypobiotic larvae that leave their cysts in the large intestine wall and then molt subsequently to adult stages.

This phenomenon may also explain the short Egg Reappearance Period (ERP) after deworming with doramectin, as this drug may have a shorter in ERP in young animals as a result of its low activity and pharmacokinetics in what concerns hypobiotic cyathostomin larvae [43].

More work must be done in an attempt to find the best anthelmintic to be used during the administration of *D. flagrans* and according to the season. In fact, studies *in vitro* with *Arthrobotrys* spp. (another kind of nematophagous fungus) showed that ivermectin, abamectin, levamisole and especially albendazole, may reduce the growth of its mycelium [44].

The average number of L3/g of feces and the percentage of development were always lower in *Duddingtonia* groups compared to control groups. On the other hand, the groups that received *D. flagrans* compared to control groups showed a higher percentage of larval population reduction in the fecal mass, from 62 to 72% over the two trials in the pasture. High percentage of reduction have been recorded in faecal cultures with the isolate *D. flagrans* CI3a reducing 90-99% of cyathostomin L3, although with a higher daily dose of $5x10^6$- $5x10^7$ spores/kg bw [22].

However, previous studies concerning the Troll A isolate of *D. flagrans* showed a more powerful nematophagous activity than other isolates, such as CI3, registering average percentage larval reduction of 98.4%. Nevertheless, these authors achieved this reduction efficacy after administration of a single dose to an individual horse, whose feces were placed in experimental plots to assess the ability of fungi in the L3reduction in fecal mass and on pasture [45].

This situation is substantially different from the administration of the fungus to a group of horses with *ad libitum* feeding, contributing to some loss of fungal material during its ingestion and different rates of intake, due to different appetites of the animals. Still, we think the values recorded in this investigation should be considered as promising for their use as a prophylactic measure of horse strongylidosis in grazing animals in different ecosystems and equine production systems, as the ones of these researches in Portugal.

Table 2.8. Hematological parameters of control group animals and *D. flagrans* in 1997/98

Leucocyte counts									
	N	EPG	Neutrophils (%)	Lynphocytes (%)	Monocytes (%)	Eosinophils (%)	Basophils (%)		
Standard values			22 - 72	17-68	0 – 7	0 - 10	0 - 4		
(Jain, 1993)			43.8 ± 7.0	47.9 ± 6.0	3.6 ± 2.0	4.1 ± 2.9	0.4 ± 0.5		
Control									
13-11-97	9	1278	42.56	50.33	3.78	3.11	0.22		
6-1-98	9	536	42.00	49.56	5.56	2.67	0.22		
3-3-98	9	2663	40.56	53.67	2.67	3.11	0.00		
D. flagrans									
13-11-97	9	1333	49.67	44.44	4.11	1.56	0.22		
6-1-98	9	759	43.22	48.56	3.78	4.11	0.22		
3-3-98	9	2212	41.67	49.78	4.00	4.11	0.44		
Hematocrit, cell counts and Hemoglobin values									
	N	EPG	Hematocrit (%)	Leucocytes ($nX10^3\mu l$)	RBC ($nX10^6\mu l$)	Hb (g/dl)			
Standard values			34.5 ± 3.8	10.81 ± 1.87	8.6 ± 0.58	11.8 ± 1.6			
(Jain, 1993)			32	9.05	9	14.4			
Control									
13-11-97	9	1278	24.83	18.72	6.18	12.71			
6-1-98	9	536	26.21	17.27	6.64	13.53			
3-3-98	9	2663	24.52	15.98	6.52	12.12			
D. flagrans									
13-11-97	9	1333	25.24	17.23	6.19	11.86			
6-1-98	9	759	25.36	16.49	6.05	12.98			
3-3-98	9	2212	24.31	15.99	6.17	12.11			
Biochemical Parameters									
	N	EPG	Total Prot. (g/dl)	Albumin (g/dl)	Globulins (g/dl)	Ratio Alb/Glob	Glucose (mg/dl)	ALT U/L 25°C	AST U/L 25°C
Standard values			5.2 – 7.9	2.60 – 3.70	2.62 – 4.04	6.2 – 14.6	75-115	3 – 23	226 - 366
(Kaneko et al 1997)			6.35 ± 0.59	3.09 ± 0.28	3.33 ± 0.71	9.6 ± 1.7	95.6± 8.5	14 ± 11	296 ± 70
Control									
13-11-97	9	1278	8.72	4.46	4.26	1.27	80.47	5.10	69.16
6-1-98	9	536	10.39	4.23	6.15	0.70	47.54	9.65	67.75
3-3-98	9	2663	4.56	3.26	1.41	2.97	*	10.2	289.4

Table 2.8. (Continued)

	N	EPG	Total Prot. (g/dl)	Albumin (g/dl)	Globulins (g/dl)	Ratio Alb/Glob	Glucose (mg/dl)	ALT U/L 25ºC	AST U/L 25ºC
D. flagrans									
13-11-97	9	1333	9.66	4.06	5.60	0.77	73.42	4.60	74.20
6-1-98	9	759	10.19	4.17	6.02	0.71	70.61	10.70	88.30
3-3-98	9	2212	4.78	3.30	1.55	2.70	*	6.71	246.50

* Not analysed.

The individual administration showed to have even greater applicability in the control of horse strongylidosis, because in horses dewormed with ivermectin and receiving *D. flagrans*, their egg-shedding remained low (<50 EPG) for 3-4 months. Besides, their intake occurs more effectively and is easily articulated with the strategic treatments of the horse farm [46].

In both groups the predominant genus of infective larvae was *Cyathostomum sl,* although were identified in all trials three and six genera, in groups D and C, respectively. However it was not possible to determine possible differences in their reduction due to the small number of specimens of some of these genera, as reported by other authors. The L3 of genus *Strongylus* were not isolated in the cultures, probably due to the long pre-patent period (6 to 11 months) and the animals used in some of the trials had less than 1 year old [24,47].

Compared to other studies, the stable trial in 2001 revealed a lower rate of L3 population reduction, possibly related to a low EPG and fewer L3 in culture, since only large numbers of infective larval stages induce a greater development of the micellar. Although not significant differences in average levels of reduction of L3 in the cultures of the treatment groups in group A (with administration of *D. flagrans*) and C (associated with ivermectin) there was an average reduction of 28 - 30% of the larvae. The lower reduction rate was achieved with *Duddingtonia* associated with moxidectin, and once again care should be taken in what concerns the best anthelmintic to be associated, since some induce more trapping structures than other, which seems to be the case of ivermectin, considered a greater stimulant of nematophagous activity [48-50].

Contamination of pastures used for animals receiving *D. flagrans* was lower than the control group, although only significant differences have been recorded in the study in 1999, a situation that may have several explanations.

The non significant reduction of L3 on pasture during the 1997/98 test may have been a result of progressive reduction of rainfall and temperature fluctuation (minimum of 7.9ºC and maximum 16-19°C) from November to February. Other authors have set a limit to the development of the micellar network, 10-35°C, below and above which the fungus is less likely to grow and reproduce, reaching the optimum point of growth at 30°C [27].

The temperature increment in February / March 1998, may have increased the nematophagous activity, contributing to the maintenance of lower levels of L3 on pasture of group D. However, the low temperatures at the beginning may have reduced the larval survival, dry weather and poor harvests of the herb in the past (especially in the paddock of group C) did not allow an effective migration of L3 into the vegetation and a more significant difference between the two groups [51, 52].

In the 1999 trial there was a different situation, with an almost constant rainfall and a progressive increase in average temperature from March to July (minimum 6.3 to 15.6ºC and maximum 16.6 to 31.7°C). These weather conditions enabled a greater activity and higher rate of migration of larval forms in the pasture, as proposed by other researchers [32, 53, 54].

Thus, there was a peak of larvae on pasture in the paddock of C1control group in April/May, which agreed with the results obtained during other epidemiological studies in horses grazing at the CL horse farm [25, 52, 55, 56].

But during this peak, the group D animals showed a remarkable reduction of L3 on pasture, which remained even after the group C1 was removed from the pasture. Thus, the temperature increase coupled with rainfall stimulated activity on the one hand and its larval migration to the grass on the other and therefore the fungal growth and nematophagous activity increased, a situation that agrees with results in Denmark [57] and partly with those in Louisiana, USA [58].

The development of the micellar network and nematophagous activity are also stimulated by higher EPG counts and a progressive contact with a growing number of infective larvae, which seems to have happened between March and July [3, 59].

Nematophagous fungi may be used more effectively during the winter and spring in the southern U.S. Our results seem to indicate better efficacy of *D. flagrans* on the larval forms of strongyles on pasture during the spring/summer and a smaller reduction in larvae during the autumn/winter, which seems to contradict the one proposed by those authors. But the autumn/winter of southern US can prove rather moisty and warm, and these climatic parameters have a positive effect on the fungi [58].

Recent investigations seem to confirm our hypothesis, because these fungi have a positive correlation between the increase in temperature (up to 35ºC) and the further development and predacious efficacy of the fungus. They also found a low and weak development of the mycelium nematophagous activity at 5 to 10ºC, which may question its use in animals grazing in the winter [60, 61].

In blood parameters, cell counts at the level of leukocytes and erythrocytes revealed no differences between the two groups. However, the leukocyte counts, neutrophils and lymphocytes of *Duddingtonia* group had values closer to the standard ones. The lymphocytosis was evident and have been responsible for the leukocytosis in both groups, although more evident at the end of the trial in control animals. This is typical of hematological picture of horse strongylidosis/cyatostominosis, mainly due to the increase of neutrophils and lymphocytes [38, 62].

In what concerns biochemical parameters there were no differences between groups and records in both a hyperproteinemia, hyperglobulinemia and inversion of the albumin/globulin ratio, with values near 0.7. This was accompanied by hypoglycemia, most evident in the control group, which seems really connected with the output of larval forms of hypobiotic cyathostomin larvae [48].

The decrease of the albumin / globulin ratio is mainly due to hypoalbuminemia caused by protein loss in the large intestine and for the hiperbetaglobulinemia, a response to cyathostomin infection. The low level of blood sugar was probably the consequence of a deficiency of absorption during the outburst of cyathostomin larvae, as evidenced by recent works on the pathophysiology of infection by these nematodes [63-65].

When tested in the autumn/winter 1997/98, the animals showed poor body condition, intermittent diarrhea and a list of albumin / globulin ratio close to 0.7, highly suspected of type II cyathostominosis syndrome diagnosed in the 1990s at the CL horse farm [48]. This condition, time of year and the use of young animals may have contributed to a lower efficacy of *D. flagrans* in grazing horses in the 1998 field trial compared with the 1999 one.

Although indirectly, this assay allowed to deduce the existence of equine cyathostomin hipobiosis in this region: a) the short egg reappearance period after deworming, b) there were EPG peaks in January when eggs will usually rise in March/April; c) the group that received the *D. flagrans* had a minimum degree of contamination of the pasture, but the pattern of elimination of eggs was similar to the control group, a sign that the determinism of this behavior must be related to exogenous cycle due to the environment stimulus that L3 receive before ingestion by the host or immunity phenomena of the infected horses [66-68].

In short, no changes were observed in blood of horses receiving fungus that could be associated with this administration, registering hematic changes characteristic of a cyatostominosis type II, a parasitic disease common in equine production system, which coincided with the seasonality of the 1997/98 trial [56].

According to available literature, the tests carried out in Portugal with *D. flagrans* were the second record of field work worldwide with naturally infected horses grazing and probably the first record in stabled horses, in which it was pretty visible the impact of the administration of the fungus in reducing their infection and contamination of pasture with strongyles.

The results proved the effectiveness of the fungus *Duddingtonia flagrans,* isolate Troll A in both reducing pasture contamination and infection of the hosts in this production system involving animals in the pasture and in the stable in a Mediterranean ecosystem.

The administration of this fungus should be understood as a complement to the strategic deworming of horses and not as a substitute for the administration of anthelmintics, which in our country may have wide application in the spring, summer and probably in early autumn. This behavior is consistent with other works done with horse strongyle larvae on pasture in regions and seasons hot and humid [57, 58].

The sustainable control of horseparasitic diseases is a combination of good hygiene of the pasture, grazing management, sensible and rational use of anthelmintics and adoption of biological control measures that have a strong acceptance by producers and a good market prospect. This form of control will allow the contact of horses with a low level of contamination in the pasture and the intake of a low concentration of L3, which allow a constant antigenic stimulus and a pre-immune status of premonition [2, 69, 70].

On the other hand, to reduce contamination of the environment, decreases the infection of the host and reduce the number of administrations of anthelmintics, thus helping to delay the onset of resistance, with special interest to the group of avermectins / milbemycin. Given the intensive use of these drugs in equine deworming, we should preserve its effectiveness and its usefulness as anthelmintics with high margin of safety, spectrum of activity and ease of administration for as long as possible [71, 72].

Studies already carried out in the region can be considered as fundamental to deworming to be carried out in March/April and September/October and if the body condition warrants, in December/January too. Administration of anthelmintics in autumn/winter must involve drugs with recognized efficacy on the larval forms in hypobiosis (L3/L4) to reduce the transmission of infective forms for grazing animals and clinical cases of cyathostominosis [51].

In this equine production farms with high grazing system, strongyle infection can be minimized from early spring until autumn with the administration of *Duddingtonia flagrans*, easily associated with conventional anthelmintic treatments.

Conclusions and Future Perspectives

The fungus *Duddingtonia flagrans,* variety Troll A, administered at a daily dose of $5x10^5$ spores / kg bw, shows a good mean efficacy in the reduction of larval population in culture, 62-72% and a significant decrease in the contamination of pasture with strongyle L3, 50-70%, especially during the spring / summer, which makes this approach a very promising means of control in horses grazing when associated with strategic treatment with anthelmintics at this time of year. This also applies to animals in stable / semi-housing systems and it is expected that this management may maintain low levels of EPG even longer, reducing to a minimum of two deworming interventions per year.

The administration of *D. flagrans* did not affect the blood parameters studied, constituting the first observation of the world of non-interference in vital parameters of the nematophagous fungus during an equine study.

The results presented here allow us to foresee the importance of implementation of biological control in the control of parasitic diseases of horses in Portugal (as well as other countries with similar equine production and management, like Spain), in particular intestinal strongylidosis, given that approach is practical, innovative and can easily be integrated with the strategic treatments conducted with avermectin/milbemycin and other forms of integrated control of parasitic diseases, such as pasture rotation and hygiene.

Acknowledgments

To CIISA / FMV / UTL and its coordinator, Professor Luis Tavares, for financing this work.

To "CHR HANSEN - BioSystems, A / S", Denmark, for the interest in this project and for providing *D.* for testing. A special thanks to Dr. Adrian Gillespie, unfortunately deceased, and without whom there would not have been possible to perform these tests and study this new approach on the control of animal parasitic diseases. The Board of Companhia das Lezírias (CL), in particular to Mr. Eng FranciscoPerestrello, for the willingness and knowledge to provide us with the animals, locations and premises where the works have made, and the Board of Coudelaria Nacional(CN) at Estação Zootécnica. The technical and auxiliary Staff of both CL and CN for invaluable input in the management and restraint of animals during the sample collections. The colleagues who worked actively in all these subsidiary works that provided the overall results, in particular Professor Isabel Fazendeiro, and Drs. Pedro Serra, Carlos Farrim, Mario Barbosa, Ricardo Agrícola and Engineer Helena Jorge. To Professor Isabel Neto of the Dept. of Animal Health, FVM/TUL, we appreciate the support in statistical analysis.

We also thank the Technical Assistant, Dr. Lidia Gomes, from Parasitology and Parasitic Diseases, for her help and collaboration in the field trials and laboratory technical support.

A final word of appreciation to all the students of 3rd and 4th years that from 1997 to 2001 accompanied the field and laboratory work, allowing more rapid processes and healthy environment of ideas exchange relating to this new type of anti-parasitic control. To everyone involved, a big thank you!

References

[1] Lyons, ET, Tolliver, SC, & Drudge, JH (1999). Historical perspective of cyathostomes: prevalence, treatment and control programs. pp. 97-111. *In* Little, SA; Moore, JN;

Dipietro, JA(Eds.) *Special issue: Equine Cyathostome Conference. Proceedings of a Conference on Equine Cyathostomes held at the University of Georgia, Athens, GA, 7-8 November 1998. Vet. Parasitol.*, 85 (2/3), 95-225 pp.

[2] Waller, PJ (1999). Biological control of parasitic nematodes of the horse; the need, practicalities and prospects. *Equine Vet. J.*, 31 (6), 449-450.

[3] Bird, J, Herd, RP (1994). Nematophagous fungi for the control of equine cyathostomes. *Comp. Cont. Educ. Pract. Veter. The Compendium, Equine,* 16 (5), 658-665.

[4] Lichtenfels, JR, Kharchenko, V.A, Krecek, RC, Gibbons, LM (1998). An annotated checklist by genus and species of 93 species level names for 51 recognized species of small strongyles (*Nematoda: Strongyloidea: Cyathostominea*) of horses, asses and zebras of the world. *Vet. Parasitol.*, 79 (1), 65-79.

[5] Lichtenfels, J.R., Kharchenko, V.A., Dvojnos, G.M. (2008) Illustrated identification keys to strongylid parasites (strongylidae: Nematoda) of horses, zebras and asses (Equidae). *Veterinary Parasitology*, Volume 156, Issues 1-2, 4-161.

[6] Lloyd, S, Soulsby, EJL (1998). Is Anthelmintic Resistance Inevitable: Back To Basics? *Equine Vet. J.*, 30 (4), 280-283.

[7] Waller, PJ (1997). Sustainable helminth control of ruminants in developing countries. *Vet. Parasitol.*, 71 (2/3), 195-207.

[8] Mota, M.A., Campos, A.K., Araújo, J.V. (2003) Controle biológico de helmintos parasitos de animais: estágio atual e perspectivas futuras. *Pesq. Vet. Bras.* 23(3):93-100.

[9] Soder, K. J. & Holden, L.A. (2005) Use of Nematode-Trapping Fungi as a Biological Control in Grazing Livestock. *The Professional Animal Scientist* 21 (2005):30–37.

[10] Araújo, J.V.; Santos, M.A.; Ferraz, S.; Maia, A.S.; Magalhães (1992) Controle de larvas infectantes de *Haemonchus placei* por fungo predadores da espécie *Monacrosporium ellypsosporum* em condições de laboratório. *Arq.Bras. Med. Vet. Zoot.*, 44 (6): 521-526.

[11] Larsen, M, Wolstrup, J, Henriksen, SA, Dackman, C, Grønvold, J, Nansen, P (1991). *In vitro* stress selection of nematophagous fungi for biocontrol of parasitic nematodes of ruminants. *J. Helminthol.*, 65 (3), 193-200.

[12] Sagüés, M.F., Purslowc, P., Fernández, S., Fusé, L., Iglesias, L. & Saumell, C. (2011) Hongos nematófagos utilizados para el control biológico de nematodos gastrointestinales en el ganado y sus formas de administración. *Rev Iberoam Micol.*. doi:10.1016/j.riam.2011.06.009.

[13] Barron, GL (1977) - *The nematode-destroying fungi*. Department of Environmental Biology, Ontario Agricultural College, University of Guelph, Guelph, Canada, 111 pp.

[14] Nordbring-Hertz, B., Jansson, H.B, Tunlid, A. (2006) Nematophagous fungi. Encyclopedia Of Life Sciences, John Wiley & Sons, Ltd. www.els.net, doi: 10.1038/npg.els.0004293, 11 pp.

[15] Waller, P.J. (1993) Nematophagous Fungi. Prospective Biological Control Agents of Animal Parasitic Nematodes. *Parasitology Today*, 9 (11): 429-431.

[16] Braga, F.R., Araújo, J.V., Carvalho, R.O., Araujo, J.M., Silva, A.R. & Campos, A.K. (2009) Controle *In Vitro* de Larvas Infectantes de Ciatostomíneos (NEMATODA: CYATHOSTOMINAE) de Equinos utilizando os Fungos Predadores *Duddingtonia flagrans, Monacrosporium thaumasium* e *Arthrobotrys robusta. Ciência Animal Brasileira,* 10 (3), 887-892.

[17] Braga, F.R., Araújo, J.V., Carvalho, R.O., Silva, A.R., Araujo, J.M., Soares, F.F., Geniêr, H.A.,, Ferreira, S.R., Queiroz, J.H. (2010) Ovicidal action of a crude enzymatic

extract of the fungus Pochonia chlamydosporia against cyathostomin eggs. *Veterinary Parasitology* 172, 264–268.

[18] Larsen, M (2000). Prospects for controlling animal parasitic nematodes by predacious micro fungi. *Parasitology*, 120, S121-S131.

[19] Grønvold, J, Wolstrup, J, Larsen, M, Henriksen, SA, Nansen, P (1993). Biological control of *Ostertagia ostertagi* by feeding selected nematode-trapping fungi to calves. *J. Helminthol.*, 67, 31-36.

[20] Larsen, M, Wolstrup, J, Henriksen, Sa, Grønvold, J, Nansen, P (1992). *In vivo* passage through calves of nematophagous fungi selected for biocontrol of parasitic nematodes. *J. Helminthol.*, 66, 137-141.

[21] Larsen, M, Faedo, M, Waller, PJ (1994). The potential of nematophagous fungi to control the free-living stages of nematode parasites of sheep: survey for the presence of fungi in fresh faeces of grazing livestock in Australia. *Vet. Parasitol.*, 53 (3-4), 275-281.

[22] Larsen, M, Nansen, P, Henriksen, SA, Wolstrup, J, Grønvold, J, Zorn, A, Wedø, E (1995). Predacious activity of the nematode-trapping fungus *Duddingtonia flagrans* against cyathostome larvae in faeces after passage through the gastrointestinal tract of horses. *Vet.Parasitol.*, 60 (3/4), 315-320.

[23] Bird, J, Herd, RP (1995). *In vitro* assessment of two species of nematophagous fungi (*Arthrobotrys oligospora and Arthrobotrys flagrans*) to control the development of infective cyathostome larvae from naturally infected horses. *Vet.Parasitol.* 56 (1/3), 181-187.

[24] Larsen, M, Nansen, P, Grondahl, C, Thamsborg, SM, Grønvold, J, Wolstrup, J, Henriksen, SA, Monrad, J (1996). The capacity of the fungus *Duddingtonia flagrans* to prevent strongyle infections in foals on pasture. *Parasitology*, 113, 1-6.

[25] Madeira de Carvalho, LM, Afonso-Roque, MM, Carvalho-Varela, M (1997). Aspectos epidemiológicos dos estrongilídeos nos cavalos em regime extensivo e dinâmica estacional das suas formas infectantes na pastagem. *"V Congresso Ibérico de Parasitologia, Universidade de Évora, 6-10 de Outubro de 1997". Resumo publicado na Acta Parasitol. Port.,* vol.4 (1/2) *: p. 41.*

[26] Araújo, J.V.; Santos, M.A.; Ferraz, S.; Maia, A.S. (1994) Biological control "in vitro" of infective *Haemonchus placei* larvae by predacious fungi *Arthrobotrys musiformis. Arq. Bras. Med. Vet. Zoot.*, 46 (3): 197-204.

[27] Grønvold, J, Nansen, P, Henriksen, SA, Larsen, M, Wolstrup, J, Bresciani, J, Rawat, H, Fribert, L (1996). Induction of traps by *Ostertagia ostertagi* larvae, chlamydospore production and growth rate in the nematode–trapping fungus *Duddingtonia flagrans. J. Helmint.*, 70: 291-297.

[28] Grønvold, J, Wolstrup, J, Nansen, P, Henriksen, SA, Larsen, M, Bresciani, J (1993a). Biological control of nematode parasites in cattle with nematode-trapping fungi: a survey of danish Studies. *Vet. Parasitol., Special Issue: Environmental impact of avermectin usage in livestock*, 48 (1/4), 311-325.

[29] Thienpont, D, Rochette, F, Vanparijs, OFJ (1986). *Diagnóstico de las helminthiasis por medio del examen coprológico.* 2ª Edición, Janssen Research Foundation, Beerse, Belgium, 205 pp.

[30] Ueno, H, Gutierres, VC (1983). *Manual para Diagnóstico das Helmintoses de Ruminantes*. 2ª Edição, Japan International Cooperation Agency, Tóquio, Japão, 176 pp.

[31] Madeira de Carvalho, LM (1993). *Relatório da Missão de estudo ao "Department of Veterinary Parasitology, Faculty of Veterinary Medicine, University of Glasgow", Glasgow ,18-31 de Outubro de 1993*. Faculdade de Medicina Veterinária, Lisboa, 46 pp.

[32] Baudena, MA; Chapman, MR; French, DD; Klei, TR (2000). Seasonal development and survival of equine cyathostome larvae on pasture in south Louisiana.*Vet. Parasitol.*, 88 (1-2), 51-60.

[33] Soulsby, EJL (1965). *Textbook of Veterinary Clinical Parasitology - Vol. I. Helminths*. 1st Edition, Blackwell Scientific Publications, Oxford, GB, 1120 pp.

[34] Arundel, JH (1985). *Parasitic Diseases of the Horse*.Veterinary Review N°28, University of Sydney.The Post-Graduate Foundation in Veterinary Science, Sydney, NSW, Australia, 150 pp.

[35] Bevilaqua, CML; Rodrigues, ML; Concordet, D (1993). Identification of infective larvae of some common nematode strongylids of horses. *Revue Méd.Vét.*, 144 (12), 989-995.

[36] Taylor, EL (1939). Technique for the estimation of pasture infestation by strongyloid larvae. *Parasitology*, 31, 473-478.

[37] Ludwig, KG, Johnstone, C (1984). Use of pasture sampling to determine equine strongyle larvae contamination.*Vet. Med.*, 1183-1189.

[38] Jain, NC (1993). *Essentials of Veterinary Hematology*.Lea & Febiger, Philadelphia, USA, 1st Edition, 417 pp.

[39] Kaneko, JJ, Harvey, JW, Bruss, ML (1997). Apendixes. pp. 885-905. *In* "KANEKO, JJ, HARVEY, JW, BRUSS, ML (Eds.) – *Clinical Biochemistry of Domestic Animals*", 5th Edition, Academic Press, 932 pp.

[40] Wolstrup, J, Grønvold, J, Henriksen, SA, Nansen, P, Larsen, M, Bogh, HO; Ilsoe, B (1994). An attempt to implement the nematode-trapping fungus *Duddingtonia flagrans* in biological control of trichostrongyle infections of first year grazing calves. *J Helminthol.* 68, 175-180.

[41] Larsen, M, Nansen, P, Wolstrup, J, Grønvold, J, Henriksen, SA, Zorn, A (1995a). Biological control of trichostrongyles in calves by the fungus *Duddingtonia flagrans* fed to animals under natural grazing conditions. *Vet. Parasitol.*, 60, 321-330.

[42] Piché, CA, Kennedy, MJ, Bauck, SW, Goonewardene, L (1990). Comparison of three anthelmintics in the control of intestinal nematodes in young horses on fall and winter pasture. *Can. Vet. J.*, 31, 841-843.

[43] Madeira de Carvalho, LM, Farrim, MC, Afonso-Roque, MM, Fazendeiro, MI (2003). Groups, efficacy and egg reappearance period of commonly used anthelmintics in equine practice in Portugal.9th Int. Cong. Eur. Ass. Vet. Pharm. Toxicol., Lisbon, Portugal, 13-18th July 2003. *J. Vet. Pharm.Therap.,* 26, Suppl.1, pp. 237-238.

[44] Araújo, JV; Santos, MA; Ferraz, S (1995). Efeito de drogas anti-helmínticas sobre o crescimento de fungos nematófagos do gênero *Arthrobotrys*. *Naturalia, São Paulo*, 20, 157-163.

[45] Silvina-Fernández, AS, Henningsen, E, Larsen, M, Nansen, P, Grønvold, J, Søndergaard, J (1999). A new isolate of the nematophagous fungus *Duddingtonia*

flagrans as biological control agent against free-living larvae of horse strongyles. *Equine Vet. J.*, 31(6), 488-491.

[46] Jorge, H, Agrícola, R, Madeira de Carvalho, LM, Gillespie, AT, Barbosa, M, Fazendeiro, IM (2000). Eficácia do fungo *Duddingtonia flagrans* no controlo dos estrongilídeos em equinos estabulados. *"V Congresso Anual da Sociedade Portuguesa de Parasitologia, Faculdade de Medicina Veterinária, Pólo Universitário do Alto da Ajuda, 23-25 de Novembro de 2000*", Resumo publicado na *Acta Parasitol. port.*, vol. 7, (1/2) : 70-71.

[47] Madeira de Carvalho, LM, Afonso-Roque, MM, Fazendeiro, MI (2001). Epidemiology of horse strongyle infections in Portugal: the rise of the cyathostomes! *1st International Symposium – Research in Veterinary Medicine, Centro de Investigação Interdisciplinar em Sanidade Animal (CIISA), Faculdade de Medicina Veterinária, Lisbon, 24-25 May 2001,* Abstract PO-13, pp. 52.

[48] Madeira de Carvalho, LM (2001). *Epidemiologia e controlo da estrongilidose em diferentes sistemas de produção equina em Portugal.* Tese de Dissertação de Doutoramento, Faculdade de Medicina Veterinária, Universidade Técnica de Lisboa, 445 + xxii pp.

[49] Madeira de Carvalho,L.M; Gillespie,A.T, Serra,P; Bernardo,F.A; Farrim,A.P; Jorge,H; Agrícola,R; Barbosa,M; Fazendeiro,M.I. (2002). Controlo biológico da estrongilidose equina em Portugal pelo fungo nematófago *Duddingtonia flagrans. Proceedings of the Veterinary Science Congress* ,Oeiras , 10-12 Outubro, Portugal.

[50] Kuzmina, T. (2001). The effect of some anthelmintics on *Duddingtonia flagrans* growth [Abstract]. *Proceedings of the18th internationalconference of the WAAVP*, Stresa – Italy, 26-30 August.

[51] Madeira de Carvalho, LM, Gillespie, A.T, Serra, PM, Farrim, AP, Bernardo, FA, Gonçalves, J (1999a) - A field trial with *Duddingtonia flagrans* to control strongyle infections in foals in Portugal. *17th International Conference of the World Association for the Advancement of Veterinary Parasitology, 15th-19th August 1999, Copenhaguen, Dinamarca,* Abstract c.7.71.

[52] Madeira de Carvalho, LM, Gillespie, AT, Serra, PM, Farrim, AP (2000). Controlo biológico dos estrongilídeos nos equinos em pastoreio pelo fungo *Duddingtonia flagrans. "V Congresso Anual da Sociedade Portuguesa de Parasitologia, Faculdade de Medicina Veterinária, Pólo Universitário do Alto da Ajuda, 23-25 de Novembro de 2000*", Resumo publicado na *Acta Parasitol. port.*, vol. 7, (1/2) : 69-70.

[53] Mfitilodze, MW, Hutchinson, GW (1988). Development of free-living stages of equine strongyles in faeces on pasture in a tropical environment. *Vet. Parasitol.*, 26 (3/4), 285-296.

[54] Hutchinson, GW, Abba, SA, Mfitilodze, MW (1989). Seasonal translation of equine strongyle infective larvae to herbage in tropical Australia. *Vet. Parasitol.*, 33 (3-4), 251-263.

[55] Madeira de Carvalho, LM, Afonso-Roque, MM, Carvalho-Varela, M (1999). Seasonal pattern of a population of horse strongyles in Portugal.*17th International Conference of the World Association for the Advancement of Veterinary Parasitology, 15th-19th August 1999, Copenhaguen, Dinamarca*, Abstract a.3.01.

[56] Madeira de Carvalho, LM, Serra, P, Afonso-Roque, MM, Fazendeiro, MI (2002). Ciatostominose equina: actualização do seu conhecimento em Portugal. *Proceedings of*

the Veterinary Sciences Congress, Congresso de Ciências Veterinárias, 100 Anos da Sociedade Portuguesa de Ciências Veterinárias, Tagus Park, Oeiras, 10-12 de Outubro de 2002, pp. 353-354.

[57] Fernández, AS, Larsen, M, Nansen, P, Grønvold, J, Henriksen, SA, Wolstrup, J (1997). Effect of the nematode-trapping fungus *Duddingtonia flagrans* on the free-living stages of horse parasitic nematodes: a pilot study. *Vet. Parasitol.,* 73 (3/4), 257-266.

[58] Baudena, MA; Chapman, MR; Larsen, M; Klei, TR (2000a) – Efficacy of the nematophagous fungus *Duddingtonia flagrans* in reducing equine cyathostome larvae on pasture in south Louisiana. *Vet. Parasitol.*, 89 (3), 219-230.

[59] Paz-Silva, A, Francisco, I., Valero-Coss, R.O., Cortiñas,F.J., Sánchez, J.A., Francisco, R, Arias, M, Suárez, M.L., López-Arellano, M.E., Sánchez-Andrade, M.R., Mendoza de Gives, P. (2011) Ability of the fungus *Duddingtonia flagrans* to adapt to the cyathostomin egg-output by spreading chlamydospores. *Vet. Parasitol.,* 179, 277–282

[60] Silvina-Fernández, AS (1998). *Duddingtonia flagrans as biological control agent against parasitic nematodes of cattle and horses*.Ph.D. Thesis, Danish Center for Experimental Parasitology, The Royal Veterinary and Agricultural University, Copenhagen, Denmark, 179 pp.

[61] Fernández, AS, Larsen, M, Wolstrup, J, Grønvold, J, Nansen, P, Bjorn, H (1999). Growth rate and trapping efficacy of nematode-trapping fungi under constant and fluctuating temperatures. *Parasitol. Res.,* 85 (8/9), 661-668.

[62] Ogbourne, CP (1978). *Pathogenesis of Cyathostome (Trichonema) infections of the horse.A review*.Commonwealth Agricultural Bureaux, Commonwealth Institute of Parasitology, Miscellaneous publication Nº 5, 25 pp.

[63] Murphy, D, Keane, MP, Chandler, KJ, Goulding, R (1997). Cyathostome associated disease in the horse: Investigation and management of four cases. *Eq. Vet. Educ.*, 9 (5), 247-252.

[64] Thamsborg, SM, Leifsson, PS, Grøndahl, C, Larsen, M, Nansen, P (1998). Impact of mixed strongyle infections in foals after one month on pasture. *Equine.Vet.J.*, 30 (3), 240-245.

[65] Smets, K, Shaw, DJ, Deprez, P, Vercruysse, J (1999) - Diagnosis of larval cyathostominosis in horses in Belgium. *Vet. Rec.*, 144 (24), 665-668.

[66] Reinemeyer, CR (1986). Small Strongyles.Recent Advances. pp. 281-312. *In* HERD, R.P. (Ed.) *Vet. Clin.North Amer., Equine Practice, Parasitology*, 2 (2), 263-463 pp.

[67] Proudman, C, Matthews, J (2000). Control of intestinal parasites in horses.*In Practice*, 22 (2), 90-97.

[68] Chapman, MR, French, DD, Taylor, HW, Klei, TR (2002). One season of pasture exposure fails to induce a protective resistance to cyathostomes but increases numbers of hypobiotic third-stage larvae. *J. Parasitol.,* 88, 678-683.

[69] Nielsen, M.K. (2008). Sustainable equine parasite control: Perspectives and research needs. *Proceeding International Equine Parasite Drug Resistance Workshop, Faculty of Life Sciences*, University of Copenhagen, July 31-August 1, 2008, pp. 13-14.

[70] Urquhart, GM, Armour, J Duncan, JL, Dunn, AM, Jennings, FW (1987). *Veterinary Parasitology.*1st Ed., Longman Scientific and Technical, Longman Group Ltd, U.K., 286 pp.

[71] Sangster, NC (1999). Pharmacology of anthelmintic resistance in cyathostomes: will it occur with the avermectin/milbemycins? pp. 189-204. *In* LITTLE, S.A.; MOORE, J.N.;

DIPIETRO, J.A.(Eds.) *Special issue: Equine Cyathostome Conference. Proceedings of a Conference on Equine Cyathostomes held at the University of Georgia, Athens, GA, 7-8 November 1998. Vet. Parasitol.*, 85 (2/3), 95-225 pp.

[72] Sangster, NC (2003). A practical approach to anthelmintic resistance. *Equine Vet. J.*, 35 (3), 218-219.

In: Fungi: Types, Environmental Impact and Role in Disease ISBN: 978-1-61942-671-9
Editors: Adolfo Paz Silva and María Sol

Chapter XV

Biological Control of Parasites in Veterinary Medicine

Silvina Fernández[1] and Carlos Saumell[2]
[1] BioNem Research Centre, ON, Canada
[2] Área de Parasitología y Enfermedades Parasitarias, Departamento de Sanidad Animal y Medicina Preventiva, Facultad de Ciencias Veterinarias, UNCPBA, Tandil, Argentina

Abstract

The capacity of microfungi known as 'nematophagous fungi' to reduce the numbers of parasitic nematodes is anything but a new subject. However, the research into the actual development of these fungi for biological control purposes and their practical application in veterinary medicine has made inroads only during the last two decades.

This chapter intends to shed light on those inroads, focusing on the application of biological control of parasites of veterinary importance. Examples of parasite-destroying fungi other than nematophagous ones are also mentioned in the chapter.

The four functional categories of nematophagous fungi are nematode-trapping, endoparasitic, egg- and cyst-parasitic fungi, and toxin-producing fungi. Although species of all categories have shown capabilities to act as biological control agents at different levels, the first group has been more extensively studied for that purpose to this date. However, the field for using the second and third groups remains wide open. While the last category is briefly mentioned and exemplified, this chapter focuses on the first three categories, including their description, biology, and behaviour in the environment.

The research into using nematophagous fungi to control parasitic nematodes of veterinary importance has not only elucidated issues such as which fungal species are the most promising candidates and their individual nematode suppressive effect under controlled conditions, but has also raised problematic questions ranging from the use of the fungi in different geographical and climate regions to the formulation, commercialisation and regulatory aspects of their use. All these and the future perspectives of this exciting field of knowledge will be covered in this chapter.

Introduction

There are several definitions of biological control. An *ad hoc* group of scientists defined biological control as the "pest suppression with biotic agents, excluding the process of breeding for resistance to pests, sterility techniques, and chemicals modifying pest behaviour" [1]. Meanwhile, deBach & Rosen [2] state that "biological control in an ecological sense can be defined as the regulation by natural enemies of another organism's population density at a lower average than would otherwise occur". Later on Van Driesche & Bellows [3] proposed biological control as "the use of parasitoid, predator, pathogen, antagonist, or competitor populations to suppress a pest population, making it less abundant and thus less damaging than it would otherwise be"; while Grønvold et al [4] defined biological control in the applied sense as "an ecological method designed by man to lower a pest or parasite population to acceptable sub-clinical densities or to keep these populations at a non-harmful level using living antagonists". All the above definitions are related in one way or another and have a common denominator, i.e. the reduction or regulation of a pest population using natural antagonists. In this chapter, the definition by Grønvold et al [4] will be adopted.

Animal parasites, whether endoparasites (e.g. nematodes, trematodes, cestodes) or ectoparasites (e.g. arthropods), have many natural antagonists, such as insects, bacteria, fungi, viruses, nematodes, acari, etc. This chapter deals exclusively with the fungal antagonists of animal parasites, and specifically with fungi attacking nematodes, i.e. nematophagous fungi. A brief mention of parasite-destroying fungi other than nematophagous ones is given in the section 'Potential fungal species for biological control in veterinary medicine' of this chapter.

The search for biological control agents against parasites of veterinary importance sprung from the need to develop alternative measures of control other than the use of chemicals. The reason for this is two-fold: the worldwide spread of parasite resistance to antiparasitic drugs [5], which can be considered chronologically the first motive, and concerns related to ecotoxicity of antiparasitic drugs and the growing public consumption of organic (ecological) foods [6]. As the tendency is to move from the exclusive use of chemical drugs to the inclusion of alternative strategies under an integrated parasite management [7], the aim of biological control using fungal antagonists is to be one of those integrated components.

Types of Nematophagous Fungi and Their Mode of Action

Nematophagous fungi have traditionally been divided into endoparasitic fungi and nematode-trapping (or predatory) fungi [8, 9] with a third group comprising of egg- and cyst-parasitic fungi [10, 11], although Barron [9] places this type of fungi within the former group.

A functional classification depending on the manner of attacking their prey divides the nematophagous fungi in four categories: 1) nematode-trapping fungi, 2) endoparasitic fungi, 3) egg- and cyst-parasitic fungi, and 4) toxin-producing fungi [12, 13, 14, 15, 16]. While the last category will be briefly mentioned and exemplified, this chapter focuses on the first three categories.

Nematode-Trapping Fungi

This group is characterised by an extensive hyphal system (mycelium) with trapping organs or devices formed along the hyphae; the fungi trap nematodes using their organs of capture, then penetrate them and consume their contents rapidly. The entire process from capture to total destruction of the nematode under laboratory conditions takes 4-7 days. The organs of capture depend on the fungal species, and are adhesive hyphae, adhesive branches, adhesive two- or three-dimensional nets, adhesive knobs, non-constricting rings, and constricting rings (Figures 1 – 3). In the case of adhesive organs, the adhesive material either covers the whole hyphae or, more commonly, is limited to the actual trapping devices.

Examples of nematode-trapping fungi featuring the different organs of capture are the genera *Stylopage* and *Cystopage* (adhesive hyphae), *Monacrosporium gephyropagum* (= *Dactylella cionopaga*) and *M. cionopagum* (adhesive branches), Arthrobotrys oligospora, *A. conoides* and *Duddingtonia flagrans* (adhesive nets), *M. robustum*, *M. ellipsosporum* (adhesive knobs), *M. haptotylum* (= *Dactylaria candida*) and *M. lysipagum* (= *D. lysipaga*) (non-constricting rings), and *A. dactyloides*, *A. anchonia* and *M. aphrobrochum* (constricting rings) [8, 9, 17].

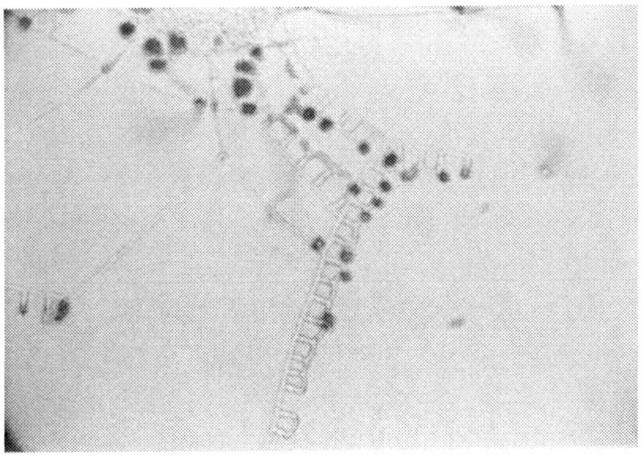

Figure 1. Two-dimensional net of *M. renderensis*, a species of Monacrosporium isolated in Brazil [100].

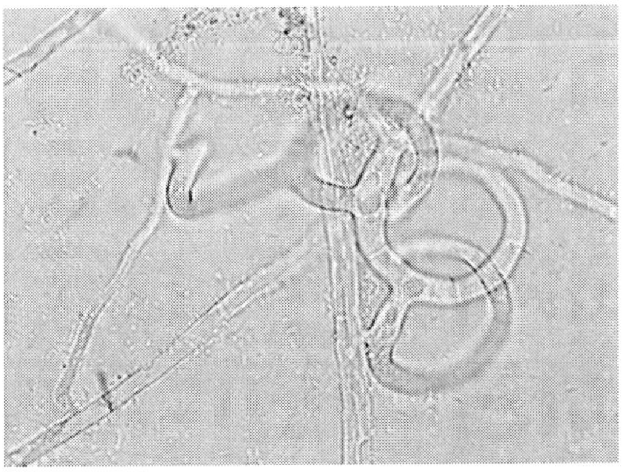

Figure 2. Three-dimensional net of *A. conoides*, which is formed by erect branches growing from the vegetative hyphae and curving on themselves to fuse with the parent hypha.

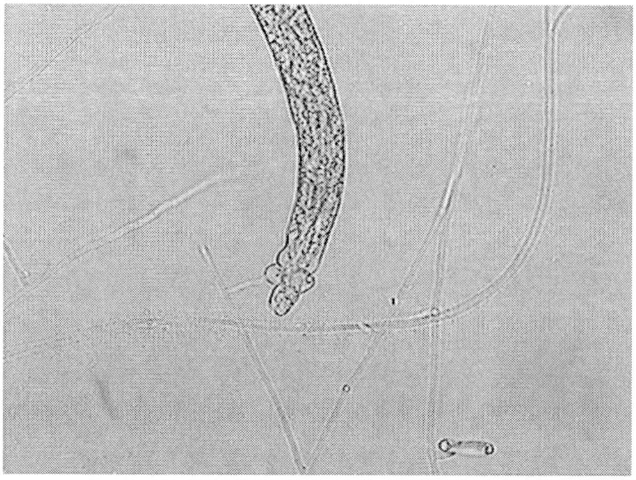

Figure 3. A nematode trapped in its cephalic region by a constricting ring of *M. aphrobrochum*.

Endoparasitic Fungi

These fungi do not grow extensive mycelia outside the nematode body; instead, they exist in the environment as zoospores or conidia (non-sexual spores) (Figure 4) and are obligated parasites. They invade nematodes through infective structures either by adhering to the their surface or being ingested by them, then they germinate quickly and develop the hyphal system exclusively in the interior of the nematode, invading and consuming the nematode body contents; development of hyphae continues and the fungus emerges to the exterior only as fertile hyphae (Figure 5). While spore adherence or ingestion can be a rapid affair, the actual effects of nematode death and consumption are usually observed after two to three weeks. Examples of endoparasitic fungi with different infective structures are *Catenaria anguillulae* and *Myzocytium* spp. (mobile flagellated zoospores), *Drechmeria coniospora* and *Verticillium* spp. (adhesive conidia), and *Harposporium* spp. (ingested conidia) [8, 9, 18, 19].

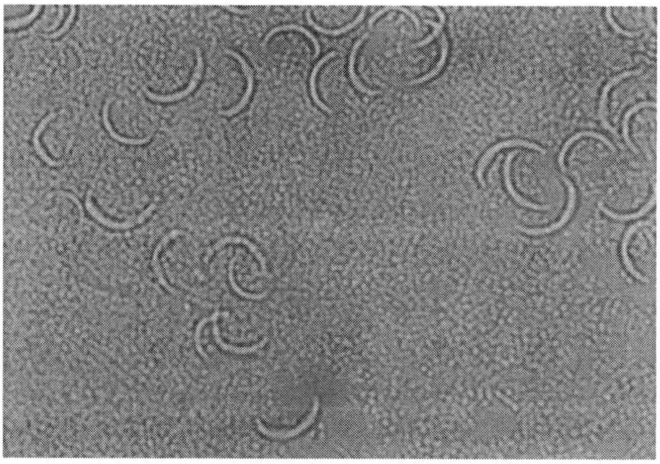

Figure 4. Infecting conidia of the endoparasitic fungus *H. anguillulae*.

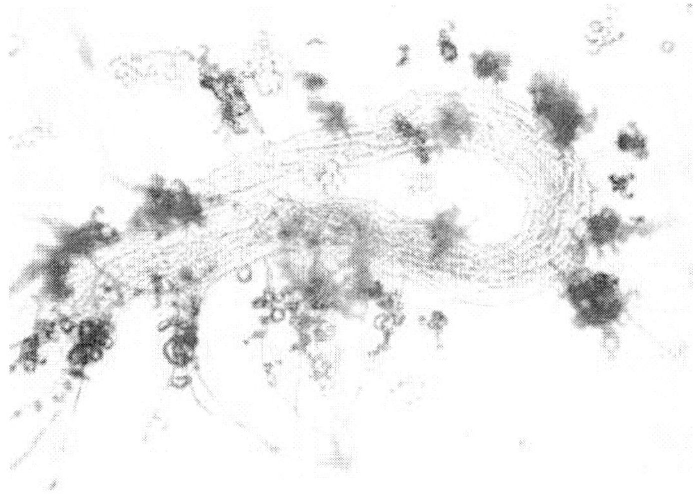

Figure 5. Fertile hyphae of the endoparasitic *H. anguillulae* emerging from the consumed body of a nematode.

The capturing of nematodes arises from the need of nutrients in a nutrient-deficient environment. *D. flagrans* does not produce trapping structures if grown on rich media [20]. *A. oligospora* seems to kill nematodes to obtain N and C; if the immediate environment surrounding the fungus contains enough N and C the fungus lives saprophytically, produces large numbers of conidiophores, and does not capture nematodes because traps are not produced. In the presence of N but absence of C, the fungus has a strong predacious activity and forms many chlamydospores (resting spores). Formation of the latter is a symptom of unsuitable conditions for growth. However, the balance between hyphal growth and trap formation is a delicate one regulated by environmental conditions. For example, under *in vitro* tropical conditions (100% humidity and 29°C) *A. oligospora* forms many traps despite the availability of both N and C [21].

Nematode Capture and Enzymes Involved

The first step of nematode infection by nematophagous fungi is the recognition stage that includes attraction, nematode chemotaxis towards the fungus, or active chemotaxis of fungi towards the nematode [22, 23]. The compounds involved during chemotaxis are not known. However, lectin or lectin-related compounds could be involved in the case of nematode-trapping fungi [24], and endoparasitic fungi [25], and a glycoprotein with mannose/glucose moieties in egg-parasitic fungi [26]. As well, carbohydrates on the nematode surface seem to take part in nematode chemotaxis [27]. Thus, the ability to capture nematodes is also related to the molecular nature of the nematode cuticle [28]. Endoparasitic fungi, because they are obligate parasites, have a stronger ability to attract nematodes than saprophytic nematode-trapping fungi [22].

Once the contact has been established, the fungus forms an adhesive extracellular material containing proteins and/or carbohydrates that keeps it attached to the nematode surface [29] and is able to penetrate the nematode – or eggshell in the case of egg-parasitic fungi, see below. All this is executed through enzymatic processes as well as physical/mechanical ones. Extracellular serine proteases have been found and partly characterized from *A. oligospora*, suggesting that these enzymes are not only key enzymes but also a virulence factor involved in the nematode penetration by fungi [30, 31, 32]. Collagenases are another type of enzyme produced by species of *Arthrobotrys* and *Monacrosporium* [33, 34, 35] and have been found to be constitutive [34]. Although it is reasonable to speculate that they could play a role in the penetration of nematode cuticles, it is not totally certain that collagenases other than serine proteases function as infectious factors [36, 37]. Besides, the key nematode stages in the interaction nematode-trapping fungi and nematodes in the case of animal parasites are the free-living larvae, i.e. first- (L1), second- (L2), and third-stage (L3) larvae; and the cuticle of these larvae contains not only collagen but cuticlin and surface proteins as well, both non-collagen proteins. Cuticlins are the main cuticle proteins in free-living infective stages, while the surface proteins are not cuticular integral structural elements and have been suggested as attachment points for microorganisms such as predatory fungi [38].

Although *D. flagrans* secretes acid phosphatase naturally, even without having formed traps, acid phosphatase activity is at its peak when the fungus has formed the three-dimensional net and nematodes are captured, thus indicating that enzyme activity is important for the penetration of the nematode by the fungus [39]. The same enzyme activity is involved during the interaction process between *A. oligospora* and nematodes [40], however *A. oligospora* does not seem to secrete acid phosphatase when not in presence of nematodes.

Some nematophagous fungi (e.g. *D. flagrans*) produce a wide range of enzymes, including enzymes thought to have distinctive properties, such as phospholipase C, and are able to degrade proteins and peptides, lecithin and oils, starch, and pectin [41]. This enzyme activity is also seen in endoparasite fungi, where acid phosphatase is present in the site of contact between conidia of *D. coniospora* and the nematode cuticle [42, 43].

Fungal Compounds and Metabolites

Linoleic acid, an aliphatic compound [44], extracted from mycelia of several species of *Arthrobotrys* (Deuteromycetes) has strong nematicidal activity [45]; its production has been shown to increase as the number of nematode-trapping structures also increase, indicating that linoleic acid is important during nematode capture. However, not all nematodes are sensitive to linoleic acid.

Three terpenoid [44] compounds with antibiotic activity have been isolated from *A. oligospora*, *A. conoides*, *A. dactyloides* and *A. brochopaga*: oligosporon, oligosporol A, and oligosporol B [46, 47]; all have antimicrobial, cytotoxic and haemolytic effects, and oligosporons show nematicidal activity against *Haemonchus contortus* [48]. Three more compounds of the oligosporon group have been detected; of these, oligosporon and 4',5'-dihydro-oligosporon have nematicidal activity [48]. Arthrobotrisins A, B, and C are metabolites of the oligosporon group [49]; arthrobotrisin C has a strong antibiotic activity against *Bacillus subtilis*, and moderate activity against *Bacillus* sp. and *Staphylococcus aureus*. However, these arthrobotrisins do not possess nematicidal effect against *Panagrellus redivivus*. Other predacious Deuteromycetes other than species of *Arthrobotrys* that produce antimicrobial and/or nematicidal metabolites are *M. haptotylum* (= *D. candida*) (both types of metabolites), *Nematoctonus concurrens* (antimicrobial), *N. robustus* (antimicrobial), and

Monacrosporium parvicolle (antimicrobial) [47]. Both species of *Nematoctonus* produced three metabolites: dihydropeurotinic acid, pleurotin, and leucopleurotin [47, 50]. The nematode-trapping fungus *D. flagrans* produces three antibiotic compounds: flagranones A, B, and C, which are structurally related to the oligosporon group produced by *A. oligospora*, and show comparable antimicrobial activity [51].

The presence of antimicrobial compounds in these fungi could explain why trapped nematodes are seldom colonised by microorganisms different than the predator itself since the antibiotic and nematicidal compounds could exert a synergistic effect on the nematode, causing its rapid colonisation and death [9].

Many secondary metabolites from different fungi have shown to have toxic effects against *Caenorhabditis elegans*, a test nematode commonly used in laboratories for *in vitro* experiments, but not against *Meloidogne incognita* (the plant-parasitic root-knot nematode), which highlights the case that, when testing nematophagous fungi for their potential use as biological control agents, the intended target/s should be used even in *in vitro* tests [47].

Egg- and Cyst-Parasitic Fungi

These are facultative fungi that are soil saprophytes and opportunistic, they only predate on sedentary stages (eggs and cysts) of nematodes [52, 53]. When a fungus grows in the environment in the vicinity of a nematode egg, a hyphal branch develops toward the egg and colonises it [9]. It can take 7-10 days for the fungi to attack the eggs, and the number of infected eggs increases progressively over few weeks [54]. However, the duration of the entire process from capture to total destruction of the egg and its content will be conditioned by the species of both fungus and nematode. Examples of this type of fungi are *Gliocladium roseum*, *Paecilomyces lilacinus* and *Pochonia chlamydosporia* [52].

The endoparasitic fungus *C. anguillulae* is also able to infect nematode eggs [55].

Parasite Egg Colonisation and Enzymes Involved

Nematode eggshells are composed of proteins, chitin and lipoproteins. The structure of nematode eggshells presents variations not only between different orders of nematodes but even between species; in the case of some species of *Ascaris*, one of the most important genera in veterinary parasitology from the point of view of biological control using egg-parasitic fungi, the eggshell presents four layers: an outer lipoprotein uterine layer; a vitelline layer also composed of lipoproteins; a chitinous layer formed of chitin and proteins; and an inner lipid layer consisting of an unique chemical nature of protein and lipids [56].

As mentioned above, a glycoprotein with mannose/glucose moieties seems to be involved in the chemotaxis process [26]. When the hyphae of egg-parasitic fungi come in contact with nematode eggs, an appressorium is formed which secretes extracellular enzymes [57]. A mucilaginous extracellular material between the appressorium and the surface of the eggshell seems to function as fungal adhesive [26, 58, 59]. The fungus then penetrates the egg by a combination of mechanical pressure and enzymatic activity and is able to infect the developing larva [60]. The formation of an appressorium is a normal occurrence, however it is not an obligatory requirement for egg infection [61, 62, 63]. It is speculated that the presence or absence of appressoria could be due to variability in the structural layers eggs of different nematode genera [64, 64]; however, this theory can be easily questioned in light of the findings of two species of *Paecilomyces* that differ in their presence/absence of appressoria when penetrating nematode eggs of the same species [63]. The presence of appressoria could be more related to the aggressiveness of the particular fungal species towards the parasite egg, given that species featuring appressoria possess higher ovicidal activity than those without appressoria, at least in *in vitro* conditions [63].

The developmental stage of nematode eggs can be a factor for fungal colonisation since earlier stages are more easily invaded than later ones, although this seems not to be relevant in the case of aggressive fungi [52].

The enzymes involved in the infection and penetration of the nematode egg are alkaline serine proteases and chitinases [35, 66, 67, 68, 69, 70, 71, 72]. It seems that both types of enzymes must be secreted by the fungi in order to degrade the different layers of the eggshell [73]. However, the inner lipid layer of the *Ascaris* eggshell is not degraded enzymatically even after the embryo/larva has been digested by the fungus [54]. The principal proteases secreted by *P. chlamydosporia* and *P. rubescens* are constitutive enzymes of conidia and germlings when nematode eggs are not present [26].

Nematotoxic Metabolites

Acetic acid with paralysing effects on nematode larvae is produced by the egg-parasitic fungus *P. lilacinus*, and is particularly effective at low pH [74]. Phomalactone and monordem (radicicol) have both has been isolated from *P. chlamydosporia*, but only the former has nematicidal activity [75].

Toxin-Producing Fungi

These fungi do not produce traps, instead they produce nematode toxins in tiny spatula-shaped secretory cells formed along the length of older hyphae and immobilise their prey prior to penetration [14]. Once the nematode is immobilised the fungal hyphae grows towards it and penetrates through body orifices [13].

Examples of toxin-producing fungi can be found in the genus *Pleurotus* [14, 15]. The oyster mushroom *P. pulmonarius* has been shown to immobilise free-living larvae of animal parasitic nematodes such as *Ostertagia ostertagi*, *Cooperia oncophora*, *Oesophagostomum quadrispinulatum* and cyathostomes species [76].

Nematotoxic Metabolites

P. ostreatus secretes trans-2-decenedioic acid ('ostreatin') [14, 77] and linoleic acid that causes shrinkage of the nematode head [78]. Linoleic acid and S-coriolic acid are the most active nematicidal compounds secreted by *P. pulmonarius* [79].

Experiments using nematode-trapping fungi [80], endoparasitic fungi [81], and toxin-producing fungi [76] indicate that the double cuticle of the third-stage, infective larvae (L3) of parasitic nematodes affecting livestock are less vulnerable to fungal attack than the pre-infective larval stages (L1 and L2). This could be either because of the double cuticle featured by the L3, which retain the L2 cuticle as it moults to third-stage, or because possible differences in layers and/or thickness of the cuticle that are dependent on the nematode developmental stages and species [38]. However, the opposite effect was observed in a laboratory study when free-living larvae of the sheep parasite *H. contortus* were subjected to the antagonistic effects of *D. flagrans* [82].

Fungal Ecology and Ecological Considerations

Fungi with restricted saprophytic capability generally show higher predatory ability than highly saprophytic ones [83]. The fact that a fungal isolate grows faster than others does not necessarily mean that is a better predator, this lack of correlation between growth rate and trapping efficacy has been noted in several fungal genera such as *Monacrosporium* and *Arthrobotrys* [84, 85], and in several isolates of *D. flagrans* [85], while an apparent inverse relation between fungal biomass and predacity has been reported [87]. It almost seems that those fungi that spend time and energy into growing their biomass do not spend much time predating and vice versa.

Nematophagous fungi are natural inhabitants of soil and are extensively found in different substrates and in all types of climatic and geographic regions, they are able to survive the most extreme climatic or nutritional conditions [88]. They are regularly found in organically rich environments such as faeces and compost; fresh animal faeces are colonised by several species of these fungi as quickly as in 1-3 days [89, 90, 91, 92].

Given that biological control using nematophagous fungi targets the free-living stages of livestock parasitic nematodes, which develop inside the faecal mass on the ground, it is clear that the faecal environment is the prime – if not the only – site of the fungal-parasite interaction; thus faeces are the substrate for fungal development and nematophagous action. Although studies on nematophagous fungi able to destroy the free-livings stages of animal parasitic nematodes around the world abound [6, 93], information on the natural behaviour of these fungi in faeces and their interaction with other faecal and soil microorganisms is limited. This information is important because laboratory studies not always mirror what happens in the environment.

Some of the elements that could alter natural fungal behaviour in faeces are: physical nature of faeces as influenced by animal species and/or diet, abiotic factors such as climate and soil, and biotic factors such as coprofauna, plant cover, and fungistasis.

The physical nature of faeces from different animal species could impact in the design of formulations to be used as vehicles for fungal administration. Factors such as moisture, pH, and organic matter are directly affected by the animal species (and diet) in question. For example, the obvious differences between cattle and sheep in water content of fresh faeces could have an impact on fungal development of the species administered or on the natural faecal colonisation by fungi from the soil. However, although water content in faeces is important for fungal development – as well as for the development of earthworms, dipterous and coleopterous [94], the large numbers and diversity of nematophagous fungi found in both cattle faeces with high water content and sheep faeces with low levels of moisture [91, 92, 95, 96, 97] would indicate that these fungi are more versatile in Nature that expected. The same seems to apply in the case of diet diversity in monogastric animals; *D. flagrans* is very effective against free-living stages of pig parasites even when pigs are subjected to different diets [98].

Climate is one of the abiotic factors that affect the colonisation of animal faeces by nematophagous fungi, and could affect as well the ability of the fungi to act as biological control agents. The population dynamics of nematophagous fungi and soil nematodes in northern Europe indicate that the highest density of both fungi and nematodes occur in autumn [99], which also happens in temperate parts of the Southern Hemisphere when compared to summer [90]. Likewise, seasonality also plays a role in the appearance of nematophagous fungi in Asia, where both number and diversity of fungal species are highest in summer, followed by autumn, spring and winter [97]. In tropical areas, the largest number of fungal species colonising animal faeces are found during the rainy season, when fungi such as *D. leptospora* and *Myzocytium* sp. are abundant, while the dry season shows a predominance of fungal species forming adhesive nets; however, fungi such as the nematode-trapping *A. oligospora* and *M. eudermatum*, and the endoparasitic *H. anguillulae* demonstrated their abundance in both wet and dry months [92, 100].

Egg-parasitic fungi such as *P. lilacinus* and *P. chlamydosporia* are well adapted to tropical temperatures, their optimal *in vitro* growth ranges from 26°C to 40°C while little growth occurs between 4°C and 10°C [101].

The ideal temperature for the growth of several species of nematode-trapping fungi is reported to be between 15 and 30-35°C [86, 102, 103, 104, 105], although several fungal species survive at temperatures between 1-18°C, including *A. oligospora*, *A. musiformis*, *M. thaumasium*, and *M. ellipsosporum* [97]. Yet, and despite earlier findings showing that the efficiency of some nematode-trapping fungi at favourable pH values is not significantly affected by temperature, as long as it is within the limits of normal growth [107], the predatory capacity of *D. flagrans* and *A. oligospora* varies with temperature. Below 10°C trapping activity is compromised, while the nematode larvae that are the fungal target are still able to develop in temperate regions. The *in vitro* predatory activity at fluctuating temperatures between 10 and 25°C is excellent, which can be attributed to the temperature oscillations *per se* and/or to a greater stimulus provided by higher number of nematode larvae [20, 104, 107, 108] developing in the faecal environment [86]. The same high predatory activity has occurred in the environment with larvae of horse cyathostomes as target when fluctuating temperatures inside the faecal mass ranged between 9°C and 27°C [109]. *In vitro* fungal development and trapping capacity seem to be seriously affected when temperature reaches 35°C [86, 107, 110, 111, 112, 113], although *D. flagrans* has shown high predatory activity against *H. contortus* at temperatures oscillating between 14°C and 39°C in 20-hour cycles (average temperature, 27.1°C) [82]. Despite all these limitations under laboratory conditions, *D. flagrans* in the environment has shown a trapping efficacy of 77% at temperatures inside the faecal mass fluctuating between 13°C and 51°C, with an average temperature of 30.5°C [109].

The nematode reduction capacity of nematophagous fungi in the faecal mass in the field at low temperatures is still unknown; if such a poor performance by the fungi is demonstrated in the environment, as it occurred in laboratory experiments, there is a need to address the basis for biological control of animal parasitic nematodes when such temperatures prevail, i.e. during the early part of the grazing season in northern Europe, or in winter time where grazing takes place all year round, such as temperate areas of South America [6]. On the other hand, the temperature in tropical regions would offer an ideal habitat for the fungi to develop their nematophagous activity. Even when fungal development is limited during the dry season, the environmental conditions throughout the rainy season favour the development of the free-living stages of parasitic nematodes; thus the fungal species administered during that time would have ample opportunities to exert their nematode-destroying action [114].

Soil is another abiotic factor that could influence the colonisation of animal faeces by nematophagous fungi and their development. The characteristics and type of soil could affect the permanence of fungi in the environment [115]. Agricultural soil, presence of animal faeces, and less acid soils have all been associated with higher incidence of nematophagous fungi in soil [116]. The presence of endoparasitic fungi – obligate parasites, unable to live saprophytically – in soil has been reported to be influenced by high soil moisture [117] and high organic matter, while the presence of predatory species seems to be influenced by soil moisture and pH [88], with some fungal species growing slowly at pH between 4-5, and others showing rapid growth at pH between 5-6 [107]. *A. oligospora* has a pH range for hyphal growth of 5-8, with adequate growth at pH 6 [104], and traps large numbers of nematodes at pH 7 in *in vitro* conditions [118]. Meanwhile, *D. flagrans* is able to grow and produce three-dimensional nets *in vitro* at pH 6.3-9.3, showing optimum net production at pH 7-8 [119]. Again, observations on cattle and sheep faeces, either freshly voided or already deposited on soil, would indicate that pH is not a conditioning factor for the colonisation by or development of a wide range of nematophagous fungi in that substrate [91, 92, 95, 96, 97, 100].

Coprofauna is one of the biotic factors that affect the colonisation of animal faeces by nematophagous fungi, and could influence as well the ability of the fungi to control animal parasitic nematodes. Diverse coprophilic organisms carry fragments of nematophagous fungi or their spores. Endoparasitic fungi, due to their obligatory parasitism, depend on soil nematodes for food and, to a certain degree, transportation. Soil nematodes seem to be also important for nematode-trapping fungi producing traps spontaneously [99] and not spontaneously [20]. Amongst the latter, *D. flagrans*, *M. megalosporoum*, and *A. oligospora* show higher predacity when nematode larval populations increase [20, 107, 108]. Adhesive knobs are produced by certain nematode-trapping fungi as organs of capture; the knobs of some of these fungal species are detachable and, once attached to the nematode, are transported to new sites for further predation [9]. As exemplified above, soil nematodes would help in the dissemination and development of nematophagous fungi in faeces. The profuse presence and the migration of those nematodes from the soil to the fresh faecal mass would be a further stimulus for the nematophagous fungi that are administered to animals and eliminated with their faeces because they would be in the faeces simultaneously with the free-living stages of animal parasitic nematodes, therefore, larger reductions of parasitic larval populations would be expected in view of this massive stimulation [114].

In terms of plant cover, it is not clear whether the nematophagous activity of fungi in faeces could be affected by the microenvironment developing in soils with different plant covers or under different grazing regimes that might modify the faecal moisture and faecal exposition to heat and UV radiation, all of which alter the composition of microorganisms colonising the faeces. Some plants would favour the appearance of different soil nematodes and, consequently, different concentrations of nematophagous fungi could appear. The same would occur with bacteria associated with fungal populations, as is the case of bacteria present in wheat rhizosphere that have a more antagonistic effect on *A. oligospora* than those existent in soy rhizosphere [120]. On the other hand, pasture species have shown not to influence the occurrence of nematophagous fungi in fresh faeces [121].

As for fungistasis, nematophagous fungi are considered poor competitors compared with other soil microorganisms, and would be affected by various antagonistic factors in the environment such as soil fungi and bacteria with fungistatic capabilities [120, 122]. For example, the presence of annelids (*Enchytraeidae*) reduces the fungal nematophagous activity in soil [123, 124, 125, 126]. The interference by other microorganisms could be an explanation for the lower efficacy of nematophagous fungi commonly observed in the environment when compared to laboratory experiments [100]. Meanwhile, field studies show the efficacy of *D. flagrans* against larval stages of parasitic nematodes in different livestock species when the fungus is deployed massively [127, 128, 129, 130, 131, 132, 133]. This efficacy, contrasted with the inefficacy of biological control of plant-parasitic nematodes, is thought to be a result of fresh faeces being a less competitive environment for the deployed fungus than the soil environment. Another point to consider is that faeces on pasture are gradually colonised by coprophilic organisms that could have the aforementioned fungistatic effect on the deployed fungi, causing them to gradually lose their nematophagous ability [134]. However, these deleterious effects by the antagonistic organisms would not occur during the first few days after faeces are deposited on pasture, since colonisation at that time is just beginning to take place, thus allowing the fungi to attack and destroy the larvae of animal parasitic nematodes during that time. Besides, nematophagous fungi are evidently equipped for surviving and competing against mycophagous microorganisms in the environment.

On the basis of all the above ecological considerations, a natural enemy will be very effective against its prey population if it shows tolerance of and good adaptation to as broad a range of environmental conditions as its prey [2].

Potental Fungal Species for Biological Control of Nematodes and other Parasites in Veterinary Medicine*

Field trials to control nematodes of economic importance in livestock have concentrated basically on two nematode-trapping fungal species, i.e. *A. oligospora* and *D. flagrans*, with the bulk of the work done on the latter [93]. The reason for this is that *D. flagrans* has proved its candidacy as biological control agent through rigorous stress selection procedures [105, 135, 136], thanks to the abundance of chlamydospores that this species produces (Figures 6 and 7).

Both *A. oligospora* and *D. flagrans* are able to destroy a wide range of animal parasitic nematodes, such as *Ostertagia* spp., *Teladorsagia circumcincta*, *Haemonchus* spp., *Cooperia* spp., *Strongyloides papillosus*, cyathosthomes, *Oesophagostomum* spp., *Strongylus* spp., etc. [6, 93]. *D. flagrans* has also shown to be effective against the bovine lungworm, *Dictyocaulus viviparus*, both *in vitro* and in plot experiments [137, 138], although its effect on the parasite is greater when also in presence of gastrointestinal nematodes [138]. However, the fungus has

*(Note: detailed information on applied biological control of parasites of different livestock species is dealt with in other chapters of this book; therefore, only a brief mention of the fungi and the parasites involved is given in this chapter).

not shown the same reduction effects on the small ruminant lungworm, *Muellerius capillaris* [139, 140].

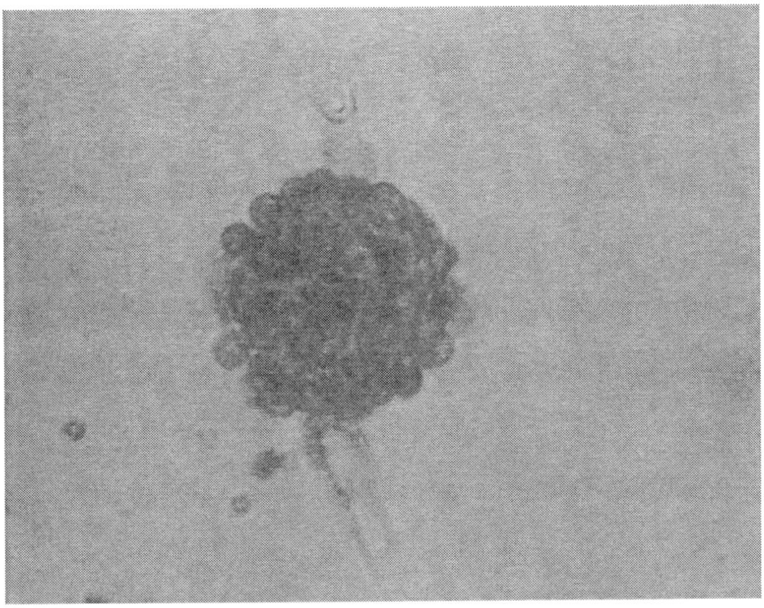

Figure 6. Mature chlamydospore of *D. flagrans*. Chlamydospores of this species resist the passage through the gastrointestinal tract of livestock animals and are voided with faeces.

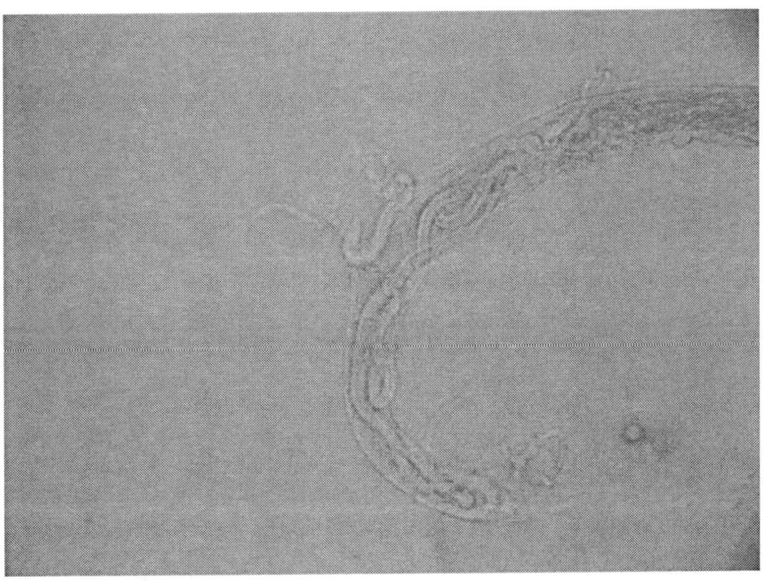

Figure 7. Three-dimensional net of *D. flagrans* with the hypha penetrating the body of a captured nematode. In the faecal environment, fungal chlamydospores germinate and develop the three-dimensional net that allows the fungus to trap free-living larvae of animal parasitic nematodes.

Endoparasitic fungi as potential biological control agents have been practically not explored despite the fact that species such as *H. anguillulae* and *Myzocytium* sp. are easily isolated from animal faeces [100]. The interest of veterinary researchers on this fungal group has been limited by different factors such as the inefficacy of few species against plant-parasitic nematodes and the difficulties in developing these fungi in pure cultures [141, 142], as well as initial results showing that although conidia of *D. coniospora* can adhere to the cuticle of third-stage larvae of gastrointestinal nematodes of ruminants, it does not infected them [81]. Nevertheless, the effect of endoparasitic fungi against free-living stages of trichonstrongylids in ruminants has been evaluated in few studies. *D. coniospora* and *H. anguillulae* were both able to reduce *in vitro* the larval population of *H. contortus* [143, 144]. *D. coniospora* infected first-stage larvae although high numbers of conidia were needed to reduce significantly the number of larvae in coprocultures [143]. A plot study revealed that *H. helicoides* is able to attack first- and second-stage larvae of *T. circumcincta* under different climatic (seasonal) conditions [145]. *In vitro* trials failed to show resistance of *H. anguillulae* to artificial digestion [135, C. Saumell, personal communication). However, under *in vivo* conditions, the fungus seemed to resist the passage through the gastrointestinal tract of calves only to a certain degree since the recovery period of chlamydospores in faeces was very short and the fungal activity against nematodes on coprocultures was scarce [146].

Studies on egg-parasitic fungi against helminths of veterinary importance only in recent years have started to gain strength. The two fungal species that currently are receiving the most attention are *P. lilacinus* and *P. chlamidosporia*.

P. lilacinus can degrade the eggshell of the pig parasite *Ascaris suum* [147], and has also shown *in vitro* activity against eggs of *Toxocara canis* [63, 148, 149, 150, 151], *Taenia hydatigena* [152], *T. saginata* [153], *Dipylidium caninum* [154], and low activity against *Moniezia* spp. [155] and the cattle trematodes *Fasciola hepatica* and *Gigantocoyle explanatum* [101].

P. chlamydosporia has shown *in vitro* activity against *Trichuris vulpis* [156], *T. canis* [151, 157], *T. vitulorum* [158], and low activity against *A. suum* [159] and *Oxiuris equi* [160]. It has also shown low activity on eggs of *T. saginata* [161], high efficacy against *T. taeniformis* [162], moderate to high activity against *Anoplocephala perfoliata* [163], and moderate efficacy against *D. caninum* [154]. The fungus is also active against eggs of *F. hepatica* and *G. explanatum* [101, 164, 165]. Crude enzyme extracts of *P. chlamydosporia* have shown to reduce the hatching of cyathostomes and *Ancylostoma* sp. eggs, leading to the speculation that the fungus could be used to control gastrointestinal parasites featuring eggs with short hatching time in the environment [166, 167]. *P. chlamydosporia* resists the passage through the gastrointestinal tract of goats [168], cattle [165] and horses [169], and retains its low ovicidal activity against *O. equi* [169].

Other fungal species that are reported to exert ovicidal activity on *T. canis* eggs are *Chrysosporium merdarium* and different species of the genera *Fusarium* and *Aspergillus* [148, 149, 170, 171, 172, 173, 174]. However, due to the effects on human health of *C. merdarium*, *F. oxysporum*, and *F. semitectun*, and of *F. moniliforme* and other species of *Fusarium* on human and animal health, only *F. sulphureum* would be recommended as biological control agent of nematode eggs, although its effects on public and animal health as well as on the environment are still unknown [149, 173].

The Ascomycete *Beauveria bassiana* is a commonly occurring entomopathogenic fungus of cosmopolitan distribution [175]. An isolate of this fungus is being used on *in vitro* experiments against the cattle tick *Rhipicephalus* (*Boophilus*) *microplus* with promising good results. Low fungal concentrations have been applied to engorged female ticks, and the fungus has been able to destroy the tick keratin, preventing egg shedding (Molento and Buzatti, unpublished results).

Environmental Impact of Deployment of Nematophagous Fungi to Control Animal Parasites

One of the main questions in relation to the use of nematophagous fungi as biological control agents against animal parasites is whether the massive deployment of fungi can have any detrimental effect on the environment, be it on faecal decomposition, on the environment below or beyond faecal depositions, or on non-target organisms. Such legitimate concerns are assuaged by several studies that have purposely looking into the matter. Because *D. flagrans* has been studied the most as biological control agent of animal parasitic nematodes, it is this fungus that features in almost all the environmental studies up to date.

Short- and long-term studies showed no adverse effect on the degradation rate of horse, cattle and goats faeces by the massive presence of *D. flagrans* [138, 176, 177, 178, 179, 180].

When fungi-containing faeces were deposited on pasture, *D. flagrans* was not detected in soil 50-60 days post-deposition [181, 182], or was scarcely detected beyond 112 days, a fact possibly related to dry environmental conditions [183]. Therefore, the fungus does not become established in soil below or adjacent to faecal mass, which has been demonstrated in different regions around the world [178, 179, 183, 184].

Non-target organisms are not impacted by the deployment of *D. flagrans* in faecal depositions. This has been corroborated for soil nematodes under different climate conditions in various geographical areas [181, 183, 185, 186, 187, 188]. Only when the fungal dose administered to animals was purportedly augmented 10 times in order to maximise eventual effects, the soil nematode population was affected [180]. Numbers of microarthropods in soil such as *Collembola* y *Acarina* do not change by the massive presence of *D. flagrans* in faeces [183]. The same lack of effect occurs on soil macrofauna, exemplified by coleopteran and lepidopteran insects [182], and earthworms [189]. The populations of faecal-colonising organisms such as dung beetles [190] and nematophagous fungi other that *D. flagrans* [181, 183] follow the same trend, i.e. are not affected by the massive deployment of this fungus.

All the evidence up to date indicates that no adverse environmental impact will occur when using *D. flagrans* to control parasitic nematodes of livestock. Regarding the environmental impact of egg-parasitic fungi, initial investigations revealed that the presence of *P. lilacinus* and *P. chlamydosporia* seem no to be a hindrance for the degradation of goat faeces, nor do they spread horizontally beyond the faecal mass [191]. As well, *P. chlamydosporia* is innocuous for three species of invertebrates, including earthworms [192].

Further Issues and Future Perspectives

As previously mentioned in this chapter, the aim of biological control in veterinary medicine is to be one more tool in the fight against animal parasites. Since another important tool are chemical drugs, the possible interaction between these two parasite control options needs to be taken into consideration.

Some benzimidazole compounds used as anthelmintics are known for their antifungal activity [193]. Benzimidazole compounds have strong inhibitory effects on *in vitro* growth and formation of trapping structures of *D. flagrans* [194, 195]. Albendazole, administered as an intraruminal slow-release capsule to sheep and buffaloes, makes the faecal recovery of *D. flagrans* impossible, while if administered as a single dose it interrupts temporarily the faecal release of the fungus [196]. However, it is possible to recover the fungus abundantly ninety-six hours after drenching goats, sheep, and buffaloes with a single dose of albendazole, and its predatory activity is displayed again [195, 196]. Both albendazole and triclabendazole temporarily interfere with the release of *P. chlamydosporia* in goat faeces when administered as a single dose [168]. Thus, given the aforementioned interactions between nematophagous fungi and benzimidazole compounds, it would be of upmost importance in an integrated parasite management system to a) timing carefully the use of both nematophagous fungi and anthelmintic treatment with benzimidazoles, or b) avoiding, when possible, the use of benzimidazoles if biological control with nematophagous fungi is to be applied.

Several aspects of biological control of gastrointestinal nematodes in livestock have been discussed in a recent international workshop. Some of the issues covered were the practical use of fungi in different geographical and climatic regions, fungal efficacy across a wide range of environmental and epidemiological conditions, the need of testing the extended use of nematophagous fungi in all grazing livestock species, efficacy differences of biological control in sheep, and dosage applications. One of the key recommendations put forward in that meeting was the development of appropriate fungal formulation/s for prompt commercial availability, and thenceforth embarking upon the extended field studies in different regions of the world and under different management conditions [197].

The formulation and administration of fungal material, whichever fungal species is considered, should be adapted to the livestock species and type of farm enterprise where biological control would be adopted. The forms of fungal administration that have been evaluated so far are mineral blocks, energy blocks, fungus-containing cereal grains, alginate pellets, and controlled-release boluses [198]. The commercially availability of *D. flagrans* in a premix feed additive form is soon to be available in Australia (M. Knox, personal communication).

The regulatory aspects of fungal use for commercialisation have been by far the most difficult obstruction. Taking the case of *D. flagrans* as example, regulatory authorities seem to have serious trouble grasping the concept that a) the animals are not actually 'treated' – as in the case of therapeutic drugs – when administered the fungal chlamydospores but are only serving as their vehicle to reach their action destination, i.e. the faecal mass, and b) a fungal product of these characteristics, and however formulated, must not be considered a drug, whether anthelmintic or any other kind. As submitted to the European Food Safety Authority [199], given that *D. flagrans* is an obligatory aerobic organism, it is unlikely that the fungus would be metabolically active in the gastrointestinal tract and is regarded as inocuous.
Another obstacle is that registration requirements are not the same in different countries, and the fungal product considered a feed additive in one country is regarded as a biological pesticide in another [200].

Final Considerations

Few considerations must be observed when it comes to biological control.

First, the fungal organisms must be safe in terms of human and animal health as well as the environment. In the case of *D. flagrans*, the European Food Safety Authority [199] states that the fungus is "unlikely to be a skin irritant or to cause dermal sensitisation" and deems it safe for the target species. As for ecotoxicity, it has been shown previously in this chapter that no negative environmental impact would take place when *D. flagrans* is used to control parasitic nematodes in livestock, either in terms of persistence, dispersal, or competitive displacement. In the case of egg-parasitic fungi, *P. chlamydosporia* has been classified as safe in toxicity and pathogenicity studies [201, 202]; but as only one preliminary evaluation on its lack of ecotoxic effects has been undertaken [192], more extensive studies on environmental health need to be carried out. Safety tests of *P. lilacinus* in rabbits and rats indicate that this egg-parasitic fungus is relatively safe; however, it could cause human infections even though the isolates used against plant-parasitic nematodes would not present a risk for human health [203]. These two egg-parasitic fungal species are commercially available for use on plant nematodes, but since they are not allowed in several countries, their potential use in veterinary medicine remains an open question.

Second, the fungus must be consistently effective. As previously noted above, detailed information on applied biological control of parasites of different livestock species is dealt with in other chapters of this book. Thus, all matters concerning this second consideration are deferred to those chapters.

Third, the fungi must be delivered to the specific site of action. Since the work on egg-parasitic fungi in veterinary medicine is relatively new, this consideration remains to be answered. As for other fungal species, the fact that free-living larvae and nematode-trapping fungi share time and space, i.e. they are present in the faecal mass and develop there at the same time, presents a unique possibility to exploit the latter as biological control agents.

Fourth, the fungal organisms must be marketable and cost-effective. There is no doubt that, given the current situation of anthelmintic resistance worldwide, the use of microfungi as biological control agents will be well received by well-informed farmers and veterinary practitioners alike. In terms of costs, it has been noted that the commercial cost needs to be competitive with current parasite control methods [200] and should not exceed that of chemical drugs in order to make it commercially effective.

Conclusion

Biological control by means of parasite-attacking fungi, and more specifically nematophagous fungi, is not 'just' an attractive idea any longer but an imminent reality. Following an apparent hiatus in research – conditioned mainly by a general marked decrease in funding for investigation, as well as regulatory issues in certain countries – the current global conditions to move forward in this area are indeed better than in the '80s and '90s, when research on biological control in veterinary medicine was at its prime. The reasons are the accelerated increase of anthelmintic resistance worldwide that affects all livestock production systems, the global growth of organic livestock production, and the heightened sense of the general population in relation to public and environmental health.

Certain level of parasitism in animal production systems has to be accepted since eradication of most parasites of economic significance is, if not possible, at least not practical. Thus, the aim of parasite control must be to ensure that parasite populations do not exceed levels compatible with economic production, which fits with the definition of biological control adopted in this chapter.

References

[1] Baker, R. (1983). State of the art: Plant diseases, In: S.L. Battenfield (Ed.), *Proc. National Interdisciplinary Biological Control Conference, CSRS/USDA*, Las Vegas, USA, pp. 14-22.

[2] DeBach, P. & Rosen, D. (1991). *Biological Control by Natural Enemies.* Second edition. Cambridge, Great Britain: Cambridge University Press.

[3] Van Driesche, R.G. & Bellows, T.S. (1996). Pest origins, pesticides and the history of biological control. In: *Biological Control*, New York, USA: Chapman & Hall.

[4] Grønvold, J., Henriksen, S. A., Larsen, M., Nansen, P. & Wolstrup, J. (1996). Biological control: Aspects of biological control, with special reference to arthropods, protozoans and helminths of domesticated animals. *Veterinary Parasitology*, *64*, 47-64.

[5] Food and Agricultural Organization (2004). Resistance management and integrated parasite control in ruminants (guidelines). CD-ROM. Animal Production and Health Division, FAO, Rome.

[6] Fernández, A.S. (1998) *Duddingtonia flagrans* as biological control agent against parasitic nematodes of cattle and horses. PhD Thesis, The Royal Veterinary and Agricultural University, Denmark, 179 p.

[7] Nari-Henrioud, A. (2011). Towards sustainable parasite control practices in livestock production with emphasis in Latin America. *Veterinary Parasitology, 180*, 2-11

[8] Gray, N.F. (1988). Fungi attacking vermiform nematodes. In: G.O. Poinar Jr. & H-B. Jansson (Eds.), *Diseases of Nematodes, Volume II* (pp. 3-38), United States of America, CRC Press.

[9] Barron, G.L. (1977) *The Nematode-Destroying Fungi. Topics in Mycobiology No. 1.* Guelph, Canada: Canadian Biological Publications.

[10] Dackman, C., Jansson, H.-B. & Nordbring-Hertz, B. (1992) Nematophagous fungi and their activities in soil. In: G. Stotzky & J. M. Bollag (Eds.), *Soil Biochemistry, Volume 7* (pp. 95-103), New York, United States of America: Marcel Dekker, Inc.

[11] Dix, N.J. & Webster, J. (1995). *Fungal Ecology*. London, Great Britain: Chapman & Hall.

[12] Mankau, R. (1980). Biocontrol: Fungi as nematode control agents. *Journal of Nematology, 12*, 244-252.

[13] Thorn, R.G. & Barron, G.L. (1984). Carnivorous mushrooms. *Science, 224*, 76-78.

[14] Barron, G.L. & Thorn, R.G. (1987). Destruction of nematodes by species of *Pleurotus. Canadian Journal of Botany, 65*, 774-778.

[15] Jansson, H.-B., Tunlid, A. & Nordbring-Hertz, B. (1997). Nematodes. In: T. Anke (Ed.), *Fungal Biotechnoloy* (pp. 38-50), Weinheim, Germany: Chapman & Hall.

[16] Li, T.F., Zhang, K.Q. & Liu, X.Z. (2000). *Taxonomy of Nematophagous Fungi*. Beijing, China: Chinese Scientific and Technological Publication.

[17] Rubner, A. (1996). Revision of predacious Hyphomycetes in the *Dactylella-Monacrosporium* complex. *Studies on Mycology, 39*, 1-134.

[18] Jansson, H.-B., Nordbring-Hertz, B., Wyss, U., Häusler, P., Hard, T. & Poloczek, E. (1995). Infection of nematodes by zoospores of *Catenaria anguillulae*. IWF Wissen und Medien, Göttingen, Germany. Film No. C 1868.

[19] Lopez-Llorca, L.V., Maciá-Vicente, J.G. & Jansson, H.-B. (2008). Mode of action and interactions of nematophagous fungi. In: A. Ciancio & K.G. Mukerji (Eds.), *Integrated and Biocontrol of Vegetable and Grain Crops Nematodes* (pp. 51-76), Dordretch, The Netherlands: Springer.

[20] Morgan, M; Behnke, J.M., Lucas, J.A. & Peberdy, J.F. (1997). *In vitro* assessment of the influence of nutrition, temperature and larval density on trapping of the infective larvae of *Heligmosomoides polygirus* by *Arthrobotrys oligospora, Duddingtonia flagrans* and *Monacrosporium megalosporum*. *Parasitology, 115*, 303-310.

[21] Scholler, M. & Rubner, A. (1994). Predacious activity of the nematode-destroying fungus *Arthrobotrys oligospora* in dependence of the medium composition. *Microbiological Research, 149*, 145-149.

[22] Jansson, H.-B. & Nordbring-Hertz, B. (1979). Attraction of nematodes to living mycelium of nematophagous fungi. *Journal of General Microbiology, 112*, 89-93.

[23] Jansson, H-B. & Thiman, L. (1992). A preliminary study of chemotaxis of zoospores of the nematodeparasitic fungus *Catenaria anguillulae. Mycologia, 84,* 109-112.

[24] Nordbring-Hertz, B. & Mattiasson, B. (1979). Action of a nematode-trapping fungus shows lectin-mediated host-microorganism interaction. *Nature, 281*, 477–479.

[25] Jansson, H.-B. & Nordbring-Hertz, B. (1984). Involvement of sialic acid in nematode chemotaxis and infection by an endoparasitic nematophagous fungus. *Journal of General Microbiology, 130*, 39-43.

[26] Lopez-Llorca, L.V., Olivares-Bernabeu, C., Salinas, J., Jansson, H.-B. & Kolattukudy, P. E. (2002). Pre-penetration events in fungal parasitism of nematode eggs. *Mycological Research, 106*, 499-506.

[27] Zuckerman, B.M. & Jansson, H.-B. (1984). Nematode chemotaxis and mechanisms of host/prey recognition. *Annual Review of Phytopathology*, *22*, 95-113.

[28] Mendoza de Gives, P., Davies, K.G., Clark, S.J. & Behnke, J.M. (1999). Predatory behaviour of trapping fungi against *srf* mutants of *Caenorhabditis elegans* and different plant and animal parasitic nematodes. *Parasitology, 119*, 95-104.

[29] Tunlid, A., Jansson, H.-B. & Nordbring-Hertz, B. (1992). Fungal attachment to nematodes. *Mycological Research, 96*, 401-412.

[30] Tunlid, A. & Jansson, S. (1991). Proteases and their involvement in the infection and immobilization of nematodes by the nematophagous fungus *Arthrobotrys oligospora. Applied and Environmental Microbiology, 57*, 2868-2872.

[31] Tunlid. A., Rosen, S., Ek, B. & Rask, L. (1994) Purification and characterization of an extracellular serine protease from the nematode-trapping fungus *Arthrobotrys oligospora. Microbiology, 140*, 1687-1695.

[32] Zhao, M.L., Mo, M.H. & Zhang, K.Q. (2004). Characterization of a neutral serine protease and its full-length cDNA from the nematode-trapping fungus *Arthrobotrys oligospora*. *Mycologia, 96*, 16-22.

[33] Schenck, S., Chase, T., Rosenzweig, W.D. & Pramer, D. (1980). Collagenase production by nematode-trapping fungi. *Applied and Environmental Microbiology, 40,* 567-570.

[34] Tosi, S., Annovazzi, L., Tosi, I., Iadarola, P. & Caretta, G. (2001). Collagenase production in an Antarctic strain of *Arthrobotrys tortor* Jarowaja. *Mycopathologia, 153,* 157-162.

[35] Yang, J., Tian, B., Liang, L. & Zhang, K.Q. (2007). Extracellular enzymes and the pathogenesis of nematophagous fungi. *Applied Microbiology and Biotechnology, 75,* 21-31.

[36] Huang, X., Zhao, N. & Zhang, K. (2004). Extracellular enzymes serving as virulence factors in nematophagous fungi involved in infection of the host. *Research in Microbiology, 155*, 811-816.

[37] Tharwat, N., Abdel-Rahman, T., Ismail, I.M.K., Elwi, E.E.A. & Abou-Elela, M. (2007). Purification and characterization of nematicidal extracellular collagenase produced by the nematophagous fungus *Arthrobotrys oligospora*. *Bulletin of the Faculty of Science, Cairo University 75*, 57-85.

[38] Fetterer, R.H. & Rhoads, M.L. (1993). Biochemistry of the nematode cuticle: relevance to parasitic nematodes of livestock. *Veterinary Parasitology, 46*, 103-111.

[39] Cruz, D.G., Silva, C.P., Carneiro, C.N., Retamal, C.A., Thiébaut, J.T., DaMatta, R.A. & Santos, C.P. (2009). Acid phosphatase activity during the interaction of the nematophagous fungus *Duddingtonia flagrans* with the nematode *Panagrellus* sp. *Journal of Invertebrate Pathology, 102*, 238-244.

[40] Toth, R., Toth, D. & Nordbring-Hertz, B. (1980). Cytochemical localization of acid phosphatase in the nematode-trapping fungus *Arthrobotrys oligospora*. *Mycologia, 72,* 813-817.

[41] Meyer, W.J. & Wiebe, M.G. (2003). Enzyme production by the nematode-trapping fungus, *Duddingtonia flagrans*. *Biotechnology Letters, 25*, 791-795.

[42] Dijksterhuis, J., Veenhuis, M. & Harder, W. (1990). Ultrastructural study of the adhesion and initial stage of infection of nematodes by conidiospores of *Drechmeria coniospora*. *Mycological Research, 94*, 1-8.

[43] Dijksterhuis, J., Harder, W., Wyss, U. & Veenhuis, M. (1991). Colonization and digestion of nematodes by the endoparasitic fungus *Drechmeria coniospora*. *Mycological Research, 95*, 873-878.

[44] Li, G., Zhang, K., Xu, J., Dong, J. & Liu, Y. (2007). Nematicidal substances from fungi. *Recent Patents on Biotechnology, 1*, 212-233.

[45] Stadler, M., Anke, H. & Sterner, O. (1993). Linoleic acid – The nematicidal principle of several nematophagous fungi and its production in trap-forming submerged cultures. *Archives of Microbiology, 160*, 401-405.

[46] Stadler, M., Sterner, O. & Anke, H. (1993). New biologically active compounds from the nematode-trapping fungus *Arthrobotrys oligospora* Fresen. *Zeitschrift für Naturforschung. C, 48*, 843-850.

[47] Anke, H., Stadler, M., Mayer, A. & Sterner, O (1995). Secondary metabolites with nematicidal and antimicrobial activity from nematophagous fungi and Ascomycetes. *Canadian Journal of Botany, 73*, S932-S939.

[48] Anderson, M.G., Jarman, T.B. & Rickards, R.W. (1995). Structures and absolute configuration of antibiotics of the oligosporon group from the nematode-trapping fungus *Arthrobotrys oligospora. Journal of Antibiotics, 48*, 391-398.

[49] Wei, L.X., Zhang, H.X., Tan, J.L., Chu, Y.S., Li, N., Xue, H.X., Wang, Y.L., Niu, X.M., Zhang, Y. & Zhang, K.Q. (2011). Arthrobotrisins A-C, oligosporons from the nematode-trapping fungus *Arthrobotrys oligospora. Journal of Natural Products, 74*, 1526-1530.

[50] Stadler, M., Sheldrick, W.S., Dasenbrock, J., Steglich, W. & Anke, H. (1994). Antibiotics from the nematode-trapping Basidiomycete *Nematoctonus robustus. Naural Product Letters, 4*, 209-216.

[51] Anderson, M.G., Rickards, R.W. & Lacey, E. (1999). Structures of flagranones A, B, and C, cyclohexenoxide antibiotics from the nematode-trapping fungus *Duddingtonia flagrans. Journal of Antibiotics (Tokyo), 52*, 1023-1028.

[52] Morgan-Jones, G. & Rodríguez-Kábana, R. (1988). Fungi colonizing cysts and eggs. In: G.O. Poinar, Jr. & H.-B. Jansson (Eds.), *Diseases of Nematodes, Vol II* (pp. 39-58), United States of America: CRC Press.

[53] Kerry, B.A. & Jaffee, B.A. (1997). Fungi as biocontrol agents for plant parasitic nematodes. In: D.T. Wicklow & B. Söderström (Eds.), *The Mycota, Vol. IV* (pp. 203-218), Berlin, Germany: Springer-Verlag.

[54] Kunert, (1992). On the mechanism of penetration of ovicidal fungi through egg-shells of parasitic nematodes. Decomposition of chitinous and ascaroside layers. *Folia Parasitologica, 39*, 61-66.

[55] Wyss, U., Voss, B. & Jansson, H.-B. (1992). *In vitro* observations on the infection of *Meloidogyne incognita* eggs by the zoosporic fungus *Catenaria anguilulae* Sorokin. *Fundamental and Applied Nematology, 15*, 133–139.

[56] Wharton, D. (1980). Nematode egg-shells. *Parasitology, 81*, 447-463.

[57] Lopez-Llorca, L.V. & Robertson, W.M. (1992). Immunocytochemical localization of a 32-kDa protease from the nematophagous fungus *Verticillium suchlasporium* in infected nematode eggs. *Experimental Mycology, 16*, 261-267.

[58] Stirling, G.R. & Mankau, R. (1979). Mode of parasitism of *Meloidogyne* and other nematode eggs by *Dactylella oviparasitica. Journal of Nematology, 11*, 282-288.

[59] Lopez-Llorca, L.V. & Claugher, D. (1990). Appressoria of the nematophagous fungus *Verticillium suchlasporium. Micron and Microscopica Acta, 21*, 125-130.

[60] Segers, R., Butt, T.M., Kerry, B.R., Beckett, A. & Peberdy, J. F. (1996). The role of the proteinase VCP1 produced by the nematophagous fungus *Verticillium chlamydosporium* in the infection process of nematode eggs. *Mycological Research, 100*, 421-428.

[61] Emmet, R.W. & Parbery, D.G. (1975). Appressoria. *Annual Review of Phytopathology 13*, 147-165.

[62] Lýsek, H. & Krajčí, D. (1987). Penetration of ovicidal fungus *Verticillium chlamydosporium* through the *Ascaris lumbricoides* egg-shells. *Folia Parasitologica, 34*, 57-60.

[63] Basualdo, J.A., Ciarmela, M.L., Sarmiento, P.L. & Minvielle, M. C. (2000). Biological activity of *Paecilomyces* genus against *Toxocara canis* eggs. *Parasitology Research, 86*, 854-859.

[64] Blaxter, M.L. & Robertson, W.M. (1998). The cuticle. In: R.N. Perry & D.J. Wright (Eds.), *The physiology and biochemistry of free-living and plant-parasitic nematodes* (pp. 25-48), Wallingford, UK: CABI Publishing.

[65] Morton, C.O., Hirsch, P.R. & Kerry, B.R. (2004). Infection of plant-parasitic nematodes by nematophagous fungi – A review of the application of molecular biology to understand infection processes and to improve biological control. *Nematology, 6*, 161-170.

[66] Lopez-Llorca, L.V. (1990). Purification and properties of extracellular proteses produced by the nematophagous fungus *Verticillium suchlasporium*. *Canadian Journal of Microbiology, 36*, 530-537.

[67] Segers, R., Butt, T. M., Kerry, B.R. & Peberdy, F. (1994). The nematophagous fungus *Verticillium chlamydosporium* Goddard produces a chymoelastase-like protease which hydrolyses host nematode proteins in situ. *Microbiology, 140*, 2715-2723.

[68] Bonants, P.J.M., Fitters, P.F.L., Thijs, H., Den Belder, E., Waalwijk, C., Willem, J. & Henfling, D.M. (1995). A basic serine protease from *Paecilomyces lilacinus* with biological activity against *Meloidogyne hapla* eggs. *Microbiology, 141*, 775-784.

[69] Olivares-Bernabéu, C. (1999). *Caracterización biológica y molecular de hongos patógenos de huevos de nematodos*. PhD Thesis, University of Alicante, Alicante, Spain, 301 p.

[70] Tikhonov, V.E., Lopez-Llorca, L.V., Salinas, J. & Jansson, H.-B. (2002). Purification and characterization of chitinases from the nematophagous fungi *Verticillium chlamydosporium* and *V. suchlasporium*. *Fungal Genetics and Biology, 35*, 67-78.

[71] Gan, Z.W., Yang, J.K., Tao, N., Yu, Z.F. & Zhang, K.Q. (2007). Cloning and expression analysis of a chitinase gene Crchi1 from the mycoparasitic fungus *Clonostachys rosea* (syn. *Gliocladium roseum*). *Journal of Microbiology, 45*, 422-430.

[72] Esteves, I., Peteira, B., Atkins, S.D., Magan, N. & Kerry, B. (2009). Production of extracellular enzymes by different isolates of *Pochonia chlamydosporia*. *Mycological Research, 113*, 867-876.

[73] Khan, A., Williams, K.L. & Nevalainen, H.K.M. (2004). Effects of *Paecilomyces lilacinus* protease and chitinase on the eggshell structures and hatching of *Meloidogyne javanica* juveniles. *Biological Control, 31*, 346-352.

[74] Djian, C., Pijarowski, L., Ponchet, M., Arpin, N. & Favre-Bonvin, J. (1991). Acetic acid: a selective nematicidal meabolite fom culture filtrates of *Paecilomyces lilacinus* (Thom) Samson and *Trichoderma longibrachiatum* Rifai. *Nematologica, 37*, 101-112.

[75] Khambay, B.P.S., Bourne, J.M., Cameron, S., Kerry, B.R. & Zaki, M.J. (2000). A nematicidal metabolite from *Verticillium chlamydosporium*. *Pest Management Science, 56*, 1098-1099.

[76] Larsen, M. & Nansen, P. (1991). Ability of the fungus *Pleurotus pulmonarius* to immobilise preparasitic nematode larvae. *Research in Veterinary Science, 51*, 146-149.

[77] Kwok, O.C.H., Plattner, R., Weisleder, D. & Wicklow, D.T. (1992) A nematicidal toxin from *Pleurotus ostreatus* NRRL 3526. *Journal of Chemical Ecology, 18*, 127-136.

[78] Satou, T., Kaneko, K., Li, W. & Koike, K. (2008). The toxin produced by *Pleurotus ostreatus* reduces the head size of nematodes. *Biological and Pharmaceutical Bulletin, 31*, 574-576.

[79] Stadler, M., Mayer, A., Anke, H. & Sterner, O. (1994). Fatty acids and other compounds with nematicidal activity from cultures of basidiomycetes. *Planta Medica, 60*, 128-132.

[80] Nansen, P., Grønvold, J., Henriksen, S.A. & Wolstrup, J. (1986). Predacious activity of the nematode-destroying fungus *Arthrobotrys oligospora* on preparasitic larvae of *Cooperia oncophora* and on soil nematodes. *Proceedings of the Helminthological Society of Washington, 53*, 237-243.

[81] Jansson, H.-B., Jeyaprakash, A. & Zuckerman, B.M. (1985). Differential adhesion and infection of nematodes by the endoparasitic fungus *Meria coniospora* (Deuteromycetes). *Applied and Environmental Microbiology, 49*, 552-555.

[82] Kahn, L.P., Norman, T.M., Walkden-Brown, S.W., Crampton, A. & O'Connor, L.J. (2007). Trapping efficacy of *Duddingtonia flagrans* at temperatures existing at lambing in Australia. *Veterinary Parasitology, 146*, 83-89.

[83] Mankau, R. (1980). Biological control of nematode pets by natural enemies. *Annual Review of Phytopathology, 18*, 415-440.

[84] Feder, W.A. (1963). A comparison of nematode-capturing efficiencies of five *Dactylella* species at four temperatures. *Mycopathologia et mycologia applicata, 19*, 99-104.

[85] Heintz, C.E. (1978). Assessing the predacity of nematode-trapping fungi *in vitro*. *Mycologia, 70*, 1086-1100.

[86] Fernández, A.S., Larsen, M., Wolstrup, J., Grønvold, J., Nansen, P. & Bjørn, H. (1999). Growth rate and trapping efficacy of nematode-trapping fungi under constant and fluctuating temperatures. *Parasitology Research, 85*, 661-668.

[87] Jansson, H.-B. (1982) Predacity of nematophagous fungi and its relation to the attraction of nematodes. *Microbial Ecology, 8*, 233-240.

[88] Gray, N.F. (1987) Nematophagous fungi with particular reference to their ecology. *Biological Review, 62*, 245-304.

[89] Hay, F.S., Niezen, J.H., Miller, C., Bateson, I. & Robertson, H. (1997). Infestation of sheep dung by nematophagous fungi and implications for the control of free-living stages of gastro-intestinal nematodes. *Veterinary Parasitology, 70*, 247-254.

[90] Hay, F.S., Niezen, J.H., Bateson, I. & Wilson, S. (1998). Invasion of sheep dung by nematophagous fungi and soil nematodes on a hill country pasture in New Zealand. *Soil Biology and Biochemistry,* 30, 1815-1819.

[91] Hay, F.S., Niezen, J.H., Leathwick, D. & Skipp, R.A. (2002). Nematophagous fungi in pasture: colonisation of sheep faeces and their potential for control of free living stages of gastro-intestinal nematode parasites of sheep. *Australian Journal of Experimental Agriculture, 42*, 7-13.

[92] Saumell, C.A. & Padilha, T. (2000). Influence of weather and time of deposition on sheep faeces colonization by nematophagous fungi in the Mata region of Minas Gerais State, Brazil. *Applied Soil Ecology, 14*, 63-70.

[93] Larsen, M. (2000). Prospects for controlling animal parasitic nematodes by predacious micro fungi. *Parasitology, 120*, S121-S131.

[94] Barth, D., Karrer, M. & Henze-Mutz, E.M (1995). Significance of moisture content of dung pats for colonisation and degradation of cattle dung. *Applied Parasitology, 36*, 11-21.

[95] Larsen, M., Faedo, M. & Waller, P. J. (1994). The potential of nematophagous fungi to control the free-living stages of nematode parasites of sheep: survey for the presence of fungi in fresh faeces of grazing livestock in Australia. *Veterinary Parasitology, 53*, 275-281.

[96] Saumell, C.A., Santos, C. de P., Padilha, T. & Roque, M.V.C. (1999). Nematophagous fungi in fresh faeces of cattle in the Mata region of Minas Gerais, Brazil. *Veterinary Parasitology, 82*, 217-220.

[97] Su, H., Hao, Y., Mo, M. & Zhang, K. (2007). The ecology of nematode-trapping hyphomycetes in cattle dung from three plateau pastures. *Veterinary Parasitology, 144*, 293-288.

[98] Petkevičius, S., Larsen, M., Bach-Knudsen, K.E., Nansen, P., Grønvold, J., Henriksen, S.A. & Wolstrup, J. (1998). The effect of the nematode-destroying fungus *Duddingtonia flagrans* against *Oesophagostoumum dentatum* larvae in faeces from pigs fed different diets. *Helminthologia, 35*, 111-116.

[99] Persmack, L., Banck, A. & Jansson, H.-B. (1996). Population dynamics of nematophagous fungi and nematodes in arable soil: vertical and seasonal fluctuations. *Soil Biology and Biochemistry, 28*, 1005-1014.

[100] Saumell, C.A. (1998). Colonização de bolos fecais bovinos e fezes ovinas por fungos nematófagos e sua ocorrĕncia em fezes frescas bovinas na Zona da Mata de Minais Gerais. PhD Thesis; Belo Horinzonte, Brazil, 175 p.

[101] De, S., Sanyal, P.K., Sarkar, A.K., Patel, N.K., Pal, S. & Mandal, S.C. (2008). Screening for Indian isolates of egg-parasitic fungi for use in biological control of fascioliasis and amphistomiasis in ruminant livestock. *Journal of Helminthology, 82*, 271-277.

[102] Cooke, R.C. (1963). Ecological characteristics of nematode-trapping Hyphomycetes. I. Preliminary studies. *Annual Applied Biology, 52*, 431-437.

[103] Pandey, V.S. (1973). Predatory activity of nematode-trapping fungi against the larvae of *Trichostrongylus axei* and *Ostertagia ostertagi*: a possible method of biological control. *Journal of Helminthology, 47*, 35-48.

[104] Grønvold, J., Korsholm, H., Wolstrup, J., Nansen, P. & Henriksen, S.A. (1985). Laboratory experiments to evaluate the ability of *Arthrobotrys oligospora* to destroy infective larvae of *Cooperia* species, and to investigate the effect of physical factors on the growth of the fungus. *Journal of Helminthology, 59*, 119-125.

[105] Pelloile, M. (1991). Selection of nematode-trapping fungi for use in biological control. In: B.R. Kerry & D.H. Crumps (Eds.), *IOBC/WPRS Bulletin 1991/XIV/2. Working Group 'Integrated control of soil pests – Methods for studying nematophagous fungi'* (pp. 13-17).

[106] Gorlenko, M. V., Kondakova, E.I. & Pletneva, O.I. (1959). On the biology of predacious fungi. *Bulletin of the Moscow Association of Experimental Biology, 64*, 89-97.

[107] Grønvold, J., Nansen, P., Henriksen, S.A., Larsen, M., Wolstrup, J., Bresciani, J., Rawat, H. & Fribert, L. (1996b). Induction of traps by *Ostertagia ostertagi* larvae,

chlamydospore production and growth rate in the nematode-trapping fungus *Duddingtonia flagrans*. *Journal of Helminthology, 70*, 1086-1100.

[108] Nansen, P., Grønvold, J., Henriksen, S.A. & Wolstrup. (1988). Interactions between the predacious fungus *Arthrobotrys oligospora* and third-stage larvae of a series of animal-parasitic nematodes. *Veterinary Parasitology, 16*, 329-337.

[109] Sagüés, F., Saumell, C., Fusé, L., Iglesias, L. & Fiel, C. (2009). Evaluación de la capacidad predadora de *Duddingtonia flagrans* en materia fecal expuesta al sol y a la sombra. *Acta de Bioquímica Clínica Latinoamericana, Suplemento 1*, 96.

[110] Olthof, H.A. & Estey, R.H. (1965). Relation of some environmental factors to growth of several nematophagous Hyphomycetes. *Canadian Journal of Microbiology, 11*, 939-946.

[111] Peloille, M. (1979). Etude des Hyphomycètes prédateurs de nématodes recontrés sur une prairie du Limousin: Morpholgie-physiologie, frequence et distribution. Thesis, l'Université de Rennes, France, 106 p.

[112] Grønvold, J. (1989). Induction of nematode-trapping organs in the predacious fungus *Arthrobotrys oligospora* (*Hyphomycetales*) by infective larvae of *Ostertagia ostertagi* (*Trichostrongylidae*). *Acta Veterinaria Scandinavica, 30*, 77-87.

[113] Naves, R.L. & Campos, V.P. (1991). Ocorrĕncia de fungos predadores du nematoídes no sul de Minas Gerais e estudo de capacidade predatória. *Nematologia Brasileira, 15*, 152-162.

[114] Saumell, C., Fusé, L., Iglesias, L., Fernández, S. & Fiel, C. (2008). Enfoque bioecológico del potencial de los hongos nematófagos en el control biológico de tricostrongilídeos de rumiantes. *Revista de Medicina Veterinaria (Buenos Aires), 89*, 45-54.

[115] Gray, N.F. (1983). Ecology of nematophagous fungi: distribution and habitat. *Annual Applied Biology, 102*, 501-509.

[116] Fowler, M. (1970). New Zealand predacious fungi. *New Zealand Journal of Botany, 8*, 283-302.

[117] Ayen, E. & Lýsek, G. (1986). Endoparasitic nematode-destroying fungi in sandy soils of a beech wood in Berlin, Germany. *FEMS Microbiology Ecology, 38*, 397-400.

[118] Nordbring-Hertz, B. (1968). The influence of medium composition and additions of animal origins on the formation of capture organs in *Arthrobotrys oligospora*. *Physiologia Plantarum, 21*, 52-65.

[119] Grønvold, J., Wolstrup, J., Nansen, P., Larsen, M., Henriksen, S.A., Bjørn, H., Kirchheiner, K., Larsen, K., Rawat, H. & Kristiasen, H.L. (1999). Biotic and abiotic factors influencing growth rate and production of traps by the nematode-killing fungus *Duddingtonia flagrans* when induced by *Cooperia oncophora* larvae. *Journal of Helminthology, 73*, 129-136.

[120] Peterson, E.A. & Katznelson, H. (1965). Studies on the relationships between nematodes and other soild microorganisms. IV. Incidence of nematode-trapping fungi in the vicinity of plant roots. *Canadian Journal of Microbiology, 11*, 491-495.

[121] Hay, F.S., Niezen, J.H., Ridley, G.S., Bateson, L., Miller, C. & Robertson, H. (1997). The influence of pasture species and time of deposition of sheep dung on infestation by nematophagous fungi. *Applied Soil Ecology, 6*, 181-186.

[122] Mankau, R. (1962). Soil fungistasis and nematophagous fungi. *Physiopathology, 52*, 611-615.

[123] Hedlund, K. & Augustsson, A. (1995). Effects of enchytraeid grazing on fungal growth and respiration. *Soil Biology and Biochemistry, 27*, 905-909.

[124] Jaffee, B.A. & Muldoon, A.E. (1997). Suppression of the root-knot nematode *Meloidogyne javanica* by alginate pellets containing the nematophagous fungi *Hirsutella rhossiliensis*, *Monacrosporium cionopagum* and *M. ellpsosporum*. *Biocontrol Science and Technology, 7*, 203-218.

[125] Jaffee, B.A., Muldoon, A.E. & Didden, W.A.M. (1997). Enchytraeids and nematophagous fungi in soil microcosmos. *Biology and Fertility of Soils, 25*, 328-388.

[126] Jaffee, B.A., Santos, P.F. & Muldoon, A.E. (1997). Suppression of nematophagous fungi by enchytraeid worms. *Oecologia, 11*, 412-423.

[127] Larsen, M., Nansen, P., Grøndahl, C., Thamsborg, S.M., Grønvold, J., Wolstrup, J., Henriksen, S.A. & Monrad, J. (1996). The capacity of the fungus *Duddingtonia flagrans* to prevent strongyle infections in foals on pasture. *Parasitology, 113*, 1-6.

[128] Nansen, P., Larsen, M., Roesptorff, A., Grønvold, J., Wolstrup, J. & Henriksen, S.A. (1996). Control of *Oesophagostomum dentatum* and *Hyostringylus rubidus* in outdoor-reared pigs by daily feeding with the microfungus *Duddingtonia flagrans*. *Parasitology Research, 82*, 580-584.

[129] Githigia, S.M., Thamsborg, S.M., Larsen, M., Kyvsgaard, N.C. & Nansen, P. (1997). The preventive effect of the fungus *Duddingtonia flagrans* on trychostrongyle infections of lambs on pasture. *International Journal for Parasitology, 27*, 931-939.

[130] Fernández, A.S., Larsen, M., Henningsen, E., Nansen, P., Grønvold, J., Bjørn, H. & Wolstrup, J. (1999). Effect of *Duddingtonia flagrans* against *Ostertagia ostertagi* in cattle grazing under different stocking rate. *Parasitology,* 119, 105-111.

[131] Šarkūnas, M., Larsen, M., Nansen, P. & Hansen, J. W. (2000). Biological control of trichostringylid infections in calves on pasture in Lithuania using *Duddingtonia flagrans,* a nematode-trapping fungus. *Journal of Helminthology, 74*, 355-359.

[132] Jobim, M.B., Santurio, J.M. & de la Rue, M.L. (2008). *Duddingtonia flagrans*: controle biológico de nematodeos de bovinos a campo. *Ciência Rural, Santa Maria, 38*, 2256-2263.

[133] Santurio, J.M., Zanette, R.A., Da Silva, A.S., Fanfa, V.R., Farret, M.H., Ragagnin, L., Hecktheuer, P.A. & Monteiro, S.G. (2011). A suitable model for the utilization of *Duddingtonia flagrans* fungus in small-flock-size sheep farms. *Experimental Parasitology, 127*, 727-731.

[134] Nansen, P., Larsen, M., Grønvold, J., Wosltrup, J., Zom, A. & Henriksen, S.A. (1995). Prevention of clinical trichostrongylidosis in calves by strategic feeding with the predacious fungus *Duddingtonia flagrans*. *Parasitology Research, 81*, 371-374.

[135] Larsen, M., Wolstrup, J., Henriksen, S.A., Dackman, C., Grønvold, J. & Nansen, P. (1991). *In vitro* stress selection of nematophagous fungi for biocontrol of pre-parasitic nematode larvae of ruminants. *Journal of Helminthology, 65*, 193-200.

[136] Larsen, M., Nansen, P., Grønvold, J., Wolstrup, J. & Henriksen, S.A. (1992). *In vivo* passage of nematophagous fungi selected for biocontrol of parasitic nematodes in ruminants. *Journal of Helminthology, 66*, 137-141.

[137] Henriksen, S.A., Larsen, M., Grønvold, J., Nansen, P. & Wolstrup, P. (1997). Nematode-trapping fungi in biological control of *Dictyocaulus viviparous*. *Acta Veterinaria Scandinavica, 38*, 175-179.

[138] Fernández, A.S., Larsen, M., Nansen, P., Grønvold, J., Henriksen, S.A., Bjørn, H. & Wolstrup, J. (1999). The efficacy of two isolates of the nematode-trapping fungus *Duddingtonia flagrans* against *Dictyocaulus viviparus* larvae in faeces. *Veterinary Parasitology, 85*, 289-304.

[139] Paraud, C. & Chartier, C. (2003). Biological control of infective larvae of a gastro-intestinal nematode (*Teladorsagia circumcincta*) and a small lungworm (*Muellerius capillaris*) by *Duddingtonia flagrans* in goat faeces. *Parasitology Research, 89*, 102-106.

[140] Paraud, C., Cabaret, J., Pors, I. & Chartier, C. (2005). Impact of the nematophagous fungus *Duddingtonia flagrans* on *Muellerius capillaris* larvae in goat faeces. *Veterinary Parasitology, 131*, 71-78.

[141] Aschner, M. & Kohn, S. (1958). The biology of *Harposporium anguillulae*. *Journal of Microbiology, 19*, 182-189.

[142] Lohmann, U. & Sikora, A. (1989). Mass production of the endoparasitic fungus *Drechmeria coniospora*, *Verticillium balanoides* and *Harposporium anguillulae* in liquid culture. *Nematologica, 35*, 97-104.

[143] Santos, C.P. & Charles, T.P. (1995). Efeito da aplição de conídios de *Drechmeria coniospora* em cultivos de fezes contend ovos de *Haemonchus contortus*. *Arquivo Brasileiro de Medicina Veterinaria e Zootecnia, 47*, 123-128.

[144] Charles, T.P., Roque, M.V.C., & Santos, C. de P. (1996). Reduction of *Haemonchus contortus* infective larvae by *Harposporium anguillulae* in sheep faecal cultures. *International Journal for Parasitology, 26*, 509-510.

[145] Waghorn, T.S., Leathwick, D.M., Chen, L.-Y., Gray, R.A.J. & Skipp, R.A. (2002). Influence of nematophagous fungi, earthworms and dung burial on development of the free-living stages of *Ostertagia* (*Teladorsagia*) *circumcincta* in New Zealand. *Veterinary Parasitology, 104*, 119-129.

[146] Roque, M.C.V., Padilha, T. & Saumell, C.A. (1996) Cattle feeding of *Harposporium anguillulae* to reduce development of trichostrongylid nematode infective larvae. *Anais XV Congresso Pan-Americano de Ciências Veterinárias (PANVET)*, Campo Grande, MS, Brazil, p. 303.

[147] Katakam, K.K., Thapa, S., Meyling Vitt, N., Dalgaard, A., Thamsborg, S.M. & Roepstorff, A. (2011). Antagonistic effect of chitin-degrading microfungi on thick-shelled pig helminth eggs. *Proceedings of the 23rd Conference of the World Association for the Advancement of Veterinary Parasitology (WAAVP) 2011*, Buenos Aires, Argentina, p. 289.

[148] Ciarmela, M.L., Minvielle, M.C. & Basualdo, J.A. (2001). Control biológico de huevos de *Toxocara canis* con hongos saprófitos del suelo. *Acta de Bioquímica Clínica Latinoamericana., 35*, 463-470.

[149] Ciarmela, M.L. (2008). Efecto ovicida de hongos del suelo sobre huevos de *Toxocara canis* (Werner) Johnston. PhD Thesis, National University of La Plata, La Plata, Argentina, 208 p.

[150] Gortari, M. C., Galarza, B. C., Cazau, M. C., & Hours, R. A. (2008). Comparison of the biological properties of two strains of *Paecilomyces lilacinus* (Thom) Samson associated to their antagonitstic effect onto *Toxocara canis* eggs. *Malaysian Journal of Microbiology, 4*, 35-41.

[151] Carvalho, R.O., Araújo, J.V., Braga, F.R., Araujo, J.M. & Alves, C.D.F. (2010). Ovicidal activity of *Pochonia chlamydosporia* and *Paecilomyces lilacinus* on *Toxocara canis* eggs. *Veterinary Parasitology, 169*, 123-127.

[152] Ciarmela, M.L., Sanchez Thevenet, P., Alvarez, H.M., Minvielle, M.C. & Basualdo, J.A. (2005). Effect of *Paecilomyces lilacinus* on the viability of oncospheres of *Taenia hydatigena*. *Veterinary Parasitology, 131*, 61-64.

[153] Braga, F.R., Araújo, J.V., Araujo, J.M., Carvalho, R.O. & Silva, A.R (2008). Efeito do fungo *Paecilomyces lilacinus* sobre ovos de *Taenia saginata*. *Revista da Sociedade Brasileria de Medicina Tropical, 41*, 686-688.

[154] Araujo, J.M., Braga, F.R., Araújo, J.V. & Carvalho, R.O. (2009). Activity of nematophagous fungi *Pochonia chlamydosporia* and *Paecilomyces lilacinus* on *Dipylidium caninum* egg capsules. *Revista del Instituto Adolfo Lutz, 68*, 488-491.

[155] Braga, F.R., Araújo, J.V., Araujo, J.M., Carvalho, R.O., Silva, A.R., Campos, A.K. & Tavela, A.O. (2008). Ovicidal activity of *Paecilomyces lilacinus* on *Moniezia* sp. eggs. *Journal of Helminthology, 82*, 241-243.

[156] Silva, A.R., Araújo, J.V., Braga, F.R., Alves, C.D.F. & Frassy, L.N. (2010). In vitro ovicidal activity of the nematophagous fungi *Duddingtonia flagrans*, *Monacrosporium thaumasium* and *Pochonia chlamydosporia* on *Trichuris vulpis* eggs. *Veterinary Parasitology, 172*, 76-79.

[157] Frassy, L.N., Braga, F.R., Silva, A.R., Araújo, J.V., Ferreira, S.R. & de Freitas, L.G. (2010). Destruição de ovos de *Toxocara canis* pelo fungo nematófago *Pochonia chlamydosporia*. *Revista da Sociedade Brasileria de Medicina Tropical, 43*, 102-104.

[158] Braga, F.R., Ferreira, S.R., Araújo, J.V., Araujo, J.M., Silva, A.R., Carvalho, R.O., Campos, A.K. & Freitas, L.G. (2010). Predatory activity of *Pochonia chlamydosporia* fungus on *Toxocara* (syn. *Neoascaris*) *vitulorum* eggs. *Tropical Animal Health and Production, 42*, 309-314.

[159] Araújo, V.V., Braga, F.R., Silva, A.R., Araujo, J.M. & Tavela, A.O. (2008). In vitro evaluation of the effect of the nematophagous fungi *Duddingtonia flagrans*, *Monacrosporium sinense* and *Pochonia chlamydosporia* on *Ascaris suum* eggs. *Parasitology Research, 102*, 787-790.

[160] Braga, F.R., Araújo, J.V., Silva, A.R., Araujo, J.M., Carvalho, R.O., Campos, A.K., Tavela, A.O., Ferreira, S.R., Frassy, L.N. & Alves, C.D.F. (2010). *Duddingtonia flagrans*, *Monacrosporium thaumasium* and *Pochonia chlamydosporia* as possible biological control agents of *Oxyuris equi* and *Austroxyuris finlaysoni*. *Journal of Helminthology, 84*, 21-25.

[161] Araujo, J.M., Araújo, J.V., Braga, F.R., Carvalho, R.O., Silva, A.R. & Campos, A.K. (2009). Interaction and ovicidal activity of nematophagous fungus *Pochonia chlamydosporia* on *Taenia saginata* eggs. *Experimental Parasitology,* 121, 338-341.

[162] Braga, F.R., Silva, A.R., Carvalho, R.O., Araújo, J.V. & Pinto, P.S.A. (2011). Ovicidal activity of different concentrations of *Pochonia chlamydosporia* chlamydospores on *Taenia taeniformis* eggs. *Journal of Helminthology, 85*, 7-11.

[163] Silva, A.R., Araújo, J.V., Braga F.R., Alves, C.D.F. & Filho, J.D.R. (2010). Destruction of *Anoplocephala perfoliata* eggs by the nematophagous fungus *Pochonia chlamydosporia*. *Journal of Equine Veterinary Science, 30*, 701-704.

[164] Braga, F.R., Araújo, J.V., Campos, A.K., Araujo, J.M., Carvalho, R.O., Silva, A.R. & Tavela, A.O. (2008). *In vitro* evaluation of the action of the nematophagous fungi

Duddingtonia flagrans, *Monacrosporium sinense* and *Pochonia chlamydosporia* on *Fasciola hepatica* eggs. *World Journal of Microbiology and Biotechnology, 24*, 1559-1564.

[165] Dias, A., Araújo, J., Braga, F., Araujo, J., Puppin, A., Fernandes, F., Ramos, R., Bertoncelli, R., da Silva, R. & Perboni, W. (2011). Biological control of *Fasciola hepatica* egss with the *Pochonia chlamydosporia* fungus after passing through the cattle gastrointestinal tract. *Parasitology Research* (19 July 2011), pp. 1-5. doi:10.1007/s00436-011-2538-6.

[166] Braga, F.R., Araújo, J.V., Carvalho, R.O., Silva, A.R., Araujo, J.M., Soares, F.E., Geniêr, H.L., Ferreira, S.R. & Queiroz, J.H. (2010). Ovicidal action of crude enzymatic extract of the fungus *Pochonia chlamydosporia* against cyathostomin eggs. *Veterinary Parasitology, 172*, 264-268.

[167] Braga, F.R., Araujo, J.M., Silva, A.R., Araújo, J.V., Carvalho, R.O., de Freitas Soares, L.G., Queiroz, J.H. & Geniêr, H.L. (2011). Ação ovicida do extrato bruto enzimático do fungo *Pochonia chlamydosporia* sobre ovos de *Ancylostoma* sp.. *Revista da Sociedade Brasileria de Medicina Tropical, 44*, 116-118.

[168] Singh, R.K., Sanyal, P.K., Patel, N.K., Sarkar, A.K., Santra, A.K., Pal, S., & Mandal, S.C. (2010). Fungus-benzimidazole interactions: a prerequisite to deploy egg-parasitic fungi *Paecilomyces lilacinus* and *Verticillium chlamydosporium* as biological agents against fascioliasis and amphistomiasis in ruminant livestock. *Journal of Helminthology, 84*, 123-131.

[169] Braga, F.R., Araújo, J.V., Silva, A.R., Carvalho, R.O., Araujo, J.M., Ferreira, S.R. & Carvalho, G.R. (2010). Viability of the nematophagous fungus *Pochonia chlamydosporia* after passage through the gastrointestinal tract of horses. *Veterinary Parasitology, 168*, 264-268.

[170] Gortari, C., Cazau, C. & Hours, R. (2007). Hongos nematófagos de huevos de *Toxicara canis* en un paseo público de La Plata, Argentina. *Revista Iberoamericana de Micología, 24*, 24-28.

[171] Ciarmela, M.L., Minvielle, M.C., Lori, G. & Basualdo, J.A. (2002). Biological interaction between soil fungi and *Toxocara canis* eggs. *Veterinary Parasitology, 103*, 251-257.

[172] Ciarmela, M., Arambarri, A., Basualdo, J. & Minvielle, M. (2008). Efecto de *Chrysosporium merdarium* sobre huevos de *Toxocara canis*. *Libro de resúmenes, III Congreso Latinoamericano de Zoonosis & VI Congreso Argentino de Zoonosis, Buenos Aires, Argentina, 18-20 June 2008.*

[173] Ciarmela, M.L., Arambarri, A.M., Basualdo, J.A. & Minvielle, M.C. (2010). Effect of saprotrophic soil fungi on *Toxocara canis* eggs. *Malaysian Journal of Microbiology, 6*, 75-80.

[174] Ciarmela, M.L. (2008). Efecto de los hongos del suelo sobre huevos de *Toxocara canis*. *Libro de resúmenes, III Congreso Latinoamericano de Zoonosis & VI Congreso Argentino de Zoonosis, Buenos Aires, Argentina, 18-20 June 2008.*

[175] Rehner, S.A. (2005). Phylogenetics of the insect pathogenic genus *Beauveria*. In: F.E. Vega & M. Blackwell (Eds.), *Insect-Fungal Associations: Ecology and Evolution* (pp. 3-27), New York, United States of America: Oxford University Press.

[176] Fernández, A.S., Henningsen, E., Larsen, M., Nansen, P., Grønvold, J. & Søndergaard, J. (1999). A new isolate of the nematophagous fungus *Duddingtonia flagrans* as

biological control agents against free-living larvae of horse strongyles. *Equine Veterinary Journal, 31* 488-491.

[177] Fernández, A.S., Larsen, M., Nansen, P., Henningsen, E., Grønvold, J., Wolstrup, J., Henriksen, S.A. & Bjørn, H. (1999). The ability of the nematode-trapping fungus *Duddingtonia flagrans* to reduce the transmission of infective *Ostertagia ostertagi* larvae from faeces to herbage. *Journal of Helminthology, 73*, 115-122.

[178] Dimander, S.-O., Höglund, J. & Waller, P.J. (2003). Disintegration of dung pats from cattle treated with the ivermectin anthelmintic bolus or the biocontrol agent *Duddingtonia flagrans*. *Acta Veterinaria Scandinavica, 44*, 171–180.

[179] Sanyal, P.K., Sarkar, A.K., Patel, N.K., Pal, S. & Mandal, S. C. (2008). Evaluation of Chhattisgarh isolate of nematode trapping fungus *Duddingtonia flagrans* for their short-term environmental impact: Studies on fungal spread and faecal decomposition. *Journal of Veterinary Parasitology, 22*, 5-8.

[180] Paraud, C., Lorrain, R., Pors, I. & Chartier, C. (2011). Administration of the nematophagous fungus *Duddingtonia flagrans* to goats: an evaluation of the impact of this fungus on the degradation of faeces and on free-living soil nematodes. *Journal of Helminthology,* doi:10.1017/S0022149X11000083.

[181] Saumell, C.A., Echeverría, F., Gonçalves, L., Iglesias, L.., Rodríguez, E. & Padilha, T. (2001). Absence of environmental impact of the nematophagous fungus *D. flagrans*. *Proceedings of the 18th Conference of the World Association for the Advancement of Veterinary Parasitology (WAAVP)*, Stressa, Italy, p. 128.

[182] Yeates, G.W., Skipp, R.A., Gray, R.A.J., Chen, L.Y. & Waghorn, T.S. (2007). Impact on soil fauna of sheep faeces containing a range of parasite control agents. *Applied Soil Ecology, 35*, 380-389.

[183] Knox, M.R., Josh, P.F. & Anderson, L.J. (2002). Deployment of *Duddingtonia flagrans* in an improved pasture system: dispersal, persistence, and effects on free-living soil nematodes and microarthropods. *Biological Control, 24*, 176-182.

[184] Faedo, M., Larsen, M., Dimander, S.O., Yeates, G.W., Höglund, J. & Waller, P.J. (2002). Growth of the fungus *Duddingtonia flagrans* in soil surrounding feces deposited by cattle or sheep fed the fungus to control nematode parasites. *Biological Control, 23*, 64-70.

[185] Yeates, G.W., Dimander, S.O., Waller, P.J. & Höglund, J. (2002). Environmental impact on soil nematodes following the use of the ivermectin sustained-release bolus or the nematophagous fungus *Duddingtonia flagrans* to control nematode parasites of cattle in Sweden. *Acta Agriculturae Scandinavica, Section A, Animal Science, 52*, 233–242.

[186] Yeates, G.W., Dimander, S.O., Waller, P.J. & Höglund, J. (2003). Soil nematode populations beneath faecal pats from grazing cattle treated with the ivermectin sustained-release bolus or fed the nematophagous fungus *Duddingtonia flagrans* to control nematode parasites. *Acta Agriculturae Scandinavica, Section A, Animal Science, 53*, 197–206.

[187] Waller, P.J., Schwan, O., Ljungström, B.-L., Rydzik, A. & Yeates, G.W. (2004). Evaluation of biological control of sheep parasites using *Duddingtonia flagrans* under commercial farming conditions on the island of Gotland, Sweden. *Veterinary Parasitology, 126*, 299-315.

[188] Yeates, G.W., Skipp, R.A., Chen, L.Y., Waghorn, T.S. & Potter, J. F. (2007). Temporal and spatial aspects of the influence on soil nematodes of depositing artificial pats of sheep faeces containing a range of parasite management agents. *Applied Soil Ecology, 37*, 106-117.

[189] Grønvold J., Wolstrup, J., Larsen, M., Nansen, P. & Bjørn, H. (2000). Absence of obvious short term impact of the nematode-trapping fungus *Duddingtonia flagrans* on survival and growth of the earthworm *Aporrectodea longa*. *Acta Veterinaria Scandinavica, 41*, 147-151.

[190] Paraud, C., Lumaret, J.-P. & Chartier, C. (2007). Lack of effect of the nematophagous fungus *Duddingtonia flagrans* on the development of the dung beetle, *Aphodius constans*. *Small Ruminant Research, 70*, 276-279.

[191] Singh, R.K. & Sanyal, P.K. (2011). Fungal spread and faecal decomposition as indicators to evaluate short term environmental impact of egg parasitic fungi *Paecylomices lilacinus* and *Verticillium chlamydosporium*. *Veterinary World, 4*, 168-170.

[192] García, L., Melchor, G., Domíngues, Y., Rodríguez, H., Pino, O. & Hidalgo, L. (2008). Ecotoxicological evaluation of *Pochonia chlamydosporia* var. *catenulata* in terrestrial invertebrates. *Revista de Toxicología en Línea (ReTeL),* 14, 39-50.

[193] Davidse, L.C. (1986). Benzimidazole fungicides: mechanism of action and biological impact. *Annual Review of Phytopathology, 24*, 43-65.

[194] Kuzmina, T. (2001). The effect of some anthelminthics in *Duddingtonia flagrans* growth. *Proceedings of the 18th Conference of the World Association for the Advancement of Veterinary Parasitology (WAAVP)*, Stressa, Italy, p. 127.

[195] Paraud, C., Pors, I. & Chartier, C. (2004). Activity of *Duddingtonia flagrans* on *Trichostrongylus colubriformis* larvae in goat faeces and interaction with a benzimidazole treatment. *Small Ruminant Research, 55*, 199-207.

[196] Sanyal, P.K., Chauhan, J.B. & Mukhopadhyaya, P.N. (2004). Implications of fungicidal effects of benzimidazole compounds on *Duddingtonia flagrans* in integrated nematode parasite management in livestock. *Veterinary Research Communications, 28*, 375-385.

[197] Saumell, C.A. & Fernández, S. (2011). Conclusions of the Workshop "Biological Control of Gastrointestinal Nematodes in Livestock". *23rd Conference of the World Association for the Advancement of Veterinary Parasitology (WAAVP) 2011*, Buenos Aires, Argentina.

[198] Sagüés, F., Purslow, P., Fernández, S., Fusé, L., Iglesias, L. & Saumell, C. (2011). Especies de hongos nematófagos utilizadas para el control biológico de nemátodes gastrointestinales en el ganado y sus formas de administración. *Revista Iberoamericana de Micología,* DOI: 10.1016/j.riam.2011.06.009.

[199] European Food Safety Authority (2006). Opinion of the Scientific Panel on Additives and Products or Substances used in Animal Feed on the safety of the micro-organism preparation of *Duddingtonia flagrans*, for use as a feed additive for calves in accordance with Council Directive 70/524/EEC. *EFSA Journal, 334*, 1-8.

[200] Gillespie, A. (2002). *Duddingtonia flagrans* for the control of parasites in farm animals: A commercial perspective. In: *Biological control of nematode parasites of small ruminants in Asia: Final proceedings of the FAO Technological co-operation project in Malaysia*, (pp. 41-43), FAO Animal Production and Health Paper, Rome: Italy, Food and Agriculture Organization of the United Nations.

[201] García, L., Bulnes, C., Gleiby, M., Vega, E., Miranda, I., Montes de Oca, N., Hidalgo, L. & Marrero, E. (2004). Safety of *Pochonia chlamydosporia* var. *catenulat* in acute oral and dermal toxicity/pathogenicity evaluations in rats and rabbits. *Veterinary and Human Toxicology, 46*, 248-250.

[202] García, L., Gleiby, M., Montes de Oca, N. & Hidalgo, L. (2004). Estudio de la irritación ocular y dérmica de *Pochonia chlamydosporia* var. *catenulata. Revista de Toxicología, 21*, 103-107.

[203] Goettel, M.S., Hajek, A.E., Siegel, J.P. & Evans, H. (2001). Safety of fungal biocontrol agents. In: T.M. Butt, C. Jackson and N. Magan (Eds.), *Fungi as Biocontrol Agents – Progress, Problems and Potential* (pp. 347-375), New York, USA: CAB International.

In: Fungi: Types, Environmental Impact and Role in Disease ISBN: 978-1-61942-671-9
Editors: Adolfo Paz Silva and María Sol

Chapter XVI

Restoration of Fungal Biota in the Soil is Essential to Prevent Infection by Endoparasites in Grazing Animals

M. Arias, J. Suárez, F. J. Cortiñas, I. Francisco, J. L. Suárez, A. Romasanta, C. Cazapal-Monteiro, R. Sánchez-Andrade and A. Paz-Silva
Equine Diseases Study Group (Epidemiology, Parasitology and Zoonoses), Animal Pathology Department, Veterinary Faculty, Santiago de Compostela University, Spain

Abstract

Grazing livestock systems ensure nutrition of the animals, provide high quality food (milk, meat) and limit the presence of unwanted vegetation in an ecologically sustainable procedure. Pasturing practices can control grass competition for moisture, nutrients, and sunlight, thereby enhancing tree growth and the benefits from timber production. The presence of grazing animals provides organic fertilization so livestock manure recycles nutrients to trees and forage.

Productivity of grazing animals can be lessened by pathogens as the gastrointestinal parasites. In infected animals, eggs from several nematode genera are passed in the feces and develop in the soil until the infective third larval stage (L3). Animals become infected when feedind the grass with the L3.

Several chemical compounds exhibiting anthelmintic activity are commercially available for their control and frequently employed in farm animals. Some troubles related to low efficacy and/or emergence of anthelmintic resistance (AR) in different animal species has been widely reported. By considering the AR, together with chemotherapy affects only to the parasites in the animals, the importance for certain measures on the environment to reduce the presence of free living and infective stages to prevent the infection of the animals is emphasized.

Duddingtonia flagrans is a nematode-trapping fungus very useful to capture nematodes from mycelia, due to its ability to produce a large quantity of chlamydospores able to resist the passage through the gastrointestinal tract of ruminants and monogastrics once they are orally administered into animals.

The possibilities for reducing the frequence of administration of dewormers to livestock by using *D. flagrans* chlamydospores in horses under a pasturing management were evaluated. Equines received ivermectin and/or fungal spores. Chemotherapy was very successful, but the presence of infective stages in the environment conferred a marked temporal solution. When using chlamydospores of the *D. flagrans* nematode-trapping fungus, a high reduction of the infective stages (L3) from the pasture was noted. It is concluded that chemotherapy and the administration of spores prevents the infection in grazing horses and helps to decrease the frequency of deworming.

1. Introduction

Management of livestock under pasturing regimes affords many benefits, being the production of high-quality products (milk and meat) the most recognized.

It is well-known that what livestock are fed has a big influence on the quality of the final alimentary products, mainly milk and meat [1-3]. Fresh cattle milk is considered to be the best tasting milk.

Recently it has been demonstrated that grazing cows produce milk which contains significantly higher beneficial fatty acids, antioxidants and vitamins than their conventional 'high input' counterparts [4-7].

Small but significant differences have been reported in chemical content and sensory quality of meat from comparable groups of lambs grazing unimproved mountain pastures compared to fertilized cultivated lowland pastures [8, 9]. Grazing resulted in higher proportions of polyunsaturated fatty acids [10, 11] as well as other compounds like carotene and tocopherol, which may improve the shelf life of meat, are also increased [12].

Other additional advantages which can be afforded are:

a) Control of unwanted vegetation, propagation of shrubs and bushes, contributing to reduce the risk of fire.

Total abandonment of agriculture results in rapid development of shrub cover with the loss of the diverse flora associated with traditional arable rotations. Agroforestry and silvopastoralism are antique ways of managing forestland that should be encouraged as they increase productivity in the short, medium and long terms (in comparison with forestland), biodiversity (in comparison with farmland) and the sustainability of land (multiproduct system) [13].

Properly managed livestock grazing contributes to control the amount and distribution of grasses and other potential fuel, and thus to reduce fire hazards.

b) Reduction of grass competition for moisture, nutrients, and sunlight, thereby enhancing tree growth and the benefits from timber production.

Grazing can control grass competition for moisture, nutrients, and sunlight, thereby enhancing tree growth. Well managed grazing provides economical control of weeds and bush, and maintains fire breaks [14, 15].

c) Increasing the possibilities for enjoying the landscape aesthetics in natural places.

Grazing regimes can increase wildlife diversity and improve water quality. The forage protects the soil from water and wind erosion, while adding organic matter to improve soil properties. These regimes provide also an attractive landscape with an aesthetically pleasing "park-like" setting. On the contrary to concentrated livestock operations, grazing systems are less probable to increase environmental affairs related to water quality, odors, dust, noise, disease problems, and animal treatment [16].

d) Allowing sustainability of vegetal and animal species.

Open grasslands and woodlands are generally dominated by non-native and/or invasive annual grasses and herbs. The vegetation, when left unmanaged, tends to inhibit the germination and growth of other plants by using up most of the available water and mineral resources in the soil and by producing large amounts of thatch [17].

Livestock grazing controls the growth of the non-native grasses and herbs so that other advantageous plants (wildflowers and native grasses) can regenerate and coexist with them. Many plants, including several scarce species require grazing to maintain viable populations [18].

By other hand, it seems clear that well-managed livestock pasturing increases the diversity of habitats available to wildlife species [19]. Many species, including several ones in danger of extinction, benefit form the vegetation management performed by livestock. This is the case of ground squirrel colonies which support in grazed areas the foraging needs of predators and at the same time, recreate underground tunnels that are used by insects, reptiles, amphibians, and many small mammals. In a similar way, the construction of ponds for livestock watering support large numbers of breeding amphibians, which also feed on the abundant insect life found in the grasslands. Proper utilization of livestock grazing endorses healthier, diverse wildlife populations in natural spaces.

1.1. Grazing Livestock Parasite Infections

It is well-known that many infections provoked by parasites affecting the gastrointestinal tract in grazing animals are responsible for a severe reduction in their production. Infection occurs when feeding grass with either resistance forms for some protozoa (oocysts), trematoda (metacercariae) or free-living stages (larvae) in the case of nematode [20]. Their negative effect can be measured either on the yielding of the animals (milk production, gain weight, reproductive indexes) or their advantageous effect on the environment (diminution of unwanted and uncontrolled vegetation, fertilization).

One of the main parasitosis in grazing livestock are caused by the ingestion of larvae belonging to several trichostrongyle nematoda, as *Strongylus, Triodontophorus, Gyalocephalus, Cyathostomum, Ostertagia, Nematodirus, Trichostrongylus, Haemonchus, Chabertia* or *Oesophagostomum*. The generalized life cycle for these parasites begins with the production of unembryonated eggs by the adult worms located at the gastrointestinal tract, which are passed by the feces to the ground. In the fecal pat, embryonation takes place, followed by the egg-hatching into an L1 larva, which moults to L2 and then to L3, the infective phase [21]. Third larval stages leave the fecal pats and move to adjacent plants, being ingested by the animals when pasturing.

Chemotherapy is the most commonly applied measure against endoparasites in domestic animals, which looks for their elimination from the definitive hosts.

1.2. Environmental Equilibrium

Under natural conditions, environment is able to maintain itself in equilibrium. For example, vegetable species serve as sustenance for herbivorous and are transformed into food (milk, meat, skin) whereas the feces are dispersed by the action of coprophagous mites. As a result, a great variety of nutrients and micronutrients are released, thus livestock manure recycles nutrients to trees and forage grows [22].

Some organisms including gastrointestinal parasites are ingested together with the herbage and reach the digestive tract of the mammals. In the soil, several bacteria (*Bacillus thuringiensis*), viruses or fungi develop a necessary parasiticide activity, preventing the infection of the herbivorous.

Figure 1. Rearing of cattle under a grazing system.

In regions under oceanic climate (typically found along the west coasts at the middle latitudes of all the world's continents, and in southeastern Australia), it is described a narrower annual range of temperatures [23], which allow that grass can grow almost all the year.

Traditionally, grazing has been the most common practice among farmers for rearing livestock, especially in areas characterized by the presence of abundant pastures [24]. Nevertheless, different systems to increase the inputs have been applied. In this way, intensive regimes were frequently developed from the 1950's among cattle, swine and poultry farmers.

From several decades, consumers began to change their diets by considering that food contributes to satisfy appetite, prevent diet-deficiency disorder or to provide the essential building blocks of nutrition for the maintenance of body tissue [14]. The need for adjusting to the new tendencies based on healthy-produced foods seems to underline the possibilities of grazing livestock by means of semi-intensive or extensive management regimes.

1.3. The Environmental Equilibrium can be disturbed

Rearing of livestock is always based on production of forage. For trying to improve this production, several agronomic procedures have been often developed on arable lands, with the help of domestic animals (cattle, horses). Although a certain level of erosion could be reached, including the lessening of active biota on the soil, some of these procedures have been also reported as beneficial for reducing the presence and survival of parasite infective stages. In this way, the environmental equilibrium could be maintained.

Machinery helps to enhance the effectiveness of land management and forage production, but the damage provoked to the soil seems irreversible for many of the organisms, especially those with parasiticide activity. In spite of their convenience, most of these procedures disturb the equilibrium in the environment, and thus the telluric biota appears to experiment a notable reduction.

Due to the absence of equilibrium, many of the infective stages belonging to different endoparasites remain in the environment, hence the possibility of infection in pasturing livestock is enhanced.

The most frequent solution consists of increasing the frequency of deworming the animals, which could be responsible for the appearance of reduced efficacy or even anthelmintic resistance (AR) [25]. By taking into account that chemotherapy reduces only the parasitic burden in the final host, it seems essential to consider certain measures on the soil to reduce the risk of animal infection.

It has been previously underlined there are several environmental components remaining in equilibrium under natural conditions. For this reason, the most useful solution for the control of parasites affecting grazing livestock appears to be the restoration of that balance by imitating natural systems. Like this, several trials for improving the presence of a trapping-nematophagous fungus, *Duddingtonia flagrans*, have been conducted.

2. Chemical and Biological Trials for the Deworming of Pasturing Horses

Among the parasites infecting horses maintained in pastures, ascarids, strongyles, oxyurids and cestoda are frequently detected, being scarce the findings of protozoa or trematoda in the feces of these animals. Chemotherapy comprises the main procedure for the control of these parasites, and several efficient drugs (chiefly benzimidazoles, macrocyclic lactones and tetrahydropyrimidines) are commercially available [26, 27].

Regardless of the availability of efficient parasiticides, several aspects associated to the deworming of horses could reduce their expected successful. The absence of periodical analysis for checking the parasitological status of the equines previous to the chemotherapy or after it for assessing the effect have been pointed as frequent and important difficulties in the action against the parasites. In many cases, treatment is only applied to the most valuable specimens, or when the visual examination of the horses seems to reveal a very poor body condition. Other consideration resides in the accurate estimation of the body weight to calculate the proper dose.

In the last decades, increasing resistance of parasites (ascarids and strongylids) to different groups of anthelmintics has been denoted [28]. Turning different classes of anthelmintics in a fast rotation scheme (eg, every few months) or a slow rotation scheme (annually) is widely practiced to prevent development of resistant parasite strains, but there is little evidence to support the utility of this procedure.

The requirement for reducing the level of pasture contamination by free-living stages has been widely indicated to ensure the successful control of a number of parasites [29]. Although several procedures have been described, focused on rotational grazing, pasture clipping, chain harrowing, manual extraction of fecal pats, or administration of chlamydospores of the nematode trapping-fungus *Duddingtonia flagrans*, their application is not very spread [30].

Herein the effect of chemical and biological procedures to enhance the control of parasites in grazing horses was analyzed.

Area of Study and Horse Management

The current investigation was carried out in Galicia (NW Spain, 42°20′–43°45′N, 6°49′–8°00′W), an agricultural region (Fig. 2). Pastures are usually composed by *Trifolium pratense*, *T. repens*, *Lolium perenne*, *L. multiflorum* and *Dactylis glomerata*.

By taking into account that climatic conditions favor the grass develops throughout February to December, grazing equines remain outdoors unless adverse weather conditions dissuade it, when they are kept into boxes.

Parasite control is performed by the annual administration (in spring) of anthelmintics, being the macrocyclic lactones or the benzimidazoles the foremost choice. Coprological analyses are seldom considered.

Experimental Design

Previous investigations showed the appropriate moment for deworming grazing horses was September-October [31]. For this reason, in September 2010 fecal samples from 16 autochthonous Pura Raza Galega (PRG) mares were collected and analyzed by using the flotation copromicroscopical technique. According to the fecal egg count (FEC) numbers of strongyles, chemotherapy was administered to 12 horses exceeding a cut-off value of 300 eggs per gram of feces (EPG) [32]. Deworming was performed by the pour on application of the macrocyclic lactone ivermectin (IVM 0.5% w/v, Noromectin®, Norbrook Laboratories Ltd, UK) at a 1 mg kg bw^{-1} dosage [33, 34]. In addition, chlamydospores of the nematode-trapping *Duddingtonia flagrans* were employed for the biological control of the free-infective larval stages (also called L3) [35].

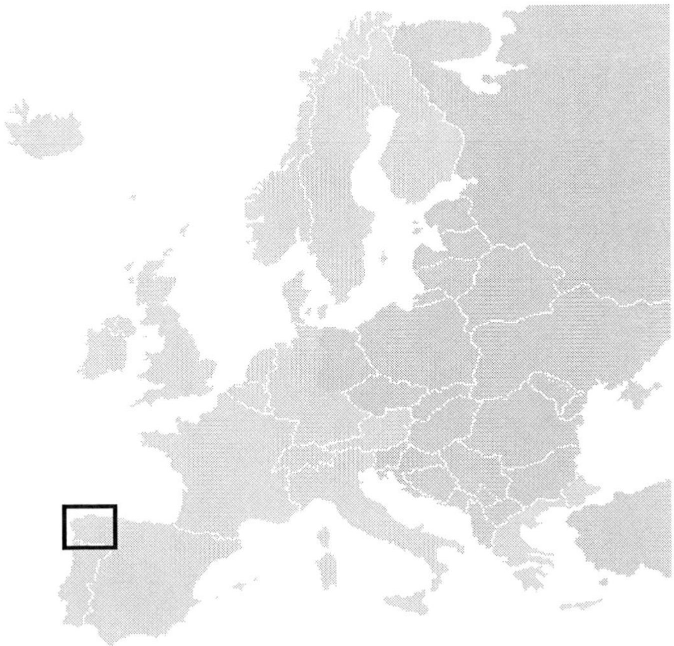

Figure 2. Localization of the area (NW Spain) where the current study was conducted.

The mares were divided into three groups and fed on a 15 Ha meadow (Fig. 3). Supplement was provided when necessary, consisting of wheat straw and barley. Water was available *ad libitum*.

The composition of the groups was the following:

- G-1: 6 PRG adult mares receiving ivermectin
- G-2: 6 PRG adult mares treated with ivermectin and grazing in a pasture where *D. flagrans* chlamydospores were directly spread onto the ground (2 x 10^6/100 m^2).
- G-3: 4 PRG adult mares remaining without chemotherapy and receiving a dose of 2 x 10^6 *D. flagrans* chlamydospores / kg bodyweight (Fig. 4).

In all cases, feces were individually collected from the rectum of each mare, by introducing them into a chute. The technician, who was blinded to treatment protocols, counted the parasites fecal eggs per 5 g by performing a McMaster modified technique with a minimum detection level of 30 EPG (Francisco et al. 2009b). The identification of the strongyles was made after the culture of the feces at 25°C for 15 days, by using morphological keys [36].

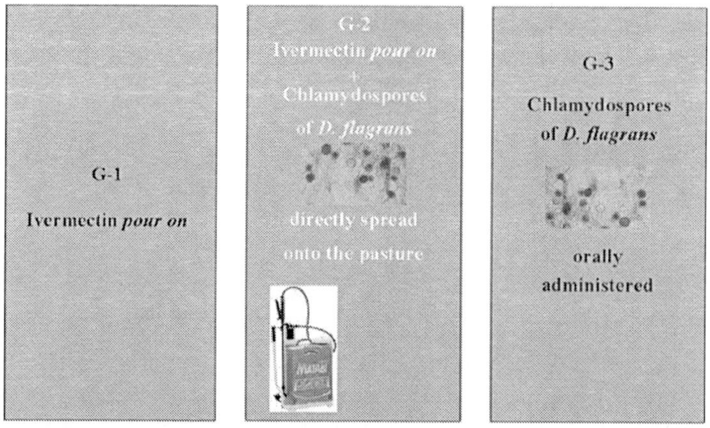

Figure 3. Distribution of the horses in the pastures.

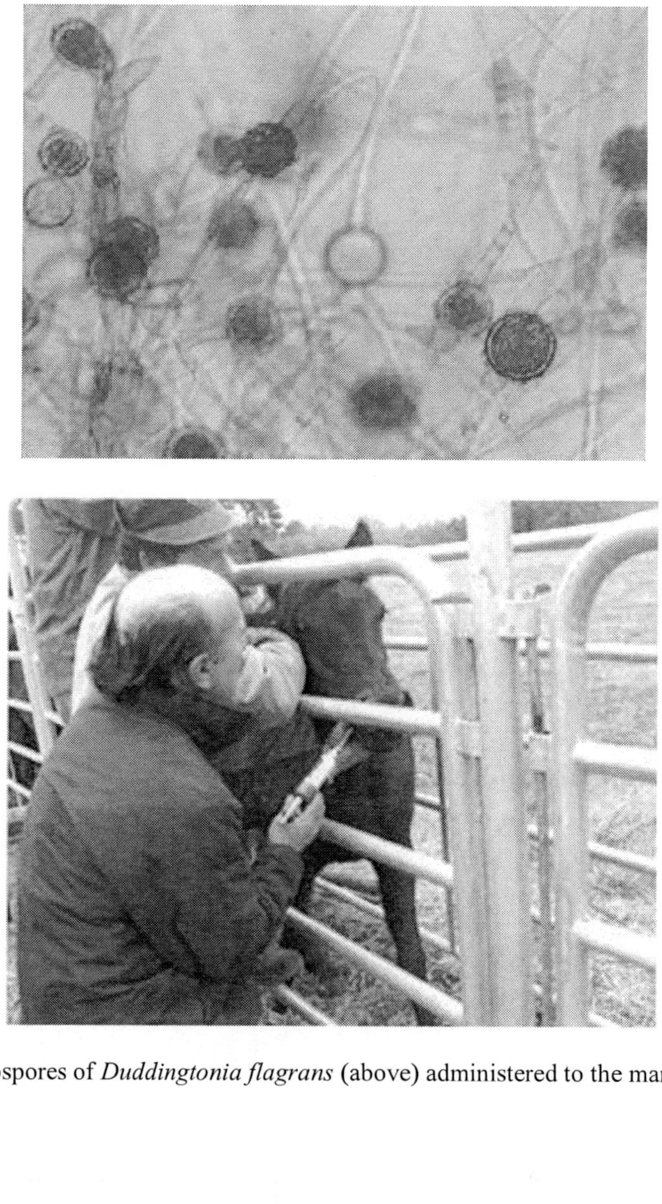

Figure 4. Chlamydospores of *Duddingtonia flagrans* (above) administered to the mares in the G-2 (below).

Efficacy of Deworming Procedures

The success of the procedures applied for the control of endoparasites in the horses was done by the analysis of fecal samples collected 2 weeks before the deworming and 2 weeks after it.

EPG values were determined on the day of treatment and at various times post-treatment as noted in Fig. 5 and in Table 1. Two parameters were calculated, the reduction in the fecal egg-count (FECR) and in the number of positive horses (PHR) to the copromicroscopical analysis, according the following formulae [34]:

$$FECR (\%) = [1 - (FEC_{post\text{-}treatment} / FEC_{pretreatment})] \times 100$$

$$PHR (\%) = [1 - (positive\ horses_{post\text{-}treatment} / positive\ horses_{pretreatment})] \times 100$$

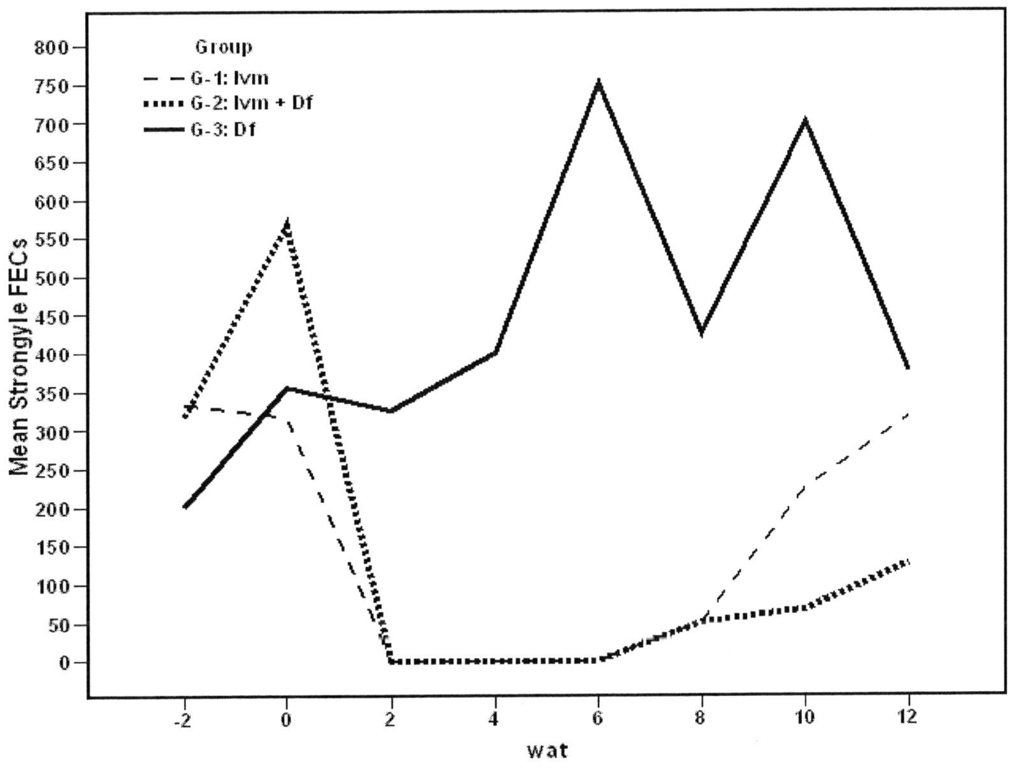

Figure 5. Kinetics of strongyles egg-output in grazing pastures expressed as average values.

Statistical Analysis

By examining the distribution of data with the Levene's probe, it was shown that the fecal egg-shedding and the FECRT values did not adjust to a normal distribution (P= 0.001 and P= 0.002, respectively), opposite to that obtained for the PHR numbers (P= 0.359). In this way, the FECs were analyzed by using the non-parametric Mann-Whitney U two-sided test (α= 0.05). Nevertheless, to make possible the discussion of the results with previous investigations, the egg-output kinetics was represented in the graphics as the mean and the standard deviation.

The values of FECR and PHR were expressed as percentages and the 95% Confidence Interval (CI) (Table 1), and analyzed by using the Mann-Whitney U test and the ANOVA, respectively.

Significant differences were considered when $P < 0.05$. All the tests were performed by using the SPSS (15.1).

2.2. Results

2.2.1. Parasitological Examinations

At the beginning of the study, eggs belonging to gastrointestinal nematode (ascarids, strongyles and oxyurids) were only observed in the feces of the equines. The coprocultures showed the presence of third larval stages belonging to the cyathostomins *Cyathostomum* and *Gyalocephalus*. No specimens of protozoa, trematoda, cestoda or lungworms were appreciated in any of the examined mares.

2.2.2. Anthelmintic Efficacy

In all the horses receiving ivermectin the overall mean FECR 14 days (2 weeks) following the treatment with ivermectin was 100% for the ascarids and the strongyles (Table 1). The same value was obtained for the PHR.

2.2.3. Trends of Strongyles Egg-Output

Kinetics

The cyathostomin egg-output was suppressed by the 2nd week after treatment (wat) with ivermectin (G-1 & G-2) (Fig. 5). Eggs reappeared in the feces of both treated groups at the 8th wat. Significant differences in the egg-output were observed only among the treated and control equines, whereas no differences between the two treated groups were obtained.

The horses belonging to the G-3 passed strongyle eggs in the feces throughout the study, achieving values higher than in the treated groups to the end of the experiment.

FECR and PHR

As summarized in Table 1, the mares receiving the macrocyclic lactone (G-1 & G-2) achieved a 100% value for the FECR between the 2nd and the 6th wat. From the 8th wat, percentages higher than 75% were maintained in the G-2, whereas reduced to 0% in the G-1 (12th wat).

Identical values for the PHR were attained both for the G-1 and G-2 groups, oscillating between 100% and 0% (Table 1).

In the horses orally administered chlamydospores of *D. flagrans* (G-3), the FECR was 0% except for the 2nd wat (8%). The PHR was 0%.

Table 1. Efficacy of strategies for the control of strongyles affecting grazing horses. FECR: fecal egg-count reduction; PHR: positive horses reduction

	FECR (%)			PHR (%)		
Weeks after treatment	G-1	G-2	G-3	G-1	G-2	G-3
0	0	0	0	0	0	0
2	100	100	8	100	100	0
4	100	100	0	100	100	0
6	100	100	0	100	100	0
8	80	91	0	83	83	0
10	10	88	0	0	0	0
12	0	77	0	0	0	0

Conclusion

Progress in the productivity of agriculture and related sectors directly increases farm and rural incomes and household food security. At the same time, agricultural growth focused on small farmers promotes overall rural and non-farm employment [17].

All the agricultural activity has to be profitable, because in the opposite situation there will be fewer grazing animals and the retention of such grassland will not be viable. There is great interest in the expansion of more sustainable agricultural systems, notably integrated farming and organic, which rely on crop rotations to add fertility and control weeds, pests and diseases [22].

When the livestock production does not achieve the required profitability, the main solution has consisted of changing the rearing systems, looking for a kind of management which ensures it. Since several decades domestic ruminants have been moved from outdoor to indoor feeding systems. As a consequence, animal nutrition with cereals and concentrated products together with introduction of machinery have been responsible to increase both yielding and economical inputs [20].

Though its elevated productivity, the indoor-based systems presents some problems which appear not easy to solve. Firstly, the acquisition of machinery reduced the people needed for working in rural areas, and thus a significant reduction in the total inhabitants was recorded. Secondly, the increment in the cost of animal feeding based on cereals and concentrates diminished the numbers of farms, and only the most productive can survive nowadays, which increased the abandonment of rural areas. Thirdly, elimination of manure constitutes a big problem due to the difficult for recycling this product, due to ground fertilization is carried out by using chemical compounds.

The agricultural activity should provide adequate inputs for ensuring an appropriate life quality. There is a very extended believe about milk yield per cow is lower on farms using grazing systems than those using confinement-feeding systems, but this situation might be reversed if additional considerations be taken into account [37]. In this way, it has been shown evidence of the high potential of *Leucaena* to improve meat and milk production from grazing cattle.

Some investigations pointed the superior taste and qualities of cheese and milk products from grazing animals when compared to dairy products from indoor fed animal [38]. In addition, ruminant milk converted into quality artisan cheeses would provide better economical and social status for the rural community [39, 40]. Artisan cheese making requires more effort, thus creating jobs which can reduce emigration of peasants. With modern technology the production of artisan food can be made attractive to urban consumers, thus improving rural conditions [41].

Other proposal could rely in the development of new milk types on the basis of the use of large quantities of legumes and herbs in the feeding ration to examine the transmission of precursors for aroma components or direct transfer of flavor components [1, 43]. It has been stated the possibility for developing products with a composition, nutritional properties and flavor that differ from conventionally ones by administering different species of legumes [44, 45].

The value for some agricultural procedures as the grazing practices is not only based on the direct economical inputs. It seems very clear to understand that grassland maintains the fertility and erosion resistance of soils by adding organic matter [46].

A well managed countryside offers an attractive environment for rural businesses which can attract inward investment, and for that reason more positive strategies to develop and market quality products from grazing livestock could stimulate new rural based businesses such as abattoirs, cutting plants and local food outlets [47]. The marketing of locally distinctive quality meat and/or milk products is an important way of re-connecting the consumer with food and the land from which it was produced.

Among the problems derived from pasturing animals, infection by parasites becomes into one of the most frequent and important for the health status and the economical affords. Internally parasitized livestock can result in significant improvement in weight gain and body condition scores, increasing overall profitability. Parasitism in cow-calf herds can appear with clinical symptoms including diarrhea, anemia, sunken flanks, rough coats and bottle jaw, being suboptimal performance, lost calf, missing milk, enlarged time to breeding and greater disease susceptibility the main subclinical manifestations. Major economic losses due to parasites can be unseen and unpredictable, and should be reduced or prevented by diminishing the risk of livestock infection.

Most of the parasite control procedures consist of deworming with successful chemical compounds. The use of pharmacological drugs in food-producing animals has raised significant concerns about microbial or parasite resistance to different compounds, and their residues can be found into the food supply.

By considering the existence of a group of parasites needing the environment for completion of their life cycle, it seems very conceivable that successful results can not be achieved if any actions on the pathogen free stages are developed, and thus pasture management looks of major importance in the prevention and transmission of parasites in grazing livestock.

There have been described different procedures for chemotherapy of the animals against the immature and mature worms, mainly focused on the deworming in summer or winter-spring [48]. With the aim to get better results, it has been proposed chemotherapy followed with strategically timed retreatments (3-4 weeks intervals) to reduce pasture contamination [25]. However, the repeated deworming can become problematic due to a) the possibility for the presence of residues in the foods; b) the development of resistance to parasiticides, and c) the increment in the cost of rearing livestock.

It is important to emphasize that the principal goal in the control of parasites affecting grazing animals should be to prevent pasture contamination by killing adult worms before pasturing, and suppressing infective present in the pasture which can infect livestock while feeding [49]. Many of the parasites infecting the gastrointestinal tract are shedding eggs and/or larvae to the ground by the feces. In this way, management practices are used to break up and scatter manure piles, with the purpose to allow worm larvae and eggs are exposed to heat and dry conditions, thus reducing contamination.

The efficacy of strategies based on the biological control of parasites by using the trapping-nematode fungus *Duddingtonia flagrans* has been widely demonstrated [50]. Most of the experiments were carried out by the oral administration of chlamydospores to the animals, which needs for their accurate immobilization and increases the needs for staff, raising the fees for the deworming [51, 52]. In the current chapter, it has been shown that a combined strategy by the pour on chemotherapy of the grazing animals in concert with the spreading of spores directly onto the meadows provides a very useful tool to reduce both the parasitim in the animals and in the environment. Like this, the cumulative effect seems to ensure an appropriate control of the parasites, as chemotherapy eliminates them from the animals, whose have a diminished possibility of reinfection due to the action of the nematophagus fungus.

It has been demonstrated the control of endoparasites in grazing livestock fails if based only on chemotherapy, and needs for the restoration of the regular biota on the soil. After an initial deworming of the animals to eliminate the parasites affecting them, parasitic infections would be controlled by means of an ecological and sustainable process consisting on the addition of chlamydospores of *Duddingtonia flagrans*.

The interest for organic farming is increasing nowadays. In essence, the basis is the simulation of natural conditions. Among the most frequent troubles under organic farming systems, control of endoparasites becomes a serious problem without possibilities for solution due to that chemotherapy is not allowed. As a consequence the inputs afforded are significantly lessened.

The described procedure could be candidate to be applied in organic farming systems, by considering a period of transition during which commercially available dewormers might be initially provided to the animals. By the administration of fungal spores to restore the soil biota, elimination of parasitic infective stages would prevent infection of the animals feeding on pastures. Although organic farming is not able for guaranteeing the alimentary requirements to all human population, but this type of farming offers very useful and attractive advantages to part of them [53].

The production of healthy and tasty products appears to be the most known, which is underlined by the possibility for transforming then into dynamic foods with high concentrations of interesting and needed components as the omega 3 fatty acids, conjugated linoleic acid (CLA), vitamin E and carotenoids, all related to a reduced risk of cardiovascular disease and cancer [7, 14, 54]. CLA is hugely popular in the US, where it is marketed as a nutritional supplement [55]. Switching to organic milk gives a natural way to increase our intake of nutritionally desirable fatty acids, vitamins and antioxidants without increasing our intake of less desirable fatty acids and synthetic forms of vitamin E. The possibility for adjusting to organic farming becomes another advantage from grazing livestock and emphasizes the interest for this agricultural practice [53].

Acknowledgements

We are in debt to Mrs B. Valcárcel for preparing and editing the manuscript. Thanks must be given to the Spanish equine associations Pura Raza Galega PURAGA, *Cabalo de Pura Raza Galega*, "Granxa do Souto", Grupoportichol and Asociación del Caballo Pura Raza Asturcón. Thanks should be given to Dr. P. Morrondo Pelayo and Dr. P. Díez-Baños (Parasitology and Parasitic Diseases (Animal Health Dept., Faculty of Veterinary, Santiago de Compostela University, Lugo, Spain) and Dr. Z. García, Director (Centro Nacional de Investigaciones Disciplinarias en Parasitología Veterinaria, Instituto Nacional de Investigaciones Forestales, Agrícolas y Pecuarias, Jiutepec, Morelos, México), for helping us in the development of the current chapter.

Conflict of Interest

All the authors declare the absence of any financial or personal relationships with other people or organizations that could inappropriately influence (bias) the current chapter. The final article has been approved for all their authors.

Policy and Ethics

The current investigation complies with the current laws for Animal Health Research in Spain and with the EC Directive 86/609/EEC for animal experiments.

Funding Source

This chapter was partly supported by the Research Funds XUGA PGIDT06RAG26102PR, 07MDS021261PR and & 10MDS261023PR (Xunta de Galicia, Spain), and no involvement in the design, collection, analysis of data, writing or decision to submit the paper has been developed. Dr. M.S. Arias was given an "Ángeles Alvariño" PostDoctoral fellowship (Xunta de Galicia, Spain).

References

[1] Moloney, A.P.; Mooney, M.T.; Kerry, J.P.; Troy, D.J. (2001). Producing tender and flavoursome beef with enhanced nutritional characteristics. *Proceedings of the Nutrition Society,* 60, 221-229.

[2] Nuernberg, K.; Dannenberg, D.; Nuernberg, G.; Ender, K.; Voigt, J.; Scollan, N.D.; Wood, J.D.; Nute, G.R.; Richardson, R.I. (2005). Effect of a grass-based and a concentrate feeding system on meat quality characteristics and fatty acid composition of

longissimus muscle in different cattle breeds. *Livestock Production Science*, 94, 137-147.

[3] Monaghan R.M. (2008). The environmental impacts of non-irrigated, pasture based dairy farming. In: R.W. McDowell (Ed), *Environmental Impacts of Pasture-based Farming*. Oxfordshire, UK: CABI; 209-231.

[4] Whitlock, L.A.; Schingoethe, D.J.; Hippen, A.R.; Kalscheur, K.F.; Abu-Ghazaleh, A.A. (2003). Milk production and composition from cows fed high oil or conventional corn at two forage concentration. *Journal of Dairy Science,* 86, 2428-2437.

[5] Banskalieva,V.; Sahlu, T.; Goetsch, A.L. (2003). Fatty acid composition of goat muscles and fat deposits: a review. *Small Ruminant Research,* 7, 255-268.

[6] Noci, F.; Monahan, F.J.; French, P.; Moloney, A.P. (2005). The fatty acid composition of muscle fat and subcutaneous adipose tissue of pasture-fed beef heifers: Influence of the duration of grazing. *Journal of Animal Science,* 83, 1167-1178.

[7] Alfaia, C.M.M.; Castro, M.L.F.; Martins, S.I.V.; Portugal, A.P.V.; Alves, S.P.A.; Fontes, C.M.G.A.; Bessa, R.J.B.; Prates, J.A.M. (2007). Effect of slaughter season on fatty acid composition, conjugated linoleic acid isomers and nutritional value of intramuscular fat in Barrosã-PDO veal. *Meat Science,* 75, 44-52.

[8] Arsenos, G.; Banos, G.; Fortomaris, P.; Katsaounis, N.; Stamataris, C.; Tsaras, L.; Zygoiannis, D. (2002). Eating quality of lamb meat: effects of breed, sex, degree of maturity and nutritional management. *Meat Science,* 60, 379-387.

[9] Ådnøy, T.; Haug, A.; Sørheim, O.; Thomassen, M.S.; Varszegi, Z.; Eik, L.O. (2005). Grazing on mountain pastures–does it affect meat quality in lambs? *Livestock Production Science,* 94, 25-31.

[10] Whale, K.; Heys, D.S.; Rotondo, D. (2004). Conjugated linoleic acid: are they beneficial or detrimental to health? *Progress in Lipid Research,* 43, 553-587.

[11] Riediger, N.D.; Othman, R.; Fitz, E.; Pierce, G.N.; Suh, M.; Moghadasian, M.H. (2008). Low n-6:n-3 fatty acid ratio, with fish- or flaxseed oil, in a high fat diet improves plasma lipids and beneficially alters tissue fatty acid composition in mice. *European Journal of Nutrition,* 47, 153-60.

[12] Lourenço, M.; Van Ranst, G.; Vlaeminck, B.; De Smet, S.; Fievez, V. (2008). Influence of different dietary forages on the fatty acid composition of rumen digesta as well as ruminant meat and milk. *Animal Feed Science and Technology*, 145, 418-437.

[13] Sharrow, S.H. (1999). Silvopastoralism: competition and facilitation between trees, livestock and improved grass-clover pastures on temperate rainfed lands. In: L.E. Buck, J.P. Lassoie & E.C.M. Fernandes (Eds), *Agroforestry in Sustainable Agricultural Systems*. Boca Raton, Florida: CRC Press LLC; 416 pp.

[14] Hasler, C. Functional Foods: their role in disease prevention and health promotion. *Food Technology*, (1998) 52, 63-70.

[15] Bradshaw, R.H.W.; Hannon, G.E.; Lister, A.M. (2003). A long-term perspective on ungulate-vegetation interactions. *Forest Ecology and Management*, 181, 267-280.

[16] White, S.L.; Benson, G.A.; Washburn, S.P.; Green, J.T.Jr. (2002). Milk production and economic measures in confinement or pasture systems using seasonally calved Holstein and Jersey cows. *Journal of Dairy Science*, 85, 95-104.

[17] Husak, A.L.; Grado, S.C. (2002). Monetary benefits in a southern silvopastoral system. *Southern Journal of Applied Forestry*, 26, 159-164.

[18] McAdam, J.H. (2004). Silvopastoral systems in North-West Europe. In M.R. Mosquera-Losada, J. McAdam & A. Rigueiro-Rodríguez (Eds), *Silvopastoralism and Sustainable Land Management*. Oxfordshire, UK: CABI; 429 pp.

[19] Grenvold, J.; Summer, C.; Holter, P.; Nansen, P. (1992). Reduced splash dispersal of bovine parasitic nematodes from cow pats by the dung beetle *Diastellipalpus queinquedens*. *The Journal of Parasitology*, 78, 845-848.

[20] Paz-Silva, A.; Arias, M.; Francisco, I.; Cortiñas, F.J.; Francisco, R.; Mochales, E.; Suárez, J.L.; Díez-Baños, P.; Morrondo, P.; Sánchez-Andrade, R. (2010). Cross-immunity and interpretation of the diagnostics of parasitic trematodosis in ruminants by means of immunoenzymatic probes. In Gregory V. LaMann (Ed.), *Veterinary Parasitology*. Hauppauge NY: Novapublishers; 271-288.

[21] Corning, S. (2009). Equine cyathostomins: a review of biology, clinical significance and therapy. *Parasites & Vectors*, Suppl 2, S1.

[22] López-Díaz, M.L.; Rigueiro-Rodríguez, A.; Mosquera-Losada, M.R. (2009). Influence of pasture botanical composition and fertilization treatments on tree growth. *Forest Ecology and Management*, 257, 1363-1372.

[23] Kottek, M.; Grieser, J.; Beck, C.; Rudolf, B.; Rubel, F. (2006). World map of the Köppen–Geiger climate classification updated. *Meteorologische Zeitschrift*, 15, 259-263.

[24] Newport, A. (1993). Fringe benefits. Oklahoma Farmer Stockman, June issue. pp. 6-7.

[25] Lyons, E.T.; Tolliver, S.C.; Collins, S.S. Study (1991 to 2001) of drug-resistant Population B small strongyles in critical tests in horses in Kentucky at the termination of a 40-year investigation. *Parasitology Research*, (2007) 101, 689-701.

[26] Monahan, C.M.; Klei, T.R. (2002). The use of macrocyclic lactones to control parasites of horses. In J. Vercruysse, R.S. Rew (Eds), *Macrocyclic lactones in antiparasitic therapy*. New York, USA: CABI.

[27] Dornbusch, P.T.; Michelotto, P.V.; Santarém, V.A.; De Carli, L.M.; Camargo, C.E.; Biava, J.S. (2006). Eficácia anti-helmíntica da ivermectina "pour-on" comparada com a formulação oral em gel nos eqüinos. *Revista da Academia de Curitiba*, 4, 21-24.

[28] Larsen, M.L.; Ritz, C.; Petersen, S.L.; Nielsen, M.K. (2011). Determination of ivermectin efficacy against cyathostomins and *Parascaris equorum* on horse farms using selective therapy. *Veterinary Journal,* 188, 44-47.

[29] Mendoza de Gives, P.; Zapata Nieto, C.; Hernández, E.L.; Arellano, M.E.; Rodríguez, D.H.; Garduño, R.G. (2006). Biological control of gastrointestinal parasitic nematodes using *Duddingtonia flagrans* in sheep under natural conditions in Mexico. *Annals of the New York Academy of Sciences*, 1081, 355-359.

[30] Casillas-Aguilar, J.A.; Mendoza de Gives, P.; López-Arellano, M.E.; Hernández, L. (2008). Evaluation of multinutritional pellets containing *Duddingtonia flagrans* chlamydospore for the control of ovine haemonchosis. *Annals of the New York Academy of Sciences*, 1149, 161-163.

[31] Francisco, I.; Arias, M.; Cortiñas, F.J.; Francisco, R.; Mochales, E.; Sánchez, J.A.; Suárez, J.L.; Morrondo, P.; Uriarte, J.; Sánchez-Andrade, R.; Díez-Baños, P.; Paz-Silva, A. (2009). Silvopastoralism and autochthonous equine livestock. Analysis of the infection by endoparasites. *Veterinary Parasitology*, 164, 357-362.

[32] Uhlinger, C.A. (2007). Evidence-based parasitology in horses. *Veterinary Clinics North America. Equine Practice*, 23, 509-517.

[33] Francisco, I.; Sánchez, J.A.; Cortiñas, F.J.; Francisco, R.; Mochales, E.; Arias, M.; Mula, P.; Suárez, J.L.; Morrondo, P.; Díez-Baños, P.; Sánchez-Andrade, R.; Paz-Silva, A. (2009). Clinical trial of efficacy of ivermectin pour-on against gastrointestinal parasitic nematodes in silvopasturing horses. *Equine Veterinary Journal*, 41, 713-715.

[34] Francisco, I.; Sánchez, J.A.; Cortiñas, F.J.; Francisco, R.; Suárez, J.; Cazapal, C.; Suárez, J.L.; Arias, M.S.; Morrondo, P.; Sánchez-Andrade, R.; Paz-Silva, A. Efficacy of ivermectin pour-on against nematodes infecting foals on pasture: coprological and biochemical analysis. *Journal of Equine Veterinary Science*, (2011), doi:10.1016/j.jevs.2011.01.012

[35] Paz-Silva, A.; Francisco, I.; Valero-Coss, R.O.; Cortiñas, F.J.; Sánchez, J.A.; Francisco, R.; Arias, M.; Suárez, J.L.; López-Arellano, M.E.; Sánchez-Andrade, R.; Mendoza de Gives, P. (2011). Ability of the fungus *Duddingtonia flagrans* to adapt to the cyathostomin egg-output by spreading chlamydospores. *Veterinary Parasitology*, 179, 277-282.

[36] Lichtenfels, J.R.; Kharchenko, V.A.; Sommer, C.; Ito, M. (1997). Key characters for the microscopical identification of *Cylicocyclus nassatus* and *Cylicocyclus ashworthi* (Nematoda: Cyathostominae) of the horse, *Equus caballus*. *Journal of the Helminthological Society of Washington*, 64, 120-127.

[37] Fraser, M.D.; Davies, D.A.; Vale, J.E.; Nute, G.R.; Hallett, K.G.; Richardson, R.I.; Wright, I.A. (2009). Performance and meat quality of native and continental cross steers grazing improved upland pasture or semi-natural rough grazing. *Livestock Science*, 123, 70-82.

[38] Eriksen, J.; Søegaard, K.; Kristensen, T.; Fretté, X.C.; Wiking, L.; Holm Nielsen, J. (2007). Herbage and milk productivity and quality when grazing different legumes and herbs. *Grassland Science in Europe*, 12, 212-214.

[39] Leiber, F.; Kreuzer, M.; Nigg, D.; Wettstein, H.R.; Scheeder, M.R.L. (2005). A study on the causes for the elevated n-3 fatty acids in cows' milk of alpine origin. *Lipids*, 40, 191-202.

[40] Galina, M.A.; Osnaya, F.; Cuchillo, H.M.; Haenlein, G.F.W. (2007). Cheese quality from milk of grazing or indoor fed Zebu cows and Alpine crossbred goats. *Small Ruminant Research*, 71, 264–272.

[41] Soryal, K.A.; Zeng, S.S.; Min, B.R.; Hart, S.P. (2004). Effect of feeding treatments and lactation stages on composition and organoleptic quality of goat milk Domiati cheese. *Small Ruminant Research*, 52, 109-116.

[42] Moloney, A.P.; Mooney, M.T.; Kerry, J.P.; Troy, D.J. (2001). Producing tender and flavoursome beef with enhanced nutritional characteristics. *Proceedings of the Nutrition Society*, 60, 221-229.

[43] Zlatanos, S.; Laskaridis, K.; Feist, C.; Sagredos, A. (2002). CLA content and fatty acid composition of Greek Feta and hard cheeses. *Food Chemistry*, 78, 471-477.

[44] Priolo, A.; Micol, D.; Agabriel, J.; Prache, S.; Dransfield, E. (2002). Effect of grass or concentrate feeding systems on lamb carcass and meat quality. *Meat Science*, 45, 169-181.

[45] Poulson, C.S.; Dhiman, T.R.; Ure, A.L.; Cornforth, D.; Olson, K.C. (2004). Conjugated linoleic acid content of beef from cattle fed diets containing high grain, CLA, or raised on forages. *Livestock Production Science*, 91, 117-128.

[46] Rotz, C.A.; Soder, K.J.; Skinner, R.H.; Dell, C.J.; Kleinman, P.J.; Schmidt, J.P.; Bryant, R.B. (2009). Grazing can reduce the environmental impact of dairy production systems. *Forage and grazinglands, DOI 10.1094/FG-2009-0916-01-RS.*

[47] Sharpley, A.N.; West, C. (2008). Pressures on beef grazing in mixed production farming. In: R.W. McDowell (Ed), *Environmental Impacts of Pasture-based Farming.* Oxfordshire, UK: CABI; 187-208.

[48] Krecek, R.C.; Guthrie, A.J.; Van Nieuwenhuizen, L.C.; Booth, L.M. A (1994). Comparison between the effects of conventional and selective antiparasitic treatments on nematode parasites of horses from two management schemes. *Journal of the South African Veterinary Association*, 65, 97-100.

[49] Gómez-Rincón, C.; Uriarte, J.; Valderrábano, J. (2006). Efficiency of *Duddingtonia flagrans* against *Trichostrongyle* infections of sheep on mountain pastures. *Veterinary Parasitology*, 141, 84-90.

[50] Llerandi-Juárez, R.D.; Mendoza de Gives, P. (1998). Resistance of chlamydospores of nematophagous fungi to the digestive processes of sheep. *Journal of Helminthology*, 72, 209-213.

[51] Baudena, M.; Chapman, M.; French, D.; Klei, T. (2000). Seasonal development and survival of equine cyathostome larvae on pasture in south Louisiana. *Veterinary Parasitology*, 88, 51-60.

[52] Braga, F.R.; Araújo, J.V.; Silva, A.R.; Araújo, J.M.; Carvalho, R.O.; Tavela, A.O.; Campos, A.K.; Carvalho, G.R. (2009). Biological control of horse cyathostomin (Nematoda: Cyathostominae) using the nematophagous fungus *Duddingtonia flagrans* in tropical southeastern Brazil. *Veterinary Parasitology*, 163, 335-340.

[53] Gray, D. (1999). Can parasitic gastro-enteritis be used as an indicator of welfare in organic sheep? UK Organic Research 2002: *Proceedings of the COR conference*, Aberystwyth. Pages 347-350.

[54] French, P.; Stanton, C.; Lawless, F.; O'Riordan, E.G.; Monahan, F.J.; Caffrey, P.J.; Moloney, A.P. (2000). Fatty acid composition, including conjugated linoleic acid, of intramuscular fat from steers offered grazed grass, grass silage, or concentrate-based diets. *Journal of Animal Science*, 78, 2849-2855.

[55] Boorton, J.; Foster, J. (2002). Alpha-tocopherol concentration and case life of lamb muscle as influenced by concentrate or pasture finishing. *Journal of Animal Science*, 80, 2513-2521.

In: Fungi: Types, Environmental Impact and Role in Disease ISBN: 978-1-61942-671-9
Editors: Adolfo Paz Silva and María Sol

Chapter XVII

Nematophagous Fungi as Control Agents of Gastrointestinal Nematodes in Small Ruminants

C. Gómez-Rincón[1*], *J. Valderrábano*[2] *and J. Uriarte*[2]
[1] Health Sciences Faculty, San Jorge University, Zaragoza, Spain
[2] Animal Health Department,
Agrifood Research and Technology Centre of Aragon (CITA)
Zaragoza, Spain

Abstract

The gastrointestinal nematodes infections are the most frequent pathological processes in the small ruminant production systems [1] and are considered to be the mainspring productive loss in small ruminants production systems based on pasture globally [2,3]. For this reason, the implementation of control methods to support the parasitic population inside a level compatible with the economic production has been consolidated as a routine practice in the managing of livestock productions.

During the last fifty years the control of the parasitic diseases and concretely the gastrointestinal nematodes, it has been based almost exclusively on the systematic use of antiparasitic drugs. Nevertheless as consequence of the improper or inadequate use of the chemical products, the anthelmintic resistance has stopped being an academic curiosity to turn into the principal threat for the developments of small ruminants throughout the world [4]. The increasing prevalence and severity of anthelmintic resistance together with increased market demand for ecological or green animal products, has led to the search for alternative methods of parasite control. The new methods of gastrointestinal nematodes control should minimize the use of drugs allow to obtain quality products, guarantee the animal welfare and reduce the environment residue, contributing to the sustainability of the livestock systems.

* E-mail address: cgomez@usj.es.

Biological Control

The utilization of natural antagonists of a not wished species, it has appeared as an effective method in the control of certain agriculture plagues [5,6]. Nevertheless, in the development of strategies of biological control it is necessary bear in mind, that the incorporation of a foreign species or the increase of natural pre-existing population, can alter the ecosystem balance and provoke unwanted effects. For this reason, the biological agents must support a highly specific relation with the species against the one that tries to use, not damaging to other organisms of the way. On the other hand, it turns out to be fundamental to understand the biological control as a regulatory measure which aim is not to finish with the organism reveille but the population controls to reduce his harmful effects [7].

The first approximations to the biological control of livestock parasites was carried out by Roaubaund and Daschies to half of last century [7]. Nevertheless, it is not up to the last decade of the mentioned century and as consequence of the need of looking for alternative methods of control that were limiting the expansion of the phenomena of resistance to the medicaments, when the biological control of the gastrointestinal nematodes has experienced a notable summit and the search of the biological "ideal" agent has been a motive of numerous works. In the prospective studies several coprophagous arthropods, some bacteria and fungi have been tested. A wide number of species of coprophagous arthropods show a relatively high efficiency against to free living stages of gastrointestinal nematodes, nevertheless, his applicability as control strategy turns out to be limited as consequence of his high seasonality variation [8,9]. On the other hand, there are several strain of *Bacillus thurigiensis* that produce nematodicidas substance that shows high efficiency under in-vitro conditions, [10]. Though, the use of these bacteria stains as biological control agents has practical limitations. Nevertheless, recent works suggest the usefulness of this bacterial origin compounds in the production of new active molecules with nematodicidas properties [11,12,13].

Finally between all the agents tested, the most promising are a small group of hyphomycetos fungi capable of catching and digesting the free living stages of soil nematodes, acquaintances as nematophagous fungi [14].

Nematophagous Fungi

The nematophagous fungus constitutes a heterogeneous and ubiquitous group of soil microfungi. Normally this fungus growth saprophytically but used free living stages of soil nematodes as secondary resource [15,16]. More than 200 species have been described, being the majority of they are included in the imperfect fungi called deuteromycetes. Depending on the strategy used for the utilization of nematodes as nutritional resource, are divides in two groups: endoparasitic and nematode – trapping fungi.

Endoparasitic fungi

These fungi are obligate parasites that invade nematode larvae and eggs from adhesive spores that stick on the cuticle, from spores that are ingested by the nematode. Once inside the nematode, the spores germinates and give rise to an infectious thalus that absorbs the nematode body content [17]. The development of hyphaes occurs only inside the nematode or his eggs, with the exception of reproductive hypha that grow outside where they release the spores [15]. Though in the majority of the species, the spores must be consumed to penetrate inside the nematode, some species as *Catenaria anguillulae*, show mobile zoospores that seem to have a positive tropism towards the nematodes, thanks to which they accumulate around to the natural orifices of the cuticle facilitating the penetration process (Figure 1).

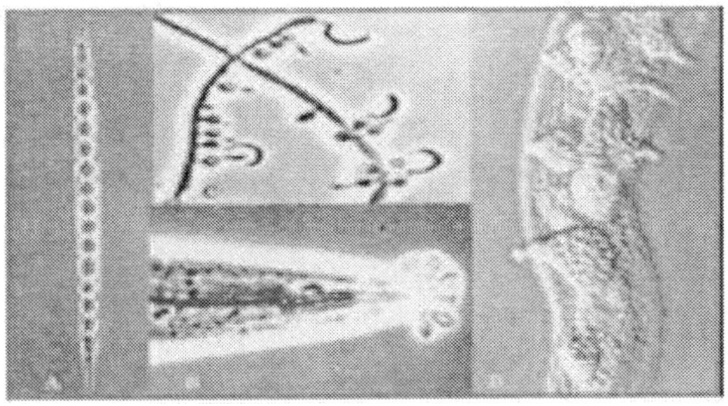

Figure 1. Zoosporas of Myzocytium spp. y Catenaria anguillulae inside of parasited nematode (A y D); Harposporium anguillulae (C) y Drenchmeria conidiospora (E) conidiophores y conidiospores. Mycologie Publications 2001.

The studies realized inside the group of endoparasitic fungi are scanty and the majority are focused on two species: *Drenchmeria coniospora y Harposporium anguillulae* (Figure 1E y 1C). The results obtained with both species show that there is necessary a highest number of spores of *Drenchmeria coniospora* (108/g of dregs) in order to have a significant effect against to *Haemonchus contortus* [18], whereas *Harposporium anguillulae* shows a marked efficiency to lower doses (3x105/g) [19]. Also between the parasitic fungi of eggs, some species like *Verticillium chlamydosporum* or *Paecilomyces lilacinus* have showed moderate *in-vitro* efficiency against to *Ascaris lumbricoides, Toxocara cannis* or *Nematodirus* spp. [20,21,22]. Nevertheless in spite of these results, the obliged parasites character of this fungi group limits his employment as biological control agents. This limitation is owed to the fact that his spread is realized only by means of direct contact between parasitized individuals, for this reason his isolation, culture on a large scale and practical application turn out to be extremely complicated.

Nematophagous fungi

The nematode-destroying fungi belong to a heterogeneous group of microfungi that utilize nematodes either as the main source of nutrients, or supplementary to a saprophytic existence [15]. These fungi are found worldwide in many different habitats, but are especially frequent in organically rich environments such as compost and faeces. Some of the most important generous inside this group are *Arthrobortrys*, *Dactylaria, Dactylella* and *Monacrosporium* [17]. These fungi produce several hyphal trapping devices, such as adhesive networks, knobs and constricting or non-constricting rings. Across these systems, these fungi are capable of catching and consuming the free living nematode stages of soil and faeces.

A specially abundant group in the nature, uses three-dimensional adhesive networks as strategy of capture. The training of three-dimensional adhesive networks is a process of a mycelial anastomosis. It begins with a lateral branch formation in the vegetative hyphae that grows and bows up to fusing with the mother hyphae. The process repetition gives place to multiple rings, which form the three-dimensional networks.

The networks are composed for more robust hyphae, whose cells present cellular walls 4-5 times thicker than those of the vegetative hyphea. The whole network is covered by a mucilaginous layer to which the nematodes remain adhered and which production seems to be stimulated by the struggle of the caught nematode [23]. This step may involve an interaction between a carbohydrate-binding protein (lectin) in the fungus and a carbohydrate receptor on the nematode. Recognition of the host is probably also important for the subsequent steps of the infection, including penetration of the nematode cuticle [24,25]. Finally, there exist evidences of the production of fungal substances that attract the nematodes towards the networks [26].

As soon as the nematode is captured, the fungus penetrates his cuticle across a mixed enzymatical - mechanical mechanism [27,28]. After the capture there are liberated of hydrolytic enzymes as proteases, chitinases and collagenase that digest nematode cuticle [29]. After the penetration an infective bulb is formed inside of nematode body [30] simultaneously, the movement of the nematode stops completely as consequence of an unknown nematodicidal substance [31,32]. From this infective bulb, the trophic hyphae is formed and grows quickly up to occupying the whole body in a few hours. Simultaneously that will digest the corporal content. After the digestion, the trophic hyphae is lysed and the fungi develops saprophytically again until the presence of nematodes in soil stimulates his predatory activity [23] (Figure 2).

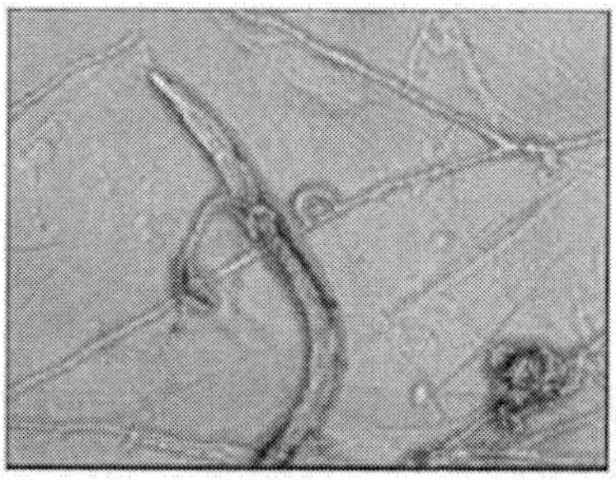

Figure 2. Three living stages of *Haemonchus contortus* nematode caught in a three-dimensional adhesive networks of *Duddintonia flagrans.*

There are several factors that concern the formation of the networks, so that it is indispensable that the fungi are stimulated by physicochemical or biological factors in order that the process of network formation begins.

Some in-vitro studies have showed that certain nematode substance can induce the morphogenetic response [33]. Likewise it has been suggested that the electrical potential generated by the nematode muscular contraction, it produces an alteration of the hyphal membrane potential that modifies his permeability to ions and peptides, unleashing the process that will give place to the formation of the networks [34]. It can be for this reason that the alive nematodes induce more rapidly the trapping network formation in-vitro that his peptides or extracts [34]. In addition, the more mobiles species of nematode are inducting more efficient [35,36] and the network formation process is highly dependent on the larval density [37,38] what might corroborate this hypothesis.

Finally, a delicate balance exists between mycelial growth and the formation of highly dependent traps on the environmental conditions [39]. Nevertheless, the ubiquity of this group does that the requirements of temperature and humidity change notably between species.

Fungal Species with Greater Potential for Biological Control

In the search of nematophagous fungi for the control of ruminants gastrointestinal nematodes, the majority of the works have been focused on predatory species since his saprophytic character allows his isolation and maintenance in pure cultures facilitating his industrial production. These fungus have been isolated in great variety of habitats specially in rich means in organic matter as compost or faeces of the pastures, which are rapidly colonized by different species [40]. On the other hand, the isolation of different fungal species from faeces excreted by grazing animals, indicates that in natural conditions they are consumed and eliminated in the faeces [41,42]. This feature makes them excellent candidates for biological control against gastrointestinal nematodes, because it is imperative that the selected species to survive passage through the digestive tract of animals and germinate in the faeces keeping intact its predatory capacity. Surveys carried out show that few species of nematophagous fungi survive the gastrointestinal tract of ruminants [43,44]. Larsen et al., (1991 and 1992) [43,45] after subjecting to 55 strains of fungi isolated from nematophagous soil and compost in-vitro conditions simulating the gastrointestinal tract, found that only two strains of genus *Arthrobotrys* and six *Duddingtonia* survive digestive juices.

Arthrobotrys oligospora

The first candidate biological control agent was *Arthrobotrys oligospora* since previous work had shown that direct inoculation of fungal conidia or mycelium in bovine faeces reduced larval evolution in cultures by 70-90%, the L3 pasture contamination by 50% for 89 and the worm burden of calves that grazed treated plots up to 37% [37,46,47,48]. However, an in vivo test, when the fungus was consumed by cattle, no significant effects on infective population of pasture was showed. Thus, it appeared that much of the fungal material was destroyed during its passage through the digestive tract [49]. For this reason the use of *A. oligospora*, at least for the strains tested to date is considered inadequate as only the ingestion of fungal material ensures direct contact fungus-parasite in the faeces.

Duddingtonia flagrans

Studies *"in vivo"* analogous to those above, showed that two strains of *Duddingtonia flagrans* were very effective after its passage through the gastrointestinal tract. These strains fell between 74-85% larval migration between faeces and grass, making it a strong candidate as a biological control agent for this reason many studies have been conducted with this species.

In recent years many authors have shown that *D. flagrans* is able to pass through the digestive tract of cattle [50,51,52], sheep [53,54,55,56], goats [57,58], pigs [59] and horses [60] and germinate in the faeces maintaining its predatory capacity. This ability is due to the production of a large amount of chlamydospores, which unlike the spores of A. oligospora are surrounded by a thick wall, which confers resistance to external conditions [14]. Recent studies show that while 90% of spores administered in sheep do not survive the digestive tract [61], once in the faeces, viable chlamydospores germinate in response to the presence of nematodes, forming three-dimensional sticky networks that trap the developing parasite larval stages. The trapping mechanism reduces the migration of infective stages of the grass and therefore the risk of animal infection.

On the other hand, the environmental requirements for the development of *D. flagrans* are undemanding in humidity and temperature [62,63]. Thus their development thermal range is between 10 and 30 °C, being the optimum range for the formation of three-dimensional networks of 25 to 28 °C. These conditions, match those required for the development of eggs and larvae of most species of gastrointestinal nematodes [64,65], which a priori, may ensure their effectiveness during periods of major epidemiological risk.

Because of the potential of *D. flagrans* as biological control agents have been numerous studies "*in vitro*" and "*in vivo*" to evaluate their effectiveness against different species of gastrointestinal nematodes. *D. flagrans* has an average efficiency of 85% in vitro against major trichostrongyle species of ruminants [43,66,67,68,69].

The tests *"in vivo"* have been carried out mostly in cold climates, mainly in northern European countries where parasite development is conditioned by heavy rains and low temperatures. Under these conditions, studies in several domestic species have shown that daily doses between 0.5 and 1 million chlamydospores of *D. flagrans* per kg of live weight (kg PV) during the grazing reduced larval evolution in the faeces and in the pasture at values ranging between 66% and 90%, reducing levels of infection and its pathogenic effects in cattle [51,70], sheep [44,53], pigs [59] and horses [71]. More recently in southern Europe, has also shown that daily application of chlamydospores 5x105 / kg PV during grazing, reduce pollution by 72% of the grass and up to 61% the parasite burden in both sheep [72] and in goats [73], improving productivity. Similar results have been reported in Australia [74] and New Zealand [58] in goats on pasture, with reductions of 44% in egg excretion and 53% in the parasite load after *D. flagrans*. These results suggest that this fungus is an effective tool in controlling gastrointestinal nematodes in small ruminants. Finally, other studies have demonstrated its efficacy against gastrointestinal nematodes of major small ruminants in tropical and subtropical conditions [76,77,78,79].

Another important aspect of *D. flagrans* in his career to become an agent of biological control of gastrointestinal nematodes of cattle, is its safety for the environment. *D. flagrans* is a bit selective predator so its use could result in the reduction of certain species of soil nematodes, which are involved in the recycling of organic matter contributing to the maintenance of fertility. However, the work of Yeates et al., (1997, 2002 and 2003) [80,81,82] have shown that treatment of cattle causes no change in size or diversity in the fauna of soil nematodes, since outside of the stool is a saprophytic poor competitor [54]. In a larger study Knox et al., (2002) [83] have shown that in addition to not disturb the soil nematode populations, treatment with *D. flagrans* in sheep does not produce negative effects on insects, arthropods, nor on other species of nematophagous fungi. In the same vein Paraud et al., (2011) [84] showed that the presence of *D. flagrans* in goat faeces does not change the rate of degradation of the same or produce significant effects on populations of free-living nematodes.

On the other hand, its ubiquitous nature combined with the low genetic variability found between strains from distant geographic areas, indicates that there may be a single clonal population of worldwide distribution [85]. This, minimize the environmental impact from the use of non-native strain in case of not having an indigenous strain that can be produced on a large scale.

Strategic Use of *Duddingtonia flagrans* in Spanish Sheep and Goats Production Systems

The establishment of biological control of gastrointestinal nematodes in livestock production systems must be based on deep understanding of the epidemiology of infection in the specific area where you want to implement. Likewise, it is essential to conduct regular coprological analysis to know the levels of egg excretion from the flock and to determine the right time to implement control measures.

The epidemiology of gastrointestinal nematode infection is determined by the interaction between parasites, climatic conditions and production system of animals. The climatic conditions and particularly temperature and moisture, regulate the development, migration and survival of infective stages of parasites. The production system and especially the reproductive system and grazing management, establish the level of contamination of grass and the possibility that the ingestion of infective forms.

The epidemiological characteristics and the periods of greatest risk of infection within each system must be taken into account in order to establish the most appropriate time to apply fungal treatments to livestock. As an example, explain the epidemiological models of production systems prevalent in Spain (non-irrigated, irrigated and upland systems) which have three critical points that promote the development cycle of the parasites:

- The overwintering larvae (larvae that survive winter in the grass, maintaining their infectivity until spring) are often regarded as a minor source of infection in areas where climatic conditions require delaying the start of the season grazing. However, in Mediterranean conditions, the overwintering larvae have an important epidemiological significance in the resumption of the spring pasture contamination.
- The increase in the excretion of parasite eggs in faeces of sheep around parturition (periparturient rise) is the main source of the high contamination of the pasture.
- The high proportion of inhibited larvae that have animals in both summer and winter contributes significantly to the resumption of pasture contamination at the next grazing season.

The recommendations made here derive from the European project WORMCOPS whose main objective is not the elimination of animals worm burden but minimize pasture contamination. Therefore, although routine deworming should be avoided, the use of antiparasitic drugs remains part of the control of animals nematodes including organic farms.

Non-Irrigated Systems

Under dry conditions, usually lambing take place in spring when grass production is sufficient to meet the needs of milk production in sheep. The lambs are finished in the fold, regardless of productive purpose (meat or milk). In winter, when there is a shortfall in the availability of green fodder, the animals have to be supplemented with excess spring fodder, while in summer cereal stubble are the main nutritional resource for animals.

In dry areas of Spain the development, migration and survival of free living stages of gastrointestinal nematodes are conditioned mainly by the humidity that the temperature remains practically throughout the year, above the minimum threshold for development [86]. In these areas, the risk of infection coincides with periods of water availability, extending from October to late spring of following year [87,88]. In these systems, animals acquire the most important infections in the middle of spring and late fall [89]. The lower risk of infection occurs when animals are eating cereal stubble in summer. At this time, however, there is a high proportion of *Teladorsagia circumcincta* larvae inhibited (> 80% from July to September) to be held in the autumn taking part in the annual resumption of pasture contamination. The increased excretion of eggs in the faeces takes place in April coinciding with the lambing period (Figure 3). In these circumstances, the application of *D. flagrans* would be limited to the spring when levels of pasture contamination are highest, as long as the coprologic analysis of flock show values upper 250 epg.

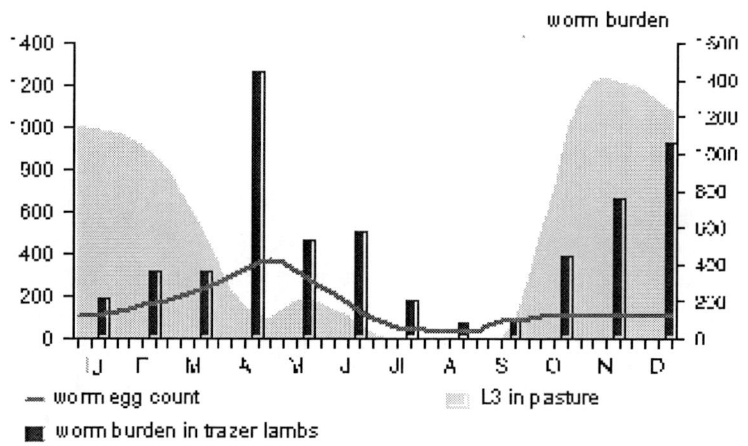

Figure 3. Epidemiological pattern of gastrointestinal nematodes in dry areas (Llorente, 1999).

In any case we must take into account that the system itself with one or no annual deworming involves the rational use of anthelmintics which a priori would be a low risk of resistance. Therefore, the need to implement strategies for biological control of gastrointestinal nematodes in dry conditions may be questionable.

Irrigated Farms

The use of irrigated land or those with high annual rainfall for forage production is an alternative for the development of ruminant production systems in Mediterranean areas. The sheep and goat production in these areas is based on a high utilization of pastures for most of the year. The integration of forage crops, whose production is highest in autumn - winter, in the farm system is often used to obviate the seasonal shortage of pasture. The profitability of these systems of production to other agricultural alternatives is only possible with high grazing pressure and grazing short intervals. These management conditions strongly favour the gastrointestinal parasite infection [88], which is the main limiting factor for the development of intensive production systems (3 births / 2 years) of sheep grazing [90].

The epidemiological model (Figure 4) is characterized by three periods of greatest risk of infection per year [91]. The first extends from February to April and is due to overwintering larvae population. The second takes place between June and July, comes from the eggs deposited during the periparturient period of spring and the reactivation of the inhibited population that consists almost exclusively of *Haemonchus contortus*. Finally, the third occurs between October and November and is caused mostly by the periparturient rise phenomenon in fall. However, the use of forage in winter means that only the last two periods constitute a real risk of infection, both the administration of chlamydospores of *D. flagrans* in those periods could contribute decisively to control the infection. In this regard, the studies of Gómez-Rincón et al., (2006 and 2007) [72,73] have shown that the application of *D. flagrans* during the fall and spring grazing significantly reduced both the population of overwintering larvae as those derived from periparturient rise and allow keeping levels of animal infection consistent with the economical production.

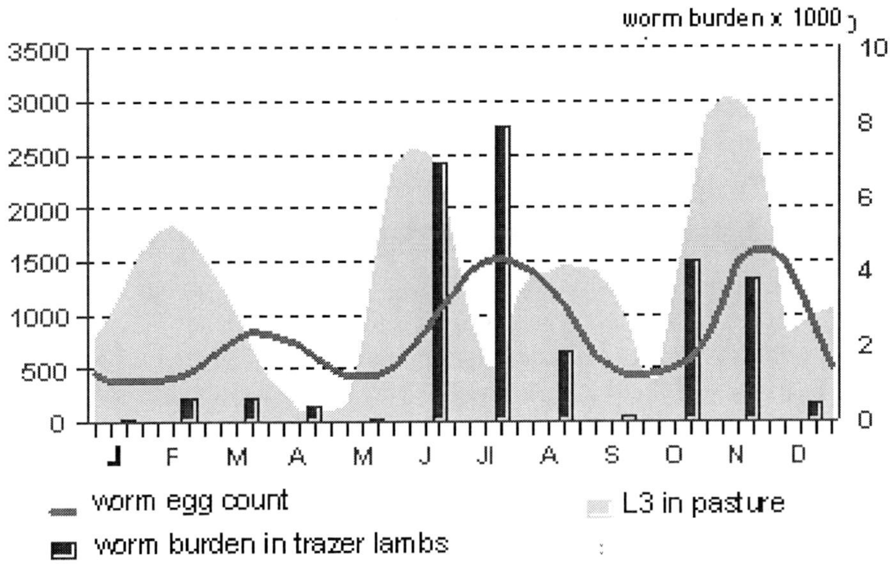

Figure 4. Epidemiological pattern of gastrointestinal nematodes in irrigated areas, three lambing /2 years (Uriarte et al., 2003).

Extensive Mountain Systems

In extensive mountain systems, lambing usually take place in early spring. At this time, the sheep and their lambs have access to high quality pastures in areas close to wining zone at the bottom of the valley. With the arrival of summer, the flock are moved to the high pasture (June-October), returning to the valley pastures in autumn, where they remain until the arrival of snow forces the winter housing. In these systems the first infection occurs at the exit to the pasture in spring due intake of the overwintering larvae population. The second wave of infection takes place from mid May until the animal movement to the high pastures and is due to the larvae derived from periparturient rise of spring (Figure 5).

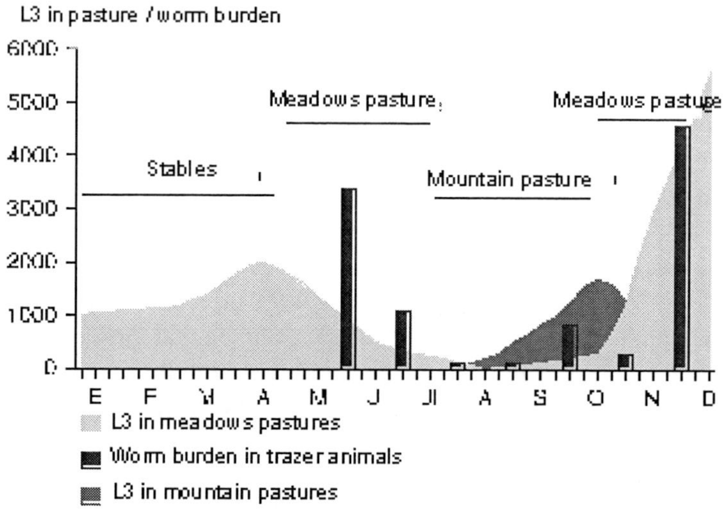

Figure 5. Epidemiological pattern of gastrointestinal nematodes in mountain areas.

In the fall there is another wave of infection with important epidemiological consequences as part of the ingested larvae remain inhibited until March of the next year, playing a definitive role in increasing pasture contamination at the beginning of spring and contaminate pastures elsewhere prior to housing. This population will constitute the future overwintering population. Under these conditions, the control of gastrointestinal nematodes should be directed to actions that minimize contamination of the pasture in early spring and the larval population derived from peripartum. Under this hypothesis the work of Gómez-Rincón et al. (2006) [72] showed that daily administration of *D. flagrans* in sheep during autumn and spring, significantly reduces both waves of infection and therefore decreases by 61% the worm burden of animals at the end of spring.

Limitations to Incorporate *Duddingtonia flagrans* as Biological Control Agent in Livestock Systems

Currently there are two limitations for the incorporation of *D. flagrans* as a measure of control of gastrointestinal nematodes in small ruminants systems. One is that it is still under investigation in small ruminants and therefore has not obtained license for marketing, not true for cattle, for which will be marketed soon. The other limitation for small ruminants is its difficulty of administration, it is necessary to ensure a daily dose for more or less longer periods of time to get a significant effect. At the moment the administration is performed by mixing the daily dose of spores necessary with small amounts of food in the form of flour, to facilitate mixing. This procedure is cumbersome as well as extensive systems impractical where animals are housed. To solve this problem are being used, successfully, the chlamydospores inclusion in mineral and salt blocks [55] pellets of sodium alginate [92] and slow release devices [55,56].

General Considerations

Returning to the essence of biological control, *D. flagrans* should be used as a prophylactic measure since their action is limited to free-living stages. For this reason must be combined with other strategies that act on the parasitic life stages. In this sense, since the short-term appearance in the market for a tool as effective, fast and economical as broad-spectrum anthelmintics seems unlikely, the integrated control scheme is more suitable for fighting gastrointestinal nematodes. Consequently new methods of control should try to combine different strategies framed in the context of sustainability and working at different levels in the parasite-host interaction, generally allows effective control of parasites. The use of biological control agents and in particular *D. flagrans* fits perfectly into this context as it can be combined with any non-pharmacological strategies are being valued today as vaccination, genetic resistance herbal or even drugs with the use of anthelmintics.

References

[1] Gray, G.D. (1991). Breeding for resistance to Trichostrongylus nematodes in sheep. Owen, J. B. y Axford, R. F. E. (Eds.), *Breeding for disease resistance in farms animals.* CAB International, Wallingfor, 139-161.

[2] Holmes, P.H. (1993). Interactions between parasites and animal production: the veterinary consequences. *Proc Nutr Soc*, 52: 113-120.

[3] Prichard, R. (1994). Anthelmintic resistance. *Vet Parasitol*, 54: 249-268.

[4] Waller, P.J. (1997). Anthelmintic resistance. *Vet Parasitol*, 72: 391-412.

[5] Albajes, R. & Alomar, O. (1999). Current and potencial use of polyphagous predators. En: R. Albajes, M. L.; Gullino, J. C.; van Lenteren, Y. (Eds), *Integrated Pest and Disease Management in Greenhouse Crops*. Kluwer Academic Publishers, Dordrecht, 265-275.

[6] Wratten, S., Lavandero, B., Scarratt, S. & Vattala, D. (2003). Conservation biological control of insect pests at the landscape scale. En: Rossing, W. A. H., Poehling, H. M., Burgio, G., (Eds). Bulletin OILB/SROP, 26: 215-220.

[7] Larsen, M. (2000). Prospects for controlling animal parasitic nematodes by predacious micro fungi. *Parasitology*, 120: 121-131.

[8] Fincher, G.T. (1973). Dung beetles as biological control agents for gastrointestinal parasites of livestock. *J Parasitol,* 59: 396-399.

[9] Bryan, R.P. & Kerr, J.D. (1989). Factors affecting the survival and migration of the free-living stages of gastrointestinal nematode parasites of cattle in central Queensland. *Vet Parasitol*, 30: 315-326.

[10] Kotze, A.C., O'Grady, J., Gough, J.M., Pearson, R., Bagnall, N.H., Kemp, D.H. & Akhurst, R.J. (2005). Toxicity of *Bacillus thuringiensis* to parasitic and free-living life-stages of nematode parasites of livestock. *Int J Parasitol*, 35: 1013-22.

[11] Bottjer, K.P., Klesius, P.H. & Bone, L.W. (1985). Effects of host serum on feeding by *Trichostrongylus colubriformis* (nematoda). *Parasite Immunol*, 7: 1-9.

[12] Meadows, J., Gill, S.S. & Bone, W.l. (1989). Factors influencing lethality of Bacillus thuringiensis kurstaki toxin for eggs and larvae of *Trichostrongylus colubriformis*. *J Parasitol*, 75: 191-194.

[13] Hu, Y., Georghiou, S.B., Kelleher, A.J. & Aroian, R.V. (2010). *Bacillus thuringiensis* Cry5B protein is highly efficacious as a single-dose therapy against an intestinal roundworm infection in mice. *PLoS Negl Trop Dis*, 4 (3): 614.

[14] Larsen, M. (1999). Biological control of helminths. *Int J Parasitol*, 29: 139-146.

[15] Barron, G.L. (1977). The nematode-destroying fungi. *Canadian Biological Publication*, Guelph.

[16] Nordbring-Hertz, B. & Jansson, H.B. (2002). Nematophagous Fungi. *Encyclopedia of Life Science (c)*. Macmillan Publishers ltd, Nature Publishing Group www.els.net

[17] Mankau, R. (1980). Biocontrol: Fungi as Nematode Control Agent. *J Nemathol*, 12: 244-252.

[18] Santos, C.P. & Charles, T.P. (1995). Efeito da aplicacao de conidios de *Drechmeria coniospora* em cultivos de fezes contendo ovos de *Haemonchus contortus*. *Arquivo Brasileiro de Medicina Veterinaria et Zootecnia*, 47: 123-128.

[19] Charles, T.P., Roque, M.Y. & Santos, C.D. (1996). Reduction of *Haemonchus contortus* infective larvae by *Harposporium anguillulae* in sheep faecal cultures. *Int J Parasitol*, 26: 509-510.

[20] Lysek, H. & Sterba, J. (1991). Colonization of *Ascaris lumbricoides* eggs by the fungus *Verticillium chlamydosporium* Goddard. *Folia Parasitol*, 38: 255-259.

[21] Araujo, J.V., Santos, M.A. & Ferraz, S. (1995). Efeito ovicida de fungos nematófagos sobre ovos embrionados de *Toxocara canis*. *Arquivo Brasileiro de Medicina Veterinaria et Zootecnia*, 47:37-42.

[22] Carvalho, R.O., Araujo, J.V., Braga, F.R., Araujo, J.M. & Alves, C.D. (2010). Ovicidal activity of *Pochonia chlamydosporia* and *Paecilomyces lilacinus* on *Toxocara canis* eggs. *Vet Parasitol*, 169: 123-7.

[23] Nordbring-Hertz, B. & Stalhammer-Carlemalm, M. (1978). Capture of nematodes by *Arthrobotrys oligospora*, an electron microscope study. *Canad J Bot*, 56: 1297-1307.

[24] Nordbring-Hertz, B. (1979). Action of nematode-trapping fungus shows lectin-mediated host microorganism interaction. *Nature*, 281: 477-479.

[25] Borrebaeck, C.A.K., Mattiasson, B. & Nordbring-Hertz, B. (1984). Isolation and partial characterization of a carbohydrate-binding protein from a nematode-trapping fungus. *J Bacteriol*, 159: 53-56.

[26] Jansson, H.B. & Nordbring-Hertz, B. (1980). Interaction betweens nematophagous fungi and plant-parasitic nematodes: Attraction, induction of trap formation and capture. *Nematologica,* 26: 383-389.

[27] Barron, G.L. (1981). Predators and parasites of microscopic animals. In G.T. Cole & K. Kendrick (Eds.), *Biology of conidial fungi*, (Vol 2, pp. 167-200). New York, Academia Press.

[28] Liang, L., Liu, S., Yang, J., Meng, Z., Lei, L. & Zhang, K. (2011). Comparison of homology models and crystal structures of cuticle-degrading proteases from nematophagous fungi: structural basis of nematicidal activity. *Faseb J*, 25: 1894-902.

[29] Yang, J.K., Huang, X.W., Tian, B.Y., Sun, H., Duan, J.X. & Wu, W.P. (2005). Characteriza-tion of an extracellular serine protease gene from the nematophagous fungus *Lecanicillium psalliotae*. *Biotechnol Lett*, 27: 1329-34.

[30] Shepherd, A. (1955). Formation of the infection bulb in *Arthrobotrys oligospora*. *Nature*, 175: 475.

[31] Olthof, T.H.A. & Estey, R.H. (1963). A nematoxin produced by the nematophagous fungus *Arthrobotrys oligospora* (Fresenius). *Nature*, 197: 514-515.

[32] Balan, J., Krizkova, L., Nemec, P. & Vollek, V. (1974). Production of nematode-attracting and nematicidal substances by predacious fungi. *Folia Microbiol,* 19: 512-519.

[33] Nordbring-Hertz, B. (1974). Qualitative characterization of some peptides inducing morphogenesis in the nematode-trapping fungus Arthrobotrys oligospora. *Physiol Plantarum*, 31: 59-63.

[34] Nordbring-Hertz, B. (1977). Nematode-induced morphogenesis in the predaciuous fungus *Arthrobotrys oligospora*. *Nematologica*, 23: 443-451.

[35] Gronvold, J., Wolstrup, J., Nansen, P. & Henriksen, S.A. (1993). Nematode-trapping fungi against parasitic cattle nematodes. *Parasitol Today*, 9: 137-140.

[36] Nansen, P., Gronvold, J., Henriksen, S.A. & Wolstrup, J. (1988). Interactions between the predacious fungus Arthrobotrys oligospora and third-stage larvae of a series of animal-parasitic nematodes. *Vet Parasitol*, 26: 329-337.

[37] Gronvold, J., Henriksen, S.A., Nansen, P., Wolstrup, J. & Thylin, J. (1989). Attempts to control infection with *Ostertagia ostertagi* (*Trichostrongylidae*) in grazing calves by adding mycelium of the nematode-trapping fungus *Arthrobotrys oligospora* (Hyphomycetales) to cow pats. *J Helminthol*, 63: 115-126.

[38] Morgan, M., Behnke, J.M., Lucas, J.A. & Peberdy, J.F. (1997). In vitro assessment of the influence of nutrition, temperature and larval density on trapping of the infective larvae of *Heligmosomoides polygyrus* by *Arthrobotrys oligospora*, *Duddingtonia flagrans* and *Monacrosporium megalosporum*. *Parasitology*, 115: 303-310.

[39] Scholler, M. & Rubner, A. (1994). Predacious activity of the nematode-destroying fungus *Arthrobotrys oligospora* in dependence of the medium composition. *Microbiol Res*, 149: 145-149.

[40] Hay, F.S., Niezen, J.H., Leathwick, D. & Skipp, R.A. (2002). Nematophagous fungi in pasture: colonisation of sheep faeces and their potential for control of free-living stages of gastro-intestinal nematode parasites of sheep. *Aus J Exp Agric*, 42: 7-13.

[41] Larsen, M., Faedo, M. & Waller, P.J. (1994). The potential of nematophagous fungi to control the free-living stages of nematode parasites of sheep: survey for the presence of fungi in fresh faeces of grazing livestock in Australia. *Vet Parasitol*, 53: 275-281.

[42] Manueli, P.R., Waller, P.J., Faedo, M. & Mahommed, F. (1999). Biological control of nematode parasites of livestock in Fiji: screening of fresh dung of small ruminants for the presence of nematophagous fungi. *Vet Parasitol*, 81: 39-45.

[43] Larsen, M., Wolstrup, J., Henriksen, S.A., Dackman, C., Gronvold, J. & Nansen, P. (1991). In vitro stress selection of nematophagous fungi for biocontrol of parasitic nematodes in ruminants. *J. Helminthol*, 65: 193-200.

[44] Waller, P.J., Larsen, M., Faedo, M. & Hennessy, D.R. (1994). The potential of nematophagous fungi to control the free-living stages of nematode parasites of sheep: in vitro and in vivo studies. *Vet Parasitol,* 51: 289-299.

[45] Larsen, M., Wolstrup, J., Henriksen, S.A., Gronvold, J. & Nansen, P. (1992). In vivo passage through calves of nematophagous fungi selected for biocontrol of parasitic nematodes. *J Helminthol*, 66: 137-141.

[46] Gronvold, J., Korsholm, H., Wolstrup, J., Nansen, P. & Henriksen, S.A. (1985). Laboratory experiments to evaluate the ability of *Arthrobotrys oligospora* to destroy infective larvae of *Cooperia* species, and to investigate the effect of physical factors on the growth of fungus. *J Helminthol*, 59: 119-125.

[47] Gronvold, J., Wolstrup, J., Henriksen, S.A. & Nansen, P. (1987). Field experiments on the ability of *Arthrobotrys oligospora* (Hyphomycetales) to reduce the number of larvae of *Cooperia oncophora* (*Trichostrongylidae*) in cow pats and surrounding grass. *J Helminthol*, 61: 65-71.

[48] Gronvold, J., Nansen, P., Henriksen, S.A., Thylin, J. & Wolstrup, J. (1988). The capability of the predacious fungus *Arthrobotrys oligospora* (Hyphomycetales) to reduce numbers of infective larvae of *Ostertagia ostertagi* (*Trichostrongylidae*) in cow pats and herbage during the grazing season in Denmark. *J Helminthol*, 62: 271-280.

[49] Gronvold, J., Wolstrup, J., Larsen, M., Henriksen, S.A. & Nansen, P. (1993). Biological control of *Ostertagia ostertagi* by feeding selected nematode-trapping fungi to calves. *J Helminthol*, 67: 31-36.

[50] Larsen, M., Nansen, P., Wolstrup, J., Gronvold, J., Henriksen, S.A. & Zorn, A. (1995) Biological control of trichostrongyles in calves by the fungus *Duddingtonia flagrans* fed to animals under natural grazing conditions. *Vet Parasitol,* 60: 321-330.

[51] Nansen, P., Larsen, M., Gronvold, J., Wolstrup, J., Zorn, A. & Henriksen, S.A. (1995). Prevention of clinical trichostrongylidosis in calves by strategic feeding with the predacious fungus *Duddingtonia flagrans*. *Parasitol Res*, 81: 371-374.

[52] Fernández, A.S., Larsen, M., Henningsen, E., Nansen, P., Gronvold, J., Bjorn, H. & Wolstrup, J. (1999). Effect of *Duddingtonia flagrans* against *Ostertagia ostertagi* in cattle grazing at different stocking rates. *Parasitology*, 119: 105-111.

[53] Githigia, S.M., Thamsborg, S.M., Larsen, M., Kyvsgaard, N.C. & Nansen, P. (1997). The preventive effect of the fungus *Duddingtonia flagrans* on trichostrongyle infections of lambs on pasture. *Int J Parasitol*, 27: 931-939.

[54] Faedo, M., Larsen, M., Dimander, S.O., Yeates, G.W., Höglund, J. & Waller, P.J. (2002). Growth of the fungus *Duddingtonia flagrans* in soil surrounding feces deposited by cattle or sheep fed the fungus to control nematodes parasites. *Biol Control*, 23: 64-70.

[55] Waller, P.J., Knox, M.R. & Faedo, M. (2001). The potential of nematophagous fungi to control the free-living stages of nematode parasites of sheep: feeding and block studies with *Duddingtonia flagrans*. *Vet Parasitol*, 102: 321-330.

[56] Waller, P.J., Faedo, M. & Ellis, K. (2001). The potential of nematophagous fungi to control the free-living stages of nematode parasites of sheep: towards the development of a fungal controlled release device. *Vet Parasitol*, 102: 299-308.

[57] Chartier, C. & Pors, I. (2003). Effect of nematophagous fungus *Duddingtonia flagrans* on larval development of goat parasitic nematodes: a plot study. *Vet Res*, 34: 221-230.

[58] Wright, D.A., Mcanulty, R.W., Noonan, M.J. & Stankiewicz, M. (2003). The effect of *Duddingtonia flagrans* on the trichostrongyle infections of Saanen goats on pasture. *Vet Parasitol*, 118: 61-69.

[59] Nansen, P., Larsen, M., Roepstorff, A., Gronvold, J., Wolstrup, J. & Henriksen, S.A. (1996). Control of *Oesophagostomum dentatum* and *Hyostrongylus rubidus* in outdoor-reared pigs by daily feeding with the microfungus *Duddingtonia flagrans*. *Parasitol Res*, 82: 580-584.

[60] Larsen, M., Nansen, P., Henriksen, S.A., Wolstrup, J., Gronvold, J., Zorn, A. & Wedo, E. (1995). Predacious activity of the nematode-trapping fungus *Duddingtonia flagrans* against cyathostome larvae in faeces after passage through the gastrointestinal tract of horses. *Vet Para*sitol, 60: 315-320.

[61] Ojeda-Robertos, N.F., Torres-Acosta, J.F., Ayala-Burgos, A.J., Sandoval-Castro, C.A., Valero-Coss, R.O. & Mendoza-de-Gives, P. (2009). Digestibility of *Duddingtonia flagrans* chlamydospores in ruminants: *in vitro* and *in vivo* studies. *BMC Vet Res*, 28 (5): 46.

[62] Gronvold, J., Nansen, P., Henriksen, S.A., Larsen, M., Wolstrup, J., Bresciani, J., Rawat, H. & Fribert, L. (1996). Induction of traps by *Ostertagia ostertagi* larvae, chlamydospore production and growth rate in the nematode-trapping fungus *Duddingtonia flagrans*. *J Helminthol*, 70: 291-297.

[63] Gronvold, J., Wolstrup, J., Nansen, P., Larsen, M., Henriksen, S.A., Bjorn, H., Kirchheiner, K., Lassen, K., Rawat, H. & Kristiansen, H.l. (1999). Biotic and abiotic factors influencing growth rate and production of traps by the nematode-trapping fungus *Duddingtonia flagrans* when induced by *Cooperia oncophora* larvae. *J Helminthol*, 73: 129-136.

[64] Rojo-Vázquez, F. (1976). Effect of temperature on the development of pre-parasitic stages of *Trichostrongylus axei* and *T. colub*riformis. *Rev Iber Parasitol*, 36: 249-255.

[65] Rose, J.H. & Small, A.J. (1984). Observations on the bionomics of the free-living stages of *Trichostrongylus vitrinus*. *J Helminthol*, 58: 49-58.

[66] Mendoza De Gives, P., Flores Crespo, J., Herrera Rodríguez, D., Vázquez Prats V., Liebano Hernández, E. & Ontiveros Fernández, G.E. (1998). Biological control of *Haemonchus contortus* infective larvae in ovine faeces by administering an oral suspension of *Duddingtonia flagrans chlamydospores* to sheep. *J Helminthol*, 72: 343-347.

[67] Peña, M.T., Miller, J.E., Fontenot, M.E., Gillespie, A. & Larsen, M. (2002). Evaluation of *Duddingtonia flagrans* in reducing infective larvae of *Haemonchus contortus* in feces of sheep. *Vet Parasitol*, 103: 259-265.

[68] Paraud, C. & Chartier, C. (2003). Biological control of infective larvae of a gastro-intestinal nematode (*Teladorsagia circumcincta*) and a small lungworm (*Muellerius capillaris*) by *Duddingtonia flagrans* in goat faeces. *Parasitol Res*, 89: 102-106.

[69] Paraud, C., Pors, I., Chicard, C. & Chartier, C. (2006). Comparative efficacy of the nematode-trapping fungus Duddingtonia flagrans against *Haemonchus contortus*,

Teladorsagia circumcincta and *Trichostrongylus colubriformis* in goat faeces: influence of the duration and of the temperature of coproculture. *Parasitol Res*, 98: 207-213.

[70] Wolstrup, J., Gronvold, J., Henriksen, S.A., Nansen, P., Larsen, M., Bogh, H.O. & Ilsoe, B. (1994). An attempt to implement the nematode-trapping fungus *Duddingtonia flagrans* in biological control of trichostrongyle infections of first year grazing calves. *J Helminthol*, 68: 175-180.

[71] Larsen, M., Nansen, P., Grondahl, C., Thamsborg, S.M., Gronvold, J., Wolstrup, J., Henriksen, S.A. & Monrad, J. (1996). The capacity of the fungus Duddingtonia flagrans to prevent strongyle infections in foals on pasture. *Parasitology*, 113: 1-6.

[72] Gómez-Rincón, C., Uriarte, J. & Valderrábano, J. (2006). Efficiency of *Duddingtonia flagrans* against Trichostrongyle infections of sheep on mountain pastures. *Vet. Parasitol*, 141: 84-90.

[73] Gómez-Rincón, C., Uriarte, J. & Valderrábano, J. (2007). Effect of nematophagous fungus *Duddingtonia flagrans* and energy supplementation on the epidemiology of naturally infected kids. *Vet. Res*, 38: 141-150.

[74] Knox, M.R. & Faedo, M. (2001). Biological control of field infections of nematode parasites of young sheep with *Duddingtonia flagrans* and effects of spore intake on efficacy. *Vet Parasitol*, 101: 155-160.

[76] Chandrawathani, P., Jamnah, O., Waller, P.J., Larsen, M., Gillespie, A.T. & Zahari, W.M. (2003). Biological control of nematode parasites of small ruminants in Malaysia using the nematophagous fungus *Duddingtonia flagrans*. *Vet Parasitol*. 117: 173-83.

[77] Chandrawathani, P., Jamnah, O., Adnan, M., Waller, P.J., Larsen, M. & Gillespie, A.T. (2004). Field studies on the biological control of nematode parasites of sheep in the tropics, using the microfungus *Duddingtonia flagrans*. *Vet. Parasitol*, 120: 177-187.

[78] Silva, A.R., Araujo, J.V., Braga, F.R., Frassy, L.N., Tavela, A.O., Carvalho, R.O. & Castejon, F.V. (2009). Biological control of sheep gastrointestinal nematodiasis in a tropical region of the southeast of Brazil with the nematode predatory fungi *Duddingtonia flagrans* and *Monacrosporium thaumasium*. *Parasitol Res*, 105: 1707-13.

[79] Silva, B.F., Carrijo-Mauad, J.R., Braga. F.R., Campos, A.K., Araujo, J.V. & Amarante, A.F. (2010). Efficacy of *Duddingtonia flagrans* and *Arthrobotrys robusta* in controlling sheep parasitic gastroenteritis. *Parasitol Res*, 106: 1343-50.

[80] Yeates, G.W., Waller, P.J. & King, K.l. (1997). Soil nematodes as indicators of the effect of management on grassland in the new England Tablelands (NSW): effect of measures for control of parasites of sheep. *Pedobiologia,* 41: 537-548.

[81] Yeates, G.W., Dimander, S.O., Waller, P.J. & Höglund, J. (2002). Environmental impact on soil nematodes following the use of either ivermectin sustained release bolus or the nematophagous fungus *Duddingtonia flagrans* to control nematode parasites of cattle in Sweden. *Ac Agric Scan Anim Sci*, 52: 233-242.

[82] Yeates, G.W., Dimander, S.O., Waller, P.J. & Höglund, J. (2003). Soil Nematode Populations Beneath Faecal Pats from Grazing Cattle Treated with the Ivermectin Sustained-release Bolus or Fed the Nematophagous Fungus *Duddingtonia flagrans* to Control Nematode Parasites. *Ac Agri Scan Anim Sci,* 53:197-206.

[83] Knox, M.R., Josh, P. & Anderson, L. (2002). Deployment of *Duddingtonia flagrans* in an improved pasture system: dispersal, persistence, and effects on free-living soil nematodes and microarthropods. *Biol Control*, 24: 176-182.

[84] Paraud, C., Lorrain, R., Pors, I. & Chartier, C.J. (2011). Administration of the nematophagous fungus *Duddingtonia flagrans* to goats: an evaluation of the impact of this fungus on the degradation of faeces and on free-living soil nematodes. *J. Helminthol,* online March 21th 2011, p 1-9. DOI: 10.1017/S0022149X11000083

[85] Skipp, R.A., Yeates, G.W., Chen, L.Y. & Glare, T.R. (2002). Occurrence, morphological characteristics and ribotyping of New Zealand isolates of *Duddingtonia flagrans*, a candidate for biocontrol of animal parasitic nematodes. *NZ J Agric Res*, 45: 187-196.

[86] Tarazona, J.M. (1980). Etiopatogenia y control de la gastroenteritis parasitaria ovina. *Comunicaciones INIA. Ser. Hig. San. Anim*, 3. 32.

[87] Uriarte, J., Tanco, J.A., Guillén, A. & Marco, M.M. (1984). Evolución y supervivencia de los estados libres de la familia *Trichostrongylidae* en los secanos de la provincia de Zaragoza. *Anales de Información Técnica de Economía Agraria. Serie Ganadera*, 20: 11-23.

[88] Tarazona, J.M., Babin, M.M., Canals, A. & Martín, M. (1985). Epizootiología de las Trichostrongylidosis ovinas en la Meseta Meridional española. *ITEA,* Extra (5): 394-396.

[89] Llorente, M.M. (1999). *Epidemiología de la gastroenteritis parasitaria ovina en sistemas extensivos del Valle Medio del Ebro: Efecto del periparto en la dinámica de infección.* Tesis Doctoral. Universidad de Zaragoza, 345 p.

[90] Uriarte, J. & Valderrábano, J. 1989. An epidemiological study of parasitic gastroenteritis in sheep under an intensive grazing system. *Vet Parasitol*, , 31: 71-81.

[91] Uriarte, J., Llorente, M.M. & Valderrábano, J. 2003. Seasonal changes of gastrointestinal nematode burden in sheep under an intensive grazing system. *Vet Parasitol*, 118: 79-92.

[92] Braga, F.R., Araujo, J.V., Silva, A.R., Araujo, J.M., Carvalho, R.O., Tavela, A.O., Campos, A.K. & Carvalho, G.R., (2009). Biological control of horse cyathostomin (Nematoda: *Cyathostominae*) using the nematophagous fungus *Duddingtonia flagrans* in tropical southeastern Brazil. *Vet Parasitol*, 163: 335-40.

In: Fungi: Types, Environmental Impact and Role in Disease
Editors: Adolfo Paz Silva and María Sol
ISBN: 978-1-61942-671-9

Chapter XVIII

Helminthosis Control of Domestic Animals, a New Approach to an Old Problem

Fabio Ribeiro Braga and Jackson Victor de Araújo
Departamento de Veterinária - Universidade Federal de Viçosa,
Viçosa Minas Gerais, Brasil

Abstract

Overall helminthic infections are an old problem and a serious barrier for cattle ranching. Typically, large portion of nematodes in fecal environment passes from egg to larval stages, reaching the infective form. Nematodiosis control in general is accomplished through drugs known as anthelmintics. This control has been secular, i.e., it is culturally held in all parts of the world for long time. However, except experiences gained in the past, currently a major problem has been its indiscriminate and by wrong way use, which has been extensively described as a major responsible for parasitic resistance installation. Thus, the uses of nematophagous fungi have been tested under controlled conditions and in natural conditions to help fighting parasitic nematodes. They are divided into three groups, endoparasites, predators, and ovicidal. In practice, the "biological control" should be understood as the use of natural antagonists, in these particular case nematophagous fungi, to reduce the target organ, which is the nematode, to an acceptable level. For over 20 years, in South America, Brazil, Professor Jackson Victor de Araújo from Universidade Federal de Viçosa (Federal University of Viçosa) led studies shown that *Duddingtonia flagrans* species destroys nematodes populations of domestic animals in laboratory and natural tests. The future prospects for using a biological control agent consider basically be virulent, be efficient, present easy and cheap production, not to change the farmer's usual cultural practices for their implementation, be compatible with chemicals, be able to quickly colonize and persist in the soil, be effective for several host species to increase the interest of industrial companies to develop it commercially, not present deleterious effects on other organisms than the target, and maintain viability during storage without special care. Although a biocontrol agent with all these attributes is not yet available, there are researches for this purpose.

Introduction

Helminth infections have been presented as an old problem and a serious barrier for livestock activity. In most cases the animals are reared for productive parasitized by a large number of endoparasites, especially here, for gastrointestinal helminth parasites that cause serious problems to their health. In the large group of helminths, stand out the phyla Platyhelminthes and Nematelminthes. In the phylum Nematelminthes stands out the class Nematoda that has parasitical importance. The gastrointestinal nematode parasites have been identified as the major responsible for global economic losses in livestock intended for production. These parasites are usually known as roundworms, due to their aspect to the cross section. The phylum Platyhelmintes houses two classes, Trematoda and Cestoda, which have the body dorsoventrally flattened up to segmented, respectively. This group also houses worms that cause damage and losses in livestock [39]. The nematodes have the body covered by a colorless layer called cuticle, a simple digestive system, a primitive excretory system, a circulatory system, a nervous system and reproductive organs. In aspect of the Veterinary Medical importance generally the nematodes are divided into two major groups, those with and without stock copulatory pouch. According to a simplified classification, the veterinary importance nematodes with copulatory pouch are part of the following superfamilies: Trichostrongyloidea, Strongyloides and Metatrosgyloidea and those without the copulatory pouch are classified as: Rhabditoidea, Ascaroidea, Oxyuroidea, Spiruroidea, Filarioidea, Trichuroidea and Dictophymatoidea [39]. A nematode, during its development, usually suffers "changes" in steps, losing their cuticle. Usually the changes occur after hatching of its eggs which are excreted in faeces, urine and more rarely during the expectoration. In the external environment can suffer the "changes" of the larval stages. In general rule the nematodes are oviparous or oviviparos, however, their eggs can vary greatly between species, both in size, shape and thickness of its shell. Also, depending on the species the eggs can hatch outside the body or after ingestion. Usually in the environment of the fecal mass, ,most nematode passes from egg to the larval stages L_1 and L_2, reaching the third stage called "infecting", ie, L_3 [39]. At that time, the large majority of animals can become infected by ingesting the L_3 contained in the external environment. In the external environment there are natural antagonists of nematodes, which are microorganisms capable of causing the death of these animals [22]. Among these natural antagonists, the highlight has been the nematophagous fungi. In relation to control of nematodioses, in general, is accomplished through the use of drugs known as anthelmintics [36, 40, 24]. This control has been secular, ie, culturally performed on all parts of the world for decades. However, unless the experiences gained in the past, currently a major problem has been the wrongly and indiscriminate use, a fact that has been extensively reported as a major responsible for the installation of parasite resistance [1]. In this aspect, the term parasite resistance regards to failures to control a population of nematodes with a prior drug effective in destroying that target population and which does not present the same efficacy when used under the same conditions and after a period of time. Briefly, the problem installed reflects the failures in the treatments, the indiscriminate use, the use of sub lethal and the rapid rotation of the bases of pharmacological compounds used [33, 32]. Thus, the use of nematophagous fungi has been tested in controlled conditions and even under natural conditions to help in combating parasitic nematodes. These fungihave attracted the attention of researchers since its function as a predator of nematodes was recognized at the end of nineteenth century by Zopf in 1888 [21]. These microorganisms are natural from environment and destroy justly the pre-parasitic contained in the external environment [5, 1, 13, 12]. They are divided into: endoparasites (1); predators (2) and ovicidal (3).

(1) Endoparasitic fungi persist mainly as spores, but sometimes as chlamydospores (resistant spores) that are released upon disintegration of the nematode. They are found in various taxonomic classes, being subdivided into three groups that follow: group I, corresponding to species that encyst, Chytridiomyces (Catenaria Sorokin) and Oomycetes (Myzocytium Schenk, Haptoglossa Drechsler and Lagenidium Schenk). Group II corresponds to the species producing adhesive conidia, Deuteromycotina (Verticillium Nees, Cephalosporium Corda, Drechmeria conospora Gans and Jansson, Hirssutella rhossiliensis Minter and Brady, Harposporium subuliforme Drechsler) and of Zygomycotina (Meristacrum asterosporium Drechsler, Zygnemomyces Miura and Gonimochaete Drechsler); and group III corresponds to species that also produce adhesive conidia regarding Basidiomycotina [21, 5].

(2) Predatory fungi produce up to six types of traps (Figure 1): Non-differentiated adhesive hyphae, branching hyphae that suffer anastomosis, forming three-dimensional adhesive networks; adhesive branches, which at times can come together to form simple two-dimensional adhesive networks; nodules adhesives; constricting rings, and non-constricting rings. The hypha is used as a trap and prey is captured by adhesion [8, 33]. It is the group most studied and therefore more widely used in biological control ofgastrointestinal nematode parasites of domestic animals, effectively reducing the population both in laboratory conditions as in field conditions. They also have the advantage of greater potential for industrialization [26, 27].

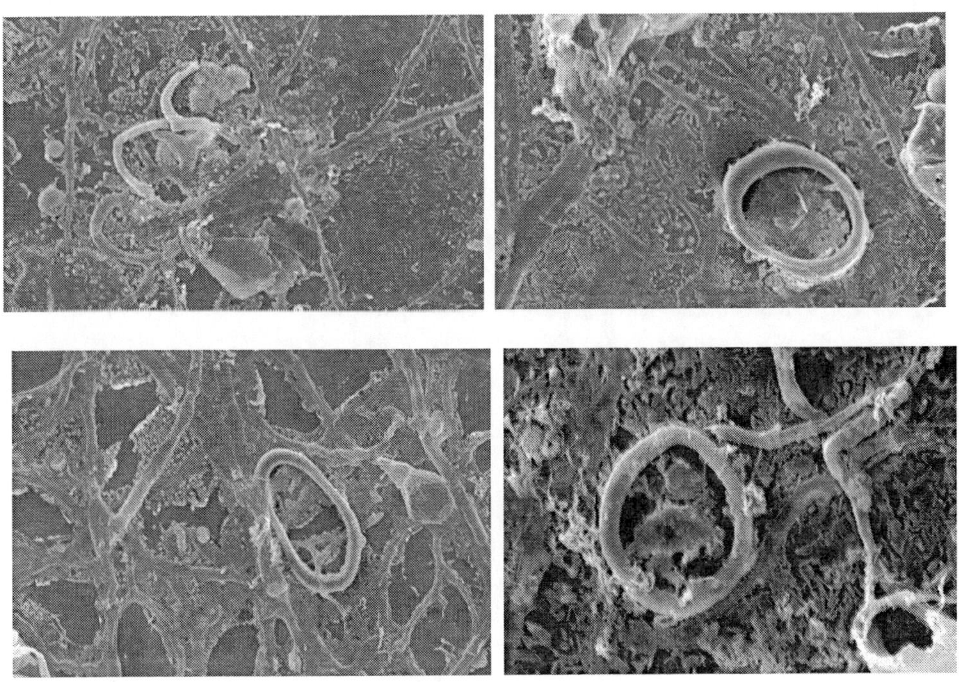

Figure 1. Traps produced by predator nematophagous fungi.

The predator fungi produce capture organs in pure culture, but the process of differentiation of hyphae into traps occur within 24 hours after the interaction of the fungus with the nematode, and the higher the motility of nematodes higher the production of traps by the fungus, seen that the stimulus will be higher in rich soil microbiota [34, 14].

(3) About the group of opportunistic fungi, the first determination of its ovicidal capacity against helminth eggs was described by Lestan in 1970 [25]. The opportunistic fungi besides of parasitize eggs, cysts and females of nematodes and helminths, are saprophytic, and, not dependent on the presence of the parasite in the soil for their survival, are therefore easily grown in laboratory (Figure 2). It is a very promising group to be used in biological control of helminths, mainly because they reduce by about 70 to 90% levels of viable eggs in the soil [9]. However, often what prevents their full effectiveness is the strategy developed by the parasites. Most of these gastrointestinal helminth parasites produce eggs that give rise to larvae rapidly, complicating the process of interaction with the eggs [23, 9, 10]. In practice the "biological control" should be understood as the use of natural antagonists, in this particular case, nematophagous fungi to reduce a level acceptable to the target organism, in this case the nematodes [22]. Moreover, it not will eliminate the use of anthelmintics but employment in a synergistic way to the chemical. However, consequently, the use of drugs may be decreased resulting in more benefits to animal health, the environment and the human population Moreover, even being reported some time ago, the study of nematophagous fungi for biological control of helminths of domestic animals can be seen as a new approach to the use of anthelmintics [8, 15]. For many years, the study of the use of predator and ovicidal fungi has been more prominent, especially through the satisfactory results that can be extrapolated to the day-to-day practice of cattle. Studies have shown that the use of predatory species such as Duddingtonia flagrans and ovicidal as *Pochonia chlamydosporia* has shown great results [26, 31, 19, 9, 10, 14]. The fungus *D. flagrans* is the most studied species and that presents a greater potential to be marketed. It is a species that produces resistant structures known as chlamydospores, and can pass through the gastrointestinal tract of domestic animals, and thus can withstand extreme conditions of "stress" [26, 4, 37, 38, 35].

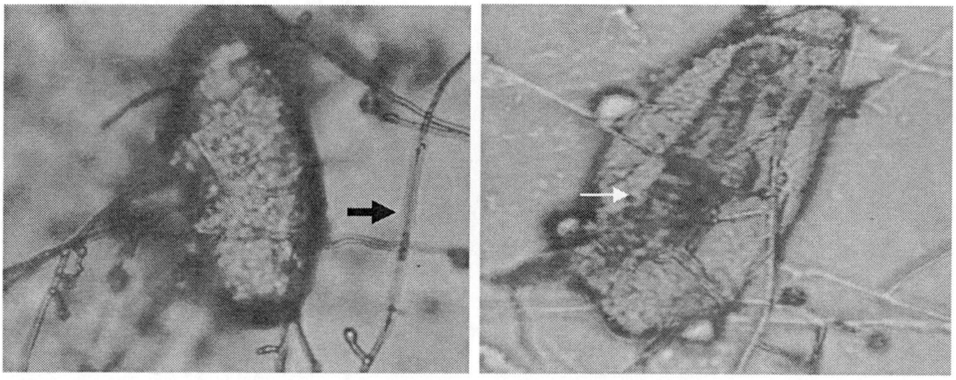

Figure 2. Interaction of the hyphae from Pochonia chlamydosporia fungus (black arrow) with Shistosoma mansoni eggs (white arrow).

However, there may be differences in the action of different isolates of this species, as some investigations have mentioned this fact in different regions [13]. On the other hand, researchers have long argued about the real possibility of their employment and then studies converged to seek a formulation that could help in environmental decontamination, and therefore in a lower helminth reinfection in animals [6, 7, 26]. As mentioned earlier, there is a wide range of "worms" that behave in most cases differently, and so the need for interaction studies in several species of parasites on various animals and have been performed. For over 20 years, investigations conducted in South America, Brazil, led by Professor Jackson Victor de Araújo from Federal University of Viçosa, Brazil, has demonstrated that the species D. flagrans destroys populations of nematodes of domestic ruminants at laboratorial assays [5, 11]. In these investigations, researchers have observed the interaction of the fungus versus L3 gastrointestinal nematode parasites after a short period of time and its complete destruction after 24 hours [37, 3]. On the other hand, under natural conditions, researchers have used a pellet formulation based on sodium alginate containing the fungus D. flagrans orally mixed in the feed [13, 37, 38]. Over many years and after several investigations, came to the conclusion that oral administration of 1g/10kg body weight of these pellets and the range of two times per week in domestic ruminants may reduce the parasite load in these animals compared to untreated animals [19, 13, 37]. Such formulation is patented and is in final stage of the manufacturing process in Latin America. On the other hand, the concomitant use of anthelmintics may also be performed jointly [5]. In equine species, has also studied the use of this formulation containing this fungus, since in the creation of these animals the problem of parasite resistance has been detected, mainly in relation to small strongyles (Figure 3). In this context, investigations have shown satisfactory and promising results with the use of 1g/10kg weight of pellets containing the fungus *D. flagrans* twice a week [13, 14]. Studies have demonstrated that animals treated with this dose and in this interval may have a lower recurrence of parasitic infection and therefore a higher feed conversion. Other investigations include the great diversity of employment of this fungus in pets such as dogs, potentially decreasing the risk of transmission of potentially zoonotic nematodes housed in these animals [17, 18, 3, 4, 16]. Regarding the use of ovicidal fungi, such as Pochonia chlamydosporia, the investigations carried out have noted a new phase of biological control, since the main approach is the destruction of L3 [11, 9, 13] . Efforts in this case have converged on the wide use of this fungus in laboratorial conditions and only after that your employment in the field. The answer so far has based is that the use of ovicidal fungi can decrease a future population of gastrointestinal nematodes, starting from the old premise that "the vast majority of the nematodes will suffer hatching in the fecal mass and then of phase change stadial" [33, 12]. The studies have shown that the fungus *P. chlamydosporia* destroys the eggs of helminths of animals, in vitro assays, by enzymatic and mechanical action [28, 29, 30, 2]. But a new approach to its use has been tested through the use of its crude extract on eggs and or even larvae of helminths potentially zoonotic [15]. However, earlier studies have mentioned the use of this fungus against *Ascaris lumbricoides*, nematode parasite of humans, and the new approach, just reported its use on the nematode parasites of domestic animals [28, 11]. According to research cited, the major challenge is to observe the great diversity in the action of various existing isolates and just from there determine the in vivo tests. In relation to results obtained, the fungus *P. chlamydosporia* destroys eggs of Nematoda, Trematoda and Cestoda, however, must be very careful in the tests with eggs which can hatch in a short period of time [17, 14]. On the other hand, the use of this fungus as a possible biological control of potentially zoonotic helminths such as Toxocara canis, Ancylostoma caninum, Dipylidium caninum and Taenia saginata has been viewed as an interesting fact, since the contamination of people have been treated for geohelminths as a public health problem [2, 3,

18, 20]. In relation to the use under natural conditions the works are in early stages, but have shown that *P. chlamydosporia* can support the passage through the gastrointestinal tract of horses, germinate in faeces and keep its viability of destruction. This fact indicates that their use in combating the nematodioses of domestic animals must obey its environmental dispersion made by the animals themselves [12]. On the other hand, the future prospects for the use of an agent in biological control are basically summarized: to be virulent, be efficient, reducing the target population levels below the economic threshold; make production easier and cheaper; not to modify cultural practices farmer's usual for its application; be compatible with chemicals; be able to quickly colonize and persist in the soil; to be effective for several species of the host to enhance the interest of industrial companies to develop it commercially; not present deleterious effects on other organisms than the target, maintaining viability during storage, without special care [33, 5, 15]. However, one does not have a biocontrol agent with all these attributes, however, the research walk for this purpose. Finally, research on epidemiology, biology and mode of action of nematophagous fungi are important for understanding the viability of using these agents in the biocontrol of helminth parasites of animals under natural conditions.

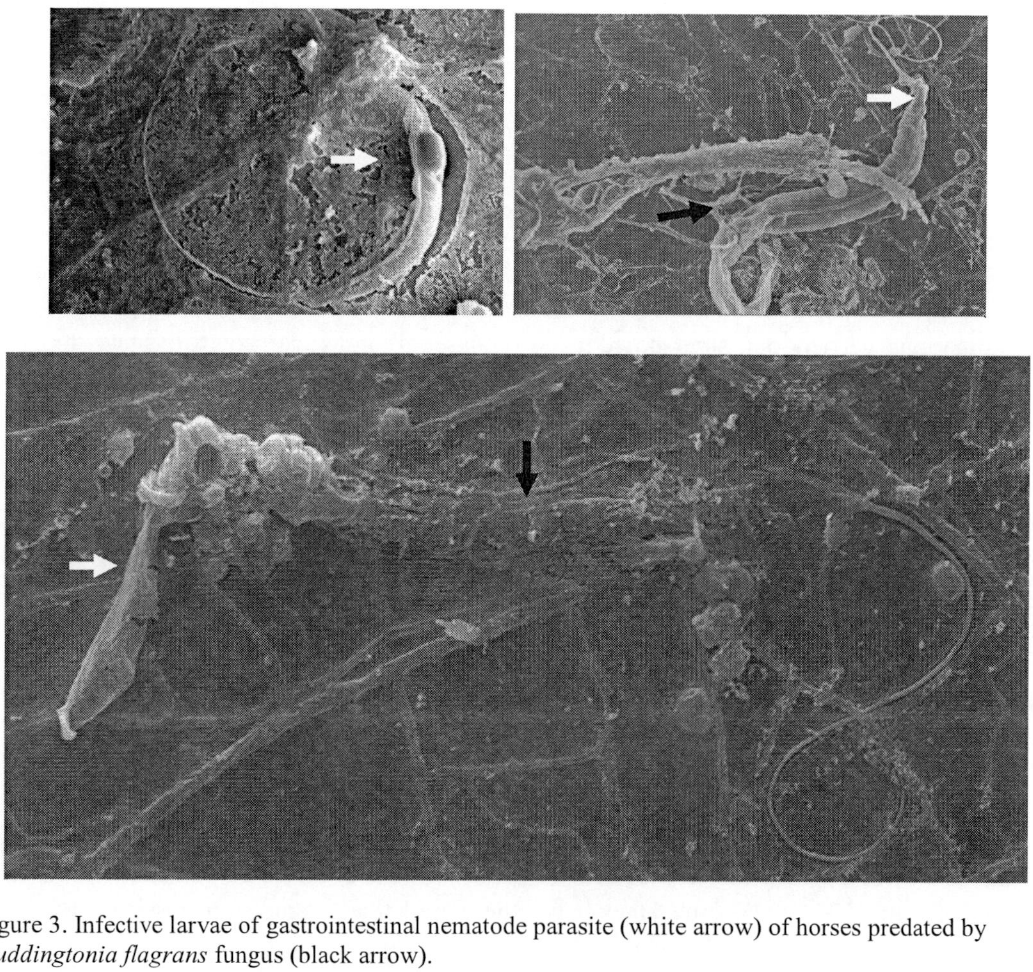

Figure 3. Infective larvae of gastrointestinal nematode parasite (white arrow) of horses predated by *Duddingtonia flagrans* fungus (black arrow).

Acknowledgments

The authors would like to thank CNPq and CAPES for the financial support and grant concession, and to the Núcleo de Microscopia Eletrônica e Microanálise – NMM of the Federal University

References

[1] Araújo, J.V., Assis, R.C.L., Campos, A.K. & Mota, M.A. (2004b). Atividade in vitro dos fungos nematófagos dos gêneros *Arthrobotrys*, *Duddingtonia* e *Monacrosporium* sobre nematóides trichostrongilídeos (Nematoda: Trichostrongyloidea) parasitos gastrintestinais de bovinos. *Revista Brasileira de Parasitologia Veterinária,* 13, 65-71.

[2] Araujo, J.M., Araújo, J.V., Braga, F.R., Carvalho, R.O., Silva, A.R. & Campos, A.K. (2009). Interaction and ovicidal activity of nematophagous fungus *Pochonia chlamydosporia* on *Taenia saginata* eggs. *Experimental Parasitology*, 121, 338-341.

[3] Araujo, J.M., Araújo, J.V., Braga, F.R. & Carvalho, R.O. (2010). *In vitro* predatory activity of nematophagous fungi and after passing through gastrointestinal tract of equine on infective larvae of *Strongyloides westeri*. *Parasitology Research*, 107,103–108.

[4] Araújo, J.V., Braga, F.R., Araujo, J.M., Silva, A.R. & Tavela, A.O. (2008) *In vitro* evaluation of the effect of the nematophagous fungi *Duddingtonia flagrans*, *Monacrosporium sinense* and *Pochonia chlamydosporia* on *Ascaris suum* eggs. *Parasitology Research* 102, 787-790.

[5] Araújo, J.V., Mota, M.A. & Campos, A.K. (2004a). Controle biológico de helmintos parasitos de animais por fungos nematófagos. *Revista Brasileira de Parasitologia Veterinária* 13, 165–170.

[6] Araújo, J.V. & Sampaio, W.M. (2000). Effects of temperature, mineral salt and passage through gastrointestinal tract of calves on alginate formulation of *Arthrobotrys robusta*. *Revista Brasileira de Parasitologia Veterinária* 9, 55-59.

[7] Araújo, J.V., Sampaio, W.M., Vasconcelos, R.S. & Campos, A.K. (2000). Effects of different temperatures and mineral salt on pellets of *Monacrosporium thaumasium* - a nematode-trapping fungus. Vet. Arhiv. 80, 181-190.

[8] Barron, G.L. (Ed.) *The nematode-destroying fungi.* Canadá: Canadian Biological Publications, 1977.

[9] Braga, F.R., Araújo, J.V., Campos, A.K., Araujo, J.M., Silva, A.R., Carvalho, R.O., Correa, D.N. & Pereira, C.A.J. (2008b). *In vitro* evaluation of the effect of the nematophagous fungi *Duddingtonia flagrans*, *Monacrosporium sinense* and *Pochonia chlamydosporia* on *Schistosoma* mansoni eggs. *World Journal of Microbiology and Biotechonology,* 24, 2713-2716.

[10] Braga, F.R., Araújo, J.V., Campos, A.K., Carvalho, R.O., Silva, A.R. & Tavela, A.O. (2008a.) *In vitro* evaluation of the effect of the nematophagous fungi *Duddingtonia flagrans*, *Monacrosporium sinense* and *Pochonia chlamydosporia* on *Fasciola hepatica* eggs. *World Journal of Microbiology and Biotechonology,* 24, 1559-1564.

[11] Braga, F.R., Araújo, J.V., Campos, A.K., Carvalho, R.O., Silva, A.R., Tavela, A.O. & Maciel, A.S. (2007). Observação in vitro da ação dos isolados fúngicos *Duddingtonia flagrans*, *Monacrosporium thaumasium* e *Verticillium chlamydosporium* sobre ovos de *Ascaris lumbricoides* (Lineu, 1758). *Revista da Sociedade Brasileira de Medicina Tropical,* 40, 356-358.

[12] Braga, F.R., Araújo, J.V., Carvalho, R.O., Silva, A.R., Araujo, J.M.., Freitas, F.E.S., Geniêr, H.L.A., Ferreira, S.R. & Queiroz, J.H. (2010b). Ovicidal action of a crude enzymatic extract of the fungus *Pochonia chlamydosporia* against cyathostomin eggs. *Veterinary Parasitology,* 172, 264-268.

[13] Braga, F.R., Araújo, J.V., Silva, A.R., Araujo, J.M., Carvalho, R.O., Tavela, A.O., Campos, A.K. & Carvalho, G.R. (2009). Biological control of horse cyathostomin (Nematoda: Cyathostominae) using the nematophagous fungus *Duddingtonia flagrans* in tropical southeastern Brazil. *Veterinary Parasitology* 163, 335-340.

[14] Braga, F.R., Araújo, J.V., Silva, A.R., Carvalho, R.O., Araujo, J.M., Ferreira, S.R., Benjamin, L.A. (2010a). Predatory activity of the nematophagous fungus Duddingtonia flagrans on horse cyathostomin infective larvae. *Tropical Animal Health Production,* 42,1161–1165.

[15] Braga, F.R., Carvalho, R.O., Silva, A.R., Araujo, J.M., Araújo, J.V., Freitas, F.E.S., Geniêr, H.L.A. & Queiroz, J.H. (2011a) Ação ovicida do extrato bruto enzimático do fungo *Pochonia chlamydosporia* sobre ovos de *Ancylostoma* sp. *Revista da Sociedade Brasileira de Medicina Tropical* 44, 116-118.

[16] Campos, A.K., Araújo, J.V. & Guimarães, M.P. (2008). Interaction between the nematophagous fungus *Duddingtonia flagrans* and infective larvae of *Haemonchus contortus* (Nematoda: Trichostrongyloidea). *Journal of Helminthology,* 82, 337-341.

[17] Carvalho, R.O., Araújo, J.V., Braga, F.R., Araujo, J.M., Silva, A.R. & Tavela, A.O. (2009) Predatory activity of nematophagous fungi on *Ancylostoma* ssp. infective larvae: evaluation *in vitro* and after passing through gastrointestinal tract of dogs. *Journal of Helminthology,* 15,1-5.

[18] Carvalho, R.O., Araújo, J.V., Braga, F.R., Araujo, J.M.& Alves, C.D.F. (2010). Ovicidal activity of *Pochonia chlamydosporia* and *Paecilomyces lilacinus* on *Toxocara canis* eggs. *Veterinary Parasitology*, 169, 123–127.

[19] Dias A.S., Araújo J.V., Campos A.K., Braga F.R. & Fonseca, T.A. (2007). Application of a formulation of the nematophagous fungus *Duddingtonia flagrans* in the control of cattle gastrointestinal nematodioses. *World Journal of Microbiology and Biotechnology*, 28, 1254-1252.

[20] Frassy, L.N., Braga, F.R., Silva, A.R., Araújo, J.V., Ferreira, S.R & Freitas, L.G. (2010) Destruição de ovos de *Toxocara canis* pelo fungo nematófago *Pochonia chlamydosporia*. Revista da Sociedade Brasileira de Medicina Tropical, 43, 102-104.

[21] Gray, N.F. Fungi attackhing vermiform nematodes. In: Poinar, O.G. & Borne, J.H. (Eds). *Diseases of nematodes*. Boca Raton: CRC Press, 1988. p. 3- 38.

[22] Gronvold, J., Henriksen, S.A., Larsen, M., Nansen, P., Wolstrp, J. (1996). Biological control Aspects of biological control, with special reference to arthropods, protozoans and helminths of domesticated animals. *Veterinary Parasitology*, 64, 47-64.

[23] Jansson, H.B. & Nordbring-Hertz, B. Infection mechanisms in the fungus nematode system. In: Poinar, G.O. & Jansson H.B. *Diseases of nematode,* v. II. CRC Press, Boca Raton, 1988. p.1-72.

[24] Kaplan, R.M. (2002). Antihelmintic resistance in nematodes of horses. *Veterinary Research*, 33, 491-507.

[25] Kerry, B.R. & Hidalgo, L. (2004). Application of *Pochonia chlamydosporia* in the integrated control of root-knot nematode on organically grown vegetable crops in Cuba. IOBC/wprs Bul. 27, 123-126

[26] Larsen, M., 1999. Biological control of helminths. *International Journal for Parasitology*, 29, 139-146.

[27] Larsen, M., Wolstrup, J., Henriksen, S.A., Gronvold, J. & Nansen, P. (1992). *In vivo* passage through calves of nematophagous fungi selected for biocontrol of parasitic nematodes. *Journal of Helminthology*, 66, 137-141.

[28] Lysek, H. (1976). Classification of ovicide fungi according to type of ovicidity. Acta University Palacki Olomue. 76, 9–13.

[29] Lysek, H., Fassatiová, O., Pineda, N.C. & Hernández, N.L. (1982). Ovicidal fungi in soils of Cuba. *Folia Parasitologica,* 29, 265-270.

[30] Lysek, H. & Sterba, J. (1991). Colonization of *Ascaris lumbricoides* eggs by the fungus *Verticillium chlamydosporium* Goddard. *Folia Parasitologica*, 38, 255–259.

[31] Mendoza-De-Gives, P., Zapata Nieto,C., Hernández, E.L., Arellano, M.E., Rodríguez, D.H. & Garduno, R.G. (2006). Biological control of gastroin-testinal parasitic nematodes using *Duddingtonia flagrans* in sheep under natural conditions in Mexico. Ann.N.Y.Acad.Sci, 1081, 355–359.

[32] Molento, M.B., Antunes, J., Bentes, R.N. & Coles, G.C. (2008). Anthelmintic resistant nematodes in Brazilian horses. *Veterinary Record* 162, 384–385.

[33] Mota, M.A., Campos, A.K. & Araújo, J.V. (2003). Controle biológico de helmintos parasitos de animais: estágio atual e perspectivas futuras. *Pesquisa Veterinária Brasileira*, 23, 93-100.

[34] Nansen, P., Gronvold, J., Henriksen, S.A. & Wolstrup, J. Predacious activity of the nematode-destroying fungus *Arthrobotrys oligospora* on preparasitic larvae of *Cooperia oncophora* and soil nematodes. Proc. Helminthology Soc. W. 53:237-243, 1986.

[35] Paz-Silva, A., Francisco, I., Valero-Coss, R.O., Cortiñas, F.J., Sánchez, J.A., Francisco, R., Arias, M., Suárez, J.L., López-Arellano, M.E., Sánchez-Andrade, R. & Mendoza-de-Gives, P. (2011). Ability ofthefungus *Duddingtonia flagrans* to adapt to the cyathostomin egg-output by spreading chlamydospores. Veterinary Parasitology, 30, 277-282.

[36] Prichard, R. (1990). Anthelmintic resistance in nematodes: extent, recent understanding and future directions for control and research. *International Journal for Parasitology*, 20, 515-523.

[37] Silva, A.R., Araújo, J.V., Braga, F.R., Frassy, L.N., Tavela, A.O., Carvalho, R.O. & Castejon, F.V. (2009). Biological control of sheep gastrointestinal nematodiasis in a tropical region of the southeast of Brazil with the nematode predatory fungi *Duddingtonia flagrans* and *Monacrosporium thaumasium*. *Parasitology Research*, 105, 1707–1713.

[38] Silva, B.F., Mauad, J.R.C., Braga,F.R., Amarante, A.F.T., Araújo, J.V. & Campos, A.K. (2010). Efficacy of *Duddingtonia flagrans* and *Arthrobotrys robusta* in controlling sheep parasitic gastroenteritis. *Parasitology Research,* 106, 1346–1350.

[39] Urquhart, G.M., Armour, J., Duncan, J.L., Dunn, A.M. & Jennings. F.W. (1998). Parasitologia Veterinária. Guanabara Koogan, Rio de Janeiro, 373 pp.

[40] Woolaston, R.R., Baker, R.L. (1996). Prospects of breeding small ruminants for resistance to internal parasites. *International Journal for Parasitology*, 26, 845–855.

Section 4: Fungi in Biotechnological Processes

In: Fungi: Types, Environmental Impact and Role in Disease ISBN: 978-1-61942-671-9
Editors: Adolfo Paz Silva and María Sol

Chapter XIX

Biotechnological Use of Fungi in the Control of Ruminant Parasitic Nematodes

P. Mendoza-de-Gives [1*] ***and J.F.J. Torres-Acosta*** [2]
[1] Laboratory of Helminthology, National Centre for Disciplinary Research in Veterinary Parasitology (CENID-PAVET-INIFAP) Morelos, Mexico
[2] Faculty of Veterinary Medicine and Animal Science, Autonomous University of Yucatan (FMVZ-UADY). Merida, Mexico

Abstract

Ruminant parasitic gastrointestinal nematodes (GIN) are organisms responsible for severe economic losses affecting grazing ruminant industry all over the world. The most common method of control of these parasites is through the regular use of antiparasitic treatments in the form of commercial anthelmintic (AH) drugs. Nevertheless, the effectiveness of the AH products is dramatically declining due to the development of resistant worms that can survive the AH treatments. As a result, animals keep showing clinical signs of nematodosis and low productivity even after treatment with conventional AH drugs. Additionally, the use of conventional AH drugs have other disadvantages. The AH molecules can remain as residual contaminants in animal products intended for human consumption. Besides, some active molecules can be eliminated in the feces of treated animals and reach the soil leaving beneficial organisms of the soil under risk of destruction.

There is an extraordinary group of microfungi living in the soil that is classified as natural antagonist of nematodes. The fungi in this group are able to form trapping devices to capture and destroy infective larvae of ruminant parasitic nematodes before they leave the feces and move to the field. Thus, they can act as biological control agents against the free living stages of GIN. In this chapter a general overview of nematophagous fungi and the perspectives of their use as tools for the biological control of GIN is presented. Special emphasis is directed to the results obtained by the group of Mexican researchers

[*] E-mail address: mendoza.pedro@inifap.gob.mx.

working with this interesting group of fungi mainly on its biotechnological potential for the control of GIN. Its successful application may help to reduce the risk of parasitic diseases affecting grazing ruminants around the world.

1. Introduction

1.1. Ruminant Parasitic Nematodes

Parasitic fauna affects animals in different production systems worldwide particularly in tropical and subtropical areas of the world [1]. Parasites may cause an important economic impact in animal production [2]. Ruminants are highly susceptible to the attack of both internal and external parasites. The presence of external parasites such as ticks or lice is easy to detect with simple observation. Thus, it is easy to quantify the abundance of these parasites and, as a result, the implementation of a treatment scheme can be made based on a rapid and direct observation of the affected hosts [3]. Also, the efficacy of antiparasitic treatments can be easily visualized with the external parasites since their presence or absence on the animal´s body is evident. In contrast, internal parasites such as gastrointestinal nematodes (GIN) are not easy to detect. Although the typical clinical signs of such parasitic disease are notorious, they can be confused with clinical signs caused by other common conditions such as under-nourishment and other diseases leading to low productivity and chronic emaciation. Subclinical helminthosis may cause a decline of up to 40% in the live weight gain [4,5] Ruminant GIN are parasites belonging to the group of helminths which are worms with a cylindrical and elongated body structure. The adult stages of these parasites inhabit the gastro-intestinal tract of ruminants and may cause serious health damages; *ie.,* anorexia, inflammation and edema of the mucosa, changes in the liver metabolism, depletion of iron reserves leading to anaemia, amongst others [6]. The result of such changes is reflected in weight loss, weakness and low productivity. In general, the health status of the animals can be dramatically compromised and may cause death in severe cases especially in young animals [7]. The most frequent GIN of ruminants in many countries around the world are *Haemonchus contortus, Mecistocirrus digitatus, Trichostrongylus colubriformis, T. axei, Bunostomum trigonocephalum, Cooperia curticei, Teladorsagia circumcincta, Nematodirus spp, Trichuris ovis, T. globulosa, Strongyloides papillosus, Gaigeria pachyscelis, Chabertia ovina and Oesophagostomum columbianum [8].*

1.2. Life Cycle of the Gastrointestinal Nematodes

The infection of the animals with GIN occurs when free-living stages of these parasites hatch from eggs to first stage larvae (L_1) in the feces. The L_1 is transformed into the second stage larvae (L_2) after a couple of days. The first two larval stages eat plant detritus, spores of bacteria or fungi, etc. After 5 to 7 days, the L_2 larvae develop to the third stage larvae (L_3). This stage is protected inside a sheath that enables them to resist desiccation outside the feces [9]. The L_3 larvae are the infective stages which abandon the feces and move towards the upper parts of grasses and herbs.

The L_3 larvae remain on the leaves (i.e., inside small droplets of water) to favour their ingestion by the ruminant hosts in the pastures. The protective sheath of the L_3 larvae allows them to survive the external environment but, once the larvae are ingested by the hosts, the L_3 larvae exsheath to allow for their establishment in the target organ [6]. Their specific target site within the host′s digestive tract depends on the genus and species of nematodes that enter the hosts. Some genus of GIN inhabit inside the abomasum which is the true stomach of ruminants (ie., H. contortus, M. digitatus, Trichostrongylus axei, Teladorsagia circumcincta).

Other worm genera like Bunostomum, Cooperia, Gaigeria, Nematodirus, Strongyloides, Toxocara and species such as Trichistrongylus colubriformis can only establish in the small intestine. Meanwhile, nematodes like Trichuris spp., Chabertia ovina and Oesophagostomum spp. establish inside the large intestine [10,11]. When the infective larvae enter the gastrointestinal tract of the hosts and reach their target organ, the nematode sheath is opened at the anterior end and the larvae escape from the sheath. The ex-sheathed larvae may enter the mucosa of the target organ and molt into the fourth stage larvae (L_4).

This stage is also known as histotropic larvae. After the L_4 stage, the fifth stage (L_5) can be found in the lumen of the gastrointestinal tract. This larval stage is also referred to as immature or juvenile worms and lead eventually to the sexually mature adult parasites. Mature male and female nematodes reproduce sexually inside the lumen of the digestive tract of the animals.

The fertilized females produce a large quantity of fertile eggs every day which are expelled to the soil in the fecal material of the ruminants [12].

1.3. How do the Ruminants Get Infected with Gastrointestinal Nematodes?

Once the feces containing the embryonated nematode eggs are eliminated by the infected animals to the soil, the nematode eggs proceed with their incubation period. The success in the development of the embryo within the egg, the larval hatching and the development process of the different free-living larval stages to reach the third stage (infective larvae) depends on the external conditions, i.e., ambient temperature, humidity and rainfall [13]. These conditions also influence the horizontal and vertical migration of nematode larvae on the plants [14].

For instance, under tropical or subtropical conditions the period from egg to infective larvae ranges from 3 to 5 days. In places where environmental conditions are drier or colder such development takes longer. Infective larvae possess the natural ability to respond to moisture and sun-light changes [15].

During the first hours of the morning, when the sun appears and the grass is wet, the larvae migrate through the water film in the leaf and they remain in the small droplets of dew at the upper end of the leaf. In this way, larvae are available to the animals that consume those pastures. When the sunlight is more intense and the ambient temperature increases, larvae respond by moving down from the top of the grass leaves and may even bury some centimeters under the soil surface [12].

The process is repeated everyday until one day larvae are successfully consumed by a susceptible host to continue their cycle previously described. If the infective larvae fail to enter a susceptible host, they will eventually finish their food reserves and die.

2. Methods of Control

2.1. Anthelmintic Chemotherapy

The AH products commercially available have been used for decades as the unique method for the control of ruminant parasitic nematodes around the world [16]. The commercial AH drugs have contributed substantially to the promotion of animal health, welfare and production for decades. Chemotherapy based on AH achieved a very high level of acceptance amongst farmers. At present, the AH drugs are used in preventive and curative programs as a good way to substantially reduce the parasitic burdens in grazing ruminants. However, only a few years after the launch of the first chemical AH drugs, the finding of parasites resistant to such drugs started to threaten the future of this control method [17]. At present, there is a worldwide concern about the inefficacy of AH compounds due to the development of GIN strains showing resistance to one or more of the three common AH groups (benzimidazoles, macrocyclic lactones and imidazothiazoles). For many years the problem of AH resistant worms was mainly considered an issue for sheep and goat farms [18]. However, recent work has shown that the problem of resistance in cattle nematodes was underestimated for a long time and now it has reached alarming levels [19,20]. International organizations are concerned about the global situation of anthelmintic resistance and its rapid spread around the world [21]. Some recommendations to manage anthelmintic resistance in livestock production systems are being suggested to governments in most countries [22]. In spite of the later, the problem of drug resistance is becoming more serious with time. Parasites are developing resistance to more than one group of AH compounds, which is known as multiple anthelmintic resistance (MAR). The MAR is a true risk for animal health and production worldwide [23,24].

Besides the problem of resistant nematodes, there are other negative issues around the use of chemical drugs ie., the potential risk to public health due to AH residues found in animal products and the potentially harmful effects of active residues from AH products excreted by the animals which can destroy beneficial organisms in the soil [25]. In spite of the problems around the use of AH drugs, at present there is no other strategy that can substitute commercial AH compounds with a similar degree of efficacy and security. Thus, a responsible use of AH against GIN is strongly recommended for preventive or curative programs [26]. Some areas within countries where grazing ruminant production used to be economically important have lost the battle against parasitic nematodes due to the dissemination of worms with MAR [27]. Therefore, there is an urgent need to find alternative methods for the control of GIN that can significantly reduce the dependence on AH drugs.

2.2. Other Strategies of Control of Ruminant Parasitic Nematodes that are Currently under Study

A number of strategies for the control of ruminant parasitic nematodes are being explored [28]. The search for new control strategies used as a single control option is changing towards the creation of different combinations of control tools leading to a more sustainable GIN control [29]. Some alternative methods of GIN control include the following:

a) Grazing management

Grazing management (*i.e.*, rotational grazing) is based on principles established by forage experts who determined that grazing areas need a resting period of a given time which may allow the grass to grow strong and healthy. Such practices aim at maximizing the use of the pasture [30]. Researchers working with the ecology of nematode larvae on the field found that larvae of parasitic nematodes have a limited survival time [9]. As we mentioned before, infective larvae actively move up and down the leaves of grasses and herbs. Such activity consumes the energy reserves of the larvae. If there are no animals grazing nearby, the chances to be ingested by a suitable host and the life expectancy are reduced. This system may help to reduce the larvae population on the grass and reduces the risk of re-infection. Under tropical conditions a resting period of 30 days is recommended to achieve the best use of tropical grasses [31] such as *Cynodon plectostachius* or *Brachiaria decumbens*. Studies performed under tropical humid conditions showed that a resting period of 30 days reduced the infectivity of the pasture significantly [32,33].

b) Improvement of the nutritional level of grazing animals

Grazing animals are confronted with two common problems under tropical conditions: the low quality of the grass and the marked seasonality of the fodder production in the fields. Most tropical grasses contain less than 8% crude protein (CP), which is the minimum level of CP needed for the correct function of the rumen. The seasonality of fodder availability in the field from the wet to the dry seasons may worsen the situation and, as a result, ruminants commonly show signs of malnourishment. Animals with a low nutritional level show poor production performance and are less capable of tolerating an infection with GIN. Studies with goats and sheep have shown that supplementary feeding (*ie.* providing a small quantity of the limiting nutrients in the pen) can help to improve the ability of these animals to endure infection and maintain production (resilience) and may also enable to implement an immune response against GIN (resistance) [4, 34]. Without any doubt, the first thing to suggest for the control of GIN in any grazing production system must be the improvement of the nutritional level of the animals.

c) Non-conventional anthelmintics

A number of non-conventional AH materials have been studied during the last decades [35]. All these AH materials share a common aspect: they have been used for reasons different to parasite control. However, they seem to have a useful degree of AH activity against GIN.

d) Copper oxide wire particles (COWP)

The COWP have shown a clear AH effect against abomasal adult nematodes (*ie., H. contortus*) of sheep and goats when administered *per os* in gel capsules [36,37]. When the needles reach the abomasum of the hosts they are retained in the folds of the abomasum for days or even weeks. The close contact between the COWP and the adult parasites takes place in the acidic environment of the abomasum and the cooper $^{+}$ions released from the needles seem responsible for their lethal effect on the parasites [38]. Its persistent AH effect was suggested by many authors and has been recently explored for up to 35 days post-treatment [39]. It is important to mention that dietary copper sulphate failed to show any AH effect [40] and the methalic copper is inert in the abomasum [41]. The COWP needles (1-2 mm length) are dosed inside gelatin capsules which are orally administered to the animals at different doses according to animal size. Growing sheep or goats are dosed with 2 g of COWP and the dose is increased in adult animals. Recent studies have shown that small doses of COWP can be offered using feed pellets with good AH effect [42]. Although this method is still under investigation, the results generated by a number of workers have created very good expectations for this parasite control tool. However, the use of COWP must avoid any toxic effect especially in sheep and will need to consider that the copper content in the liver of treated animals will be increased (limiting its quality for human consumption). Furthermore, the presence of copper in the feces and its effect on the normal microbiota in the soil needs to be evaluated.

e) Traditional medicine

This method is based on indigenous knowledge of the local pharmacopeia which has been an important source of information to test interesting plant materials that attack many infectious agents during the last decades. Ancestral indigenous knowledge from different cultures used medicinal plants in the form of plant extracts, infusions and other methodologies [43]. In many cultures around the world, plant extracts are being used as the only method to cure a number of diseases in humans and animals [44]. So far, there is a long list of bioactive compounds obtained from plants which are potentially active against animal parasitic nematodes. Several substances contained in different parts of the plants (leafs, roots, bark, etc.) have been explored. Several chemical compounds found in those materials have shown correlation with an AH effect: condensed tannins, flavonoids, terpens, sesquiterpens, etc. These and other bioactive compounds are frequently found in many plants of the local flora in many countries. Even more, some of those substances are present in the leaves eaten by ruminants in the field [45]. In recent years, the possibility of using plant materials to control GIN has taken great importance as a natural way to reduce the dependence on commercial AH drugs [46]. The number of plants and bioactive compounds obtained from plants around the world with potential use in the control of animal parasitic nematodes is increasing [47,48,49,50,51). The use of plant materials in the form of plant extracts or simply in the form of medicinal or nutraceutical fodder will hopefully be available in the near future and will contribute to the control of animal parasitic nematodes [52].

3. Free-Living Stages of Ruminant Parasitic Nematodes as a Vulnerable Target

In the last decades researchers have focused their attention on the study of free-living stages of ruminant parasitic nematodes in the field [53]. Some interesting facts that occur once the nematode eggs are deposited in the soil have been discovered. After the hatching process, the larvae face a number of natural enemies that attack and kill them. This phenomenon can explain why a single adult female (i.e., *H. contortus*) in the abomasum of a ruminant needs to produce several thousands eggs every day. The free-living stages of ruminant parasitic nematodes include the eggs and the three larval stages (L_1, L_2 and L_3). Similar to the situation with soil nematodes affecting agriculture, the free-living larval stages of ruminant parasitic nematodes are very fragile and susceptible to the attack of natural enemies like bacteria, virus, mites, other nematodes and macro and microfungi [54]. Researchers around the world are searching for a strategy to control ruminant parasitic nematodes using natural antagonists of nematodes as an additional tool of control. So far, the most widely studied natural enemies of ruminant parasitic nematodes are the nematophagous fungi. In the next section this important topic will be consider including the general aspects of this group of fungi, their important role in nature and their potential biotechnological use for the control of ruminant parasitic nematodes.

4. Nematophagous Fungi

4.1. General Aspects

Nematophagous fungi are microorganisms that possess the ability to act as predators (or parasites) of nematodes in nature. They are considered as one of the main bio-regulators of the nematode population in the soil. These microorganisms form part of the soil myco-biota and are common inhabitants of various substrates in nature such as soil, decaying plant material, plant roots and animal feces [55,56]. This group of fungi is saprophytic. However, in the presence of nematodes, such fungi become parasitic or predatory as they are facultative saprophyte fungi [57,58]. Nematophagous fungi can be isolated from the soil by sprinkling a small sample of soil in a Petri dish containing water agar [56]. After some days of incubation at room temperature, the nematophagous fungi can be visualized on a microscope (X50 and X75 magnifications). Nematophagous fungi produce two growing phases on agar plates. One phase grows under the agar surface. Once the fungi germinates on the agar, filament mycelia branches are disseminated under the agar surface. After some time the growing process generates aerial structures which include conidiophores which are a sort of stems that form the structures where spores (conidia) or chlamydospores will be produced (Figure 1 c,d) through a process known as conidio-genesis [59]. Other aerial structures can be seen as chlamydospores or resting spores produced by some fungi and trapping devices or even trapped nematodes that will be mentioned later in this chapter [56,60]. The growing behavior of the nematophagous fungi under *in vitro* conditions varies depending on the quality and quantity of the nutrients in the agar. Some fungi respond with abundant growth of mycelia and only a few conidiophores/conidia when media culture contains a rich source of protein and energy in the agar. In contrast, when a poor source of nutrients is supplied into the agar, some fungi respond with a large production of conidiophores/conidia [61]. According to the authors' own experience, using a poor source of nutrients in the agar is highly recommended to achieve the isolation of nematophagous fungi. Once conidiophores/conidia are visualized on the agar's surface with the stereo-microscope, a single conidia or conidiophore can be collected with a sterile fine needle and transferred to a sterile water agar plate. After one week of incubation at room temperature, the agar surface of plates is reviewed to verify if the nematophagous fungi were successfully isolated. The isolation process of nematophagous fungi can require two or more passes into sterile agar plates to achieve a pure culture.

4.2. Nematophagous Fungi: Mode of Action

Nematophagous fungi are classified for their study into two groups, related to their mode of action against nematodes. One of these groups is the endozoic or endoparasitic fungi group [56]. These fungi possess adhesive spores that adhere to the nematode cuticle and dissolve the cuticle wall. After a mechanical and a biochemical process, they penetrate and colonize the nematode´s body to consume its content. In this group of micro-fungi there are also some species which produce small spores which are ingested by the nematodes. Once the spores get inside the nematode´s intestine, they germinate and eat the internal tissues of the nematode, destroying its body from the inside [62].

There is another group of nematophagous fungi classified as nematode-trapping fungi. This group of fungi is considered as saprophytic. However, in the presence of nematodes, such fungi modify their mycelia by transforming them into trapping devices especially designed to capture and destroy nematodes, as well as feed on them (Figure 1 a,b). This is the way that nematophagous fungi become facultative saprophytes fungi [57]. Early evidence showed that, during the transformation from the saprophytic to the predatory state of the fungi, they respond to the stimulus provoked by the nematode cuticle protein called "nemin" [63]. The most common trapping devices produced by nematode-trapping fungi are tree-dimensional adhesive rings, adhesive branches or single and constricting rings.

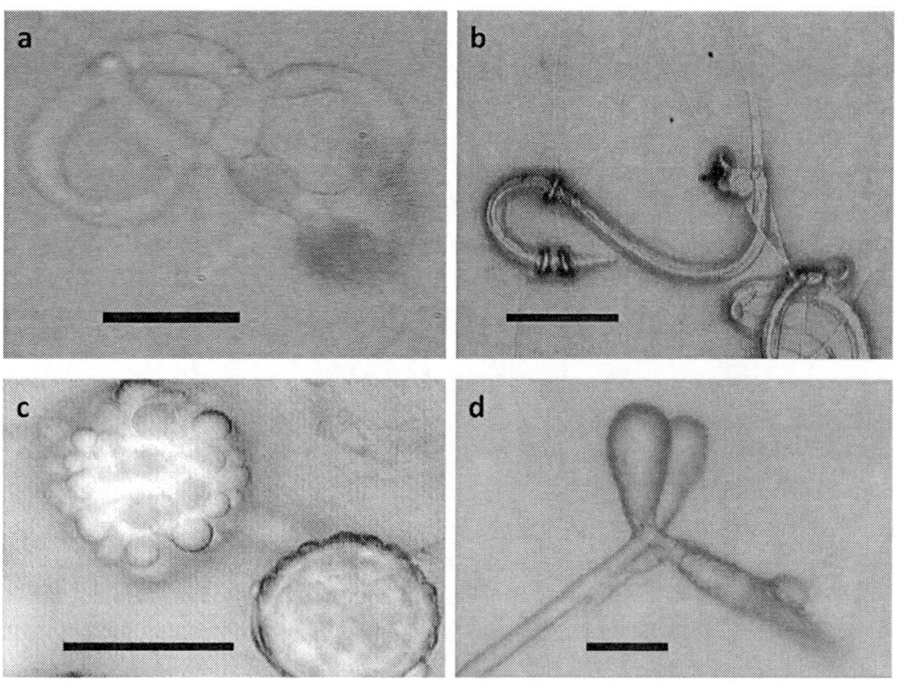

Figure 1. Microphotographs showing different aspects of the nematode-trapping fungi *Duddingtonia flagrans*: a) three-dimensional adhesive rings (bar=10µ); b*) Haemonchus contortus* infected larvae (L_3) captured by the *D. flagrans* trapping devices (bar=100 µ); c) chlamydospores (bar=10 µ); d) spores (bar=10µ).

Whatever the kind of trapping device available, the fungi establish a point of contact with the nematode cuticle wall and an infective bulb is developed from the fungus cell that makes mechanical pressure on the nematode′s cuticle [64]. Such mechanical process is complemented with a biochemical process produced by enzymes that hydrolyze the cuticle wall. The latter is eventually broken, helping the infective bulb to penetrate the nematode body. Likewise, some enzymes produced by the nematode trapping fungi are being identified as responsible for the lethal or paralyzing effect against nematodes that also contribute to the degradation of the nematode body [62]. Once the nematode is penetrated by the fungus, hyphae are differentiated in specialized infective hyphae that grow through the nematode body. Infective hyphae grow and invade the whole body before they become an assimilative hyphae with a clear biological activity: taking the nutrients from the nematode internal tissues [65]. The predation process of nematodes by the nematode-trapping fungi ends when the nematode body is totally degraded and the nematode carcass is completely full of fungal hyphae [64].

Nematophagous fungi have been explored as possible agents for the control of plant [66,67,68] and animal parasitic nematodes [69,70,71,72,73,74]. Studies about the predatory or parasitic behavior of nematophagous fungi against parasitic nematodes of importance in agriculture have shown that some specie of fungi *ie., Arthrobotrys conoides* and *A. oligospora* can produce enzymes with proteoilitic activity [75,76,77]. Other fungi like *Verticillium chlamydosporium* (*Pochonia*) are being classified as ovicidal fungi. The latter produces enzymes that contribute to the invasion of eggs of *Meloidogyne incognita,* which is one of the major agriculture plagues affecting the root system in tomato plants [78,79,80]. Other authors studying the activity of *Verticillium* spp. against *Ascaris lumbricoides* eggs obtained from pigs found that after exposure to these fungi species the egg shells were degraded by the enzyme activity followed by the infection of the eggs [81,82,83,84]. Nematophagous fungi have been evaluated as possible agents to control parasites in dogs *ie., Ancylostoma* sp., with good results [85].

4.3. Nematophagous Fungi in the Control of Ruminant Parasitic Nematodes

For many years researchers have looked for a practical way to use nematophagous fungi for the biological control of ruminant parasitic nematodes. From the beginning, they faced an almost impossible task: finding a practical method to establish the close contact between the nematophagous fungi and the free living stages of the GIN in the soil. In the early forties (during the twentieth century), a group of researchers in France performed the first attempts to establish the possible use of nematophagous fungi for the control of GIN in the livestock industry with good results [86,87,88]. The time of such studies coincided with the second World War and after that period no more work was published by this group. In the mid-seventies, Dr. Pryadko and colleagues from the National Academy of Sciences of Khazakstan (former USSR) had the smart idea of administering spores of selected nematophagous fungi (*Arthrobotrys flagrans* and *A. arthrobotryoides*) to the animals through an oral suspension. By doing this the spores were expected to pass through the digestive tract of the hosts to be expelled in the feces to the soil. By doing this, the spores and the recently hatched larvae of GIN established close contact in the feces. This allowed the fungi to display their powerful predatory effect against the nematode free-living larvae before they could abandon the fecal material. The results obtained by Pryadko and his group in reducing the nematode larval population of gastrointestinal and lung parasites in feces were spectacular. The recovered larvae in the feces of treated sheep were significantly lower than those in the non-treated sheep feces [89]. Such results renewed expectations for a new strategy to control ruminant parasites. During the eighties and nineties, Dr. Peter Nansen and his team from the Danish Centre for Experimental Parasitology, Royal Veterinary and Agricultural (University in Copenhagen, Denmark), continued the study of nematophagous fungi and searched for its possible commercial use to control ruminant parasitic nematodes [69,90]. This research group found that *D. flagrans* was capable of surviving after passing the gastrointestinal tract of ruminants both under *in vitro* and *in vivo* conditions. Their results confirmed Dr. Pryadko′s results: the oral administration of spores of the same specie *D. flagrans* (*A. flagrans*) fed to sheep resulted in a substantial reduction of the nematode larvae population in sheep feces. The research group of Dr Michael Larsen in Denmark intensified their studies and successfully established the basis for the control of ruminant parasitic nematodes based on the oral administration of spores using their own autochthonous Danish *D. flagrans* strain [90]. Likewise, Dr. Peter Waller from the CSIRO Livestock Industry, Armidale, Australia, also explored the use of *D. flagrans* in the control of sheep parasites and incorporated *D. flagrans* chlamydospores into nutritional blocks. Their results showed that the consumption of either barley grains or blocks containing *D. flagrans* chlamydospores virtually eliminated the larval numbers from fecal cultures [91]. Meanwhile, in Mexico, a local strain of *D. flagrans* was found (FTHO-8) by the team of the National Centre for Disciplinary Research in Veterinary Parasitology from the National Institute of Forestry, Agriculture and Livestock. It was obtained from sheep feces at a sheep farm in Fierro del Toro, Morelos, Mexico. The strain has been studied since the early nineties, searching for its potential use for the control of ruminant parasitic nematodes with encouraging results under controlled *in vivo* assays with sheep [60,92,93] and goats [94]. Further studies have attempted to combine the efficacy of the combined use of Albendazol and *D. flagrans* chlamydospores for the control of sheep haemonchosis. The results showed that this method reduced from 69 to 79% the larval population in fecal cultures [72]. A field trial was performed in the cool high plateau of Tres Marías, Mexico, using "tracer" sheep. A group of grazing sheep was supplemented with *D. flagrans* chlamydospores mixed with oat grains and molasses. The control group remained without fungi. This method resulted in a substantial reduction in the larvae population on the grass: 51.1% and 100% for *H. contortus* and *Cooperia* sp. infective larvae, respectively compared to the control. Likewise, the number of adult nematodes recovered at necropsy from both groups showed an important reduction of 56% in the parasitic burden of *T. circumcincta* [71]. More recently, a new strategy is being developed to administer *D. flagrans* chlamydospores to sheep. Such strategy consists in the design of a nutritional mix containing a source of protein and energy, which can help to improve the resilience of sheep and to stimulate the immune system of lambs against GIN, plus an amount of chlamydospores of the Mexican isolate of *D. flagrans* (FTHO-8). These ingredients are also being integrated into nutritional pellets (also called "cookies for sheep") which are easily consumed by sheep. The idea is to substitute part of the commercial feed for the use of nutritional pellets containing *D. flagrans* chlamydospores. This system has showed 82% reduction of the nematode larvae population in fecal cultures [95]. Recent trials showed that the spores contained in the nutritional pellets possess a long shelf-life (at least 80 days) under different environmental conditions (*i.e.*, out-doors, refrigeration and in-doors) [96]. Even more, nutritional pellets were equally useful to administer the spores when compared to the administration using a suspension prepared with oat grains and molasses [97]. These nutritional pellets are expected to help improve the nutritional status of the animals, achieving a better immune response against adult parasites, and also to "clean" the infective larvae in the pasture by the action of the fungi incorporated in the nutritional pellets [98]. The Mexican group has also investigated means to identify the relationship between the dose of spores offered to the animals *per os* and the number of spores being recovered in the feces. For this purpose the McMaster technique was adapted to quantify the spores in the feces [93]. Different densities were tested and, fortunately, the same density used for counting nematode eggs was useful for the quantification of chlamydospores per gram of feces. A clear and strong correlation was observed between the spores offered to the animals and the spores counted in the feces. Once a technique was available to count the number of spores in the feces, a new opportunity was found: testing the apparent digestibility of the chlamydospores offered to the animals *per os* [99]. Previous attempts to determine the survival/presence of spores after passing the gastrointestinal tract were qualitative (the feces were either positive or negative). Now it was possible to count the number of spores surviving the gastrointestinal tract of ruminants. The *in vivo* and *in vitro* results showed that a large proportion of spores are destroyed in the gastrointestinal tract (more than 89% under the conditions of the *in vivo* experiment). Furthermore, the *in vitro* work showed that spores are not destroyed in the rumen but are destroyed when exposed to the acidic environment of the abomasum. No information is available on the effect of the intestinal enzymes on the integrity of the spores. Information on the apparent digestibility of spores may help to explain variation in efficacy results between *in vivo* trials using different basal diets [99]. A basal diet with more volume (bulky) might cause lower digestion of spores but a basal diet with small volume (concentrates) will cause a higher digestion (destruction) of spores.

In spite of the initial doubts about the possible practicality of this tool in large scale commercial application farms, teams of workers from different parts of the world continued attempting to find a possible practical application of the nematophagous fungi. For example, in Brazil a trial was conducted to evaluate the effect of *D. flagrans* chlamydospores (produced in sorghum) mixed with a maintenance daily ration when grazing a paddock during eight months. Their results showed a fecal egg count reduction of around 56.8% in the last three months of the experiment, with a range between 10.1 and 67.1% in the treated group compared to the control. The faecal cultures demonstrated that the main nematodes found in both groups were *Cooperia* and *Haemonchus*. Larvae counts in the pasture showed a reduction percentage of around 77.1% in the treated group at the end of experiment [73]. More recently, a field experiment was conducted in Australia administering a supplement containing chlamydospores of an autochthonous *D. flagrans* strain everyday for 60 days. The control group was kept under the same conditions except for the presence of *D. flagrans* in the supplements. After this period, sheep were moved out from the experiment and replaced by worm-free tracer sheep. After 22 days sheep were transferred to pens with slatted floors for 20 days. Ten sheep of each group were slaughtered for worm burden estimation. Their results showed that the numbers of larvae in feces were reduced more than 60% and the parasitic burdens in the animals reduced by 57% [100]. Better results have been recently obtained in Uruguay, with the use of energy blocks containing *D. flagrans* for the control of GIN of sheep using "tracer" animals. Their results showed reductions in the parasitic burdens at necropsy of 100% against *H. contortus* and *T. circumcincta,* 89.9% against *T. colubriformis*, 87.5% against *Cooperia oncophora* and 90% against *T. axei* [101].

If *D. flagrans* is going to be used for the control of GIN under field conditions, it is important to consider its use only when the environmental conditions are favorable for the hatching of GIN eggs in the feces. If the eggs cannot hatch (during the dry season, cold winter or any other adverse time), the spores will have a reduced efficacy against the larvae in the feces. Also, when the spores are fed to animals with low egg counts in the feces, the efficacy of *D. flagrans* might be reduced as it cannot achieve its predatory activity with the same high efficacy achieved in feces from an animal with high worm egg counts (leading to high numbers of larvae). Therefore, any control tool that reduces the number of eggs in the feces may be contra-indicated when using *D. flagrans* spores. Furthermore, it might be sensible to use the spores only in those animals shedding high worm egg counts only (selective application during the peri-parturient period). By doing this, the control effort will be concentrated in those animals that contaminate the field with larvae and there will be no waste of biological material in those animals with low egg counts.

In most of the trials described above with *D. flagrans* spores, the reduction percentages were variable and depended on the individual experimental conditions of each research protocol. Nevertheless, the results obtained with *D. flagrans* chlamydospores, whatever the means of administration, showed reductions in the population of larvae either in fecal cultures or in pasture. Also the parasitic burden estimated from tracer animals at necropsy ranged between 60 and 80% lower than the control group (as a unique measure of control). The use of nematophagous fungi should not be considered as the only tool for the control of GIN. It is important to consider this tool as part of an integrated control program either combining or alternating it with other measures of control, *ie.,* the use of plant extracts with AH effects, COWP needles and grazing rotation. Even other biological agents of GIN control that are currently under study *ie.,* "cannibal nematodes" [102], bacteria like *Bacillus thuringiensis* [103] or nematode-predatory mites [104] could be combined with nematophagous fungi in order to achieve the best results in the control of ruminant parasitic nematodes.

Conclusions

The use of *D. flagrans* as a biological control agent to control GIN of ruminants is still a promising option. The success of this tool of control will depend on the understanding of the complex relationships between the fungi strain used, the type of diet consumed by the animals and the factors affecting the survival of the spores and the nematode eggs in the feces, as well as the relationship between the fungi and the larvae in the feces. Although many aspects of these interactions are yet to be discovered, the field trials with *D. flagrans* performed in different countries clearly indicate that this tool must not be discarded from the basket of options for the sustainable control of GIN of ruminants. Clear advances have been achieved with different administration protocols and long term field studies so far. The use of *D. flagrans* as a biological control tool against GIN is still at the stage of implementation, and many other biological control tools are yet to be included in our basket of options.

References

[1] FAO, (2001). Sustainable approaches for managing haemonchosis in sheep and goats. *FAO Animal Production and Health Paper.* Final Report of a FAO Technical Co-operation Project in South Africa.

[2] Poglaven, G. and Battelli, G. (2006) An insight into the epidemiology and economic impact of gastro-intestinal nematodes in small ruminants. *Parasitologia.* 48: 409-413.

[3] Guglielmone, A.A., Castelli, M.E, Mangold, A.J., Aguirre, D.H., Alcaraz, E., Cafrune, M.M., Cetrá, B., Luciani, C.A. and Suárez, V.H. (2007). El uso de acaricidas para el control de Rhipicephalus (Boophilus) microplus (Canestrini) (Acari: Ixodidae) en la Argentina. *Revista de Investigaciones Agropecuarias.* 36 (abril).

[4] Torres-Acosta J.F.J., Jacobs D.E., Aguilar-Caballero A., Sandoval-Castro C., May-Martinez M. and Cob-Galera L.A. (2004). The effect of supplementary feeding on the resilience and resistance of browsing Criollo kids against natural gastrointestinal nematode infections during the rainy season in tropical Mexico. *Veterinary Parasitology.* 124: 217-238.

[5] West, D.M., Pomroy, W.E., Kenyon, P.R., Morris, S.T., Smith, S.L. and Burnham, D.L. (2008). The effect of subclinical parasitism on a ewe flock 2008. In: *Proceedings of the Society of Sheep and Beef Cattle Veterinarians of the New Zealand Veterinary Association, Annual Seminar.* pp. 97-102.

[6] Hidalgo-Arguello, R.M. and Cordero del Campillo, M. (1999). Estrongiloidosis. In: Parasitología Veterinaria. M. Cordero del Campillo & Rojo Vázquez, A.F. (Eds) McGraw-Hill. 234-248.

[7] Macedo-Barragán, R., Arredondo-Ruiz, V., Ramírez-Rodríguez, J. and García-Márquez, L.J. (2009). Grazing sheep poisoned by milkweed Asclepias curassavica or gastrointestinal nematosis? *A case report findings Veterinaria México.* 40: 275-281.

[8] Torres-Acosta, J.F.J. and Aguilar-Caballero, A.J. (2005). Epidemiología, prevención y control de nematodos gastrointestinales en rumiantes. En: Rodríguez, V.I. (Editor). *Enfermedades de importancia económica en producción animal.* Editorial McGraw-Hill Interamericana / UADY. pp. 145-173.

[9] Ellenby, C. (1968). Desiccation survival of the infective larva of Haemonchus contortus. *Journal of Experimental Biology.* 49: 469-475.

[10] Valcárcel-Sancho, F., Rojo-Vázquez, F.A., Olmeda-García, A.S., Arribas-Novillo, B., Márquez-Sopeña, L., Fernando-Pat, N. (2009). Atlas de Parasitología Ovina. *Editorial Servet. Zaragoza,* España, 137p.

[11] González-Garduño, R., Córdoba-Pérez, C., Torres-Hernández, G., Mendoza-de-Gives, P. and Arece-García, J. (2011). Prevalence of gastrointestinal parasites in slaughtered sheep at a slaughterhouse in Tabasco, Mexico. *Veterinaria México.* 42: 125-135.

[12] Olsen, W.O. (1974). Animal Parasites. Their Life Cycles and Ecology. Third Edition. *University Park Press.* London, UK. pp. 430-433.

[13] Fernández-Ruvalcaba, M, Vázquez-Prats, V., and Liébano-Hernández, E. (1994). Development and recovery of Haemonchus contortus first larval stages on experimental plots in Mexico. *Veterinary Parasitology.* 51: 263-269.

[14] Onyali, I.O., Onwuliri, C.O.E., and Ajayi, J.A. (1990). Development and survival of Haemonchus contortus larvae on pasture at Von, Plateau State, Nigeria. *Veterinary Research Communications*. 14: 211-216.

[15] Chaudary, F.R., Qayyum, M., and Miller, E.J. (2008). Development and survival of Haemonchus contortus infective larvae derived from sheep faeces under sub-tropical conditions in the Potohar region of Pakistan. *Tropical Animal Health and Production*. 40: 85-92.

[16] Keyyu, J.D., Kyvsgaard,N.C., Kassuku, A.A. and Willingham, A.L. (2003). Worm control practices and anthelmintic usage in traditional and dairy cattle farms in the southern highlands of Tanzania. *Veterinary Parasitology*. 114: 51–61.

[17] Love, S., (2011). Drench resistance and sheep worm control. http://www.dpi.nsw.gov.au/__data/assets/pdf_file/0009/111060/Drench-resistance-and-sheep-worm-control.pdf

[18] Kaplan, R.M. (2004). Drug resistance in nematodes of veterinary importance: a status report. *Trends in Parasitology*. 20: 477-481.

[19] Canul-Ku, H.L., Rodriguez-Vivas, R.I., Torres-Acosta J.F.J., Aguilar-Caballero, A.J., Perez-Cogollo, L.C. and Ojeda-Chi, M.M. (2011). Prevalence of cattle herds with ivermectin resistant nematodes in the hot sub-humid tropics of Mexico. *Veterinary Parasitology*. doi:10.1016/j.vetpar.2011.07.029.

[20] Sutherland, A.I. and Leathwick, M.D. (2011). Anthelmintic resistance in nematode parasites of cattle: a global issue? *Trends in Parasitology*. 27: 176-181.

[21] FAO, (2009). Conferencia Electrónica 2009. Uso de alternativas para el control de los parásitos" Red de Helmintología para América Latina y el Caribe. 14 Octubre – 14 Noviembre 2009. http://cnia.inta.gov.ar/helminto/Guidelines/hel3.htm

[22] Pech, C.L., Doole, G.J. and Pluske, J.M. (2009). Economic management of anthelmintic resistance: model and application. *The Australian Journal of Agricultural and Resource Economics*. 53: 585-602.

[23] Sargison, N., Jackson, F., Bartley, D., Wilson, D., Stenhouse, L., and Penny, C. (2007). Observations on the emergence of multiple anthelmintic resistance in sheep flocks in the south-east of Scotland. *Veterinary Parasitology*. 145: 65-76.

[24] Taylor, M.A., Learmount, J., Lunn, E., Morgan, C. and Craig, B.H. (2009). Multiple resistance to anthelmintics in sheep nematodes and comparison of methods used for their detection. *Small Ruminant Research*. 86: 67-70.

[25] Martínez, M.I. and Cruz, R.M. (2009). The use of agricultural and livestock chemical products in the cattle-ranching area of Xico, central Veracruz, Mexico, and their possible environmental impact. *Acta Zoológica Mexicana*. 25: 637-681.

[26] Sargison, N. (2011). Responsible use of anthelmintics for nematode control in sheep and cattle. *In Practice*. 33: 318-327.

[27] Van Wyk, J.A. (2003) Think refugia or lose the battle against drug resistance. In: *Proceedings of the 5th International Seminar of Animal Parasitology*. Mérida, Mexico, p.p. 39–47.

[28] Padula, A. (2011). Gastrointestinal nematodes of sheep and cattle: Biology and control. *Australian Veterinary Journal*. 89: 54.

[29] Hoste, H., and Torres-Acosta, J.F.J. (2011). Non chemical control of helminths in ruminants: adapting solutions for changing worms in a changing world. *Veterinary Parasitology*. 180: 144-54.

[30] Aljoe, H. (2006). Rotation improves grazing efficiency. Southwest Farm Press. http://southwestfarmpress.com/rotation-improves-grazing-efficiency

[31] Waller, P., Chandrawathani, P. (2005). Haemonchus contortus: Parasite problem No. 1 from Tropics - Polar Circle. Problems and prospects for control based on epidemiology. *Tropical Biomedicine*. 22: 131–137.

[32] Barger, I.A. (1996a). Grazing management for the control of nematode parasites. In: Sustainable parasite control in small ruminants. An international workshop sponsored by ACIAR and held in Bogor, Indonesia 22-25 April 1996. *ACIAR Proceedings,* No. 74: 146-150.

[33] Barger, I.A. (1996b). Prospects for integration of novel parasite control options into grazing systems. *International Journal for Parasitology*. 26: 1001-1007.

[34] Retama-Flores, C., Torres-Acosta, J.F.J., Sandoval-Castro, C.A., Aguilar-Caballero, A.J., Cámara-Sarmiento, R. and Canul-Ku, H.L. (2012). Effect of maize supplementation on the fodder selection of growing sheep and the metabolic cost of gastrointestinal parasite infection. *Animal*. 6: 145–153.

[35] Stear, J.M., Doligalska, M. and Donskow-Schmelter, K. (2007). Alternatives to anthelmintics for the control of nematodes in livestock. *Parasitology,* 134: 139–151.

[36] Chartier, C., Etter, E., Hoste, H., Pors, I., Koch, C., Dellac, B. (2000). Efficacy of copper oxide needles for the control of nematode parasites in dairy goats. *Veterinary Research Communications*. 24: 389–399.

[37] Knox, M.R., 2002. Effectiveness of copper oxide wire particles for Haemonchus contortus control in sheep. *Australian Veterinary Journal*. 80: 224-227.

[38] Miller, E.J., Burke, M.J., Terrill, H.T. and Kearney, T.M. (2011). A comparison of two integrated approaches of controlling nematode parasites in small ruminants. *Veterinary Parasitology*. 178: 300-310.

[39] Galindo-Barboza, A.J., Torres-Acosta, J.F.J., Cámara-Sarmiento, R., Sandoval-Castro, C.A., Aguilar-Caballero, A.J., Ojeda-Robertos, N.F., Reyes-Ramírez, R. and España-España, E., (2011). Persistence of the efficacy of copper oxide wire particles against Haemonchus contortus in sheep. *Veterinary Parasitology*. 176: 201–207.

[40] Burke, J.M. and Miller, J.E. (2008). Dietary copper sulfate for control of gastrointestinal nematodes in goats. *Veterinary Parasitology*. 154: 289-293.

[41] Langlands, J.P., Bowles, J.E., Donald G.E. and Smith, A.J. (1986). Trace element nutrition of grazing ruminants. I. Degree of oxidation, diet, frequency of dosing and location in the gastro-intestinal tract as factors affecting the ability of oxidized copper wire to promote hepatic copper storage. *Australian Journal of Agricultural Research*. 37: 179-188.

[42] Burke, J.M., Soli, F., Miller, J.E., Terrill, T.H., Wildeus, S., Shaik, S.A., Getz, W.R. and Vanguru, M. (2010). Administration of copper oxide wire particles in a capsule or feed for gastrointestinal nematode control in goats. *Veterinary Parasitology*. 168: 346–350.

[43] Yadav, N., Vasudeva, N., Sing, S. and Sharma, K.S. (2007). Medicinal properties of genus Chenopodium Linn. *Natural Product Radiance*. 6: 131-134.

[44] Willcox, M., Bodeker, G. and Rasoanaivo, P. (2005). Traditional medicinal plants and malaria: Volume 4 of the Traditional Herbal Medicine for Modern Times Series. *The Journal of Alternative and Complementary Medicine*. 11: 381-382.

[45] Martínez-Ortíz-de-Montellano, C., Vargas-Magaña, J.J., Canul-Ku, H.L., Miranda-Soberanis, R., Capetillo-Leal, C., Sandoval-Castro, C.A., Hoste, H. and Torres-Acosta, J.F.J. (2010). Effect of a tropical tannin-rich plant Lysiloma latisiliquum on adult populations of Haemonchus contortus in sheep. *Veterinary Parasitology*. 172: 283-90.

[46] Hoste, H., Torres-Acosta, J.F., Alonso-Díaz, M.A., Brunet, S., Sandoval-Castro, C. and Adote, S.H. (2008). Identification and validation of bioactive plants for the control of gastrointestinal nematodes in small ruminants. *Tropical Biomedicine*. 25: 56-71.

[47] Githiori, B.J., Athanasiadou, S. and Thamsborg, M.S. (2006). Use of plants in novel approaches for control of gastrointestinal helminths in livestock with emphasis on small ruminants. *Veterinary Parasitology*. 139: 308-20.

[48] Al-Shaibani, I.R.M., Phulan, M.S., Arijoand, A. and Qureshi, A. (2008). Ovicidal and larvicidal properties of Adhatoda vasica (L.) extracts against gastrointestinal nematodes of sheep in vitro. *Pakistan Veterinary Journal*. 28: 79-83.

[49] López-Aroche, U., Salinas-Sánchez, D.O., Mendoza-de-Gives, P., López-Arellano, M.E., Liébano-Hernández, E., Valladares-Cisneros, G., Arias-Ataide, D.M. and Hernández-Velázquez, V. (2008). In vitro nematicidal effect of medicinal plants from the Sierra de Huautla Biosphere Reserve, Morelos, Mexico against Haemonchus contortus infective larvae. *Journal of Helminthology*. 82: 25-31.

[50] Eguale, T. and Mirutse, G. (2009). In vitro anthelmintic activity of three medicinal plants against Haemonchus contortus. *International Journal of Green Pharmacy*. 3: 29-34.

[51] Abdel-Ghaffar, F., Semmler, M., Khaled, A., Al-Rasheid, S., Strassen, B., Fischer, K., Aksu, G., Klimpel, S. and Mehlhorn, H. (2011). The effects of different plant extracts on intestinal cestodes and on trematodes. *Parasitology Research*. 108: 979–984.

[52] Alonso-Díaz, M.A., Torres-Acosta, J.F.J., Sandoval-Castro, C.A. and Hoste, H. (2010). Tannins in tropical tree fodders fed to small ruminants: a friendly foe? *Small Ruminant Research*. 89: 164-173.

[53] O'Connor, L.J., Walkden-Brown, S.W. and Kahn, L.P. (2006). Ecology of the free-living stages of major trichostrongylid parasites of sheep. *Veterinary Parasitology*. 142: 1–15.

[54] Mankau, R. (1980). Biological control of nematode pests by natural enemies. *Annual Review of Phytopathology*. 18: 415-440.

[55] Zwirn-Hirsch, H.E. (1947). Nematode destroying fungi isolated from sheep dung. *Palestine Journal Botany*. SER 4: 56-57.

[56] Barron, G.L. (1977). The nematode-destroying fungi. Topics in Mycobiology. No 1, Canadian Biological Publications, LTD, Guelph, Canada. 140 pp.

[57] López-Llorca, L.V., Maciá-Vicente, J.G. and Janson, H.B. (2007). Mode of action and interactions of nematophagous fungi. *Integrated Managenent and Biocontrol of Vegetable and Grain Crops Nematodes*. 2: 51-76.

[58] Mendoza-de-Gives, P. and Valero-Coss, R.O. (2009). Uso de hongos nematófagos: Una herramienta biotecnológica para el control de nematodos parásitos del ganado. *Folleto Técnico No. 2 CENID-Parasitología Veterinaria del INIFAP*. 39 pp.

[59] Rubner, A. (1996). Revision of predacious hyphomycetes in the Dactylella-Monacrosporium complex. *Studies in Mycology*. 39:1–134.

[60] Llerandi-Juarez, J.R.D. and Mendoza-de-Gives, P. (1998). Resistance of nematophagous fungi chlamydospores to the digestive processes of sheep in Mexico. *Journal of Helminthology*. 72: 155-158.

[61] Saxena, G., Dayal, R. and Mukerji, K.G. (1989). Nutritional studies on nematode-trapping fungi. *Folia Microbiologica*. 34: 42-48.

[62] Nordbring-Hertz, B., Zunke, U., Wyss, U. and Veenhuis, M. (1986). Trap formation and capture of nematodes by Arthrobotrys oligospora. *Film* (C1622). Produced by Institut fur den Wissenschaftlichen Film, Gottingen. Germany.

[63] Pramer, D. and Kuyama, S. (1963). Nemin and the nematode-trapping fungi. *Bacteriological Reviews*. 27: 282–292.

[64] Duddington, C.L. (1955). Fungi that attack microscopic animals. *Botanical Review*. 21: 377-439.

[65] Grønvold, J., Korsholm, H., Wolstrup, J., Nansen, P. and Henriksen, S.A. 1985. Laboratory experiments to evaluate the ability of Arthrobotrys oligospora to destroy infective larvae of Cooperia species, and to investigate the effect of physical factors on the growth of the fungus. *Journal of Helminthology*. 59: 119-126.

[66] Bordallo, J.J., Lopez-Llorca, V., Jansson, H-B, Salinas, J., Persmark, L. and Asensio, L. 2002. Colonization of plant roots by egg-parasitic and nematode-trapping fungi. *New Phytologist*. 154: 491–499.

[67] Kerry, B. (2004). Biological Control of Nematodes. In: Encyclopedia of Plant and Crop Science. (Edited by Robert M Goodman). Nematode interaction Unit. *Rothamsted Experimental Station. Marcel Decker,* London UK 1300p.

[68] Flores-Camacho, R., Atkins, D.S., Manzanilla-López, R.E., Cid-del-Prado-Vera, I. and Martínez-Garza, A. 2008. Caracterización de Aislamientos Mexicanos de Pochonia chlamydosporia var. chlamydosporia (Goddard) Gams y Zare para el control biológico de Nacobbus aberrans (Thorne) Thorne y Allen. *Revista Mexicana de Fitopatología*. 26: 93-104.

[69] Larsen, M. (2000). Prospects for controlling animal parasitic nematodes by predacious micro fungi. *Parasitology*. 120: S121-S131.

[70] Mendoza-de-Gives, P. (1998) Perspectives of the use of nematode-trapping fungi in the livestock industry in Mexico. In: International Workshop on Biological Control of Gastro-intestinal nematodes of ruminant using predacious fungi. Veterinary Research Institute, Ipoh, Malaysia. 5-12 October 1997. Published by FAO. *FAO Animal Production and Health Paper* No 141. Pp. 82-85.

[71] Mendoza-de-Gives, P. Zapata-Nieto, C., Liébano-Hernández, E., López-Arellano, M.E., Herrera-Rodríguez, D. and González-Garduño, R. (2006). Biological control of gastro-intestinal parasitic nematodes using Duddingtonia flagrans in sheep under natural conditions in Mexico. *Annals of the New York Academy of Science*. 1081: 355-359.

[72] Arroyo-Balán, F.L., Mendoza-de-Gives, P., López-Arellano, M.E., Liébano-Hernández, E. Vázquez-Prats, V.M., Miranda-Miranda, E. and Ortiz-de-Montellano-Nolasco, A.M. (2008). Evaluación de un método combinado de control de la hemoncosis ovina bajo condiciones controladas. *Técnica Pecuaria en México*. 46: 217-223.

[73] Bañolas-Jobim, M., Morais-Santurio, J. and De-la-Rue, M.L. (2008) Duddingtonia flagrans: biological control of cattle nematodes in the field. *Ciência Rural, Santa Maria*. 38: 2256-2263.

[74] Soto-Barrientos, N., Oliveira, J.B, Vega-Obando, R., Montero-Caballero, D., Vargas, B., Hernández-Gamboa, J. and Orozco-Solano, C., (2011). In-vitro predatory activity of nematophagous fungi from Costa Rica with potential use for controlling sheep and goat parasitic nematodes. *International Journal of Tropical Biology.* 59: 37-52.

[75] Tunlid, A. and Jansson, S. (1991). Proteases and their involvement in the infection and immobilization of nematodes by the nematophagous fungus Arthrobotrys oligospora. *Applied and Environmental Microbiology.* 57: 2868-2872.

[76] Aswani, V.H. and Jaffar, M.B. (1992). Protease from two nematophagous fungi, Arthrobotrys oligospora and A. conoides. *Indian Journal of Experimental Biology.* 30: 881-884.

[77] Persson, Y. and Friman, E. (1993). Intracellular proteolytic activity in mycelia of Arthrobotrys oligospora bearing mycoparasitic or nematode trapping structures. *Experimental Mycology.* 17: 182-190.

[78] Segers, R., Butt, M.T., Kerry, R.B., Beckett, A. and Peberdy, F.J. (1994). The nematophagous fungus Verticillium chlamydosporium produces a chymoelastase-like protease which hydrolyses host nematode proteins in situ. *Microbiology.* 140: 2715-2723.

[79] Segers, R., Butt, M.T., Kerry, R.B., Beckett, A. and Peberdy, F.J. (1996). The role of proteinase VCP1 produced by the nematophagous fungus Verticillium chlamydosporum in the infection process of nematode eggs. *Mycological Research.* 100: 421-428.

[80] Segers, R., Butt, M.T., Kerry, R.B., Carder, H.J., Keen, N.J., Kerry, R.B. and Peberdy, F.J. (1999). The subtilisins of fungal pathogens of insects, nematodes and plants: distribution and variation. *Mycologycal Research.* 103: 395-402.

[81] Lýsek, H. and Bacovskí, J. (1979). Penetration of ovicidal fungi into altered eggs of Ascaris lumbricoides. *Folia Parasitologica.* 26: 139-142.

[82] Lýsek, H. and Krajcí, D. (1987). Penetration of ovicidal fungus Verticillium chlamydosporium through the Ascaris lumbricoides egg-shells. *Folia Parasitologica.* 34: 57-60.

[83] Kunert, J., Zemek, J., Augustin, J. and Kuniak, L. (1987). Proteolytic activity of ovicidal soil fungi. *Biologia* (Bratislava). 42: 695-705.

[84] Lýsek, H. and Sterba, J. (1991). Colonization of Ascaris lumbricoides eggs by the fungus Verticillium chlamydosporium Goddard. *Folia Parasitologica.* 38: 255-259.

[85] Carvalho R.O., Araújo J.V., Braga F.R., Araujo J.M., Silva A.R. and Tavela A.O. (2009). Predatory activity of nematophagous fungi on infective larvae of Ancylostoma sp.: evaluation in vitro and after passing through the gastrointestinal tract of dogs. *Journal of Helminthology.* 83: 231-236.

[86] Descazeaux, J. (1939). Action des champignons hyphomycetes prédateurs sur les larves de certains nématodes parasites des Ruminants. *Bulletin of the Society of Pathologic Exotics.* 32:457-459.

[87] Descazeaux, J. and Capelle, N. (1939). Contributions á l´étude des Champignons prédateurs de larves de Nématodes parasites des animaux domestiques. *Bulletin de l'Academie Veterinaire de France.* 12: 284-288.

[88] Deschiens, R. (1939). Capture et destruction de larves de strongylidés du singe et du boeuf par des hypomycetes. *Bulletin of the Exotic Pathology Society.* 32: 394-398.

[89] Pryadko, E.I. and Osipov, P.P. (1986). Trials of nematophagous fungi in field conditions. *Biologicheskaya.* 1: 30-33.

[90] Larsen, M. (1999). Biological control of helminths. *International Journal for Parasitology.* 29: 139-146.

[91] Waller, J.P., Knox, R.M. and Faedo, M. (2001). The potential of nematophagous fungi to control the free-living stages of nematode parasites of sheep: feeding and block studies with Duddingtonia flagrans. *Veterinary Parasitology.* 102: 321–330.

[92] Ojeda-Robertos, N.F., Torres-Acosta, J.F.J., Ayala-Burgos, A., Aguilar-Caballero, A.J., Cob-Galera, L.A. and Mendoza–de-Gives, P. (2008a). A technique for the quantification of Duddingtonia flagrans chlamydospores in sheep faeces. *Veterinary Parasitology.* 152: 339-343.

[93] Ojeda-Robertos, N.F., Torres-Acosta, J.F.J., Aguilar-Caballero, A.J., Ayala-Burgos, A., Cob-Galera, L.A., Sandoval-Castro, C.A., Barrientos-Medina, R.C., and Mendoza–de-Gives, P. (2008b). Assessing the efficacy of Duddingtonia flagrans chlamydospores per gram of faeces to control Haemonchus contortus larvae. *Veterinary Parasitology.* 158: 329-335.

[94] Ojeda-Robertos, F.N., Mendoza-de Gives, P., Torres-Acosta, J.F.J., Rodríguez-Vivas, R.I. and Aguilar-Caballero A.J. (2005). Evaluating the effectiveness of a Mexican strain of Duddingtonia flagrans as a biological control agent against gastrointestinal nematodes in goat faeces. *Journal of Helminthology.* 79: 151-157.

[95] Casillas-Aguilar, J.A., Mendoza-de-Gives, P., López-Arellano, M.E. and Liébano-Hernández. E. (2008). Evaluation of multi-nutritional bio-pellets containing Duddingtonia flagrans chlamydospore for the control of ovine haemonchosis. *Annals of the New York Academy of Science.* 1149: 161-163.

[96] Fitz-Aranda, J.A., Mendoza-de-Gives, P., Liébano-Hernández, E., Valero-Coss, R.O., Quiróz-Romero, H. and Torres-Acosta, J.F.J. (2011a). Duddingtonia flagrans in nutritional pellets: effect of different storage conditions on the viability of chlamydospores. *Proceedings of the 23rd WAAVP Conference. Towards good management Practices in Parasite Control.* Buenos Aires Argentina. p. 300.

[97] Fitz-Aranda, J.A., Mendoza-de-Gives, P., Quiróz-Romero, H., Torres-Acosta, J.F.J., Sandoval-Castro, C.A. and Ojeda-Robertos, N.F. (2011b). Effect of storage time on the viability of Duddingtonia flagrans chlamydospores when fed to sheep using nutritional pellets. *Proceedings of the 23rd WAAVP Conference. Towards good management Practices in Parasite Control.* Buenos Aires, Argentina, p. 301.

[98] Aguilar-Marcelino, L., Mendoza-de-Gives, P., Torres-Hernández, G., Liébano-Hernández, E., López-Arellano, Ma. E., Orihuela, A., González-Camacho, J.M. and Torres-Acosta, J.F.J. (2011a). Effect of feeding Santa Cruz sheep with nutritional pellets containing Duddingtoinia flagrans chlamydospores in the reduction of Haemonchus contortus (L3) in feces. *Proceedings of the 23rd WAAVP Conference. Towards good management Practices in Parasite Control.* Buenos Aires, Argentina. p. 42.

[99] Ojeda-Robertos, N.F., Torres-Acosta, J.F.J., Ayala-Burgos A.J., Sandoval-Castro C.A., Valero-Coss R.O. and Mendoza-de-Gives, P. (2009). Digestibility of Duddingtonia flagrans chlamydospores in ruminants: in vitro and in vivo studies. *BMC Veterinary Research* 5: 46.

[100] Knox, M., Healey, K., Lawlor, C., and Chambers, M. (2011). Field assessment of the efficacy of Duddingtonia flagrans against infective stages of sheep, goat and horse nematodes. *Proccedings of the 23rd International Conference of the WAAVP*. Buenos Aires, Argentina. p. 290.

[101] Sagués, M.F., Fusé, L.A., Fernández, A.S., Iglesias, L.E., Moreno, F.C. and Saumel, A.C. (2011). Efficacy of an energy block containing Duddingtonia flagrans in the control of gastrointestinal nematodes of sheep. *Proccedings of the 23rd International Conference of the WAAVP*. Buenos Aires, Argentina. p. 27.

[102] Rodríguez-Esparza, J.R., Mendoza-de-Gives, P., Liébano-Hernández, E., López-Arellano, M.E., Cid-del-Prado, I., Franco-Navaro, F. and López-Martínez, V. (2007). In vitro predatory activity of "cannibal" nematodes against Haemonchus contortus and Rhabditis sp. *Proceedings of the 21st International Conference of the World Association for the Advancement of Veterinary Parasitology*. Gent, Belgium. 19th – 23rd August 2007. p. 273.

[103] López-Arellano, ME., Bravo-de-la-Parra, A., Mendoza-de-Gives, P., Liébano-Herández, E. and Aguilar-Marcelino, L. (2011). Bacillus thuringiensis toxicity against ruminant parasitic nematodes. Revista Mexicana de Ciencias Pecuarias. *Revista Mexicana de Ciencias Pecuarias*. 3(1):77-88.

[104] Aguilar-Marcelino, L., Quintero-Martínez, MT, Mendoza de Gives, P., López-Arellano, Ma E., Liébano-Hernández, E., Torres-Hernández, G., González-Camacho' JM, and Cid-del-Prado, I. (2011b). *Haemonchus contortus infective larvae (L_3) (Nematoda: Trichostrongylidae) and free-living nematodes are devoured by the mite Lasioceius penicilliger* (In preparation).

In: Fungi: Types, Environmental Impact and Role in Disease
Editors: Adolfo Paz Silva and María Sol
ISBN: 978-1-61942-671-9

Chapter XX

Fundamentals, Diversity and Application of White-Rot Fungi

G. M. Maciel[1], A. Bracht[1], C. G. M. Souza[1], A. M. Costa[2] and R. M. Peralta[1*]

[1] Laboratory of Biochemistry of Microorganisms, Department of Biochemistry, Universidade Estadual de Maringá, PR, Brazil
[2] Universidade Estadual de Santa Cruz, BA, Brazil

Abstract

White-rot fungi (WRF) are essentially basidiomycetes that exhibit singular biochemical abilities and potential applications. These fungi are efficient carbon recyclers in nature, can degrade lignin completely to carbon dioxide, act as ecological indicators and are important components of ecosystems.

Their unique ability to degrade lignin is due to their capability to synthesize a set of oxidative (ligninolytic) extracellular enzymes, mainly peroxidases and laccases. The ligninolytic enzymes of WRF have broad substrate specificity and have been implicated in the transformation and mineralization of organopollutants, including pesticides, poly-aromatic hydrocarbons and dyes. The most studied WRF is *Phanerochaete chrysosporium*. Also quite extensively studied are species of the genera *Trametes*, *Bjerkandera*, *Ganoderma* and *Pleurotus*.

The biotechnological potential of ligninolytic enzymes has stimulated the search for other promising producers. The WRF biodiversity in temperate environments has been more widely studied than in tropical environments.

This review focus especially on the biodiversity of WRF of tropical environments, mainly of Brazilian biomes, their impact on rain forest ecosystems, their economical impact as plant pathogens, their possible uses in green chemistry and the state of the art of research on this exceptional group of fungi.

[*] E-mail address: rmperalta@uem.br.

1. Introduction

Fungi are the major decomposers of lignocellulosic material in several ecosystems and play an essential role in the cycling of carbon and other nutrients. The wood-decaying fungi comprise mainly basidiomycetes and ascomycetes and usually are saprotrophic, utilizing dead wood, or parasitic, attacking living trees.

A wide variety of wood types and trees in different stages of decomposition can be colonized by these fungi. Many wood-decaying fungi have specific preferences for certain host species and stages of wood decay. Besides, wood-decaying fungi that are plant pathogens are specialized to survive plant defences with particular abilities to defeat antifungal substances such as phenolics, tannins and alkaloids.

Wood is a carbon-rich substrate with low contents of nitrogen and other essential nutrients. Lignocellulose, the main component of wood, is a complex mixture of polymers which mostly include cellulose, hemicellulose and lignin. The proportions and location of these polymers are quite variable among plant groups or species. In order to use woody materials as substrate and to have access to its limited nutrients the species of wood-decaying fungi have evolved distinct mechanisms of growth as well as metabolic and enzymatic abilities which are dependent on environmental conditions such as temperature, humidity and availability of food resources.

A diverse set of enzymes (Fig. 1) can be produced by the wood-decaying fungi to enable them to breakdown wood components. The rate and extent of wood decay is closely related to the wood structure (i.e. chemical composition) and to the characteristics of the different hydrolytic and oxidative enzymatic machinery as well as to the hyphae morphology of particular groups and species of the wood-decaying fungi. Usually, the wood-decaying fungi are distributed into three physiological groups (Fig. 2) based on their decay capabilities and morphological changes they cause in the wood: the brown-rot, soft-rot and white-rot fungi. Brief descriptions of the brown-rot and soft-rot fungi groups are presented in the following sections as well as a detailed description of the white-rot fungi group which is the main subject of this chapter.

2. Brown-Rot Fungi

The brown-rot fungi causing wood decay are predominantly *Basidiomycota* commonly detected on live and dead trees of coniferous forests and attacking indoor wood in buildings. The most frequent cause of deterioration in wooden structures is an aggressive decay by brown-rot basidiomycetes.

Typical species of brown-rot fungi can be found in the families *Coniophoraceae* (*Coniophora puteana*, known as the "cellar fungus"), *Fomitopsidaceae* (*Fomitopsis pinicola*, *Laetiporus sulphureus*, *Piptoporus betulinus*, *Postia placenta*), *Gloeophyllaceae* (*Gloeophyllum trabeum*) and *Serpulaceae* (*Serpula lacrymans*, known as "dry rot fungus").

During growth on wood, the brown-rot fungi may develop mycelial cords for exploitation of distant food resources and efficient transportation of water and nutrients as well as conservation of nitrogen. The formation of cords in fungal growth is seen as an adaptation to a spatially heterogeneous carbon and mineral nutrient environment [1-3]. Some brown-rot fungi, such as *L. sulphureus* and *P. placenta*, can also produce chlamydospores as a mechanism to survive unfavorable environmental conditions [4,5].

The degradation targets of the brown-rot fungi in wood are cellulose and hemicelluloses which are mainly located in the secondary cell wall layers. Lignin, which protects these polysaccharides from enzymatic degradation, is oxidized by the fungi and partially removed.

The possibility of partial degradation of lignin by these fungi (and not only oxidative modification) has recently been investigated. The growth of the brown-rot fungus *G. trabeum* on spruce wood resulted in non-selective depletion of all intermonomer side-chain linkages in lignin and formation of an aromatic polymer no longer recognizable as lignin [6].

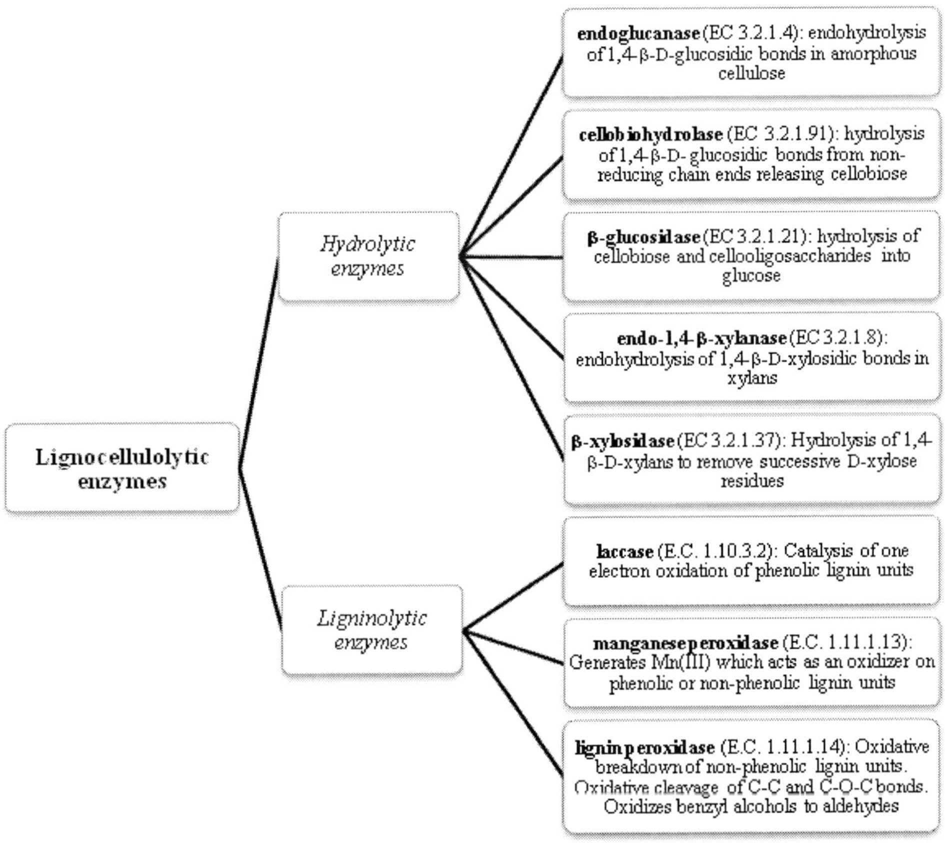

Figure 1. The main enzymes involved in the degradation of lignocelluloses. Additional hemicellullases (e.g. acetylesterase, α-glucuronidase, endo-1,4-β-mannanase, α-galactosidase) and oxidoreductases (e.g. aryl alcohol oxidase, glucose-1-oxidase, glyoxal oxidase, pyranose-2-oxidase) may also participate in the degradation process.

The initial stage of wood decay by brown-rot fungi has been mainly considered as a nonenzymatic process since most cellulases and hemicellulases are larger than wood pores and lignin-degrading enzymes usually are not observed in brown-rot species. In this stage, there is extensive depolymerization of structural polysaccharides in wood with substantial strength loss but low weight loss. In an advanced state of decay, when structural polysaccharides are depleted leaving an aromatic polymer derived from lignin, the wooden cells lose their rigidity and shrink resulting in wood breaking into brown cubical fragments which is characteristic of brown-rot decay.

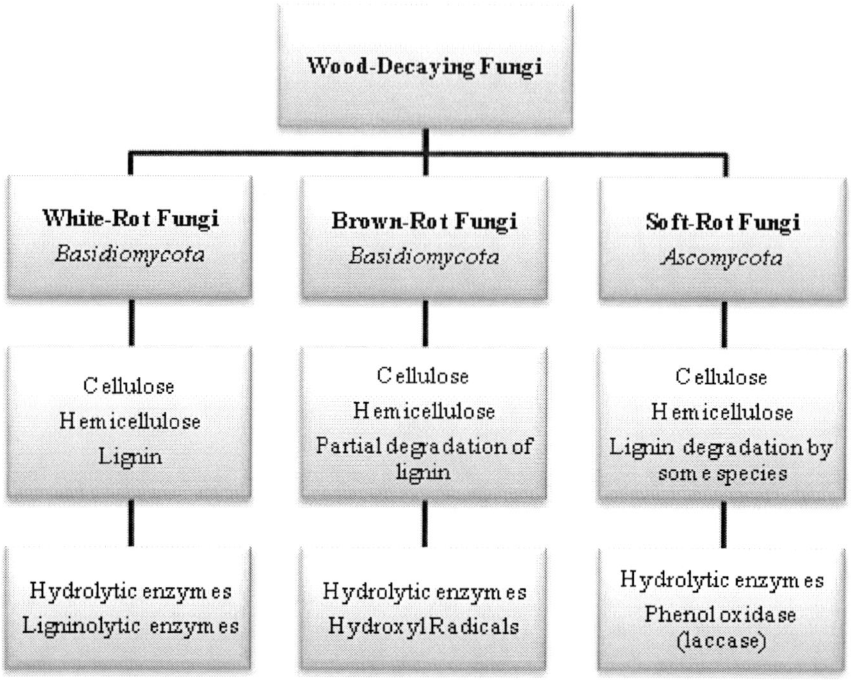

Figure 2. Distribution of wood-decaying fungi into three physiological groups according to their abilities to degrade the cell wall components and main mechanisms involved in wood degradation.

Brown-rot basidiomycetes seem to depend on the attack of highly reactive extracellular hydroxyl radicals produced by the fungi for depolymerization of lignocelluloses, at least in an early stage [6-8]. The structural polysaccharides are depolymerized by these radicals through the abstraction of hydrogen atoms from the sugar subunits as lignin is attacked by electrophilic addition to its aromatic rings [6,9,10].

The formation of hydroxyl radicals (•OH) by the brown rot fungi has been based on a Fenton reaction which involves hydrogen peroxide (H_2O_2) and iron (Fe): $H_2O_2 + Fe^{2+} + H^+ \rightarrow H_2O + Fe^{3+} + \bullet OH$. Many mechanisms have been proposed for the generation of H_2O_2 and Fe^{2+} which ultimately reacts to produce hydroxyl radicals [7,11-15]. The participation of a cellobiose dehydrogenase in Fe^{3+} oxalate reduction to produce Fe^{2+} and the autoxidation of Fe^{2+} oxalate as a source of H_2O_2 have been suggested for *C. puteana* [14]. The involvement of a hydroquinone-driven Fenton reaction to reduce Fe^{3+} and produce H_2O_2 has been proposed for *G. trabeum* [10,15,16]. A mechanism to oxidize hydroquinones in an environment with high oxalate concentrations has recently been investigated with *P. placenta* [7]. In this process, the brown rot basidiomycete secretes a laccase (which is not usually detected in brown-rot fungi species) which oxidizes a metabolic hydroquinone (2,5-dimethoxyhydroquinone) producing H_2O_2 and catalyzing the production of reactive oxygen species resulting in a complete Fenton system without the need of the hydroquinone to react with Fe3+ (which is strongly chelated by oxalic acid) and contributing to incipient brown-rot.

Most of the brown-rot fungi produce large amounts of oxalic acid which accumulates in the cultures [14,17,18]. The production of high concentrations of oxalic acid mainly depends on carbon/nitrogen ratio of the substrate and the absence of the enzyme oxalate decarboxylase [18]. The presence of sufficient oxalic acid is fundamental for the performance of cellobiose dehydrogenase mechanism [14] as at low pH the autooxidation of Fe^{2+} is slow and the hyphae are protected from oxidative damage. Other than protection from oxidative damage and formation of soluble iron complexes, oxalic acid has an important function in the chelation of metal cations such as calcium which might be toxic to the fungi in large amounts. Calcium oxalate crystals are commonly found in decayed wood of *C. puteana* [18,19].

The cellulolytic enzyme system of most brown-rot fungi is tipically composed of endoglucanases and β-glucosidases [20-23] and plays an important role in the degradation and utilization of cellulose and hemicellulose sugar units. It is believed that, during wood decay, when the pores in the lignocellulose network are adequate for the admission of cellulolytic enzymes [9] cellulose in the amorphous regions of wood is rapidly degraded. The absence of an active cellobiose dehydrogenase in most brown-rot fungi (except *C. puteana*) leave unclear the mechanism and rate achieved in crystalline cellulose decomposition and cellobiose production. Despite this, processive endoglucanases exhibiting the ability to hydrolyse crystalline cellulose have recently been reported for *G. trabeum*, *Fomitopsis palustris* and *F. pinicola* [20,24,25]. The action of hydroxyl radicals can be mainly responsible for breaking the structure of crystalline cellulose [26] and the activity of cellulases might occur in cooperation with hydroxyl radicals to achieve sufficient degradation of cellulose [27].

Several differences in the cellullolytic enzyme system can be observed among species of the brown-rot fungi and among this group and the other groups of wood-decaying fungi. In general, cellulases of the brown-rot fungi are not as well studied and characterized as those of soft-rot and white-rot fungi. Additionaly, many species are still unexplored and more research outside model species of brown-rot fungi is needed to clarify the chemical and biochemical mechanisms involved in degradation of lignocellulosic material. Recently, the degradative capabilities of brown-rot fungi have received increased attention because of their application in saccharification processes and biofuels production.

3. Soft-Rot Fungi

Fungi that cause soft-rot on wood belong to the phylum *Ascomycota* and usually occur in environmental conditions which are detrimental for growth of most common basidiomycetes: very high or very low moisture content. Although most ascomycetes inhabit soil, many of them have the ability to colonize wood and contribute to decay. The *Ascomycota* capable of lignocelluloses degradation are in general taxonomically classified in the classes of *Eurotiomycetes* and *Sordariomycetes* (Table 1).

Resistance to low availability of oxygen (e.g. in a condition of excessive moisture), to the presence of wood preservatives, the ability to survive in extreme temperatures (thermophilic and thermotolerant fungi) and to support high pH ranges are particular characteristics of some species of soft-rot fungi [28-30]. As a result, these fungi have been found in several adverse circumstances such as colonizing timbers in water cooling towers [31], decaying wood at the historic huts on Ross Island in Antarctica [32] and at archeological woods in very dry environments (e.g. wooden objects in ancient tombs) [33,34].

Table 1. List of some *Ascomycota* that cause soft-rot on wood

Class	Family	Species
Eurotiomycetes	*Trichocomaceae*	*Aspergillus flavus* *Paecilomyces inflatus*
Sordariomycetes	*Chaetomiaceae*	*Chaetomium cellulolyticum* *Chaetomium globosum* *Humicola grisea* *Humicola insolens*
	Coniochaetaceae	*Coniochaeta ligniaria*
	Hypocreaceae	*Trichoderma reesei*
	Microascaceae	*Petriella setifera* *Trichurus spiralis*
	Xylariaceae	*Daldinia concentrica* *Daldinia eschscholzii* *Hypoxylon atropunctatum* *Hypoxylon mammatum* *Xylaria hypoxylon* *Xylaria polymorpha*

Species names and taxonomic position according to the Index Fungorum (IFS) database.

When wood is attacked by this group of fungi in an environment with elevated moisture, the surface layer of the wood becomes soft, an effect that originated the term soft-rot [35]. Independently of the environmental conditions, a usual characteristic that identifies soft-rot fungi colonization is the presence of cavities on wood resulting from hyphal growth in the secondary wall (S2 layer) [36]. Tipically, two different decay patterns caused by soft-rot fungi can be distinguished: Type 1 and Type 2. In Type 1, the fungal hyphae form a chain of cylindrical longitudinal cavities in the secondary cell wall as wood colonization proceeds. The middle lamella is not degraded [34,37]. In Type 2, the formation of cavities within the cell wall is absent as a progressive erosion occurs from the cell lumen to the middle lamella [29,34,37]. In advanced stages of decay, the entire secondary wall may be degraded and the middle lamella is either slightly modified or not degraded (on the contrary white-rot fungi degrade the middle lamella) [37-39].

Cellulose and hemicelluloses are preferentially degraded when soft-rot fungi are attacking wood, though some fungi (especially Xylariaceae) have also shown the ability to cause extensive degradation of lignin. *Xylaria hypoxylon* and *Xylaria polymorpha* exhibited the ability to degrade a synthetic lignin in solid and liquid media; however, the pattern of the lignocellulose fragments produced by these fungi differed considerably from those produced by white-rot fungi [40]. The same soft-rot ascomycetes were grown on beech-wood microcosms and found to degrade wood lignin [41]. Total wood mass loss was moderate and laccase activity was correlated with the formation of high-molecular mass fragments [41].

Soft-rot fungi of the family *Xylariaceae* that exhibit the activity of phenol oxidases and a bleach effect during wood decay were once recognized as white-rot fungi, though the classification of their mode of decay as Type 2 soft-rot is more appropriate [4,37,38].

Although presenting the ability to oxidize and degrade lignin, most soft-rot ascomycetes are not as efficient as white-rot basidiomycetes: lignin modification is more limited or achieved at a slower rate [40,42,43]. The amount and type of lignin usually affect wood decay by ascomycetes [4,37,44]. Soft-rot fungi predominantly decay hardwoods rather than softwoods mainly because of the lower content of guaiacyl units in lignin of hardwoods [37]. The reduced ability to degrade guaiacyl lignin might be due to the production of either a low amount of ligninolytic oxidants or weak ligninolytic oxidants [45-47]. Nevertheless, syringyl lignin is oxidized and degraded by the soft-rot fungi [37]. The erosion of the secondary cell wall and the decrease in the content of acid-insoluble material are characteristics of lignin degradation in hardwoods by some soft-rot fungi [45,48].

The enzymatic system of soft-rot fungi, specially the ligninolytic system, is still unexplored in many species. The hydrolytic and oxidative enzymes detected in some species of soft-rot ascomycetes are listed in Table 2. The description and production of cellulases by this group of fungi have received most of the attention.

Trichoderma reesei (syn. *Hypocrea jecorina*) is a sof-rot fungus able to produce a complete set of cellulases. This fungus is a common source of cellulases and hemicellulases for applications in industrial processes and the genome sequencing of this fungus with the analysis of its enzymes has already been performed [49,50]. However, ligninolytic enzymes such as laccase have not been found in this fungus, and without a system for removing the lignin, this fungus cannot initiate degradation of wood or of the cellulose component of wood [26]. Consequently, cellulolytic ascomycetes incapable of degrading lignin possibly function as secondary wood-decomposing fungi, decaying wood after delignification by other fungi [51]. Despite the fact that soft-rot is considered the first decay type observed in a general sequence of wood colonization [29,52], several species of soft-rot fungi, including *Trichoderma* spp., have been isolated from later stages of wood decomposition [53].

Table 2. Hydrolytic and Oxidative enzymes detected in some species of soft-rot ascomycetes

Enzymes	Fungal species	References
β-glucosidase	*Chaetomium* sp.	[54]
	Daldinia eschscholzii	[55]
	Trichoderma reesei	[56,57]
Cellobiohydrolase	*Chaetomium cellulolyticum*	[58]
	Humicola insolens	[59,60]
	Trichoderma reesei	[50,56,61]
Endoglucanase	*Coniochaeta ligniaria*	[62]
	Daldinia eschscholzii	[63]
	Paecilomyces inflatus	[64]
	Trichoderma reesei	[50,56]
Laccase	*Paecilomyces inflatus*	[65]
	Xylaria hypoxylon	[40,41]
	Xylaria polymorpha	[40,41]
Xylanase	*Coniochaeta ligniaria*	[62]
	Xylaria polymorpha	[40]

Species names and taxonomic position according to the Index Fungorum (IFS) database.

4. White-Rot Fungi

White-rot fungi are a diverse and abundant group of filamentous fungi taxonomically classified into the phylum *Basidiomycota* in the *Agaricomycetes* class. Most commonly studied species of white-rot fungi are distributed over five orders: *Agaricales*, *Auriculariales*, *Hymenochaetales*, *Polyporales* and *Russulales* (Table 3).

These fungi can be found colonizing either living trees (e.g. heart rot) or dead wood (e.g. logs, stumps) from temperate to tropical climates, presenting a variety of morphologies such as caps, brackets or resupinaceous (corticioid) basidiomes. The term white-rot is related to the bleached (white) appearance frequently observed on wood attacked by these fungi.

A common characteristic among white-rot fungi is the ability to degrade all main components of plant cell wall: cellulose, hemicelluloses and lignin. The decomposition of lignocelluloses is achieved by a series of enzymatic (hydrolases and oxidoreductases) and nonenzymatic mechanisms and the production of ligninolytic enzymes is typical of this group of fungi. The degradation of lignin, which is extremely recalcitrant, is a critical step for efficient carbon cycling, thus white-rot fungi, as organisms capable of completely metabolize lignin, are an essential part of forest ecosystems.

Besides, the ability of this group of fungi to degrade lignin has drawn attention to the application of ligninolytic microorganisms and their degradative systems in a wide range of processes, including: biopulping and biobleaching, bioremediation and bioconversion of agroindustrial residues.

Table 3. Most commonly studied species of white-rot fungi

Order	Family	Species
Agaricales	***Agaricaceae***	*Agaricus bisporus*
		Agaricus brasiliensis
	Marasmiaceae	*Lentinula edodes*
	Physalacriaceae	*Armillaria mellea*
	Pleurotaceae	*Pleurotus djamor*
		Pleurotus eryngii
		Pleurotus ostreatus
		Pleurotus pulmonarius
Auriculariales	***Auriculariaceae***	*Auricularia auricula-judae*
		Auricularia polytricha
Hymenochaetales	***Hymenochaetaceae***	*Inonotus hispidus*
		Inonotus obliquus
		Phellinus igniarius
		Phellinus linteus
		Phellinus ribis
Polyporales	***Ganodermataceae***	*Ganoderma applanatum*
		Ganoderma lucidum
	Meruliaceae	*Bjerkandera adusta*
		Irpex lacteus
		Phlebia radiata
	Phanerochaetaceae	*Ceriporiopsis subvermispora*
		*Phanerochaete chrysosporium**
	Polyporaceae	*Pycnoporus cinnabarinus*

	Family	Species
		Pycnoporus sanguineus
		Trametes hirsuta
		Trametes versicolor
Russulales	***Bondarzewiaceae***	*Heterobasidion annosum*

*Phanerochaete chrysosporium is the most studied white-rot fungus and is considered a model organism

Species names and taxonomic position according to the Index Fungorum (IFS) database.

4.1. Lignocelluloses Biodegradation Patterns

The pattern and rate of wood and woody materials degradation vary among white-rot fungi. Usually, two degradation systems can be distinguished: simultaneous rot and selective delignification (Table 4).

In simultaneous rot, lignin and structural polysaccharides are concomitantly degraded by the fungi, at similar rates, leading to progressive erosion and thinning of wood cell walls [66,67]. This is considered the normal or common process of wood degradation that is used by most white-rot fungi [34,67]. In some cases, a slight preference for the removal of lignin in advance of carbohydrates may occur, but extensive loss of carbohydrates usually appears simultaneously or immediately after lignin removal [68,69]. In selective delignification, white-rot fungi preferentially remove lignin from wood, causing moderate losses of hemicelluloses and leaving cellulose practically intact or slightly degraded [66,70]. Selective decayed wood presents white pockets consisting entirely of cellulose [66]. The possible applications of lignin primary removal from lignocellulosic materials by these fungi have received increasing attention. The list of applications includes industrial processes such as biopulping and bioethanol production as well as in cellulose enriched products in animal feed [71-75]. However, the application of white rot fungi in selective delignification may depend on strain, cultivation methods and lignocellulosic material composition [73,76-78], whereas considerable variation exists in the ability of certain fungi to cause selective delignification and/or simultaneous rot. Some white-rot fungi cause either simultaneous rot or selective delignification as others may produce a combination of both patterns causing a simultaneous rot in one part of the substrate and a selective delignification in another [94,95]. Usually, environmental conditions and the type of lignocelllulosic material greatly affect the pattern of decay caused by the fungi [66,91]. Reproducibility of decay patterns in laboratory and field conditions is frequently low [81]. Variations in wood degradation patterns among species of the same genera and between strains of the same species have been described for *Ganoderma* spp. [83,96] and *Pleurotus* spp. [91]. A great variation has been reported for strains of *Phanerochaete chrysosporium* [69,87,88]. Among selective decayers, *Ceriporiopsis subvermispora* has shown the greatest potential for biopulping [97-99].

4.2. Cellulose Degradation

The main goal of wood biodegradation by white-rot fungi is to gain access to cellulose: the principal component of plant cell wall and the primary energy source for most fungi. Cellulose is a polymer composed of over 10,000 D-glucose residues linked by β-1,4-

glycosidic bonds [100] and naturally occurs in insoluble particles or imbedded within hemicelluloses and lignin polymers [101].

Table 4. White-rot fungi species causing selective lignification (SL) or simultaneous rot (SR) on lignocellulosic material

Fungal Species	Decay pattern	References
Ceriporiopsis subvermispora*	SL	[70,74,75]
Dichomitus squalens	SL	[44]
Echinodontium taxodii	SL	[79,80]
Ganoderma australe	SL	[81]
Ganoderma boninense	SR	[82]
Ganoderma lucidum	SR, SL	[83]
Irpex lacteus	SL	[71,84]
Perenniporia medulla-panis	SL	[85]
Phanerochaete chrysosporium	SR, SL	[69,86-88]
Phellinus pini	SL	[44]
Phlebia brevispora	SL	[72,73]
Phlebia fascicularia	SL	[72,73]
Phlebia floridensis	SL	[72,73]
Phlebia radiata	SL	[72,73]
Phlebia subserialis	SL	[89,90]
Pleurotus eryngii	SL	[87]
Pleurotus ostreatus	SL, SR	[87,91]
Trametes hirsuta	SR	[92]
Trametes versicolor**	SR	[92,93]

*C. subvermispora is a model fungus for selective lignin degradation
**Trametes versicolor is a model fungus for simultaneous rot
Species names and taxonomic position according to the Index Fungorum (IFS) database.

The hydrolysis of the β-1,4-glycosidic bonds in cellulose is achieved by a set of extracellular cellulases with different specificities, including endoglucanases (4-β-D-glucan 4-glucanohydrolases), cellobiohydrolases (4-β-D-glucan cellobiohydrolases) and β-glucosidases (β-D-glucoside glucohydrolase). Endoglucanases initiate cellulose hydrolysis randomly by attacking internal amorphous regions of cellulose generating new chain ends and opening sites for the attack of cellobiohydrolases [102,103]. Cellobiohydrolases (known as exoglucanases) act in a progressive mode by hydrolyzing β-1,4-glycosidic bonds from the cellulose chain ends, releasing cellobiose as the major product [102]. Cellobiose and soluble cellodextrins are hydrolyzed by β-glucosidases producing glucose [102]. Cellobiohydrolases can hydrolyze crystalline cellulose [104,105].

Cellulase systems act in a coordinated manner exhibiting higher collective activity than the sum of the activities of the individual enzymes [102]. Endoglucanases and cellobiohydrolases usually work synergistically in the hydrolysis of cellulose [106].

The cellulolytic system of white-rot fungi is complete and many cellulases of these fungi have been purified and characterized and the respective genes have been cloned, sequenced and expressed [107-112].

Selective white-rot fungi can inhibit cellulose depolymerization by suppression of the Fenton reaction during lipid peroxidation catalyzed by manganese peroxidase with metabolites such as alkylitaconates and, consequently attenuate the production of hydroxyl radicals [113,114].

4.3. Hemicellulose Degradation

Hemicelluloses, the second most common polysaccharides in nature, are heterogeneous polymers of branched and linear polysaccharides such as xylans, mannans, mixed linkage β-glucans and xyloglucans [67,115,116]. The main component of hemicellulose, xylan, is a linear polymer of β-D-xylopyranosyl units linked by β-1,4-glycosidic bonds [117]. Besides xylose, xylans may contain arabinose, glucuronic acid or its 4-O-methyl ether, and acetic, ferulic, and p-coumaric acids. The frequency and composition of branches are dependent on the source of xylan [118-120]. The branches render hemicellulose non-crystalline because they prevent efficient packing of hemicelluloses [121].

The complete degradation of hemicelluloses, as a heterogenic polymer, is more complicated than cellulose and requires the cooperative action of a variety of hydrolytic enzymes [23,122]. The major enzymes involved in xylan degradation are endo-1,4-β-xylanases (4-beta-D-xylan xylanohydrolases), which produce xylooligomers, and β-xylosidases (4-beta-D-xylan xylohydrolases), which hydrolyze the xylooligomers into xylose. It is believed that endo-1,4-β-xylanase is a key enzyme for hemicellulose degradation [123]. Depending on the composition of hemicelluloses, additional enzymes are necessary such as xylan esterases, ferulic and *p*-coumaric esterases, α-1-arabinofuranosidases and α-4-O-methyl glucuronosidases [122]. These enzymes act synergistically to achieve efficient hydrolysis of wood xylans and mannans [122]. Feruloyl esterases have been shown to act in synergy with cellulases, xylanases and pectinases to break down complex plant cell wall carbohydrates [124]. A possible synergy occurs between xylanases and feruloyl esterases to disrupt the hemicellulose–lignin complex without mineralization of lignin *per se* [76,125]. Hemicellulose degradation can be a necessary step for efficient lignin removal [76].

Enzymatic systems that are active on cellulose might also be active on hemicelluloses, and enzymes with specific activity on hemicelluloses are usually co-produced by cellulolytic microorganisms [102,122]. However, xylanases are frequently more expressed in wood-grown than in cellulose-grown cultures [121].

Xylanases with high specificity or lack of significant cellulase activity can be obtained with *C. subvermispora* [126]. High levels of α-glucuronidase can be produced with *Phanerochaete chrysosporium* [127]. *Trametes versicolor* (syn. *Coriolus versicolor*) produces a complex set of xylanolytic enzymes [128]. Low activities of endo-1,4-β-xylanase have been found for species of *Pleurotus* [129] strains of *Irpex lacteus* [71].

4.4. Lignin Structure and Degradation

Lignins are three dimensional complex polymers resulting from oxidative coupling of 4-hydroxyphenylpropanoids that occur in woody and vascular plants. They are mainly biosynthesized by combinatorial polymerization reactions of *p*-coumaryl, coniferyl and sinapyl alcohol monomers (Figure 3), but other components (phenolic monomers and conjugates) may also be a part of lignin building blocks [130,131]. The heterogeneous and complex structure of lignin difficults its degradation by either chemical or biological systems and protects plant tissues from the attack of many microorganisms [35,132]. The only microorganisms capable of degrading lignin completely to carbon dioxide are the white-rot fungi.

The nature of the lignin has a great influence on wood degradation by fungi, and the exact composition of lignin varies widely within species [37]. In general, lignins can be divided into two major groups: guaiacyl lignins and guaiacyl-syringyl lignins [37,133]. Guaiacyl lignins are predominantly polymerization products of coniferyl alcohol while guaiacyl-syringyl lignins are composed of varying parts of aromatic nuclei guaiacyl and syringyl (originated from sinapyl alcohol), together with small amounts of p-hydroxyphenyl units (originated from p-coumaryl alcohol) [37,134]. Gymnospermous wood consists almost exclusively of guaiacyl monomers, whereas angiospermous wood consists of approximately equal ratios of guaiacyl and syringyl monomers [37,135]. White rot fungi can be found colonizing both types of wood but usually are more associated with angiospermous wood [37,96,136,137].

Lignin is not used as an energy source by white-rot fungi, thus it is not a substrate of primary metabolism. The biodegradation of lignin is an oxidative process that occurs during secondary metabolism to access energy-rich cellulose and hemicelluloses, which are chemically bound to lignin in plant cell wall [138,139].

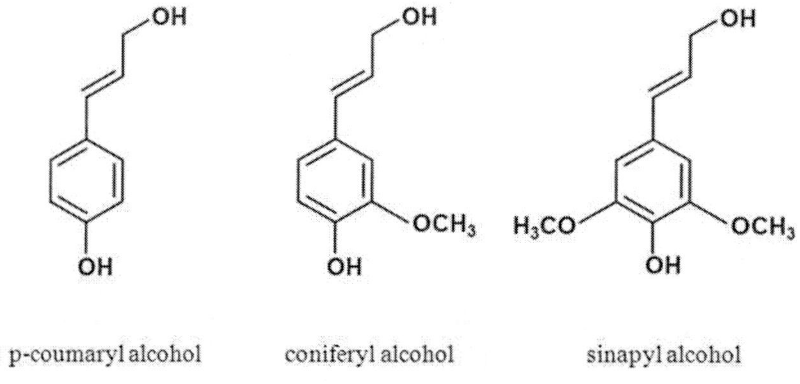

Figure 3. Structure of predominant lignin monomers: p-coumaryl, coniferyl and synapyl alcohol.

White-rot fungi apparently initiate lignin depolymerization by secreting a set of enzymes including laccases (LAC), lignin peroxidases (LiP), manganese peroxidases (MnP) and versatile peroxidases (VP), but other smaller oxidative agents may be simultaneously involved [8,140,141]. These enzymes generate reactive and nonspecific free radicals which spontaneously break the covalent bonds in lignin [8,142,143]. The lignin degrading system of the white-rot fungi is also composed of H_2O_2-generating oxidases (e.g. aryl alcohol oxidase, glyoxal oxidase and glucose oxidase) and low molecular weight cofactors (Mn^{2+}, veratryl alcohol, oxalate and chlorinated anysil metabolites) [130].

Mechanisms of lignin biodegradation by basidiomycetes have been well described and reviewed [130,132,144-146].

In general, when attacking wood (Fig. 4), the white-rot fungus secretes ligninolytic enzymes (such as LAC, LiP, MnP and VP) which oxidize lignin by producing aromatic cation radical intermediates [132,146]. These are very unstable and advance into different non-enzymatic reactions, which were classified into various groups by Martinez et al. [146] as illustrated by Figure 4. A brief description of the most important groups is given below.

(A) The first group corresponds to C-4-ether breakdown. Phenoxy radicals are released which can repolymerize on lignin if they are not reduced by oxidases to phenolic compounds (these can be reoxidized by ligninolytic enzymes). Phenoxy radicals can also be converted into quinones by Cα-Cβ cleavage, and the quinones can contribute to oxygen activation in redox cycling reactions via quinone reductases (QR) and peroxidases/laccases;

(B) Aromatic ring cleavage with the formation of muconate-type structures;

(C) Cα-Cβ cleavage with the release of aromatic aldehydes which are also synthesized *de novo* by the fungus. These are the main products of the enzymatic depolymerization of lignin and are the substrates of aryl alcohol oxidases (AAO) in redox cyclic reactions involving aryl alcohol dehydrogenase (AAD) for generation of H_2O_2 [147]. As a result of AAO/AAD reactions, benzyl alcohols can be subjected to the attack of peroxidases/laccases generating quinones, which can participate with quinones from phenoxy radicals, in redox cycling reactions and contribute to oxygen activation;

(D) Demethoxylation which results in the release of methanol.

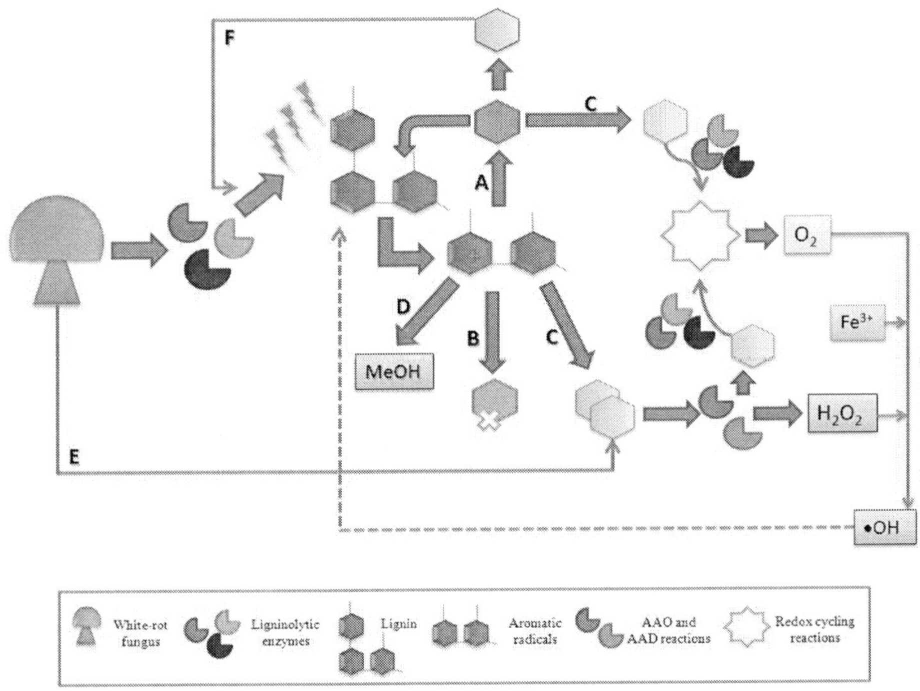

Figure 4. Mechanism of lignin degradation (adapted from Martinez et al. [146]). The main nonenzymatic reactions are represented as (A) C-4-ether breakdown, (B) aromatic ring cleavage, (C) Cα-Cβ cleavage and (D) demethoxylation. Other reactions: (E) *de novo* synthesis of aromatic aldehydes, (F) re-oxidation of phenolic compounds by ligninolytic enzymes.

Hydroxyl radicals (•OH) are formed from the quinone redox cycling reaction involving O_2, Fe^{3+} (naturally present in wood), a semiquinone radical and H_2O_2 [8]. Hydroxyl radicals are highly mobile and strong oxidizers that can attack lignin by abstracting aliphatic hydrogens and by adding to aromatic rings in early stages of wood decay, when the small size of pores in the still-intact cell wall prevents the penetration of ligninolytic enzymes [8,146]. As lignin biodegradation proceeds by the continuous attack of ligninolytic enzymes, the simple products from degradation are incorporated by fungal hyphae and used in intracellular catabolic reactions [146].

4.5. Special Mechanisms of Lignin Degradation

Selective white-rot fungi or those unable to produce LiP possess alternative mechanisms to oxidize lignin: lipid peroxidation and the formation of peroxyl radicals [148]. A mixture of unsaturated fatty acids, Mn^{2+} and MnP cause lipid peroxidation generating carbon-centered lipid radicals that react with oxygen-forming peroxyl radicals [148,149]. Peroxyl radicals apparently act in a different manner from that of the unselective hydroxyl radicals. They abstract electrons only from activated carbons leading to lignin oxidation instead of cellulose degradation [8], which is a characteristic of selective white-rot fungi.

4.6. Ligninolytic Enzymes of White-Rot Fungi

The ability of white-rot fungi to produce ligninolytic i.e. lignin-modifying enzymes (LMEs) is well known. Most LMEs are secreted as multiple isoforms by many different species of white-rot fungi under varying conditions. The set of LMEs comprise a phenoloxidase, laccase, and three peroxidases, lignin peroxidase (LiP), manganese peroxidase (MnP) and versatile peroxidase (VP). Laccase was firstly described over 128-years ago becoming one of the oldest known enzymes [150]. LiP and MnP were firstly discovered in *Phanerochaete chrysosporium* [151-153].Versatile peroxidase (VP) is a new addition to the group of LMEs and was discovered in a strain of *Pleurotus eryngii* [154,155].

The genomes of some species of white-rot fungi such as *Phanerochaete chrysosporium* [156] and *Schizophyllum commune* [157] have already been sequenced and are useful tools for understading the enzymatic systems used by different fungal species in lignocelluloses degradation. *P. chysosporium* has genes encoding peroxidases but no laccases, however *S. commune* has genes encoding laccases but no peroxidases. Therefore, these fungi probably have different mechanisms to degrade lignin.

The LMEs are highly non-specific and, thus, capable of degrading a wide range of recalcitrant compounds structurally similar to lignin. White-rot fungi usually secrete one or more of the LMEs in different combinations. Distribution of white-rot fungi into groups according to their enzymatic systems have been undertaken [137,158,159]. General classification is based on the capacity of different fungi to produce one or a combination of three ligninolytic enzymes: LiP, Laccase and MnP. However, the absence or non-detection of these enzymes in some white-rot fungi, the sequencing of white-rot fungal genomes and the discovery of new enzymes has led to the establishment of different groups [160]. Generically, white-rot fungi can be distributed into four groups, according to their ability to produce laccases and peroxidases (LiP, MnP and VP):

(A) laccase and MnP and LiP (*Trametes versicolor*, *Bjerkandera adusta*)
(B) laccase and at least one of the peroxidases (*Lentinus edodes, Pleurotus eryngii, Ceriporiopsis subvermispora*)
(C) only laccase (*Schizophyllum commune*)
(D) only peroxidases (*Phanerochaete chrysosporium*)

The most frequently observed LMEs among white-rot fungi species are laccases and MnP and the least are LiP and VP. A considerable amount of reviews detailing the numerous characteristics and applications of ligninolytic enzymes have been published in the last years [132,143,149,150,161-164]. A general description of the mechanisms and functions of the main ligninolytic enzymes of the white-rot fungi is given in Table 5. The LMEs may function cooperatively or separately from each other, however auxiliary enzymes (unable to degrade lignin on their own) are necessary to complete the process of lignin and/or xenobiotics degradation: aryl alcohol oxidase (AAO), aryl alcohol dehydrogenase (AAD), glyoxal oxidase (GLOX), quinone reductase (QR), cellobiose dehydrogenase (CDH), superoxide dismutase (SOD), glucose 1-oxidase (GOX), pyranose 2-oxidase (P2Ox) and methanol oxidase. These are mostly oxidases generating H_2O_2 and dehydrogenases. Cytochrome P450 monooxygenases are also significant components involved in the degradation of lignin and many xenobiotics. Recent additions to the enzymatic system of white-rot fungi include dye-decolorizing peroxidases or DyP involved in the oxidation of synthetic high redox-potential dyes and nonphenolic lignin model compounds [165] and aromatic peroxygenases or APOs that catalyze diverse oxygen transfer reactions which can result in the cleavage of ethers [41,166]. Other important components of the lignin degradative system of white-rot fungi are low molecular mass oxidants such as hydroxyl radicals (•OH) and chelated Mn^{3+} (Mn^{3+} mainly chelated by oxalic acid), produced through the action of the ligninolytic enzymes. These are particularly important during the early stages of wood decay and can also act on the degradation of xenobiotics [8,113,130,149].

4.7. Production of Lignin-Modifying Enzymes

The environmental or cultivation conditions (i.e. temperature, carbon/nitrogen ratio, agitation, cultivation system) have a major impact on the expression and production of LMEs by different fungal strains. Some species of white-rot fungi have their ligninolytic abilities regulated by nitrogen and oxygen levels. *P. chrysosporium* usually degrades lignin only under low levels of nitrogen. However, ligninolytic activity of many species and strains of white-rot fungi, such as *Trametes versicolor*, are not nitrogen regulated [130,186,187]. An increase in oxygen tension (e.g. static cultures) has shown to stimulate the production of LiP and MnP [130,144] though laccase production can be enhanced by agitation [161]. The presence of a wide variety of phenolic or aromatic compounds and also copper can induce the production of laccase [188,189]. Interestingly, the production of LMEs in synthetic media is not always reproduced in field conditions or soil cultivation [190].

Table 5. Lignin-modifying enzymes of the white-rot fungi

Enzyme	Active site/ Prosthetic group	Catalytic Mechanism	Functions	Mediators	Fungal sources	References
Laccase	Four copper atoms	Catalysis of the one electron oxidation of phenolics, aromatic amines and other aromatic hydrogen donors with simultaneous reduction of O_2 to H_2O. Reaction: 4 benzenediol + O_2 = 4 benzosemiquinone + 2 H_2O	Removal of polyphenols, degradation of pesticides and textile dyes, bleaching, removal of lignin, synthesis of melanin, formation of basidiomes, generation of hydroxyl radicals, stress defense	Requires the presence of redox mediators. 3-HAA ABTS HBT RBB	*Ganoderma lucidum*	[167,168]
					Pleurotus eryngii	[169]
					Pleurotus ostreatus	[170]
					Rigidoporus lignosus	[171]
					Trametes versicolor	[172,173]
LiP Lignin peroxidase	Heme (Iron protoporphyrin IX)	Oxidation of lignin side chains by one electron abstraction in the presence of H_2O_2, generating reactive radicals. Cleavage of C-C and ether bonds in lignin. Cleavage of aromatic rings. Reaction: 1,2-bis(3,4-dimethoxyphenyl)propane-1,3-diol + H_2O_2 = 3,4-dimethoxybenzaldehyde + 1-(3,4-dimethoxyphenyl)ethane-1,2-diol + H_2O	Degrades lignin (non-phenolic lignin units), mineralize recalcitrant aromatic compounds.	VA 2Cl-14DMB	*Bjerkandera adusta*	[76]
					Phanerochaete chrysosporium	[174-176]
					Trametes cervina	[177]
					Trametes versicolor	[178]
MnP Manganese peroxidase	Heme (Iron protoporphyrin IX)	Oxidize Mn^{2+} into Mn^{3+}. Mn^{3+} acts as a diffusible oxidizer on phenolic or non-phenolic lignin units and amino-aromatic compounds Reaction: 2 Mn^{2+} + 2 H^+ + H_2O_2 = 2 Mn^{3+} + 2 H_2O	Oxidation and depolymerization of lignin, degradation of dyes, phenols and other recalcitrant compounds	Chelated Mn^{3+} Co-oxidants: unsaturated fatty acids and thiols	*Ceriporiopsis subvermispora*	[99]
					Irpex lacteus	[179]
					Lentinus edodes	[180]
					Pleurotus ostreatus	[180,181]
					Trametes versicolor	[172]
VP Versatile peroxidase	Heme (Iron protoporphyrin IX) Hybrid molecular architecture with multiple binding sites	Combination of the catalytic properties of LiP and MnP; Oxidation of Mn^{2+} to Mn^{3+} and also phenolic and nonphenolic aromatic compounds. Reaction: donor + H_2O_2 = oxidized donor + 2 H_2O	Oxidize phenols, hydroquinones, dyes, amines, aromatic alcohols and xenobiotics.	Chelated Mn^{3+}	*Bjerkandera adusta*	[182,183]
					Pleurotus eryngii	[184,185]

3-HAA (3-Hydroxyanthranilic acid), ABTS (2,2′-azinobis-(3-ethylbenzothiazoline-6-sulphonate)), HBT (1-hydroxybenzotriazole), RBB (Remazol brilliant blue), VA (veratryl alcohol), 2Cl-14DMB (2-Chloro-1,4-dimethoxybenzene).

The use of the most diverse agricultural residues worldwide for the production of ligninolytic and hydrolytic enzymes from white-rot fungi in submerged cultivation or solid-state cultivation has proven to be a low cost, efficient and environmental correct alternative [168,191,192].

4.8. Impact on Forest Ecosystems

The white-rot fungi are indispensable for the functioning of forest ecosystems. These fungi maintain equilibrium between formation and breakdown of wood (i.e. wood turnover), contribute to carbon recycling and soil formation and establish symbiotic associations with many organisms [193]. Some insects feed on the fungal basidiomes and the fungal spores can be spread by these animals. Apparently, insects which feed on bracket fungi are more host specific than insects feeding on mushrooms [194]. Some white-rot fungi exhibit associations only with certain families or genera of plants. Some fungal species are more incident and abundant as others are rarer. The production of basidiomes is generally climate dependent. As a result of all the previously mentioned characteristics, many species of white-rot fungi can be used as indicators of forests conservation status.

White-rot fungi are potential bioindicators of heavy metals in the environment. The sensitivity towards metals varies among species as well as the mechanism involved in the protection [195]. Physiological and ecological effects of heavy metals on white-rot fungi include: altering the expression of fungal oxidases which influences the energy flux in the ecosystem and affects the growth rate or reproduction leading to changes in the community structure [172,195]. Heavy metals can also be immobilized in the basidiome. Usually the amount of heavy metals in fungal basidiomes is indicative of the concentration available in their environment [195]. As the concentration of heavy metals in wood is very low, white-rot fungi can be used as biomarkers of atmospheric pollution which would in this case be the main source of the heavy metals accumulated by the fungi [195,196]. In view of the ability to accumulate heavy metals from the environment, the white-rot fungi are also applied in biosorption strategies.

4.9. Economical Impact

The use of white-rot fungi in many industrial processes has long been considered as an eco-friendly, efficient and economic alternative to physico-chemical techniques. The main applications of these fungi include biopulping, biofuel production and bioremediation of industrial effluents and other xenobiotics. Other than these applications, white rot fungi are important sources of bioactive components such as polysaccharides and phenolic compounds with beneficial effects on human health [197-201]. The nutritional and pharmacological properties of these fungi have stimulated the global market of edible mushrooms which is predominantly composed by species of white-rot fungi in the genera *Agaricus*, *Lentinus* and *Pleurotus* [71,201].

Besides the beneficial and valuable properties of the white-rot fungi, these fungi can attack timber in service and also act as plant pathogens causing relevant economical impact.

Most of these fungi decay dead trees and logs or stumps contributing in a great extent to the dynamics and survival of forest ecosystems. However some are parasites attacking living trees as an extended form of biodeterioration [202] and leading to losses of productivity, structural weakening or death of the trees. In most cases, the white-rot fungi are parasites of already injured trees. The establishment of white-rot fungi in living trees involves the ability to tolerate or detoxify the secondary metabolites produced by the trees, such as phenolics, and to degrade lignin. The type of decay caused by wood-decaying fungi in living trees varies among root-, butt-, or heart-rot.

Some white-rot fungi that are common pathogens of living trees include *Armillaria mellea*, *Ganoderma* spp., *Heterobasidion annosum* and *Phellinus* spp. [203-206]. *Schizophyllum commune* can be occasionally found in living trees but mostly attack damp wood or timbers of deciduous trees [157,207]. Other species may be important plant pathogens in particular sites.

Species of *Armillaria* are among the most significant and widespread plant fungal parasites. *Armillaria mellea* spreads by rhizomorphs from one tree to another and is quite difficult to control. *Armillaria luteobubalina* is an aggressive and economically important plant parasite in Australia [208]. Species of the genera *Ganoderma* usually are not aggressive pathogens but are known for attacking many species of living trees. A serious economic loss of oil palm is caused worldwide by species of *Ganoderma* [82,202,209]. Stem Rot disease which is caused by *Ganoderma boninense* is the most destructive disease in oil palm plantations [82,210].

Wood-decaying fungi that are tree parasites in forests and plantations can also cause a detrimental impact on urban tree populations. Trees in the city are exposed to several stress factors that can increase the possibility of parasitism by wood-decaying fungi which can decrease the stability of the trees and increase the chance of mechanical failure [37,211-213].

The attack of living trees by white-rot fungi can occur in the most diverse ecosystems, but are especially significant in humid tropics where the environmental conditions are frequently adequate for the growth of these fungi. In Brazil, a list of phytopathogenic fungi, including white-rot fungi (Table 6), has been produced in an effort to map local plant parasites [214]. Some white-rot fungi have been observed as plant pathogens of more than one plant species. The most frequent white-rot fungi in many different types of plant species in Brazil are: *Rigidoporus microporus* (23 species), *Schizophyllum commune* (17 species), *Polyporus zonalis* (16 species), *Ganoderma philippii* (8 species), *Armillaria* sp. (7 species), *Ganoderma* sp. (6 species) and *Armillaria mellea* (5 species). The majority of the listed fungi are associated to angiosperms (hardwood) except the species of *Armillaria* which are mostly parasites of gymnosperms (softwood), especially *Pinus* sp. The most attacked species of plants in Brazil are in the genera *Pinus*, *Araucaria*, *Theobroma*, *Hevea*, *Coffea* and *Malus*.

Table 6. White-rot fungi as plant pathogens in Brazil

Fungal species	Plant hosts	Pathology
Amylostereum ferreum	*Podocarpus lamberti*	NS
Armillaria mellea	*Pinus elliotti*	NS
	Pinus patula	NS
	Pinus sp.	NS

Fungal species	Plant hosts	Pathology
	Prunus persica	Root rot
	Vitis vinifera	Root rot
***Armillaria* sp.**	*Araucaria angustifolia*	Root rot, Tree death
	Camellia sinensis	Root rot
	Pinus caribaea Morelet var. *hondurensis*	NS
	Pinus maximinoi	NS
	Pinus oocarpa	NS
	Pinus sp.	Tree death, Root rot, Armillaria root disease
	Pinus tecunumanii	NS
Auricularia auriculare	*Torresea cearensis*	NS
Auricularia delicate	*Citrus aurantium*	NS
Collybia apiahyna	*Araucaria brasiliensis*	NS
Collybia aquosa	*Pinus* sp.	NS
***Conocybe* sp.**	*Elaeis guineensis*	NS
***Coriolopsis occidentalis* (syn. *Coriolus occidentalis*)**	*Prunus persica*	NS
Crepidotus albidus* var. *boliviae	*Syagrus romanzoffiana*	NS
Crepidotus alpiniae	*Alpinia aromática*	NS
Crepidotus croceitinctus* var. *cystidiophilus	*Syagrus romanzoffiana*	NS
Crepidotus epibryus	*Araucaria angustifolia*	NS
Crepidotus euterpicola	*Euterpe edulis*	NS
Crepidotus martinii	*Eucalyptus* sp.	NS
***Elfvingia* sp.**	*Araucaria angustifolia*	White-rot
Ganoderma australe (syn. Fomes annularis)	*Hevea brasiliensis*	NS
Ganoderma lucidum	*Cocos nucifera*	Root rot
Ganoderma oerstedii	*Delonix regia*	NS
Ganoderma philippii	*Bixa orellana*	Red root-rot
	Caryophyllus aromaticus	Red rot disease
	Coffea Arabica	NS
	Hevea brasiliensis	Root rot
	Hevea sp.	Red rot disease
	Paullinia cupana	Red root-rot
	Theobroma cacao	Red rot disease, Red root-rot
	Theobroma grandiflorum	Red rot disease
***Ganoderma* sp.**	*Caesalpinia echinata*	NS
	Caesalpinia leiostachya	NS
	Cariniana legalis	NS

Table 6. (Continued)

Fungal species	Plant hosts	Pathology
	Delonix regia	NS
	Holocalyx glaziovii	NS
	Poinciana regia	NS
Inonotus rheades	*Eucalyptus grandis*	NS
	Eucalyptus sp.	NS
Irpex lacteus	*Prunus persica*	NS
Lentinus villosus	*Araucaria angustifolia*	White-rot
	Areca oleracea	NS
	Aspidosperma polyneuron	NS
Mycena citricolor	*Bryophyllum* sp.	NS
	Coffea Arabica	NS
	Inga edulis	NS
***Phellinus* sp.**	*Araucaria angustifolia*	Root rot, Heart rot
	Bertholletia excels	Post-harvest fungus
	Pinus sp.	Root rot, Tree death
Pleurotus ostreatus	*Araucaria angustifolia*	White-rot
Polyporus zonalis	*Artocarpus altilis*	NS
	Artocarpus heterophyllus	NS
	Bombax munguba	NS
	Calophyllum sp.	NS
	Calycophyllum spruceanum	NS
	Carapa guianensis	NS
	Cecropia sp.	NS
	Ceiba pentandra	NS
	Cocos nucifera	NS
	Coffea Arabica	NS
	Ficus sp.	NS
	Hevea brasiliensis	NS
	Hura crepitans	NS
	Mangifera indica	NS
	Pouteria macrocarpa	NS
	Theobroma cacao	NS
Polystictus pinsitus	*Mimosa bracaatinga*	NS
Polystictus sector	*Theobroma bicolor*	NS
Pycnoporus cinnabarinus	*Malus domestica*	NS
	Araucaria angustifolia	White-rot
Rigidoporus microporus (syn. Fomes lignosus)	*Araucaria angustifólia*	White-rot
	Artocarpus altilis	NS

Fungal species	Plant hosts	Pathology
	Artocarpus heterophyllus	NS
	Bombax munguba	NS
	Calophyllum sp.	NS
	Calycophyllum spruceanum	NS
	Carapa guianensis	NS
	Cecropia sp.	NS
	Ceiba pentandra	NS
	Cocos nucifera	NS
	Coffea arabica	NS
	Ficus sp.	NS
	Hevea brasiliensis	White root-rot
	Hevea sp.	White-rot, White root-rot
	Hura crepitans	NS
	Khaya ivorensis	NS
	Mangifera indica	NS
	Manihot esculenta	White-rot
	Paullinia cupana	NS
	Piper nigrum	White root-rot
	Pouteria macrocarpa	NS
	Theobroma cacao	White-rot
	Theobroma grandiflorum	White root-rot
Schizophyllum commune	*Achras sapota*	NS
	Bambusa sp.	NS
	Betula papyrifera	NS
	Carya sp.	NS
	Castanea sp.	NS
	Coffea arabica	NS
	Diospyros sp.	NS
	Elaeis guineensis	Loss of germinating viability
	Eucalyptus sp.	NS
	Gleditsia sp.	NS
	Gossypium hirsutum	NS
	Jatropha curcas	NS
	Liquidambar sp.	NS
	Manihot esculenta	NS
	Prosopanche burmeisteri	NS
Fungal species	**Plant hosts**	**Pathology**
	Saccharum officinarum	NS
	Theobroma cacao	NS
Stereum hirsutum	*Malus domestica*	NS

Table 6. (Continued)

Fungal species	Plant hosts	Pathology
***Stereum* sp.**	*Baccharis oxyodonta*	NS
Trametes ochracea	*Malus domestica*	NS
Trametes versicolor	*Malus domestica*	NS

NS – not specified
Data obtained from: Mendes & Urben [214]. Database produced from technical books and scientific journals from 1933 to 2002 and collected and published online by Brazilian Agricultural Research Corporation (EMBRAPA) as an initial mapping of phytopathogenic fungi in Brazil.
Available at: http://pragawall.cenargen.embrapa.br/aiqweb/michtml/fgbanco01.asp.

An important step to develop efficient and eco-friendly mechanisms to control plant parasitism by fungi is to investigate the interactions between plants and fungi and understand the biological and ecological processes of these associations that lead fungi to become pathogens. The control of moisture content on wood is an important control measure against the attack of these fungi on wood in service, as they usually do not decay wood that is too dry or too wet. The identification of wood-decaying fungi in initial stages of decay is also significant for parasitism control. Recently, techniques based on the identification of fungal DNA in wood samples have been developed [215,216].

4.10. Biodiversity of Agaricomycetes in Tropical Environments

The number of fungi distributed worldwide in most diverse ecosystems is estimated at 3.5 to 5.1 million species [217]. However, only 97,330 species of fungi have been described so far [218]. Most of these species are *Ascomycota* (64,163 species) and *Basidiomycota* (31,515) [218].

The largest diversity of most groups of organisms, including fungi, is distributed in tropical environments [219,220]. Despite this, the tropical biodiversity of basidiomycetes is widely unexplored and less described than the biodiversity of this group of fungi in temperate environments.

The climate and predominant vegetation of a particular environment is greatly associated to the number and species of wood-decaying fungi present in that habitat. In South America, particularly in Brazil, there are two of the most important biomes in the world in terms of tropical biodiversity: the Amazon rainforest (Amazônia) and the Atlantic forest (Mata Atlântica). Other important and extent tropical biomes present in Brazil are Cerrado which is characterized by tropical savannas [221,222] and Caatinga, a seasonally dry tropical forest [223]. Recently, updated lists from literature records of lignocellulolytic *Agaricomycetes* observed in these biomes in Brazil have been published [224-228]. However, in view of the wide territorial extent of Amazon, Atlantic forest and Cerrado biomes as well as other tropical biomes, additional work is necessary not only in field to find and catalogue new species but also in laboratory to characterize the species.

Basidiomycota are widely distributed in Brazil and can be found in all phytogeographic domains. A total of 1730 species are taxonomically accepted and distributed in 376 genera. Among these species, 417 are endemic [229]. Most studied white wood-decaying fungi can be commonly found in the orders *Agaricales, Auriculariales*, *Hymenochaetales, Polyporales* and *Russulales*. A total of 255 species (34 endemic) of *Agaricales*, 28 species of *Auriculariales*, 136 species (9 endemic) of *Hymenochaetales*, 333 species (50 endemic) of *Polyporales* and 46 species (6 endemic) of *Russulales* are described in Brazil [230-234]. Many of the species are native and endemic (Table 7). The *Polyporales* present the highest diversity and distribution in the Brazilian biomes (Figure 5). The Atlantic forest has the highest diversity and endemic species of *Agaricomycetes*. The genus *Phellinus* is the most common. Most of the recorded species in the biomes cause white-rot. In Caatinga, 97.3% of the listed *Agaricomycetes* are white-rot fungi and, interestingly, a higher species richness of *Hymenochaetaceae* is observed on live hosts than on dead substrates, being a distinct characteristic from other tropical forests [227,235].

The composition and richness of fungal species communities can be affected by natural disturbances, but anthropogenic actions such as discharge of pollutants in the air, water and soil and wood removal are the main causes of loss of fungal biodiversity. Therefore, the research of fungi biodiversity and ecology is of vital importance to understand the dynamics and equilibrium of forest ecosystems as well as to develop conservation strategies and find novel natural resources with biotechnological potential.

Table 7. Endemic and native species of wood-decaying basidiomycetes of the most species rich orders in the largest tropical biomes in Brazil

Biome	Characteristics	Order	Endemic and Native Species
Amazon	Area in Brazil: 4,196,943 km^2 ; Total area: approximately 7,000,000 km^2; Tropical forest with diverse vegetation. The climate is equatorial humid or tropical rainforest and the average temperature is around 25°C.	*Agaricales*	NO
		Auriculariales	NO
		Hymenochaetales	*Coltricia globispora*
		Polyporales	*Amauroderma unilatereum* *Gloeoporus cystidiatus* *Perenniporia martia* *Perenniporia sinuosa* *Podoscypha bubalina* *Rigidoporus amazonicus* *Tyromyces singeri*
		Russulales	*Lachnocladium brasiliense*
Atlantic Forest	Area: 1,110,182 km^2 Diverse vegetation from tropical and subtropical moist forests to tropical dry forests. The climate is mainly tropical warm and humid with an average temperature from 22 to 25°C in most areas.	*Agaricales*	*Campanella elongatispora* *Lycoperdon marginatum* *Morganella benjaminii* *Stropharia araucariae* *Tetrapyrgos nigripes*
		Auriculariales	NO

Table 7. (Continued)

Biome	Characteristics	Order	Endemic and Native Species
		Hymenochaetales	*Clavariachaete peckoltii* *Hymenochaete pratensis* *Hymenochaete proxima* *Phellinus macrospores* *Phellinus neocallimorphus* *Phellinus piptadeniae* *Phellinus rickii* *Phellinus spinescens*
		Polyporales	*Amauroderma unilatereum* *Antrodiella luteocontexta* *Antrodiella mollis* *Antrodiella multipileata* *Auriporia brasílica* *Ceriporiopsis cystidiata* *Ceriporiopsis latermarginata* *Cerrena cystidiata* *Diacanthodes fluminensis* *Diplomitoporus dilutabilis* *Diplomitoporus marianoi-rochae* *Diplomitoporus navisporus* *Diplomitoporus taquarae* *Flaviporus subhydrophilus* *Fomes paulensis* *Gloeoporus guerreroanus* *Grammothelopsis bambusicola* *Grammothelopsis puiggarii* *Junghuhnia polycystidifera* *Microporellus brasiliensis* *Mycoacia subconspersa* *Navisporus terrestris* *Perenniporia martia* *Perenniporia neofulva* *Phaeolus subbulbipes* *Phlebiopsis galochroa* *Podoscypha bubalina* *Podoscypha moelleri* *Podoscypha replicata* *Podoscypha semiresupinata* *Podoscypha viridans* *Polyporus aquosus* *Polyporus lentinoides* *Polyporus puiggarianus* *Polyporus sapurema* *Rigidoporus amazonicus* *Rubroporus carneoporis* *Skeletocutis roseola* *Steccherinum hydneum* *Trametes farcta* *Trametes ochroflava*

Biome	Characteristics	Order	Endemic and Native Species
			Tyromyces atroalbus *Tyromyces crassisporus*
		Russulales	*Asterostroma olivaceum* *Asterostroma sordidum* *Lachnocladium brasiliense* *Xylobolus frustulatus*
Caatinga	Area: 844,453km^2 Seasonally dry tropical forest: dry/rainy seasons. Vegetation mainly composed of thorny trees and xerophytes. The climate is semi-arid with an average temperature of 25.5°C.	*Agaricales*	*Tulostoma excentricum*
		Auriculariales	NO
		Hymenochaetales	*Phellinus piptadeniae*
		Polyporales	*Navisporus terrestris*
		Russulales	NO
Cerrado	Area: 2,036,448 km^2 There are extensive formations of tropical savannas and some gallery forests. The climate is tropical wet and dry. The average temperature is 25°C.	*Agaricales*	*Crucibulum laeve* *Montagnea haussknechtii* *Tulostoma exasperatum*
		Auriculariales	NO
		Hymenochaetales	NO
		Polyporales	*Polyporus lentinoides* *Steccherinum hydneum*
		Russulales	NO

NO – not observed

Data obtained from: Capelari, et al. [230]; Gibertoni, et al. [231]; Gomes-Silva, et al. [225]; Gugliotta, et al. [232]; IBGE, 2004 [236]; Loguercio-Leite & Campos-Santana [233]; Neves, et al. [234].

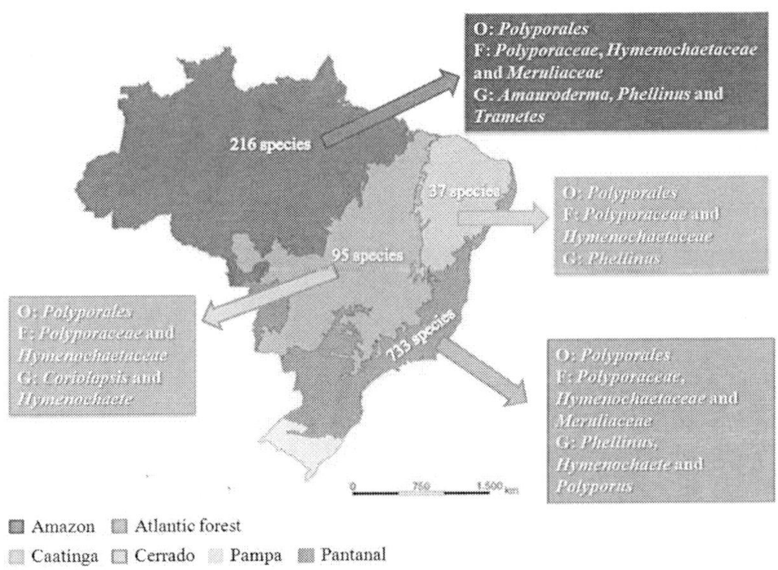

Figure 5. Number of *Agaricomycetes* species in the largest Brazilian biomes. The order (O), families (F) and genera (G) with the highest number of species are displayed for each biome. Data obtained from: Baltazar & Gibertoni [226]; Drechsler-Santos, et al. [227]; Gibertoni & Drechsler-Santos [228]; Gomes-Silva & Gibertoni [224].

4.11. Bioluminescence in White-Rot Fungi

Many white-rot fungi are bioluminescent, i.e. they have the special ability to emit light from the basidiomes and/or mycelium at a determined wavelength. Fungal bioluminescence has been well reported, but the mechanism of light emission by luminous fungi and the ecological impact or biological function has not yet been fully understood.

The biochemical mechanism of fungal luminescence has been characterized as the luciferin-luciferase type [237] and is an oxygen-dependent metabolic process [238]. Recently, the mechanism of bioluminescence has been reproduced enzymatically *in vitro* [239].

It has been proposed that bioluminescence is involved in lignin degradation by basidiomycetes [240]. Bioluminescence has also been suggested as an advantageous biochemical process providing protection against reactive oxygen species produced mainly by mitochondria during respiration [238].

A total of 64 species of bioluminescent fungi are recognized so far, and are generally distributed in the genera *Armillaria*, *Mycena*, *Pleurotus*, *Omphalotus* and *Panellus* [241,242]. The Atlantic forest in Brazil is probably the site with the greatest diversity of bioluminescent basidiomycetes, and many species of white-rot fungi in the genera *Mycena*, including endemic species, have been detected in this tropical forest [238].

Recently, the bioluminescent fungus *Agaricus gardneri* was rediscovered in Brazil and taxonomically classified as *Neonothopanus gardneri*. This is one of the largest bioluminescent agaric fungus in the world and the biggest in Brazil [243].

4.12. Research and Recent Applications in Green Technologies

The research regarding the most diverse characteristics of the white-rot fungi has evolved considerably over the last twenty years (Figure 6).

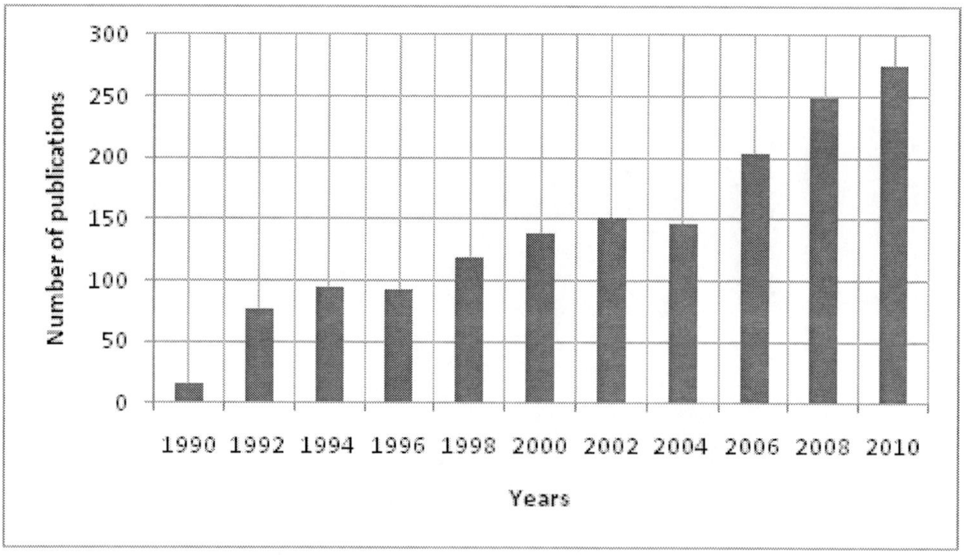

Figure 6. Number of publications on white-rot fungi produced in two decades. Data Source: Thomson Reuters Web of Knowledge.

The progressive increase in the numbers of publications (17-fold from 1990 to 2010) considering white rot fungi as the main topic mostly reflects the great interest of using the species or enzymes of this group of fungi in green chemistry (environmentally benign technologies) (Table 8). Nowadays, the ten most cited articles in the topic "white rot fungi" (according to Thomson Reuters Web of Knowledge) mainly describe the applications of this group of fungi in bioremediation strategies, especially in the removal of dyes from textile effluents by laccases and biosorption. Interestingly, the second most cited article reports the possible treatments of lignocelluloses for ethanol production [244] demonstrating a tendency of research in this subject.

Table 8. Some of the most recent applications of white-rot fungi in green technologies

Application	Fungus	Substrates/Effluents/Compounds	References
Biopretreatment of agroindustrial residues for biofuel production	*Ceriporiopsis subvermispora*	Sugarcane bagasse	[245]
	Echinodontium taxodii	Corn stover	[246]
	Ceriporiopsis subvermispora and *Phellinus* sp.	Softwood (*Cryptomeria japonica*)	[247]
	Trametes hirsuta	Starch, wheat bran and rice straw	[248]
Bioremediation of pharmaceuticals	*Trametes versicolor*	Naproxen, Ketoprofen, Diclofenac, Ibuprofen, Atenolol, Propanolol and Carbamazepine	[249-253]
	Trametes versicolor	Sewage sludge contaminated with many pharmaceuticals. Complete removal of phenazone, bezafibrate, fenofibrate, cimetidine, clarithromycin, sulfamethazine and atenolol	[254]
Bioremediation of pesticides	*Ganoderma lucidum*	Bentazon and Diuron	[167]
	Anthracophyllum discolor and *Phanerochaete chrysosporium*	Pentachlorophenol in soil	[255]
	Phanerochaete chrysosporium, *Trametes versicolor* and *Pleurotus ostreatus*	Straw extract medium and soil extract medium contaminated with thiabendazole, imazalil, thiophanate methyl, ortho-phenylphenol, diphenylamine and chlorpyrifos	[256]
Bioremediation of dyes	*Trametes trogii*, *T. villosa* and *Coriolus versicolor* var. *antarcticus*	Gentian Violet, Xylidine, Congo Red, Malachite Green, Remazol Brilliant Blue R, Indigo Carmine and Anthraquinone Blue	[257]
	Pycnoporus coccineus and *Pycnoporus sanguineus*	Poly R-478, Remazol brilliant blue R, Azure B, Bromocresol purple, Reactive black 5, Acid yellow 17, Nitrosulfonazo III	[258]

4.13. Research in Brazil

Brazil has a rich diversity of white-rot fungi and research on this group of fungi has expanded over the years in the country (Fig.7). The discovery of new species and bioprospection of novel applications of the known species, the generation of agroindustrial residues and industrial effluents have encouraged the research on this group of fungi.

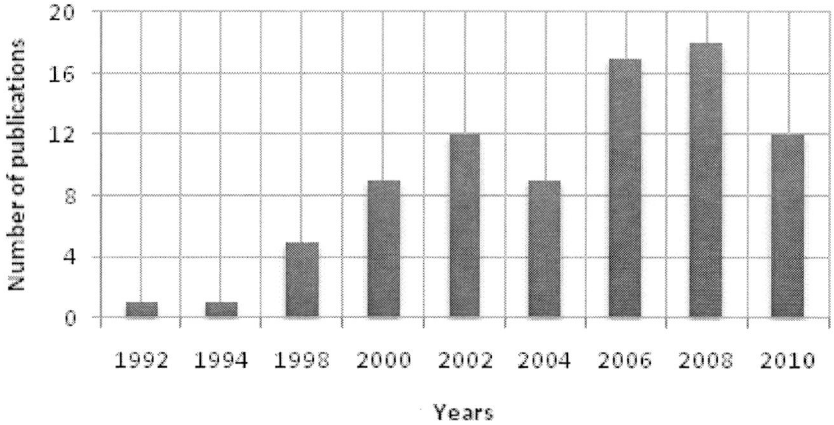

Figure 7. Number of publications produced in Brazil on white-rot fungi. Data source: Thomson Reuters Web of Knowledge.

Brazil is among the ten countries with the highest percentage of total publications on white-rot fungi (approximately 4% of the 4,465 publications produced worldwide so far) (Fig. 8).

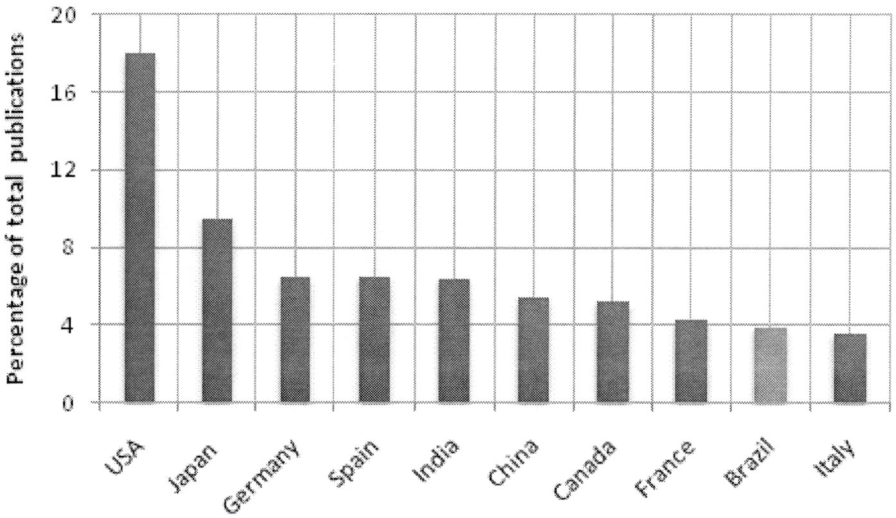

Figure 8. Brazil´s share of world publication about white-rot fungi. Data source: Thomson Reuters Web of Knowledge.

The ten most cited articles (according to Thomson Reuters Web of Knowledge) from Brazilian researchers on the topic "white-rot fungi" have mainly described the use of solid-state fermentation [259] and sugarcane bagasse [260], the applications of laccases [261,262], the decolorization of dyes [263,264] and the biodegradation of wood [265,266].

The research on white-rot fungi in Brazil over the last five years has focused on:

(A) diversity, ecology and bioprospection
- *Agaricomycetes* – Brazilian semi-arid region, Cerrado biome, Atlantic rain forest in northeast Brazil and Brazilian Amazonia [205,224,227,228];
- *Mycena* – bioluminescent species [238];
- Polypores – south of Brazil [267];
- *Trametes* – Brazilian Amazonia [268];
- Polyporoid fungi – Brazilian Amazonia [269];
- White-rot fungi from *Eucalyptus* spp. Forests [270];
- Basidiomycetes from Amazonia [271];

(B) biopulping
- *Ceriporiopsis subvermispora* – *Eucalyptus grandis* and *Pinus taeda* [126];
- *Ceriporiopsis subvermispora* – *Pinus taeda* [272];
- *Ceriporiopsis subvermispora* – *Eucalyptus grandis* [273];
- Technological advances [274];

(C) biopretreatment, enrichment and saccharification of lignocellulosic materials
- *Pleurotus sajor-caju* – sugarcane bagasse [275];
- *Pleurotus ostreatus* and *Pleurotus sajor-caju* – soft rush and rice residues [276];
- *Lentinula edodes* - effluent from wood laminate industry [277];

(D) production of ligninolytic and hydrolytic enzymes from agroindustrial residues
- *Pleurotus ostreatus*, *Pleurotus pulmonarius*, *Macrocybe titans*, *Ganoderma lucidum*, and *Grifola frondosa* – laccases, pectinases, and aryl-β-D-glycosidases (β-glucosidases, β-xylosidases, and β-galactosidases). Substrate: passion fruit waste [168];
- *Coriolopsis byrsina* SXS16, *Lentinus strigellus* SXS355 and *Lentinus* sp. SXS48, *Pycnoporus sanguineus* SXS 43 and *Phellinus rimosus* SXS47 – Laccase, MnP and LiP. Substrates: wheat straw and rice straw [278];
- *Agaricus brasiliensis* and *Pleurotus ostreatus* – holocellulase. Substrates: sugarcane bagasse, banana stems, dirty cotton, corn and soybean residues [279];

(E) detection and application of bioactive compounds
- *Agaricus brasiliensis* (*Agaricus blazei* Murril) – antioxidant capability and total phenolic contents of methanolic extracts [198];
- *Agaricus brasiliensis* – effect of hydroalcoholic extract on liver function [199];

(F) bioremediation of contaminated soil, pesticides, synthetic dyes and industrial effluents
- *Pleurotus sajor-caju* CCB 20, *P. chrysosporium*, *Pleurotus* sp. 068, *P. shimeji* and *Ganoderma* sp. – degradation of vinasse from industrial processing of sugarcane [280,281];
- *Ganoderma lucidum* – dye decolorization by laccase [282];

- *Lentinula edodes* CCB 554, *Perenniporia medulla-panis* ATCC 42463 and *Trametes versicolor* – degradation of Azure B [283];
- *Caripia montagnei, Datronia caperata, Ganoderma stiptatum, Hexagonia hydnoides, Pleurotus ostreatus, Pycnoporus sanguineus, Trametes membranacea* and *Earliella scabrosa* – decolorization of synthetic dyes [284];
- *Agaricus subrufescens* (syn. *A. brasiliensis*) and *Lentinula* edodes – spent mushroom substrates for lettuce cultivation and soil bioremediation [285];
- *Ganoderma lucidum* – degradation of diuron and bentazon [167,286];
- *Pleurotus pulmonarius* – degradation of pentachlorophenol [287];

(G) biocatalysis of chemical compounds

- *Irpex lacteus* CCB 196, *Pycnoporus sanguineus* CCB 196, *Trametes rigida* CCB 285, *Trametes versicolor* CCB 202 and *Trichaptum byssogenum* CCB 203 - racemic selenocyclohexanone and (±)-2-(phenylthio)cyclohexanone [288,289];
- *Lentinus strigellus* - aromatic and aliphatic ketones [290];

Conclusion

The diversity and importance of white-rot fungi in the most diverse ecosystems is undeniable. Much remains to be investigated about the biology and ecology of this group of fungi, especially in tropical environments. It must be emphasized that the group as a whole offers excellent opportunities in aiding to reduce pollution worldwide. This is a field which is still in its beginnings. Furthermore, the industrial applications, especially as enzyme producers, by far the most advanced branch, will certainly show considerable expansion in the next years, as it has not yet exhausted its multifaceted possibilities.

Acknowledgments

This work was supported by grants from Fundação Araucária and Conselho Nacional de Desenvolvimento Científico e Tecnológico (CNPq, Edital 047/2010-SISBIOTA, proc. 563260/2010-6). R.M. Peralta and A. Bracht are fellowship holders of CNPq. G.M. Maciel is a fellowship holder of the Coordenação de Aperfeiçoamento de Pessoal de Ensino Superior (CAPES).

References

[1] Tlalka, M., Fricker, M. & Watkinson, S. (2008). Imaging of long-distance alpha-aminoisobutyric acid translocation dynamics during resource capture by *Serpula lacrymans*. *Applied and Environmental Microbiology*, 74, 2700-2708.

[2] Lindahl, B.O., Taylor, A.F.S. & Finlay, R.D. (2002). Defining nutritional constraints on carbon cycling in boreal forests - towards a less 'phytocentric' perspective. *Plant and Soil*, 242, 123-135.

[3] Boddy, L. (1999). Saprotrophic cord-forming fungi: meeting the challenge of heterogeneous environments. *Mycologia*, 91, 13-32.

[4] Schwarze, F.W.M.R., Mattheck, C., & Engels, J. (2000). *Fungal strategies of wood decay in trees* (1st ed.). Germany: Springer-Verlag Berlin Heidelberg.

[5] Powell, M.R. (2002). A model for probabilistic assessment of phytosanitary risk reduction measures. *Plant Disease*, 86, 552-557.

[6] Yelle, D.J., Ralph, J., Lu, F.C. & Hammel, K.E. (2008). Evidence for cleavage of lignin by a brown rot basidiomycete. *Environmental Microbiology*, 10, 1844-1849.

[7] Wei, D.S., Houtman, C.J., Kapich, A.N., Hunt, C.G., Cullen, D. & Hammel, K.E. (2010). Laccase and Its Role in Production of Extracellular Reactive Oxygen Species during Wood Decay by the Brown Rot Basidiomycete *Postia placenta. Applied and Environmental Microbiology*, 76, 2091-2097.

[8] Hammel, K.E., Kapich, A.N., Jensen, K.A. & Ryan, Z.C. (2002). Reactive oxygen species as agents of wood decay by fungi. *Enzyme and Microbial Technology*, 30, 445-453.

[9] Goodell, B., Jellison, J., Liu, J., Daniel, G., Paszczynski, A., Fekete, F., et al. (1997). Low molecular weight chelators and phenolic compounds isolated from wood decay fungi and their role in the fungal biodegradation of wood. *Journal of Biotechnology,* 53(2-3), 133-162.

[10] Suzuki, M.R., Hunt, C.G., Houtman, C.J., Dalebroux, Z.D. & Hammel, K.E. (2006). Fungal hydroquinones contribute to brown rot of wood. *Environmental Microbiology* 8, 2214-2223.

[11] Schmidt, C.J., Whitten, B.K. & Nicolas, D.D. (1981). A proposed role for oxalic acid in non-enzymatic wood decay by brown-rot fungi. *Proceedings of American Wood Preservation Association* 77, 157-164.

[12] Enoki, A., Hirano, T. & Tanaka, H. (1992). Extracellular substance from the brown-rot basidiomycete *Gloeophyllum trabeum* that produces and reduces hydrogen peroxide. *Material Und Organismen* 27, 247-261.

[13] Hirano, T., Tanaka, H. & Enoki, A. (1995). Extracellular substance from the brown-rot basidiomycete *Tyromyces palustris* that reduces molecular oxygen to hydroxyl radicals and ferric iron to ferrous iron. Mokuzai Gakkaishi 41, 334-341.

[14] Hyde, S.M. & Wood, P.M. (1997). A mechanism for production of hydroxyl radicals by the brown-rot fungus *Coniophora puteana*: Fe(III) reduction by cellobiose dehydrogenase and Fe(II) oxidation at a distance from the hyphae. *Microbiology-Uk* 143, 259-266.

[15] Kerem, Z., Jensen, K.A. & Hammel, K.E. (1999). Biodegradative mechanism of the brown rot basidiomycete *Gloeophyllum trabeum*: evidence for an extracellular hydroquinone-driven fenton reaction. *Febs Letters* 446, 49-54.

[16] Jensen, K.A., Houtman, C.J., Ryan, Z.C. & Hammel, K.E. (2001). Pathways for extracellular fenton chemistry in the brown rot basidiomycete *Gloeophyllum trabeum. Applied and Environmental Microbiology* 67, 2705-2711.

[17] Shimada, M., Akamtsu, Y., Tokimatsu, T., Mii, K. & Hattori, T. (1997). Possible biochemical roles of oxalic acid as a low molecular weight compound involved in brown-rot and white-rot wood decays. *Journal of Biotechnology* 53, 103-113.

[18] Makela, M.R., Hilden, K. & Lundell, T.K. (2010). Oxalate decarboxylase: biotechnological update and prevalence of the enzyme in filamentous fungi. *Applied Microbiology and Biotechnology* 87, 801-814.

[19] Dutton, M.V., Evans, C.S., Atkey, P.T. & Wood, D.A. (1993). Oxalate production by basidiomycetes, including the white-rot species *Coriolus versicolor* and *Phanerochaete chrysosporium*. *Applied Microbiology and Biotechnology* 39, 5-10.

[20] Yoon, J.J., Cha, C.J., Kim, Y.S. & Kim, W. (2008). Degradation of cellulose by the major endoglucanase produced from the brown-rot fungus *Fomitopsis pinicola*. *Biotechnology Letters* 30, 1373-1378.

[21] Okamoto, K., Sugita, Y., Nishikori, N., Nitta, Y. & Yanase, H. (2011). Characterization of two acidic beta-glucosidases and ethanol fermentation in the brown rot fungus *Fomitopsis palustris*. *Enzyme and Microbial Technology* 48, 359-364.

[22] Valaskova, V. & Baldrian, P. (2006). Degradation of cellulose and hemicelluloses by the brown rot fungus *Piptoporus betulinus* - production of extracellular enzymes and characterization of the major cellulases. *Microbiology-Sgm* 152, 3613-3622.

[23] Duncan, S.M. & Schilling, J.S. (2010). Carbohydrate-hydrolyzing enzyme ratios during fungal degradation of woody and non-woody lignocellulose substrates. *Enzyme and Microbial Technology* 47, 363-371.

[24] Cohen, R., Jensen, K.A., Houtman, C.J. & Hammel, K.E. (2002). Significant levels of extracellular reactive oxygen species produced by brown rot basidiomycetes on cellulose. *Febs Letters* 531, 483-488.

[25] Yoon, J.J. & Kim, Y.K. (2005). Degradation of crystalline cellulose by the brown-rot basidiomycete *Fomitopsis palustris*. *Journal of Microbiology*, 43, 487-492.

[26] Kaneko, S., Yoshitake, K., Itakura, S., Tanaka, H. & Enoki, A. (2005). Relationship between production of hydroxyl radicals and degradation of wood, crystalline cellulose, and a lignin-related compound or accumulation of oxalic acid in cultures of brown-rot fungi. *Journal of Wood Science*, 51, 262-269.

[27] Martinez, D., Challacombe, J., Morgenstern, I., Hibbett, D., Schmoll, M., Kubicek, C. P., et al. (2009). Genome, transcriptome, and secretome analysis of wood decay fungus *Postia placenta* supports unique mechanisms of lignocellulose conversion. *Proceedings of the National Academy of Sciences of the United States of America,* 106, 1954-1959.

[28] Daniel, G., & Nilsson, T. (1998). Developments in the study of soft rot and bacterial decay. In A. Bruce & J. W. Palfreyman (Eds.), *Forest Products Biotechnology* (pp. 37-62). London: Taylor and Francis.

[29] Eaton, R. A., & Hale, M. D. C. (1993). *Wood: Decay, Pests and Protection*. London: Chapman & Hall.

[30] Daniel, G. (2003). Microview of Wood under Degradation by Bacteria and Fungi. In B. Goodell, D. D. Nicholas & T. P. Schultz (Eds.), *Wood Deterioration and Preservation: Advances in our Changing World* (Vol. 845, pp. 34-72). Washington: American Chemical Society.

[31] Udaiyan, K. & Manian, S. (1991). Fungi colonizing wood in the cooling tower water system at the Madras Fertilizer Company, Madras, India. *International Biodeterioration*, 27, 351-371.

[32] Blanchette, R.A., Held, B.W., Jurgens, J.A., McNew, D.L., Harrington, T.C., Duncan, S.M. & Farrell, R.L. (2004). Wood-destroying soft rot fungi in the historic expedition huts of Antarctica. *Applied and Environmental Microbiology* 70, 1328-1335.

[33] Nilsson, T. & Daniel, G. (1990). Structure and the aging process of dry archaeological wood. *Advances in Chemistry Series*, 67-86.

[34] Blanchette, R.A. (2000). A review of microbial deterioration found in archaeological wood from different environments. *International Biodeterioration & Biodegradation* 46, 189-204.

[35] Eriksson, K.E. (1981). Fungal degradation of wood components. *Pure and Applied Chemistry*, 53, 33-43.

[36] Khalili, S., Nilsson, T. & Daniel, G. (2001). The use of soft rot fungi for determining the microfibrillar orientation in the S2 layer of pine tracheids. *Holz Als Roh-Und Werkstoff*, 58, 439-447.

[37] Schwarze, F.W.M.R. (2007). Wood decay under the microscope. *Fungal Biology Reviews*, 21, 133-170.

[38] Blanchette, R.A., Nilsson, T., Daniel, G. and Abad, A. (1990). Biological degradation of wood. *Advances in Chemistry Series*, 141-174.

[39] Schwarze, F.W.M.R., Lonsdale, D. & Mattheck, C. (1995). Detectability of wood decay caused by *Ustulina deusta* in comparison with other tree-decay fungi. *European Journal of Forest Pathology*, 25, 327-341.

[40] Liers, C., Ullrich, R., Steffen, K.T., Hatakka, A. & Hofrichter, M. (2006). Mineralization of C-14-labelled synthetic lignin and extracellular enzyme activities of the wood-colonizing ascomycetes *Xylaria hypoxylon* and *Xylaria polymorpha*. *Applied Microbiology and Biotechnology*, 69, 573-579.

[41] Liers, C., Arnstadt, T., Ullrich, R. & Hofrichter, M. (2011). Patterns of lignin degradation and oxidative enzyme secretion by different wood- and litter-colonizing basidiomycetes and ascomycetes grown on beech-wood. *FEMS Microbiology Ecology*, 1-12.

[42] Regalado, V., Rodriguez, A., Perestelo, F., Carnicero, A., delaFuente, G. & Falcon, M.A. (1997). Lignin degradation and modification by the soil-inhabiting fungus *Fusarium proliferatum*. *Applied and Environmental Microbiology,* 63, 3716-3718.

[43] Pointing, S.B. (2001). Feasibility of bioremediation by white-rot fungi. *Applied Microbiology and Biotechnology*, 57, 20-33.

[44] Eriksson, K. E., Blanchette, R. A., & Ander, P. (1990). *Microbial and enzymatic degradation of wood and wood components*. Berlin: Springer.

[45] Nilsson, T. & Daniel, G. (1989). Chemistry and microscopy of wood decay by some higher ascomycetes. *Holzforschung*, 43, 11-18.

[46] Hammel, K.E., Jensen, K.A., Mozuch, M.D., Landucci, L.L., Tien, M. & Pease, E.A. (1993). Ligninolysis by a purified lignin peroxidase. *Journal of Biological Chemistry*, 268, 12274-12281.

[47] Shary, S., Ralph, S.A. & Hammel, K.E. (2007). New insights into the ligninolytic capability of a wood decay ascomycete. *Applied and Environmental Microbiology*, 73, 6691-6694.

[48] Worrall, J.J., Anagnost, S.E. & Zabel, R.A. (1997). Comparison of wood decay among diverse lignicolous fungi. *Mycologia*, 89, 199-219.

[49] Ortega, N., Busto, M.D. & Perez-Mateos, M. (2001). Kinetics of cellulose saccharification by *Trichoderma reesei* cellulases. *International Biodeterioration & Biodegradation*, 47, 7-14.

[50] Martinez, D., Berka, R. M., Henrissat, B., Saloheimo, M., Arvas, M., Baker, S. E., et al. (2008). Genome sequencing and analysis of the biomass-degrading fungus *Trichoderma reesei* (syn. *Hypocrea jecorina*). *Nature Biotechnology*, 26, 553-560.

[51] Tanaka, H., Enoki, A., Fuse, G. & Nishimoto, K. (1988). Interactions in sucessive exposure of wood to varying wood-inhabiting fungi. *Holzforschung*, 42, 29-35.

[52] Raberg, U., Terziev, N. & Land, C.J. (2009). Early soft rot colonization of Scots sapwood pine in above-ground exposure. *International Biodeterioration & Biodegradation*, 63, 236-240.

[53] Fukasawa, Y., Osono, T. & Takeda, H. (2009). Microfungus communities of Japanese beech logs at different stages of decay in a cool temperate deciduous forest. *Canadian Journal of Forest Research-Revue Canadienne De Recherche Forestiere*, 39, 1606-1614.

[54] Vasil'chenko, L.G., Khromonygina, V.V., Karapetyan, K.N., Vasilenko, O.V. & Rabinovich, M.L. (2005). Cellobiose dehydrogenase formation by filamentous fungus *Chaetomium* sp. INBI 2-26(-). *Journal of Biotechnology*, 119, 44-59.

[55] Karnchanatat, A., Petsom, A., Sangvanich, P., Piaphukiew, J., Whalley, A.J.S., Reynolds, C.D. & Sihanonth, P. (2007). Purification and biochemical characterization of an extracellular beta-glucosidase from the wood-decaying fungus *Daldinia eschscholzii* (Ehrenb.: Fr.) Rehm. *FEMS Microbiology Letters,* 270, 162-170.

[56] Rahman, Z., Shida, Y., Furukawa, T., Suzuki, Y., Okada, H., Ogasawara, W. & Morikawa, Y. (2009). Evaluation and characterization of *Trichoderma reesei* cellulase and xylanase promoters. *Applied Microbiology and Biotechnology* 82, 899-908.

[57] Chandra, M., Kalra, A., Sharma, P.K. & Sangwan, R.S. (2009). Cellulase production by six *Trichoderma* spp. fermented on medicinal plant processings. *Journal of Industrial Microbiology & Biotechnology*, 36, 605-609.

[58] Ankudimova, N.V., Baraznenok, V.A., Becker, E.G. & Okunev, O.N. (1999). Cellulase complex from *Chaetomium cellulolyticum*: Isolation and properties of major components. *Biochemistry-Moscow*, 64, 1068-1073.

[59] Boisset, C., Petrequin, C., Chanzy, H., Henrissat, B. & Schulein, M. (2001). Optimized mixtures of recombinant *Humicola insolens* cellulases for the biodegradation of crystalline cellulose. *Biotechnology and Bioengineering*, 72, 339-345.

[60] Schulein, M. (1997). Enzymatic properties of cellulases from *Humicola insolens*. *Journal of Biotechnology,* 57, 71-81.

[61] Miettinen-Oinonen, A., Paloheimo, M., Lantto, R.A. & Suominen, P. (2005). Enhanced production of cellobiohydrolases in *Trichoderma reesei* and evaluation of the new preparations in biofinishing of cotton. *Journal of Biotechnology*, 116, 305-317.

[62] Lopez, M.J., del Carmen Vargas-Garcia, M., Suarez-Estrella, F., Nichols, N.N., Dien, B.S. & Moreno, J. (2007). Lignocellulose-degrading enzymes produced by the ascomycete *Coniochaeta ligniaria* and related species: Application for a lignocellulosic substrate treatment. *Enzyme and Microbial Technology*, 40, 794-800.

[63] Karnchanatat, A., Petsom, A., Angvanich, P.S., PiapukieW, J., Whalley, A.J.S., Reynoldsc, C.D., Gadd, G.M. & Sihanonth, P. (2008). A novel thermostable endoglucanase from the wood-decaying fungus *Daldinia eschschokii* (Ehrenb.: Fr.) Rehm. *Enzyme and Microbial Technology*, 42, 404-413.

[64] Kluczek-Turpeinen, B., Maijala, P., Tuomela, M., Hofrichter, M. & Hatakka, A. (2005). Endoglucanase activity of compost-dwelling fungus *Paecilomyces inflatus* is

stimulated by humic acids and other low molecular mass aromatics. *World Journal of Microbiology & Biotechnology*, 21, 1603-1609.

[65] Kluczek-Turpeinen, B., Maijala, P., Hofrichter, M. & Hatakka, A. (2007). Degradation and enzymatic activities of three *Paecilomyces inflatus* strains grown on diverse lignocellulosic substrates. *International Biodeterioration & Biodegradation*, 59, 283-291.

[66] Blanchette, R.A. (1991). Delignification by wood decay fungi. *Annual Review of Phytopathology*, 29, 381-398.

[67] Kirk, T. K., & Cullen, D. (1998). Enzymology and molecular genetics of wood degradation by white rot fungi. In R. A. Young & M. Akhtar (Eds.), *Environmentally friendly technologies for the pulp and paper industry* (pp. 273-308). New York: John Wiley & Sons.

[68] Blanchette, R.A., Abad, A.R., Farrell, R.L. & Leathers, T.D. (1989). Detection of lignin peroxidase and xylanase by immunocytochemical labeling in wood decayed by basidiomycetes. *Applied and Environmental Microbiology*, 55, 1457-1465.

[69] Pandey, K.K. & Pitman, A.J. (2003). FTIR studies of the changes in wood chemistry following decay by brown-rot and white-rot fungi. *International Biodeterioration & Biodegradation*, 52, 151-160.

[70] Guerra, A., Mendonca, R. & Ferraz, A. (2003). Molecular weight distribution of wood components extracted from *Pinus taeda* biotreated by *Ceriporiopsis subvermispora*. *Enzyme and Microbial Technology*, 33, 12-18.

[71] Dias, A.A., Freitas, G.S., Marques, G.S.M., Sampaio, A., Fraga, I.S., Rodrigues, M.A.M., Evtuguin, D.V. & Bezerra, R.M.F. (2010). Enzymatic saccharification of biologically pre-treated wheat straw with white-rot fungi. *Bioresource Technology*, 101, 6045-6050.

[72] Arora, D.S., Sharma, R.K. & Chandra, P. (2011). Biodelignification of wheat straw and its effect on in vitro digestibility and antioxidant properties. *International Biodeterioration & Biodegradation*, 65, 352-358.

[73] Sharma, R.K. & Arora, D.S. (2011). Biodegradation of paddy straw obtained from different geographic locations by means of *Phlebia* spp. for animal feed. *Biodegradation* 22, 143-152.

[74] Wan, C. & Li, Y. (2010). Microbial delignification of corn stover by *Ceriporiopsis subvermispora* for improving cellulose digestibility. *Enzyme and Microbial Technology*, 47, 31-36.

[75] Wan, C. & Li, Y. (2010). Microbial pretreatment of corn stover with *Ceriporiopsis subvermispora* for enzymatic hydrolysis and ethanol production. *Bioresource Technology*, 101, 6398-6403.

[76] Dinis, M.J. et al. (2009). Modification of wheat straw lignin by solid state fermentation with white-rot fungi. *Bioresource Technology*, 100, 4829-4835.

[77] Shi, J., Chinn, M.S. & Sharma-Shivappa, R.R. (2008). Microbial pretreatment of cotton stalks by solid state cultivation of *Phanerochaete chrysosporium*. *Bioresource Technology,* 99, 6556-6564.

[78] Shi, J., Sharma-Shivappa, R.R. & Chinn, M.S. (2009). Microbial pretreatment of cotton stalks by submerged cultivation of *Phanerochaete chrysosporium*. *Bioresource Technology*, 100, 4388-4395.

[79] Zhang, X.Y., Yu, H.B., Huang, H.Y. & Liu, Y.X. (2007). Evaluation of biological pretreatment with white rot fungi for the, enzymatic hydrolysis of bamboo culms. *International Biodeterioration & Biodegradation,* 60, 159-164.

[80] Yu, H.B., Guo, G.N., Zhang, X.Y., Yan, K.L. & Xu, C.Y. (2009). The effect of biological pretreatment with the selective white-rot fungus *Echinodontium taxodii* on enzymatic hydrolysis of softwoods and hardwoods. *Bioresource Technology*, 100, 5170-5175.

[81] Martinez, A.T., Rencoret, J., Nieto, L., Jimenez-Barbero, J., Gutierrez, A. & del Rio, J.C. (2011). Selective lignin and polysaccharide removal in natural fungal decay of wood as evidenced by in situ structural analyses. *Environmental Microbiology*, 13, 96-107.

[82] Rees, R.W., Flood, J., Hasan, Y., Potter, U. & Cooper, R.M. (2009). Basal stem rot of oil palm (*Elaeis guineensis*); mode of root infection and lower stem invasion by *Ganoderma boninense*. *Plant Pathology*, 58, 982-989.

[83] Luna, M.L., Murace, M.A., Keil, G.D. & Otano, M.E. (2004). Patterns of decay caused by *Pycnoporus sanguineus* and *Ganoderma lucidum* (*Aphyllophorales*) in poplar wood. *Iawa Journal*, 25, 425-433.

[84] Xu, C.Y., Ma, F.Y., Zhang, X.Y. & Chen, S.L. (2010). Biological Pretreatment of Corn Stover by *Irpex lacteus* for Enzymatic Hydrolysis. *Journal of Agricultural and Food Chemistry,* 58, 10893-10898.

[85] Arantes, V., Milagres, A.M.F., Filley, T.R. & Goodell, B. (2011). Lignocellulosic polysaccharides and lignin degradation by wood decay fungi: the relevance of nonenzymatic Fenton-based reactions. *Journal of Industrial Microbiology & Biotechnology*, 38, 541-555.

[86] Flournoy, D.S., Paul, J.A., Kirk, T.K. & Highley, T.L. (1993). Changes in the size and volume of pores in sweetgum wood during simultaneous rot by *Phanerochaete chrysosporium* burds. *Holzforschung*, 47, 297-301.

[87] Martinez, A. T., Camarero, S., Guillen, F., Gutierrez, A., Munoz, C., Varela, E., et al. (1994). Progress in biopulping of non-woody materials - chemical, enzymatic and ultrastructural aspects of wheat-straw delignification with ligninolytic fungi from the genus *Pleurotus*. *FEMS Microbiology Reviews* 13, 265-274.

[88] Blanchette, R.A., Burnes, T.A., Leatham, G.F. & Effland, M.J. (1988). Selection of white-rot fungi for biopulping. *Biomass,* 15, 93-101.

[89] Akhtar, M., Attridge, M.C., Myers, G.C. & Blanchette, R.A. (1993). Biomechanical pulping of loblolly-pine chips with selected white-rot fungi. *Holzforschung*, 47, 36-40.

[90] Fatehi, P., Ates, S. & Ni, Y.H. (2009). Fungal pretreatment of wheat straw and its effect on the soda-AQ pulps. *Nordic Pulp & Paper Research Journal*, 24, 193-198.

[91] Brozzoli, V., Bartocci, S., Terramoccia, S., Conto, G., Federici, F., D'Annibale, A. & Petruccioli, M. (2010). Stoned olive pomace fermentation with *Pleurotus* species and its evaluation as a possible animal feed. *Enzyme and Microbial Technology*, 46, 223-228.

[92] Srebotnik, E. & Messner, K. (1994). A simple method that uses differential staining and light-microscopy to assess the selectivity of wood delignification by white-rot fungi. *Applied and Environmental Microbiology*, 60, 1383-1386.

[93] Mohebby, B. (2005). Attenuated total reflection infrared spectroscopy of white-rot decayed beech wood. *International Biodeterioration & Biodegradation*, 55, 247-251.

[94] Blanchette, R.A. (1984). Selective delignification of wood by white-rot fungi. *Applied Biochemistry and Biotechnology*, 9, 323-324.

[95] Otjen, L., Blanchette, R., Effland, M. & Leatham, G. (1987). Assessment of 30 white rot basidiomycetes for selective lignin degradation. *Holzforschung*, 41, 343-349.

[96] Adaskaveg, J.E., Gilbertson, R.L. & Blanchette, R.A. (1990). Comparative studies of delignification caused by *Ganoderma* species. *Applied and Environmental Microbiology*, 56, 1932-1943.

[97] Ferraz, A., Rodriguez, J., Freer, J. & Baeza, J. (2001). Biodegradation of *Pinus radiata* softwood by white- and brown-rot fungi. *World Journal of Microbiology & Biotechnology*, 17, 31-34.

[98] Yaghoubi, K., Pazouki, M. & Shojaosadati, S.A. (2008). Variable optimization for biopulping of agricultural residues by *Ceriporiopsis subvermispora*. *Bioresource Technology,* 99, 4321-4328.

[99] Mendonca, R.T., Jara, J.F., Gonzalez, V., Elissetche, J.P. & Freer, J. (2008). Evaluation of the white-rot fungi *Ganoderma australe* and *Ceriporiopsis subvermispora* in biotechnological applications. *Journal of Industrial Microbiology & Biotechnology*, 35, 1323-1330.

[100] Baldrian, P. & Valaskova, V. (2008). Degradation of cellulose by basidiomycetous fungi. *FEMS Microbiology Reviews*, 32, 501-521.

[101] Tomme, P., Warren, R.A.J., Miller, R.C., Kilburn, D.G. & Gilkes, N.R. (1995). Cellulose-binding domains: Classification and properties. *Enzymatic Degradation of Insoluble Carbohydrates*, 618, 142-163.

[102] Lynd, L.R., Weimer, P.J., van Zyl, W.H. & Pretorius, I.S. (2002). Microbial cellulose utilization: Fundamentals and biotechnology. *Microbiology and Molecular Biology Reviews*, 66, 506-77.

[103] Pérez, J., Muñoz-Dorado, J., de la Rubia, T. & Martínez, J. (2002). Biodegradation and biological treatments of cellulose, hemicellulose and lignin: an overview. *International Microbiology*, 5, 53-63.

[104] Teeri, T.T. (1997). Crystalline cellulose degradation: New insight into the function of cellobiohydrolases. *Trends in Biotechnology*, 15, 160-167.

[105] Liu, Y.S., Baker, J.O., Zeng, Y.N., Himmel, M.E., Haas, T. & Ding, S.Y. (2011). Cellobiohydrolase Hydrolyzes Crystalline Cellulose on Hydrophobic Faces. *Journal of Biological Chemistry*, 286, 11195-11201.

[106] Rabinovich, M.L., Melnick, M.S. & Bolobova, A.V. (2002). The structure and mechanism of action of cellulolytic enzymes. *Biochemistry-Moscow*, 67, 850-871.

[107] Toda, H., Nagahata, N., Amano, Y., Nozaki, K., Kanda, T., Okazaki, M. & Shimosaka, M. (2008). Gene cloning of cellobiohydrolase II from the white rot fungus *Irpex lacteus* MC-2 and its expression in *Pichia pastoris*. *Bioscience Biotechnology and Biochemistry*, 72, 3142-3147.

[108] Covert, S.F., Bolduc, J. & Cullen, D. (1992). Genomic organization of a cellulase gene family in *Phanerochaete chrysosporium*. *Current Genetics*, 22, 407-413.

[109] Vanden Wymelenberg, A., Denman, S., Dietrich, D., Bassett, J., Yu, X. C., Atalla, R., et al. (2002). Transcript analysis of genes encoding a family 61 endoglucanase and a putative membrane-anchored family 9 glycosyl hydrolase from *Phanerochaete chrysosporium*. *Applied and Environmental Microbiology*, 68, 5765-5768.

[110] Vanden Wymelenberg, A., Sabat, G., Martinez, D., Rajangam, A.S., Teeri, T.T., Gaskell, J., Kersten, P.J. & Cullen, D. (2005). The *Phanerochaete chrysosporium* secretome: Database predictions and initial mass spectrometry peptide identifications in cellulose-grown medium. *Journal of Biotechnology*, 118, 17-34.

[111] Vanden Wymelenberg, A., Gaskell, J., Mozuch, M., BonDurant, S. S., Sabat, G., Ralph, J., et al. (2011). Significant Alteration of Gene Expression in Wood Decay Fungi *Postia placenta* and *Phanerochaete chrysosporium* by Plant Species. *Applied and Environmental Microbiology*, 77, 4499-4507.

[112] Hamada, N., Fuse, N., Shimosaka, M., Kodaira, R., Amano, Y., Kanda, T. & Okazaki, M. (1999). Cloning and characterization of a new exo-cellulase gene, cel3, in *Irpex lacteus*. *FEMS* Microbiology Letters 172, 231-237.

[113] Watanabe, T., Teranishi, H., Honda, Y. & Kuwahara, M. (2002). A selective lignin-degrading fungus, *Ceriporiopsis subvermispora*, produces alkylitaconates that inhibit the production of a cellulolytic active oxygen species, hydroxyl radical in the presence of iron and H_2O_2. *Biochemical and Biophysical Research Communications*, 297, 918-923.

[114] Rahmawati, N., Ohashi, Y., Watanabe, T. & Honda, Y. (2005). Ceriporic acid B, an extracellular metabolite of *Ceriporiopsis subvermispora*, suppresses the depolymerization of cellulose by the Fenton reaction. *Biomacromolecules*, 6, 2851-2856.

[115] Saha, B.C. (2003). Hemicellulose bioconversion. *Journal of Industrial Microbiology & Biotechnology*, 30, 279-291.

[116] Ebringerova, A., Hromadkova, Z., & Heinze, T. (2005). Hemicellulose. *Polysaccharides 1: Structure, Characterization and Use,* 186, 1-67.

[117] Polizeli, M., Rizzatti, A.C.S., Monti, R., Terenzi, H.F., Jorge, J.A. & Amorim, D.S. (2005). Xylanases from fungi: properties and industrial applications. *Applied Microbiology and Biotechnology*, 67, 577-591.

[118] Aspinall, G. O. (1980). The Biochemistry of Plants (A Comprehensive Treatise). In J. Preiss (Ed.), *Carbohydrates: Structure and Function* (Vol. 3, pp. 473-500). New York: Academic Press.

[119] Kormelink, F.J.M. & Voragen, A.G.J. (1993). Degradation of different (glucurono)arabino xylans by a combination of purified xylan-degrading enzymes. *Applied Microbiology and Biotechnology*, 38, 688-695.

[120] Ebringerova, A., & Heinze, T. (2000). Xylan and xylan derivatives - biopolymers with valuable properties, 1 - Naturally occurring xylans structures, procedures and properties. *Macromolecular Rapid Communications,* 21, 542-556.

[121] Sato, S., Liu, F., Koc, H. & Tien, M. (2007). Expression analysis of extracellular proteins from *Phanerochaete chrysosporium* grown on different liquid and solid substrates. *Microbiology-Sgm*, 153, 3023-3033.

[122] Sanchez, C. (2009). Lignocellulosic residues: Biodegradation and bioconversion by fungi. *Biotechnology Advances*, 27, 185-194.

[123] Belien, T., Van Campenhout, S., Robben, J. & Volckaert, G. (2006). Microbial endoxylanases: Effective weapons to breach the plant cell-wall barrier or, rather, triggers of plant defense systems? *Molecular Plant-Microbe Interactions*, 19, 1072-1081.

[124] Hermoso, J.A., Sanz-Aparicio, J., Molina, R., Juge, N., Gonzalez, R. & Faulds, C.B. (2004). The crystal structure of feruloyl esterase a from *Aspergillus niger* suggests evolutive functional convergence in feruloyl esterase family. *Journal of Molecular Biology*, 338, 495-506.

[125] Wong, D.W.S. (2006). Feruloyl esterase - A key enzyme in biomass degradation. *Applied Biochemistry and Biotechnology,* 133, 87-112.

[126] Heidorne, F.O., Magalhaes, P.O., Ferraz, A.L. & Milagres, A.M.F. (2006). Characterization of hemicellulases and cellulases produced by *Ceriporiopsis subvermispora* grown on wood under biopulping conditions. *Enzyme and Microbial Technology*, 38, 436-442.

[127] Castanares, A., Hay, A.J., Gordon, A.H., McCrae, S.I. & Wood, T.M. (1995). D-xylan-degrading enzyme system from the fungus *Phanerochaete chrysosporium*: Isolation and partial characterisation of an alpha-(4-O-methyl)-D-glucuronidase. *Journal of Biotechnology*, 43, 183-194.

[128] El-Nasser, N.H.A., Helmy, S.M. & El-Gammal, A.A. (1997). Formation of enzymes by biodegradation of agricultural wastes with white rot fungi. *Polymer Degradation and Stability,* 55, 249-255.

[129] de Menezes, C.R., Silva, I.S., Pavarina, E.C., Dias, E.F.G., Dias, F.G., Grossman, M.J. & Durrant, L.R. (2009). Production of xylooligosaccharides from enzymatic hydrolysis of xylan by the white-rot fungi *Pleurotus*. *International Biodeterioration & Biodegradation*, 63, 673-678.

[130] ten Have, R. & Teunissen, P.J.M. (2001). Oxidative mechanisms involved in lignin degradation by white-rot fungi. *Chemical Reviews*, 101, 3397-3413.

[131] Ralph, J., Lundquist, K., Brunow, G., Lu, F., Kim, H., Schatz, P. F., et al. (2004). Lignins: Natural polymers from oxidative coupling of 4-hydroxyphenyl- propanoids. *Phytochemistry Reviews,* 3, 29-60.

[132] Ruiz-Duenas, F.J. & Martinez, A.T. (2009). Microbial degradation of lignin: how a bulky recalcitrant polymer is efficiently recycled in nature and how we can take advantage of this. *Microbial Biotechnology,* 2, 164-177.

[133] Gibbs, R. D. (1958). The Maule reaction, lignins, and the relationships between woody plants. In K. V. Thimann, W. D. Critchfield & M. H. Zimmermann (Eds.), *The physiology of forest trees* (pp. 269-312). New York: Ronald Press.

[134] Fengel, D., Wegener, G. & Greune, A. (1989). Studies on the delignification of spruce wood by organosolv pulping using SEM-EDXA and TEM. *Wood Science and Technology*, 23, 123-130.

[135] Whetten, R. & Sederoff, R. (1995). Lignin biosynthesis. *Plant Cell*, 7, 1001-1013.

[136] Schwarze, F.W.M.R. & Fink, S. (1998). Host and cell type affect the mode of degradation by *Meripilus giganteus*. *New Phytologist,* 139, 721-731.

[137] Tuor, U., Winterhalter, K. & Fiechter, A. (1995). Enzymes of white-rot fungi involved in lignin degradation and ecological determinants for wood decay. *Journal of Biotechnology*, 41, 1-17.

[138] Jeffries, T.W. (1990). Biodegradation of lignin-carbohydrate complexes. *Biodegradation*, 163-176.

[139] Blanchette, R.A., Krueger, E.W., Haight, J.E., Akhtar, M. & Akin, D.E. (1997). Cell wall alterations in loblolly pine wood decayed by the white-rot fungus, *Ceriporiopsis subvermispora*. *Journal of Biotechnology*, 53, 203-213.

[140] Robertson, S.A., Mason, S.L., Hack, E. & Abbott, G.D. (2008). A comparison of lignin oxidation, enzymatic activity and fungal growth during white-rot decay of wheat straw. *Organic Geochemistry* 39, 945-951.

[141] Fackler, K., Gradinger, C., Hinterstoisser, B., Messner, K. & Schwanninger, M. (2006). Lignin degradation by white rot fungi on spruce wood shavings during short-time solid-state fermentations monitored by near infrared spectroscopy. *Enzyme and Microbial Technology*, 39, 1476-1483.

[142] Rabinovich, M.L., Bolobova, A.V. & Vasil'chenko, L.G. (2004). Fungal decomposition of natural aromatic structures and xenobiotics: A review. *Applied Biochemistry and Microbiology*, 40, 1-17.

[143] Kersten, P. & Cullen, D. (2007). Extracellular oxidative systems of the lignin-degrading Basidiomycete *Phanerochaete chrysosporium. Fungal Genetics and Biology*, 44, 77-87.

[144] Kirk, T.K. & Farrell, R.L. (1987). Enzymatic combustion - the microbial degradation of lignin. *Annual Review of Microbiology*, 41, 465-505.

[145] Leonowicz, A., Matuszewska, A., Luterek, J., Ziegenhagen, D., Wojtas-Wasilewska, M., Cho, N.S., Hofrichter, M. & Rogalski, J. (1999). Biodegradation of lignin by white rot fungi. *Fungal Genetics and Biology,* 27, 175-185.

[146] Martinez, A. T., Speranza, M., Ruiz-Duenas, F. J., Ferreira, P., Camarero, S., Guillen, F., et al. (2005). Biodegradation of lignocellulosics: microbial chemical, and enzymatic aspects of the fungal attack of lignin. *International Microbiology,* 8, 195-204.

[147] Guillen, F. & Evans, C.S. (1994). Anisaldehyde and veratraldehyde acting as redox cycling agents for H_2O_2 production by *Pleurotus eryngii. Applied and Environmental Microbiology*, 60, 2811-2817.

[148] Kapich, A.N., Jensen, K.A. & Hammel, K.E. (1999). Peroxyl radicals are potential agents of lignin biodegradation. *Febs Letters*, 461, 115-119.

[149] Hofrichter, M. (2002). Review: lignin conversion by manganese peroxidase (MnP). *Enzyme and Microbial* Technology 30, 454-466.

[150] Mayer, A.M. & Staples, R.C. (2002). Laccase: new functions for an old enzyme. *Phytochemistry*, 60, 551-565.

[151] Tien, M. & Kirk, T.K. (1984). Lignin-degrading enzyme from *Phanerochaete chrysosporium* - purification, characterization and catalytic properties of a unique H_2O_2-requiring oxygenase. *Proceedings of the National Academy of Sciences of the United States of America-Biological Sciences,* 81, 2280-2284.

[152] Tien, M. & Kirk, T.K. (1988). Lignin peroxidase of *Phanerochaete chrysosporium. Methods in Enzymology*, 161, 238-249.

[153] Glenn, J. K., & Gold, M. H. (1985). Purification and characterization of an extracellular Mn(II)-dependent peroxidase from the lignin-degrading basidiomycete *Phanerochaete chrysosporium. Archives of Biochemistry and Biophysics,* 242, 329-341.

[154] Martinez, A.T. (2002). Molecular biology and structure-function of lignin- degrading heme peroxidases. *Enzyme and Microbial Technology*, 30, 425-444.

[155] Martinez, M.J., Ruiz-Duenas, F.J., Guillen, F. & Martinez, A.T. (1996). Purification and catalytic properties of two manganese peroxidase isoenzymes from *Pleurotus eryngii. European Journal of Biochemistry*, 237, 424-432.

[156] Martinez, D., Larrondo, L. F., Putnam, N., Gelpke, M. D. S., Huang, K., Chapman, J., et al. (2004). Genome sequence of the lignocellulose degrading fungus *Phanerochaete chrysosporium* strain RP78. *Nature Biotechnology,* 22, 695-700.

[157] Ohm, R. A., de Jong, J. F., Lugones, L. G., Aerts, A., Kothe, E., Stajich, J. E., et al. (2010). Genome sequence of the model mushroom *Schizophyllum commune. Nature Biotechnology,* 28, 957-965.

[158] Hatakka, A. (1994). Lignin-modifying enzymes from selected white-rot fungi - production and role in lignin degradation. *FEMS Microbiology Reviews*, 13, 125-135.

[159] Nerud, F. & Misurcova, Z. (1996). Distribution of ligninolytic enzymes in selected white-rot fungi. *Folia Microbiologica*, 41, 264-266.

[160] Kuhar, S., Kapoor, M., Kapoor, R., Sharma, K. K., Singh, A., & Kuhad, R. C. (2007). Biodiversity of ligninolytic fungi. In R. C. Kuhad & A. Singh (Eds.), *Lignocellulose biotechnology: future prospects* (pp. 37-53). New Delhi: I. K. International.

[161] Wesenberg, D., Kyriakides, I. & Agathos, S.N. (2003). White-rot fungi and their enzymes for the treatment of industrial dye effluents. *Biotechnology Advances*, 22, 161-187.

[162] Baldrian, P. (2006). Fungal laccases - occurrence and properties. *FEMS Microbiology Reviews*, 30, 215-242.

[163] Lundell, T.K., Makela, M.R. & Hilden, K. (2010). Lignin-modifying enzymes in filamentous basidiomycetes - ecological, functional and phylogenetic review. *Journal of Basic Microbiology*, 50, 5-20.

[164] Conesa, A., Punt, P.J. & van den Hondel, C. (2002). Fungal peroxidases: molecular aspects and applications. *Journal of Biotechnology*, 93, 143-158.

[165] Liers, C., Bobeth, C., Pecyna, M., Ullrich, R. & Hofrichter, M. (2010). DyP-like peroxidases of the jelly fungus *Auricularia auricula-judae* oxidize nonphenolic lignin model compounds and high-redox potential dyes. *Applied Microbiology and Biotechnology*, 85, 1869-1879.

[166] Hofrichter, M., Ullrich, R., Pecyna, M.J., Liers, C. & Lundell, T. (2010). New and classic families of secreted fungal heme peroxidases. *Applied Microbiology and Biotechnology*, 87, 871-897.

[167] Coelho, J.S., de Oliveira, A.L., Marques de Souza, C.G., Bracht, A. & Peralta, R.M. (2010). Effect of the herbicides bentazon and diuron on the production of ligninolytic enzymes by *Ganoderma lucidum. International Biodeterioration & Biodegradation*, 64, 156-161.

[168] Zilly, A., dos Santos Bazanella, G., Helm, C., Araújo, C., de Souza, C., Bracht, A. & Peralta, R. (2011). Solid-state bioconversion of passion fruit waste by white-rot fungi for production of oxidative and hydrolytic enzymes. *Food and Bioprocess Technology*, 1-8.

[169] Rodriguez, E., Ruiz-Duenas, F.J., Kooistra, R., Ram, A., Martinez, A.T. & Martinez, M.J. (2008). Isolation of two laccase genes from the white-rot fungus *Pleurotus eryngii* and heterologous expression of the pel3 encoded protein. *Journal of Biotechnology*, 134, 9-19.

[170] Faraco, V., Pezzella, C., Miele, A., Giardina, P. & Sannia, G. (2009). Bioremediation of colored industrial wastewaters by the white-rot fungi *Phanerochaete chrysosporium* and *Pleurotus ostreatus* and their enzymes. *Biodegradation*, 20, 209-220.

[171] Cambria, M.T., Ragusa, S., Calabrese, V. & Cambria, A. (2011). enhanced laccase production in white-rot fungus *Rigidoporus lignosus* by the addition of selected phenolic and aromatic compounds. *Applied Biochemistry and Biotechnology*, 163, 415-422.

[172] Lebrun, J.D., Lamy, I. & Mougin, C. (2011). Favouring the bioavailability of Zn and Cu to enhance the production of lignin-modifying enzymes in *Trametes versicolor* cultures. *Bioresource Technology* 102, 3103-3109.

[173] Lopez, M., Loera, O., Guerrero-Olazaran, M., Viader-Salvado, J.M., Gallegos-Lopez, J.A., Fernandez, F.J., Favela-Torres, E. & Viniegra-Gonzalez, G. (2010). Cell growth and *Trametes versicolor* laccase production in transformed *Pichia pastoris* cultured by solid-state or submerged fermentations. *Journal of Chemical Technology and Biotechnology,* 85, 435-440.

[174] Cohen, S., Belinky, P.A., Hadar, Y. & Dosoretz, C.G. (2009). Characterization of catechol derivative removal by lignin peroxidase in aqueous mixture. *Bioresource Technology*, 100, 2247-2253.

[175] Matityahu, A., Hadar, Y. & Belinky, P.A. (2010). Involvement of protein kinase C in lignin peroxidase expression in oxygenated cultures of the white rot fungus *Phanerochaete chrysosporium. Enzyme and Microbial Technology*, 47, 59-63.

[176] Wen, X., Jia, Y. & Li, J. (2009). Degradation of tetracycline and oxytetracycline by crude lignin peroxidase prepared from *Phanerochaete chrysosporium* - A white rot fungus. *Chemosphere*, 75, 1003-1007.

[177] Miki, Y., Ichinose, H. & Wariishi, H. (2010). Molecular characterization of lignin peroxidase from the white-rot basidiomycete *Trametes cervina*: a novel fungal peroxidase. *FEMS Microbiology Letters*, 304, 39-46.

[178] Takamiya, M., Magan, N. & Wamer, P.J. (2008). Impact assessment of bisphenol A on lignin-modifying enzymes by basidiomycete *Trametes versicolor*. *Journal of Hazardous Materials*, 154, 33-37.

[179] Sklenar, J., Niku-Paavola, M.L., Santos, S., Man, P., Kruus, K. & Novotny, C. (2010). Isolation and characterization of novel pI 4.8 MnP isoenzyme from white-rot fungus *Irpex lacteus*. *Enzyme and Microbial Technology,* 46, 550-556.

[180] Elisashvili, V., Penninckx, M., Kachlishvili, E., Tsiklauri, N., Metreveli, E., Kharziani, T. & Kvesitadze, G. (2008). *Lentinus edodes* and *Pleurotus* species lignocellulolytic enzymes activity in submerged and solid-state fermentation of lignocellulosic wastes of different composition. *Bioresource Technology* 99, 457-462.

[181] Elisashvili, V., Kachlishvili, E. & Penninckx, M.J. (2008). Lignocellulolytic enzymes profile during growth and fruiting of *Pleurotus ostreatus* on wheat straw and tree leaves. *Acta Microbiologica Et Immunologica Hungarica*, 55, 157-168.

[182] Mohorcic, M., Bencina, M., Friedrich, J. & Jerala, R. (2009). Expression of soluble versatile peroxidase of *Bjerkandera adusta* in *Escherichia coli*. *Bioresource Technology*, 100, 851-858.

[183] Taboada-Puig, R., Lu-Chau, T., Teresa Moreira, M., Feijoo, G., Jesus Martinez, M. & Manuel Lema, J. (2011). A new strain of *Bjerkandera* sp. production, purification and characterization of versatile peroxidase. *World Journal of Microbiology & Biotechnology*, 27, 115-122.

[184] Chen, M., Yao, S., Zhang, H. & Liang, X. (2010). Purification and characterization of a versatile peroxidase from edible mushroom *Pleurotus eryngii*. *Chinese Journal of Chemical Engineering*, 18, 824-829.

[185] Ruiz-Duenas, F.J., Morales, M., Mate, M.J., Romero, A., Marinez, M.J., Smith, A.T. & Martinez, A.T. (2008). Site-directed mutagenesis of the catalytic tryptophan environment in *Pleurotus eryngii* versatile peroxidase. *Biochemistry*, 47, 1685-1695.

[186] Collins, P. J., & Dobson, A. D. W. (1995). Extracellular lignin and manganese peroxidase production by the white-rot fungus *Coriolus versicolor*-290. *Biotechnology Letters, 17*(9), 989-992.

[187] Tien, M. & Myer, S.B. (1990). Selection and characterization of mutants of *Phanerochaete chrysosporium* exhibiting ligninolytic activity under nutrient-rich conditions. *Applied and Environmental Microbiology,* 56, 2540-2544.

[188] Tychanowicz, G.K., de Souza, D.F., Souza, C.G.M., Kadowaki, M.K. & Peralta, R.M. (2006). Copper improves the production of laccase by the white-rot fungus *Pleurotus pulmonarius* in solid state fermentation. *Brazilian Archives of Biology and Technology*, 49, 699-704.

[189] de Souza, C.G.M., Tychanowicz, G.K., de Souza, D.F. & Peralta, R.M. (2004). Production of laccase isoforms by *Pleurotus pulmonarius* in response to presence of phenolic and aromatic compounds. *Journal of Basic Microbiology*, 44, 129-136.

[190] Novotny, C., Erbanova, P., Cajthaml, T., Rothschild, N., Dosoretz, C. & Sasek, V. (2000). *Irpex lacteus*, a white rot fungus applicable to water and soil bioremediation. *Applied Microbiology and Biotechnology*, 54, 850-853.

[191] Reddy, G.V., Babu, P.R., Komaraih, P., Roy, K. & Kothari, I.L. (2003). Utilization of banana waste for the production of lignolytic and cellulolytic enzymes by solid substrate fermentation using two *Pleurotus* species (*P. ostreatus* and *P. sajor-caju*). *Process Biochemistry* 38, 1457-1462.

[192] Gassara, F., Brar, S.K., Tyagi, R.D., Verma, M. & Surampalli, R.Y. (2010). Screening of agro-industrial wastes to produce ligninolytic enzymes by *Phanerochaete chrysosporium*. *Biochemical Engineering Journal*, 49, 388-394.

[193] Lonsdale, D., Pautasso, M. & Holdenrieder, O. (2008). Wood-decaying fungi in the forest: conservation needs and management options. *European Journal of Forest Research*, 127, 1-22.

[194] Jonsell, M. & Nordlander, G. (2004). Host selection patterns in insects breeding in bracket fungi. *Ecological Entomology*, 29, 697-705.

[195] Baldrian, P. (2003). Interactions of heavy metals with white-rot fungi. *Enzyme and Microbial Technology*, 32, 78-91.

[196] Gabriel, J., Baldrian, P., Rychlovsky, P. & Krenzelok, M. (1997). Heavy metal content in wood-decaying fungi collected in Prague and in the National Park Sumava in the Czech Republic. *Bulletin of Environmental Contamination and Toxicology*, 59, 595-602.

[197] Yu, Z., Yin, L., Qian, Y. & Yan, L. (2009). Effect of *Lentinus edodes* polysaccharide on oxidative stress, immunity activity and oral ulceration of rats stimulated by phenol. *Carbohydrate Polymers*, 75, 115-118.

[198] Soares, A.A., Marques de Souza, C.G., Daniel, F.M., Ferrari, G.P., Gomes da Costa, S.M. & Peralta, R.M. (2009). Antioxidant activity and total phenolic content of

Agaricus brasiliensis (*Agaricus blazei* Murril) in two stages of maturity. *Food Chemistry*, 112, 775-781.

[199] de Oliveira, A.L., Eler, G.J., Bracht, A. & Peralta, R.M. (2010). Purinergic effects of a hydroalcoholic *Agaricus brasiliensis* (*A. blazei*) extract on liver functions. *Journal of Agricultural and Food Chemistry*, 58, 7202-7210.

[200] Smiderle, F. R., Olsen, L. M., Carbonero, E. R., Baggio, C. H., Freitas, C. S., Marcon, R., et al. (2008). Anti-inflammatory and analgesic properties in a rodent model of a (1 -> 3),(1 -> 6)-linked beta-glucan isolated from *Pleurotus pulmonarius*. *European Journal of Pharmacology,* 59(1-3), 86-91.

[201] Smiderle, F.R., Carbonero, E.R., Sassaki, G.L., Gorin, P.A.J. & Iacomini, M. (2008). Characterization of a heterogalactan: Some nutritional values of the edible mushroom *Flammulina velutipes*. *Food Chemistry,* 108, 329-333.

[202] Paterson, R.R.M. (2007). *Ganoderma* disease of oil palm - A white rot perspective necessary for integrated control. *Crop Protection*, 26, 1369-1376.

[203] Hood, I. A., Redfern, D. B., & Kile, G. A. *Armillaria* in planted hosts. In C. G. Shaw & G. A. Kile (Eds.), *Armillaria root disease* (Vol. 691, pp. 122-149). Washington: USDA Agricultural Handbook.

[204] Filip, G. M., & Morrison, D. J. (1998). Impact, control and management of *Heterobasidion annosum* root and butt rot in North America. In S. Woodward, J. Stenlid, R. Karjalainen & A. Huttermann (Eds.), *Heterobasidion annosum: Biology, Ecology, Impact and Control* (pp. 589). Wallingford: CAB International.

[205] Gibertoni, T.B., Santos, P.J.P. & Cavalcanti, M.A.Q. (2007). Ecological aspects of *Aphyllophorales* in the Atlantic rain forest in northeast Brazil. *Fungal Diversity*, 25, 49-67.

[206] Guglielmo, F., Gonthier, P., Garbelotto, M. & Nicolotti, G. (2008). A PCR-based method for the identification of important wood rotting fungal taxa within *Ganoderma*, *Inonotus* s.l. and *Phellinus* s.l. *FEMS Microbiology Letters*, 282, 228-237.

[207] Schmidt, O. & Liese, W. (1980). Variability of wood degrading enzymes of *Schizophyllum commune*. *Holzforschung* 34, 67-72.

[208] Falk, S.P. & Parbery, D.G. (1995). *Armillaria luteobubalina* population structure in horticultural plantings in Victoria, Australia. *Mycological Research*, 99, 216-220.

[209] Corley, R. H. V. & Tinker, P. B. (2003). *The Oil Palm* (4 ed.). Oxford: John Wiley and Sons Ltd.

[210] Susanto, A., Sudharto, P.S. & Purba, R.Y. (2005). Enhancing biological control of basal stem rot disease (*Ganoderma boninense*) in oil palm plantations. *Mycopathologia*, 159, 153-157.

[211] Robles, C.A., Carmaran, C.C. & Lopez, S.E. (2011). Screening of xylophagous fungi associated with *Platanus acerifolia* in urban landscapes: Biodiversity and potential biodeterioration. *Landscape and Urban Planning*, 100, 129-135.

[212] Schwarze, F.W.M.R. & Baum, S. (2000). Mechanisms of reaction zone penetration by decay fungi in wood of beech (*Fagus sylvatica*). *New Phytologist*, 146, 129-140.

[213] Terho, M., Hantula, J. & Hallaksela, A.M. (2007). Occurrence and decay patterns of common wood-decay fungi in hazardous trees felled in the Helsinki City. *Forest Pathology*, 37, 420-432.

[214] Mendes, M. A. S., & Urben, A. F. (2011). Fungos relatados em plantas no Brasil. 2011, Aug 10. Available from EMBRAPA:

http://pragawall.cenargen.embrapa.br/aiqweb/michtml/fgbanco01.asp
[215] Guglielmo, F., Bergemann, S.E., Gonthier, P., Nicolotti, G. & Garbelotto, M. (2007). A multiplex PCR-based method for the detection and early identification of wood rotting fungi in standing trees. *Journal of Applied Microbiology*, 103, 1490-1507.
[216] Guglielmo, F., Gonthier, P., Garbelotto, M. & Nicolotti, G. (2010). Optimization of sampling procedures for DNA-based diagnosis of wood decay fungi in standing trees. *Letters in Applied Microbiology*, 51, 90-97.
[217] O'Brien, H.E., Parrent, J.L., Jackson, J.A., Moncalvo, J.M. & Vilgalys, R. (2005). Fungal community analysis by large-scale sequencing of environmental samples. *Applied and Environmental Microbiology,* 71, 5544-5550.
[218] Kirk, P. M., Cannon, P. F., Minter, D. W. & Stalpers, J. A. (2008). *Dictionary of the Fungi* (10 ed.). Wallingford: CAB International.
[219] Hillebrand, H. (2004). On the generality of the latitudinal diversity gradient. *American Naturalist*, 163, 192-211.
[220] Arnold, A.E. & Lutzoni, F. (2007). Diversity and host range of foliar fungal endophytes: Are tropical leaves biodiversity hotspots? *Ecology*, 88, 541-549.
[221] Eiten, G. (1982). Brazilian Savannas. *Ecological Studies*, 42, 25-47.
[222] Ratter, J.A., Ribeiro, J.F. & Bridgewater, S. (1997). The Brazilian cerrado vegetation and threats to its biodiversity. *Annals of Botany*, 80, 223-230.
[223] Pennington, R.T., Prado, D.E. & Pendry, C.A. (2000). Neotropical seasonally dry forests and Quaternary vegetation changes. *Journal of Biogeography*, 27, 261-273.
[224] Gomes-Silva, A.C. & Gibertoni, T.B. (2009). Checklist of the aphyllophoraceous fungi (*Agaricomycetes*) of the Brazilian Amazonia. *Mycotaxon*, 108, 319-322.
[225] Gomes-Silva, A.C., Ryvarden, L. & Gibertoni, T.B. (2009). New and interesting species of *Hymenochaetaceae* from the Brazilian Amazonia. *Mycological Progress*, 8, 273-279.
[226] Baltazar, J.M. & Gibertoni, T.B. (2009). A checklist of the aphyllophoroid fungi (*Basidiomycota*) recorded from the Brazilian Atlantic Forest. *Mycotaxon*, 109, 439-442.
[227] Drechsler-Santos, E.R., Gibertoni, T.B., Goes-Neto, A. & Cavalcanti, M.A.Q. (2009). A re-evaluation of the lignocellulolytic Agaricomycetes from the Brazilian semi-arid region. *Mycotaxon*, 108, 241-244.
[228] Gibertoni, T.B. & Drechsler-Santos, E.R. (2010). Lignocellulolytic Agaricomycetes from the Brazilian Cerrado biome. *Mycotaxon*, 111, 87-90.
[229] Maia, L. C., & Carvalho Jr., A. A. (2010). Fungos *in* Lista de Espécies da Flora do Brasil. 2011, Aug 10. Available from: *http://floradobrasil.jbrj.gov.br/2010/FB120181*
[230] Capelari, M., Baseia, I. G., Cortez, V. G., Sulzbacher, M. A., Silveira, R. M. B., Loguercio-Leite, C., et al. (2010). Agaricales *in* Lista de Espécies da Flora do Brasil. 2011, Aug 10. Available from: *http://floradobrasil.jbrj.gov.br/2010/FB000012*
[231] Gibertoni, T. B., Baltazar, J. M., Gomes-Silva, A. C., Drechsler-Santos, E. R., Loguercio-Leite, C., Campos-Santana, M., et al. (2010). Hymenochaetales *in* Lista de Espécies da Flora do Brasil. 2011, Aug 10. Available from: *http://floradobrasil.jbrj.gov.br/2010/FB000017*
[232] Gugliotta, A. M., Silveira, R. M. B., Loguercio-Leite, C., Campos-Santana, M., Gibertoni, T. B., Drechsler-Santos, E. R., et al. (2010). Polyporales *in* Lista de Espécies

da Flora do Brasil. 2011, Aug 10. Available from: *http://floradobrasil.jbrj.gov.br/2010/FB000023*

[233] Loguercio-Leite, C., & Campos-Santana, M. (2010). Auriculariales *in* Lista de Espécies da Flora do Brasil. 2011, Aug 10. Available from: *http://floradobrasil.jbrj.gov.br/2010/FB000013*

[234] Neves, M. A., Silveira, R. M. B., Loguercio-Leite, C., Campos-Santana, M., & Drechsler-Santos, E. R. (2010). Russulales *in* Lista de Espécies da Flora do Brasil. 2011, Aug 10. Available from: *http://floradobrasil.jbrj.gov.br/2010/FB000024*

[235] Drechsler-Santos, E.R., Santos, P.J.P., Gibertoni, T.B. & Cavalcanti, M.A.Q. (2010). Ecological aspects of *Hymenochaetaceae* in an area of Caatinga (semi-arid) in Northeast Brazil. *Fungal Diversity*, 42, 71-78.

[236] IBGE. (2004). Mapa de Biomas e de Vegetação. Aug 10, 2011. Available from: *http://www.ibge.gov.br/home/presidencia/noticias/noticia_visualiza.php?id_noticia=169*

[237] Mihail, J.D. & Bruhn, J.N. (2007). Dynamics of bioluminescence by *Armillaria gallica*, *A. mellea* and *A. tabescens*. *Mycologia* 99, 341-350.

[238] Desjardin, D.E., Capelari, M. & Stevani, C. (2007). Bioluminescent *Mycena* species from Sao Paulo, Brazil. *Mycologia*, 99, 317-331.

[239] Oliveira, A.G. & Stevani, C.V. (2009). The enzymatic nature of fungal bioluminescence. *Photochemical & Photobiological Sciences*, 8, 1416-1421.

[240] Bermudes, D., Petersen, R.H. & Nealson, K.H. (1992). Low level bioluminescence detected in *Mycena haematopus* basidiocarps. *Mycologia*, 84, 799-802.

[241] Desjardin, D.E., Oliveira, A.G. & Stevani, C.V. (2008). Fungi bioluminescence revisited. *Photochemical & Photobiological Sciences*, 7, 170-182.

[242] Shimomura, O. (1992). The role of superoxide dismutase in regulating the light emission of luminescent fungi. *Journal of Experimental Botany*, 43, 1519-1525.

[243] Capelari, M., Desjardin, D.E., Perry, B.A., Asai, T. & Stevani, C.V. (2011). *Neonothopanus gardneri*: a new combination for a bioluminescent agaric from Brazil. *Mycologia*, 11-097.

[244] Sun, Y. & Cheng, J.Y. (2002). Hydrolysis of lignocellulosic materials for ethanol production: a review. *Bioresource Technology*, 83, 1-11.

[245] Sasaki, C., Takada, R., Watanabe, T., Honda, Y., Karita, S., Nakamura, Y. & Watanabe, T. (2011). Surface carbohydrate analysis and bioethanol production of sugarcane bagasse pretreated with the white rot fungus, *Ceriporiopsis subvermispora* and microwave hydrothermolysis. *Bioresource Technology*, In Press, Accepted Manuscript.

[246] Yang, X.W., Zeng, Y.L., Ma, F.Y., Zhang, X.Y. & Yu, H.B. (2010). Effect of biopretreatment on thermogravimetric and chemical characteristics of corn stover by different white-rot fungi. *Bioresource Technology*, 101, 5475-5479.

[247] Baba, Y., Tanabe, T., Shirai, N., Watanabe, T. & Honda, Y. (2011). Pretreatment of Japanese cedar wood by white rot fungi and ethanolysis for bioethanol production. *Biomass & Bioenergy*, 35, 320-324.

[248] Okamoto, K., Nitta, Y., Maekawa, N. & Yanase, H. (2011). Direct ethanol production from starch, wheat bran and rice straw by the white rot fungus *Trametes hirsuta*. *Enzyme and Microbial Technology*, 48, 273-277.

[249] Marco-Urrea, E., Perez-Trujillo, M., Vicent, T. & Caminal, G. (2009). Ability of white-rot fungi to remove selected pharmaceuticals and identification of degradation products of ibuprofen by *Trametes versicolor*. *Chemosphere*, 74, 765-772.

[250] Marco-Urrea, E., Perez-Trujillo, M., Blanquez, P., Vicent, T. & Caminal, G. (2010). Biodegradation of the analgesic naproxen by *Trametes versicolor* and identification of intermediates using HPLC-DAD-MS and NMR. *Bioresource Technology*, 101, 2159-2166.

[251] Marco-Urrea, E., Perez-Trujillo, M., Cruz-Morato, C., Caminal, G. & Vicent, T. (2010). Degradation of the drug sodium diclofenac by *Trametes versicolor* pellets and identification of some intermediates by NMR. *Journal of Hazardous Materials*, 176, 836-842.

[252] Marco-Urrea, E., Radjenovic, J., Caminal, G., Petrovic, M., Vicent, T. & Barcelo, D. (2010). Oxidation of atenolol, propranolol, carbamazepine and clofibric acid by a biological Fenton-like system mediated by the white-rot fungus *Trametes versicolor*. *Water Research* 44, 521-532.

[253] Marco-Urrea, E., Perez-Trujillo, M., Cruz-Morato, C., Caminal, G. & Vicent, T. (2010). White-rot fungus-mediated degradation of the analgesic ketoprofen and identification of intermediates by HPLC-DAD-MS and NMR. *Chemosphere*, 78, 474-481.

[254] Rodriguez-Rodriguez, C.E., Jelic, A., Llorca, M., Farre, M., Caminal, G., Petrovic, M., Barcelo, D. & Vicent, T. (2011). Solid-phase treatment with the fungus *Trametes versicolor* substantially reduces pharmaceutical concentrations and toxicity from sewage sludge. *Bioresource Technology*, 102, 5602-5608.

[255] Rubilar, O., Tortella, G., Cea, M., Acevedo, F., Bustamante, M., Gianfreda, L. & Diez, M.C. (2011). Bioremediation of a Chilean Andisol contaminated with pentachlorophenol (PCP) by solid substrate cultures of white-rot fungi. *Biodegradation*, 22, 31-41.

[256] Karas, P.A., Perruchon, C., Exarhou, K., Ehaliotis, C. & Karpouzas, D.G. (2011). Potential for bioremediation of agro-industrial effluents with high loads of pesticides by selected fungi. *Biodegradation*, 22, 215-228.

[257] Levin, L., Melignani, E. & Ramos, A.M. (2010). Effect of nitrogen sources and vitamins on ligninolytic enzyme production by some white-rot fungi. Dye decolorization by selected culture filtrates. *Bioresource Technology* 101, 4554-4563.

[258] Uzan, E., Nousiainen, P., Balland, V., Sipila, J., Piumi, F., Navarro, D., et al. (2010). High redox potential laccases from the ligninolytic fungi *Pycnoporus coccineus* and *Pycnoporus sanguineus* suitable for white biotechnology: from gene cloning to enzyme characterization and applications. *Journal of Applied Microbiology*, 108, 2199-2213.

[259] Pandey, A., Soccol, C.R. & Mitchell, D. (2000). New developments in solid state fermentation: I-bioprocesses and products. *Process Biochemistry* 35, 1153-1169.

[260] Pandey, A., Soccol, C.R., Nigam, P. & Soccol, V.T. (2000). Biotechnological potential of agro-industrial residues. I: sugarcane bagasse. *Bioresource Technology*, 74, 69-80.

[261] Duran, N., Rosa, M.A., D'Annibale, A. & Gianfreda, L. (2002). Applications of laccases and tyrosinases (phenoloxidases) immobilized on different supports: a review. *Enzyme and Microbial Technology*, 31, 907-931.

[262] Minussi, R.C., Pastore, G.M. & Duran, N. (2002). Potential applications of laccase in the food industry. *Trends in Food Science & Technology*, 13, 205-216.

[263] Chagas, E.P. & Durrant, L.R. (2001). Decolorization of azo dyes by *Phanerochaete chrysosporium* and *Pleurotus sajor-caju*. *Enzyme and Microbial Technology*, 29, 473-477.

[264] Boer, C.G., Obici, L., de Souza, C.G.M. & Peralta, R.M. (2004). Decolorization of synthetic dyes by solid state cultures of *Lentinula* (*Lentinus*) *edodes* producing manganese peroxidase as the main ligninolytic enzyme. *Bioresource Technology*, 94, 107-112.

[265] Ferraz, A., Baeza, J., Rodriguez, J. & Freer, J. (2000). Estimating the chemical composition of biodegraded pine and eucalyptus wood by DRIFT spectroscopy and multivariate analysis. *Bioresource Technology*, 74, 201-212.

[266] Ferraz, A., Cordova, A.M. & Machuca, A. (2003). Wood biodegradation and enzyme production by *Ceriporiopsis subvermispora* during solid-state fermentation of *Eucalyptus grandis*. *Enzyme and Microbial Technology*, 32, 59-65.

[267] Westphalen, M.C., Reck, M.A. & da Silveira, R.M.B. (2010). Polypores from Morro Santana, Rio Grande do Sul, Brazil. *Hoehnea*, 37, 647-662.

[268] Gomes-Silva, A.C., Ryvarden, L. & Gibertoni, T.B. (2010). Notes on *Trametes* from the Brazilian Amazonia. *Mycotaxon,* 113, 61-71.

[269] Gibertoni, T.B. (2008). Polyporoid fungi (*Agaricomycetes*, *Basidiomycota*) in the Estação Científica Ferreira Penna (State of Pará, Brazilian Amazonia): diversity and ecological aspects. *Scientifica Acta*, 2, 70-74.

[270] Alonso, S.K.D., Silva, A.G.D., Kasuya, M.C.M., Barros, N.F.D., Cavallazzi, J.R.P., Bettucci, L., Lupo, S. & Alfenas, A.C. (2007). Isolamento e seleção de fungos causadores da podridão-branca da madeira em florestas de *Eucalyptus* spp. com potencial de degradação de cepas e raízes. *Revista Árvore*, 31, 145-155.

[271] de Souza, H.Q., de Oliveira, L.A. & Andrade, J.S. (2008). Screening of basidiomycetes from Amazonia for the production of biotechnological interest enzymes. *Ciencia E Tecnologia De Alimentos*, 28, 116-124.

[272] Aguiar, A., Mendonca, R., Rodriguez, J. & Ferraz, A. (2010). Behavior of *Ceriporiopsis subvermispora* during *Pinus taeda* biotreatment in soybean-oil-amended cultures. *International Biodeterioration & Biodegradation*, 64, 588-593.

[273] Vicentim, M.P. & Ferraz, A. (2007). Enzyme production and chemical alterations of *Eucalyptus grandis* wood during biodegradation by *Ceriporiopsis subvermispora* in cultures supplemented with Mn^{2+}, corn steep liquor and glucose. *Enzyme and Microbial Technology*, 40, 645-652.

[274] Ferraz, A., Guerra, A., Mendonca, R., Masarin, F., Vicentim, M.P., Aguiar, A. & Pavan, P.C. (2008). Technological advances and mechanistic basis for fungal biopulping. *Enzyme and Microbial Technology* 43, 178-185.

[275] Camassola, M. & Dillon, A.J.P. (2009). Biological pretreatment of sugar cane bagasse for the production of cellulases and xylanases by *Penicillium echinulatum*. *Industrial Crops and Products*, 29, 642-647.

[276] Fonseca, G.G., Gandra, E.A., Sclowitz, L.F., Antunes Correa, A.P. & Vieira Costa, J.A. (2009). Protein enrichment and digestibility of soft rush (*Juncus effusus*) and rice residues using edible mushrooms *Pleurotus ostreatus* and *Pleurotus sajor-caju*. *World Journal of Microbiology & Biotechnology*, 25, 449-456.

[277] Barreto-Rodrigues, M., Aguiar, C.M. & da Cunha, M.A.A. (2009). Biotreatment of an effluent from a wood laminate industry using *Lentinula edodes* UEC 2019. *Journal of Hazardous Materials*, 164, 1556-1560.

[278] Gomes, E., Aguiar, A.P., Carvalho, C.C., Bonfa, M.R.B., da Silva, R. & Boscolo, M. (2009). Ligninases production by basidiomycetes strains on lignocellulosic agricultural residues and their application in the decolorization of synthetic dyes. *Brazilian Journal of Microbiology*, 40, 31-39.

[279] de Siqueira, F. G., de Siqueira, A. G., de Siqueira, E. G., Carvalho, M. A., Pinto Peretti, B. M., Duque Jaramillo, P. M., et al. (2010). Evaluation of holocellulase production by plant-degrading fungi grown on agro-industrial residues. *Biodegradation,* 21, 815-824.

[280] Ferreira, L.F., Aguiar, M., Pompeu, G., Messias, T.G. & Monteiro, R.R. (2010). Selection of vinasse degrading microorganisms. *World Journal of Microbiology & Biotechnology*, 26, 1613-1621.

[281] Ferreira, L.F.R., Aguiar, M.M., Messias, T.G., Pompeu, G.B., Lopez, A.M.Q., Silva, D.P. & Monteiro, R.T. (2011). Evaluation of sugar-cane vinasse treated with *Pleurotus sajor-caju* utilizing aquatic organisms as toxicological indicators. *Ecotoxicology and Environmental Safety*, 74, 132-137.

[282] Zilly, A., Coelho-Moreira, J.d.S., Bracht, A., Marques de Souza, C.G., Carvajal, A.E., Koehnlein, E.A. & Peralta, R.M. (2011). Influence of NaCl and Na_2SO_4 on the kinetics and dye decolorization ability of crude laccase from *Ganoderma lucidum*. *International Biodeterioration & Biodegradation*, 65, 340-344.

[283] Arantes, V. & Milagres, A.M.F. (2007). The synergistic action of ligninolytic enzymes (MnP and Laccase) and Fe^{3+}-reducing activity from white-rot fungi for degradation of Azure B. *Enzyme and Microbial Technology*, 42, 17-22.

[284] Lyra, E. S., Moreira, K. A., Porto, T. S., Carneiro da Cunha, M. N., Paz Junior, F. B., Neto, B. B., et al. (2009). Decolorization of synthetic dyes by basidiomycetes isolated from woods of the Atlantic Forest (PE), Brazil. *World Journal of Microbiology & Biotechnology, 25*(8), 1499-1504.

[285] Ribas, L.C.C., de Mendonca, M.M., Camelini, C.M. & Soares, C.H.L. (2009). Use of spent mushroom substrates from *Agaricus subrufescens* (syn. *A. blazei*, *A. brasiliensis*) and *Lentinula edodes* productions in the enrichment of a soil-based potting media for lettuce (*Lactuca sativa*) cultivation: Growth promotion and soil bioremediation. *Bioresource Technology*, 100, 4750-4757.

[286] Coelho, J.S., de Souza, C.G.M., de Oliveira, A.L., Bracht, A., Costa, M.A.F. & Peralta, R.M. (2010). Comparative Removal of Bentazon by *Ganoderma lucidum* in Liquid and Solid State Cultures. *Current Microbiology*, 60, 350-355.

[287] de Souza, D.F., da Costa, S.C., Dacome, A.S., Marques de Souza, C.G., Bracht, A. & Peralta, R.M. (2011). Pentachlorophenol Removal by *Pleurotus Pulmonarius* in Submerged Cultures. *Brazilian Archives of Biology and Technology,* 54, 357-362.

[288] .Piovan, L., Capelari, M., Andrade, L.H., Comasseto, J.V. & Porto, A.L.M. (2007). Biocatalytic reduction of a racemic selenocyclohexanone by Brazilian basidiomycetes. *Tetrahedron-Asymmetry*, 18, 1398-1402.

[289] Piovan, L., Kagohara, E., Ricci, L.C., Keppler, A.F., Capelari, M., Andrade, L.H., Comasseto, J.V. & Porto, A.L.M. (2008). Chemoselective screening for the reduction of a chiral functionalised (+/-)-2-(phenylthio)cyclohexanone by whole cells of Brazilian micro-organisms. *Tetrahedron-Asymmetry*, 19, 2385-2389.

[290] Barros-Filho, B.A., de Oliveira, M.d.C.F., Lemos, T.L.G., de Mattos, M.C., de Gonzalo, G., Gotor-Fernandez, V. & Gotor, V. (2009). *Lentinus strigellus*: a new versatile stereoselective biocatalyst for the bioreduction of prochiral ketones. *Tetrahedron-Asymmetry*, 20, 1057-1061.

In: Fungi: Types, Environmental Impact and Role in Disease ISBN: 978-1-61942-671-9
Editors: Adolfo Paz Silva and María Sol

Chapter XXI

Contributions to Research on White-Rot Fungi, Lignin-Degrading Enzymes, and Their Industrial and Environmental Applications

L. Levin and F. Forchiassin*
Laboratorio de Micología Experimental,
Departamento de Biodiversidad y Biología Experimental,
Facultad de Ciencias Exactas y Naturales,
Universidad de Buenos Aires,
Ciudad Universitaria, Pab. II, C1428EHA CABA,
Argentina

Abstract

White-rot fungi (WRF), the most frequent wood-rotting organisms, are so far unique in their ability to completely degrade all components of our principal renewable resource, lignocellulose. They produce an impressive array of enzymes (cellulases, hemicellulases, pectinases and ligninases) with potential in a wide range of biotechnological applications. WRF and their ligninolytic enzymes (such as laccase and Mn-peroxidase) are able to degrade many pollutants. In pulp and paper manufacturing they are applied in biopulping, biobleaching and effluent treatment. Recently, there has been a growing interest in studying the ligninolytic enzymes of a wider array of WRF, with the expectation of finding better lignin-degrading systems for use in various biotechnological applications. To be able to use the white-rotters in these processes it is necessary to learn how their enzymes are secreted, how they operate and under what conditions they are active. Our current research is focused on these subjects and in the screening of new isolates for their ability to degrade xenobiotics, in the search of more efficient strains. Reducing the cost for enzyme production is still needed in order to develop enzymatic treatment processes for different industrial and environmental applications, competitive with other treatment

* E-mail address: lale@bg.fcen.uba.ar.

technologies. At present, one of the objectives of our work is reducing this cost by means of process optimization using statistical experimental designs, and the use of cheaper growth substrates such as agricultural and food wastes. We identified new strains of *Trametes versicolor, Trametes villosa, Stereum hirsutum* and *Fomes sclerodermeus* which demonstrated an outstanding ability in dye degradation. *Trametes trogii* BAFC 463 almost completely removed several dyes as well as high priority pollutants such as polycyclic aromatic hydrocarbons and polychlorinated biphenyls. *Pycnoporus sanguineus* BAFC 2126 proved to be a potential candidate for use in softwoods biopulping processes. Efficient bleaching of loblolly pine kraft pulp could be achieved in a chlorine-free sequence, with culture supernatants from *T. trogii*, followed by a peroxide stage under mild conditions. Its high potential purified laccase, was recently applied as biocatalyst for oxygen electro-reduction in biofuel cells. With our results we wish to contribute to expand the knowledge on this important group of fungi and their biotechnological applications.

1. Introduction

1.1. White-Rot Fungal Decay of Wood

White-rot fungi (WRF), a filamentous group of fungi common inhabitants of forest litter and fallen trees, are so far unique in their ability to completely degrade all components of our principal renewable resource, lignocellulose. Lignocellulose degradation is a central step for carbon recycling in land ecosystems [1]. The use of lignocellulose as a source of fuel, food and fiber is of major economic significance. WRF, the most frequent wood-rotting organisms, are a physiological rather than taxonomic group comprising those fungi characterized by their ability to degrade lignin, hemicelluloses, and cellulose often giving rise to a cellulose-enriched white material. The name white-rot derives from the appearance of wood attacked by these fungi, in which lignin removal results in a bleached appearance to the substrate [2]. Most known WRF are basidiomycetes, although a few ascomycete genera within the Xylariaceae are also capable of white-rot decay [3-5]. Due to the ability of WRF to degrade lignin selectively or simultaneously with cellulose, two white-rot patterns have been described in different types of wood, namely selective delignification and simultaneous rot. The fungal isolate, solid substrate and ecophysiological factors affect the fungal growth and wood degradation pattern [1]. WRF have two types of extracellular enzymatic systems; the hydrolytic system, which produces hydrolases that are responsible for polysaccharide degradation and a unique oxidative and extracellular ligninolytic system, which degrades lignin and opens phenyl rings [6]. The ability to catabolize cellulose and hemicellulose, the polysaccharides forming the other main components of lignocellulose, is fairly common as a primary metabolic process among fungi and other organisms, and occurs under a range of environmental conditions. As a result, it is not regarded as a rate-limiting step in carbon flux. Lignin (a polymer of phenylpropane units connected by different C–C and C–O–C linkages) however, is extremely recalcitrant and is mineralized in an obligatory aerobic oxidative process, carried out appreciably only by the WRF, which possess the unique ability of efficiently degrading lignin to carbon dioxide and water. This degradative ability of WRF is due to the strong oxidative activity and low substrate specificity of their ligninolytic enzymes. Lignin is oxidized and degraded by a ligninase system made up of at least three enzyme activities: lignin peroxidase (EC 1.11.1.14) (LiP), Mn-peroxidase (EC 1.11.1.13) (MnP) and laccase (EC 1.10.3.2), which function together with H_2O_2-producing oxidases and secondary metabolites [2,7]. WRF secrete one or more of the three enzymes essential for lignin degradation. Ligninases act nonspecifically, generating free radicals on a vast number of phenolic and non-phenolic compounds. In addition, several findings provided information about low weight molecules that participate in the initial attack of lignocellulose. Fenton reaction produces $\cdot OH$ [8], this radical is the strongest oxidant found in WRF; another reactive molecules involved in lignin attack are the peroxyl radicals rendered from lipid peroxidation [9].

1.2. Environmental and Industrial Applications of White-Rot Fungi and Their Enzymes

The same unique nonspecific mechanisms that give these fungi the ability to degrade lignin also allow them to degrade a wide range of pollutants, which are resistant to degradation by most microorganisms, including: polycyclic aromatic hydrocarbons (PAHs), organochlorines, polychlorinated biphenyls (PCBs), explosives, pesticides and dyes [10-18]. Interest in WRF and their lignocellulolytic enzyme system has increased during the past two decades spurred by their biotechnological and industrial applications. Cellulosic biofuel has attracted growing interests as the concerns raises over the fact that grain-based biofuel presents competition with food supply. A major challenge to full realization of the potential of lignocellulosic biomass is to overcome their high resistance to degradation to become fermentable sugars. Often the first step in lignocellulosic biomass utilization is to remove lignin or hemicelluloses in a process called pre-treatment so that the cellulose polymer can be exposed to cellulolytic enzyme's attack to be broken down to glucose. The current thermochemical methods for deconstruction of lignocellulosic material, however, are mainly energy intensive, environmentally harsh, and produce chemicals toxic to ethanol producing strains. Therefore, research and development are needed to explore sustainable approaches to degrade lignocellulosics biologically with enzymes including lignin-degrading peroxidases and hemicellulase [19-21].

1.3. Bioprospection for New Sources of Ligninolytic Enzymes

Although the ability of WRF and their ligninolytic enzymes to transform or mineralize a wide range of organopollutants has been confirmed, there are still unanswered questions about the catabolism of these compounds, particularly the precise role and regulation of the enzymes involved in these processes. All three ligninolytic enzymes (MnP, LiP, laccase) are produced in multiple isoforms and encoded by gene families with complex regulation. Nutrient levels, mediator compounds and required metal ions [Mn (II) for MnP, Cu (II) for laccase] affect transcription of respective genes [12]. The utilization of fungi for bioremediation requires an understanding of the factors that enhance their degrading ability, and clarification of the enzyme mechanisms used [18]. The majority of information so far available concerns the lignin-degrading enzyme systems of *Phanerochaete chrysosporium* and *Trametes versicolor* [10,13,22]. Since the enzymatic system secreted by basidiomycetes depends even on the strain and culture conditions [23], a growing interest has raised in studying the ligninolytic enzymes of a wider array of WRF, with the expectation of finding better lignin-degrading systems for use in various biotechnological applications. The importance, properties and applications of ligninolytic enzymes were described by several review papers [10-21,24-36]. The ligninolytic activity of various Argentinean white-rot strains is presently under investigation in our laboratory. The present chapter reviews our current research, focused in learning how their enzymes are secreted, how they operate and under what conditions they are active; in the screening of new isolates for their ability to degrade xenobiotics, meant to identify more efficient strains; and in reducing the cost for enzyme production in order to develop enzymatic treatment processes for different industrial and environmental applications, competitive with other treatment technologies.

2. Results

2.1. Enzyme Production through the Bioconversion of Lignocellulosic Residues

Reducing the cost for enzyme production is still needed in order to develop enzymatic treatment processes for different industrial and environmental applications, which might be more competitive than conventional and other novel treatment technologies. This might be achieved by enzyme overproduction in a suitable host. Over expression of recombinant enzymes in heterologous systems, such as yeasts and bacteria, has been actively studied to enhance the production of fungal peroxidases and laccases, as well as to improve the catalytic activity and stability of these enzymes, during the last several years. Gene amplification and expression in appropriate hosts could be promising for abundant production and affordable price of ligninolytic enzymes, as is already the case with laccases used commercially in the pulp and paper industry [14,17]. A lot of research has been done in the field of genetic engineering, mainly on the most investigated fungus *P. chrysosporium* [22,37-41]. However, ligninolytic enzymes are difficult to overexpress heterologously in an active form.

In alternative to recombinant DNA techniques, enzyme overproduction might be attained by means of process optimization using statistical experimental designs and the use of cheap growth substrates [14,42-45]. The fungal production of ligninolytic enzymes through the bioconversion of lignocellulosic residues has been widely investigated in recent years. This approach is attractive because of foreseeable effects on cost reduction, waste reuse and enhanced enzyme production. Solid state fermentation (SSF) processes have shown to be particularly suitable for the production of enzymes by filamentous fungi, since they reproduce the natural habitats of such fungi.

Trametes trogii BAFC 463, an Argentinean white-rot strain, besides efficiently degrading lignin in wood [46], has been tested successfully in biomechanical pulping experiments [47] and also proved to be a good producer of xylanases [48] and ligninases (laccase and MnP) [49]. Its ability to almost completely remove high priority pollutants such as PCBs, PAHs, nitrobenzene, anthracene, and a wide range of textile dyes, has been recently demonstrated [50-54]. The fact that *T. trogii* produces high levels of both laccase and MnP is a further advantage with respect to their possible synergism in bioremediation processes [55]. The simultaneous presence in this fungus of high ligninolytic and hydrogen peroxide producing activities (such as glyoxal oxidase), essential for peroxidase activity and rate limiting for pollutant degradation, make it an attractive microorganism on which to base future biotechnological applications. On the other hand, in contrast to the well studied model organism *P. chrysosporium*, *T. trogii* is N-unregulated. Sufficient or excess N-nutrients stimulate high MnP and laccase titers in parallel with high biomass production. This characteristic of the fungus makes it an outstanding candidate for large-scale fermentation to produce ligninolytic enzymes in bulk for bioremediation [49]. We applied response surface optimization techniques to maximize *T. trogii* ligninolytic enzyme production in SSF [56] and in submerged fermentation [54,57]. In order to reduce the effects of the crude culture fluids on the cellulose fraction of lignocellulose, with the expectation of applying them in biobleaching, optimization was aimed at simultaneously minimizing cellulase activity and maximizing ligninolytic enzyme production. Various inexpensive agro-industrial wastes, among them wood chips, cereal grains, straw and bran, were evaluated during the past few years by our research group for ligninolytic enzyme production [56,58]. *Fomes sclerodermeus* BAFC 2752 another Argentinean WRF, produced high levels of laccase and MnP when cultured in SSF in a natural medium based on wheat bran [58]. Moreover, the system wheat bran- *F. sclerodermeus* proved to be effective in the removal and degradation of the fungicide dye Malachite Green from aqueous solution [59]. Several uses have been suggested for biodegraded lignocellulosic wastes; such as raw material for the production of ethanol [19-21]. Some lignin-degrading fungi, notably *Pleurotus* sp. and *Lentinus edodes* produce edible mushrooms and can directly convert lignocellulose into human food [6,60]. The production of lignocellulolytic enzymes during growth and fruiting of the edible fungi *Lentinus tigrinus* [61] and *Grifola frondosa* [62] on lignocellulosic wastes was the subject of other of our investigations. An important problem in bioremediation processes involving microorganisms is the amount of time required for their growth. The use of solid spent substrates (SFS) from mushroom cultivation in wastewater treatment solves the problem of waiting for microorganisms to grow. We employed solid residues from *Pleurotus ostreatus* production in dye removal from aqueous solutions. These residues constitute a material that could be useful in wastewater bioremediation. Decolorization mediated by *P. ostreatus* ligninolytic enzymes and adsorption to SFS were both involved in this process [63]. These residues are composed not only of lignocellulosic materials which can be used for biosorption but also by chitin, which possesses a high adsorption capacity [59,64].

2.2. Application of White-Rot Fungi and Their Enzymes in the Pulp and Paper Industry

The importance of microbial enzymes in pulp and paper manufacturing has grown significantly in the last two decades. Microbial enzymes: cellulases, xylanases and ligninases, create new technologies for pulp and paper processing. Xylanases reduce the amount of chemicals required for bleaching; cellulases smooth fibers, enhance drainage and promote ink removal. Ligninolytic enzymes remove lignin from pulps, and may be applied in biopulping, biobleaching and effluent treatment, although currently their most important application is in the prebleaching of Kraft pulp [65-67]. WRF and their enzymes (particularly ligninases and xylanases) are considered for the treatment of wood chips prior to pulping. While ligninases attack the lignin, xylanases degrade hemicelluloses and make the pulp more permeable for the removal of residual lignin [68]. Termed ''biopulping'', this process removes not only lignin but also some of the wood extractives, thus reducing the pitch content and effluent toxicity. When biological treatment is followed by mechanical pulping, there is as much as 30% energy saving, whereas when it is followed by sulfite or kraft pulping, the cooking time is dramatically reduced. The paper strength properties have also been found to improve after biopulping. However, this process is still in its beginnings and no full-scale biopulping mills are in operation at the moment [66,69,70]. In the search of new strains suitable for softwood biopulping processes, several argentinean WRF were screened for their ability to selectively delignify loblolly pine (*Pinus taeda*) chips [71]. They were selected among 33 basidiomycetes from Argentina because of their capacity to decolorize Poly R-478 and Azure B. Fungal pretreatment caused changes in wood chemical composition as well as in its physical structure. The study allowed the identification of a new strain, *Pycnoporus sanguineus* BAFC 2126 which was able to reduce lignin content in 11% in 14 days of treatment. *P. taeda* wood suffered notable structural changes of lignin and hemicelluloses during the treatment, as revealed from ^{13}C CP-MAS NMR spectra. An increase of 15% in porosity of decayed wood confirmed physical changes due to fungal attack. In addition to the chemical modification caused by selective delignification, physical changes, as the increase in pore size, affects directly the penetration of cooking chemicals. Both combined effects increase reaction's rate, leading to shorter cooking times or the possibility of decreasing chemical charge reducing in consequence the environmental impact of the process [70]. The removal of lignin from chemical pulps is called bleaching, and it is necessary for aesthetic reasons and for improvement of paper properties. Present-day bleaching of kraft pulp uses large amounts of chlorine and chlorine chemicals. Byproducts from using these chemicals are chlorinated organic substances, some of which are toxic and mutagenic. Nowadays peroxide, oxygen and ozone constitute environmentally friendly alternatives to develop totally chlorine-free (TCF) sequences. However, most TCF chemical reagents, due to their lower delignification power, are less efficient than chlorine reagents in attaining high and stable pulp brightness degrees. The use of enzymes appears as a promising approach for chlorine-free pulp bleaching process. Xylanase prebleaching technology is now in use at several mills worldwide [65-68]. Some ligninolytic enzymes, such as laccases and MnPs, also showed potential to perform biobleaching reactions by specific lignin oxidation and removal [72-77]. The results obtained with MnP in kraft pulp bleaching have been promising and the enzyme treatment has had only a minor effect on paper strength or yield. This enzyme can use unsaturated fatty acids as mediators through lipid peroxidation to degrade synthetic lignins efficiently [75-77]. Due to its rather low redox potential (0.5-0.8 V), laccase is able to attack only the phenolic moieties in the lignin polymer, thus being less efficient in delignification and bleaching of pulp. However, the substrate range of laccases can be expanded to include non-phenolic compounds in the presence of synthetic or natural mediators such as HBT (1-hydroxybenzotriazole), ABTS [2,2-azino-bis-(3-ethylbenzothiazoline-6-sulfonic acid)] and HBA (4-hydroxybenzoic acid) [72-74,78,79]. Although the biological significance of the laccase-mediator system in nature is to be established, it presents high biotechnological interest. Pulp bleaching with a laccase-mediator system has reached pilot plant stage and is expected to be commercialized soon, but several aspects need to be solved before mill implementation of a ligninolytic enzymatic stage in TCF bleaching. The main constraints in the use of laccases in industrial processes are the large-scale availability required, and their production costs. Synthetic laccase mediators have also a high cost and a growing concern exists about their possible toxicity [67,68]. Crude culture filtrates from *T. trogii* BAFC 463, recently demonstrated their potential in biobleaching. We obtained a noteworthy increase in pulp Luminance [L* = 94.45, compared with 94.5 of the white reference standard (titanium oxide), L* = 69.94 of the initial pulp, and 83.11 of the peroxide-bleached control], when loblolly pine kraft pulp was bleached in a TCF sequence that involved treatment with this low-cost source of enzyme, followed by a peroxide stage. The whole process was performed under essentially mild conditions (at 28 ºC), thus reducing energy consumption, and did not require mediator addition in the enzymatic stage [80]. Crude culture filtrates offer several advantages over the use of purified enzymes: their obtention process is not as expensive, they may include natural laccase-mediators secreted by the fungus, and in addition, proteins, residual macronutrients, soluble extracellular metabolites or other factors present in the medium may stabilize crude enzymes [81,82]. The fact that *T. trogii* produces high levels of both laccase and MnP is another advantage with respect to their possible synergism in biobleaching [83]. Taking into account that one of the most promising approaches to improve the economics of kraft pulp production consists in increasing overall pulp yields [84], an additional advantage of the current process was its minor pulp yield loss (less than 5%). Moreover, the cost of the mediator is the determining factor for the commercialization of the laccase-mediator bleaching system, therefore the fact that crude culture filtrates from *T. trogii* did not require the addition of synthetic mediators for biobleaching (with the concomitant undesired effect of enzyme deactivation) seems to be a promising alternative [85]. WRF and their ligninolytic enzymes have also potential for the treatment of pulp and paper industry effluents: for their decolorization as well as the reduction of adsorbable organic halides and the chemical oxygen demand [66,86,87]. Black liquor, a colored and toxic effluent, is a residue from the most widely used process in the industrial production of cellulosic pulps, kraft pulping. During pulping treatments, chromophoric and highly oxidized polymeric lignin/chlorolignin derivatives are formed, giving rise to the characteristic dark color. Its intense brown color along with its recalcitrance to biodegradation makes this residue an issue of environmental concern. Traditional technologies used for the treatment of black liquor are not always effective and may not to be an environmentally friendly process. Several strains of Argentinean WRF were evaluated for their ability to decolorize black liquor on plates and on SSF using vermiculite as solid support and a liquid phase with malt extract. One of this strains *Steccherinum* sp. BAFC 1171 reduced up to 92% the color of the black liquor in malt extract medium; this percentage is equivalent to 5.2 g/l of decolorized black liquor, the highest value reported to date. Based on decolorization percentages achieved, vermiculite-WRF systems could be an attractive alternative or a complement of the existing technologies such as adsorption, precipitation and membrane filtration, for black liquor treatment [88].

2.3. Decolorization of Dyes by White-Rot Fungi

WRF are the most intensively studied dye decolorizing microorganisms. Their potential so far, however, has not yet found a real application, mainly because of the difficulty in selecting organisms capable of growing and degrading in the very variable and restrictive conditions of the textile industry wastewaters. In addition, most of the research studies on dye degradation are conducted at Erlenmeyer scale. Nevertheless, before an industrial application can be implemented, fungal bioreactors which can be operated under industrial conditions, using real wastewaters must be developed [89,90]. Two strategies for WRF application in dye degradation have been pursued: (1) direct transformation of pollutants by active cultures of WRF, and (2) use of enzymes extracted from the culture media. The advantage of the second strategy is a higher independence from influences of wastewater composition that may affect the survival of the fungi, e.g. high salinity [91]. A disadvantage is the higher costs resulting from the need for extraction and cleaning steps. Considering the high volume of effluents generated from dyeing units, the use of purified and/or immobilized enzymes increases the cost of degradation processes. But, to target a wider range of chemical structures in waste effluents, the use of crude culture filtrates containing more than one ligninolytic enzyme would be advantageous, since each one may attack different chemical structures. As the use of whole fungal cultures for decolorization is not easily applicable in a large scale, at present the studies of *in vitro* decolorization of synthetic dyes by crude enzymes become more important [92]. Several methods for improving the industrial application of enzymes exist, these include less expensive production strategies, improvement in yield from current producer organisms, or isolation of more industrially applicable forms of the enzyme that exhibit enhanced stability and superior catalytic properties through bio-prospecting [93]. The ability to produce ligninolytic enzymes and decolorize industrial dyes was recently studied in various Argentinean WRF [94,95]. Comparing the isolates with the well-known dye-degrader *P. chrysosporium*, a new fungus was identified: *Coriolus antarcticus* BAFC 266. Based on the magnitude of the decoloration obtained (whole cultures of this fungus were able to decolorize in an hour 28-100% of five different dyes added (Malachite Green, Remazol Brilliant Blue R; Xylidine, Indigo Carmine and Poly R-478), this strain seems potentially competitive in the bioremediation of textile processing effluents [95]. Its efficiency in Malachite Green decolorization using a two-phase bioreactor was investigated [96]. In the first phase the decolorization ability of this fungus, growing under SSF, was proved; in the second phase the capacity of the enzymes present in extracts from the solid residues was exploited. *T. trogii* BAFC 463 is another promising fungal strain for dye decolorization purposes. Laccase mediated dye decolorization has been described with crude and purified forms from many fungi; however, most dyes are only transformed in the presence of redox mediators [97]. Laccase from *T. trogii* was reported to have one of the highest redox potential among laccases (E° 0.79 at pH 5.0) [98], which makes this laccase particularly interesting since high redox potentials correlate with high laccase activity [99]. The effect of amino acids, complex nitrogen sources and vitamin addition on *T. trogii, Trametes villosa* and *C. antarcticus* ligninolytic enzyme production, was evaluated. Dye decolorization by their culture filtrates was compared. Glutamic acid followed by peptone, were the best N sources for laccase and MnP production. *T. trogii* grown in glutamic acid produced culture filtrates with the highest laccase (188.3 U/ml) and MnP activities (4.5 U/ml), rendering the best results in decolorization [53]. The broad-spectrum decolorization efficiency of *T. trogii* BAFC 463 was assessed using chemically different dyes. The study showed the ability of *T. trogii* culture fluids, as well as its purified laccase of decolorizing a variety of synthetic dyes, including azoic, indigoid, triphenylmethane, anthraquinonic and heterocyclic dyes, in the absence of redox mediators [52]. The similar results obtained with crude and purified laccase confirmed the role of the enzyme in dye decolorization. Nevertheless, using this crude filtrate without purification would reduce the cost of enzymatic-based decolorization process and provide more stability for the enzyme. *T. trogii* culture fluids and its purified laccase also demonstrated an outstanding ability to decolorize the heterocyclic dyes Azure B and Methylene Blue, when comparing with laccases from other WRF (30% after 24 h). With HBT addition they decolorized more than 90% of Methylene Blue (15 μM) and Azure B (18 μmM) in 1 h. Azure B is among the most recalcitrant dyes, and is used to detect LiP activity because it is not oxidized by laccase or MnP alone [100-101]. LiP from culture fluids of *P. chrysosporium* supplemented with 4 mM H_2O_2 removed 54% of Methylene Blue (15.6 μM) after 1 h [102]. *T. trogii* laccase with HBT has a comparable effectiveness in Methylene Blue decolorization, but presents an advantage over peroxidases in utilizing oxygen as cofactor. As industrial wastes are generally complex mixtures of various xenobiotics including heavy metal salts, their inhibitory effects must be taken into consideration during their bioremediation [103]. The resistance of the catalytically active protein structure toward high temperature, pH and other potentially denaturing factors is one of the chief criteria for commercialization and for industrial applications of the enzymes [82]. Culture fluids of *T. trogii* were able to decolorize in the presence of high concentrations of heavy metals, and not only pure solutions of synthetic dyes, but also synthetic dyebaths containing inorganic salts and EDTA, and showed efficient decolorization even at 70 °C and pH 7.0. These results show the potential of using *T. trogii* culture fluids for textile-effluent degradation.

An inherent limitation of bioremediation by active cultures of WRF arises from the tolerance limits in which the active microorganism can degrade the pollutant [104]. Members of our research group recently developed an assembly for bioremediation of polluted water based on the immobilization of alginate beads loaded with filamentous fungus *Stereum hirsutum* BAFC 2234 inside nanoporous silica hydrogels [105]. The resulting hybrid device exhibited good physical, chemical and biological stability, being effective in the removal and degradation of Malachite Green, even in solutions with a high concentration of the dye. This fact was a consequence of adsorption and regulated transport of the dye, as well as the retention of dye degradation enzymes inside the hydrogel. To the best of our knowledge, this is the first proposal of bioremediation of organic pollutants with microorganisms encapsulated in mineral cages. The results obtained open the possibility of bioremediation without dissemination of exotic organisms to the environment.

2.4. Laccase-Based Biofuel Cells

Enzyme-based biofuel cells are attracting attention rapidly partially due to the promising advances achieved lately. Many attempts have been made to immobilize enzymes using various nanostructures, since the large surface area afforded by these nanostructures usually leads to high enzyme loading, which can be used in improving the power density of enzyme-based biofuel cells [106]. The catalytic ability of laccases to activate the four-electron reduction of oxygen at unprecedented high electrode potentials (between 0.35 and 0.50V vs. Ag/AgCl) has promoted their study in cathodes for biofuel cells. High potential purified *T. trogii* laccase has been studied as a biocatalyst for oxygen cathodes composed of layer-by-layer self-assembled thin films by sequential immersion of mercaptopropane sulfonate-modified Au electrode surfaces in solutions containing laccase and osmium-complex bound to poly(allylamine), (PAH-Os) [107,108].

Conclusions

The majority of information so far available concerns the ligninolytic enzyme systems of *P. chrysosporium* and *T. versicolor*. Since the enzymatic system secreted by basidiomycetes depends even on the strain and culture conditions, in recent years a growing interest has raised worldwide, in studying the ligninolytic enzymes of a wider array of WRF, from both a basic and applied viewpoint. The ligninolytic enzyme systems of several Argentinean WRF and their potential biotechnological applications are currently under investigation in our laboratory. This chapter underlines the importance of exploring the potential of new strains in the process of bioremediation in order to expand the pool of existing biocleaning and biobleaching WRF. More research and development is still needed for cost-effective and sustainable application of WRF and their ligninolytic enzymes. Degradation of many hazardous chemicals and wastes has been demonstrated on a laboratory-scale, especially under sterile conditions [13]. Pilot- and full-scale applications are essential to evaluate the effectiveness and economy of the fungal technology. Many technical and engineering challenges remain in the application area.

References

[1] Blanchette, R.A. (1995). Degradation of the lignocellulose complex in wood. *Canadian Journal of Botany*, 73 (Suppl. 1), S999-S1010.

[2] Martínez, T., Speranza, M., Ruiz-Dueñas, F.J., Ferreira, P., Camarero, S., Guillén, F., Martínez, M.J., Gutierrez, A. & del Río, J.C. (2005). Biodegradation of lignocellulosics: microbiological, chemical and enzymatic aspects of fungal attack to lignin. *International Microbiology*, 8, 195-204.

[3] Liers, C., Ullrich, R., Steffen, K.T., Hatakka, A. & Hofrichter, M. (2006). Mineralization of ^{14}C-labelled synthetic lignin and extracellular enzyme activities of the wood-colonizing ascomycetes *Xylaria hypoxylon* and *Xylaria polymorpha. Applied Microbiology and Biotechnology*, 69, 573-579.

[4] Pointing, S.B., Parungao, M.M. & Hyde, K.D. (2003). Production of wood-decay enzymes, mass loss and lignin solubilization in wood by tropical Xylariaceae. *Mycological Research*, 107, 231-235.

[5] Shary, S., Ralph, S.A. & Hammel, K.E. (2007). New insights into the ligninolytic capability of a wood decay Ascomycete. *Applied and Environmental Microbiology*, 73, 6691-6694.

[6] Sánchez, C. (2009). Lignocellulosic residues: Biodegradation and bioconversion by fungi. *Biotechnology Advances*, 27, 185-194.

[7] Wong, D. (2009). Structure and action mechanism of ligninolytic enzymes. *Applied Biochemistry and Biotechnology*, 157, 174-209.

[8] Arantes, V., Milagres, A.M.F., Filley, T.R. & Goodell, B. (2011). Lignocellulosic polysaccharides and lignin degradation by wood decay fungi: the relevance of nonenzymatic Fenton-based reactions. *Journal of Industrial Microbiology and Biotechnology*, 38, 541-555.

[9] Kapich, N., Jensen, K.A. & Hammel, K.E. (1999). Peroxyl radicals are potential agents of lignin biodegradation. *FEBS Letters*, 461, 115-119.

[10] Pointing, S.B. (2001). Feasibility of biorremediation by white-rot fungi. *Applied Microbiology and Biotechnology*, 57, 20-33.

[11] Singh H (2006) *Mycoremediation, Fungal Bioremediation.* New Jersey, USA: John Wiley and Sons.

[12] Wesenberg, D., Kyriakides, I. & Agathos, S.N. (2003). White-rot fungi and their enzymes for the treatment of industrial dye effluents. *Biotechnology Advances*, 22, 161-187.

[13] Gao, D., Du, L., Yang, J., Wu, W. & Liang, H. (2010). A critical review of the application of white rot fungus to environmental pollution control. *Critical Reviews in Biotechnology*, 30, 70-77.

[14] Ikehata, K., Buchanan, I.D. & Smith, D.W. (2004). Recent developments in the production of extracellular fungal peroxidases and laccases for waste treatment. *Journal of Environmental Engineering and Science*, 3, 1-19.

[15] Fu, Y. & Viraraghavan, T. (2001). Fungal decolorization of dye wastewaters: a review. *Bioresource Technology*, 79, 251-262.

[16] Levin, L. & Forchiassin, F. (2007). White-rot fungi, new biotechnological tools for a cleaner environment. *Current Topics in Biotechnology*, 3, 49-56.

[17] Gianfreda, L. & Rao, M.A. (2004), Potential of extracellular enzymes in remediation of polluted soils: a review. *Enzyme and Microbial Technology*, 35, 339-354.

[18] Cerniglia, C.E., Sutherland, J.B. & Crow, S.A. Fungal metabolism of aromatic hydrocarbons. In: Winkelmann, G., *Microbial degradation of natural products*, Weinheim: VCH Press; 1992; 193-217.

[19] Sun, Y. & Cheng, J. (2002). Hydrolysis of lignocellulosic materials for ethanol production: a review. *Bioresource Technology*, 83, 1-11.

[20] Martin, C., López, Y., Plasencia, Y. & Hernández, E. (2006). Characterization of agricultural and agroindustrial residues as raw materials for ethanol production. *Chemical and Biochemical Engineering Quarterly*, 20, 443-447.

[21] Ward, P.O. & Singh, A. (2002). Bioethanol technology: developments and perspectives. *Advances in Applied Microbiology*, 51, 53-80.

[22] Kersten, P. & Cullen, D. (2007). Extracellular oxidative systems of the lignin-degrading basidiomycete *Phanerochaete chrysosporium*. *Fungal Genetics and Biology*, 44, 77-87.

[23] Zouari-Mechichi, H., Mechichi, T., Dhouib, A., Sayadi, S., Martínez, A.T. & Martínez, M.J. (2006). Laccase purification and characterization from *Trametes trogii* isolated in Tunisia: decolorization of textile dyes by the purified enzyme. *Enzyme and Microbial Technology*, 39, 141-148.

[24] Klaus, P. (2002). New insights into lignin peroxidase. *Indian Journal of Chemistry*, 41, 46-53.

[25] Dashtban, M., Schraft, H., Syed, T.A. & Qin, W. (2010). Fungal biodegradation and enzymatic modification of lignin. *International Journal of Biochemistry and Molecular Biology*, 1, 36-50.

[26] Hamid, M. & Rehman, K. (2009). Potential applications of peroxidases. *Food Chemistry*, 115, 1177-1186.

[27] Hofrichter, M. (2002). Review: lignin conversion by manganese peroxidase (MnP). *Enzyme and Microbial Technology*, 30, 454-466.

[28] Lundell, T.K., Mäkelä, M.R. & Hildén, K. (2010). Lignin modifying enzymes in filamentous basidiomycetes – ecological, functional and phylogenetic review. *Journal of Basic Microbiology*, 50, 5-20.

[29] Mayer, M. & Staples, R.C. (2002). Laccase: new function for an old enzyme. *Phytochemistry*, 60, 551-555.

[30] Rodríguez-Couto, S. & Toca-Herrera, J.L. (2006). Industrial and biotechnological application of laccase: A review. *Biotechnology Advances*, 24, 500-513.

[31] Giardina, P., Faraco, V., Pezzella, C., Piscitelli, A., Vanhulle, S. & Sannia, G. (2010). Laccases: a never-ending story. *Cell Molecular Life Science*, 67, 369-385.

[32] Pérez, J., Muñoz-Dorado, J., De-la-Rubia, T. & Martínez, M.J. (2002). Biodegradation and biological treatments of cellulose, hemicellulose and lignin: an overview. *International Microbiology*, 5, 53-63.

[33] Riva, S. (2006). Laccases: blue enzymes for green chemistry. *Trends in Biotechnology*, 24, 219-226.

[34] Hammel, K.E. & Cullen, D. (2008). Role of fungal peroxidases in biological ligninolysis. *Current Opinion in Plant Biology*, 11, 349-355.

[35] Baldrian, P. (2006). Fungal laccases-occurrence and properties. *FEMS Microbiology Reviews*, 30, 215-242.

[36] Martínez, T., Ruiz-Dueñas, F.J., Martínez, M.J., del Río, J.C. & Gutiérrez, A. (2009). Enzymatic delignification of plant cell wall: from nature to mill. *Current Opinion in Biotechnology*, 20, 348-357.

[37] Bouws, H., Wattenberg, A. & Zorn, H. (2008). Fungal secretomes—nature's toolbox for white biotechnology. *Applied Microbiology and Biotechnology*, 80, 381-388.

[38] Piscitelli, A., Pezzella, C., Giardina, P., Faraco, V. & Giovanni, S. (2010). Heterologous laccase production and its role in industrial applications. *Bioengineering Bugs*, 1, 252-262.

[39] Vanden Wymelenberg, A., Gaskell, J., Mozuch, M., Sabat, G., Ralph, J., Skyba, O., Mansfield, S.D., Blanchette, R.A., Martinez, D., Grigoriev, I., Kersten, P.J. & Cullen, D. (2010). Comparative transcriptome and secretome analysis of wood decay fungi *Postia placenta* and *Phanerochaete chrysosporium. Applied and Environmental Microbiology*, 76, 3599-3610.

[40] Martinez, D., Larrondo, L.F., Putman, N., Gelpke, M.D.S., Huang, K., Chapman, J., Helfenbein, K.G., Ramaiya, P., Detter, J.C., Larimer, F., Coutinho, P.M., Henrissat, B., Berka, R., Cullen, D. & Rokhsar, D. (2004). Genome sequence of the lignocellulose degrading fungus *Phanerochaete chrysosporium* strain RP78. *Nature Biotechnology*, 22, 1-6.

[41] Colao, M.C., Lupino, S., Garzillo, A.M., Buonocore, V. & Ruzzi, M. (2006). Heterologous expression of *lcc1* gene from *Trametes trogii* in *Pichia pastoris* and characterization of the recombinant enzyme. *Microbial Cell Factories*, 5, 31-41.

[42] Rodríguez Couto, S. & Toca-Herrera, J.L. (2007). Laccase production at reactor scale by filamentous fungi. *Biotechnology Advances*, 25, 558-569.

[43] Rodríguez Couto, S. & Sanromán, M.A. (2005). Application of solid-state fermentation to ligninolytic enzyme production. *Biochemical Engineering Journal*, 22, 211-219.

[44] Pandey, A., Selvakumar, P., Soccol, C.R. & Nigam P. (1999). Solid-state fermentation for the production of industrial enzymes. *Current Science*, 77, 149-162.

[45] Viniegra-González, G., Favela-Torres, E., Aguilar, C.N., Romero-Gómez, S.J., Díaz-Godínez, G. & Augur, C. (2003). Advantages of fungal enzyme production in solid state over liquid fermentation systems. *Biochemical Engineering Journal*, 13, 157-167.

[46] Levin, L. & Castro, M.A. (1998). Anatomical study of the decay caused by the white rot fungus *Trametes trogii* (Aphyllophorales) in wood of *Salix* and *Populus*. *IAWA Journal*, 19, 169-180.

[47] Planes, E., Bassi, M., Bensignor, J. & Burachik, M. (1986). Pulpado biológico de maderas de sauce. *ATIPCA*, 25, 44-46.

[48] Levin, L. & Forchiassin, F. (1998). Influence of growth conditions on the production of xylanolytic enzymes by *Trametes trogii*. *World Journal of Microbiology and Biotechnology*, 14, 443-446.

[49] Levin, L., Ramos, A.M. & Forchiassin, F. (2002). Copper induction of lignin modifying enzymes in the white rot fungus *Trametes trogii*. *Mycologia*, 94, 377-383.

[50] Haglund, C., Levin, L., Forchiassin, F., Lopez, M. & Viale, A. (2002). Degradation of environmental pollutants by *Trametes trogii*. *Revista Argentina de Microbiología*, 34, 157-162.

[51] Levin, L., Viale, A. & Forchiassin, F. (2003). Degradation of organic pollutants by the white rot basidiomycete *Trametes trogii*. *International Biodeterioration and Biodegradation*, 52, 1-5.

[52] Grassi, E., Scodeller, P., Filiel, N., Carballo, R. & Levin, L. (2011). Potential of *Trametes trogii* culture fluids and its purified laccase for the decolorization of different types of recalcitrant dyes without the addition of redox mediators. *International Biodeterioration and Biodegradation*, 65, 635-643.

[53] Levin, L., Melignani, E. & Ramos, A.M. (2010). Effect of nitrogen sources and vitamins on ligninolytic enzyme production by some white-rot fungi. Dye decolorization by selected culture filtrates. *Bioresource Technology*, 101, 4554-4563.

[54] Levin, L., Forchiassin, F. & Viale, A. (2005). Ligninolytic enzyme production and dye decolorization by *Trametes trogii*. Application of Plackett-Burman experimental design to evaluate nutritional requirements. *Process Biochemistry*, 40, 1381-1387.

[55] Champagne, P.P. & Ramsay, J.A. (2005). Contribution of manganese peroxidase and laccase to dye decoloration by *Trametes versicolor*. *Applied Microbiology and Biotechnology*, 69, 276-285.

[56] Levin, L., Herrmann, C. & Papinutti, L. (2008). Optimization of lignocellulolytic enzyme production by the white-rot fungus *Trametes trogii* in solid-state fermentation using response surface methodology. *Biochemical Engineering Journal*, 39, 207-214.

[57] Trupkin, S., Levin, L., Forchiassin, F. & Viale, A. (2003). Optimization of a culture media for ligninolytic enzyme production and synthetic dye decolorization using response surface methodology. *Journal of Industrial Microbiology and Biotechnology*, 30, 682-690.

[58] Papinutti, V.L., Diorio, L.A. & Forchiassin, F. (2003). Production of laccase and manganese peroxidase by *Fomes sclerodermeus* grown on wheat bran. *Journal of Industrial Microbiology and Biotechnology*, 30, 157-160.

[59] Papinutti, V., Mouso, N. & Forchiassin, F. (2006). Removal and degradation of the fungicide dye malachite green from aqueous solution using the system wheat bran-*Fomes sclerodermeus*. *Enzyme and Microbial Technology*, 39, 848-853.

[60] Rani, P., Kalyani, N. & Prathiba, K. (2008). Evaluation of lignocellulosic waste for production of edible mushroom. *Applied Biochemistry and Biotechnology*, 151, 151-159.

[61] Lechner, B.E. & Papinutti, V.L. (2006). Production of lignocellulosic enzymes. *Process Biochemistry*, 41, 594-598.

[62] Orrego, C.E., Montoya, S. & Levin, L. (2010). Lignocellulolytic enzyme production by *Grifola frondosa* during solid state fermentation on two formulations based on coffee spent-ground and oak-sawdust. The 32nd Symposium on Biotechnology for Fuels and Chemicals (FL,USA) at http://sim.confex.com/sim/32nd/webprogram/Paper14036.html

[63] Papinutti, L. & Forchiassin, F. (2010). Adsorption and decolorization of dyes using solid residues from *Pleurotus ostreatus* mushroom production. *Biotechnology and Bioprocess Engineering*, 15, 1102-1109.

[64] Denizli, A., Cihangir, N., Tüzmen, N. & Alsancak, G. (2005). Removal of chlorophenols from aquatic systems using the dried and dead fungus *Pleurotus sajor caju*. *Bioresource Technology,* 96, 59-62.

[65] Mai, C., Kues, U. & Militz, H. (2004). Biotechnology in the wood industry. *Applied Microbiology and Biotechnology*, 63, 477-494.

[66] Bajpai, P. (1999). Application of enzymes in the pulp and paper industry. *Biotechnology Progress*, 15, 147-157.

[67] Bajpai, P. (2004). Biological bleaching of chemical pulps. *Critical Reviews in Biotechnology*, 24, 1-58.

[68] Valls, C. & Roncero, M.B. (2009). Using both xylanase and laccase enzymes for pulp bleaching. *Bioresource Technology,* 100, 2032-2039.

[69] Ali, M. & Sreekrishnan, T.R. (2001). Aquatic toxicity from pulp and paper mill effluents: a review. *Advances in Environmental Research*, 5, 175-181.

[70] Messner, K., Koller, K., Wall, M.B., Akhtar, M. & Scott, G.M. Fungal treatment of wood chips for chemical pulping. In: Young, R. & Akhtar, M., *Environmentally friendly technologies for the Pulp and Paper Industry*, New York: Wiley; 1998; 385-419.

[71] Levin, L., Villalba, L., Da Re, V., Forchiassin, F. & Papinutti, L. (2007). Comparative studies of loblolly pine biodegradation and enzyme production by Argentinean white rot fungi focused on biopulping processes. *Process Biochemistry*, 42, 995-1002.

[72] Ibarra, D., Camarero, S., Romero, J., Martínez, M.J. & Martínez, A.T. (2006). Integrating laccase mediator treatment into an industrial-type sequence for totally chlorine-free bleaching of eucalypt kraft pulp. *Journal of Chemical Technology and Biotechnology*, 81, 1159-1165.

[73] Sigoillot, C., Record, E., Belle, V., Robert, J.L., Levasseur, A., Punt, P.J., van den Hondel, C.A.M.J.J., Fournel, A., Sigoillot, J.C. & Asther, M. (2004). Natural and recombinant fungal laccases for paper pulp bleaching. *Applied Microbiology and Biotechnology*, 64, 346-352.

[74] Moldes, D., Díaz, M., Tzanov, T. & Vidal, T. (2008). Comparative study of the efficiency of synthetic and natural mediators in laccase-assisted bleaching of eucalyptus kraft pulp. *Bioresource Technology,* 99, 7959-7965.

[75] Moreira, M.T., Feijoo, G., Canaval, J. & Lema, J.M. (2003). Semipilot-scale bleaching of kraft pulp with manganese peroxide. *Wood Science and Technology*, 37, 117-123.

[76] Bao, W., Fukushima, Y., Jensen Jr, K.A., Moen, M.A. & Hammel, K.E. (1994). Oxidative degradation of non-phenolic lignin during lipid peroxidation by fungal manganese peroxidase. *FEBS Letters*, 354, 297-300.

[77] Bermek, H., Li, K. & Eriksson, K.E. (2002). Studies on mediators of manganese peroxidase for bleaching of wood pulps. *Bioresource Technology,* 85, 249-252.

[78] Morozova, O.V., Shumakovich, G.P., Shleev, S.V. & Yaropolov, Y.I. (2007). Laccase-mediator systems and their applications: a review. *Applied Biochemical Microbiology,* 43, 523-535.

[79] Moldes, D. & Vidal, T. (2008). Laccase–HBT bleaching of eucalyptus kraft pulp: influence of the operating conditions. *Bioresource Technology,* 99, 65-85.

[80] Da Re, V., Papinutti, L., Forchiassin, F. & Levin, L. (2010). Biobleaching of loblolly pine kraft pulp with *Trametes trogii* culture fluids followed by a peroxide stage. Application of Doehlert experimental design to evaluate process parameters. *Enzyme and Microbial Technology*, 46, 281-286.

[81] Johannes, C. & Majcherczyk, A. (2000). Natural mediators in the oxidation of polycyclic aromatic hydrocarbons by laccase mediator systems. *Applied and Environmental Microbiology*, 66, 524-528.

[82] Papinutti, L., Dimitriu, P. & Forchiassin, F. (2008). Stabilization studies of *Fomes sclerodermeus* laccases. *Bioresource Technology,* 99, 419-424.

[83] Galliano, H., Gas, G., Seris, J.L. & Boudet, A.M. (1991). Lignin degradation by *Rigidoporus lignosus* involves synergistic action of two oxidizing enzymes: Mn peroxidase and laccase. *Enzyme and Microbial Technology*, 13, 478-482.

[84] Ragauskas, A. Mill designed biobleaching technologies. In: Technical Report Number: DE-FC36-99GO10374. 2004. DOI 10 2172/828216 2004.

[85] Da Re, V., Papinutti, L., Villalba, L., Forchiassin, F. & Levin, L. (2008). Preliminary studies on the biobleaching of loblolly pine Kraft pulp with *Trametes trogii* crude extracts. *Enzyme and Microbial Technology*, 43, 164-168.

[86] Wu, J., Xiao, Y.Z. & Yu, H.Q. (2005). Degradation of lignin in pulp mill wastewaters by white-rot fungi on biofilm. *Bioresource Technology,* 96, 1357-1363.

[87] Font, X., Caminal, G., Gabarrell, X., Romero, S. & Vicent, M.T. (2003). Black liquor detoxification by laccase of *Trametes versicolor* pellets. *Journal of Chemical Technology and Biotechnology*, 78, 548-554.

[88] Da Re, V. & Papinutti, L. (2011). Black liquor decolorization by selected white-rot fungi. *Applied Biochemistry and Biotechnology*, DOI 10.1007/s12010-011-9260-9.

[89] Blanquez, P., Sarrà, A. & Vicent, T. (2008). Development of a continuous process to adapt the textile wastewater treatment by fungi to industrial conditions. *Process Biochemistry*, 43, 1-7.

[90] Anastasi, A., Spina, F., Prigione, V., Tigini, V., Giansanti, P. & Varese, G.C. (2010). Scale up of a bioprocess for textile wastewater treatment using *Bjerkandera adusta. Bioresource Technology,* 101, 3067-3075.

[91] Moreira, M.T., Palma, C., Mielgo, I., Feijoo, G. & Lema, J.M. (2001). *In vitro* degradation of a polymeric dye (Poly R-478) by manganese peroxidase. *Biotechnology and Bioengineering*, 75, 362-368.

[92] Kokol, V., Doliska, A., Eichlerova, I., Baldrian, P. & Nerud, F. (2007). Decolorization of textile dyes by whole cultures of *Ischnoderma resinosum* and by purified laccase and Mn-peroxidase. *Enzyme and Microbial Technology*, 40, 1673-1677.

[93] Jordaan, J., Pletschke, B.I. & Leukes, W.D. (2004). Purification and partial characterization of a thermostable laccase from an unidentified basidiomycete. *Enzyme and Microbial Technology*, 34, 635-641.

[94] Levin, L., Papinutti, L. & Forchiassin, F. (2004). Evaluation of Argentinean white rot fungi for their ability to produce lignin-modifying enzymes and decolorize industrial dyes. *Bioresource Technology,* 94, 169-176.

[95] Papinutti, L. & Forchiassin, F. (2004). Modification of malachite green by *Fomes sclerodermeus* and reduction of toxicity to *Phanerochaete chrysosporium. FEMS Microbiology Letters*, 231, 205-209.

[96] Diorio, L., Mercuri, A., Nahabedian, D. & Forchiassin, F. (2008). Development of a bioreactor system for the decolorization of dyes by *Coriolus versicolor f. antarcticus. Chemosphere*, 72, 150-156.

[97] Baldrian, P. (2004). Purification and characterization of laccase from the white rot fungus *Daedalea quercina* and decolorization of synthetic dyes by the enzyme. *Applied Microbiology and Biotechnology*, 63, 560-563.

[98] Garzillo, M., Colao, M.C., Buonocore, V., Oliva, R., Falcigno, L., Saviano, M., Santoro, A.M., Zappala, R., Bonomo, R.P., Bianco, C., Giardina, P., Palmieri, G. & Sannia, G. (2001). Structural and kinetic characterization of native laccases from *Pleurotus ostreatus*, *Rigidoporus lignosus* and *Trametes trogii. Journal of Protein Chemistry*, 20, 191-201.

[99] Li, K., Xu, F. & Eriksson, K.E.L. (1999). Comparison of fungal laccases and redox mediators in oxidation of a nonphenolic lignin model compound. *Applied and Environmental Microbiology*, 65, 2654-2660.

[100] Archibald, F.S. (1992). A new assay for lignin-type peroxidases employing the dye Azure B. *Applied and Environmental Microbiology*, 58, 3110-3116.

[101] Camarero, S., Ibarra, D., Martínez, M.J. & Martínez, A. (2005). Lignin-derived compounds as efficient laccase mediators for decolorization of different types of recalcitrant dyes. *Applied and Environmental Microbiology*, 71, 1775-1784.

[102] Alam, M.Z., Mansor, M.F. & Jalal, K.C.A. (2009). Optimization of decolorization of methylene blue by lignin peroxidase enzyme produced from sewage sludge with *Phanerochaete chrysosporium. Journal of Hazardous Materials*, 162, 708-715.

[103] Hatvani, N. & Mecs, I. (2003). Effects of certain heavy metals on the growth, dye decolorization, and enzyme activity of *Lentinula edodes. Ecotoxicology and Environmental Safety*, 55, 199-203.

[104] Boopathy, R. (2000). Factors limiting bioremediation technologies. *Bioresource Technology,* 74, 63-67.

[105] Perullini, M., Jobbagy, M., Mouso, N., Forchiassin, F. & Bilmes, S.A. (2010). Silica-alginate-fungi biocomposites for remediation of polluted water. *Journal of Materials Chemistry*, 20, 6479-6483.

[106] Kim, J., Jia, H. & Wang, P. (2006). Challenges in biocatalysis for enzyme-based biofuel cells. *Biotechnology Advances*, 24, 296-308.

[107] Szamocki, R., Flexer, V., Levin, L., Forchiassin, F. & Calvo, E.J. (2009). Oxygen cathode based on a layer-by-layer self-assembled laccase and osmium. *Electrochimica Acta*, 53, 6740-6752.

[108] Scodeller, P., Carballo, R., Szamocki, R., Levin, L., Forchiassin, F. & Calvo, E. (2010). Layer by-layer self assembled osmium polymer-mediated laccase oxygen cathodes for biofuel cells: The role of hydrogen peroxide. *Journal of the American Chemical Society*, 132, 11132-11140.

In: Fungi: Types, Environmental Impact and Role in Disease
Editors: Adolfo Paz Silva and María Sol
ISBN: 978-1-61942-671-9

Chapter XXII

White-Rot Fungi as Promising Bioremediation Agents

***Margarida S. Roriz*[1] *and Susana Rodríguez-Couto*[2, 3*]**
[1] Francisco Vaz da Costa Marques, Filhos & Ca., S.A., Rua do Pombal, Guimarães, Portugal
[2] CEIT, Unit of Environmental Engineering, Paseo Manuel de Lardizábal 15, San Sebastian, Spain
[3] IKERBASQUE, Basque Foundation for Science, Bilbao, Spain

Abstract

Environmental pollution with hazardous xenobiotic compounds has become one of the major ecological problems world-wide. In addition, the implementation of more stringent environmental regulations, especially in the most developed countries, has impelled the necessity of finding innovative and environmentally-friendly treatment technologies to complement or substitute the conventional ones.

White-rot fungi are the only micro-organisms able to degrade the whole wood components and are by far the most efficient ligninolytic micro-organism described to date. This ability to degrade lignin is related to the secretion of a non-specific and non-stereo selective extracellular enzymatic system. This enzymatic system consists mainly of oxidative enzymes like lignin peroxidases (LiP; E.C. 1.11.1.14), manganese-dependent peroxidases (MnP; E.C.1.11.1.13), laccases (EC 1.10.3.2) and hydrogen peroxide-generating oxidases. The same unique non-specific mechanisms that give these fungi the ability to degrade lignin also allow them degrading a wide range of recalcitrant pollutants such as polycyclic aromatic hydrocarbons (PAHs), pesticides, fuels, alkanes, polychlorinated biphenyls (PCBs), explosives and synthetic dyes. This chapter deals with the developments made for the use of white-rot fungi in the treatment of pollutants and their future perspectives.

---

* E-mail address: srodriguez@ceit.es; phone: +34943212800 (ext. 2239); fax: +34943213076.

Introduction

Environmental pollutants are a serious concern world-wide because of the hazards they pose to the health of humans and animals. Around 80 billion pounds of hazardous organo-pollutants are produced annually by the chemical, agricultural, oil, paper, textile, aerospace and other industries [1]. Only about 10% of these wastes are believed to be disposed of in an environmentally safe manner [2]. Traditional methods of disposing of hazardous wastes such as physical, chemical, and thermal treatments and land filling are not always efficacious because of the high cost, sludge formation and disposal problems. One method that is becoming increasingly popular for environment decontamination is bioremediation.

Recent research on bioremediation of environmentally relevant chemicals has centred on four important aspects: the characterisation of the biodegradation processes useful for the treatment of xenobiotic compounds, the development of technical protocols for increasing the degradation rates and substrate ranges of enzymes from micro-organisms, the design and engineering of bioreactor systems and bio-treatment strategies to optimise biodegradation processes and development of information on the ecological and human health risks associated with exposure to the chemicals [3].

The use of indigenous or suitable introduced micro-organisms at contamination sites often provides an efficient and economically attractive solution to the pollution problem. One of the earliest reports on the above-mentioned topic indicated that lignin-degrading white-rot fungi, as exemplified by *Phanerochaete chrysosporium*, can degrade an extremely diverse group of environmental pollutants [4]. Since then, there has been intense world-wide research to assess the true potential of white-rot fungi in bioremediation. The ability of white-rot fungi to degrade a wide spectrum of environmental pollutants sets them apart from many other microbes used in bioremediation.

At present bioremediation conducted on a commercial scale utilises prokaryotes, with comparatively few recent attempts to use white-rot fungi. These filamentous fungi, however, offer advantages over bacteria in the diversity of compounds they are able to oxidise [5]. In addition, they are robust organisms and are generally more tolerant to high concentrations of pollutants than bacteria [6].

This chapter will focus on the developments made for biodegradation of real industrial wastewater and real polluted soils by white-rot fungi.

White-Rot Fungi

Three types of fungi are involved in the biodegradation of wood polymers named white-rot, brown-rot and soft-rot fungi, depending on the type of decay they cause. The two formers belong mostly to the basidiomycetes, whereas the latter are ascomycetes. White-rot fungi are able to decompose all wood polymers, including lignin, which leaves the wood with a white, fibrous appearance. Brown-rot fungi efficiently degrade wood polysaccharides and are capable to modify lignin only to a certain extent, mostly by demethoxylation, which leaves the wood brown, dry and with poor strength [7]. Soft-rot fungi typically attack wood in wet environments. As a result, cavities or complete erosion in the secondary cell walls of the wood attacked is detected and the decayed wood loses its mechanical strength due to cellulose breakdown [8, 9]. Soft-rot fungi can degrade lignin to some extent, especially in hardwood. They are able to cause demethylation of lignin and possibly degrade the side chains and aromatic rings, although the extent of degradation is low. Therefore, the brown and soft-rot fungi cause only limited degradation of lignin.

In nature, white-rot fungi live on woody tissues that are composed mainly of three biopolymers: cellulose, hemicellulose and lignin. Lignin is, after cellulose, the second most abundant renewable biopolymer in nature. It is an essential part of the cell wall in plants, imparting rigidity and protecting the easily degradable cellulose from attack by pathogens. Lignin is a heterogeneous, complex and stable polymer composed of various aromatic monomers connected to each other by various bonds carbon-carbon and ethers [10] (Figure 1). Due to its complicated structure and non-hydrolysable bonds, lignin is more difficult to break down than cellulose or hemicellulose. The molecular mass of lignin is high, about 100 kDa or more, which prevents its uptake inside the microbial cell [11]. Thus, the biological degradation of macromolecular lignin must occur through the activity of extracellular enzymes. The physiological importance of lignin biodegradation is the destruction of the matrix it forms, so that the micro-organism can gain better access to the real substrates: hemicellulose and cellulose [12], from where they obtain energy.

Due to the similarity between the lignin structure and the chemical structure of several recalcitrant pollutants, the use of white-rot fungi and/or their enzymes for the degradation of such pollutants have been considered [13, 14]. Thus, the same mechanisms that give lignin-degrading fungi the ability to degrade lignin can also be used to degrade a wide range of recalcitrant pollutants such as polycyclic aromatic hydrocarbons (PAHs), pesticides, fuels, alkanes, polychlorinated biphenyls (PCBs), explosives and synthetic dyes. This characteristic is the greatest advantage of the use of white-rot fungi in bioremediation, since a mixture of different pollutants are usually found in most contaminated sites [15].

Along with their ability to degrade complex chemicals, white-rot fungi have different advantages that are not associated with other bioremediation systems:

- Their extracellular enzymatic system allows them tolerating high concentration of xenobiotics as well as to degrade very insoluble components as lignin and several hazardous environmental pollutants such as PAHs.
- The non-specific nature of their enzymatic system enables them to degrade complex mixtures of pollutants such as creosote to carbon dioxide.
- Their degrading system is usually induced by nutrient depletion and not by a particular pollutant. This is important because repression of enzyme synthesis does not occur when the concentration of a chemical is too low for effective enzyme induction [16].
- They can effectively degrade very low concentrations of a pollutant to non-detectable or nearly non-detectable levels. The effective decontamination of very low concentrations of chemicals is important because government regulatory agencies are continually lowering the maximum permissible levels of the environmental pollutants [16].
- very inexpensive growth substrates such as corncobs or other agro-food and forestry wastes can be used to cultivate white-rot fungi in soil. The lignin degrading system of the white-rot fungi allows them to access the limited carbon source from such

substrates. Other soil microbes are less able to use these substrates. Therefore, white-rot fungi maintain a competitive advantage [16].

- Fungal treatment requires a long treatment time, but the energy requirement of the process is low and the treatment conditions are mild [17].

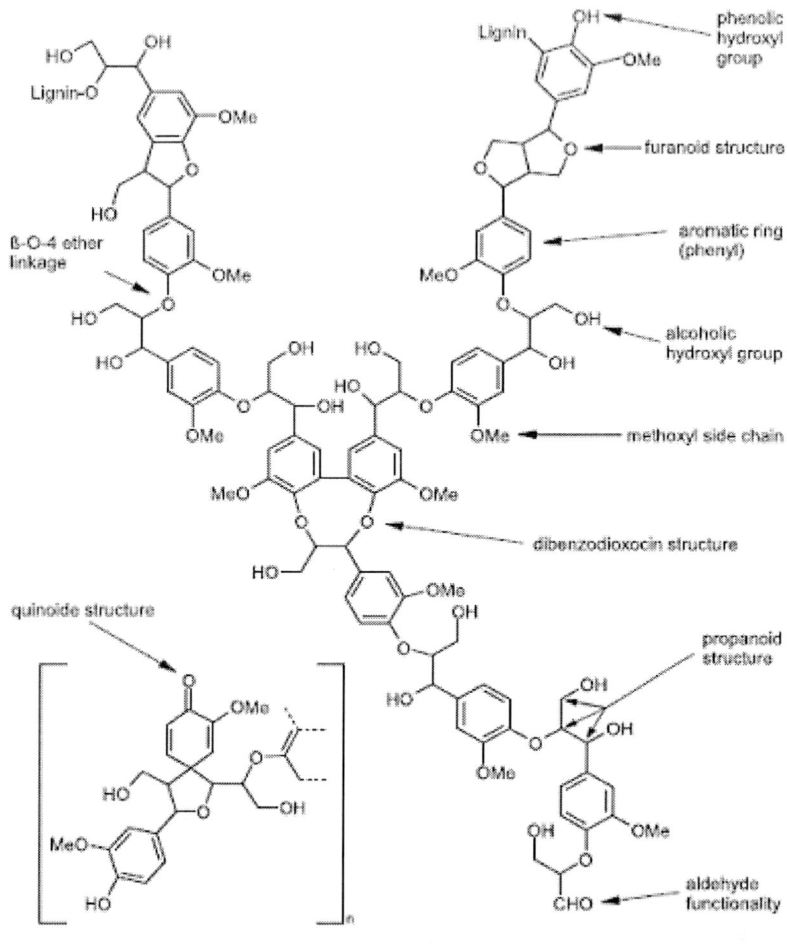

Figure 1. Structural model of lignin by Brunow [144] (Brunow, G. In M. Hofrichter & A. Steinbüchel (Eds.): Methods to reveal the structure of lignin, pp. 89-116, 2001. Copyright Wiley-VCH Verlag GmbH & Co. KGaA. Reproduced with permission).

Enzymatic System of White-Rot Fungi

White-rot fungi secrete extracellular enzymes that are essential for lignin degradation. Such enzymes are often referred to as lignin-modifying enzymes (LMEs). Table 1 summarises the ligninolytic enzymes and their substrates and reactions [18]. The main extracellular enzymes secreted by white-rot fungi comprise two glycosylated heme-containing peroxidases, lignin peroxidase (LiP, E.C. 1.11.1.14) and manganese-dependent peroxidase (MnP, E.C. 1.11.1.13) [19], and a copper-containing phenoloxidase, laccase (E.C.

1.10.3.2) [20]. Some authors have also reported Mn-independent peroxidase (MiP) and versatile peroxidase (VP) activities in some white-rot fungi [21-23]. In addition, there are several enzymes associated with LMEs in lignin breakdown, although they are unable to degrade lignin alone, such as glyoxal oxidase (E.C. 1.2.3.5), superoxide dismutase (E.C. 1.15.1.1), glucose oxidase (E.C. 1.1.3.4), aryl alcohol oxidase (E.C. 1.1.3.7) and cellobiose dehydrogenase (E.C. 1.1.99.18). They produce H_2O_2 required by peroxidases or serve to link lignocellulose degradation pathways [24].

Table 1. Ligninolytic enzymes and their main reactions (extracted from Hatakka [18]. In M. Hofrichter & A. Steinbüchel (Eds.): Biopolymers, pp. 129-180, 2001. Copyright Wiley-VCH Verlag GmbH & Co. KGaA. Reproduced with permission)

Enzyme and abbreviation	Cofactor	Substrate, mediator	Reaction
Lignin peroxidase, LiP	H_2O_2	Veratryl alcohol	Aromatic ring oxidised to cation radical
Manganese peroxidase, MnP	H_2O_2	Mn, organic acids as chelators, thiols, unsaturated fatty acids	Mn(II) oxidised to Mn(III); chelated Mn(III) oxidises phenolic compounds to phenoxyl radicals; other reactions in the presence of additional compounds
Laccase	O_2	Phenols, mediators, e.g., HBT or ABTS	Phenols are oxidised to phenoxyl radicals; other reactions in the presence of mediators

LME production by white-rot fungi occurs during secondary metabolism and is induced by limited nutrient levels, particularly nitrogen. Some taxa have, however, been shown to produce LiP, MnP and laccase under conditions of nitrogen sufficiency [25]. Also, growth conditions such as temperature and agitation of fungal cultures significantly affect the appearance and levels of activity of LMEs in white-rot fungi [26-28]. Additionally, other authors have reported that LME production by white-rot fungi is affected by mediator compounds, various other chemicals and required-metal (Mn^{2+}, Cu^{2+}) concentrations [29-31].

Lignin peroxidase

Lignin peroxidase (LiP, E.C. 1.11.1.14) is a glycosylated heme-containing peroxidase with a molecular mass of 40-45 kDa. It was first found in the white-rot fungus *P. chrysosporium*. In the presence of endogenously generated hydrogen peroxide LiP catalyses the oxidation of non-phenolic aromatic structures in lignin and generates aryl cation radicals. LiP is secreted by the white-rot fungi extracellularly, mostly at the onset of secondary metabolism, triggered by nitrogen limitation [32]. However, it can be also produced in nitrogen-rich conditions [33].

During its catalytic cycle (Figure 2) LiP is oxidised by H_2O_2 to form a two-electron oxidised intermediate (compound I) which in turn oxidises substrates by removing one electron and producing a more reduced enzyme intermediate (compound II) and a substrate radical. This intermediate can then oxidise substrates by one electron, returning the enzyme to its initial state. However, compound II has a very high reactivity with H_2O_2, therefore in the presence of a poor substrate and excess H_2O_2 is instead converted to an inactive form of the enzyme (compound III) [34, 35]. Compounds such as veratryl alcohol and tryptophane have been shown to exert a protective effect against the enzyme inactivation by excess of H_2O_2 [36] and when present, they are more favourable substrates for compound II and convert it into the resting enzyme, completing the catalytic cycle. An additional role for veratryl alcohol and tryptophane as diffusible mediators in the LiP-catalysed oxidation of environmental contaminants has been proposed [33, 37].

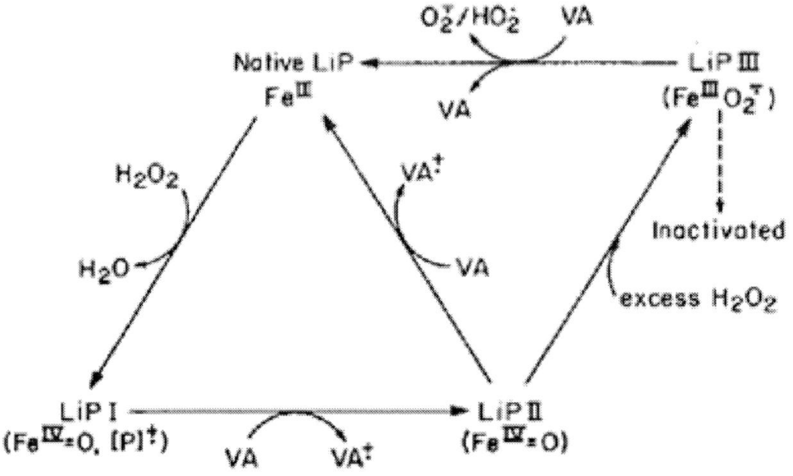

Figure 2. The catalytic cycle of lignin peroxidase (Wariishi & Gold [35], reproduced with kind permission of Elsevier Ltd.).

Manganese-dependent peroxidase

Manganese-dependent peroxidase (MnP, E.C. 1.11.1.13) is an extracellular heme-containing peroxidase that catalyses an H_2O_2-dependent oxidation of Mn^{2+} to highly reactive Mn^{3+}, which is stabilised by fungal chelators such as oxalic acid and oxidises phenolic components of lignin resulting in the formation of free radicals. MnP is often produced in multiple isoforms exhibiting molecular masses of 45-55 kDa [38-40]. These isoforms differ mostly in their isoelectric points, which are usually rather acidic (pH 3-4) [41, 42], but neutral MnPs have also been found. The expression of MnP in fungal cultures is regulated at the level of gene transcription by H_2O_2 and various chemicals including ethanol, 2,4-dichlorophenol as well as by Mn^{2+} concentration and heat shock [30, 43, 44].

The catalytic cycle of MnP (Figure 3) is similar to that of other heme peroxidases like LiP. It includes the native ferric enzyme as well as the reactive intermediates (compound I and compound II) [45]. In contrast to other peroxidases, MnP uses Mn^{2+} as the preferred substrate. The cycle starts by binding of H_2O_2 to the native ferric enzyme and the formation of an ironperoxide complex [45]. The subsequent cleavage of the peroxide oxygen-oxygen bond requires a two-electron transfer from the heme resulting in the formation of MnP compound I. Afterwards one water molecule is released. A subsequent reduction proceeds through MnP compound II. A mono-chelated Mn^{2+} ion acts as the one-electron donor for this enzyme intermediate and is oxidised to Mn^{3+}. The reduction of MnP compound II proceeds in a similar way and another Mn^{3+} is formed from Mn^{2+} leading to the generation of the native enzyme and the release of a second molecule of water [45].

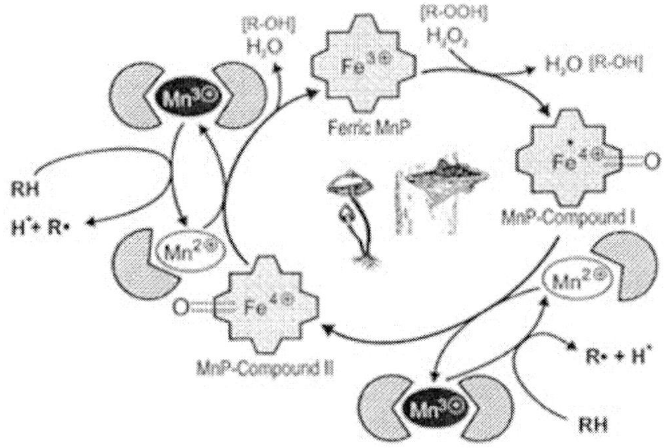

Figure 3. The catalytic cycle of manganese-dependent peroxidase (MnP) (Hofrichter [45], reproduced with kind permission of Elsevier Ltd.).

Whereas MnP compound I resembles that of LiP and can, besides Mn^{2+}, be reduced by other electron donors (ferrocyanide, phenolics), MnP compound II is reduced by other substrates only very slowly and requires Mn^{2+} to complete the catalytic cycle. The Mn^{3+} formed during the cycle is stabilised by carboxylic acids (oxalate, malonate, malate, tartrate or lactate). The chelates of Mn^{3+} cause one-electron oxidations of various substrates. MnP is, similarly to LiP, sensitive to high concentrations of H_2O_2 that cause reversible inactivation of the enzyme by forming MnP compound III.

Laccase

Laccase (E.C. 1.10.3.2; benzendiol:oxygen oxidoreductase) is one of the oldest known enzymes. It was originally found in the exudate of the Japanese lacquer tree *Rhus vernicifera* more than one hundred years ago [46]. Laccases are known to be produced by many fungi, particularly in white-rot fungi, in multiple isoforms with a typical molecular mass of 60-80 kDa and an acidic p*I* value of 3-6 [20, 47]. Additionally to lignin degradation, fungal laccases are also involved in other processes such as the formation of fruiting bodies, synthesis of melanin and other pigments, sporulation, conidiation and plant pathogenesis [48, 49].

Typical fungal laccases contain four copper atoms located in two centres (T1, T2/T3). The T1 site contains the mononuclear "blue" copper, while the T2/T3 site contains one T2 copper and two T3 copper ions [50, 51]. The copper atoms are coordinated by ten conserved histidine residues and one conserved cysteine residue. Some "non-blue" fungal laccases (*Pleurotus ostreatus* POXA1w, *Phellinus ribis* laccase) are reported to harness other metals (Zn, Fe, Mn) instead of some of the mentioned copper atoms [52, 53]. It is assumed that four electrons are transferred from the T1 reducing substrate-binding site to the T2/T3-copper site during redox reactions [50]. The electron migration between the substrate and the enzyme is regulated by the T1 site, also known as blue copper site [50], and, thus, depends on its redox potential. In fungal laccases, oxidation potentials of the blue copper vary between 0.46 V and 0.79 V. Figure 4 shows the catalytic cycle of laccase enzyme [54]. The first step is the reduction of the reducing substrate at the T1 site [55-57]. The electrons extracted from the reducing substrate are transferred to the T2/T3 trinuclear site, resulting in the conversion of the resting form (fully oxidised) of the enzyme to a fully reduced state. A successive four electron oxidation (from four substrate molecules) is required to fully reduce the enzyme.

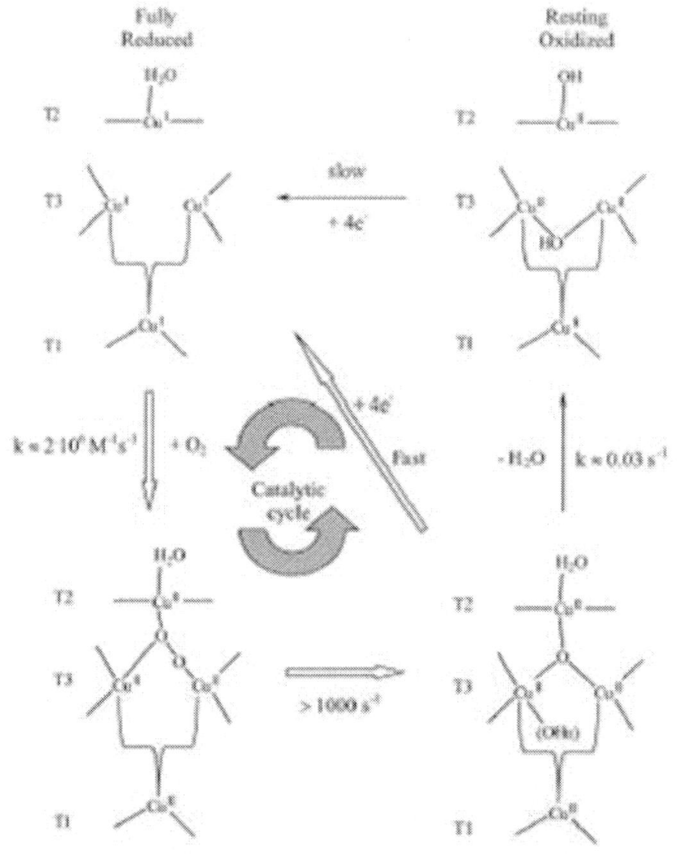

Figure 4. Catalytic cycle of a four-copper laccase (Shleev et al.[54], reproduced with kind permission of Elsevier Ltd.).

Laccases are remarkably non-specific as to their reducing substrates. Thus, they oxidise different compounds such as phenols, polyphenols and aromatic amines with the concomitant reduction of molecular oxygen to water. This rather broad substrate spectrum can be additionally extended by the presence of redox mediators [58, 59]. The basis of the laccase mediator system (LMS) is the use of low-molecular weight compounds that are oxidised by laccase to organic radicals or positively charged intermediates, which, in turn, act as redox mediators [60]. The charged mediators are capable of oxidising compounds that are not substrates for laccase and, furthermore, it is assumed that they can diffuse away from the fungal hyphae and oxidise larger molecules like lignin moieties in wood [61]. Examples of laccase mediators are 2,2'-azinobis-(3-ethylbenzthiazoline- 6-sulphonate (ABTS), 1-hydroxybenzotriazole (HBT), and 3-hydroxyanthranilate (3-HAA), of which the last one is found to occur naturally in the cultures of *Pycnoporus cinnabarinus* [60, 62]. Also some phenolic lignin precursors and degradation products might act as laccase mediators in nature [63].

The production of laccase activity by white-rot fungi can be improved by the addition of Cu^{2+}, which regulates laccase production at the level of gene transcription [64-66], to the cultures. Apart from copper, other heavy metal ions (Cd^{2+}, Ag^{2+} and Hg^{2+}) have been reported as strong inducers of laccase activity or expression of laccase transcripts [31, 67, 68]. Also Mn^{2+} has been shown to regulate the laccase transcript levels [69]. Several other chemicals such as 2,5-xylidine, veratryl alcohol and guaiacol have also been shown to have an inducing effect on laccase production [62, 70].

Bioremediation Potential of White-Rot Fungi

The industrialisation of agriculture, the rapid growth of the chemical industry, military activities and the necessity to generate economical forms of energy have caused the continuous release of organic chemicals into the biosphere [71]. Soil and aquatic environment are natural and preferential sinks for contamination, their pollution representing an important concern for the environment and human health [71].

The finding that the ligninolytic enzymes secreted by white-rot fungi were able to transform a wide variety of organic compounds with structures similar to lignin has impelled research oriented towards the practical application of such fungi in bioremediation [72].

Soil bioremediation

There is growing public concern about the wide range of xenobiotic compounds being introduced inadvertently or deliberately into soil. Many xenobiotic compounds have medium to long-term stability in soil and their persistence results in significant impact on the soil ecosystem as well as pose a threat to human health. Figure 5 compares the persistence of several xenobiotic compounds in soil [73].

Soil contamination is classified as either localised or diffuse contamination [74]. The sources of localised contamination are frequently related to industrial discharges, improper waste disposal or accidental spills during transportation or handling of hazardous compounds. Diffuse contamination is produced mostly by intensive agriculture and forestry practices, transportation and industrial activities.

Bioremediation of soil is considered a low-cost alternative to other remediation techniques. Since degradation of xenobiotic compounds in soil by bacteria has shown to be difficult, new techniques for degrading them are required. One possibility is the use of white-rot fungi. However, a successful development of this technique requires that other matters are taken into account such as fungal colonisation of the soil and effects of the fungi on indigenous soil micro-organisms.

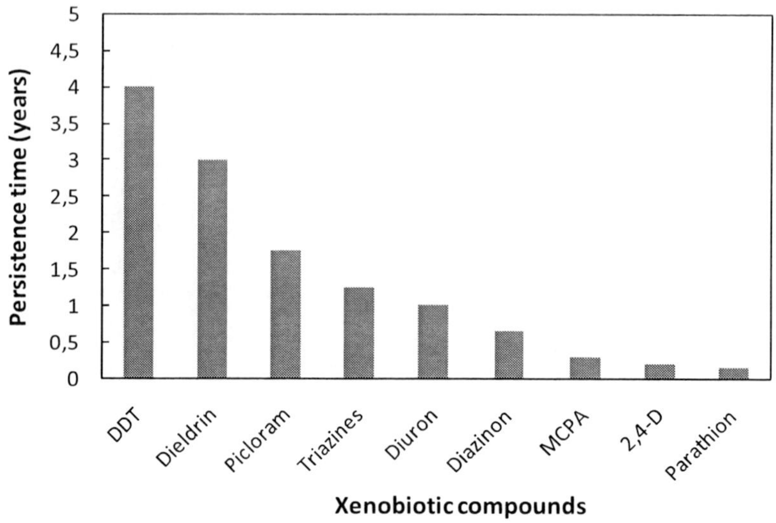

Figure 5. Relative persistence levels of different xenobiotic compounds in soil (extracted from Magan et al. [73]).

Table 2. Use of the white-rot fungi in pollutant degradation in field soil

White-rot fungus	Pollutant	Removal	Reference
P. chrysosporium *P. ostreatus* *C. versicolor*	Oil	68.7% in 12 months 53.1% in 12 months 78.1% in 12 months	[75]
P. ostreatus	Creosote	86% of total 16 PAHs; 89% of 3-ring PAHs; 87% of 4-ring PAHs; 48% of 5-ring PAHs.	[76]
P. ostreatus	Creosote	Reference soil and soil pretreated without fertiliser: 1-10% Soil pretreated with fertiliser: 18-30%	[77]
P. chrysosporium *T. versicolor* *P. ostreatus*	Delor 103	0% 0% 40% in 2 months	[78]
P. tuberregium	Crude oil	---	[79]
T. versicolor	PCP	Decreased from 800 to 50 mg/Kg after 74 weeks	[80]
T. versicolor	PCP	70% after 7 weeks	[81]
P. ostreatus *I. lacteus*	PAHs	3- and 4-ring PAHs removal: 73% and 58% depending on the soil higher or similar to *P. ostreatus* depending on the soil	[82]
P. ostreatus *I. lacteus*	Creosote	55-67% 27-36%	[83]
T. versicolor	Athrazine	98%	[84]

Table 2 shows several papers on the treatment of real polluted soils with different white-rot fungi. Yateem et al. [75] tested the ability of the white-rot fungi *P. chrysosporium, P.*

ostreatus and *Coriolus versicolor* to degrade oil in contaminated soil. They studied the effect of the strain used, the inoculum concentration and the addition of nitrogen on bioremediation efficiency and found that *C. versicolor* showed the highest degradation rate under nitrogen non-limited conditions. Thus, the total petroleum hydrocarbon (TPH) concentration was reduced by 78.1% in 12 months in the soil treated with *C. versicolor*. *P. chrysosporium* showed a similar degradability but under nitrogen-limiting conditions.

Eggen [76] studied the degradation of PAHs in aged-creosote polluted soil by two commercial *P. ostreatus* species. They found that both the source of inoculum and the type of PAHs to treat were important variables. The two commercial fungi degraded 3-ring compounds similarly. However, for 4- and 5-ring compounds there was significant difference, spent mushroom compost being more effective than fungal substrate generated before mushroom production. They also found that the 4- and 5-ring compounds degraded better if soil and substrate were mixed rather than layered. Furthermore, the addition of fish oil led to the best degradation of PAHs. Thus, after 7 weeks of treatment the following removal was obtained: 86% of total 16 PAHs, 89% of 3-ring PAHs, 87% of 4-ring PAHs and 48% of 5-ring PAHs.

Eggen & Sveum [77] investigated the effect of inoculation of white-rot fungus (*P. ostreatus*), temperature and two pre-treatment methods on PAH degradation in aged-creosote contaminated soil. It was shown that *P. ostreatus* had an overall positive effect on PAH degradation and that degradation was enhanced by increased temperature and pre-treatment. In general, adding bark and incubating at 22°C before inoculation with *P. ostreatus* increased PAH degradation compared to no pre-treatment or pre-treatment with fertiliser. At low temperature (8°C) fungal inoculation had the best effect when fertiliser was not added and a significant effect on degradation on different groups of PAH compounds, except for the more easily degradable compounds, 3-ring PAHs and heterocyclic compounds was obtained. Pre-treatment with fertiliser stimulated microbial activity at low temperature and enhanced PAH degradation even without the addition of fungi.

Kubátová et al. [78] studied the ability of six strains of white-rot fungi to degrade low PCB commercial mixture (Delor 103) in a real soil system. They found that *P. chrysosporium* and *T. versicolor* were not able to degrade PCBs in soil. On the contrary, the four strains of *P. ostreatus* tested degraded about 40% of Delor 103 in 2 months. In addition, all *P. ostreatus* strains degraded PCBs selectively showing preference for congeners with chlorine atoms in orto>metha>para position. They also observed that degradation efficiency decreased with increasing number of chlorination.

Isikhuemhen et al. [79] examined the ability of the white-rot fungus *Pleurotus tuberregium* to ameliorate crude oil polluted soil. Their results showed that bioremediation of soil contaminated with crude oil was possible, especially when the fungus had been allowed to establish and fully colonise the substrate mixed with the soil.

Walter et al. [80] designed engineered soil cells to develop proof-of-concept biopiles for bioremediation of aged pentachlorophenol (PCP) contaminated soil with the New Zealand white-rot fungus *T. versicolor* isolate HR131. PCP degradation and fungal survival were monitored at regular intervals for 2.5 years. They reported that the experiments were reproducible and there was no effect of fungal inoculum concentration. PCP residues declined from 800-1000 mg/kg to 4 (0-9.4 mg/kg) according to first-order kinetics. The engineered soil cells proved to be functional and required little to no maintenance during the trial period.

Ford et al. [81] performed a study to improve the ability of indigenous New Zealand white-rot fungi (*T. versicolor* isolate HR131 and *Trametes* sp. isolate HR577) to remove PCP from contaminated field soil. The tested conditions were fungal growth substrate and co-substrate composition, culture age and Tween 80 addition to the contaminated soil. Maximum PCP removal was 70% after 7 weeks from a 1043 mg/kg PCP-contaminated soil inoculated with an 11-day-old fungal culture of *T. versicolor* isolate HR131. There was minimal production of undesirable pentachloroanisole by the fungi. The results showed that PCP biodegradation was affected by inoculum culture age, by the presence of a co-substrate amendment and by growth substrate composition after white-rot fungal bioaugmentation into PCP-contaminated field soils.

Leonardi et al. [82] reported the degradation of PAHs in historically contaminated soils, differing in PAH content and bioavailability, by the white-rot fungi *Irpex lacteus* and *P. ostreatus*. They found that both fungi performed better in the soil with higher pollutant bioavailability.

Byss et al. [83] determined the efficacy of selected basidiomycetes in the removal of PAH from creosote-contaminated soil. They found that *P. ostreatus* and *I. lacteus* degraded PAH with 4–6 aromatic rings more efficiently than the microbial community initially present in the soil. In general, PAH removal was higher in *P. ostreatus* treatments (55–67%) than in *I. lacteus* ones (27–36%).

Bastos & Magan [84] examined the feasibility of *T. versicolor* to actively degrade atrazine (0.5 mg/g) in non-sterile calcareous clay soil microcosms for up to 24 weeks (20ºC) under low water availability. They found that up to 98% and 85% (at -0.7 and -2.8 MPa, respectively) of atrazine was degraded in soil containing both the atrazine and the inoculant, compared to 96% and 50% in soil containing atrazine only. Thus, the contribution of *T. versicolor* to atrazine degradation was only significant under the driest soil treatment conditions.

In conclusion, white-rot fungi potentially represent a powerful tool for soil bioremediation with some species already patented [85]. Interestingly, only a few companies have incorporated ligninolytic fungi for soil remediation into their programme, e.g., "EarthFax Development Corp." in USA or "Gebruder Huber Bodenrecycling" in Germany [86].

Wastewater bioremediation

In view of the extensive contamination of the environment by recalcitrant pollutants originating from industrial wastewater, it is imperative to develop efficient methods for their remediation. In recent years, many researchers have indicated that white-rot fungi are promising micro-organisms in wastewater treatment, *P. chrysosporium* being the most investigated species.

Agro-industrial effluents

Industries of olive oil, tapioca starch, distillery (molasses), cotton bleaching and pulp and paper processing produce several billion litres of coloured, often toxic and harmful effluents world-wide annually. Such effluents have strong concentrations of chemical oxygen demand

(COD) (10-200 g/L), phenol and its derivatives (0.5-8 g/L) and often contain proteins, cyanides, chlorinated-lignin compounds and dyes [87-94].

Table 3 shows the use of different white-rot fungi in the treatment of real agro-industrial effluents. Martirani et al. [95] investigated the ability of the white-rot fungus *P. ostreatus* in the detoxification of phenol compounds from olive oil mill wastewater (OOMW). They found that OOMW considerably lost its colour and the concentration of phenol compounds decreased by 90% after 100 h of treatment. In addition, the toxicity towards *Bacillus cereus* was reduced 7-fold when compared with that of untreated wastewater. Moreover, the fungal biomass could be used for 3 consecutive cycles with the same efficiency.

Table 3. Use of white-rot fungi in the treatment of different real industrial wastewater

White-rot fungus	Type of wastewater	Removal	Reference
Agro-industrial effluents			
P. ostreatus	OOMW	71% decolouration, 90% phenol reduction	[95]
C. gallica	Alkaline paper effluents from a paperboard industry	64% decolouration in 7 days	[96]
L. edodes	OMW	decolouration: 75% (1st batch), 72% (2nd batch), 34% (3rd batch); TOC reduction: 73% (1st batch), 88% (2nd batch), 75% (3rd batch); phenol reduction: 83.5% (1st batch), 88.5% (2nd batch), 78% (3rd batch)	[97]
C. versicolor *P. chrysosporium*	DMSW effluent	71.5% decolouration, 90%COD reduction 53.5% decolouration, 73.0%COD reduction:	[98]
T. versicolor	Bagasse-based paper mill effluent	60% decolouration in 4 days	[99]
C. versicolor *F. trogii*	OOMW	63% COD reduction, 90% phenol reduction, 65% decolouration (within 6 days) 70% COD reduction, 93% phenol reduction, 81% decolouration (within 6 days)	[93]
C. subvermispora	First extraction stage effluent from kraft pulp bleaching	90% decolouration, 45% COD reduction, 62% lignin removal, 32% AOX removal, 36% EOX removal	[100]
P. flavido alba	OMW	70% decolouration and ≈100% phenol reduction in 48 h	[101])
P. chrysosporium *P. ostreatus*	OMW	>50% decolouration and phenol reduction in 6 days >50% decolouration and phenol reduction in more than 12 days	[102]
C. versicolor	Bleach plant effluents	53% decolouration, 40% AOX removal and 59% COD reduction (72 h)	[103]
P. ostreatus	OMW	50-78.3% phenol reduction depending on the conditions	[104]
Pleurotus spp.	OOMW	69–76% phenol reduction	[105]
T. versicolor	Black liquors from a soda pulping mill	70–80% reduction in colour and aromatic compounds, 60% COD reduction	[106]
Pleurotus spp.	ADW (vinasses)	39% decolouration in 10 h	[107]

Table 3. (Continued)

White-rot fungus	Type of wastewater	Removal	Reference
P. tigrinus	OMW	depended on wastewater composition	[108]
F. flavus	MSW	60% decolouration in 5 days	[109]
P. coccineus	ADW	17% COD reduction (free), 19% COD reduction (immobilised)	[110]
P. chrysosporium, P. ostreatus, L. edodes, T. versicolor, S22	Black liquor from a pulp and paper mill	71% lignin removal, 48% COD reduction	[111]
P. tigrinus	OMW	60.9% COD reduction, 97.2 % dephenolisation, 75% decolouration (bubble column bioreactor)	[112]
P. ostreatus	OMW	70% polyphenol reduction in 4–7 days (without nutrients); 95% polyphenol reduction in 4–7 days (with nutrients)	[113]
T. versicolor	Phenolic wastewater	removal of 0.125 g phenol/g biomass and 0.231 g o-cresol/g biomass per day.	[114]
P. chrysosporium, T. versicolor, C. polyzona, P. coccineus	OOMW	depended on cultivation mode	[115]
Trametes sp. I-62	Molasses	---	[116]
T. pubescens	Wine-related distillery wastewater	88% decolouration, 83% COD reduction, 87% phenol reduction	[117]
T. trogii *P. chrysosporium*	LFL	79% COD reduction 68% COD reduction	[118]
T. versicolor	OMW	phenol reduction: 78% (shake flasks); 39% (static culture); 70% (CSTR)	[119]
P. sajor caju, R. oryzae	Final effluent from the secondary treatment of a bleached kraft pulp mill processing *Eucaliptus globulus*	74-81% COD reduction in 10 days	[120]
P. chrysosporium	Coking wastewater	84% phenolic reduction and 80% COD reduction in 3 days	[121]
T. versicolor	Anaerobically-treated weak black liquor	30%- 32% COD reduction, 69% decolouration, 54% ligninoid removal	[122]
L. edodes (strain Le119)	OMW	65% decolouration, 75% phenol removal	[123]
T. pubescens	Wastewater from Amarula Cream production	71-77% COD reduction, 87-92% phenol removal	[124]
T. versicolor	OWW	64% decolouration in 216 h, 87% COD reduction in 216 h and 88% phenol removal in 144h (batch); 65% decolouration, 73% COD reduction and 89% phenol removal in 192 h (continuous)	[125]
P. sajor-caju	Sugar-cane vinasse effluents	82.8% COD reduction, 75.3% BOD reduction, 99.2% decolouration, 99.7% turbidity reduction	[126]
P. chrysosporium	Bagasse-based paper mill effluent	98.7% BOD reduction in 9 days, 98.5 COD reduction in 9 days	[127]
White-rot fungus	Type of wastewater	Removal	Reference

Marine fungus NIOCC #2a	Molasses-based raw effluents (2nd and 3rd step)	60–80% decolouration, 50–70% COD reduction, 60–70% phenol removal (hybrid technology)	[128]
Dyed effluents			
C. dusenii	Raw mixed-dye wastewater from a textile dye-producing plant	87% decolouration in 20 days	[131]
Thelephora sp	Dye industry effluent	61% decolouration in 3 days (batch); 50% decolouration in 7 days (continuous)	[132]
T. versicolor	Textile effluent	92% decolouration	[133]
T. versicolor	Carpet dye effluent	95% decolouration in 10 h	[134]
P. sajor-caju; *P. ostreatus*	Municipal effluent	---	[135]
P. sajor caju	Textile industry	100% decolouration in 14 days	[136]
P. flabellatus	Textile industry	60-70% decolouration	[137]
C. versicolor RC3	Batik dyeing process	80% decolouration, 67% COD reduction	[138]
T. versicolor	Spent dyeing bath from a textile industry	60% decolouration in 4-6 days	[139]
C. versicolor IBL-04	Textile industry	84.4% decolouration in 3 days	[140]
P. chrysosporium	Antiviral and antibiotical drugs	---	[141]
B. adusta	Dyeing industry	84% decolouration	[142]
B. adusta	Textile industry	---	[143]

Calvo et al. [96] studied the decolouration of alkaline paper effluents by *Coriolopsis gallica*. They found a colour reduction of 64% on day 7 which was maintained until the end of the experiment (14 days). Also, the alkali lignin fraction experienced a notable reduction, with a weight loss of 77% at day 14.

D'Annibale et al. [97] reported the biodegradation of olive mill wastewater (OMW) by *Lentinula edodes* immobilised on cubes of polyurethane sponge. After three consecutive treatment cycles of the effluent, its contents in total organic carbon (TOC), total phenols and total ortho-diphenol were significantly reduced together with a considerable decolouration.

Kumar et al. [98] tested the ability of four white-rot fungi, *P. chrysosporium, Mycelia steriliu, C. versicolor* and *Geotrichum candidum,* to decolourise and bioremediate anaerobically digested molasses spent wash (DMSW) generated by biomethanation plants. They found that *C. versicolor* and *P. chrysosporium* were able to decolourise and reduce the COD of diluted DMSW (12.5% v/v). However both fungi needed an additional carbon source, which together with the need to dilute the DMSW were two major shortcomings that needed further research.

Modi et al. [99] studied the decolouration of a bagasse-based paper mill effluent by the white-rot fungus *Trametes versicolor*. They found that maximum colour removal (60%) was achieved on the fourth day of cultivation after addition of 0.175% (w/v) ammonium nitrate and 1.0% (w/v) glucose.

Yesilada et al. [93] studied the biodegradation of OOMW by *C. versicolor* and *Funalia trogii*. They found that initial COD concentration, agitation and inoculum size affected significantly biodegradation. On the contrary, glucose addition, sulphate or nitrogen had no effect on biodegradation. During growth in optimum conditions, *C. versicolor* removed approximately 63% COD, 90% phenol and 65% colour within 6 days and *F. trogii* removed approximately 70% COD, 93% phenol and 81% colour of the OOMW used. The fungi also excreted large amounts of extracellular laccase into the medium. High biodegradation yields were also obtained when the fungi were immobilised in calcium alginate gels.

Nagarathnamma et al. [100] showed the abilty of *Ceriporiopsis subvermispora* to decolourise and degrade the first extraction stage effluent from kraft pulp bleaching. With glucose at 1 g/L, *C. subvermispora* removed up to 90% colour, 45% COD, 62% lignin, 32% adsorbable organic halide (AOX) and 36% extractable organic halide (EOX) whereas in the absence of glucose, the fungus removed up to 62% of the colour. In addition, toxicity to zebra fish was totally removed and chlorinated aromatic compounds were considerably reduced.

Pérez et al. [101] reported that *Phanerochaete flavido-alba* was able to decolourise OMW by 70% and that this decolouration was associated with a decrease in the content of free phenol of nearly 100%.

Kissi et al. [102] compared the ability of a *P. chrysosporium* strain isolated from Moroccan OMW and a commercial *P. ostreatus* strain to degrade OMW in different culture conditions. They found that *P. chrysosporium* was more efficient than *P. ostreatus* in decolourising and detoxifying OMW in the presence of added nutrients. Thus, *P. chrysosporium* was able to remove more than 50% of the colour and phenols from OMW within 6 days of incubation, whereas *P. ostreatus* needed more than 12 days to reach similar results in the same conditions.

Van Driessel &Christov [103] studied the decolouration of bleach plant effluents from the pulp and paper industry generated during bleaching with chlorine-containing chemicals in a rotating biological contactor (RBC) reactor by the white-rot fungus *C. versicolor* and the mucoralean fungus *Rhizomucor pusillus.* They found that these two fungi showed different mechanisms for decolouration. Thus, decolouration by *C. versicolor* was due to adsorption *plus* biodegradation whereas it was only due to adsorption for *R. pusillus*.

Fountoulakis et al. [104] investigated the capability of *P. ostreatus* to degrade phenols of OMW in different conditions such as in sterilised and thermally-processed (at 100ºC) wastewater, with and without dilution. It was found that *P. ostreatus* removed phenols from the culture medium, under all different conditions that were examined. Thus, the degradation of phenols reached up to 78.3% for the sterilised and 50% diluted OMW, 66.7% and 64.7% for the thermally-processed OMW, with and without dilution, respectively. They also studied the effect of OMW pre-treatment with *P. ostreatus* on the performance of anaerobic digestion and found that this pre-treatment enhanced the anaerobic digestion of OMW.

Tsioulpas et al. [105] studied the ability of several *Pleurotus* spp. strains to remove phenolic compounds from an OOMW. They found that all the strains tested were able to grow in OOMW without any addition of nutrients and any pre-treatment, except sterilisation. In addition, high laccase activities were detected in the growth medium and 69-76% of the initial phenolic compounds were removed. Moreover, the black colour of OOMW became yellow-brown and brighter as the strains grew and a decrease of the phytotoxicity, as described by the parameter Germination Index, was noticed in the OOMW treated with some *Pleurotus* spp strains, although this decrease was not proportional to the phenolic removal. They considered a new parameter namely Phenol-toxicity Index and using such parameter it was found that the remaining phenolics and/or some of the oxidation products of the laccase reaction in the treated OOMW were more toxic than the original phenolic compounds.

Font et al. [106] treated black liquors from a soda pulping mill with the white-rot fungus *T. versicolor* in aerated reactors. Reductions in colour and aromatic compounds of 70-80% and in COD of 60% were achieved.

Rodríguez et al. [107] reported that the treatment of fermentation residues (vinasse) and liquid extract of coffee pulp with the fungus *P. ostreatus* reduced both COD and colour.

D'Annibale et al. [108] assessed the potential of the white-rot fungus *Panus tigrinus* in removing organic load, colour and toxic phenols from OMW. They found that 4-hydroxy-substituted simple phenols were predominantly removed and that the polymeric aromatic fraction underwent simultaneous polymerization and depolymerization.

Raghukumar et al. [109] reported the decolouration of molasses spent wash (MSW) by the white-rot fungus *Flavodon flavus* immobilised on polyurethane foam. They found a decolouration of about 60 and 73% by days 5 and 7, respectively, for a 10% diluted MSW. The immobilised fungus could be effectively used for a minimum of three successive cycles. They also found that toxicity and PAH content of the MSW was reduced after fungal treatment.

Chairattanamanokorn et al. [110] reported that *Pycnoporus coccineus* strains showed a higher potential for alcohol distillery wastewater (ADW) decolouration both on agar media and liquid media. They also reported that, in general, mycelia immobilised on polyurethane foam cubes performed better than free mycelia.

Wu et al. [111] tested the ability of the white-rot fungi *P. chrysosporium*, *P. ostreatus*, *Lentinus edodes*, *T. versicolor* and S22, grown on a porous plastic media, to treat black liquor from a pulp and paper mill. They found that around 71% of lignin and 48% COD were removed from wastewater. In addition, *P. chrysosporium*, *P. ostreatus* and S22 showed high capacity to degrade lignin even under strong alkaline conditions.

D'Annibale et al. [112] investigated the ability of *P. tigrinus* to reduce the polluting load of OMW in bioreactors with different agitation modes (mechanical and pneumatic). They found that both aeration and agitation considerably affected the treatment efficiency, the best pollutant degradation being achieved in the pneumatically agitated bioreactor.

Olivieri et al. [113] studied the growth of *P. ostreatus* and the polyphenol conversion in OMW. It was found that *P. ostreatus* grew on OMW at polyphenol concentrations as large as 1.4 g/L without added nutrients and exhibited a remarkable ability to catalyse polyphenol bioconversion. Thus, bioconversion of polyphenols was 70% over 4-7 days, which increased to 95% over the same time period when nutrients were supplied. No appreciable decolouration took place along with remediation. Also, bioconversion was satisfactorily tested in an internal loop airlift bioreactor.

Ryan et al. [114] studied the bioremediation of a phenolic wastewater by *T. versicolor* and found that 0.125 g phenol/g biomass and 0.231 g o-cresol/g biomass were removed from solution per day.

Alaoui et al. [115] tested the efficiency of four white-rot fungi, *P. chrysosporium*, *T. versicolor*, *Coriolopsis polyzona* and *P. coccineus*, to treat OOMW regarding their cultivation mode (free mycelium, mycelium immobilised in alginate beads and solid-state cultivation on Petri dishes). They found that degradation by *C. polyzona* and *P. coccineus* was slightly affected by the cultivation mode. However, *P. chrysosporium* and *T. versicolor* showed better degradation performances for solid-state and alginate immobilisation procedures, respectively.

González et al. [116]reported the induction of laccases by molasse wastewater and molasse melanoidins in the white-rot-fungus *Trametes* sp. I-62. In addition, they correlated effluent decolouration and laccase activity.

Strong & Burgess [117] studied the treatment of wine-related wastewater by *Trametes pubescens*. The fungal treatment resulted in reduction in COD, phenolic compounds and colour of 83%, 87% and 88%, respectively.

Ellouze et al. [118] studied the detoxification of landfill leachates (LFL) by *Trametes trogii* and *P. chrysosporium*. COD removal efficiencies for *T. trogii* and *P. chrysosporium* were 79% and 68%, respectively, for a 50% diluted effluent. In addition, enzyme secretion and reduction in phenols and hydrocarbons was obtained.

Ergül [119] showed that *T. versicolor* FPRL-28A-INI was able to remove 78% of total phenolics in shake flask experiments and 39% in static ones of undiluted OMW without any addition of nutrients or pre-treatment. Also, in continuously stirred tank reactor (CSTR) conditions, 70% of total phenolics removal was achieved.

Freitas et al. [120] tested the potential of the white-rot fungi *Pleurotus sajor caju*, *T. versicolor* and *P. chrysosporium* for the biological treatment of effluents derived from the secondary treatment of a bleached kraft pulp mill processing *Eucalyptus globulus*. *P. sajor caju* was the most effective in the biodegradation of organic compounds present in the effluent and, in addition to this, it was the only specie able to perform the biological treatment without promoting toxicity of the final effluent.

Lu et al. [121] used the white-rot fungus *P. chrysosporium* immobilised on wood for biodegradation of phenolic compounds in coking wastewater. The removal rates of phenolic compounds and COD by the immobilised fungus were 87.05% and 72.09% in 6 days and operating under optimal conditions over 84% for phenolics and 80% for COD in 3 days.

Ortega-Clemente et al. [122] evaluated the post-treatment of anaerobically-treated weak black liquor in an aerobic, up-flow reactor packed with *T. versicolor* immobilised onto small cubes of holm oak wood. Organic matter removals of 30% and 32% COD were obtained during an experimental run of 60 days at 5-day hydraulic retention time (HRT) and 35 days at 2.5-day HRT, respectively. Colour and ligninoids contents were removed at higher percentages (69% and 54% respectively, average of both HRT). The bioreactor could sustain its removal activity for 95 days of continuous operation, without supplementation of soluble carbon sources.

Lakhtar et al. [123] evaluated the ability of 4 strains of *L. edodes* (Le118, Le119, Le121, Le122) to remove total phenols from OMW in liquid culture without nutritional supplements. The strain Le119 was the most efficient for OMW treatment showing 65% decolouration and 75% removal of total phenols.

Strong [124] tested the ability of the white-rot fungi *T. pubescens*, *Ceriporiopsis subvermispora*, *P. cinnabarinus* and *P. chrysosporium*) to treat distillery wastewater generated from the production of the beverage Amarula Cream. They found that *T. pubescens* led to the highest reductions in phenolic compounds, COD and colour.

Cerrone et al. [125] reported the depollution of OWW by *T. versicolor* grown in a bubble-column bioreactor operating in continuous mode. A reduction in colour, COD and phenols of 65%, 73% and 89%, respectively, was attained. A small bioreactor (10 m^3) could treat daily the amount of OWW produced by a standard olive washing machine. Therefore, the proposed process could be regarded as a simple, efficient and low-cost OWW treatment.

Romanholo-Ferreira et al. [126] reported the biodegradation of sugar-cane vinasse effluents by *P. sajor-caju* with reductions of 82.8% in COD, 75.3% in BOD, 99.2% in colouration and 99.7% in turbidity. In addition, the toxicity was reduced.

Sharari et al. [127] showed that *P. chrysosporium* reduced by 98.7% BOD and by 98.5% COD of a bagasse effluent from a pulp and paper mill. Also, total dissolved solids were reduced from 3950 to 575 mg/L.

Verma et al. [128] reported a novel three-step technology for treatment of molasses-based raw industrial effluents: (1) sonication, (2) whole-fungal treatment by a ligninolytic marine fungus and (3) biosorption of the residual colour with heat-inactivated biomass of the same fungus. After the third step, a reduction of 60–80% in colour, 50–70% in COD and 60–70% in total phenolics was achieved. In addition, toxicity against *Artemia larva*e was reduced by two to five folds.

Dyed effluents

The effluents of pharmaceutical industries, dyeing, printing, photographs, textile and cosmetics contain dyes [129]. For example, about 800,000 tons of dyes are produced annually world-wide, of which about 10% are lost in industrial effluents [130]. Effluents from textile industries are a complex mixture of many polluting substances such as organochlorine-based pesticides, heavy metals, pigments and dyes (Table 4). The majority of these dyes are slowly removed by the wastewater treatment plants (WWTP), because of their toxicities to indigenous micro-organisms. To overcome these difficulties, fungi are being investigated for their potential to decolourise effluents.

Table 4. Characteristics of wastes produced by textile industries (extracted from Kaushik & Malik [145], with kind permission of Elsevier Ltd.)

Waste	Process in which it is released	Concentration
Process wastewater		pH alkaline; BOD: 700-2000 mg/L; COD: 2-5 × BOD
Toxic organics including phenols	Dyeing and finishing	
Halogenated organics	Bleaching	
Dyes	Dyeing, printing	Below 1g/L [146]
Heavy metals		Cr: 0.35-170 mg/L; Cu: 0.12-4.20 mg/L; Ni: 0.10-0.96 mg/L; Pb: 0.10-4.00 mg/L; Zn: 0.10-58 mg/L; Mg: 0.50-10 mg/L [147]
Pesticides (preservation of natural fibres)	Washing and scouring	
Brominate flame retardants (synthetic fibres)		
Isocyanates	Lamination	
Volatile organic compounds (VOC)	Finishing and drying	10-350 mg carbon/m^3
Oils		

Table 3 shows studies on the treatment of real dyed effluents with different white-rot fungi. Thus, Wesenberg et al. [131] reported the partly decolouration of a raw mixed-dye wastewater, consisting of a complex mixture of azo and anthraquinonic dyes, from a textile dye-producing plant by the white-rot fungus *Clitocybula dusenii*. Operating under the best conditions up to 87% of the dyes in the effluent were decolourised in 20 days.

Selvam et al. [132] used the white rot fungus *Thelephora* sp. to treat a dye industry effluent in batch and continuous modes. They obtained a maximum decolouration of 61% on the third day in the batch mode and of 50% on the seventh day in the continuous mode.

Amaral et al. [133] reported a decolouration of 92% for an effluent from a textile-dyeing industry (50 mg dye/L) by the white-rot fungus *T. versicolor*.

Ramsay and Goode [134] reported the decolouration of a non-sterile carpet dye effluent containing two anthraquinonic dyes by *T. versicolor* immobilised on jute twine in a rotating biological contacting reactor for four successive batches.

Dellamatrice et al. [135] evaluated the decolouration of an effluent of sludge from domestic and textile origin by three strains of *Pleurotus* (*P. sajor-caju* F2, F6 and *P. ostreatus*). They found that the strains of *P. sajor caju* F2 and F6 were able to decolourise the sludge, while *P. ostreatus* was less efficient. They also found that effluent toxicity decreased but not that of its sludge.

Kamida et al. [136] studied the biodegradation of a textile effluent containing indigo dye by *P. sajor-caju* grown on cane bagasse. A total decolouration of the effluent in 14 days was found.

Nilsson et al. [137] investigated the decolouration of wastewater from a textile industry by *Pleurotus flabellatus* growing on luffa sponge packed in a continuous reactor. The reactor was operated at an HRT of 25 h and a decolouration of 60-70% was achieved.

Srikanlayanukul et al. [138] investigated the decolouration of wastewater from Batik dyeing process with the white-rot fungus *C. versicolor* RC3 immobilised in cubes of polyurethane foam in repeated-batch system. It was found that the immobilised fungus decolourised up to 80% and reduced 67% of COD in 48 h.

Blánquez et al. [139] reported the continuous treatment of real industrial textile wastewater under non-sterile conditions by pellets of *T. versicolor* in a bioreactor at pilot scale operating at an HRT of 48 h. Wastewater was treated during 15 days with colour reduction levels between 40 and 60%.

Asgher et al. [140] assessed the decolouration potential of the local white-rot fungus *C. versicolor* IBL-04 for practical industrial effluents collected from five different textile industries of Faisalabad (Pakistan). The best decolouration percentage was 36.3% in 6 days for one of the effluents tested. The decolouration potential of this fungus was further enhanced to 84.4% in 3 days by optimising pH and temperature and using 1% starch as a supplementary carbon source.

Accinelli et al. [141] assessed the capacity of *P. chrysosporium* entrapped in a granular bioplastic formulation to remove four selected pharmaceuticals from wastewater samples of a WWTP. It was shown that the removal of the antiviral drug oseltamivir (Tamiflu) and the antibiotics erythromycin, sulfamethoxazol and ciprofloxacin was significantly increased.

Anastasi et al. [142] shown that *Bjerkandera adusta* was able to decolourise efficiently large volumes of a real wastewater, containing acid, direct, reactive and disperse dyes, in a fixed-bed reactor during 10 cycles under non-sterile conditions. In addition, dilution of the wastewater, addition of nutrient sources or control of chemical parameters, such as pH was not necessary. More recently, Anastasi et al. [143] compared fungal treatment (*B. adusta*) and ozonation of wastewater collected from a treatment plant for dyed effluents. The fungal treatment, although less efficient than ozonation, caused a good decolouration of the effluent.

Future Perspectives

Removal of recalcitrant pollutants from environment is an important challenge. For this, several biotechnological procedures have been developed but they are not always very efficient, produce sludge or require high energy costs. The use of white-rot fungi appears to be a better alternative since they are able to remove xenobiotic compounds without formation of toxic sub-products or accumulation of sludge.

The use of white-rot fungi in environmental biotechnology has been studied for more than 20 years. Most research work has been performed with particular fungal species, especially with the strain *P. chrysosporium*, which has become the model strain for this type of fungi. White-rot fungi have been successfully applied in treatment and decomposition of different hazardous compounds at a laboratory scale, especially under sterile conditions. However, much work is needed to explore the potential of such fungi in the removal of xenobiotics in real situation.

On one hand, recent advances in enzyme immobilisation may help in enhancing the efficiency of the process. On the other hand, search for robust, alkaline and heat tolerant strains might make the process more economical and viable. Future work should combine both aspects.

Another approach that deserves further research is the investigation of the technical feasibility of combining physico-chemical (advanced oxidation processes, AOPs) and biological treatments (fungal or enzymatic), which can be complementary. The total oxidation and the mineralisation of organic compounds by AOPs is not feasible due to the enormous amount of energy and reagents needed for the total mineralisation. Integrated AOPs and biological treatments would provide a feasible alternative and could help to remove economically and effectively xenobiotic compounds from industrial effluents.

Acknowledgments

This research was financed by the Spanish Ministry of Education and Science (Project CTM2008-02453/TECNO).

References

[1] Aust, S. D. (1990). Degradation of environmental pollutants by *Phanerochaete chrysosporium*. *Microbial Ecology*, *20*, 197–209.

[2] Fernando, T. & Aust, S. D. (1994). Biodegradation of toxic chemicals by white rot fungi. In G. R. Chaudhry (Ed.), *Biological Degradation and Bioremediation of Toxic Chemicals* (pp. 386 402). London: Chapman & Hall.

[3] Mueller, J. G., Cerniglia, C. E. & Pritchard, P. H. (1996). Bioremediation of environments contaminated by polycyclic aromatic hydrocarbons. In R. L. Crawford & D.L. Crawford (Eds.), *Bioremediation: Principles and Applications* (pp. 125–194). Cambridge, UK: Cambridge University Press.

[4] Bumpus, J. A., Tien, M., Wright, D. & Aust, S. D. (1985). Oxidation of persistent environmental pollutants by a white-rot fungus. *Science*, *228*, 1434–1436.

[5] Pointing, S. (2001) Feasibility of bioremediation by white-rot fungi. *Applied Microbiology and Biotechnology*, 57, 20-33.

[6] Evans, C. &Hedger, J. (2001). Degradation of cell wall polymers. In G. Gadd (Ed.), *Fungi in Bioremediation*. Cambridge, UK: Cambridge University Press.

[7] Blanchette, R.A. (1995) Degradation of the lignocellulose complex in wood. *Canadian Journal of Botany, 73*, S999-S1010.

[8] Akhtar M., Blanchette R.A., Kirk T.K. (1997) Fungal delignifi cation and biomechanical pulping of wood. In: K. E. L. Eriksson (Ed.), *Advances in Biochemical Engineering/Biotechnology* (pp. 159-196). Berlin, Germany: Springer Verlag.

[9] Carlile, M. J., Watkinson, S. C. & Gooday, G. W. (2001). *The Fungi*. London, UK: Academic Press.

[10] Gellerstedt, G. & Northy, R. A. (1989). Analysis of birch wood lignin by oxidative degradation. *Wood Science Technology, 23*, 75-83.

[11] Eriksson K.-E. L., Blanchette R.A. & Ander P. (1990) Microbial and enzymatic degradation of wood and wood components. Berlin, Germany: Springer-Verlag.

[12] Canet, R., Birnstingl J., Malcolm, D., Lopez-Real, J. & Beck, A. (2001) Biodegradation of polycyclic aromatic hydrocarbons (PAHs) by native microflora and combinations of white-rot fungi in a coal-tar contaminated soil. *Bioresource Technology*, *76*, 113-117.

[13] Paszczynski, A., Pasti-Grigsby, M. B., Goszczynski, S., Crawford, D. L., Crawford, R. L. (1991). New approach to improve degradation of recalcitrant azo dyes by *Streptomyces* spp. and *Phanerochaete chrysosporium*. *Enzyme and Microbial Technology*, *13*, 378-384.

[14] Zhou W., Zimmermann W. (1993). Decolourisation of industrial effluents containing reactive dyes by actinomycetes. *FEMS Microbiology Letters, 107*, 157-162.

[15] Mester, T.& Tien, M. (2000) Oxidation mechanism of ligninolytic enzymes involved in the degradation of environmental pollutants. *International Biodeteoration & Biodegradation*, *46*, 51-59.

[16] Barr, D. P. & Aust, S. D. (1994). Pollutant degradation by white rot fungi. *Reviews in Environmental Contamination and Toxicology*, *138*, 49-72.

[17] Sun, Y. & Cheng, J. (2002) Hydrolysis of lignocellulosic material for ethanol production: a review. *Bioresource Technology, 83*, 1-11.

[18] Hatakka, A. (2001). Biodegradation of lignin. In: M. Hofrichter & A. Steinbüchel (Eds.), *Biopolymers. Vol 1: Lignin, humic substances and coal* (pp. 129-180). Weinheim, Germany: Wiley-VCH.

[19] Orth, A. B. & Tien, M. (1995). Biotechnology of Lignin Degradation. In K. Esser & P. A, Lemke (Eds.). *The Mycota. Vol II. Genetics and Biotechnology* (pp. 287-302). Berlin: Springer-Verlag.

[20] Thurston, C. (1994). The structure and function of fungal laccases. *Microbiology, 140*, 19-26.

[21] Heinfling A., Martínez, M. J., Martínez A. T., Bergbauer, M. & Szewzyk, U. (1998). Purification and characterization of peroxidases from the dye-decolorizing fungus *Bjerkandera adusta*. *FEMS Microbiology Letters, 165*, 43-50.

[22] Camarero S., Sarkar, S., Ruiz-Dueñas, F. J., Martínez, M. J. & Martínez, A. T. (1999). Description of a versatile peroxidase involved in the natural degradation of lignin that has both manganese peroxidase and lignin peroxidase substrate interaction sites. *Journal of Biological Chemistry, 274*, 10324-10330.

[23] Ruiz-Dueñas, F. J., Camarero, S., Pérez-Boada, M., Martinez M., Martinez A.T., (2001). A new versatile peroxidise from *Pleurotus*. *Biochemical Society Transactions, 29*, 116-122.

[24] Leonowicz, A., Cho, N. S., Luterek, J., Wilkolazka, A., Wojtas-Wasilewska, M., Matuszewska, A., Hofrichter, M., Wesenberg, D. & Rogalski J. (2001). Fungal laccase: properties and activity on lignin. *Journal of Basic Microbiology, 41*, 185-227.

[25] Buswell, J. A., Mollet, B. & Odier, E. (1984). Ligninolytic enzyme production by *Phanerochaete chrysosporium* under conditions of nutrient sufficiency. *FEMS Microbiology Letters, 25*, 295-299

[26] Vyas, B. R. M., Volc, J. & Sasek, V. (1994). Ligninolytic enzymes of selected white rot fungi cultivated on wheat straw. *Folia Microbiologica, 39*, 235–240.

[27] Darah, I. & C. O. Ibrahim, 1996. Effect of agitation on lignin-degrading enzyme production by Phanerochaete chrysosporium grown in shake-flask cultures. *Asia Pacific Journal of Molecular Biology and Biotechnology, 4*, 174-182.

[28] Podgornik, H., Podgornik, A., Milavec, P. & Perdih, A. (2001). The effect of agitation and nitrogen concentration on lignin peroxidase (LiP): isoform composition during fermentation of *Phanerochaete chrysosporium*. *Journal of Biotechnology, 88*, 173-176.

[29] Dittmer, J. K., Patel, N. J., Dhawale, S. W. & Dhawale, S. S. (1997). Production of multiple laccase isoforms by *Phanerochaete chrysosporium* grown under nutrient sufficiency. *FEMS Microbiology Letters, 149*, 65-70.

[30] Scheel, T., Hofer, M., Ludwig, S. & Holker, U. (2000). Differential expression of manganese peroxidase and laccase in white-rot fungi in the presence of manganese or aromatic compounds. *Applied Microbiology & Biotechnology, 54*, 686–691.

[31] Galhaup C., Wagner H., Hinterstoisser B., Haltrich D. (2002). Increased production of laccase by the wood-dergrading basidiomycete *Trametes pubescens*. *Enzyme & Microbial Technology, 30*, 529-536.

[32] Tien, M. & Kirk, T. K. (1988). Lignin peroxidase of *Phanerochaete chrysosporium*. *Methods in Enzymology, 161*, 238–249.

[33] Collins, P. J. & Dobson, A. D. W. (1995). Extracellular lignin and manganese peroxidase production by the white rot fungus *Coriolus versicolor* 290. *Biotechnology Letters, 17*, 989-992.

[34] Gold, M. H., Wariishi, H. & Valli, K. (1989). Extracellular peroxidases involved in lignin degradation by the white-rot basidiomycete *Phanerochaete chrysosporium*. In J. F. Whitaker & P. E. Sonnet (Eds.). *Biocatalysis in agricultural biotechnology. ACS symposium series* (pp. 127–140), Washington, DC: American Chemical Society.

[35] Wariishi, H. & Gold, M. H. (1989). Lignin peroxidase Compound III: formation, inactivation and conversion to the native enzyme. *FEBS Letters, 243*, 165-168.

[36] Collins, P. J., Field J. A., Teunissen, P. & Dobson, A. D.W . (1997). Stabilization of lignin peroxidases in white rot fungi by tryptophan. *Applied & Environmental Microbiology, 63*, 2543-2548.

[37] Paszczynski, A. & Crawford, R. L. (1991). Degradation of azo compounds by ligninase from *Phanerochaete chrysosporium*: involvement of veratryl alcohol. *Biochemical and Biophysical Research Communications, 178*, 1056-1063.

[38] Lobos, S., Larraín, J., Cullen, D. & Vicuña, R. (1994). Isoenzymes of manganese dependent peroxidase and laccase produced by the lignin-degrading basidiomycete *Ceriporiopsis subvermispora*. *Microbiology, 140*, 2691-2698.

[39] de la Rubia, T., Linares, A., Perez, J., Munoz-Dorado, J., Romera, J. & Martinez, J. (2002). Characterization of manganese-dependent peroxidase isoenzymes from the ligninolytic fungus *Phanerochaete flavido-alba*. *Research in Microbiology, 153*, 547-554.

[40] Martínez, A. T. (2002). Molecular biology and structure-function of lignin-degrading heme peroxidases. *Enzyme & Microbial Technology, 30*, 425-444.

[41] Palma C., Martínez A. T., Lema J. M. & Martínez M. J. (2000). Different fungal manganese-oxidizing peroxidases: a comparison between *Bjerkandera sp.* and *Phanerochaete chrysosporium*. *Journal of Biotechnology, 77*, 235-245.

[42] Ha, H. C., Honda, Y., Watanabe, T. & Kuwahara, M. (2001). Production of manganese peroxidase by pellet culture of lignin-degrading basidiomycete, *Pleurotus ostreatus*. *Applied Microbiology & Biotechnology, 55*, 704–711.

[43] Li, D., Alic, M., Brown, A. & Gold, M.H. (1995). Regulation of manganese peroxidase gene transcription by hydrogen peroxide, chemical stress, and molecular oxygen. *Applied & Environmental Microbiology, 61*, 341-345.

[44] Gettemy, J. M., Ma, B., Alic, M. & Gold, M. H. (1998). Reverse transcription-PCR analysis of the regulation of the manganese peroxidase gene family. *Applied & Environmental Microbiology, 64*, 569-574.

[45] Hofrichter, M. (2002). Review: lignin conversion by manganese peroxidase (MnP). *Enzyme & Microbial Technology, 30*, 454-466.

[46] Yoshida, H. (1883). Chemistry of lacquer (Urushi). Part 1. *Journal of Chemical Society, 43*, 472–486.

[47] Baldrian, P. (2006). Fungal laccases – occurrence and properties. *FEMS Microbiology Reviews, 30*, 215-242

[48] Alexandre, G. & Zhulin, I.B. (2000). Laccases are widespread in bacteria. *Trends in Biotechnology, 18*, 41–42.

[49] Mayer, A. M. & Staples, R. C. (2002). Laccase: new functions for an old enzyme. *Phytochemistry, 60*, 551–565.

[50] Solomon, E. I., Sundaram, U. M. & Machonkin, T. E. (1996). Multicopper oxidases and oxygenases. *Chemical Reviews, 96*, 2563-2606.

[51] Morozova, V., Shumakovich, G.P., Gorbacheva, M.A., Shleev, S.V. & Yaropolov, A.I. (2007a). "Blue" laccases. *Biochemistry (Moscow), 72*, 1136–1150.

[52] Palmieri, G., Giardina, P., Bianco, C., Scaloni, A., Capasso, A. & Sannia, G. (1997). A novel white laccase from *Pleurotus ostreatus*. *Journal of Biological Chemistry, 272*, 31301–31307.

[53] Min, K. L., Kim, Y. H., Kim, Y. W., Jung, H. S. & Hah, Y. C. (2001). Characterization of a novel laccase produced by the woodrotting fungus *Phellinus ribis*. *Archives of Biochemistry & Biophysics, 392*, 279–286.

[54] Shleev, S., Reimann, C. T., Serezhenkov, V., Burbaev, D., Yaropolov, A. I., Gorton, L. & Ruzgas, T. (2006). Autoreduction and aggregation of fungal laccase in solution phase: possible correlation with a resting form of laccase. *Biochimie, 88*, 1275-1285.

[55] Messerschmidt, A., Ladenstein, R., Huber, R., Bolognesi, M., Avigliano, L., Petruzzelli, R., Rossi, A. & Finazzi-Agró, A. (1992). Refined crystal structure of ascorbate oxidase at 1.9Å resolution. *Journal of Molecular Biology, 224*, 179–205.

[56] Zoppellaro, G., Sakurai, T., & Huang, H.-W. (2001). A novel mixed valence form of Rhus vernicifera laccase and its reaction with dioxygen to give a peroxide intermediate bound to the trinuclear center. *Journal of Biochemistry, 129*, 949–953.

[57] Huang, H.-W., Zopellaro, G., & Sakurai, T. (1999). Spectroscopic and kinetic studies on the oxygencentered radical formed during the four-electron reduction process of dioxygen by Rhus vernicifera laccase. *Journal of Biological Chemistry, 274*, 32718–32724.

[58] Bourbonnais, R., Paice, M. G., Reid, I. D., Lanthier, P. & Yaguchi, M. (1995). Lignin oxidation by laccase isozymes from *Trametes versicolor* and role of the mediator 2,2'-Azinobis(3-Ethylbenzthiazoline-6-Sulfonate): in kraft lignin depolymerization. *Applied and Environmental Microbiology, 61*, 1876-1880.

[59] Johannes, C. & Majcherczyk, A. (2000). Natural mediators in the oxidation of polycyclic aromatic hydrocarbons by laccase mediator systems. *Applied and Environmental Microbiology*, 66, 524–528.

[60] Morozova, O. V., Shumakovich, G. P., Shleev, S. V., Yaropolov, Y. I. (2007b). Laccase–mediator Systems and their applications: a review. *Applied Biochemistry & Microbiology*, *43*, 523–535.

[61] Hatakka, A., Lundell, T., Hofrichter, M. & Maijala, P. (2003). Manganese peroxidase and its role in the degradation of wood lignin. In S. D. Mansfield & J. N. Saddler (Eds.), *ACS Symposium series 855, Applications of Enzymes to Lignocellulosics* (pp. 230-243). Washington, DC: American Chemical Society.

[62] Eggert, C., Temp, U. & Eriksson, K. E. L. (1996). The ligninolytic system of the white rot fungus *Pycnoporus cinnabarinus*: Purification and characterization of the laccase. *Applied and Environmental Microbiology, 62*, 1151-1158.

[63] Camarero, S., Ibarra, D., Martínez, M. J. & Martínez, A. T. (2005). Lignin-derived compounds as efficient laccase mediators for decolorization of different types of recalcitrant dyes. *Applied and Environmental Microbiology*, *71*, 1775-1784.

[64] Palmieri, G., Giardina, P., Bianco, C., Fontanella, B. & Sannia, G. (2000). Copper induction of laccase isoenzymes in the ligninolytic fungus *Pleurotus ostreatus*. *Applied and Environmental Microbiology, 66*, 920-924.

[65] Galhaup, C. & Haltrich, D. (2001). Enhanced formation of laccase activity by the white-rot fungus *Trametes pubescens* in the presence of copper. *Applied Microbiology and Biotechnology, 56*, 225-232.

[66] Saparrat, M .C. N., Guillen, F., Arambarri, A. M., Martinez, A. T. & Martinez, M. J. (2002). Induction, isolation, and characterization of two laccases from the white rot basidiomycete *Coriolopsis rigida*. *Applied and Environmental Microbiology, 68*, 1534-1540.

[67] Karahanian, E., Corsini G., Lobos, S. & Vicuña, R. (1998). Structure and expression of a laccase gene from the ligninolytic basidiomycete *Ceriporiopsis subvermispora*. *Biochimica et Biophysica Acta*, *1443*, 65-74.

[68] Baldrian, P. & Gabriel, J. (2002). Copper and cadmium increase laccase activity in *Pleurotus ostreatus*. *FEMS Microbiology Letters, 206*, 69-74.

[69] Hildén, K. S., Bortfeldt, R., Hofrichter, M., Hatakka, A. & Lundell, T. K. (2008). Molecular characterization of the basidiomycete isolate *Nemataloma frowardii* b19 and its manganese peroxidase places the fungus in the corticioid genus *Phlebia*. *Microbiology, 154*, 2371-2379.

[70] Quaratino, D., Federici, F., Petruccioli, M., Fenice, M. & D'Annibale, A. (2007). Production, purification and partial characterisation of a novel laccase from the white-rot fungus *Panus tigrinus* CBS 577.79. *Antonie van Leeuwenhoek, 91*, 57-69.

[71] Gianfreda, L. & Bollag, J. M. (2002). Isolated enzymes for the transformation and detoxification of organic pollutants. In R. G. Burns & R. P. Dick (Eds.), *Enzymes in the environment: Activity, ecology, and applications* (pp. 495–538). New York: Marcel Dekker Inc.

[72] Baldrian, P. (2008). Mini-review: Wood-inhabiting ligninolytic basidiomycetes in soils: Ecology and constraints for applicability in bioremediation. *Fungal Ecology, 1*, 4-12.

[73] Magan, N., Fragoeiro, S. & Bastos, C. (2010). Environmental Factors and Bioremediation of Xenobiotics Using White Rot Fungi. *Mycobiology, 38*, 238-248.

[74] Jones, A., Montanarella, L. & Jones, R. (2005). Soil Atlas of Europe. European Soil Bureau Network, European Commission. Luxembourg. Accessed 08/2010. http://eusoils.jrc.ec.europa.eu/projects/soil_atlas/.

[75] Yateem, A., Balba, M. T., AI-Awadhi, N. & EI-Nawawy, A. S. (1998). White rot fungi and their role in remediating oil-contaminated soil. *Environment International, 24*, 181-187.

[76] Eggen, T. (1999). Application of fungal substrate from commercial mushroom production - *Pleurotus ostreatus* - for bioremediation of creosote contaminated soil. *International Biodeterioration and Biodegradation, 44*, 117-126

[77] Eggen, T. & Sveum, P. (1999). Decontamination of aged creosote polluted soil: the influence of temperature, white rot fungus *Pleurotus ostreatus*, and pretreatment. *International Biodeterioration and Biodegradation*, *43*, 125–133.

[78] Kubatova, A., Erbanova, P., Eichlerova, I., Homolka, L., Nerud, F. & Sasek, V. (2001). PCB Congener selective biodegradation by the *white-rot* fungus *Pleurotus ostreatus* in contaminated soil. *Chemosphere, 43*, 207-215.

[79] Isikhuemhen, O. S., Anoliefo, G. O. & Oghale, O. I. (2003). Bioremediation of crude oil polluted soil by the white rot fungus, *Pleurotus tuberregium. Environmental Science and Pollution Research, 10*, 108-112.

[80] Walter, M., Boyd-Wilson, K., Boul, L., Ford, C., McFadden, D., Chong, B. & Pinfold, J. (2005). Field-scale bioremediation of pentachlorophenol by *Trametes versicolo*r. *International Biodeterioration & Biodegradation, 56*, 51–57.

[81] Ford, C. I., Walter, M., Northcott, G. L., Di, H. J., Cameron, K. C. & Trower, T. (2007). Fungal inoculum properties: extracellular enzyme expression and pentachlorophenol removal in highly contaminated field soils. *Journal of Environmental Quality, 36*, 1599–1608.

[82] Leonardi, V., Šašek, V., Petruccioli, M., D'Annibale, A., Erbanova, P. & Cajthaml, T. (2007). Bioavailability modification and fungal biodegradation of PAHs in aged industrial soils. *International Biodeterioration and Biodegradation*, *60*, 165-170.

[83] Byss, M., Elhottová, D., Tlíska, J. & Baldrian, P. (2008). Fungal bioremediation of the creosote-contaminated soil: Influence of *Pleurotus ostreatus* and *Irpex lacteus* on polycyclic aromatic hydrocarbons removal and soil microbial community composition in the laboratory-scale study. *Chemosphere, 73*, 1518–1523.

[84] Bastos, A. C. & Magan N. (2009). *Trametes versicolor*: potential for atrazine bioremediation in calcareous clay soil, under low water availability conditions. *International Biodeterioration and Biodegradation, 63*, 389-394.

[85] Sasek, V. (2003). Why mycoremediations have not yet come to practice. In: V. Sasek, (Ed.), *The Utilization of Bioremediation to Reduce Soil Contamination: Problems and Solutions* (pp. 247–276). Kluwer Academic Publishers.

[86] Fragoeiro, S. (2005). Use of fungi in bioremediation of pesticides. Ph.D. thesis, Bedford, UK: Cranfield University.

[87] Borja, R., Martin, A., Maestro, R., Alba, J. & Fiestas, J. A. (1992). Enhancement of the anaerobic digestion of olive mill wastewater by the removal of phenolic inhibitors. *Process Biochemistry, 27*, 231-237.

[88] Borja, R., Alba, J. & Banks, C. J, (1997). Impact of the main phenolic compounds of olive mill wastewater (OMW) on the kinetics of acetoclastic methanogenesis. *Process Biochemistry, 32*, 121-133.

[89] Nieto, L. M., Cormenzana, A. R., Pareja, P. G. & Hoyos, S. E. G. (1992). Phenolic compounds biodegradatio of olive mill wastewater with *Aspergillus terreus*. *Grasas y Aceites, 43*, 75- 81.

[90] Bengtsson, B. E. & Triet, T. (1994). Tapioca starch wastewater toxicity characterized by Microtox and Duckweed tests. *Ambio, 23*, 473-477.

[91] Garcia, I. G., Venceslada, J. L. B., Pena, P. R. J., Gomez, E. R. (1997). Biodegradation of phenol compounds in vinasse using *Aspergillus terreus* and *Geotrichum candidum*. *Water Research, 31*, 2005-2011.

[92] Jiménez, A. M. & Borja, R. (1997). Influence of aerobic pretreatment with *Penicillium decumbens* on the anaerobic digestion of beet molasses alcoholic fermentation wastewater in suspended and immobilized cell bioreactors. *Journal of Chemical Technology & Biotechnology, 69*, 193-202.

[93] Yesilada, O., Sik, S. & Sam, M. (1998). Biodegradation of olive oil mill wastewater by Coriolus versicolor and *Funalia trogii*: effects of agitation, initial COD concentration, inoculum size and immobilization. *World Journal of Microbiology and Biotechnology, 14*, 37–42.

[94] Kahmark, K. A. & Unwin, J. P. (1999). Pulp and paper effluent management. *Water & Environmental Research*, 71, 836–852.

[95] Martirani, L., Giardina, P., Marzullo, L. & Sannia, G. (1996). Reduction of phenol content and toxicity in olive oil mill waste waters with the ligninolytic fungus *Pleurotus ostreatus*. *Water Research, 30*, 1914–1918.

[96] Calvo, A. M., Copa-Patiño, J. L., Alonso, O. & González, A. E. (1998). Studies of the production and characterization of laccase activity in the basidiomycete C*oriolopsis gallica*, an efficient decolorizer of alkaline effluents. *Archives of Microbiology*, *171*, 31-36.

[97] D'Annibale, A., Crestini, C., Vinciguerra, V. & Sermanni, G. G. (1998). The biodegradation of recalcitrant effluents from an olive mill by a white-rot fungus. *Journal of Biotechnology*, *61*, 209-218.

[98] Kumar, V., Wati, L., Nigam, P., Banat, I. M., Yadav, B. S., Singh, D. & Marchant, R. (1998). Decolorization and biodegradation of anaerobically digested sugarcane molasses spent wash effluent from biomethanation plants by white-rot fungi. *Process Biochemistry*, *33*, 83-88.

[99] Modi, D. R., Chandra, H. & Garg, S. K. (1998). Decolourization of Bagasse-based Paper Mill Effluent by the White-rot Fungus *Trametes versicolor*, *Bioresource Technology, 66,* 79-81.

[100] Nagarathnamma, R., Bajpai, P., Bajpai, P. K. (1999). Studies on decolourization, degradation and detoxification of chlorinated lignin compounds in kraft bleaching effluents by *Ceriporiopsis subvermispora. Process Biochemistry*, *34*, 939-948.

[101] Pérez, J., de la Rubia, T. & Martínez, J. (2000). Degradación del alpechín por hongos de la podredumbre blanca de la madera. III Jornadas Mediterréneas de Olivar Ecológico-ECOLIVA.

[102] Kissi, M., Mountadar, M., Assobhei, O., Gargiulo, E., Palmieri, G., Giardina, P. & G. Sannia, G. (2001). Roles of two White-rot basidiomycete fungi in decolorisation and detoxification of olive mill waste water. *Applied Microbiology & Biotechnology*, 57, 221–226.

[103] van Driessel, B. & Christov, L. (2002). Adsorption of colour from a bleach plant effluent using biomass and cell wall fractions from *Rhizomucor pusillus*. *Journal of Chemical Technology & Biotechnology, 77*, 155-158.

[104] Fountoulakis, M.S., Dokianakis, S.N., Kornaros, M.E., Aggelis, G.G. & Lyberatos, G. (2002). Removal of phenolics in olive mill wastewaters using the white-rot fungus *Pleurotus ostreatus*. *Water Research*, 36, 4735–4744.

[105] Tsioulpas, A., Dimou, D., Iconomou, D., & Aggelis, G. (2002). Phenolic removal in olive mill waste water by strains of *Pleurotus ostreatus spp.* in respect to their phenol oxidase (laccase) activity. *Bioresource Technology*, *84*, 251–257.

[106] Font, X., Caminal, G., Gabarrell, X., Romero, S. & Vicent, M T. (2003). Black liquor detoxification by laccase of *Trametes versicolor* pellets. *Journal of Chemical Technology & Biotechnology, 78*, 548–554.

[107] Rodrıguez, S., Fernandez, M., Bermudez, R.C. & Morris, H. (2003). Tratamiento de efluentes industriales coloreados con *Pleurotus* spp. *Revista Iberoamericana de Micología*, 20, 164–168.

[108] D'Annibale, A., Ricci, M., Quaratino, D., Federici, F. & Fenice, M. (2004). *Panus tigrinus* efficiently removes phenols, colour and organic load from olive mill wastewater. *Research in Microbiology*, 155, 596–604.

[109] Raghukumar, C., Mohandass, C., Kamat, S. & Shailaja, M. S. (2004). Simultaneous detoxification and decolorization of molasses spent wash by the immobilized white-rot fungus Flavodon flavus isolated from a marine habitat. Enzyme & Microbial Technology, 35, 197–202.

[110] Chairattanamanokorn, P., Imai, T., Kondo, R., Sekine, M., Higuchi, T. & Ukita, M. (2005). Decolorization of alcohol distillery wastewater by thermotolerant white rot fungi. *Applied Biochemistry & Microbiology*, 41, 583-588.

[111] Wu, J., Xiao, Y.-Z. & Yu, H.-Q. (2005). Degradation of lignin in pulp mill wastewaters by white-rot fungi on biofilm. *Bioresource Technology, 96*, 1357–1363,

[112] D'Annibale, A., Quaratino, D., Federici, F. & Fenice, M. (2006). Effect of agitation and aeration on the reduction of pollutant load of olive mill wastewater by the white-rot fungus *Panus tigrinus*. *Biochemical Engineering Journal, 29*, 243–249.

[113] Olivieri, G., Marzocchella, A., Salatino, P., Giardina, P., Cennamo, G. & Sannia, G. (2006). Olive mill wastewater remediation by means of *Pleurotus ostreatus*. *Biochemical Engineering Journal, 31*, 180-187.

[114] Ryan, D., Leukes, W. & Burton, S. (2007). Improving the bioremediation of phenolic wastewaters by *Trametes versicolor*. *Bioresource Technology*, 98, 579–587.

[115] Alaoui, S. M., Merzouki, M., Penninckx, M. J. & Benlemlih, M. (2008). Relationship between cultivation mode of white rot fungi and their efficiency for olive oil mill wastewaters treatment. *Electronic Journal of Biotechnology, 11*, 1–8.

[116] González, T., Terrón, M.C., Yagüe, S., Junca, H., Carbajo, J. M., Zapico, E. J., Silva. R., Arana-Cuenca, A., Téllez, A., González, A. E. (2008). Melanoidin-containing wastewaters induce selective laccase gene expression in the white-rot fungus *Trametes* sp. I-62. Research in Microbiology, 159, 103-109.

[117] Strong, P. J. & Burgess, J. E. (2008). Fungal and enzymatic remediation of a wine lees and five wine-related distillery wastewaters. *Bioresource Technology*, *99*, 6134–6142.

[118] Ellouze, M., Aloui, F. & Sayadi, S. (2009). Effect of high ammonia concentrations on fungal treatment of Tunisian landfill leachates. *Desalination*, *246*, 468–477.

[119] Ergul, F. E., Sargin, S., Ongen, G. & Sukan, F. V. (2009). Dephenolisation of olive mill wastewater using adapted Trametes versicolor. *International Biodeterioration and Biodegradation*, *63*, 1–6.

[120] Freitas, A. C., Ferreira, F., Costa, A. M., Pereira, R., Antunes, S. C., Gonçalves, F. Rocha-Santos, T. A. P., Diniz, M. S., Castro, L., Peres, I. & Duarte, A. C. (2009). Biological treatment of the effluent from a bleached kraft pulp mill using basidiomycete and zygomycete fungi. *Science of the Total Environment, 407*, 3282–3289.

[121] Lu, Y., Yan, L., Wang, Y., Zhou, S., Fu, J. & Zhang, J. (2009). Biodegradation of phenolic compounds from coking wastewater by immobilized white rot fungus *Phanerochaete chrysosporium*. *Journal of Hazardous Materials, 165*, 1091–1097.

[122] Ortega-Clemente, A., Caffarel-Mendez, S., Ponce-Noyola, M. T., Barrera-Cortes, J. & Poggi-Varaldo, H. M. (2009). Fungal post-treatment of pulp mill effluents for the removal of recalcitrant pollutants. *Bioresource Technology, 100*, 1885-1894.

[123] Lakhtar, H., Ismaili-Alaoui, M., Philippoussis, A., Perraud-Gaime, I. & Roussos, S. (2010). Screening of strains of *Lentinula edodes* grown on model olive mill wastewater in solid and liquid state culture for polyphenol biodegradation. *International Biodeterioration & Biodegradation*, 64, 167–172.

[124] Strong, P. J. (2010). Fungal remediation of Amarula distillery wastewater. *World Journal of Microbiology & Biotechnology, 26*, 133–144.

[125] Cerrone, F., Barghini, P., Pesciaroli, C. & Fenice, M. (2011). Efficient removal of pollutants from olive washing wastewater in bubble-column bioreactor by *Trametes versicolor*. *Chemosphere*, 84, 254–259.

[126] Romanholo Ferreira, L. F., Aguiar, M. M., Messias, T. G., Pompeu, G. B., Queijeiro Lopez, A. M., Silva, D. P. & Monteiro R. T. (2011). Evaluation of sugar-cane vinasse treated with Pleurotus sajor-caju utilizing aquatic organisms as toxicological indicators. *Ecotoxicology and Environmental Safety, 74*, 132–137.

[127] Sharari, M., Jahan Latibari, A., Guillet, A., Aurousseau, M., Mouhamadou, B., Rafeiee, Gh., Mirshokraei, A. & Parsapaghouh, D. (2011). Application of the white rot fungus *Phanerochaete chrysosporium* in biotreatment of bagasse effluent. *Biodegradation, 22*, 421–430.

[128] Verma, A. K., Raghukumar, C. & Naik C. G. (2011). A novel hybrid technology for remediation of molasses-based raw effluents. *Bioresource Technology, 102*, 2411–2418.

[129] McMullan, G., Meehan, C., Conneely, A., Kirby, N., Robinson, T., Nigam, P., Banat, I. M., Marchant, R. & Smyth, W. F. (2001). Microbial decolourisation and degradation of textile dyes. *Applied Microbiology & Biotechnology, 56*, 81-87.

[130] Vaidya, A. A. & Datye, K. V. (1982). Environmental pollution during Chemicals processing of synthetic fibers. *Colourate, 14*, 3-10.

[131] Wesenberg, D., Buchon, F. & Agathos, S.N. (2002). Degradation of dye-containing textile effluent by the agaric white-rot fungus *Clitocybula dusenii*. *Biotechnology Letters*, 24, 989–993.

[132] Selvam, K., Swaminathan, K. & Chae, K.-S. (2003). Decolourization of azo dyes and a dye industry effluent by a white rot fungus *Thelephora* sp. *Bioresource Technology, 88*, 115–119.

[133] Amaral, P. F. F., Fernandes, D. L. A., Tavares, A. P. M., Xavier, A. B. M. R., Cammarota, M. C., Coutinho, J. A. P. & Coelho, M. A. Z. (2004). Decolorization of Dyes from textile wastewater by *Trametes versicolo*r. *Environmental Technology, 25*, 1313-1320.

[134] Ramsay, J. A. & Goode, C. (2004). Decoloration of a carpet dye effluent using *Trametes versicolor*. *Biotechnology Letters, 26*, 97-201.

[135] Dellamatrice, P. M., Monteiro, R. T. R., Kamida, H. M., Nogueira, N. L., Rossi, M. L. & Blaise, C. (2005). Decolourization of municipal effluent and sludge by *Pleurotus sajor-caj*u and *Pleurotus ostreatus*. *World Journal of Microbiology and Biotechnology, 21*, 1363-1369.

[136] Kamida, H. M., Durrant, L. R., Rosim Monteiro, R. T., Dutra de Armas, E. (2005). Biodegradação de efluente têxtil por *Pleurotus sajor-caju*. *Quimica Nova, 28*, 629-632.

[137] Nilsson, I., Moller, A., Mattiasson, B., Rubindamayugi, M. S. T. & Welander, U. (2006). Decolorization of synthetic and real textile wastewater by the use of white-rot fungi. *Enzyme and Microbial Technology, 38*, 94-100.

[138] Srikanlayanukul, M., Khanongnuch, C., Lumyong, S. (1996). Decolorization of textile wastewater by immobilized *Coriolus versicolor* RC3 in repeated-batch system with the effect of sugar addition. *Chiang Mai University Journal* (Thailand), 5, 301-306.

[139] Blánquez, P., Sarra, M. & Vicent, T. (2008). Development of a continuous process to adapt the textile wastewater treatment by fungi to industrial conditions. *Process Biochemistry*, 43, 1–7.

[140] Asgher, M., Azim, N. & Bhatti, H. N. (2009). Decolorization of practical textile industry effluents by white rot fungus *Coriolus versicolor* IBL-04. *Biochemical Engineering Journal, 47*, 61–65.

[141] Accinelli, C., Saccà, M. L., Batisson, I., Fick, J., Mencarelli, M. & Grabic, R. (2010). Removal of oseltamivir (Tamiflu) and other selected pharmaceuticals from wastewater using a granular bioplastic formulation entrapping propagules of *Phanerochaete chrysosporium*. *Chemosphere, 81*, 436-443.

[142] Anastasi, A., Spina, F., Prigione V., Tigini, V., Giansanti, P. & Varese, G. C. (2010). Scale-up of a bioprocess for textile wastewater treatment using *Bjerkandera adusta*. *Bioresource Technology, 101*, 3067–3075.

[143] Anastasi, A., Parato, B. Spina, F., Tigini, V., Prigione, V. &Varese. G. C. (2011). Decolourisation and detoxification in the fungal treatment of textile wastewaters from dyeing processes. *New Biotechnolgy*. In Press. Available from doi:10.1016/j.nbt. 2011.08.006.

[144] Brunow, G. (2001). Methods to reveal the structure of lignin. In M. Hofrichter & A. Steinbüchel (Eds.), Biopolymers (1, pp. 89-116). Weinheim, Germany: Wiley-VCH.

[145] Kaushik, P. & Malik, A. (2009). Fungal dye decolourization: Recent advances and future potential. *Environment International, 35*, 127–141.

[146] O'Neill, C., Hawkes, F. R., Hawkes, D. L., Lourenco, N. D., Pinheiro, H. M. & Delee, W. (1999). Colour in textile effluents — sources, measurement, discharge consents and simulation: a review. Jo*urnal of Chemical Technology & Biotechnology, 74,* 1009-1018.

[147] Netzer, A., Miyamato, H. K., Wilkinson, P. (1975). Heavy metal concentration of dye bath effluents. *Bulletin of Environmental Contamination & Toxicology, 14*, 301-303.

Index

A

B

C

D

E

F

G

H

I

J

K

L

M

N

O

P

Q

R

S

T

U

V

W

X

Y

Z